Self-Organising Polymers

University of Leeds
July 19-21, 2004

FARADAY DISCUSSIONS
Volume 128, 2005

RS•C

Self-Organising Polymers

A General Discussion on Self-Organising Polymers was held at the University of Leeds, UK on 19th, 20th and 21st July 2004.

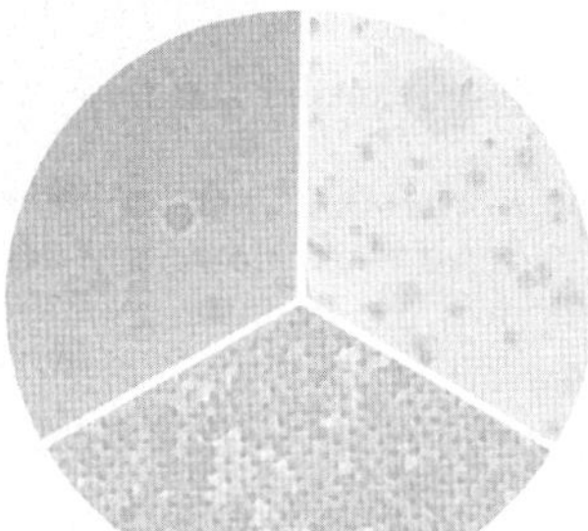

Cover

See T. P. Lodge, J. Bang, Z. Li, M. A. Hillmyer and Y. Talmon, p. 1 and H. Mao, P. L. Arecchea, T. S. Bailey, B. J. S. Johnson and M. A. Hillmyer, p. 149.

The self-assembly of block copolymers is a versatile and powerful strategy for preparing nanostructured materials. The image shows electron micrographs of three novel structures described in this issue: segmented worm-like micelles from aqueous solutions of mikto arm star ABC triblocks (left), core/shell/corona disk-like micelles from aqueous solutions of linear ABC triblocks (right), and nanoporous materials with hydrophilic pore walls from AB/AC diblock blends (bottom).

Image kindly supplied by Professor T. P. Lodge and Professor M. A. Hillmyer, Department of Chemistry, University of Minnesota, USA.

Chemical biology articles published in this journal also appear in the *Chemical Biology Virtual Journal*: **www.rsc.org/chembiol**

contents

INTRODUCTORY LECTURE

PAPERS AND DISCUSSIONS

CONCLUDING REMARKS

ADDITIONAL INFORMATION

Electronic supplementary information (ESI) is available *via* the online article (see http://www.rsc.org/esi for general information).

Introductory Lecture

Strategies for controlling intra- and intermicellar packing in block copolymer solutions: Illustrating the flexibility of the self-assembly toolbox

Timothy P. Lodge,*[ab] Joona Bang,[b] Zhibo Li,[a] Marc A. Hillmyer[a] and Yeshayahu Talmon[c]

[a] *Department of Chemistry, University of Minnesota, Minneapolis MN 55455, USA. E-mail: lodge@chem.umn.edu*
[b] *Department of Chemical Engineering & Materials Science, University of Minnesota, Minneapolis MN 55455, USA*
[c] *Department of Chemical Engineering, Technion-Israel Institute of Technology, Haifa 32000, Israel*

Received 18th August 2004, Accepted 18th August 2004
First published as an Advance Article on the web 28th October 2004

Block copolymers constitute a class of self-assembling macromolecules that offer remarkable flexibility for controlling nanostructure, both in discrete objects and in bulk materials. Block copolymer micelles may be formed with multiple compartments by judicious choice of ingredients in an ABC triblock copolymer. For example, we have shown that a poly(ethylene oxide–*b*–styrene–*b*–fluorinated butadiene) triblock assembles in dilute aqueous solution into large, flat core/shell/corona disks, with the fluorine containing block forming the core. In contrast, the unfluorinated precursor generates large spherical micelles. A numerical analysis suggests that the disk-like motif is characteristic of the so-called superstrong segregation regime, whereby the interfacial tension becomes so large as to overwhelm the conformational entropy of the core blocks. For ABC miktoarm stars comprising polyethylene oxide, polyethylethylene, and polyhexafluoropropylene oxide arms, a much richer variety of micellar structures are observed. Prominent amongst these is a "segmented worm", in which alternating layers (5–7 nm thick) of hydrocarbon and fluorocarbon blocks form disks (6–10 nm in radius) that stack into cylindrical aggregates. The disk radii suggest almost fully stretched blocks, again consistent with the superstrong segregation regime. These structures are rationalized in terms of the constraints imposed by the star architecture, combined with the extremely strong interfacial tensions. In contrast, for lipids, surfactants, and aqueous diblock copolymers, increasing the interfacial tension can induce a transition from spheres to cylinders to flat bilayers; the disk-like motif is not usually seen. Interestingly, in aqueous diblocks both worm-like micelles and vesicles have been well-documented, whereas in "simple" organic systems they have not. We have shown that by suitable choice of block composition and solvent selectivity, the curvature sequence sphere/cylinder/vesicle can also be observed in poly(styrene–*b*–isoprene) diblocks in dialkyl phthalates. In more concentrated solutions such spherical micelles assemble onto body-centered or face-centered cubic lattices. In some cases a thermoreversible fcc/bcc transition has been noted. We have recently demonstrated that this transition corresponds to a particular aggregation number; higher aggregation numbers, found at lower temperatures, favor the "hard-sphere-like" fcc packing. All of these results are based on a combination of dynamic light scattering, small angle X-ray and neutron scattering, and cryogenic transmission electron microscopy.

Introduction

A central goal of nanoscience is to exert control over the structure of materials, and thus over their function, over the range of about 5–200 nm. One general strategy that is enjoying tremendous attention in the research community is self-assembly, whereby the component molecules within some material spontaneously select a structure with a predictable degree and type of ordering. Such structures might comprise discrete objects (*e.g.*, micelles, vesicles, nanoparticles, *etc.*), or they might correspond to materials that are continuous in either two (*e.g.*, thin films) or three dimensions. Self-assembly is often considered the epitome of "bottom-up" engineering, as the processing of nanostructures may be very economical in terms of labor or energy input. Within the realm of self-assembly, an emerging paradigm is that of hierarchical structure, *i.e.*, materials with designed organization on at least two distinct length scales. Success in this endeavor would presumably enable many more, relatively sophisticated applications.

Among the various classes of molecules that undergo self-assembly, block copolymers are particularly versatile and rich in promise.[1,2] Consider the simplest case of an AB diblock copolymer, in the bulk state. Fig. 1 illustrates the seven known structures (four symmetries) that such materials adopt.[3] The periodicities of the structures are dictated primarily by molecular weight, and as such are naturally in the 5–50 nm range. They may also be tuned continuously, by blending, dilution, and temperature, for example. The self-assembly process itself is usually quite tolerant of component heterogeneity, so that commercially viable materials are perfectly acceptable. Furthermore, different functional groups or additives may be selectively incorporated into one domain. The four symmetries shown in Fig. 1 span the range of possible domain connectivities, *i.e.* the minor component microdomains are connected in zero (spheres), one (cylinders), two (lamellae) or three (double gyroid) dimensions. As these various morphologies are equilibrium phases, it is reasonable to expect, and indeed found, that theory can predict the relative free energies of possible structures rather reliably.[3,4] If a greater wealth of possible structures are sought, then straightforward extension to 3 component systems, such as ABC triblock copolymers, offers morphological richness that has only begun to be explored.[1]

Block copolymers are but one class of structured-directing materials, but they can also illustrate the great design flexibility that self-assembly affords. It is perhaps helpful to consider four general ways in which the experimenter has access to control over structure. First is the choice of molecules themselves. For block copolymers, the design variables include number and chemical identity of ingredients, molecular weight, composition, and architecture (*i.e.*, linear *versus* star *versus* branched, diblock *versus* triblock *versus* multiblock, *etc.*). The chemical ingredients may be chosen on the basis of the dominant interactions (ionic, dipolar, hydrogen-bonding, dispersive, *etc.*), because they are responsive to some external trigger, or because they have a particular structure-directing character

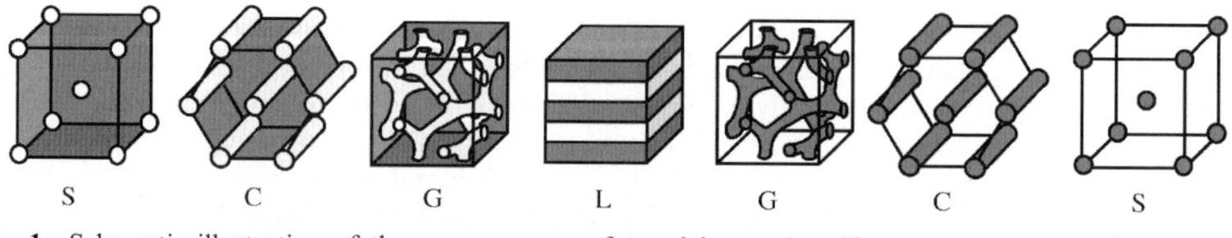

Fig. 1 Schematic illustration of the nanostructures formed by an AB diblock copolymer in the bulk. The various phases are S (bcc spheres), C (hexagonally-packed cylinders), G (double gyroid), and L (lamellae). The white and grey domains represent the A and B blocks, respectively, as the fraction of A in the copolymer increases from left to right.

(mesogens, helix formers, crystallizable blocks, *etc.*) Second, we may consider the composition of the mixture. Here important variables include the selectivity of an added solvent or diluent, the miscibility or immiscibility of different polymeric components, the use of complementary groups such as hydrogen bond donors and acceptors, or particular recognition elements. Third, external stimuli such as temperature, pressure, pH, and ionic strength may be used to tune structure. Finally we may consider the use of external fields to produce the desired organization, particularly for anisotropic structures. Examples include shear and extensional flow, electric and magnetic fields, and even gradients in temperature. In short, there are many dials that may be turned to affect the final structure, and the papers in this volume illustrate many of the examples just listed.

In this introductory paper, we provide four examples of block copolymer self-assembly from our recent work that highlight some of the important issues and possibilities. In the first, we demonstrate that the sequence of interfacial curvature sphere/cylinder/vesicle, which is well documented in lipids, surfactants and aqueous block copolymers, can also be achieved in a rational and systematic way in standard nonaqueous block copolymer solutions.[5] Second, we show how a linear ABC triblock copolymer, with suitably chosen block lengths and mutual interaction strengths, can form core/shell/corona micelles in water.[6] Furthermore, we argue that the formation in one case of a flattened oblate disk-like micelle is indicative of a new arena in block copolymer assembly,[7] the so-called superstrong segregation regime.[8,9] In the third example, the introduction of an ABC "miktoarm" star architecture[10] leads to a fascinating variety of multicompartment micelles, in which separate fluoropolymer and hydrocarbon blocks form multiple distinct nanodomains within a micellar core.[11,12] Finally, we discuss recent progress in understanding the packing of spherical micelles onto superlattices at higher concentrations,[13–19] and in particular identify the control variable that dictates whether the chosen cubic lattice is body-centered or face-centered.

Experimental section

Polymers

Poly(styrene-*b*-isoprene) diblock copolymers were prepared by sequential living anionic polymerization, following standard protocols as previously described.[19] Molecular weights and polydispersity indices were determined by size exclusion chromatography (SEC) with light scattering detection, and compositions were determined by ^{1}H NMR spectroscopy. Samples are designated SI(x-y), where x and y denote the styrene and isoprene block molecular weights in kDa.

A poly(ethylene oxide-*b*-styrene-*b*-1,2-butadiene) linear triblock terpolymer was synthesized by anionic polymerization, as previously described.[6] The styrene block was initiated with a *tert*-butyldimethylsiloxylithium protected initiator, and polymerized in cyclohexane. The butadiene block was then polymerized, with addition of THF as a polar modifier to produce almost complete 1,2 addition; the resulting SB diblock was terminated with isopropanol. The protected initiator was cleaved off with tetrabutyl ammonium fluoride, followed by acid treatment. The resulting hydroxyl group was used to initiate polymerization of the ethylene oxide, using potassium naphthalenide in THF. The resulting "OSB" triblock had block molecular weights of 13,200, 4700, and 1300 Da, respectively. The butadiene block was reacted with perfluorohexyliodide in the presence of triethylborane and air as previously described,[20] to produce a fluorinated butadiene derivative in an "OSF" triblock. The molecular weight of the resulting "F" block was 4600 Da, and the O and S blocks were essentially unaffected by this free radical chemistry.

ABC triblock miktoarm stars were prepared as follows.[12] Butadiene was polymerized in THF at low temperature, and terminated with a protected bifunctional terminator 2-methoxymethoxymethyloxirane. The functional 1,2 polybutadiene was fully hydrogenated over Pd/$CaCO_3$ to produce polyethylethylene, without compromising the functional groups. The terminal hydroxyl group was then used to initiate ethylene oxide polymerization, using diphenyl methylpotassium in THF. The resulting midfunctional EO diblock was terminated with bromoethane. The methoxymethyl protecting group was cleaved off with HCl, to provide a pendant hydroxyl group at the EO junction. Acid chloride terminal poly(perfluoropropylene oxide) (Krytox® from DuPont) was then conjugated to the EO diblock in a mixture of fluorinated solvents, to produce the desired triblock miktoarm stars, designated μ-EOF(x-y-z).

Solutions

Aqueous solutions were prepared by direct dissolution of OSB, OSF, and various μ-EOF terpolymers, assisted by stirring and, in some cases, mild sonication. All solutions appeared clear to the eye. Solutions were filtered through 0.2 μm filters prior to dynamic light scattering analysis. Solutions of SI copolymers in various dialkyl phthalates (dimethyl phthalate, DMP; diethyl phthalate, DEP; di-*n*-butyl phthalate, DBP) and tetradecane were prepared by codissolution in methylene chloride, which was subsequently stripped off under flowing nitrogen.

Dynamic light scattering

DLS measurements were taken on a homebuilt goniometer in the homodyne mode, using a BI-9000 correlator. Temperatures were controlled to within ±0.2 °C using a flow-through heater immersed in the silicon oil index-matching fluid. Correlation functions were obtained at several scattering angles to confirm the diffusive character of the decay modes. Individual correlation functions were analyzed with single exponential, double exponential, cumulant expansion, or Laplace inversion techniques, as appropriate.

Small angle X-ray scattering

SAXS measurements were performed either at the University of Minnesota Characterization Facility, or at the Argonne National Laboratory at the DND-CAT beamline 5-D-ID. Scattered X-rays were incident on area detectors, and the (isotropic) intensities were azimuthally averaged to give intensity, I, *versus* scattering wave vector, q $(= (4\pi/\lambda) \sin(\theta/2))$. No attempt was made to place the intensities on an absolute scale, but background/solvent subtractions were performed in those cases where $I(q)$ was fitted to specific form factor expressions.

Small angle neutron scattering

Measurements were performed at the National Institute of Standards and Technology NSF/ExxonMobil/University of Minnesota beamline NG7. A neutron wavelength of 6 Å with spread $\Delta\lambda/\lambda = 0.1$ was selected, and various detector positions employed to achieve a suitable q range. Scattering patterns were corrected for transmission, background, empty cell, and incoherent contributions, and placed on an absolute scale using the direct beam method.

Cryotransmission electron microscopy

A small droplet of each solution was placed on a lacey carbon film supported on microperforated cryoTEM grid. Careful blotting produced thin films (typical thickness 200 ± 100 nm) within the pores of the film. The aqueous solutions were vitrified by rapid immersion in liquid ethane. Solutions in dialkyl phthalates were frozen in liquid nitrogen. All samples were examined using low dose conditions on a JEOL 1210 at 120 kV. The high resolution cryoTEM of μ-EOF(2-7-2), μ-EOF(2-13-2), and μ-EOF(2-13-3) aqueous solutions was performed at the Hannah and George Krumholz Advanced Microscopy Laboratory of the Technion Project on Complex Fluids, Microstructure and Macromolecules. The other measurements were performed in the IT Characterization Facility at the University of Minnesota.

Results and discussion

1. Sphere/cylinder/vesicle sequence in a non-aqueous diblock copolymer/solvent system

In small molecule surfactants, the observed micellar shape can often be well correlated with a packing parameter that balances the effective size of the headgroup with that of the hydrophobic tail.[21] The larger the former, the larger the spontaneous curvature, and spherical micelles result. As the tail size increases, the curvature changes to form first cylinders, or worm-like micelles, and then flat bilayer sheets, which usually wrap around to form closed vesicles or liposomes. The same curvature sequence, *i.e.* spheres/cylinders/bilayers, has been recently documented in aqueous solutions of both nonionic[22–25] and ionic block copolymers.[26–29] To the best of our knowledge, the equivalent sequence has not been clearly established in a particular system for non-aqueous block copolymers, which is perhaps surprising, given that micellization in such systems (*e.g.*, PS–PI, PS–PB, PS–PMMA) has been studied for over 40 years. There have been some intriguing precedents, however. Canham *et al.* found long, worm-like micelles in PS–PB–PS triblocks dissolved in ethyl acetate.[30] Pedersen *et al.* found PS–PI and PS–PI–PS copolymers that formed prolate ellipsoidal micelles, with an aspect ratio of about 3.[31] Most recently, LaRue *et al.* observed long worm-like micelles in PS–PI in heptane.[32] Clear evidence of vesicles is described in a paper in this volume, where Borsali and coworkers found both worms and vesicles in solutions of cyclic PS–PI diblocks.[33,34]

There are at least three variables that one might tune in order to induce this curvature sequence: relative block length, solvent, and temperature. The latter two are more convenient, and are qualitatively equivalent in that they both act through the interfacial tension. In our study we selected SI(13-71) in a series of styrene-selective solvents.[5] The larger isoprene core block accesses the "crew-cut" micelle regime, where the penalty for compressing the relatively short corona blocks as the curvature decreases is mitigated. The dialkyl phthalates form a well-studied homologous series of styrene-selective solvents, which may be blended as desired to tune the selectivity. For example, for SI(15-13), which forms spherical micelles exclusively, the critical micelle temperature was found to be well below room temperature, 83 °C, and 156 °C for 1% solutions in DBP, DEP, and DMP, respectively.[35] We examined 1% solutions of SI(13-71) in DBP, DEP, and DMP, and in 3 : 1, 1 : 1, and 1 : 3 mixtures of DBP : DEP and DEP : DMP, all at room temperature.[5] We found spherical micelles in DBP and 3 : 1 DBP : DEP; short cylindrical micelles in 1 : 1 and 1 : 3 DBP : DEP; vesicles in DEP, 3 : 1 DEP : DMP, and 1 : 1 DEP : DMP. Solutions in 1 : 3 DEP : DMP and pure DMP phase separated. Evidence for these structural assignments is presented in Figs. 2 and 3.[5] The former shows typical SAXS form factors, along with the minima associated with the assigned structures. The relative spacings of the minima are consistent with the assigned structures, and the values of the core dimensions thus extracted agree well with both cryoTEM and full fits of the SAXS data. Further discussion of the form factors, and the full fits to the appropriate mathematical

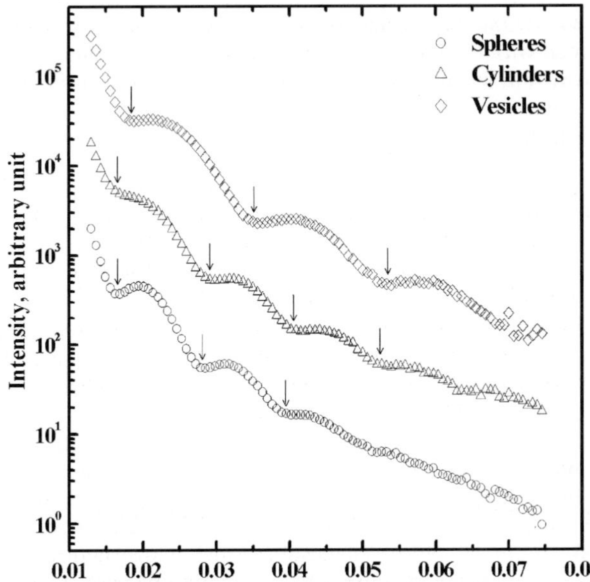

Fig. 2 Typical SAXS form factors for SI(13-71) showing spheres (in 3 : 1 DBP : DEP), cylinders (in 1 : 1 DBP : DEP), and vesicles (in 1 : 1 DMP : DEP). The vertical arrows indicate the positions of the characteristic minima of the associated form factors, from which the core dimensions (spheres: radius 28 ± 1 nm; cylinders: radius 25 ± 2 nm; vesicles: half-layer thickness 17 ± 1 nm) may be obtained.

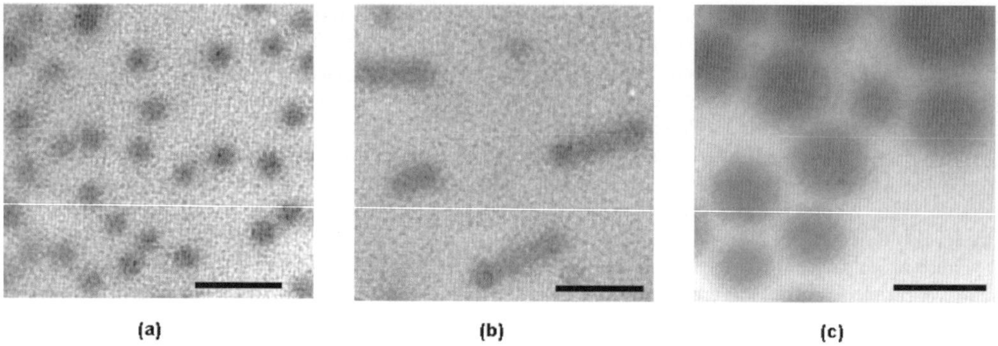

Fig. 3 CryoTEM images of SI(13-71) micelles; the scale bars are 200 nm. (a) spherical micelles in 3 : 1 DBP : DEP; (b) short cylinders in 1 : 1 DBP : DEP; (c) vesicles in 1 : 1 DEP : DMP. Larger fields of view, and further images of these solutions, may be found elsewhere.[5]

functions,[36,37] will be provided elsewhere.[5] Fig. 3 provides representative cryoTEM images of the three structures. It is worth noting that the cryoTEM images are rather overexposed, but when underexposed, the micelles appear lighter than the solvent, due to the relative electron densities. The glass-forming nature of the dialkyl phthalates is actually an advantage for cryoTEM, as crystallization does not compete with vitrification. The sequence of curvatures is completely consistent with expectation, namely as the solvent selectivity increases, the interfacial tension goes up. This is accommodated by the change in curvature, which reduces the area per chain (from roughly 13 to 11 to 8 nm^2 $chain^{-1}$ in this case). These results confirm the universality of this curvature sequence in nonaqueous block copolymer systems.

2. Transformation from sphere to flat core/shell/corona disk

As a first example of micelles formed by ABC triblock terpolymers, we examined 1% aqueous solutions of linear OSB and OSF by a variety of techniques.[6] In the case of OSB, the relatively short S and B blocks, combined with the relatively weak interaction parameter χ_{SB}, meant that these two blocks could be mixed within an overall hydrophobic micelle core. In block copolymer language, the product χN for this pair of blocks is on the order of 5, well short of the critical value needed for microphase separation in the bulk ($\chi_{SB}N > 10$). The small angle scattering curves could be well described by a spherical form factor, containing a single S/B core, with a core radius about 20 nm.[6] This is actually a rather large value, given the low molecular weights, implying that the core SB chains are significantly stretched. This is consistent with the notion that the interfacial tension in this system is very high, driving the micelles to adopt a large radius and therefore a large aggregation number (*ca.* 3000 in this case). Dynamic light scattering on this system also indicated very large micelles, with a hydrodynamic radius of about 55 nm, and the ability of single exponential decays to fit the correlation functions indicates a narrow distribution of micellar sizes.[6]

Dynamic light scattering on the OSF system gave similar results, namely a narrow distribution of sizes and a mean hydrodynamic radius of 52 nm.[6] However, the more detailed characterization experiments revealed a much more interesting micellar morphology. Both SAXS and SANS curves were well described by a flattened disk (oblate ellipsoid) core/shell corona structure, whereby the F block was segregated into a core with principal radii of 25, 25, and 5 nm, and the S block formed a shell approximately 5 nm thick.[6] Definitive confirmation of this unique morphology was obtained *via* cryoTEM;[7] an example is shown in Fig. 4. Most of the micellar cores appear circular in the image, but with a uniform optical density suggesting a disk-like structure. The invisible O corona chains are nevertheless evidenced by the empty halo surrounding each micelle. Fortuitously, some of the micelles were captured standing on edge in the field of view (see inset for an expanded image), in which the electron rich F core is clearly resolved from the lighter gray S shell.

The question of immediate interest is why this micelle is a flattened disk. The fact that the F and S blocks are segregated is not a surprise *per se*; the S/F interaction is expected to be significantly stronger than the S/B case.[38] Similarly, the S and F blocks occupy about the same volume, so that

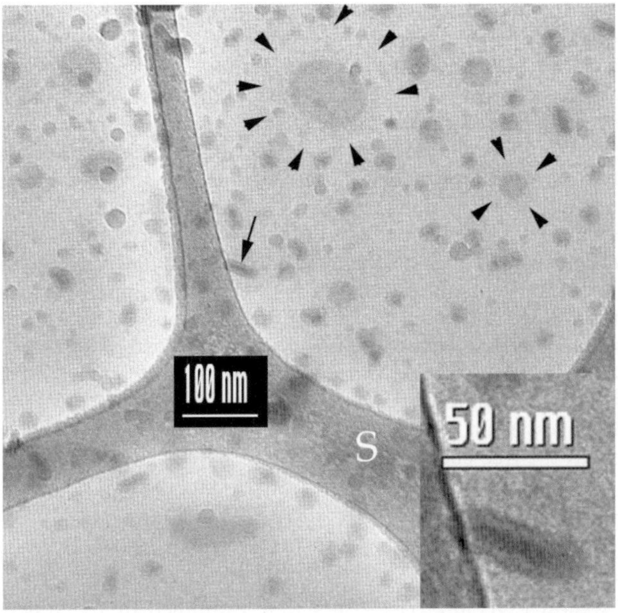

Fig. 4 CryoTEM images of OSF core/shell/corona disk-like micelles. The circles correspond to disks exposed along the short axis, and the arrow heads indicate the extent of the (almost invisible) PEO corona. The inset shows an expanded view of a micellar core viewed edge-on; the dark F core and the grey S shell are clearly resolved. Another image from this system has been presented elsewhere;[7] the inset is reproduced with permission. The label 'S' denotes the perforated support film.

all other things being equal a flat interface would be preferred. However, the interfacial tension at the shell/corona interface cannot be significantly different from the OSB case. Furthermore, the O block is still significantly larger than the combined S and F blocks. Accordingly, a spherical micelle might have been expected, or, perhaps, a cylindrical structure in order to acknowledge the increase in the size of the core block. Instead, a flat disk is seen. This is therefore in contrast to the curvature sequence discussed in the previous section. A schematic illustration of the situation is given in Fig. 5. For "typical" micelle systems, increasing the interfacial tension and/or increasing the size of the core block induces transitions from spheres to cylinders to flat sheets; in the OSB to OSF case, the intermediate structure is a flattened disk.

We propose that the difference resides in the fact that the OSF system reflects a new regime of block copolymer phase behavior, designated the superstrong segregation regime (SSSR).[7] The

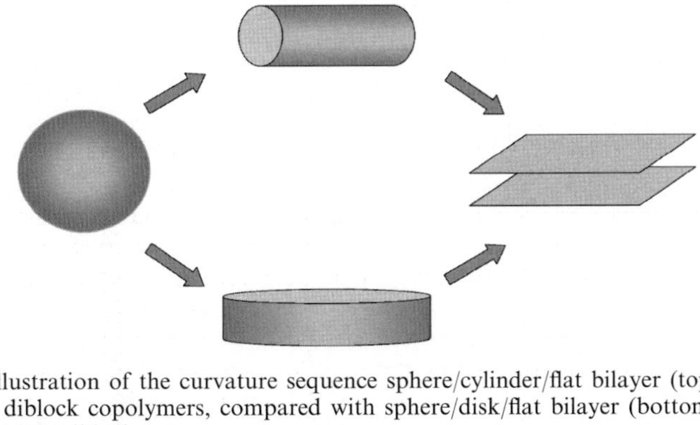

Fig. 5 Schematic illustration of the curvature sequence sphere/cylinder/flat bilayer (top) adopted by typical surfactants and AB diblock copolymers, compared with sphere/disk/flat bilayer (bottom), adopted by super-strongly segregated ABC triblocks.

SSSR was proposed and analyzed theoretically by Semenov *et al.* a decade ago,[8] but as yet we are unaware of any nonionic chemical system that appears to satisfy the necessary criteria. The underlying physical concept is straightforward. In a normal "hairy" micellar system, *i.e.*, a diblock in which the corona block is longer than the core block, an increase in interfacial tension will lead to an increase in aggregation number and core radius, at the price of increased crowding of the corona chains.[39] Depending on conditions, this will either just increase the size of a spherical micelle, or induce the transition to a cylinder. However, if the system is very strongly segregated, yet the corona block is long enough to suppress the transition to a cylinder, the sphere will increase in size until the core blocks are essentially fully extended. This, in fact, is apparently the case for OSB. Any further increases in interfacial tension must be accommodated by a change in structure, but as the core block is fully stretched, the dominant parameter is the interfacial area per chain. This quantity is minimized by a flattened disk, rather than by a cylindrical shape. We have recently made numerical estimates of the various contributions to the free energy, after extending the analysis of Semenov *et al.*[8] to the triblock case, and find that indeed the OSB system is on the brink of accessing the SSSR, and the OSF system achieves it.[7] An important aspect of the problem is the fact that the presence of two interfaces in an ABC copolymer micelle significantly enhances the interfacial tension contributions to the free energy balance.[7]

3. Multicomponent micelles: constraint of star architecture

Dilute solutions of several EOF miktoarm stars were examined by dynamic light scattering, and cryoTEM.[11] In all cases the DLS correlation functions could be reasonably well fit with the sum of two exponential decays, suggesting the presence of two (or more) distinct micellar species. However, it should be noted that because the ratio of the decay constants for the faster and slower modes was typically only 3–5, we cannot discount a distribution of particle sizes centered about these two modes. Both modes were diffusive, and hydrodynamic radii were extracted *via* the Stokes–Einstein relation. The values of R_h are reproduced in Table 1. The smaller values (typically *ca.* 20–30 nm) are much too large for single chains, and can be assigned to discrete micelles. The larger values are too big for spherical micelles, given the estimated radii of gyration of the various blocks (also listed in Table 1).

Identical or equivalent solutions were examined by cryoTEM,[11] and representative images are presented in Figs. 6a–c. In Fig. 6a, micelles formed by μ-EOF(2-13-3) are shown. In this case there are discrete cores, approximately 15 nm in diameter. The cores are roughly uniformly distributed in space, with the distance of closest approach reflecting the invisible O corona chains. The cores themselves have a speckled appearance, which we attribute to the segregation of small F domains within an E matrix. In Fig. 6b, discrete micelles and strings of micelles are both visible for μ-EOF (2-7-2). These results are initially rather surprising, in that one might expect the O corona to provide steric stabilization for the individual micelles, and thus a repulsive intermicellar potential. The image shows, however, that there is some kind of attraction between the cores, and furthermore that this attraction is not spherically symmetric. Finally, in Fig. 6c, one can see that μ-EOF(2-9-3) forms long, segmented worms. In this case the dark stripes reflect the F domains, which are separated by E layers. Again the O coronas are not directly visible.

Table 1 Dimensions of micelles formed by μ-EOF polymers in water by dynamic light scattering[11]

Sample	^aR_g/nm (PEE)	^aR_g/nm (PEO)	^aR_g/nm (PFPO)	R_h/nm (small)	R_h/nm (large)
μ-EOF(2-6-2)	1.3	2.8	0.9	34	97
μ-EOF(2-7-2)	1.3	2.9	0.9	24	111
μ-EOF(2-9-2)	1.3	3.4	0.9	17	68
μ-EOF(2-9-3)	1.3	3.4	1.0	27	98
μ-EOF(2-9-5)	1.3	3.4	1.3	33	118
μ-EOF(2-13-2)	1.3	4.1	0.9	19	81
μ-EOF(2-13-3)	1.3	4.1	1.0	19	82

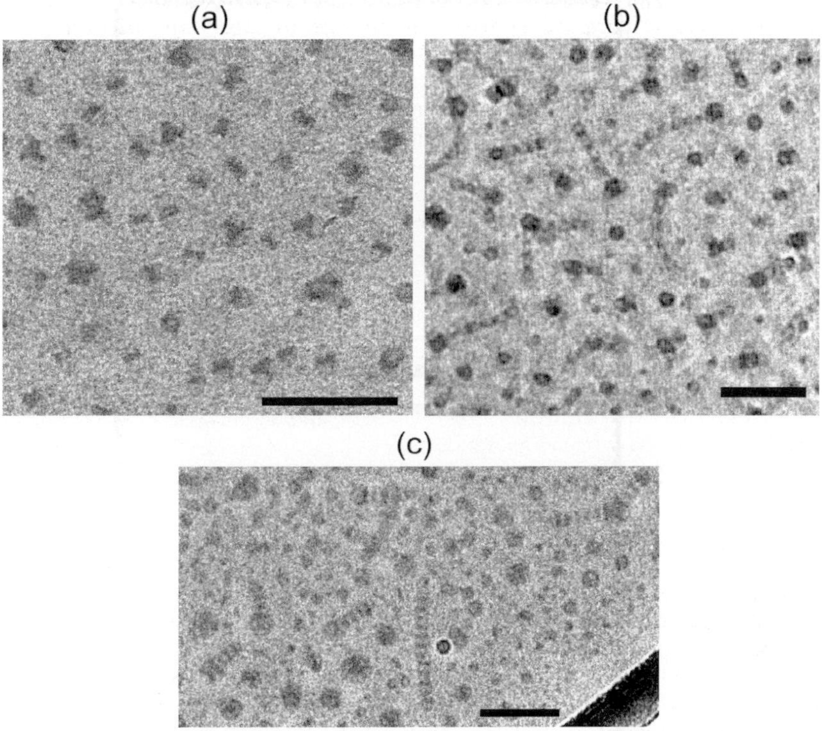

Fig. 6 CryoTEM images of micelles formed from (a) μ-EOF(2-13-3); (b) μ-EOF(2-7-2); (c) μ-EOF(2-9-3); these images are reprinted with permission from a larger series which have been reported elsewhere.[11] The scale bar is 100 nm.

These images, and many others on the μ-EOF series,[11] may be understood at least qualitatively in the following terms. First, all three blocks are extremely strongly segregated from one another, and as the E and F blocks are very hydrophobic, the micellar cores all involve separate compartments of E and F. The very strong segregation causes very high interfacial tensions, which encourages flat interfaces. At the same time, the O block is the longest of the three in all cases, and thus osmotic crowding in the corona favors higher curvature. This interplay is illustrated by the fact that in Fig. 6a the stars with the longest O blocks form discrete micelles, whereas the micelles in Figs. 6b and c show more aggregation of multiple core domains. Although E and F are both hydrophobic, the latter is presumably much more so, so that within the μ-EOF(2-13-3) cores the F nanodomains are surrounded by an E matrix. Note, however, that the star architecture requires the E and F domains to intersect on the outer surface of the core, in contact with the O corona; the F blocks cannot form a separate inner core as was observed in linear triblocks.[6] The formation of multiple small domains of such a strongly segregating group within an overall hydrophobic core has been anticipated theoretically.[40]

In the case of μ-EOF(2-7-2), the attraction between micelles is proposed to arise as follows. The F blocks aggregate as an ellipsoidal layer, with the E blocks on top and bottom, somewhat in the manner of a hamburger, where E chains constitute the buns.[11] The O corona chains are anchored at the E/F interface, and must then wrap around the top and bottom of the buns. The origin of the attraction between micelles is therefore clear: the O corona can fluctuate and expose the hydrophobic domains, so that when two micelles aggregate, the E blocks come into contact and the O chains can relax and stabilize a roughly cylindrical aggregate. The same packing motif is apparent in Fig. 6c, except that in this case the E and F domains have each segregated into flat disks. The diameters of these "segmented worm" micelles are comparable to the fully extended contour lengths of the E and F blocks, once again suggesting that the system is in the superstrong segregation regime.[8]

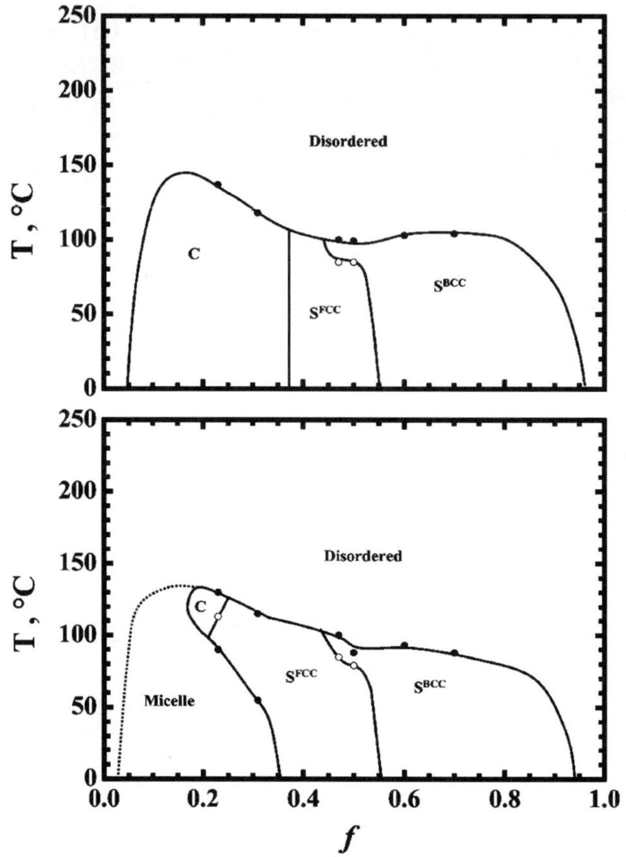

Fig. 7 Phase behavior of SI diblocks as a function of composition in DEP at the indicated polymer concentrations; figure updated with permission from ref. 19.

4. Lattice transition in ordered spherical micelles

It is well established that spherical micelles will order on a lattice at sufficiently high concentrations (typically above 10–20% polymer). At still higher concentrations the spheres may transform to cylinders, and then possibly also to gyroid or to lamellae, depending on the copolymer composition and temperature. Detailed phase diagrams have recently been presented for a variety of SI diblocks, in a range of solvents of varying selectivity.[19,41,42] An interesting observation is that the spherical micelles can pack onto either an fcc or a bcc lattice, depending on conditions. This should be contrasted with the bulk, where only the bcc phase is reported.[3] Furthermore, in several solutions a thermoreversible fcc/bcc transition is found. This experimentally convenient order–order transition offers a means to ascertain the control parameter that dictates the choice of lattice symmetry. Examples of the location of the fcc, bcc, and cylinder phases are shown in Fig. 7, in the temperature–composition plane.[19]

The pioneering work of McConnell and Gast,[13,14] and the extensive studies of Hamley, *et al.*,[15,43] provide some experimental guidance as to the location of the fcc/bcc boundary in SI and in aqueous PEO–PPO and PEO–PBO systems, respectively. The former authors related the micellar results to previous studies on charged colloids, and computer simulations, in which the range and steepness of the intermicellar potential is the key factor. Qualitatively, steeper, shorter-ranged interactions favor fcc, whereas softer, longer ranged potentials favor bcc. The authors did not study the effect of temperature, and thus the phase boundary was crossed by changing polymer composition. The authors proposed that the boundary corresponded to a value of *ca.* 1.5 for the ratio of the corona block to core block dimensions.

Here we take the following experimental approach.[17,44,45] An SI(16-15) copolymer with a perdeuterated styrene block was dissolved in a mixture of normal and perdeuterated tetradecane, an isoprene selective solvent, to mask out the SANS from the corona block. Similarly, SI(15-14) with a perdeuterated isoprene block was dissolved in the styrene-selective solvent DEP. Low concentrations (0.5, 1, and 2%) were examined as a function of temperature, from room temperature up to the critical micelle temperature.[44,45] The resulting scattering intensities were placed on an absolute basis, and the angular dependence fit to the appropriate spherical form factor.[37] The higher concentration solutions also required a term accounting for the structure factor, as previously described.[44,45] In this way it was possible to quantify the temperature dependence of the micellar characteristics, including the aggregation number, the core radius, the fraction of solvent in the core, and the associated equivalent hard-sphere radius. The results between the two systems were very similar, in that at room temperature and up to about 40 °C, the core was devoid of solvent, and the aggregation number and micellar size remained almost constant. However, at higher temperatures the diminishing solvent selectivity led to appreciable solvent uptake in the core. This was actually accompanied by a *decrease* in mean aggregation number and in micellar dimensions, and an associated *increase* in the number density of micelles. These changes are a direct consequence of the lowering of the interfacial tension at the core/corona interface, an effect that is completely consistent with the results for SI(13-71) referred to above.[5]

Solutions of the same polymers in tetradecane and DEP were examined by SAXS at higher concentrations, in the ordered state.[45] Both systems adopted the fcc phase at room temperature, but converted to bcc at a higher temperature (but below the dilute solution CMT). The position of the principal scattering peak, q^*, and therefore the lattice parameter, were independent of temperature up to the point where the dilute solution micelles began to shrink. Above that temperature the lattice parameter began to decrease, consistent with a denser packing of progressively smaller but more numerous objects. On the assumption that the temperature dependence of the spherical micellar dimensions is essentially the same in both dilute and concentrated solutions, an assumption that is justified at greater length elsewhere,[45] it is possible to assign a particular aggregation number as the critical value for the fcc/bcc transition. Then, direct comparison can be made to recent simulations and density functional theory calculations for highly branched star polymers.[46] The agreement between our micellar data and the predictions for stars is remarkable. Experimentally, aggregation numbers of about 90–100 correspond to the transition, whereas in the simulations the critical number of arms is about 60–70. This small difference is actually consistent with the presence of some loose chains, which in this case act to further stabilize the bcc phase.[47] Qualitatively, these results are in accord with the previous interpretation in terms of the steepness of the potential.[14] Specifically the lower the aggregation number, the lower the penalty for the overlap of corona chains; this overlap is required in the less efficiently packed bcc lattice. However, we do find that the corona/core length ratio at the transition may vary by up to a factor of two from system to system, but is independent of temperature for a given system, thereby indicating that this is not a reliable parameterization.

Summary

The self-assembly of block copolymers represents a versatile platform for producing nanostructures in the bulk, in thin films, and as discrete objects. In this paper we have illustrated several interesting aspects of micelle formation in block copolymer systems. First, we have demonstrated that the curvature sequence sphere/cylinder/flat bilayer may be observed in "conventional" styrene–isoprene diblock copolymer micelles dispersed in organic solvents, confirming the universality of this phenomenon. The key step in this case is to find a sufficiently selective solvent system, given the rather similar solubility parameters of the two blocks, and furthermore to adjust the composition of the copolymer to minimize osmotic crowding in the corona. Second, the self-assembly of a linear ABC terpolymer leads to an interesting new morphology, a core/shell/corona flattened disk in water. This was achieved by combining a hydrophilic block, a lipophilic block, and a hydrophobic/lipophobic fluorocarbon block. We propose that the observation of a flattened disk, as opposed to a spherical or cylindrical core/shell/corona structure, reflects entry into the superstrong segregation regime. Third, when the architecture of the ABC terpolymer is changed from linear to a miktoarm star, the core/shell/corona motif is suppressed in favor of multicompartment micelles. A fascinating array of micellar structures is observed in cryoTEM, depending on the relative lengths of the three blocks. Finally, we discuss how the thermoreversible fcc/bcc transition observed in higher concentration solutions of spherical micelles may be directly related to changes in the mean aggregation number of the micelles.

Acknowledgements

This work was supported primarily by the MRSEC Program of the National Science Foundation under Award Number DMR-0212302.

References

1 F. S. Bates and G. H. Fredrickson, *Phys. Today*, 1999, **52**, 32–38.
2 T. P. Lodge, *Macromol. Chem. Phys.*, 2003, **204**, 265–273.
3 I. W. Hamley, *The Physics of Block Copolymers*, Oxford University Press, Oxford, 1998.
4 M. W. Matsen and F. S. Bates, *Macromolecules*, 1996, **29**, 1091–1098.
5 J. Bang, S. Jain, Z. Li, T. P. Lodge, J. S. Pedersen, E. Kesselman and Y. Talmon, in preparation.
6 Z. L. Zhou, Z. B. Li, Y. Ren, M. A. Hillmyer and T. P. Lodge, *J. Am. Chem. Soc.*, 2003, **125**, 10182–10183.
7 T. P. Lodge, M. A. Hillmyer, Z. L. Zhou and Y. Talmon, *Macromolecules*, 2004, **37**, 6680–6682.
8 A. N. Semenov, I. A. Nyrkova and A. R. Khokhlov, *Macromolecules*, 1995, **28**, 7491–7500.
9 I. A. Nyrkova, A. R. Khokhlov and M. Doi, *Macromolecules*, 1993, **26**, 3601–3610.
10 H. Iatrou and N. Hadjichristidis, *Macromolecules*, 1992, **25**, 4649–4651.
11 Z. Li, E. Kesselman, Y. Talmon, M. A. Hillmyer and T. P. Lodge, *Science (Washington, D. C.)*, 2004, **306**, 98–101.
12 Z. Li, M. A. Hillmyer and T. P. Lodge, *Macromolecules*, 2004, in press.
13 G. A. McConnell, A. P. Gast, J. S. Huang and S. D. Smith, *Phys. Rev. Lett.*, 1993, **71**, 2102–2105.
14 G. A. McConnell and A. P. Gast, *Macromolecules*, 1997, **30**, 435.
15 I. W. Hamley, C. Daniel, W. Mingvanish, S.-M. Mai, C. Booth, L. Messe and A. J. Ryan, *Langmuir*, 2000, **16**, 2508–2514.
16 I. W. Hamley, J. A. Pople and O. Diat, *Colloid Polym. Sci.*, 1998, **276**, 446.
17 J. Bang, T. P. Lodge, X. H. Wang, K. L. Brinker and W. R. Burghardt, *Phys. Rev. Lett.*, 2002, **8921**, 18.
18 J. Bang and T. P. Lodge, *J. Phys. Chem. B*, 2003, **107**, 12071–12081.
19 T. P. Lodge, B. Pudil and K. J. Hanley, *Macromolecules*, 2002, **35**, 4707–4717.
20 Y. Ren, T. P. Lodge and M. A. Hillmyer, *Macromolecules*, 2001, **34**, 4780–4787.
21 J. N. Israelachvili, *Intermolecular and Surface Forces*, 1991.
22 Y.-Y. Won, A. K. Brannan, H. T. Davis and F. S. Bates, *J. Phys. Chem. B*, 2002, **106**, 3354–3364.
23 Y.-Y. Won, H. T. Davis and F. S. Bates, *Science (Washington, D. C.)*, 1999, **283**, 960–963.
24 S. Jain and F. S. Bates, *Science (Washington, D. C.)*, 2003, **300**, 460–464.
25 S. Jain and F. S. Bates, *Macromolecules*, 2004, **37**, 1511–1523.
26 H. Shen and A. Eisenberg, *Macromolecules*, 2000, **33**, 2561–2572.
27 H. Shen and A. Eisenberg, *J. Phys. Chem. B*, 1999, **103**, 9473–9487.
28 L. Zhang and A. Eisenberg, *Science (Washington, D. C.)*, 1995, **268**, 1728–1731.
29 L. Zhang and A. Eisenberg, *Macromolecules*, 1999, **32**, 2239–2249.
30 P. A. Canham, T. P. Lally, C. Price and R. B. Stubbersfield, *J. Chem. Soc., Faraday Trans. I.*, 1980, **76**, 1857–1867.
31 J. S. Pedersen, I. W. Hamley, C. Y. Ryu and T. P. Lodge, *Macromolecules*, 2000, **33**, 542–550.
32 I. LaRue, M. Adam, M. da Silva, S. S. Sheiko and M. Rubinstein, *Macromolecules*, 2004, **37**, 5002–5005.
33 R. Borsali, E. Minatti, J.-L. Putaux, M. Schappacher, A. Deffieux, P. Viville, R. Lazzaroni and T. Narayanan, *Langmuir*, 2003, **19**, 6–9.
34 J.-L. Putaux, E. Minatti, C. Lefebvre, R. Borsali, M. Schappacher and A. Deffieux, *Faraday Discuss.*, 2004, **128**, DOI: 10.1039/b403157a.
35 K. J. Hanley, PhD Thesis, University of Minnesota, 2001.
36 J. S. Pedersen and M. C. Gerstenberg, *Macromolecules*, 1996, **29**, 1363.
37 J. S. Pedersen, *J. Appl. Crystallogr.*, 2000, **33**, 637–640.
38 Y. Ren, T. P. Lodge and M. A. Hillmyer, *Macromolecules*, 2002, **35**, 3889–3894.
39 A. Halperin, M. Tirrell and T. P. Lodge, *Adv. Polym. Sci.*, 1991, **100**, 33.
40 E. E. Dormidontova and A. R. Khokhlov, *Macromolecules*, 1997, **30**, 1980–1991.
41 T. P. Lodge, K. J. Hanley, B. Pudil and V. Alahapperuma, *Macromolecules*, 2003, **36**, 816–822.
42 K. J. Hanley, T. P. Lodge and C.-I. Huang, *Macromolecules*, 2000, **33**, 5918.
43 I. W. Hamley, J. A. Pople and O. Diat, *Colloid Polym. Sci*, 1998, **276**, 446.
44 J. Bang, K. Viswanathan, T. P. Lodge, M. J. Park and K. Char, *J. Chem. Phys.*, 2004, in press.
45 T. P. Lodge, J. Bang, M. J. Park and K. Char, *Phys. Rev. Lett.*, 2004, **92**, 145501.
46 M. Watzlawek, C. N. Likos and H. Löwen, *Phys. Rev. Lett.*, 1999, **82**, 5289.
47 E. Stiakakis, D. Vlassopoulos, C. N. Likos, J. Roovers and G. Meier, *Phys. Rev. Lett.*, 2002, **89**, 208302.

Aggregation across the length-scales in β-lactoglobulin

Elizabeth H. C. Bromley,* Mark R. H. Krebs and Athene M. Donald

Cavendish Laboratory, University of Cambridge, Madingley Road, Cambridge CB3 0HE. E-mail: ehca2@phy.cam.ac.uk

Received 1st March 2004, Accepted 20th April 2004
First published as an Advance Article on the web 14th September 2004

The protein β-lactoglobulin (BLG) has been widely studied, in large part because of its importance to the food industry. Following denaturation during heating, under different conditions of pH it has been found to form either particulate (around the isoelectric point at pH 5.1) or fibrillar gels. The nature of the fibrils has recently been suggested to be the same as that identified with amyloid fibrils known for a wide-range of different proteins and implicated in many disease states. We confirm that the BLG fibrils show all the classical signatures of amyloid fibrils. In addition, the fibrils are capable themselves of aggregating further to form large-scale (many microns in size) spherulites. Polarized light microscopy and Environmental scanning electron microscopy (ESEM) have been used to explore the internal structure of these spherulites under conditions in which the solvent has not been dried off. The factors which determine whether or not the spherulites form have also been considered, together with implications for other amyloid-containing systems.

Introduction

Bovine β-lactoglobulin is a key member of the lipocalin family, which has been studied extensively with a wide range of techniques, both biochemical and physical.[1] There are two reasons for this level of interest. Firstly it is a protein of great commercial importance to the milk industry, which is concerned with improving the processing of whole milk and milk derivatives as well as searching for new uses for the protein.[2] As much of milk processing involves heat treatment, the behaviour of BLG upon heating is of particular interest as the aggregation properties of the protein are widely thought to be responsible for fouling and loss of efficiency in heat exchangers. Secondly it is a robust protein, which can be isolated easily and as such has become a standard protein for the testing of biochemical and physical techniques.

BLG gels fall into two broad categories: fine stranded gels composed of strands of various lengths (crossing the nanometer and micrometer ranges) and diameters (in the nanometer range), and particulate gels composed of much coarser particles (of micrometer range diameters). Fine stranded gels are found away from the isoelectric point and particulate gels near to the isoelectric point.[3] In between the two regimes is a region in which inhomogeneous gels (pH 5.8–7.5, the range depending on heating rate) are found.[4] Over this pH range the charge on each individual molecule varies from zero at the isoelectric point of pH 5.1 to around +20 at pH 2. This change in the net charge per molecule leads to variation in the amount of oligomerisation present in solution, with large net charges leading to an equilibrium shift towards the monomer.[5] There is a noticeable correlation between the equilibrium state of oligomerisation as a function of pH and the formation of particulate and fine stranded gels. Fine stranded gels form under conditions in which the monomer is preferred and particulate gels under conditions where the dimer (and higher oligomers) are favoured. This leads to the suggestion that particulate gels form directly from the dimer upwards whereas fine stranded gels are built *via* dissociation into the monomer.[6]

This paper is mainly focused on the behaviour of BLG in the pH range which leads to fine stranded gels, which has been studied extensively in recent years with a variety of techniques.

At the molecular level, the protein does not appear to undergo a major structural transformation on aggregation as seen using WAXS, however a slight increase in the level of structured inter-molecular β-sheet was found from FTIR.[7] Fine stranded gels formed at pH 2.5 have been found to be held together entirely by non-covalent bonds, indicating that the inter-molecular disulfide bonding, which may occur in the isoelectric aggregation process, is not significant in this regime.[8]

Scattering patterns taken from fine stranded samples at pH 2 show a fractal dimension of one (corresponding to rigid rods) for low ionic strengths.[9,10] As the ionic strength is increased the fractal dimension rises to around two (corresponding to random coil). These morphological assignments are supported by TEM images which show long straight aggregates at low ionic strength and much shorter more flexible fibres at high ionic strength.[7] Generally, at pH 2 the strands are found, from scattering experiments, to have the same mass per unit length as a chain of monomers, but a slightly larger radius.[9]

The mesoscopic gel structures formed at pH 7, while still sometimes classified as fine stranded, are much coarser (15 nm strands rather than 2 nm for pH 2) and particulate-like.[3,10] An intermediate aggregate state has been found at this pH, which consists of particles containing around 85 monomers.[11] This is proposed to be the building block of the larger fractal aggregates, with a fractal dimension of 2, formed under these conditions. These intermediate state particles have been imaged using AFM.[12] Whereas the disulfide linked intermediate particles are produced at a faster rate as the pH is increased from pH 6.7 to 8.7, the reverse is true for the association of the particles into larger scale structures.[13] This implies that disulfide bonding does not play a key role from the intermediate particle formation upwards.[13]

In the neutral pH case, the aggregation kinetics can be modelled using two steps, first an increase in the number of intermediate aggregate particles, followed by a step in which these particles themselves aggregate into strands.[14] This secondary aggregation begins at a critical concentration of intermediate particles which decreases with increasing ionic strength.

The formation of fibrillar structures may not be the end point of the aggregation process. The strands may further associate either non-specifically into random cross-linked gels or in a more ordered fashion into spherulites as described below. The fine stranded gels have been investigated extensively both in terms of their kinetics of formation[8,9,15] and their rheological properties.[15–17]

Higher order semi-crystalline aggregates in the form of spherulites have been observed in BLG systems.[17] The process and mechanisms of spherulite formation have most extensively been studied in synthetic polymer systems such as polyethylene.[18] In such a system lamellar crystals, in which the polymer chain is folded such that the backbone is oriented along the smallest dimension of the crystal, form out of the melt and assemble into a structure resembling a wheat sheaf. The lamellae are prevented from lying parallel to one another by the presence of cilia which are parts of individual polymer chains, not contained within the lamellae. This leads to a preferential angle between adjacent lamellae which, when extended outwards, creates a spherical shape in which the lamellae are aligned radially and the polymer chains are aligned tangentially.[18] Under a polarizing microscope, spherulites show a typical Maltese cross pattern of light extinction which is caused by the difference in refractive index between the axes in the plane of the lamellae and the perpendicular axis. Higher order semi-crystalline assemblies have also been observed in naturally occurring polymer systems including cellulose,[19] chitin[20] and amylose.[21]

Recently a connection between the fine stranded gels formed by BLG and the amyloid fibril form of protein aggregation has begun to emerge.[22,23] Amyloid fibrils are a form of aggregated protein of great medical significance, with its appearance in the form of amyloid plaques characterizing various pathological conditions including Alzheimer's disease and transmissible spongiform encephalopathies.[24] Traditionally the term amyloid fibril only applies to protein aggregates found *in vivo*. However it has recently become apparent that many non-disease related proteins can form the same fibrillar structure *in vitro*[25,26] and it is now commonplace to use the term for any structure exhibiting certain characteristic properties.

The defining characteristics of amyloid fibrils are as follows.[27] They have a predominantly cross β-sheet (antiparallel) structure which has characteristic distances between neighbouring β-strands. This is detected using X-ray diffraction, which will show bands at 4.7–4.8 Å and 10–13 Å. These two distances are perpendicular to one another within the structure. In a sufficiently oriented sample the X-ray pattern will therefore show a 90 degree shift between the brightest regions of the two bands. The dominant secondary structure is β-sheet even if the native state of the fibril forming protein is predominantly α-helical. This can be detected by any of the methods which are sensitive to secondary structure including FTIR, CD and Raman spectroscopy.

Fibrils bind the amyloid specific dye congo red (CR), causing a red shift and increase in absorption in the circular dichroism (CD) spectrum of the dye as well as a shift in the wavelength of the absorption peak. Fibrils also cause a large increase in the fluorescence intensity of Thioflavin T (ThT). Observation using TEM (transmission electron microscopy) or AFM (atomic force microscopy) should result in long, straight (but sometimes twisted), unbranching fibrils with diameters in the region of a few nanometres.

The existing evidence that fine stranded BLG gels may be composed of amyloid fibrils consists mostly of TEM images which show a striking resemblance to those taken of fibrils from known amyloidogenic proteins.[7] It has also been shown that BLG solutions incubated at pH 7, 37 °C in the presence of 3–5 M urea bind ThT.[28]

Spherulite formation has also been seen in other amyloid fibril forming systems both *in vitro*, including a pathogenic immunoglobulin light chain[29] and bovine insulin,[30] and *in vivo*, in the form of amyloid plaques which are associated with Alzheimer's disease[31] and in amyloid tumours.[32]

This current work aims to investigate two aspects of structure formation and aggregation by *β*-lactoglobulin. Firstly it is not yet clear how complete the correspondence is between the fine stranded gels formed over a wide range of pH and the amyloid fibril structure. The present work addresses this issue by testing gels formed *via* heating for the characteristics of the amyloid structure over a wide range of pH. All of the gels were tested for binding with Congo Red and Thioflavin T and specific samples were examined using X-ray fibre diffraction. Secondly the factors influencing the formation of spherulites, and the structure of these spherulites, will be investigated.

Experimental techniques

The protein used in this work was bovine β-lactoglobulin obtained from Sigma (product number L3908) lot number 119H7008. The protein is 90% pure and contains a mixture of genetic types A and B. The protein was used without further purification. The main components found by mass spectrometry had molecular weights of 18282.8 Da and 18368.4 Da which correspond to the main chain weights of the B and A genetic variants respectively. All other chemicals used were of analytical grade.

Congo Red (CR) is the dye traditionally used to characterise *in vivo* amyloid formation. On binding to fibrils, the dye undergoes a shift in its absorption spectra, with the usual maximum at around 480 nm moving to 490 nm and a shoulder developing at 540 nm.[33]

The binding of CR to samples was measured by first recording the absorption spectra of 995 μL of the buffer (pH 7.4, 150 mM NaCl, 10 mM Na-phosphate) then re-recording after adding 5 μL of filtered (0.2 μm filter) 1 mM CR solution, dissolved in the same buffer and mixed. The spectrum resulting from 985 μL of buffer with 10 μL of protein sample added was recorded and finally the spectrum after addition and mixing of 5 μL of CR. The spectra were recorded using a Cary 300 Bio UV-Visible Spectrophotometer.

On binding to amyloid fibrils, Thioflavin T undergoes a large increase in fluorescence, which can be used to confirm the presence of amyloid and to follow the kinetics of formation.[34] ThT stock solution was prepared by dissolving ThT (2.5 mM) in buffer (pH 7.4 150 mM NaCl, 10 mM Na^+-phosphate) and filtering through a 0.2 μm filter. The stock was then diluted 50-fold in the same buffer. 5 μL of protein sample was added to 995 μL of the diluted stock and stirred in the dark for 5 min to allow time for the binding to occur. The fluorescence was measured by exciting at 440 nm and monitoring the subsequent emission at 482 nm over a period of a minute. The fluorescence of the diluted stock solution and the native protein was also recorded. The data were recorded using a Cary Eclipse Fluorescence Spectrophotometer.

Wide angle X-ray patterns were collected for dehydrated, low pH gels which had been dried into fibres. X-ray stalks were formed using a stretch drying method where droplets of solution are suspended between capillary tubes sealed with wax. As the solution dries it both concentrates and is stretched between the ends of the tubes (which are a few millimetres apart). This can encourage alignment of fibres within the sample. The data was collected at the Biochemistry Department, Cambridge using a Rigaku Cu Kα rotating anode source, with a R-Axis IV image-plate X-ray detector. The wavelength was 1.55 Å and the sample detector distance was 300 mm. The images were interpreted using the Bragg law with the standard calibration performed using the Java software *Crystal Clear*.

Optical microscopy was performed using a Zeiss Axioplan optical microscope. The polarizer and analyser were in a fixed position (east–west and north–south on the image shown in Fig. 7 later) and a 551 nm full wave retardation plate was inserted at a 45° angle to the polarisors with the slow axis (high refractive index) running north–east to south–west.

Environmental scanning electron microscopy (ESEM) images were obtained by placing aliquots of incubated solutions on a copper stub inside the chamber of an FEI 2010 ESEM. The samples were left to equilibrate at 2 °C, maintained by a Peltier device under the stub, before being sealed in the chamber and the pressure reduced to around 6 Torr. The air in the chamber was then replaced with water vapour by cycling between 6 and 9 Torr. The pressure is then reduced to around 5 Torr, at which point small variations in pressure will cause the addition or removal of excess water from the sample surface without causing the loss of hydration in the bulk of the sample.

Results

Binding results

Samples were examined for binding of the amyloid specific dye Congo Red (CR). The absorption spectra of the buffer and the CR mixed with the buffer were taken and revealed the expected peak in absorption at 490 nm. For each sample the background spectrum resulting from the sample mixed with buffer before the addition of the CR was subtracted from the spectrum taken after the CR was added.

The spectra fall into two categories, those which exhibit a significant shift of the peak to longer wavelengths (a larger shift than the 6–7 nm shift of the native protein) and those that do not. Fig. 1 shows the dependence of the position of the absorption maximum of each spectrum on the pH for BLG samples formed after heating at 60 °C for 24 h. In this case we would categorize samples with pH values between 4.05 and 5.55 inclusive as well as the native sample as not binding CR, and

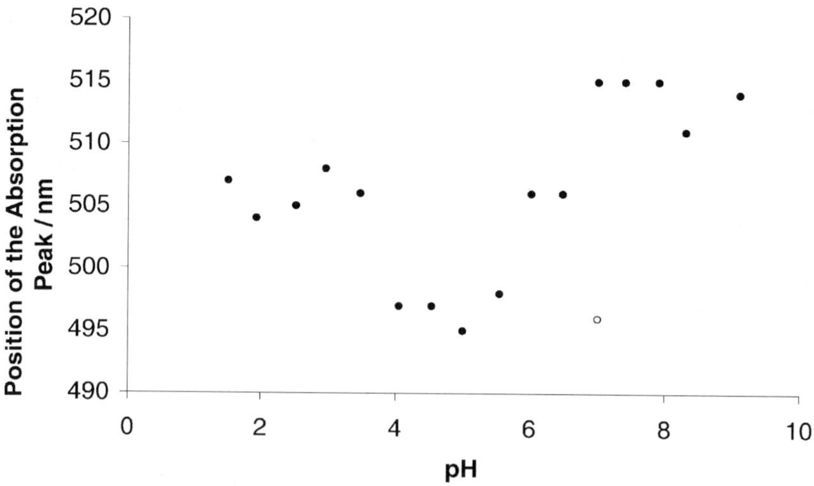

Fig. 1 This graph shows the position of the maximum of the CR absorption spectrum as a function of the pH of each sample (●). The peak in the absorption spectrum for the native protein sample is shown for comparison (○).

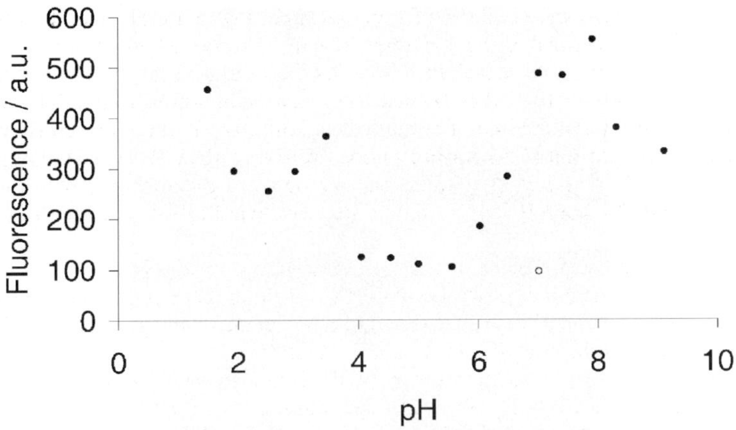

Fig. 2 This graph shows the fluorescence of ThT on samples heated at 60 °C, with 4% w/w protein and no NaCl as a function of the solution pH conditions (●). The fluorescence of the native sample is shown for comparison. All samples show more fluorescence than the native sample (○), but the increase is very small for the range between pH 4.05 and 5.55.

therefore not exhibiting amyloid structure. The rest of the pH range has bound CR and therefore is positively identified with the amyloid structure.

In order to build on the identifications made by CR, the ThT fluorescence was also measured as shown in Fig. 2. This graph shows a significant increase in fluorescence for all samples except those in the 4.05–5.55 pH range confirming the identifications made by the CR assay.

The influence of other parameters was also tested including varying salt concentration up to 0.1 M NaCl, altering concentration between 1 and 10% w/w and varying the holding temperature between 60 and 80 °C (all non-varying parameters were kept at 0 NaCl, 4% w/w, 60 °C and pH 2.5). All of these samples showed positive binding of CR and ThT (data not shown).

These results suggest that the CR and ThT binding structure is a default structure which will form on heating except under circumstances where less ordered non-specific aggregation can take place. This non-specific aggregation, which leads to the formation of particulate gels, occurs near the isoelectric point where the charge per molecule, and hence the inter-molecular repulsive force, is less significant. The aggregation in this regime is thought of as less specific because there is no long range order present within the particles.

X-ray fibre diffraction

X-ray patterns were collected for samples formed by incubating 4% w/w samples at pH 2.5 at 60 °C for 24 h in the absence of NaCl and in the presence of 50 mM NaCl. Figs. 3 and 4 show the X-ray diffraction patterns. Both patterns show bright rings corresponding to 4.6 Å and 10.6 Å. The pH 2.5, 50 mM NaCl sample also shows a third ring at 29 Å. The rings in the pH 2.5–0 NaCl sample pattern are different in that they exhibit visible anisotropy. The ring at 29 Å is not usually present in amyloid X-ray patterns and is very sharp which may indicate it is caused by a contaminant. A pH 7–0 NaCl sample X-ray pattern was also obtained (data not shown) which was very similar to the pattern found for the pH 2.5–50 mM NaCl sample both in terms of the ring positions and the isotropy of the rings.

The pH 2.5–no NaCl and 50 mM NaCl sample patterns were analysed by determining the intensity as a function of the azimuthal angle for each of the rings. The intensity as a function of azimuthal angle is shown in Figs. 5 and 6. The X-ray pattern for the pH 2.5–no NaCl sample demonstrates the presence of the two characteristic distances 4.6 Å and 10.6 Å (inter-strand and inter-sheet) and the position of the maximum intensity within these rings demonstrates that these two distances are perpendicular to one another. This strongly supports the assignment of the structure as amyloid. The lack of anisotropy in the 50 mM NaCl sample has two possible causes. Firstly the orientational order may not be present within the fibres formed under these conditions.

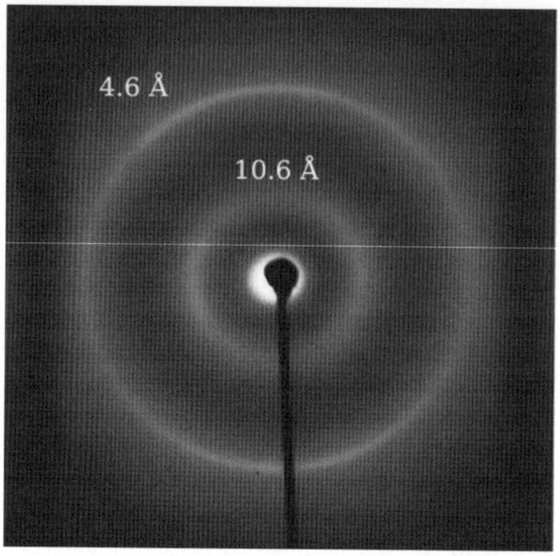

Fig. 3 This pattern is from a pH 2.5 sample with no NaCl. Only two rings are visible and they are anisotropic and at 90° to one another.

This would imply that the fibre structure does not have the same underlying symmetry as that of the amyloid structure. Secondly it may be that the structures are the same but that the fibres do not align on drying. This would also remove the anisotropy from the WAXS pattern.

In summary, the results in this section have shown that BLG positively binds the amyloid indicating dyes CR and ThT over the pH range in which fine stranded gels are formed. However, the more specific structural information revealed by WAXS only confirms the positive result for the pH 2.5 no NaCl sample.

Spherulite formation

The propensity for spherulite formation was investigated using optical microscopy over the pH range tested for amyloid specific dye binding, with a protein concentration of 4% w/w in the absence of NaCl and samples being heated at 60 °C for 24 h. Spherulites were found in varying number densities for all pH samples except at the extreme of pH 9.1, and the near isoelectric pH samples of 6.5 and 6.0. The other near isoelectric samples (pH 5.55 down to pH 4.05) showed very

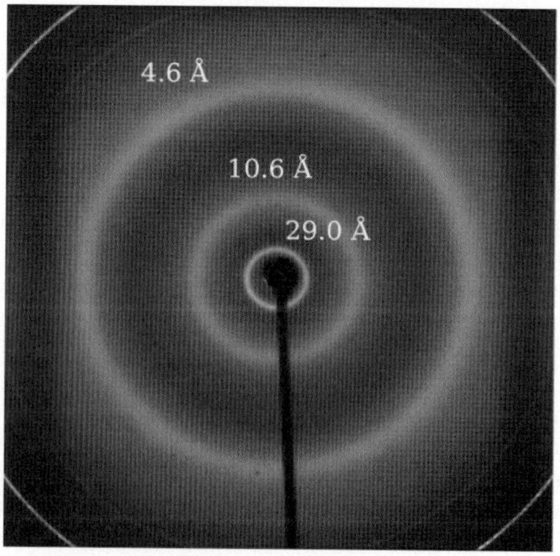

Fig. 4 This is a pattern from a pH 2.5, 50 mM NaCl sample. Once again there are three isotropic rings visible.

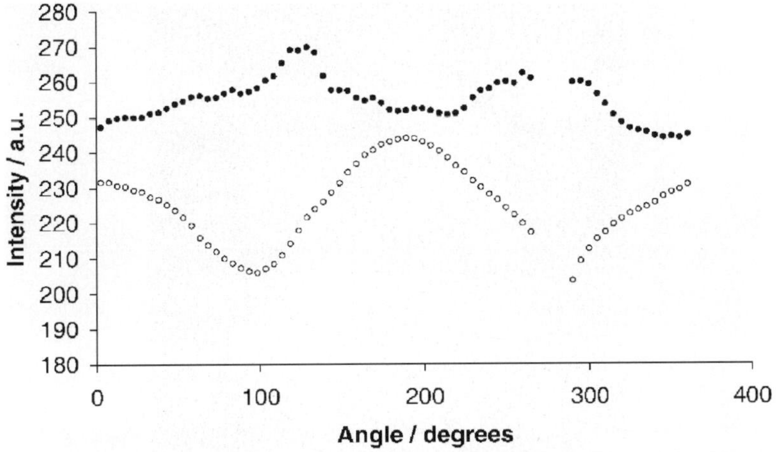

Fig. 5 A graph of the intensity of the β-sheet bands as a function of angle for the pH 2.5 no NaCl sample (10.6 Å ring (○) and 4.6 Å ring (●)). Here there are two peaks in each band and they are 90° out of phase with each other. This indicates the anisotropy expected for an amyloid structure.

little spherulite formation. This is the pH region in which amyloid formation as indicated by the binding of CR and ThT was not found, suggesting that it is the amyloid like structure which assembles into spherulites. Although the data was not collected quantitatively, the highest number densities of spherulites were seen at around pH 7.5 and at the low pH extreme of pH 1.5 (data not shown). Further experiments showed no spherulite formation at incubation temperatures higher than 70 °C or NaCl concentrations higher than 50 mM.

Spherulites formed at pH 2.5 were examined using polarised light microscopy. Under crossed polarizing conditions the spherulites appear as circular objects exhibiting a Maltese cross pattern. Rotation of the polarizer and analyser leads to a rotation of the regions of extinction in the same sense. The addition of a 551 nm wave-plate highlights regions of the sample where the fast (low refractive index) direction of the wave-plate is aligned with the fast direction of the sample in blue, and regions where the two axes are aligned at right angles appears orange. Fig. 7 shows a photograph of a spherulite under these conditions with the relevant colours and axis shown. All of the spherulites seen have this same basic appearance. It should also be noted that non-spherulitic

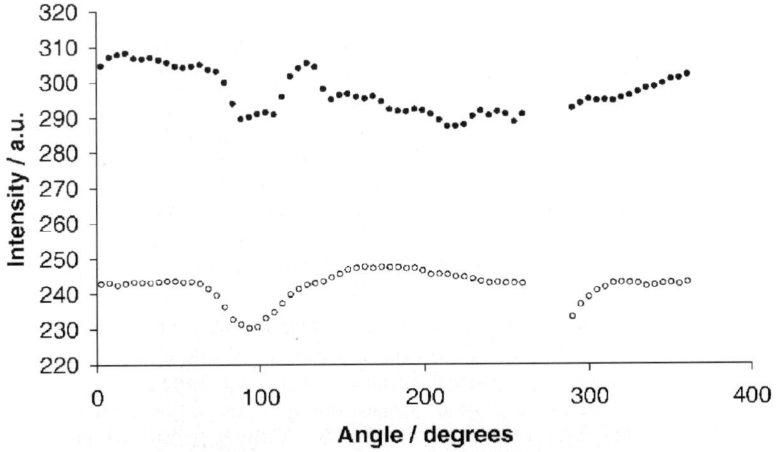

Fig. 6 A graph showing how the intensity of the bands varies as a function of angle for the β-sheet bands for the pH 2.5 50 mM NaCl sample (10.6 Å ring (○) and 4.6 Å ring (●)). There is no obvious anisotropy. Where the signal does fluctuate it does so in the same place for both bands. This is likely to be due to shadowing.

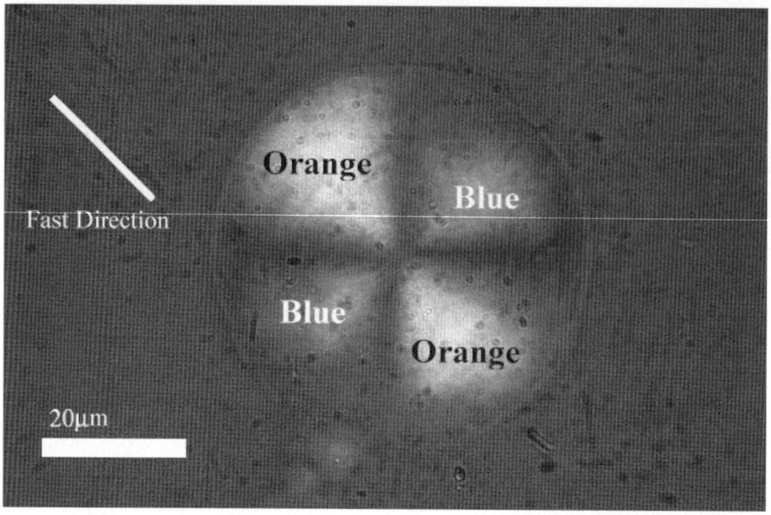

Fig. 7 A digital image taken under crossed polarising conditions with the addition of a 551 nm waveplate, of a BLG spherulite formed at pH 2.5 in the absence of NaCl by incubation at 60 °C for 24 h. The low refractive index (fast) direction of the waveplate is indicated along with the colour of the segments of the spherulite.

areas of the sample, which may contain fibrils which have not assembled into spherulites, are not visibly birefringent. Previous work has indicated that the higher refractive index (slow) direction for an amyloid fibril is along its long axis.[35] Fig. 7 shows that the slow direction within the spherulite is the radial direction and therefore indicates that the fibrils lie radially within the spherulites. Spherulites formed under similar conditions using bovine insulin are found to exhibit the same optical response.[30]

The spherulites from the same sample were also examined in the ESEM. The number density of spherulites found in these samples was small in comparison with samples prepared under similar conditions using insulin.[30] This results in a large amount of protein in the system which is either unaggregated or has formed isolated fibrils. This protein forms an obscuring film when water is removed from the sample in order to allow imaging. In order to generate images of the BLG spherulites it was therefore necessary to wash the sample after it had been deposited on the sample stub. This was done by adding a drop of water and then removing the excess *via* absorption onto filter paper. Fig. 8 shows the result of this procedure. The film of unaggregated protein and fibrils can be seen in the left of the picture. However sufficient material has been removed such that the film does not obscure the entire sample leaving the surface of a spherulite clearly exposed. Fig. 9, show a higher magnification image of this surface. The surface appears to have texture on the length scale of a few microns but shows no obvious orientational order.

The spherulites seen here do not appear to demonstrate the radial internal order which was implied by the light microscopy. However, these micrographs only show the surface of the spherulites. The internal structure of the spherulites can occasionally be observed by deliberately inducing fracture *via* fast dehydration (see Figs. 10 and 11). In both cases the irregular texture of the surface appears to be localized to the surface with the internal structure being far more ordered. What features are visible in this internal region appear to indicate that the sample has cracked along radial planes.

The spherulites formed in these experiments have been shown to have a structure consistent with a central region constructed from radially aligned amyloid fibrils, and a surface region which is not aligned although it is still thought to contain amyloid fibrils. The appearance of spherulites under neutral pH conditions is a further argument suggesting that the amyloid structure is also found under these conditions. The lack of anisotropy (in the X-ray pattern) which cast the amyloid assignment into doubt cannot be the result of random alignment of the molecules in fibres because when these fibrils are aligned themselves, within the structure of a spherulite, they exhibit birefringence.

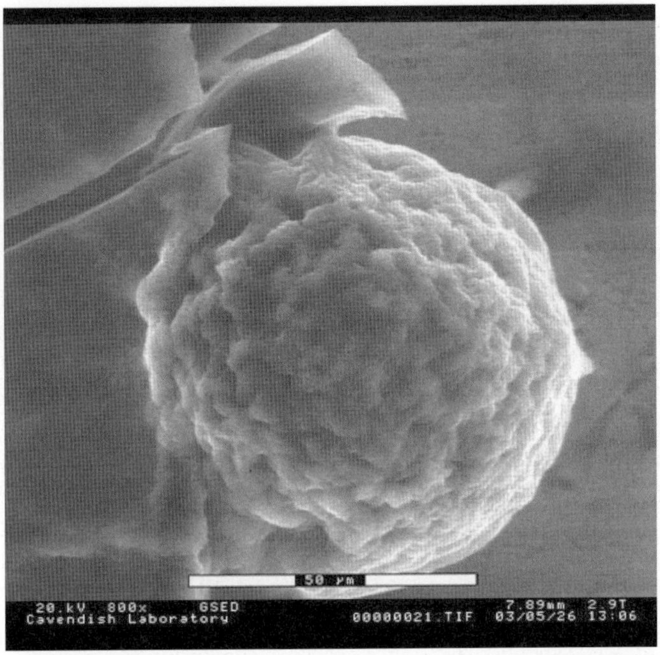

Fig. 8 ESEM image of a 'washed' sample (scale bar is 50 µm). Repeated wetting of the sample removes some of the unaggregated protein from the system, making imaging of a spherulite without interference more likely. This image shows a spherulite caught at the edge of the receding film. The surface is now much more visible.

Once formed, BLG spherulites are stable entities which are not destroyed either by dilution of the samples for imaging or by the modest shear forces involved in pipetting the sample. This suggests that under these experimental conditions the spherulitic structure is not reversible. Although a

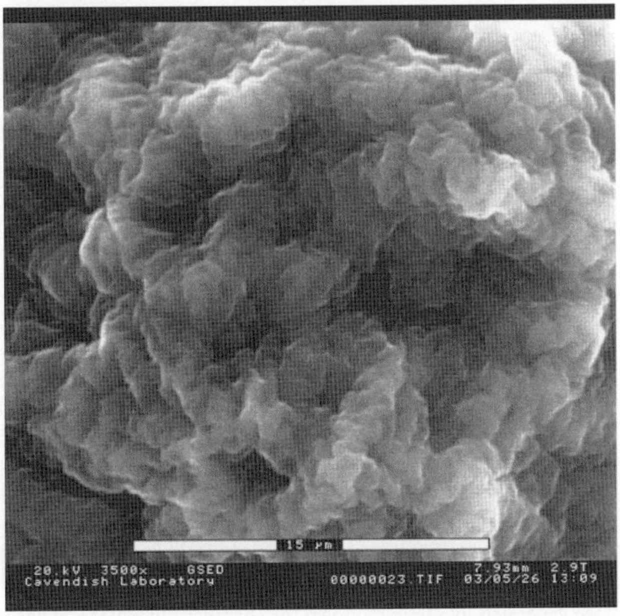

Fig. 9 A higher magnification image showing the rough surface of the spherulite (scale bar is 15 µm). This surface texture has a dominant length scale of about 2–3 µm.

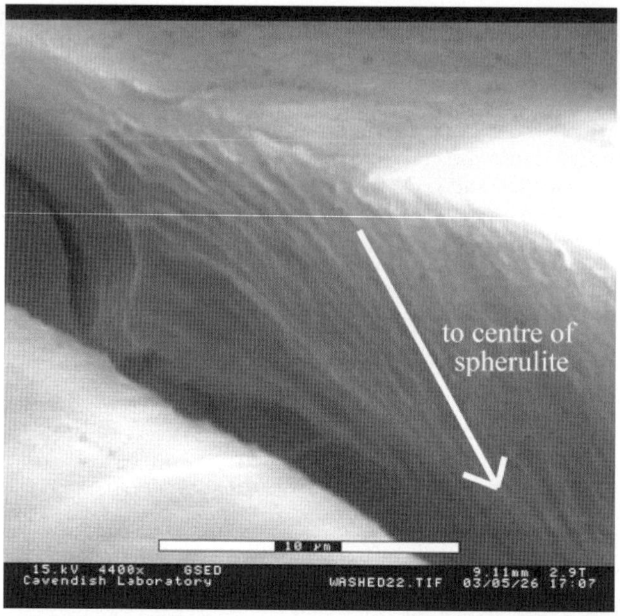

Fig. 10 This micrograph was taken of an unwashed sample where the stresses generated in the film have created a crack which has run through a spherulite (scale bar is 10 μm). Radially aligned texture is clearly visible.

phase diagram describing conditions under which either the random gel structure or the spherulitic structure are produced on heating, once formed there is no way to move the system from one region to another. This lack of reversibility also implies that the spherulitic structure may appear for kinetic rather than thermodynamic reasons and may not in fact be the equilibrium structure at all. It appears that BLG spherulites have more in common with the spherulites formed by synthetic

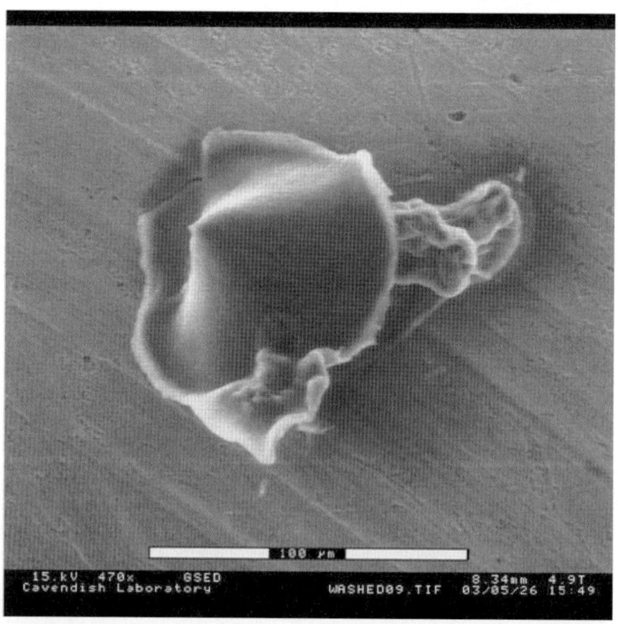

Fig. 11 An ESEM image generated by pumping down the chamber quickly (scale bar is 100 μm). The skin of the spherulites is thought to contract on drying and hence a fast drying rate can cause them to explode. Once again, radial texture is apparent and can be seen here to be in contrast with the surface layer.

polymers from the melt than the disclinations formed by liquid crystal systems suggested by some workers.[36] Firstly, spherulites formed by quenching synthetic polymers from the melt are kinetically trapped rather than thermodynamically stable. Secondly, the lamellae formed in synthetic systems have some similarities of structure with that of amyloid fibrils, with the main chain running across the length of both structures. There is also a strong possibility that the regions of the protein chain which are not incorporated into the main extended β-sheet core of the fibrils may perform the same role as the cilia in the lamellae and introduce a preferred splay angle between fibrils.

It is currently unknown whether the fibrils that compose the spherulites form individually and then associate once mature or grow *in situ*. Either way, the random gel structure is achieved by samples in which individual fibrils are stable against association and the spherulitic state by samples whose fibrils are not sufficiently stable. Therefore in order to understand the process of spherulite formation, the competing process of isolated fibril formation must also be considered.

Fibril formation kinetics

Both increasing NaCl concentration and increasing incubation temperature were found to inhibit spherulite formation. However both of these factors may also influence the production of fibrils. The effect of temperature and NaCl concentration on the kinetics of fibril formation was therefore investigated using ThT binding as an indication of the total quantity of protein engaged in fibrillar structures.

Samples containing 4% w/w protein and either 0 or 50 mM NaCl at pH 2.5 were incubated at 70 °C. This incubation temperature was chosen for two reasons. Firstly the measurement of fibril formation will be less accurate if spherulites are actually present in large quantities in solution due to interference with the dye binding, and secondly, the higher temperature encourages fibrillation to proceed in a time scale compatible with measurement, which at the lower temperature of 60 °C takes several days. 70 °C was chosen as a balance between faster kinetics, and minimal higher order structure formation and the likelihood of still finding a key difference between the kinetics in 0 and 50 mM NaCl samples.

Fig. 12 shows the fluorescence of ThT as a function of time for a 4% w/w pH 2.5–0 NaCl sample and a 4% w/w pH 2.5–50 mM NaCl sample heated at 70 °C. The 0 NaCl curve shows the generic features of amyloid formation kinetics. The fluorescence present at the beginning is taken as the background level, and the system spends a 'lag-time' at this level. There is then a sharp increase in fluorescence which levels off once the free protein is depleted. In this case the lag-time is around 250 min. The 50 mM NaCl sample shows no lag-time, instead fluorescence increases directly from $t = 0$. The growth phase part of the curve is much less steep than the equivalent region of the 0 NaCl curve.

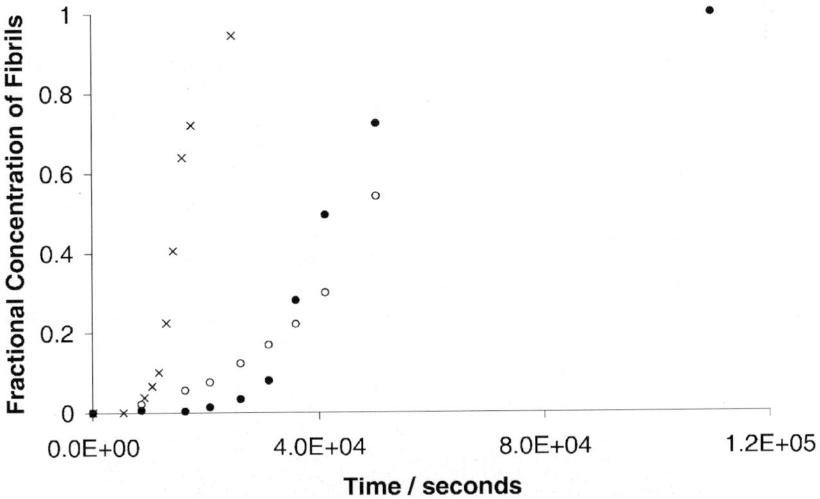

Fig. 12 A graph of ThT fluorescence as a function incubation time for a 4% w/w sample at pH 2.5 heated at 70 °C in the absence of NaCl (●), with 50 mM NaCl added (○) and a 4% w/w sample at pH 2.5 heated at 75 °C in the absence of NaCl (×).

It was found for the 4% w/w pH 2.5–0 NaCl sample, that if heating was interrupted during the lag-time, the sample only needed to be returned to 70 °C up to a total heating time of 250 min in order for fibril formation to begin. This indicates that in this case a reaction is occurring within the lag-time which leads to an actively amyloid forming species. This confirms the evidence of soluble oligomers found in many fibril forming systems[37] including BLG.[14,13] Together these facts strongly indicate that phase separation (from a uniform solution of protein into regions of high and low protein concentration) in BLG, and in amyloid systems in general, may proceed *via* a nucleation and growth based mechanism. Based on this information a model can be proposed in which the native protein is converted in a first order reaction to an 'active species'. This 'active species' then undergoes a nucleation reaction which is of order n, where n is the number of active molecules required to form a critical nucleus. Growth has been modelled as a first order reaction involving all of the protein not currently aggregated (including both native and 'active' species. If the depletion of the active state by the nucleation and growth processes is taken into account, a set of three coupled differential equations, describing active species formation, nucleus formation and growth of existing fibrils, which could not be solved analytically are generated. However, if this depletion effect is removed the equation determining the number density of the active species S simplifies to

$$\frac{\mathrm{d}S}{\mathrm{d}t} = k_1 \tag{1}$$

where k_1 is the rate constant for the reaction. The assumption that the depletion of the active species due to nucleation can be ignored is equivalent to saying that the nucleation rate is always small in comparison to the rate of the reaction towards the active species. This is not necessarily physically justified in this case, as there is no reason to suspect that the nucleation rate must be small, however the resulting model fits the data reasonably. Integrating eqn. (1) and taking the nucleation rate as being proportional to S^n we find:

$$\frac{\mathrm{d}N}{\mathrm{d}t} = k_2(k_1 t)^n. \tag{2}$$

Here k_2 is the rate constant for the nucleation reaction and N is the number density of nuclei. Integrating eqn. (2) and setting the rate of fibril formation to be the growth rate (rate constant k_3) multiplied by the number of nuclei gives,

$$\frac{\mathrm{d}F}{\mathrm{d}t} = k_3(1 - F)\left[\frac{k_2(k_1 t)^{n+1}}{k_1(n+1)}\right] \tag{3}$$

where F is the number density of protein molecules bound up in aggregates. This assumes that there is no contribution to the fluorescence of the ThT due to the nuclei themselves. This will be a reasonable approximation as long as the size of the nuclei is small in comparison to the fibrils. It has also been assumed that (unlike the nucleation step) growth may occur through the addition of any unaggregated protein molecule in the system including both native and active species. The other possibility, that growth occurs from the active state exclusively, will lead to significantly larger depletion effects and hence invalidate the assumptions already made. A more complete numerical model, which explicitly conserves protein, has been formulated and tested for both growth scenarios. The preliminary results indicate that the non-specific growth model (growth from all non-aggregated protein) gives a significantly better fit for the data then the specific growth model (data not shown). Eqn. (3) can be integrated to give the fractional concentration of fibrillar matter as a function of time,

$$F = 1 - \exp\left[-k_3 \frac{k_2(k_1 t)^{n+2}}{k_1^2(n+1)(n+2)}\right]. \tag{4}$$

The data can be re-plotted in order to give a linear relationship, by plotting $\ln[-\ln(1-F)]$ against $\ln t$. Fig. 13 shows the 0 and 50 mM NaCl heated at 70 °C data sets re-plotted in this way (it should be noted that not all of the data can be plotted in this graph because ln 0 is undefined). The gradient

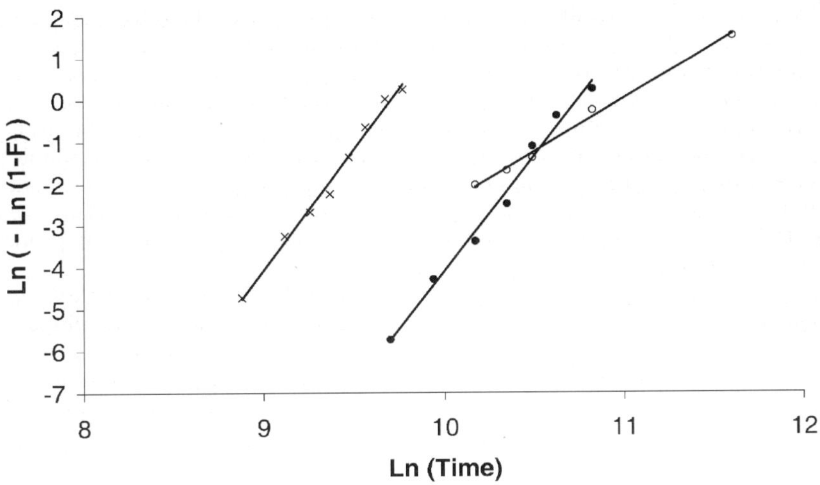

Fig. 13 A graph of $\ln[-\ln(1-F)]$ against $\ln t$ for the three data sets shown in Fig. 12 (4% w/w sample at pH 2.5 heated at 70 °C in the absence of NaCl (●), 50 mM NaCl added (○) and 4% w/w sample at pH 2.5 heated at 75 °C in the absence of NaCl (×)). The gradients of the linear fits give $n+2$.

of a linear fit on Fig. 13 gives an estimate of n. This gives $n=3.5$ for the 0 NaCl 70 °C data, $n=0.4$ for the 50 mM NaCl 70 °C data and $n=3.8$ for the 0 mM NaCl 75 °C data.

The predicted values of $n=3.5$ for the no NaCl 70 °C sample and $n=3.8$ for the no NaCl 75 °C sample are certainly reasonable and match well with the estimate of the critical nucleus size of 4 generated by the modelling of critical gel concentrations.[16] Furthermore the increase in n as the temperature increases may be interpreted as a raising of the energy of the aggregated state with respect to the monomeric state which is expected due to the increasing contribution to the free energy of entropy as the temperature rises. The value of $n=0.4$, for the 50 mM NaCl sample, is not reasonable for a critical nucleus size since it is less than unity. However the value may be interpreted as indicating that the active state is only slightly more likely to nucleate than the folded state. This system evolves with a time dependence which only just exceeds that which could be described by a nucleation and growth model that did not include an activation step (this would predict a power law exponent of 2). It is also possible that the forward reaction generating the active species is not first order.

The significantly lower value of n in the the 50 mM NaCl sample kinetics may indicate that it assembles into fibrils from the monomer upwards whereas the 0 NaCl sample requires several monomers before the fibrillar state is stable. This may explain the difference in appearance in fibrils formed under these differing conditions. Fibrils formed in higher NaCl conditions are found to be narrower, shorter and more flexible than those formed without NaCl.[15]

Therefore the model offers two possible explanations for the reduction of spherulite formation upon the addition of NaCl. Firstly, the fibrils formed are shorter and more flexible and therefore less likely to form the radially ordered structure found in the spherulites. Secondly, the maximum rate of nucleation of fibrils is much lower with the addition of NaCl.

Conclusions

This work has indicated that in the case of pH 2.5 and no added NaCl, the aggregates formed by BLG have all of the characteristics of the amyloid fibril structure. Under other conditions in which BLG forms fine stranded gels, the situation is less conclusive. The aggregates bind the amyloid indicative dyes CR and ThT, however the X-ray patterns have not been shown to demonstrate the amyloid specific symmetries which could be due to a failure in aligning the fibrils. The fact that spherulites, which require at least some of the molecular orientational order present in amyloid fibrils in order to exhibit birefringence, form across the fine stranded regime lends some weight to the argument that the amyloid structure can be formed under these conditions also.

The spherulites formed under fine stranded gel conditions were deduced from optical microscopy to be assembled from radially aligned fibrils. ESEM images of the spherulites indicated that they also have a disordered surface layer. The spherulites found in this work were irreversible aggregates with the same order as a nematic, rather than part of a reversible liquid crystal-like system.

The kinetics of fibril formation were examined using ThT binding under conditions in which the propensity for spherulite formation differed. The kinetic curves were fitted using a simple nucleation and growth model based on a growing population of the nucleating species. This model fitted the experimental curves well, giving reasonable predicted values for the critical nucleus size for the no NaCl samples. The model also offers explanations for the observed different propensities for spherulite formation.

The definite characterization of amyloid fibrils formed by BLG under at least some of the conditions described in this work, and the fibrils' ability to adopt spherulitic phases, as well as forming random gels, supports the notion that amyloid fibrils can exhibit behaviours normally associated with synthetic polymer systems.[30] The application of polymer physics to these systems has the potential to open up new ways of understanding and exploiting amyloid fibril forming systems.

Acknowledgements

We are grateful to S. Meehan for assistance with the assays in this work, and to S. Gras and B. F. Luisi for help with the WAXS experiment. Thanks go also to A. Squires and S. Rogers for helpful discussions concerning birefringence. We acknowledge the BBSRC and the EPSRC for funding.

References

1 L. Sawyer and G. Kontopidis, *Biochim. Biophys. Acta*, 2000, **1482**, 136.
2 M. Langton, A. Astrom, M. Stading and A. Hermansson, *Food Hydrocolloids*, 1997, **11**, 217.
3 A. H. Clark, C. D. Lee-Tuffnell and A. Hermansson, *Functional Properties of Food Macromolecules*, Elsevier Applied Science Publishers, 1986.
4 M. Stading, M. Langton and A. Hermansson, *Food Hydrocolloids*, 1992, **6**, 455.
5 M. Verheul, J. S. Pedersen, S. P. Roefs and K. G. de Kruif, *Biopolymers*, 1999, **49**, 11.
6 T. Lefevre and M. Subirade, *Biopolymers*, 2000, **54**, 578.
7 G. M. Kavanagh, A. H. Clark and S. B. Ross-Murphy, *Biological Macromolecules*, 2000, **28**, 41.
8 E. P. Schokker, H. Singh, D. N. Pinder and L. K. Creamer, *Int. Dairy J.*, 2000, **10**, 233.
9 P. Aymard, T. Nicolai and D. Durand, *Macromolecules*, 1999, **32**, 2543.
10 S. Takata, T. Norisuye, N. Tanaka and M. Shibayama, *Macromolecules*, 2000, **33**, 5470.
11 C. Le Bon, T. Nicolai and D. Durand, *Int. J. Food Sci. Technol.*, 1999, **34**, 451.
12 S. Ikeda and V. J. Morris, *Biomacromolecules*, 2002, **3**, 382.
13 R. Bauer, R. Carrotta, C. Rischel and L. Ogendal, *Biophys. J.*, 2000, **79**, 1030.
14 M. Verheul, S. P. Roefs and K. G. de Kruif, *J. Agric. Food Chem.*, 1998, **46**, 896.
15 G. H. Kavanagh, A. H. Clark and S. B. Ross-Murphy, *Langmuir*, 2000, **16**, 9584.
16 A. H. Clark, G. M. Kavanagh and S. B. Ross-Murphy, *Food Hydrocolloids*, 2001, **15**, 383.
17 L. M. Sagis, C. Veerman, R. Ganzevles, M. Ramaekers, S. G. Bolder and E. van der Linden, *Food Hydrocolloids*, 2002, **16**, 207.
18 D. C. Bassett, *J. Macromol. Sci. Phys.*, 2003, **B42**, 227.
19 S. Kobayashi, L. J. Hobson, J. Sakamoto, S. Kimura, J. S. abd T. Imai and T. Itoh, *Biomacromolecules*, 2000, **1**, 168.
20 S. B. Murray and A. C. Neville, *Int. J. Biol. Macromol.*, 1998, **22**, 137.
21 S. G. Ring, M. J. Miles, V. J. Morris, R. Turner and P. Colonna, *Int. J. Biol. Macromol.*, 1987, **9**, 158.
22 W. S. Gosal and S. B. Ross-Murphy, *Curr. Opin. Colloid Interface Sci.*, 2000, **5**, 188.
23 R. Carrotta, R. Bauer, R. Waninge and C. Rischel, *Protein Sci.*, 2001, **10**, 1312.
24 J. D. Sipe and A. S. Cohen, *J. Struct. Biol.*, 2000, **130**, 88.
25 C. M. Dobson, *Trends Biochem. Sci.*, 1999, **24**, 329.
26 C. M. Dobson, *Nature (London)*, 2003, **426**, 884.
27 R. Kisilevsky, *J. Struct. Biol.*, 2000, **130**, 99.
28 D. Hamada and C. M. Dobson, *Protein Sci.*, 2002, **11**, 2417.
29 R. Raffen, L. J. Dieckman, M. Szpunar, C. Wunschl, P. R. Pokkuluri, P. Dave, Wilkins, P. Stevens, X. Cai, M. Schiffer and F. J. Stevens, *Protein Sci.*, 1999, **8**, 509.
30 M. R. Krebs, C. E. MacPhee, A. F. Miller, I. Dunlop, C. M. Dobson and A. M. Donald, *Proc. Natl. Acad. Sci. U.S.A.*, 2004, submitted.

31 L. W. Jin, K. A. Claborn, M. Kurimoto, M. A. Geday, I. Maezawa, F. Sohraby, M. Estrada, W. Kaminsky and B. Kahr, *Proc. Natl. Acad. Sci. U.S.A.*, 2003, **100**, 15 294.
32 E. Acebo, M. Mayorga and J. F. Val-Bernal, *Pathology*, 1999, **31**, 8.
33 W. E. Klunk, J. W. Pettegrew and D. J. Abraham, *J. Histochem. Cytochem.*, 1989, **37**, 1273.
34 H. Naiki, K. Higuchi, M. Hosokawa and T. Takeda, *Anal. Biochem.*, 1989, **177**, 244.
35 E. J. Ambrose and A. Elliott, *Proc. R. Soc. London*, 1950, **58**, 75.
36 A. Aggeli, I. A. Nyrkova, M. Bell, R. Harding, L. Carrick, T. C. McLeish, A. N. Semenov and N. Boden, *Proc. Natl. Acad. Sci. U.S.A.*, 2001, **98**, 11 857.
37 A. J. Modler, K. Gast, G. Lutsch and G. Damaschun, *J. Mol. Biol.*, 2003, **325**, 135.

Peptide mediated formation of hierarchically organized solution and solid state polymer nanostructures

Harm-Anton Klok,*[a] Guido W. M. Vandermeulen,[b] Harald Nuhn,[a] Annette Rösler,[b] Ian W. Hamley,[c] Valeria Castelletto,[c] Hui Xu[d] and Sergei S. Sheiko[d]

[a] *École Polytechnique Fédérale de Lausanne (EPFL), Institut des Matériaux, Laboratoire des Polymères, Bâtiment MX-D, CH-1015, Lausanne, Switzerland. E-mail: harm-anton.klok@epfl.ch.; Fax: ++41 21 693 5650*

[b] *Max Planck Institute for Polymer Research Ackermannweg 10, D-55128, Mainz, Germany*

[c] *Department of Chemistry, University of Leeds, Leeds LS2 9JT, UK*

[d] *Department of Chemistry, CB# 3290 University of North Carolina at Chapel Hill, Chapel Hill NC 27599-3290, USA*

Received 16th February 2004, Accepted 2nd April 2004
First published as an Advance Article on the web 13th September 2004

Biologically-inspired peptide sequences have been explored as auxiliaries to mediate self-assembly of synthetic macromolecules into hierarchically organized solution and solid state nanostructures. Peptide sequences inspired by the coiled coil motif and "switch" peptides, which can adopt both amphiphilic α-helical and β-strand conformations, were conjugated to poly(ethylene glycol) (PEG). The solution and solid state self-assembly of these materials was investigated using a variety of spectroscopic, scattering and microscopic techniques. These experiments revealed that the folding and organization properties of the peptide sequences are retained upon conjugation of PEG and that they provide the driving force for the formation of the different nanoscale structures which were observed. The possibility of using defined peptide sequences to direct structure formation of synthetic polymers together with the potential of peptide sequences to induce a specific biological response offers interesting prospects for the development of novel self-assembled and biologically active materials.

Introduction

Block copolymers composed of two or more amorphous, chemically incompatible segments are well-known for their ability to self-assemble into a variety of phase separated nanoscale morphologies.[1] The phase behaviour of such block copolymers has been intensively investigated over the past decades and is now relatively well understood. Block copolymers are not only of scientific interest due to their fascinating phase behaviour and appealing morphologies, but also find application, *e.g.* as thermoplastic elastomers. Recently, there has been an increasing interest in block copolymers for nanotechnology applications,[2] such as lithographic masks[3] or templates for the fabrication of metallic nanostructures.[4]

The chemical incompatibility of the constituent blocks represents the major driving force for block copolymer self-assembly. For polystyrene–polyisoprene, for example, this can lead to the formation of seven different nanoscale morphologies.[5] The number of nanoscale structures and their complexity can be increased in a number of ways. One possibility is to increase the number of blocks,[6] a second option is to introduce blocks that are not amorphous but (liquid)crystalline.[7,8] Increasing the number of blocks in a block copolymer from two to three increases the number of independent system variables; for a linear ABC triblock copolymer these are three binary interaction parameters χ, two independent composition variables and three possible sequences. This larger number of system variables increases the number of possible morphologies.[6] The phase behaviour of (liquid-)crystalline–amorphous block copolymers is governed by the competition between phase separation of the constituent blocks and (liquid-)crystal formation of the rigid segments.[7,8] These competing interactions can lead to the formation of unconventional morphologies.

Despite the advances that have been made over the past years, the level of control that polymer chemists have over block copolymer structure formation is still far away from Nature's capabilities to generate complex, hierarchically organized and functional nanostructures (*i.e.* proteins) from polypeptides. An interesting strategy to further enhance control over the nanoscale structure formation of synthetic polymers would be to use peptide sequences inspired by or derived from the protein folding process as auxiliaries to mediate the self-assembly of synthetic macromolecules. Large proteins can consist of several domains, or folding motifs, which are well-defined assemblies generated from a defined number of α-helical or β-strand secondary structural units.[9,10] The discrete size and shape of these folding motifs makes the corresponding peptide sequences attractive candidates as auxiliaries to promote self-assembly of synthetic macromolecules into unprecedented hierarchically organized structures. In this contribution, the solution and solid state structure formation of two classes of hybrid block copolymers, in which peptide sequences derived from protein folding motifs are conjugated to synthetic polymers, will be discussed. In the examples that will be presented, poly(ethylene glycol) (PEG) is used as the synthetic polymer. PEG is attractive because it is water soluble and a well known biocompatible polymer, which may be advantageous in view of possible biomedical applications.[11]

Experimental

Synthesis

Peptides and block copolymers were prepared and purified as described previously.[12,13]

Circular dichroism spectroscopy

Circular dichroism (CD) spectra were recorded on a Jasco J-715 spectropolarimeter equipped with a Jasco PTC-348WI temperature-controlled cell.

Small-angle X-ray scattering (SAXS)

SAXS experiments were performed at the Synchrotron Radiation Source (SRS) at Daresbury Laboratory, Warrington, UK. Experiments were performed on beamline 2.1 configured with an X-ray wavelength $\lambda = 1.5$ Å. The data from the two-dimensional area detector were converted into one-dimensional intensity profiles by integration in a vertical triangular sector. The resulting corrected intensity curves are denoted $I(q)$. Samples were mounted in sealed 1 mm thick brass cells, with an inner spacer ring to hold liquids, sealed between mica windows. The cells were connected to a water bath for temperature control. A typical heating ramp was performed from 25 °C to 75 °C or 85 °C at a rate of 2 °C min^{-1} or 3 °C min^{-1}.

Fourier transform infrared spectroscopy (FTIR)

Transmission FTIR spectra were recorded on a Nicolet Magna IR 850 spectrometer using a triglycinesulfate (TGS) detector (resolution: 4 cm^{-1}, number of scans: 128). Spectra were recorded at room temperature. The sample and detector chamber were purged with dry air before and during the measurement. Samples were prepared by spincoating a DMF solution on a silicon wafer.

Attenuated total reflection fourier transform infrared spectroscopy (ATR-FTIR)

ATR-FTIR was performed on a Bruker Vector 22 spectrometer with a Golden Gate 531 ATR unit. Samples were slightly pressed with a sapphire against the diamond of the ATR unit to obtain a thin film. Spectra were recorded at room temperature in the range 8000–200 cm^{-1} with a resolution of 1 cm^{-1}. Typically, 16 scans were averaged for a single spectrum.

Transmission electron microscopy (TEM)

Peptide solutions were deposited onto clean glass microscopy slides. Copper electron microscopy grids (300 mesh size, carbon coated and glow discharged) were placed in contact with the peptide solutions for 1 min. Excess peptide sample was drained off the grids. The grids were left to dry overnight at room temperature. The grid was then introduced on top of a droplet of 4% uranyl acetate negative staining solution for 20 s and excess staining solution was drained off the grid before the stained sample was left to dry overnight. TEM was performed using a Philips CM10 electron microscope.

Atomic force microscopy (AFM)

AFM was performed using a MultiMode AFM with Nanoscope IIIa controller (Veeco Instruments Ltd, Santa Barbara) in tapping mode using Si tips. The image shown in Fig. 7 (see later) was obtained by spincasting a dilute peptide solution on freshly cleaved mica at room temperature and a spinning velocity of 2000 rpm. The images shown in Fig. 11 (see later) were also obtained by deposition of the peptide solution onto mica.

Results and discussion

Design and synthesis

The structures of the peptides and block copolymers that have been studied are summarized in Table 1. The synthesis of samples **1–9** is outlined in Scheme 1. The amino acid sequence of these samples is derived from a *de novo* designed coiled coil, which was reported by Hodges *et al.* in 1994.[14] Coiled coils are among the most abundant folding motifs and consist of 2–5 helices that are wrapped around each other in a superhelical fashion.[15] The primary structure of coiled coil forming proteins is characterized by a heptad periodicity, *–a–b–c–d–e–f–g–*. In this sequence, positions *a* and *d* are typically occupied by hydrophobic amino acids. The aggregation number of the coiled coil superstructure and the relative orientation of the constituent helices depend on the nature of the amino acids at positions *a* and *d*, and *e* and *g*, respectively.[16] Based on these guiding principles for the *de novo* design of coiled coils, the peptide sequences of samples **1–9** were anticipated to promote the formation of parallel, tetrameric coiled coil structures.

Samples **10–14** were prepared following a procedure described in a previous publication.[13] The amino acid sequences of these samples are based on a design reported earlier by Mutter *et al.*[17] The peptide sequences in **10–14**, however, contain an additional glycine residue (Gly, G) at the peptide's N- and C-terminus to minimize possible steric hindrance for the attachment of the PEG segments. The peptide sequences are patterned with hydrophobic and hydrophilic substituents in such a way that they can form both amphiphilic α-helices and amphiphilic β-strands. The design principle is illustrated in Fig. 1. This figure also shows that the helical structures contain one hydrophilic and one hydrophobic face, which cover approximately equal areas of the helical surface. Circular dichroism (CD) experiments indicated that depending on the pH these peptides could indeed adopt both α-helical and β-strand conformations.[17] Due to the amphiphilic character of the secondary structures, these pH-induced conformational changes may be used to modulate the solution supramolecular organization of these so-called "switch" peptides. It is interesting to note that the original peptide sequences were found to adopt predominantly or exclusively β-strand structures under a number of experimental conditions, in spite of the fact that all of the constituent α-amino acids have an intrinsic propensity to form α-helices.[18] This illustrates that the formation of peptide secondary structure is not only dictated by the intrinsic folding propensities of the individual

Table 1 Amino acid sequence and composition of the investigated peptide and block copolymer samples. Amino acid sequences are indicated with the N-terminus at the left and the C-terminus at the right-end (Ac = acetyl, Am = amide).

Sample	Structure
1	Ac–G–E–A–K–L–A–E–I–Y–Am
2	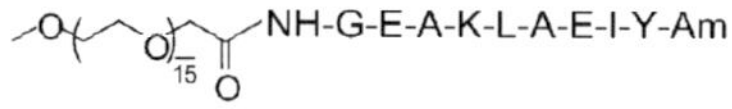 NH-G-E-A-K-L-A-E-I-Y-Am
3	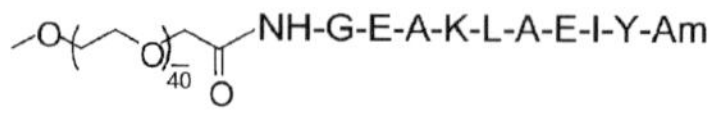 NH-G-E-A-K-L-A-E-I-Y-Am
4	Ac–G–E–A–K–L–A–E–I–E–A–K–L–A–E–I–Y–Am
5	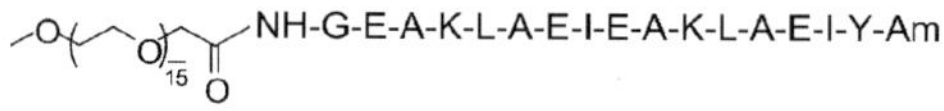 NH-G-E-A-K-L-A-E-I-E-A-K-L-A-E-I-Y-Am
6	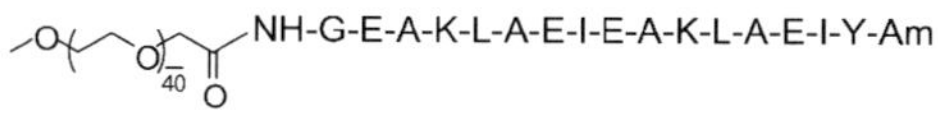 NH-G-E-A-K-L-A-E-I-E-A-K-L-A-E-I-Y-Am
7	Ac–G–E–A–K–[L–A–E–I–E–A–K]$_2$–L–A–E–I–Y–Am
8	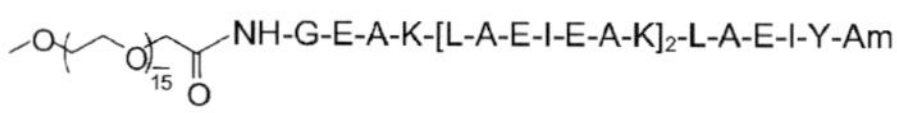 NH-G-E-A-K-[L-A-E-I-E-A-K]$_2$-L-A-E-I-Y-Am
9	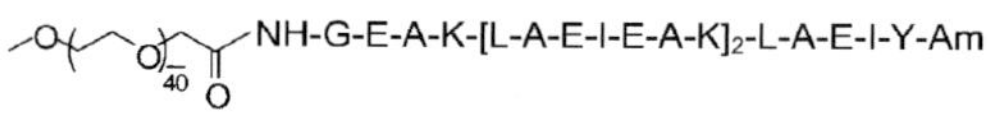 NH-G-E-A-K-[L-A-E-I-E-A-K]$_2$-L-A-E-I-Y-Am
10	H_2N–G–A–A–L–E–A–A–L–K–L–A–A–E–L–A–A–K–G–CO_2H
11	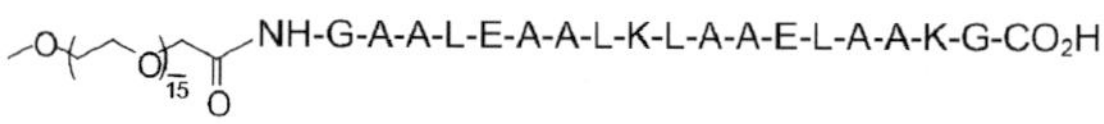 NH-G-A-A-L-E-A-A-L-K-L-A-A-E-L-A-A-K-G-CO_2H
12	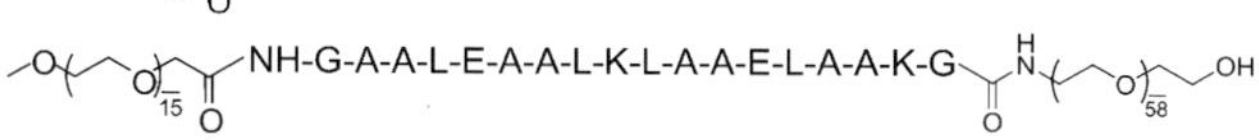 NH-G-A-A-L-E-A-A-L-K-L-A-A-E-L-A-A-K-G
13	H_2N–G–A–A–L–K–A–A–L–E–L–A–A–K–L–A–A–E–G–CO_2H
14	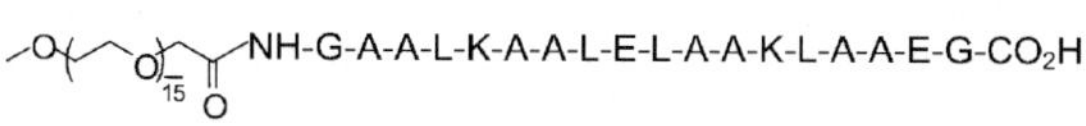 NH-G-A-A-L-K-A-A-L-E-L-A-A-K-L-A-A-E-G-CO_2H

α-amino acids, but also by other factors, such as the distribution of hydrophilic and hydrophobic substituents along the peptide chain.

Our interest in peptide sequences such as those illustrated in Table 1 is driven by the question whether they can be used as auxiliaries to mediate the self-assembly of synthetic macromolecules, such as PEG, into unprecedented, hierarchically organized assemblies. In the following, we will subsequently discuss the solution and solid state organization of the samples listed in Table 1.

Structure formation: Coiled coil based samples

The solution self-assembly of samples **1–9** has been studied extensively by circular dichroism (CD) spectroscopy. Figs. 2 and 3 show results obtained for samples **1–3** and **4–6**, respectively. The negative CD band around 195 nm in the spectra of **1–3** (Fig. 2) is indicative of the absence of any regular secondary structure in these samples.[19] The CD spectra of samples **4–6** (Fig. 3) reveal a negative CD band around 222 nm, which suggests the presence of some α-helical secondary structure. The spectra in Fig. 3, however, lack a second negative minimum at 208 nm, as well as a positive band at 195 nm, which would be expected for a perfect α-helix. From the mean residue ellipticities at 222 nm, the helix content of samples **4–6** can be estimated to 10–25%.[20] As a result, the major fraction of the peptide segments of these samples lacks any regular secondary structure.

Introduction of a second heptad repeat unit (samples **7–9**) results in CD spectra characteristic for α-helical secondary structures, and, depending on sample concentration, in helix contents of up to 90%

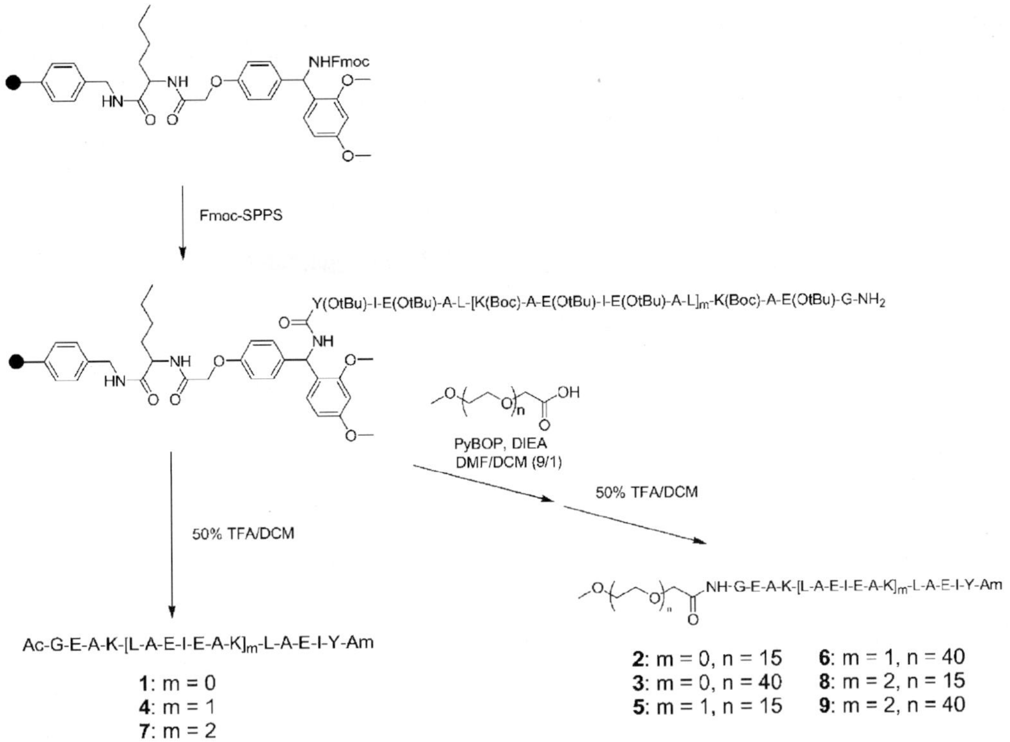

Scheme 1

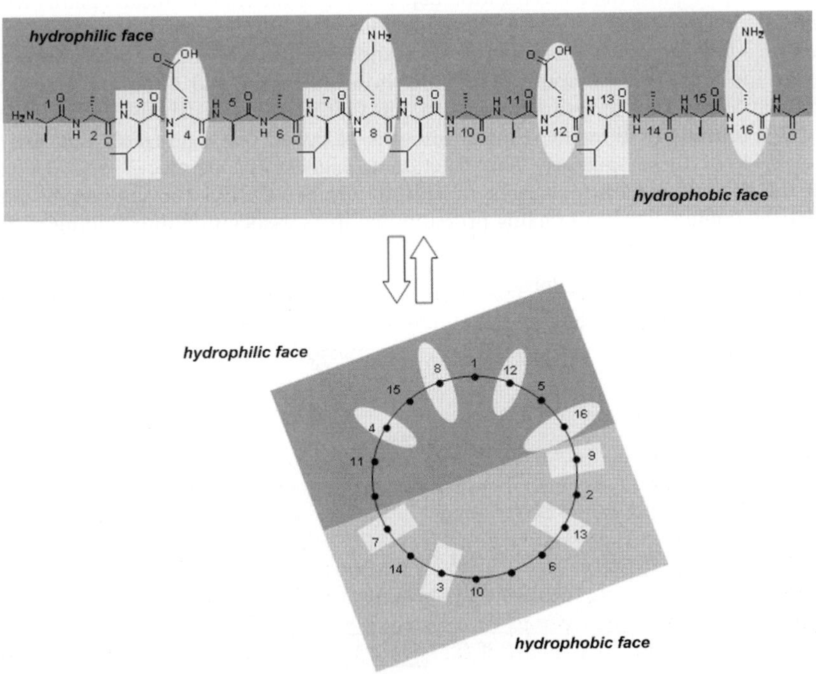

Fig. 1 Peptide sequence of samples **10–14** shown in β-strand conformation (top) and as α-helical wheel representation (bottom). Rectangles and ovals mark hydrophobic and hydrophilic amino acids, respectively.

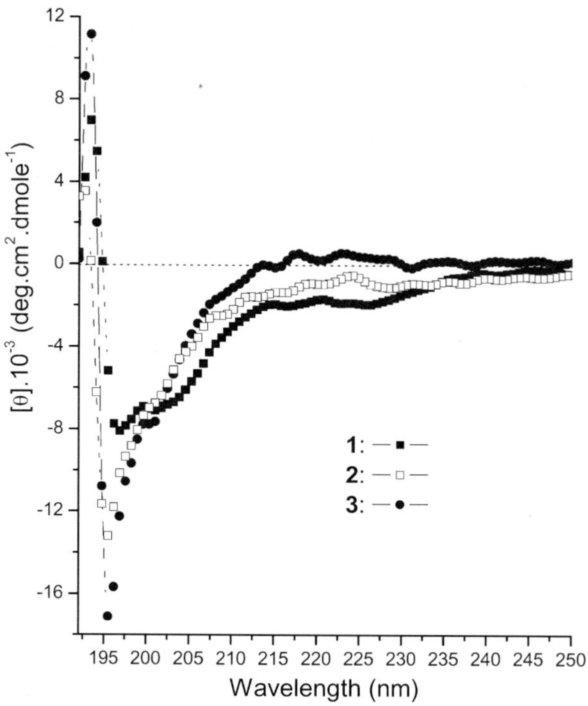

Fig. 2 CD spectra of peptide **1** and diblock copolymers **2** and **3** measured in 0.1 M PBS at pH 7.4 and a sample concentration of 750 μM.

(Figs. 4 and 5). Fig. 5b plots the concentration dependence of the ratio of mean residue ellipticities at 222 nm and 208 nm ($[\theta]_{222}/[\theta]_{208}$). At a concentration of ~100 μM, the ellipticity ratio reaches a plateau level of ~0.95, characteristic for coiled coil superstructures,[14] for all three samples. The data summarized in Fig. 4 are in great contrast to those shown in Figs. 2 and 3 and illustrate the influence of peptide chain length on the formation and stability of coiled coil structures. The different helix contents of **7–9** reflect the different relative amounts of coiled coil aggregates that are present in solutions of each of these samples. This fraction is highest for peptide **7** and decreases upon conjugation of PEG and with increasing molecular weight of PEG for samples **8** and **9**. Steric hindrance of the conjugated PEG chains probably hampers association and folding of the peptide

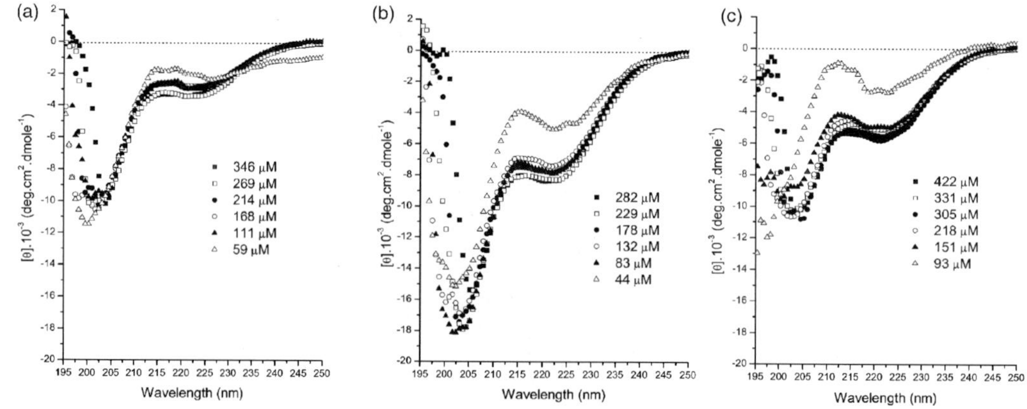

Fig. 3 Concentration-dependent CD spectra of peptide **4** (a) and diblock copolymers **5** (b) and **6** (c) measured in 0.1 M PBS at pH 7.4.

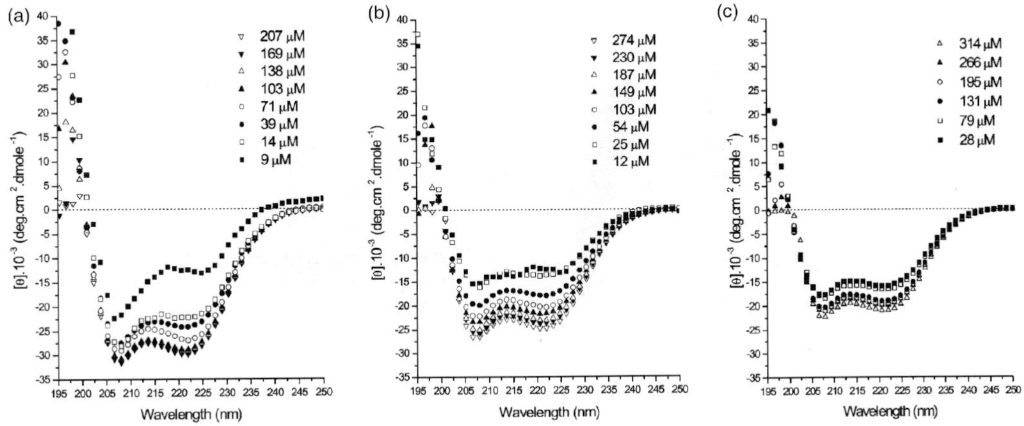

Fig. 4 Concentration-dependent CD spectra of peptide **7** (a) and diblock copolymers **8** (b) and **9** (c) measured in 0.1 M PBS at pH 7.4.

sequences in samples **8** and **9** in comparison with peptide **7**. As described in an earlier publication, analytical ultracentrifugation experiments revealed that the solution self-assembly of **7–9** can be described in terms of an equilibrium between partially unordered monomers and dimeric and tetrameric coiled coil aggregates (Fig. 6).[12] More recently, further support for the model in Fig. 6 was obtained from EPR measurements on spin labelled samples, which suggested a parallel alignment of the constituent helices and a close wrapping of the PEG chains around the peptide shell.[21]

The solid state organization of **7–9** was investigated by means of FTIR spectroscopy and atomic force microscopy (AFM). FTIR spectra recorded from thin, spincoated films of **7** and **9** revealed amide I and amide II bands at 1650 cm^{-1} and 1550 cm^{-1}, respectively, indicating that the peptide sequences maintain their α-helical conformation also in the solid state.[22] AFM studies on spin-coated layers of the samples did not reveal any discrete features for peptide **7**, except for several large agglomerates, which may be due to unspecific association of the coiled coil aggregates. AFM images of **8** and **9**, however, showed distinct nanosized features.[21] These were particularly evident for sample **9**, which contains the highest molecular weight PEG segment, and had dimensions of approximately 1×10 nm (H × D) corresponding roughly with the estimated size of a tetrameric PEGylated coiled coil aggregate (Fig. 7). The AFM studies also provided indirect support for the model shown in Fig. 6. The PEG shell that is closely wrapped around the coiled coil core in the case of the PEGylated block copolymers results in the formation of discrete nano-sized supramolecular objects and prevents unspecific association of the peptide cores.

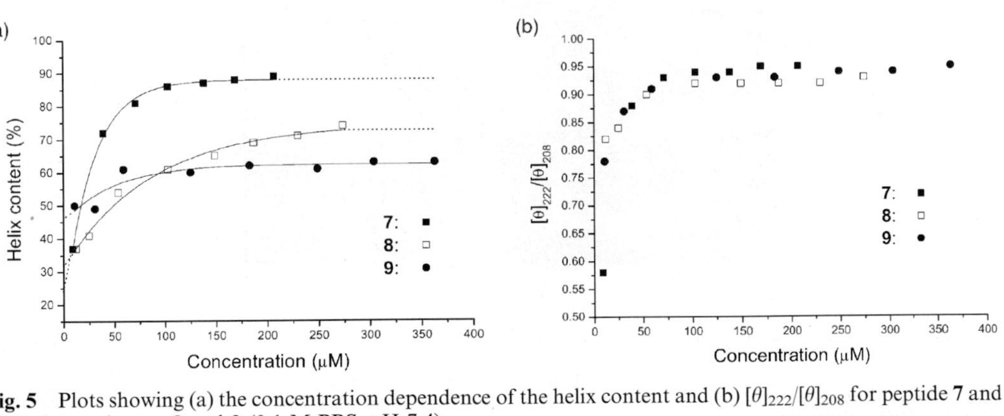

Fig. 5 Plots showing (a) the concentration dependence of the helix content and (b) $[\theta]_{222}/[\theta]_{208}$ for peptide **7** and diblock copolymers **8** and **9** (0.1 M PBS, pH 7.4).

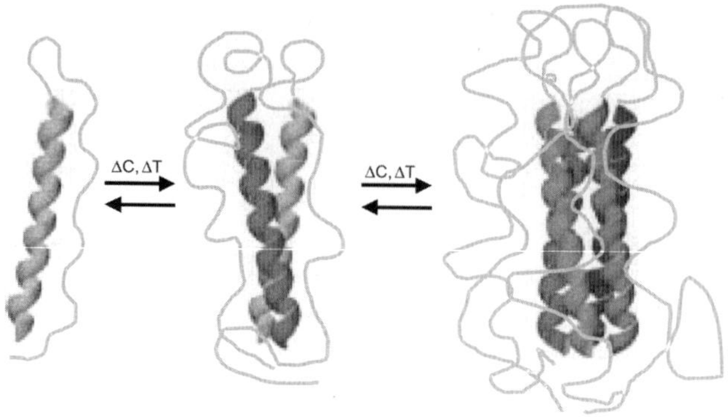

Fig. 6 Proposed model for the self-organization of the PEG-*b*-peptide diblock copolymers **8** and **9**.

Structure formation: "Switch" peptide based samples

The supramolecular organization of samples **10–14** was studied using a variety of spectroscopic, scattering and microscopy techniques. CD experiments were performed to obtain insight into the secondary structure of the peptide segments. As an example, Fig. 8 shows CD spectra recorded for samples **10**, **11** and **13** under various conditions. Fig. 8a indicates that at pH 7 peptide **10** has predominantly a β-strand secondary structure. This sample, however, adopts a partial α-helical secondary structure at pH 2 and pH 11. Temperature dependent CD experiments on this peptide at pH 3 revealed an increase in mean residue ellipticity with increasing temperature. Fitting of the CD spectra indicated that the increase in temperature from 20 °C to 80 °C resulted in an increase in the fraction of random coil content, from ~45% to 53%.[23] This increase in random coil content with increasing temperature is accompanied by a decrease in α-helix and β-strand content. The pH dependent CD spectra of **13** resemble those of **10**; at pH 7 the peptide has predominantly β-strand character, whereas the CD spectra at pH 2 and 11 indicate a partial α-helical secondary structure. In contrast to **10** and **13**, the secondary structure of the peptide sequence in **11** is essentially independent of pH. The spectra shown in Fig. 8d suggest that the peptide segment in this block copolymer has partial α-helical character, which does not change much with pH.

The peptide sequences of **10–14** have been designed in such a way that they can fold into amphiphilic α-helical or β-strand secondary structures. Due to their amphiphilic character, these

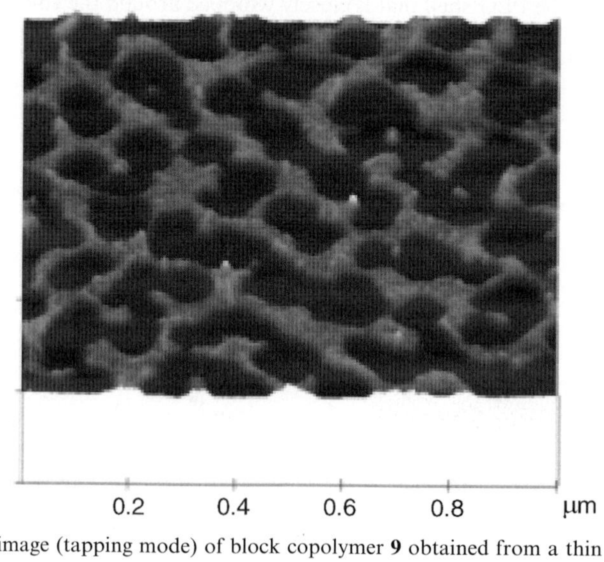

Fig. 7 AFM phase image (tapping mode) of block copolymer **9** obtained from a thin spincast film on mica.

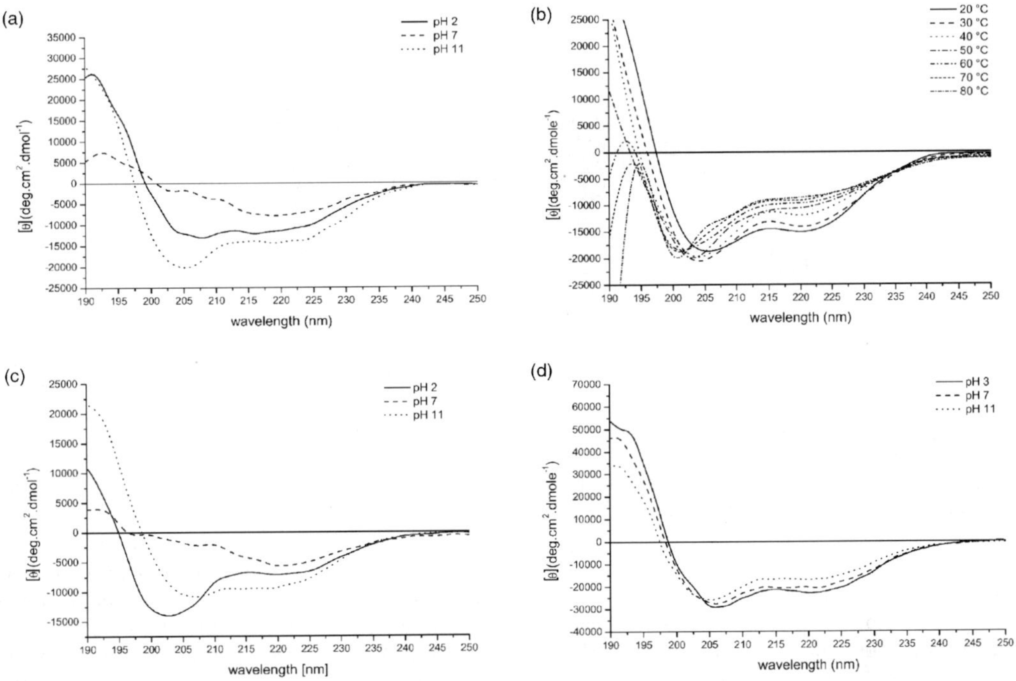

Fig. 8 (a) CD spectra of peptide **10** at pH 2, 7 and 11 ($c = 32$ μM); (b) Temperature-dependent CD spectra of **10** at pH 3 ($c = 32$ μM, sample heated from 20 °C to 80 °C); (c) CD spectra of peptide **13** measured at pH 2, 7 and 11 ($c = 58$ μM); (d) CD spectra from block copolymer **11** measured at pH 3, 7 and 11 ($c = 29$ μM).

peptide secondary structures are expected to aggregate in aqueous solutions of appropriate pH. The supramolecular organization of the samples in dilute aqueous solution was probed by SAXS. As an example, Fig. 9 shows temperature dependent SAXS intensity profiles as well as the corresponding modified Guinier plots for a 0.5 wt.% solution of **10** at pH 3. The linear regions in the Guinier plots indicate that this peptide forms rod-like supramolecular assemblies. From the scattering data, a cylindrical cross sectional radius of gyration (R_g) of ~17 Å can be determined, which slightly decreases with increasing temperature. CD experiments had shown that **10** adopts a partial α-helical conformation at pH 3. In a regular α-helix, each α-amino acid residue contributes 1.5 Å to the total length of the α-helix, measured along the α-helix axis. For sample **10**, which is composed of 18 α-amino acid residues, this gives an approximate helix length of 27 Å, which is in fair agreement with

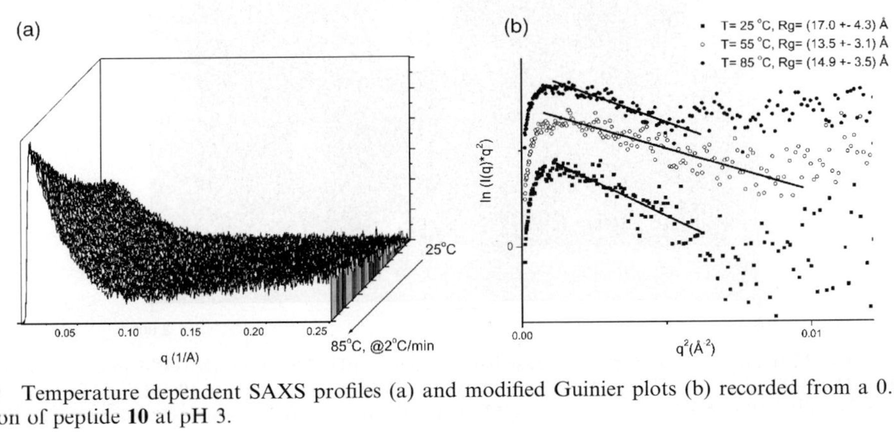

Fig. 9 Temperature dependent SAXS profiles (a) and modified Guinier plots (b) recorded from a 0.5 wt.% solution of peptide **10** at pH 3.

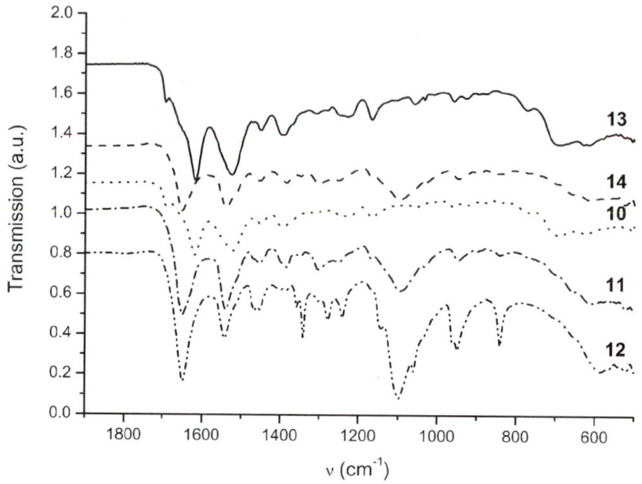

Fig. 10 ATR-FTIR spectra from samples **10–14**.

the 30–35 Å rod diameter determined from the SAXS experiments. This suggests that the α-helical peptides may be assembled side by side into strands in which the helical axis is perpendicular to the strand direction (see Fig. 12a). An alternative model, which seems less likely but cannot be completely ruled out based on the present data, is shown in Fig. 12b and involves lateral assembly

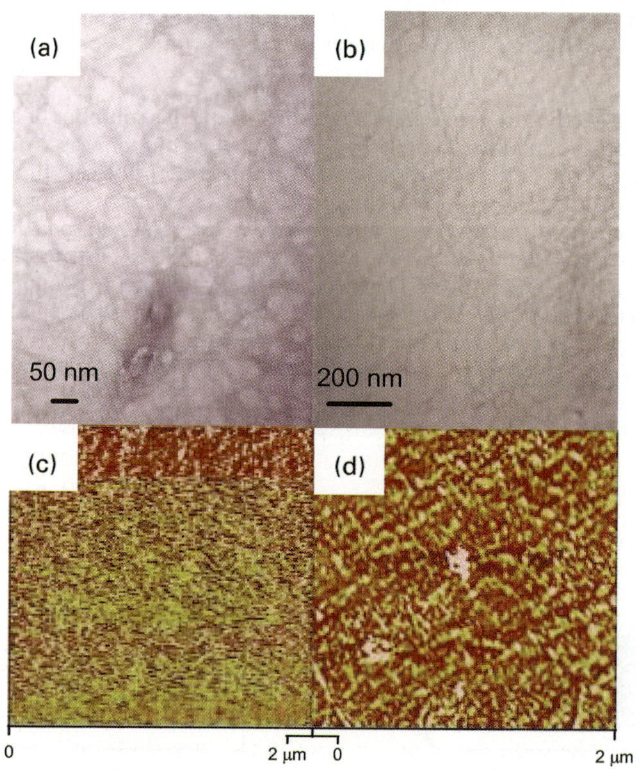

Fig. 11 (a) TEM micrograph of peptide **13** cast from 0.1 wt.% pH 3 solution; (b) TEM micrograph of block copolymer **11** cast from a 0.1 wt.% pH 3 solution; (c) AFM phase image of peptide **10** obtained from a 0.5 wt.% pH 3 solution; (d) AFM phase image of peptide **13** obtained from a 0.01 wt.% pH 3 solution.

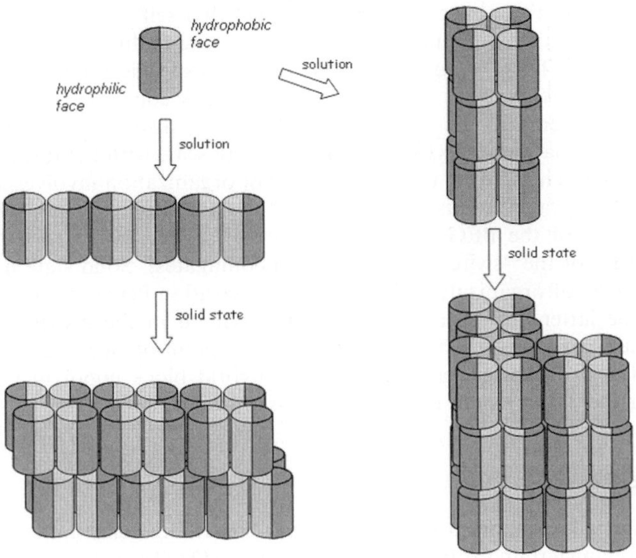

Fig. 12 Proposed models for solution and solid state nanostructures generated from "switch" peptides or "switch" peptide based block copolymers that contain α-helical peptide sequences.

of a small number of α-helical peptides into aggregates that stack on top of each other into longer rod-like structures.

ATR-FTIR spectra recorded from thin films of **10–14** are shown in Fig. 10. These spectra provide insight into the secondary structure of the peptide sequences in the solid state. The amide I bands at $\sim 1620\ cm^{-1}$ and $\sim 1690\ cm^{-1}$ in the spectra of **10** and **13** suggest an antiparallel β-sheet.[22] This is further supported by the presence of an amide V band around $700\ cm^{-1}$.[24] Interestingly, for the PEG containing block copolymers **11**, **12** and **14**, the amide I band is shifted to $1650\ cm^{-1}$, indicating an α-helical secondary structure. In addition, the spectra of the block copolymers contain a weak amide V band around $\sim 600\ cm^{-1}$, which is also characteristic for an α-helical conformation.

In an attempt to visualize the supramolecular structures that are formed in aqueous solution, samples **10–14** were investigated by means of TEM and AFM. Fig. 11a shows a TEM micrograph obtained after depositing a drop of a pH 3 solution of **13** on a TEM grid. The image reveals a network of fibrils with a diameter in the range of 10–20 nm. Since these diameters are much larger than the dimensions of the peptide, the fibrils are probably bundles of rod-like aggregates. The image in Fig. 11b, which was obtained under identical conditions from **11**, shows a similar fibrillar morphology. CD experiments had demonstrated that at pH 3, which is the pH of the solution from which the TEM sample was prepared, the peptide segments of **11** have a partial α-helical secondary structure. The ATR-FTIR spectra in Fig. 10 indicate that the α-helical secondary structure is retained in the solid state. Two possible models for the formation of a fibrillar superstructure mediated by the amphiphilic α-helical peptide secondary structures are shown in Fig. 12. The fibrillar superstructures were also found upon AFM analysis of **10–14** on mica substrates. Fig. 11c and 11d show AFM images obtained from a 0.5 wt.% pH 3 solution of **10** and a 0.01 wt.% pH 3 solution of **13**, respectively.

Conclusions

In this contribution, we have discussed two classes of peptide sequences, that can act as auxiliaries to mediate the self-assembly of synthetic polymers into hierarchically organized solution or solid state nanostructures. The first class of peptide sequences is derived from a *de novo* designed coiled coil and can self-assemble into discrete tetrameric aggregates. The second class is based on so-called "switch" peptides, which can fold into amphiphilic β-strand or α-helical secondary structures. In both cases, self-assembly of these peptides can lead to the formation of extended structures.

The results presented in this paper have demonstrated that self-assembly of conjugates of these peptide sequences with PEG is determined by the very specific folding and organization properties of the peptide segment. This is in great contrast to the unspecific interactions that provide the driving force for the self-assembly of common synthetic block copolymers. Folding and organization of the peptide segments of the PEG block copolymers also leads to hierarchical structure formation. Structure formation starts at the smallest length scale with the formation of α-helical or β-strand conformations. The second, next higher level of organization involves self-assembly of the peptide secondary structural elements, which can lead to either discrete nano-sized supramolecular aggregates (in the case of the PEG modified coiled coil sequences) or to extended ribbon-type assemblies (in the case of the "switch" peptide based conjugates). Solid state nanoscale structures may reveal a third level of organization due to phase separation between the peptide segment and the PEG block. The latter, however, has not been observed in the present contribution. Since peptide sequences are not only attractive from a structural point of view, but also offer possibilities to interact with biological matter, peptide–synthetic hybrid block copolymers may be attractive building blocks for the development of novel, self-assembled biomaterials.

Acknowledgements

This work was financially supported by the Deutsche Forschungsgemeinschaft (KL 1049/3–1 and 3–2) and the Bundesministerium für Bildung und Forschung (MaTech–Neue Materialien, 03N8641). Dr Imtiyaz Ansari is gratefully acknowledged for his contributions to the TEM experiments.

References

1 Recent reviews (*a*) S. Förster and T. Plantenberg, *Angew. Chem. Int. Ed. Engl.*, 2002, **41**, 688; S. Förster and T. Plantenberg, *Angew. Chem.*, 2002, **114**, 712; (*b*) I. W. Hamley, *Angew. Chem. Int. Ed. Engl.*, 2003, **42**, 1692; I. W. Hamley, *Angew. Chem.*, 2003, **115**, 1730; (*c*) C. Park, J. Yoon and E. L. Thomas, *Polymer*, 2003, **44**, 6725.
2 For a recent review see: I. W. Hamley, *Nanotechnology*, 2003, **14**, R39.
3 See *e.g.*(*a*) M. Park, C. Harrison, P. M. Chaikin, R. A. Register and D. H. Adamson, *Science*, 1997, **276**, 1401; (*b*) C. Harrison, M. Park, P. M. Chaikin, R. A. Register and D. H. Adamson, *J. Vac. Soc. Technol. B*, 1998, **16**, 544; (*c*) R. R. Li, P. D. Dapkus, M. E. Thomson, W. G. Jeong, C. Harrison, P. M. Chaikin, R. A. Register and D. H. Adamson, *Appl. Phys. Lett.*, 2000, **76**, 1689.
4 See *e.g.*(*a*) T. Thurn-Albrecht, J. Schotter, G. A. Kästle, N. Emley, T. Shibauchi, L. Krusin-Elbaum, K. Guarini, C. T. Black, M. T. Tuominen and T. P. Russell, *Science*, 2000, **290**, 2126; (*b*) J. Y. Cheng, C. A. Ross, V. Z.-H. Chan, E. L. Thomas, R. G. H. Lammertink and G. J. Vancso, *Adv. Mater.*, 2001, **13**, 1174.
5 A. K. Khandpur, S. Förster, F. S. Bates, I. W. Hamley, A. J. Ryan, W. Bras, K. Almdal and K. Mortensen, *Macromolecules*, 1995, **28**, 8796.
6 V. Abetz and T. Goldacker, *Macromol. Rapid Commun.*, 2000, **21**, 16.
7 (*a*) M. Walther and H. Finkelmann, *Progr. Polym. Sci.*, 1996, **21**, 951; (*b*) G. Mao and C. K. Ober, *Acta Polym.*, 1997, **48**, 405; (*c*) H.-A. Klok and S. Lecommandoux, *Adv. Mater.*, 2001, **13**, 1217; (*d*) M. Lee, B. K. Cho and W. C. Zin, *Chem. Rev.*, 2001, **101**, 3869.
8 I. W. Hamley, *Adv. Polym. Sci.*, 1999, **148**, 113.
9 C. Branden and J. Tooze, *Introduction to Protein Structure*, 2nd edn., Garland Publishing, New York, 1999.
10 For a recent discussion on protein folding: C. M. Dobson, *Nature*, 2003, **426**, 884.
11 *Poly(ethylene glycol)–Chemistry and Biological Applications*, eds. J.M. Harris and S. Zalipsky, *ACS Symp. Ser.*, 1997, **680**.
12 G. W. M. Vandermeulen, C. Tziatzios and H.-A. Klok, *Macromolecules*, 2003, **36**, 4107.
13 A. Rösler, H.-A. Klok, I. W. Hamley, V. Castelletto and O. O. Mykhaylyk, *Biomacromolecules*, 2003, **4**, 859.
14 J. Y. Su, R. S. Hodges and C. M. Kay, *Biochemistry*, 1994, **33**, 15 501.
15 For some reviews see *e.g.*(*a*) A. Lupas, *Trends Biochem. Sci.*, 1996, **21**, 375; (*b*) W. D. Kohn and R. S. Hodges, *Trends Biotechnol.*, 1998, **16**, 379; (*c*) P. Burkhard, J. Stetefeld and S. V. Strelkov, *Trends Cell Biol.*, 2001, **11**, 82.
16 J. M. Mason and K. M. Arndt, *ChemBioChem*, 2004, **5**, 170.
17 M. Mutter, R. Gassmann, U. Buttkus and K.-H. Altmann, *Angew. Chem. Int. Ed. Engl.*, 1991, **30**, 1514.
18 T. E. Creighton, *Proteins: Structures and Molecular Properties*, W.H. Freeman, New York, 2nd edn.,1992.
19 N. Sreerama and R. W. Woody, Circular Dichroism of Peptides and Proteins, in *Circular Dichroism: Principles and Applications*, eds. N. Berova, K. Nakanishi and R.W. Woody, John Wiley & Sons, Inc., New York, 2nd edn., 2000, ch. 21.

20 (*a*) Y.-H. Chen, J. T. Yang and H. M. Martinez, *Biochemistry*, 1972, **11**, 4120; (*b*) Y.-H. Chen, J. T. Chang and K. H. Chau, *Biochemistry*, 1974, **13**, 3350.
21 G. W. M. Vandermeulen, D. Hinderberger, H. Xu, S. S. Sheiko, G. Jeschke and H.-A. Klok, *ChemPhysChem*, 2004, **5**, 488.
22 T. Miyazawa and E. R. Blout, *J. Am. Chem. Soc.*, 1961, **83**, 712.
23 J. Reed and T. A. Reed, *Anal. Biochem.*, 1997, **254**, 36.
24 T. Miyazawa, Y. Masuda and K. Fukushima, *J. Polym. Sci.*, 1962, **62**, S62.

Metal–ligand induced supramolecular polymerization: A route to responsive materials

Stuart J. Rowan* and J. Benjamin Beck

Department of Macromolecular Science and Engineering, Case Western Reserve University, 2100 Adelbert Road, Cleveland, USA. E-mail: stuart.rowan@case.edu; Fax: 216 368 4202; Tel: 216 368 4242

Received 1st March 2004, Accepted 21st April 2004
First published as an Advance Article on the web 19th August 2004

The use of metal–ligand binding as the driving force for the self-assembly polymerizations of a ditopic ligand offers a facile route to the preparation of organic/inorganic hybrid materials. Such metallo-supramolecular polymers potentially offer the functionality of the metal ion along with the processibility of a polymer. We report, herein, the preparation of gel-like metallo-supramolecular polymers prepared from a monomer unit, which consists of a 2,6-bis-(benzimidazolyl)-4-hydroxypyridine unit attached to either end of a polyether chain, mixed with a lanthanoid metal (*e.g.* La(III), Eu(III)) and a transition metal ion (*e.g.* Co(II) or Zn(II)). Such materials show dramatic reversible responses to a variety of stimuli, including thermal, mechanical, chemical and light. The nature of the response can be controlled by the nature of the combination of transition metal ion and lanthanoid metal ion used.

1. Introduction

Stimuli-responsive polymers (SRPs) exhibit an ideally dramatic change in properties upon application of an external stimulus, such as a change in temperature, ionic strength, pH, electric or magnetic fields or by chemical or biological analytes. Examples include liquid crystal polymers,[1] polymer solutions and gels,[2] which can undergo a change in phase morphology,[3] electro- and magnetorheological fluids[4] and electro-active polymers (EAPs).[5] SRPs could play a role in a wide range of potential applications, including smart films, sensors, actuators, electro-optic devices, *etc.* Supramolecular chemistry,[6] with its use of weak reversible non-covalent interactions, has the potential to be a powerful tool in building SRPs which respond to environmental conditions that affect the degree of the molecular interactions. In order to maximise the effect that the environmental stress will have on the material's properties, systems which utilize non-covalent interactions in the construction of polymeric aggregates can be designed. One example of such a supramolecular polymer is the attachment of supramolecular motifs onto two ends of a monomer unit. These monomer units will then aggregate together, self-assembling into polymeric architectures which have a mixture of covalent bonds and noncovalent interactions along the polymeric backbone. Fig. 1 shows a schematic representation of two homoditopic monomer units which have complementary binding sites that drive the self-assembly of an A–A/B–B type supramolecular polymer. The use of such a supramolecular polymerization process,[7–15] means that any change in the strength of the intermolecular non-covalent interactions can and will result in dramatic modification of the

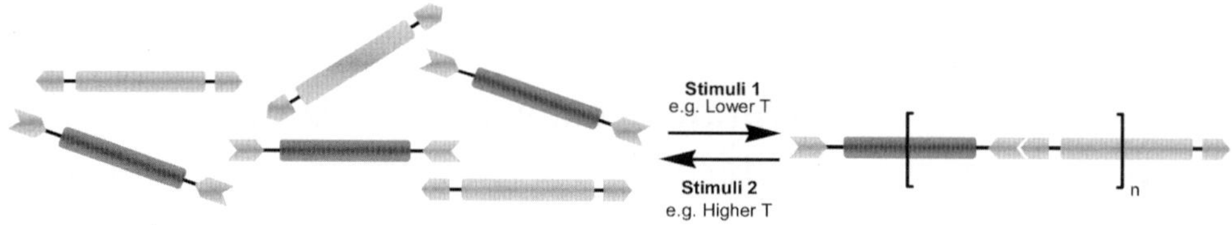

Fig. 1 A schematic representation of an A–A/B–B homoditopic supramolecular polymerization process.

supramolecular structure (*e.g.* degree of polymerization, DP) and thus drive significant changes in the properties of the material.

Supramolecular polymers have a number of interesting properties which help to make them attractive from the point of view of developing SRPs: (1) they form spontaneously, without the need for an initiation process (or catalyst); (2) they are 'dynamic' *i.e.*, are formed under reversible conditions; and (3) termination processes during the self-assembly (polymerization) process are limited. As a result, the degree of polymerization depends, to a large extent, on the strength of the supramolecular interaction between the monomers (K_a) and the total monomer (repeat unit) concentration [M]. If growth of the supramolecular polymer operates through a *Multi-stage Open Association mechanism*, which is, simply put, a standard reversible step–growth process where the binding constant is independent of the molecular weight, then the size of the aggregate (DP) will be proportional to $2(K_a[M])^{\frac{1}{2}}$. This suggests that in order to obtain significant molecular weights of such self-assembled, reversible polymers then supramolecular motifs with large binding constants and/or large total monomer concentrations need to be employed.

There are a wide range of noncovalent interactions that can be utilized to build supramolecular polymers, from hydrophobic interactions to hydrogen bonding and metal–ligand interactions. We have focused our attention in this area on two classes of noncovalent binding motifs, namely nucleobase interactions[16] and metal–ligand interactions. Metal–ligand interactions are of particular interest for a number of reasons. There are a myriad of different types of metal–ligand interactions, which offer a broad range in both their thermodynamic and kinetic stabilities, that can be employed in the construction of supramolecular polymers. This diversity of possible metal/ligand combinations allows the researcher to tune the properties of these self-assembled materials by simply varying the metal ion and/or ligand. The thermodynamic stability of the metal/ligand interaction will influence the size of the aggregate (or degree of polymerization), while its kinetic stability should influence how responsive the material is to environmental factors. For example, kinetically labile metal–ligand systems should produce 'dynamic' polymers which are under continuous equilibrium (*c.f.* supramolecular polymers), while kinetically 'inert' metal–ligand combinations will produce polymers similar in nature to standard covalent systems. Furthermore, the incorporation of metal ions into polymeric systems opens up the possibility of imparting the functional properties of the metal ion, *e.g.* catalysis, light-emitting, conducting, gas binding *etc.*, into the polymer. In recent years a number of groups have started to investigate the potential that this type of non-covalent bond has in the formation of such organic/inorganic polymer hybrids.[17]

We have recently shown that the tridentate ligand 2,6-bis(1′-methyl-benzimidazolyl)-4-alkoxy-pyridine (RO-BIP) can be used for the construction of polymeric aggregates.[18] This ligand is easy to prepare from commercial starting materials and exhibits a wide range of metal ion binding capabilities. For example, the similar 2,6-bis(benzimidazolyl)pyridine (BIP) ligands have been shown to bind to a variety of transition metal ions in a ratio of 2 : 1, as well as being able to bind to the larger lanthanoid ions in a ratio of 3 : 1.[19] Our initial studies have shown that the simple addition of lanthanoid metal ions (<5 mol% per ligand) followed by transition metal ions (>95 mol% per ligand) to a solution of the ditipoic monomer **1** results in the formation of metallo-supramolecular gels (Fig. 2).[20] These gels exhibit responses to a number of different stimuli, including thermo, chemo (with formic acid), and mechanical, in addition to being photoluminescent (Fig. 3). The mechano-responsive **1**:Zn/La gel swollen with 800% (by wt.) of acetonitrile, exhibits a thixotropic (shear-thinning) behaviour. Shaking the gel results in the formation of a free-flowing liquid, which upon standing for *ca.* 20 s results in the reformation of the gel-like material. Initial rheological studies[21] on this system demonstrate that this material has a yield point (under the conditions studied) of approximately 155 Pa. The recovery time is naturally very dependent on the

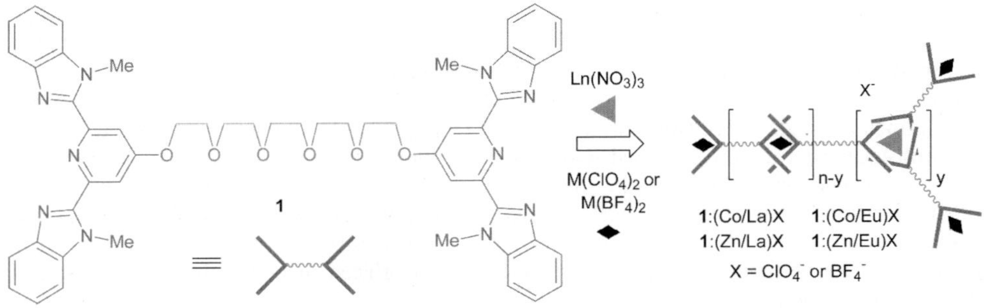

Fig. 2 Schematic representation of the formation of a metallo-supramolecular gel using a combination of lanthanoid and transition metal ions mixed with monomer **1**.

amount of acetonitrile present in the system (material with less solvent recovers more quickly). In this paper we will focus on a more detailed examination of these gels in an attempt to understand the molecular mechanism of the thermo and chemo gel response.

2. Experimental

2.1 Materials

Acetonitrile (spectrophotometric grade), zinc(II) perchlorate hexahydrate, cobalt(II) perchlorate hexahydrate, cobalt(II) tetrafluoroborate, lanthanum(III) nitrate, europium(III) nitrate, chelidamic acid, *N*-methyl-1,2-phenylene diamine, penta(ethylene glycol)ditosylate, sodium iodide, and ammonium chloride were purchased from Aldrich. Phosphoric acid (85%), hydrochloric acid (conc.), potassium carbonate, dimethyl sulfoxide, and chloroform (spectrophotometric grade) were purchased from Fisher Scientific. All the chemicals were used without further purification except for formic acid which was obtained by standard distillation.

2.2 General instrumentation

NMR spectra were recorded on a Varian Gemini-300 NMR spectrometer. Mass spectrometry was carried out on a Bruker BIFLEX III MALDI time-of-flight mass spectrometer using 2-(hydroxyphenylazo)-benzoic acid), HABA as the matrix.

2.3 UV-visible spectroscopy and fluorescence spectroscopy

UV-visible spectra of the gels were obtained by a Perkin Elmer Lambda 800 UV-VIS spectrometer. Titration experiments were performed in quartz cuvets scanning in the range 200–700 nm with an

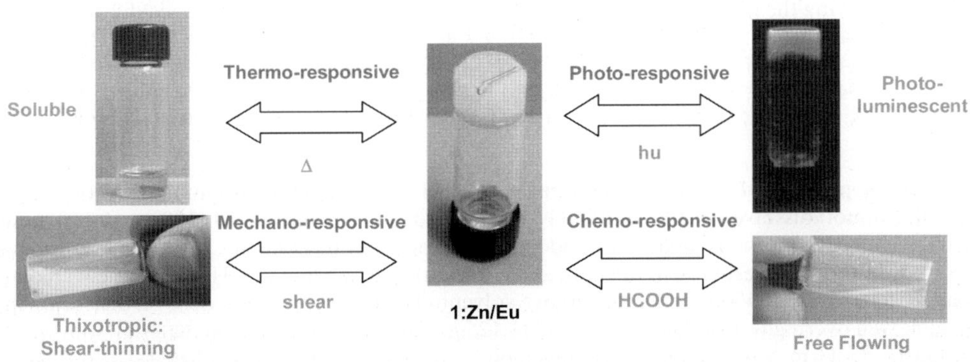

Fig. 3 Mutli-stimuli responsive metallo-supramolecular gels.

integration time of 0.24 s. Variable temperature experiments used a Perkin Elmer PTP-1 Peltier System temperature programmer. In a typical VT experiment, a gel was heated to sol and then transferred *via* pipet to a quartz cuvet. The temperature controller thermocouple was inserted into the sol and the gel reformed as it cooled in the cuvet. Spectra were obtained measuring % transmittance through the gel/solution at different temperatures ranging from 15–80 °C.

Fluorescence spectra were obtained with a SPEX Fluorolog 3 (Model FL3-12); corrections for the spectral dispersion of the Xe-lamp, the instrument throughput, and the detector response were applied. Titration experiments were performed in right-angle mode with quartz fluorescence cuvets. Gel samples were placed between glass slides and run in front-face mode. Variable temperature experiments were performed using a fiber optic detector placed in line with solution containing sealed vials atop a hot stage. The entire hot stage was enclosed so as to prevent ambient light from interfering. Temperatures reported are estimates based on the setting of the hot stage. VT fluorescence spectra described in this publication are the average of five scans using 1.000 s integration time with an interval of 1 nm. All emission spectra were excited at 340 nm.

Addition of 0.85% by weight formic acid to a **1 : Zn/Eu** (97 : 3) gel followed by shaking, causes a breakdown of the gel to a turbid sol. The fluorescence of this is measured in the same fashion as the VT experiments but without the use of the hot stage.

2.4 Synthesis

Synthesis of HOBIP. Chelidamic acid (6.75 g, 33.5 mmol) and *N*-methyl-1,2-phenylenediamine (8.4 mL, 74.0 mmol) were added to 25 mL of phosphoric acid (85%) and stirred at 220 °C for 10 h. After removing heat, the dark blue mixture was poured into 600 mL of water and stirred until it yielded a light blue precipitate. The precipitate that formed was filtered and then resuspended in 1 L of 10% K_2CO_3 solution. The mixture was stirred until all the solids had become pink in color at which time they were once again filtered off. The solids were washed twice with water and then dissolved in a stirring flask containing 1 L of hot methanol. The dark pink methanolic solution was neutralized dropwise with 1.0 M HCl until it turned a dark blue color and then was allowed to stand and cool overnight. The product was collected as gray feathery crystals filtered from the dark blue solution, yielding 11.0 g of 2,6-bis(1′-methyl-benzimidazolyl)-4-hydroxypyridine (HO-BIP) (92%).^{1}H NMR (DMSO-d$_6$): δ 4.32 (6H, s) 7.54 (4H, dd) 7.91 (2H, d) 7.95 (2H, d) 8.01 (2H, s); ^{13}C NMR (DMSO-d$_6$): δ 165.8, 151.6, 150.5. 142.6, 137.7, 123.9, 123.1, 120.2, 113.2, 111.6 (Ar), 33.4 (CH_3); MALDI-TOF MS (matrix: HABA) 356 *m*/*z* [M + H].

Synthesis of monomer 1. HO-BIP (4.85 g, 13.7 mmol) and bisiodopenta(ethylene glycol) (2.5 g, 5.5 mmol) were dissolved into a solution of K_2CO_3 (5.2 g) in 50 mL of DMSO and stirred at 90 °C for 12 h. After removing the heat, the mixture was poured into 200 mL of half-saturated NH_4Cl aqueous solution and washed with 100 mL of dichloromethane (3×). The organic fractions were collected and reduced to 5 mL, then extracted again from a mixture of water and dichloromethane. The organic fractions were then evaporated and the residue dried under vacuum. The material was purified *via* column chromatography (100 : 0 to 97 : 3 dichloromethane : methanol) and precipitation, by dissolving the material dichloromethane and adding to diethyl ether, to yield 2.7 g of **1** as a white solid (54% yield). ^{1}H NMR (DMSO-d$_6$): δ 7.92 (4H, s) δ 7.77 (4H, d) δ 7.68 (4H, d) δ 7.36 (4H, dd) δ 7.34 (4H, dd) δ 4.43 (4H, t) δ 4.25 (12H, s) δ 3.86 (4H, t) δ 3.65–3.55 (12H, m). ^{13}C NMR ($CDCl_3$): δ 164.8, 151.9, 149.9, 142.7, 137.5, 124.0, 123.2, 120.5, 111.9, 110.3, 71.4, 70.9, 69.5, 68.1, 32.5. UV-VIS (CH_3CN): $\lambda_{max} = 312$ nm; MALDI-TOF MS (matrix: HABA) 914 *m*/*z* [M + H].

Sample preparation of the metallo-supramolecular gel-like material. Europium(III) nitrate (0.938 mg, 0.002 mmol) dissolved in 100 μL of CH_3CN was added to a solution of **1** (100 mg, 0.110 mmol) in 500 μL of $CHCl_3$. To this mixture was added zinc(II) perchlorate (39.5 mg, 0.106 mmol) dissolved in 500 μL of CH_3CN, resulting in the immediate formation of a white gel. The gel was covered and placed into an oven (70 °C) for 1 h to remove solvent. The off-white solids were further dried in a vacuum oven overnight. Addition of CH_3CN (1.125 g, 800% w/w) to the solid material, followed by heating to 85 °C in a sealed vial causes the solids to dissolve. A fully swollen gel forms as the solution is allowed to cool to room temperature.

3. Results and discussion

3.1 Synthesis of the monomer and gel self-assembly

Our initial studies have focused on the use of the bis-ligand functionalized compound **1** as the monomer unit in conjunction with either Co(II) or Zn(II) ions as the linear chain extension binding units, and La(III) or Eu(III) ions as the crosslinking components. Monomer **1** was prepared by reacting 2,6-bis(1′-methyl-benzimidazolyl)-4-hydroxypyridine (HO-BIP),[22] prepared *via* a Philips condensation of chelidamic acid and *N*-methyl-1,2-phenylenediamine heated in phosphoric acid (85%) at 220 °C for 10 h, with di-iodopentaethylene glycol under basic conditions. The self-assembly of the gel-like materials occurred spontaneously upon addition of the lanthanoid(III) nitrate (3 mol% based on the total number of BIP ligands) followed by the transition metal ion perchlorate (97 mol% based on the total number of BIP ligands) to a solution of **1** in chloroform/acetonitrile. Removal of the organic solvent left a solid which could be reswollen in pure acetonitrile. Conceptually the preparation of the gels could be achieved in a number of ways which differ in the order in which the two metal ions are added to a solution of the monomer unit. The first method (as outlined above) is to add the lanthanoid ion to the monomer followed by subsequent addition of the transition metal ion. Addition of a small percentage of the lanthanoid ion to the ditopic ligand **1** should result in the formation of a small amount of a self-assembled trifunctional crosslinking unit, assuming a 3 : 1 binding of the monomer unit around the lanthanoid ion. Subsequent addition of the transition metal ion, which can only bind in a 1 : 1 ratio with the ditopic monomer, results in the formation of the gel. These gels show thermo-responsive behaviour exhibiting a reversible gel–sol transition. An alternative method of gel preparation would be the addition of the transition metal ion first followed by addition of the lanthanoid ion. However, when this is done no gel is observed and only a precipitate is formed. Repeated heating and cooling of this sample does result in the formation of a gel. This suggests that, initially, two different molecular architectures are being formed the nature of which depends on the order of the metal ion addition. The fact that altering the order of metal ion addition has such a dramatic effect on the nature of the material suggests that at least one of the metal/ligand interactions is not completely reversible on the time-scale of the self-assembly process. Alternatively, the formation of a kinetic product could be the result of kinetic trapping by the formation of the gel or precipitate. These observations lead to a number of questions which we hoped to try and address in order to gain a better insight into these materials. (1) Do either of the metal ions form reversible interactions with the ligand under the conditions investigated? (2) What are the self-assembled molecular architectures that are being formed? and (3) what is the mechanism of the response?

3.2 Lanthanoid metal ion and transitional metal ion binding to 1

With the goal of attaining insight into the mechanism of the gel self-assembly process and how the tridentate 2,6-bis(1′-methyl-benzimidazolyl)-4-alkoxypyridine (RO-BIP) ligand binds to the different metal ions used in the construction of the gel, a series of studies were carried out on acetonitrile solutions of a 1 : 1 mixture of $Zn(ClO_4)_2$ and the ditopic monomer **1** and a 1 : 3 mixture of $Eu(NO_3)_3$ and **1**.

An examination of Zn(II) ion binding to the monomer was carried out and Fig. 4 shows the UV spectra obtained upon titration of $Zn(ClO_4)_2$ into a solution of **1**. The formation of the 2 : 1 ligand to Zn(II) complex is a two step process: binding of the first ligand to the free metal ion followed by binding of the second ligand to this 1 : 1 complex. Fig. 4 (inset a) confirms that formation of a 2 : 1 ligand to metal complex and curve fitting the data to an appropriate 2 : 1 binding model allows an estimation of the binding constants ($K_{overall} > 10^6\ M^{-1}$). Fig. 4 (inset b) shows the Scatchard plot of this data which confirms that significant cooperativity is observed upon the binding of the second ligand to the 1 : 1 complex. A strongly cooperative binding process encourages the formation of 2 : 1 complexes and this along with a large $K_{overall}$ is favorable for the formation of supramolecular polymers.

A photo-luminescent (PL) study (excitation at 340 nm) was also undertaken on the 1 : 1 **1:Zn(ClO$_4$)$_2$** system in order to examine how temperature affects the metal ligand complex. Fig. 5 shows the PL spectra of **1** and **1:Zn(ClO$_4$)$_2$** at both 25 °C and 100 °C. Upon addition of $Zn(ClO_4)_2$

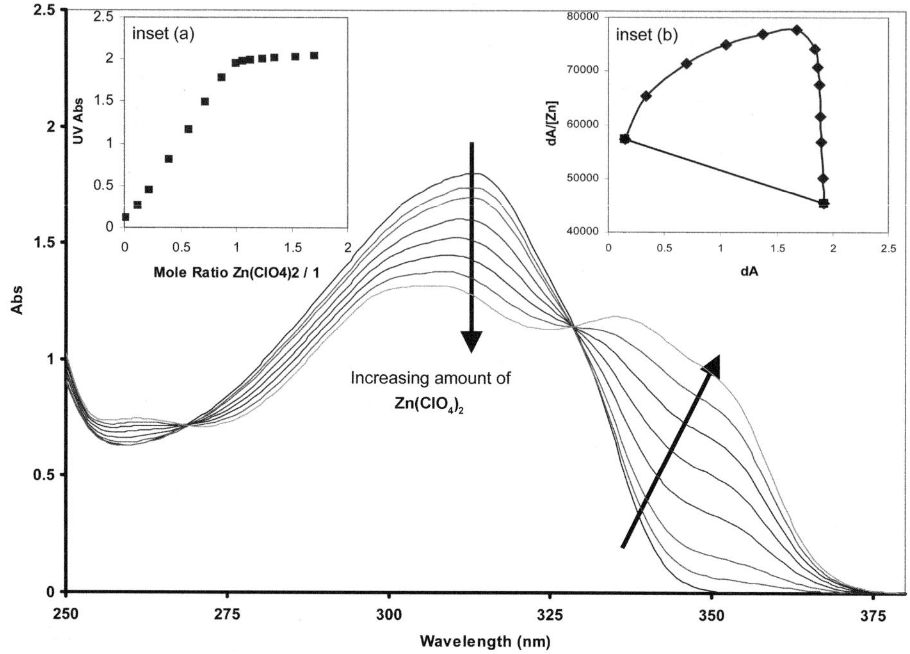

Fig. 4 UV titration experiment of $Zn(ClO_4)_2$ with **1**. An acetonitrile solution of $Zn(ClO_4)_2$ (0.107 mM) and **1** (0.0125 mM) was titrated into a 0.0125 mM solution of **1** in acetonitrile. The data shows the titration from pure **1** to a 1 : 1 mole ratio of **1** to $\mathbf{Zn(ClO_4)_2}$, which shows the presence of isosbestic points. After the 1 : 1 mole ratio of $\mathbf{1{:}Zn(ClO_4)_2}$ is reached the UV curves cease to intersect with the isosbestic points (not shown). Inset (a) shows the effect on the absorbance (350 nm) of varying the number of equivalents of $Zn(ClO_4)_2$ to 1, indicating a 1 : 1 binding of $\mathbf{1{:}Zn(ClO_4)_2}$ and therefore the formation of 2 : 1 ligand/metal ion complexes. Insert (b) shows the Scatchard plot of the titration data which indicates the presence of strong cooperative binding.

to **1** at 25 °C a shift in the emission band from $\lambda_{max} = 368$ nm to 393 nm is observed. Heating $\mathbf{1{:}Zn(ClO_4)_2}$ to 100 °C does not result in a shift in the emission band or even an increase in emission at 368 nm which would occur if the system is decomplexing and free ligand was being formed. This

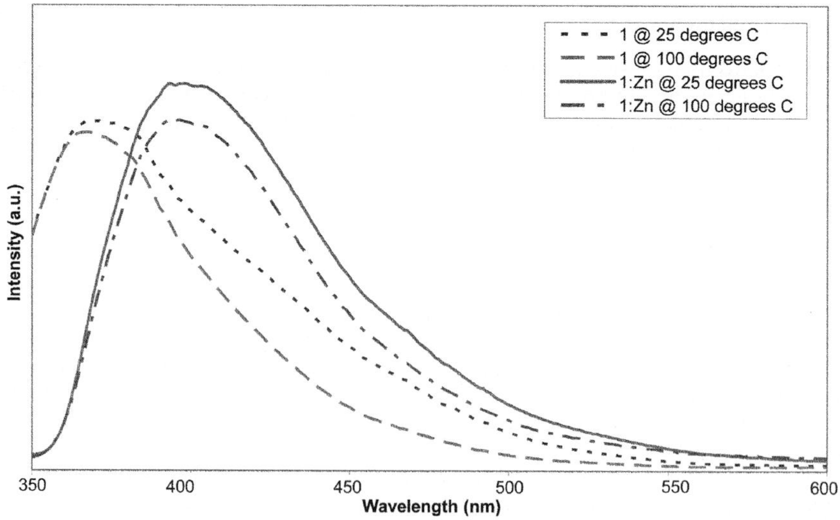

Fig. 5 Photo-luminescence spectra of **1** and $\mathbf{1{:}Zn(ClO_4)_2}$ in acetonitrile at 25 °C and 100 °C.

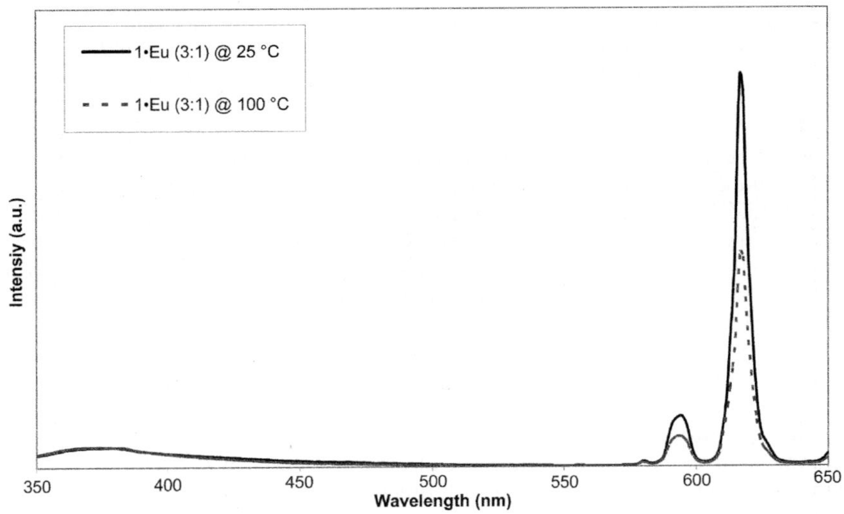

Fig. 6 Photo-luminescent spectra of the 3 : 1 mixture of **1** · **Eu(NO$_3$)$_3$** in acetonitrile at 25 °C and 100 °C.

suggests that at this temperature and concentration (0.1 mM, which is more dilute than is present in the gels) no significant decomplexation of the **1:Zn(ClO$_4$)$_2$** complex is occurring.

We then turned our attention to the Ln(III)/ligand interaction. The Eu(III) containing materials offer the possibility of utilizing the interesting spectroscopic properties of this lanthanide ion.[23] For example, Eu(III) ions can show an intense metal-centered luminescence in the presence of an appropriate UV absorbing ligand *via* the so-called "antenna effect".[24] This is in effect a light conversion process which occurs by absorption of the light by the ligand, followed by a ligand-to-metal energy transfer process finally resulting in the metal ion-based emission. The BIP ligand has been shown[25] to act as an "antenna" for Eu(III) ions. Furthermore, we can use this "antenna effect" to demonstrate whether the ligand is binding to the Eu(III) ion. A PL study was therefore carried out on a 3 : 1 solution of **1**:Eu(NO$_3$)$_3$ in acetonitrile (Fig. 6). In addition to the ligand emission at 368 nm we see a number of peaks indicative of the europium metal centered emission (581, 594, 616, 652 nm). Upon heating this sample to 100 °C we see a significant decrease in the intensity of the europium emission bands and no significant change in the intensity of the ligand based emission. This suggests that at this temperature the ligand and the Eu(III) ions are in fact, at least partially, decomplexing.

The results carried out in the two model systems suggest that the Eu(III) ions form weaker more labile complexes with the RO-BIP ligand than the Zn(II) ions. This is consistent with the thermal responsive nature of the gels being caused by the decomplexation of the crosslinking Ln(III) ions and not the decomplexation of the chain extending transition metal ions.

3.3 Thermal response of the gels

To further test the idea that it might be the Ln(III)/ligand interaction that is responsible for the thermal response of the gels we have carried out a series of spectroscopic investigations on the gels themselves. All the mixed metal systems in which either 3 mol% (based on ligand) La(NO$_3$)$_3$ or Eu(NO$_3$)$_3$ are added to **1** first, followed by the addition 97 mol% (based on ligand) of either Co(ClO$_4$) or Zn(ClO$_4$), result in the formation of thermal responsive gels. For example, simply heating the **1**:**(Co/La)ClO$_4$** gel swollen with 800% solvent (by wt.) to *ca.* 100 °C results in a reversible gel–sol transition. The resulting solution is orange in color indicative of the persistence of the Co(II) ion complex in solution at these temperatures. This, again, is consistent with the idea that decomplexation of the transition metal ions is not the cause of the thermo-solubilization of the gel. It is interesting to note that while it appears that the nature of the transition metal ion (in this case at least) does not affect the $T_{G\text{–}S}$ the nature of its counterions can. For example, while **1:(Zn/La)ClO$_4$** and **1:(Co/La)ClO$_4$** swollen with 700% acetonitrile both exhibit $T_{G\text{–}S}$ at *ca.* 100 °C, the **1:(Co/La)BF$_4$** gel (which was prepared by mixing 3 mol% $La(NO_3)_3$ and 97 mol% $Co(BF_4)_2$, based on the total number of BIP ligands, with **1**) swollen with the same amount of acetonitrile has a $T_{G\text{–}S}$ at *ca.* 45 °C. The exact reason for the difference in this thermo-response is unclear at this point; however it could be due to the counterions from the transition metal ion affecting the lanthanoid/ligand binding in different ways. Furthermore, it could also be expected that the gel–sol transition temperature would be sensitive to the amount of solvent with which the gel is swollen. The transmittance of the **1:(Co/La)BF$_4$** gel, which exhibited a lower $T_{G\text{–}S}$ than its perchlorate analogue, was utilized to better elucidate the gel–sol transition behavior. The transmittance of **1:(Co/La)BF$_4$** swollen with 500% acetonitrile remains low at low temperatures (Fig. 7). However upon heating, the transmittance shows a dramatic increase as the gel–sol transition occurs ($T_{G\text{–}S}$ *ca.* 65 °C). Furthermore, if the gel is swollen with more solvent there is a drop in the temperature at which this dramatic increase in transmittance occurs, indicating that a gel swollen with more acetonitrile exhibits a much lower $T_{G\text{–}S}$. For example, when the **1:(Co/La)BF$_4$** gel is swollen with 600% acetontrile the $T_{G\text{–}S}$ drops to *ca.* 55 °C.

As mentioned above, the photoluminescence of the Eu(III) ion can be useful in trying to further elucidate the molecular processes which occur during the thermal response of the gels. Furthermore, the light-emitting behavior of these systems opens up the possibility of utilizing them as photo- or electroluminescent materials.[26] **1:Zn/Eu** gel was indeed photoluminescent showing the peaks indicative of the europium metal centered emission (581, 594, 616, 652 nm) as well as a ligand centered emission (397 nm). Heating **1:Zn/Eu** gel to *ca.* 100 °C shows a substantial reduction in the lanthanide-based emission (Fig. 8) but no significant change in the ligand emission, further supporting the fact that it is mainly the Eu(III)/ligand interactions that are decomplexing at the higher temperatures and not the transition metal ion/ligand interactions. The nature of the different metal ions within the gels can be used to alter the PL of the material. For example, as La(III) does not show any metal centered based emission the **1:Zn/La** gel displays only the metal-bound ligand based emission (at 397 nm). The **1:Co/Eu** gel on the other hand shows no fluorescent behavior, probably on account of the Co(II) ions having low energy metal centered levels which facilitate radiationless decay processes.[27]

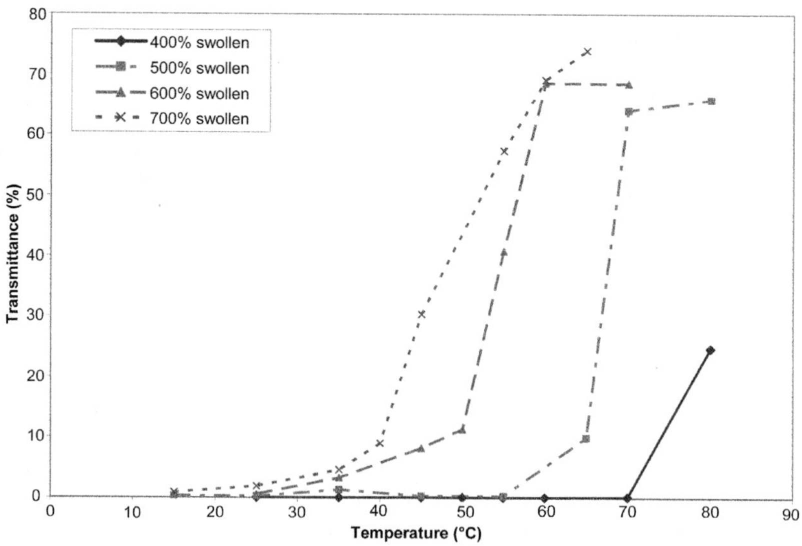

Fig. 7 The effect that the amount of swelling solvent (acetonitrile) has on the temperature dependent transmittance of the **1:(Co/La)BF$_4$** gel (measured at a wavelength of 675 nm). The amount of solvent is the weight percent solvent to dry gel.

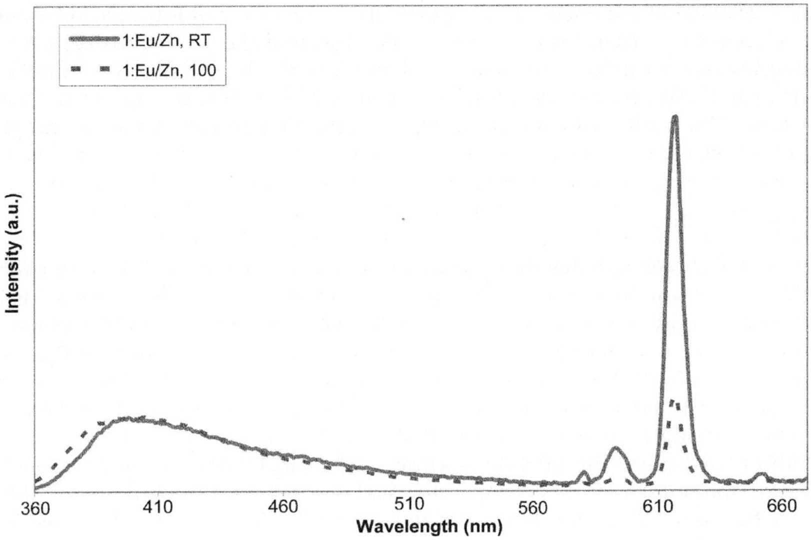

Fig. 8 Photo-luminescent spectra of the **1:Eu/Zn** swollen with 800% acetonitrile at room temperature and 100 °C.

3.5 Chemo-response of the gels

Lanthanides are known to be oxophilic and bind well to carboxylic acids.[28] Therefore, it was reasoned that these lanthanide-containing gels should also be chemo-responsive to such molecules. Any ligand exchange between the Ln(III) bound tridentate aza-ligand in **1** with a mono-functional carboxylic acid should result in a decrease in crosslinking of the system and a subsequent breakdown of the gel. Indeed, addition of a small amount of formic acid (0.85% by weight) to the **1**:**Zn**/**Eu** gel results in not only the loss in the mechanical stability of the gel but also quenching of the Eu(III) emission (Fig. 9). It should be noted that 0.85% formic acid by weight is based on the

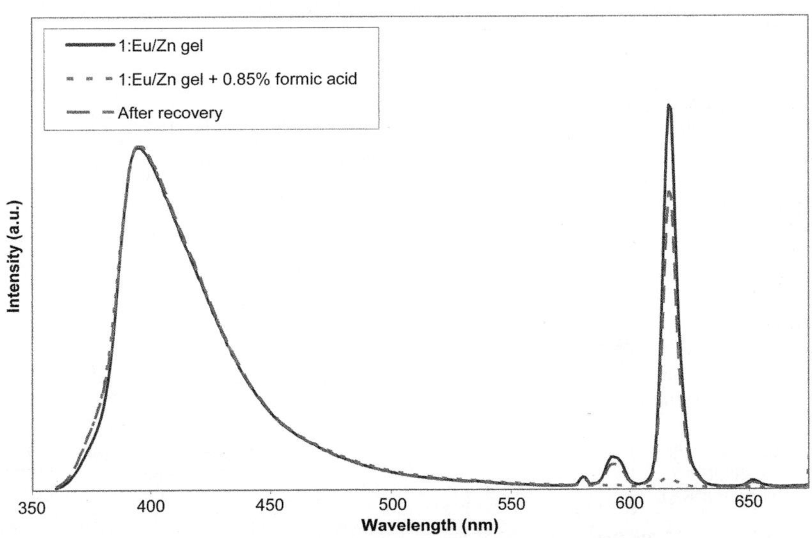

Fig. 9 Photo-luminescent spectra of the **1:Eu/Zn** gel swollen with 800% acetonitrile, **1:Eu/Zn** gel + 0.85% formic acid before and after recovery. The recovery of the gel occurred by placing in a vacuum oven at 40 °C overnight and reswelling with 800% acetonitrile. The intensities of the data have been normalized to the emission peak at 393 nm.

total weight of gel plus solvent and that this equates to a large excess of formic acid (about 100 mole equivalents) compared to the amount of the lanthanide ion in the gel. This is consistent with the formate anion displacing the RO-BIP ligand on the Eu(III) cation, resulting in a "switching off" of the aromatic ligands antenna effect. This process can be reversed by drying out the material in a vacuum oven for 24 h at 70 °C. Upon reswelling the material with acetonitrile an increased Eu(III) emission is observed.

3.6 Conclusions

In summary, we have used a combination of metal ions in conjunction with the bis-ligand monomer **1** to produce a supramolecular polyelectrolyte gel which can exhibit thermo-, chemo- and mechanical responses, as well as light-emitting properties. The nature of the response displayed by these systems depends upon the metal ion, counter-ion and the amount of swelling solvent. We have used a combination of UV and PL spectroscopic data to try and elucidate the mechanism of these responses. This initial data is consistent with a thermal-response mechanism in which there is at least partial decomplexation of the RO-BIP ligand from the more labile, weaker bound lanthanoid ions upon heating (or addition of formic acid), which in turn results in a de-crosslinking of the system. It appears that at these temperatures all of, or at least the majority of, the transition metal ions are bound to the ligand thus resulting in a solution which is mainly a mixture of linear transition metal ion bound oligomers and lanthanoid ions bound to 0, 1 or 2 ligands (Fig. 10).

The fact that little or no Zn(II)/ligand decomplexation occurs, even at the higher temperature, may hint as to why it is important for the preparation of gels to add the Ln(III) ion first to **1**. Addition of Ln(III) ions to a large excess of the ditopic ligand **1** should result in the self-assembly of a supramolecular tri-functional unit, which will act as a crosslinking moiety (Fig. 10). Subsequent addition of the Zn(II) metal ion results in a crosslinking of the system. It should be noted that it is also possible for some percentage of the Zn(II) ions binding to **1** to form macrocycles which are trapped within the gel. However, if the Zn(II) ions are added to the flexible monomer **1** first then this could result in the formation of metallo-macrocycles and a small amount of linear species which are kinetically inert under the self-assembly conditions. Addition of a small amount of the weaker binding Ln(III) ions, which can not displace the more strongly bound Zn(II) ions, to this system means that the Ln(III) ions will predominately bind to the free ligands. As the concentration of free ligand is very low (assuming all the Zn(II) ions are bound then there should only be 3 mol% free

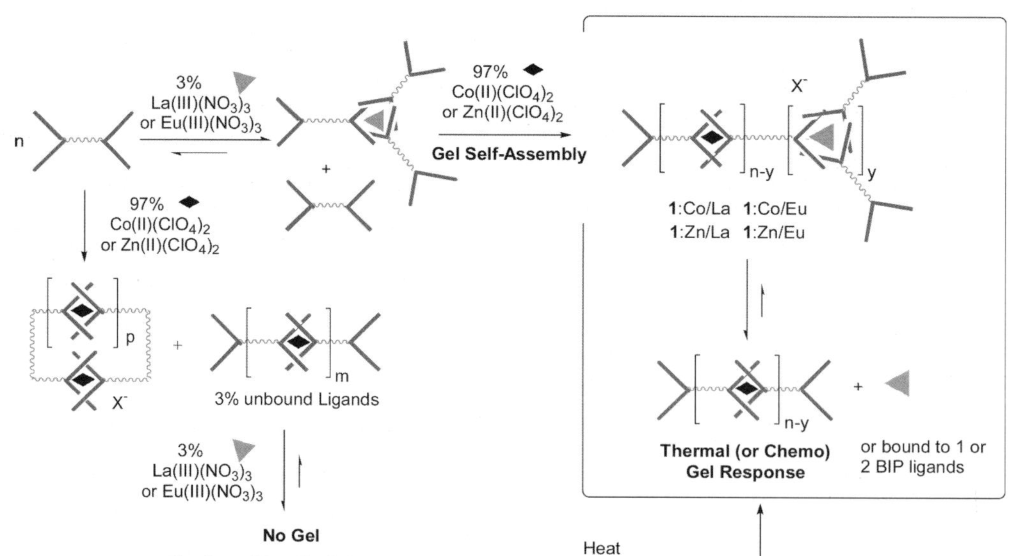

Fig. 10 Possible mechanism for the self-assembly and thermal response of the gels formed by the interaction of the ditopic ligand **1** with a mixture of lanthanoid and transition metal ions.

ligand left) then the likelihood of Ln(III) ion participating in crosslinking is reduced. Therefore, the formation of Ln(III) containing macrocycles becomes more likely and as a result no gel is formed.

It appears that the combination of metal ions and their different ligand binding strengths and kinetics is important to the formation of these multi-stimuli responsive supramolecular gels. Given the wide variety of metal ions (with different metal/ligand kinetic stabilities and functional properties), counter ions and possible monomer cores, a wide variety of environmentally responsive metallo-supramolecular materials can be envisaged.

Acknowledgements

This material is based upon work supported by the National Science Foundation under Grant No. CAREER: CHE-0133164 and by the Case School of Engineering.

References

1 S. J. Picken, *Macromol. Symp.*, 2000, **154**, 95.
2 For example, see *Stimuli-Responsive Water Soluble and Amphiphilic Polymers*, ACS Symposium Series 780, ed. C. L. McCormick , American Chemical Society, Washington, DC, 2001.
3 R. A. Siegel and B. A. Firestone, *Macromolecules*, 1988, **21**, 3254; I. C. Kwon, Y. H. Bae and S. W. Kim, *Nature*, 1991, **354**, 291; K. Kontturri, S. Mafe, J. A. Manzanares, B. L. Svarfvar and P. Viinikka, *Macromolecules*, 1996, **29**, 5740; J. H. Holtz and S. A. Asher, *Nature*, 1997, **389**, 829; Y. Miyata, N. Asami and T. Uragami, *Nature*, 1999, **399**, 766.
4 A. Lengalova, V. Pavlinek, P. Saha, O. Quadrat and J. Stejskal, *Colloid Surf. A*, 2003, **227**, 1; J. H. Park, B. D. Chin and O. O. Park, *J. Colloid Interface Sci.*, 2001, **240**, 349.
5 For example, see *Electroactive Polymers [EAP] Actuators as Artificial Muscles: Reality, Potential, and Challenges*, ed. Y. Bar-Cohen, SPIE Press, Bellingham, 2001.
6 For example, see(*a*) J.-M. Lehn, *Science*, 1985, **227**, 849; (*b*) J.-M. Lehn, *Angew. Chem. Int. Ed. Engl.*, 1988, **27**, 89; (*c*) M. Mascal, *Contemp. Org. Synth.*, 1994, **1**, 31; (*d*) J.-M. Lehn, *Supramolecular Chemistry*, VCH, Weinheim, 1995; (*e*) *Comprehensive Supramolecular Chemistry*, ed. J. L. Atwood, J. E. D. Davies, D. D. MacNicol and F. Vögtle, Pergamon, Oxford, 1996, **11**; (*f*) H.-J. Schneider, A. Yatsimirsky, *Principles and Methods in Supramolecular Chemistry*, Wiley, Chichester, 2000.
7 L. Brunsveld, B. J. B. Folmer, E. W. Meijer and R. P. Sijbesma, *Chem. Rev.*, 2001, **101**, 4071.
8 A. Ciferri, *Macromol. Rapid Commun.*, 2002, **23**, 511.
9 *Supramolecular Polymers*, A. Credi, Marcel Dekker, New York, 2000.
10 M. Kotera, J. -M. Lehn and J.-P. Vigneron, *J. Chem. Soc., Chem. Commun.*, 1994, 197.
11 C. St Pourcain and A. C. Griffin, *Macromolecules*, 1995, **28**, 4116.
12 T. Shimizu, R. Iwaura, M. Masuda, T. Hanada and K. Yase, *J. Am. Chem. Soc.*, 2001, **123**, 5947.
13 K. Yamaguchi, J. R. Lizotte and T. E. Long, *Macromolecules*, 2002, **35**, 8745.
14 R. Iwaura, K. Yoshida, M. Masuda, K. Yase and T. Shimizu, *Chem. Mater.*, 2002, **14**, 3047.
15 E. A. Fogleman, W. C. Yount, J. Xu and S. L. Craig, *Angew. Chem. Int. Ed Engl.*, 2002, **41**, 4026.
16 S. J. Rowan, P. Suwanmala and S. Sivakova, *J. Polym. Sci.: Part A: Polym. Chem.*, 2003, **41**, 3589; S. Sivakova and S. J. Rowan, *Chem. Commun.*, 2003, 2428.
17 H. Hofmeier, S. Schmatloch, D. Wouters and U. S. Schubert, *Macromol. Chem. Phys.*, 2003, **204**, 2197; B. Lahn and M. Rehahn, *Polymers*, 2002, **1**, 1; D. G. Kurth, A. Meister, A. F. Thuenemann and G. Foerster, *Langmuir*, 2003, **19**, 4055; W. C. Yount, H. Juwarker and S. L. Craig, *J. Am. Chem. Soc.*, 2003, **125**, 15 302.
18 J. B. Beck and S. J. Rowan, *J. Am. Chem. Soc.*, 2003, **125**, 13 922.
19 C. Piguet, J.-C. G. Bünzli, G. Bernardinelli, G. Hopfgartner and A. F. Williams, *J. Alloys Comp.*, 1995, **225**, 324.
20 For a review on low molecular weight gels see P. Terech and R. G. Weiss, *Chem. Rev.*, 1997, **97**, 3133–3159.
21 Y. Zhao, J. B. Beck, A. M. Jamieson and S. J. Rowan, *Macromolecules*, 2004, **37**, 3529.
22 HO-BIP was prepared using a modified literature procedure, P. Froidevaux, J. M. Harrowfield and A. N. Sobolev, *Inorg. Chem.*, 2000, **39**, 4678.
23 J. L. Bender, P. S. Corbin, C. L. Fraser, D. H. Metcalf, F. S. Richardson, E. L. Thomas and A. M. Urbas, *J. Am. Chem. Soc.*, 2002, **124**, 8526.
24 S. Lis, *J. Alloys Comp.*, 2002, **341**, 45; N. Sabbatini, M. Guardigli and J.-M. Lehn, *Coord. Chem. Rev.*, 1993, **123**, 201.
25 S. Petoud, J.-C. G. Bünzli, T. Glanzman, C. Piguet, Q. Xiang and P. Thummel, *J. Lumin.*, 1999, **82**, 69–79.
26 J. Kido and Y. Okamoto, *Chem. Rev.*, 2002, **102**, 2357.
27 N. Armaroli, L. De Cloa, V. Balzani, J.-P. Sauvage, C. O. Dietrich-Buchecker, J.-M. Kern and A. Bailal, *J. Chem. Soc., Dalton Trans.*, 1993, 3241.
28 For example see L. Ma, O. R. Evans, B. M. Foxman and W. Lin, *Inorg. Chem.*, 1999, **38**, 5837–5840; M. Kawa and J. M. J. Frechet, *Chem. Mater.*, 1998, **10**, 286.

Responsive brushes and gels as components of soft nanotechnology

A. J. Ryan,[a] C. J. Crook,[a] J. R. Howse,[a] P. Topham,[a] R. A. L. Jones,[b] M. Geoghegan,[b] A. J. Parnell,[b] L. Ruiz-Pérez,[b] S. J. Martin,[b] A. Cadby,[b] A. Menelle,[c] J. R. P. Webster,[d] A. J. Gleeson[e] and W. Bras[f]

[a] *Department of Chemistry, University of Sheffield, Sheffield, UK*
[b] *Department of Physics and Astronomy, University of Sheffield, Sheffield, UK*
[c] *Laboratoire Léon Brillouin, CEA-Saclay, F-91191, Gif-sur-Yvette Cédex, France*
[d] *ISIS Facility, Rutherford Appleton Laboratory, Chilton, Didcot, Oxfordshire, UK, OX11 OQX*
[e] *CCLRC Daresbury Laboratory, Warrington, UK, WA4 4AD*
[f] *DUBBLE CRG, ESRF, 6 rue Jules Horowitz, BP 220, F-38043, Grenoble Cédex 9, France*

Received 15th April 2004, Accepted 10th May 2004
First published as an Advance Article on the web 14th September 2004

Progress in the development of generic molecular devices based on responsive polymers is discussed. Characterisation of specially synthesised polyelectrolyte gels, "grafted from" brushes and triblock copolymers is reported. A Landolt pH-oscillator, based on bromate/sulfite/ferrocyanide, with a room temperature period of 20 min and a range of $3.1 < pH < 7.0$, has been used to drive periodic oscillations in volume in a pH responsive hydrogel. The gel is coupled to the reaction and changes volume by a factor of at least 6. A continuously stirred, constant volume, tank reactor was set-up on an optical microscope and the reaction pH and gel size monitored. The cyclic force generation of this system has been measured directly in a modified JKR experiment. The responsive nature of polyelectrolyte brushes, grown by surface initiated ATRP, have been characterised by scanning force microscopy, neutron reflectometry and single molecule force measurements. Triblock copolymers, based on hydrophobic end-blocks and either polyacid or polybase mid-block, have been used to produce polymer gels where the deformation of the molecules can be followed directly by SAXS and a correlation between molecular shape change and macroscopic deformation has been established. The three systems studied allow both the macroscopic and a molecular response to be investigated independently for the crosslinked gels and the brushes. The triblock copolymers demonstrate that the individual response of the polyelectrolyte molecules scale-up to give the macroscopic response of the system in an oscillating chemical reaction.

Introduction

The ability of individual polymer molecules to respond to changes in temperature and chemical environment with drastic changes in size and conformation has been appreciated for many years.

The volume transition of responsive gels represents a direct, macroscopic manifestation of the conformational response of the individual molecules making up the gel.[1] Responsive gels have been enthusiastically greeted as candidates for a new generation of intelligent materials with sensor, processor and actuator functions. A wide variety of different stimuli-responsive gels have been developed for specific applications such as drug release and actuators for artificial muscle.[2] However, macroscopic applications of stimulus responsive gels have not developed as fast as was initially anticipated. Part of the reason for this is the fact that diffusion limits the rate at which the gel can respond to changes in environment. Sometimes this simply leads to impractically long response times, but it can also give rise to internal stresses within the gel, as one part attempts to shrink while nearby regions stay swollen, which ultimately can lead to cracking and total mechanical failure.

The scaling of diffusion with size suggests that these undesirable features of gel response will become less severe with decreasing size. Examples of microscopic applications of responsive gels include "flow-sorter" valves for fluidic systems,[3] where the volume transition of the gels open and close 200 μm channels. The ultimate in miniaturisation occurs in polymer brushes, where a single layer of polymer molecules is end-grafted to a solid substrate. For polymer brushes, changes in the environment translate directly into changes in the thickness of the polymer layer.[4] One way in which these thickness changes have been exploited is to line nanoscale pores in a membrane with a pH responsive brush, resulting in a membrane whose permeability alters with the acidity of the environment.[5] Another recent application exploits the fact that a collapsed polymer brush offers a much more favourable surface for protein adsorption than an expanded one to create a programmable surface to hold and release proteins on demand.[6]

One feature of the collapse transition in responsive polymers, whether in the form of a macroscopic gel or as a polymer brush, is that it represents a direct coupling between the chemical environment and a mechanical response. A device which converts chemical energy directly into mechanical energy at a molecular level is known as a molecular motor. Molecular motors are ubiquitous in nature, but synthetic analogues are almost unknown. Although the variety of biological molecular motors is huge, in many cases the basic mode of operation is the same-the "power stroke" of the motor is produced by a *molecular conformational change* induced by a *cyclic chemical reaction*.[7] This mechanism is profoundly different from any used in a macroscopic motor, because it depends fundamentally on the random Brownian motion of the components of the motor. Not only is the "piston" of a molecular motor not stiff as we would try to achieve in a macroscopic engine, its very operation depends on its flexibility, because it is the random Brownian motions within the motor macromolecule that lead to the conformational changes that drive the motor.

At this fundamental level, nanodevices that exploit chain collapse have a broad analogy with the natural nanodevices of cell biology. They exploit the different physics that pertains at the nanoscale in aqueous solution. In this environment, the presence of large dissipation and the dominance of Brownian motion and strong surface forces means that the engineering principles we employ to make nanoscale devices should be very different from the world of engineering at the macroscale. In our experimental program, we attempt to use chain collapse as the basis of nanodevices, that although crude, employ some of the same design principles as cell biology. We hope that these efforts will provide the beginnings of what we might call *biomimetic nanotechnology* or *soft nanotechnology*.

An obvious target for such a program is the development of an autonomously cycling molecular motor, in which cycles of mechanical work are continuously repeated for as long as an input of chemical fuel is maintained. At the macroscopic level, a prototype of such a free-running chemical motor is provided in pioneering studies carried out by Yoshida and coworkers.[8–12] They have coupled the collapse phenomenon in macroscopic responsive gels with oscillating chemical reactions (such as the Belousov–Zhabotinsky, BZ, reaction) to create conditions where "pseudo" non-equilibrium systems which maintain rhythmical oscillations can be demonstrated, in both quiescent[9] and continuously stirred[10] reactors. The ruthenium complex of the BZ reaction was introduced as a functional group into poly(*N*-isopropyl acrylamide) (PNIPAM), which is a temperature-sensitive polymer.[8] The ruthenium group plays its part in the BZ reaction, and the oxidation state of the catalyst changes the collapse temperature of the gel.[10] The result is, at intermediate temperature, a gel whose shape oscillated (by a factor of 2 in volume) in a BZ reaction, providing an elegant demonstration of oscillation in a polymer gel. This system, however, is limited by the concentration of the catalyst which has to remain relatively small.[12]

In our program, we concentrate on weak polyacids and polybases, which respond to changes in pH. At the macroscopic level, these systems, when coupled with a chemical reaction that cause spontaneous pH oscillations, allow us to make a free-running chemical motor. To miniaturise this to the single molecule level, we need to develop chemical routes to create well characterised polyelectrolyte brushes. We have characterised the response of these brushes to changes in pH by using scanning force microscopy (SFM) and neutron reflectometry. The latter technique, in particular, allows us to follow the collapse transition in some detail. Having confirmed that our polyelectrolyte brushes do indeed exhibit a substantial conformational change in response to changes in pH, our next step was to characterise the way the mechanical properties of a single chain change in response to pH changes. This we have achieved by carrying out single molecule force measurements in a molecular force probe, an atomic force microscope optimised for carrying out force measurements. These results give us the information we need to analyse a motor cycle in our target single molecule synthetic motor. In addition, we are able to quantify the degree to which our polyelectrolyte brushes act as surfaces with switchable adhesion. Finally, we explicitly link the macroscopic behaviour by studying the response of triblock copolymers to pH oscillations. The active chains are the mid-blocks and are effectively crosslinked by the domains of the hydrophobic end block. The expansion and contraction of the grafted polyelectrolyte chains causes the distance between hydrophobic domains to change, this separation can be monitored by SAXS and the molecular and macroscopic effects correlated.

Experimental

Synthesis section

Experimental studies are reported on four related polymer systems. Crosslinked polyacid gels have been described previously[13,14] and the results are included here for completeness. Triblock copolymers were synthesized by living anionic polymerization. A precursor polymer was made with a mid-block of poly(*tert*-butyl methacrylate) and end blocks of polymethylmethacrylate (PMMA). The tertiary butyl ester was subsequently hydrolysed to give the parent polymethacrylic acid (PMAA). Polymer brushes, comprising polyacid (PMAA) and polybase chains, poly((diethylamino)ethyl methacrylate) (PDEAMA), were grown from functionalized surfaces by atom transfer radical polymerization (ATRP). The polyacid was formed by hydrolyzing a poly(*tert*-butyl methacrylate) brush whereas the polybase brush was grown directly.

Polyacid brushes

The substrates used were silicon wafers (Mitsubishi Research and Crystran Ltd.) polished to the (100) face and having a native oxide thickness of 15 Å. The silicon surface was rendered hydrophilic by first cleaning in an oxygen plasma for 10 min followed by the RCA method stage 1[15] (pure water, ammonia (37%) and hydrogen peroxide (20% v/v) with a volume ratio of 5 : 1 : 1, respectively at 80 °C for 10 min). The sample was then washed repeatedly with clean water. The cleaned silicon was dried under a nitrogen atmosphere for several minutes to remove all traces of water. The sample was then transferred to a refluxing environment of 5% (v/v) 3-aminopropyltriethoxysilane in toluene. The sample was held in place for 16 h using a specially designed PTFE holder at a height just above the condensing region. Once removed the sample was placed in a reaction vessel containing a solution of 2-bromo-2-methylpropionic acid (33.4 mg, 0.20 mmol) and *p*-(dimethyl amino) pyridine (6 mg, 0.05 mmol) in 10 ml of dry dichloromethane.[16] The solution was cooled to 0 °C before adding dicyclohexylcarbodiimide (51.6 mg, 0.25 mmol). The wafers were left at room temperature overnight before being washed with toluene and acetone. Before being transferred to the polymerisation medium the wafers were dried with nitrogen.

The chemicals employed in the synthesis were purified as follows: *tert*-butyl methacrylate (tBuMA)(Aldrich, 98%) was stirred over CaH_2 overnight before being distilled under reduced pressure. *N*,*N*,*N*′,*N*′,*N*″-pentamethyldiethylene-triamine (PMDETA) (Aldrich 99%) was used as received. Cu(I)Br (Aldrich, 98%) was stirred with glacial acetic acid overnight, before being filtered and thoroughly washed with ethanol and acetone, and then dried under vacuum. Benzonitrile (Aldrich, 98%) was used as supplied.

Methyl-2-bromoprionate (0.103 ml, 9.25×10^{-4} mol), Cu(I)Br (133 mg, 0.925 mmol) and 1 ml of benzonitrile, were introduced to a reaction tube. The mixture was stirred vigorously under an inert atmosphere for 20 min before the addition of PMDETA (0.579 ml, 2.78×10^{-3} mol), resulting in the development of an intense green colour. After another 20 min, tBuMA (15 ml, 9.2×10^{-2} mol) was added and stirred until homogeneous. An initiator-coated silicon wafer, held in a wire cradle was then lowered into the mixture, and the temperature raised to 100 °C. The polymerisation mixture rapidly increased in viscosity and, after a period of an hour, the silicon wafer was removed from the vessel and was washed clean in tetrahydrofuran (THF). The free polymer was dissolved in THF and recovered by precipitation in methanol/water (95/5) mixture. The polymer was then filtered and cleaned by further dissolving/precipitating cycles. Following polymerisation the brushes were hydrolysed to remove the *t*-butyl group by refluxing in a solution *p*-toluene sulfonic acid in 1,4-dioxane. Sacrificial free polymer was used in order to reach ideal concentrations, *i.e.* 20% polymer in solution and 5% acid with respect to polymer. The brushes were then removed and placed in a water/methanol mixture followed by several wash steps with deionised water.

Polybase brushes

The substrates were the same as those used for polyacid brush synthesis, 30 × 30 × 10 mm pieces of silicon polished to the (100) face and having a native oxide thickness of 15 Å. These samples were required for neutron reflectivity experiments and so were prepared on large silicon blocks to provide a large surface area and suitable thickness for the reflection geometry. The initiator surface was generated by a 2-step process and differs slightly from that used in the generation of PMAA brushes as described above. Here we attached the initiator to the surface in a single step transfer using (11-(2-bromo-2-methyl)propionyloxy)undecyl trichlorosilane as described by Beers *et al.*[17] and shown in Fig. 1. The silicon surface was rendered hydrophilic as described above and dried in a vacuum oven at 120 °C for 30 min before placing in a PTFE beaker with a tight fitting lid containing 20 ml of a 1.5 μl ml^{-1} solution of trichlorosilane initiator in dry toluene. After the required period the silicon block was removed and subjected to washings with toluene (×2), acetone, and ethanol before finally dried under nitrogen. The blocks were stored under vacuum until required for the brush synthesis.

(Diethylamino)ethyl methacrylate (DEAMA) (Aldrich, 99%) was stirred over calcium hydride overnight before being distilled under high vacuum at 70 °C. PMDETA (Aldrich 99%) was used as received. Cu(I)Br (Aldrich, 98%) was stirred with glacial acetic acid overnight, before being filtered and thoroughly washed with ethanol and acetone, and then dried under vacuum. THF was distilled over sodium/potassium benzophenone and stored under vacuum before use.

Copper (I) bromide (163 mg, 1.14 mmol) and PMDETA (270 μl, 1.29 mmol) were added to 15 ml of stirred THF. The reagents were purged with nitrogen and thoroughly mixed, to which the DEAMA monomer (25 ml, 124 mmol) was subsequently introduced. After 20 min of further stirring, whilst maintaining the flow of nitrogen, the suspended initiators were lowered into the bright blue solution at 75 °C. After 17 h, the silicon-based polymer brushes were withdrawn from the solution (now dark green) and purified using a THF soxhlet reflux system for 24 h. A reference wafer with a saturated initiator surface of dimensions 10 mm × 10 mm × 0.625 mm was also used for each polymerisation bath to assess the consistency of polymer chain growth. With the absence of any direct means of measuring the polymer chain molecular weight the use of a reference wafer allows for some assessment of the molecular weight of the polymer chains on the dilute brush under study.

Block copolymers

Methyl methacrylate (Aldrich, 98%) and *tert*-butyl methacrylate (Aldrich, 98%) were stirred overnight with CaH_2 and then distilled under reduced pressure prior to use. THF was dried using sodium/potassium alloy, and then stored over sodium benzophenone, using the deep purple colour as an indicator of dryness. Double diphenyl ethylene (DDPE) was synthesised by the method previously reported.[18]

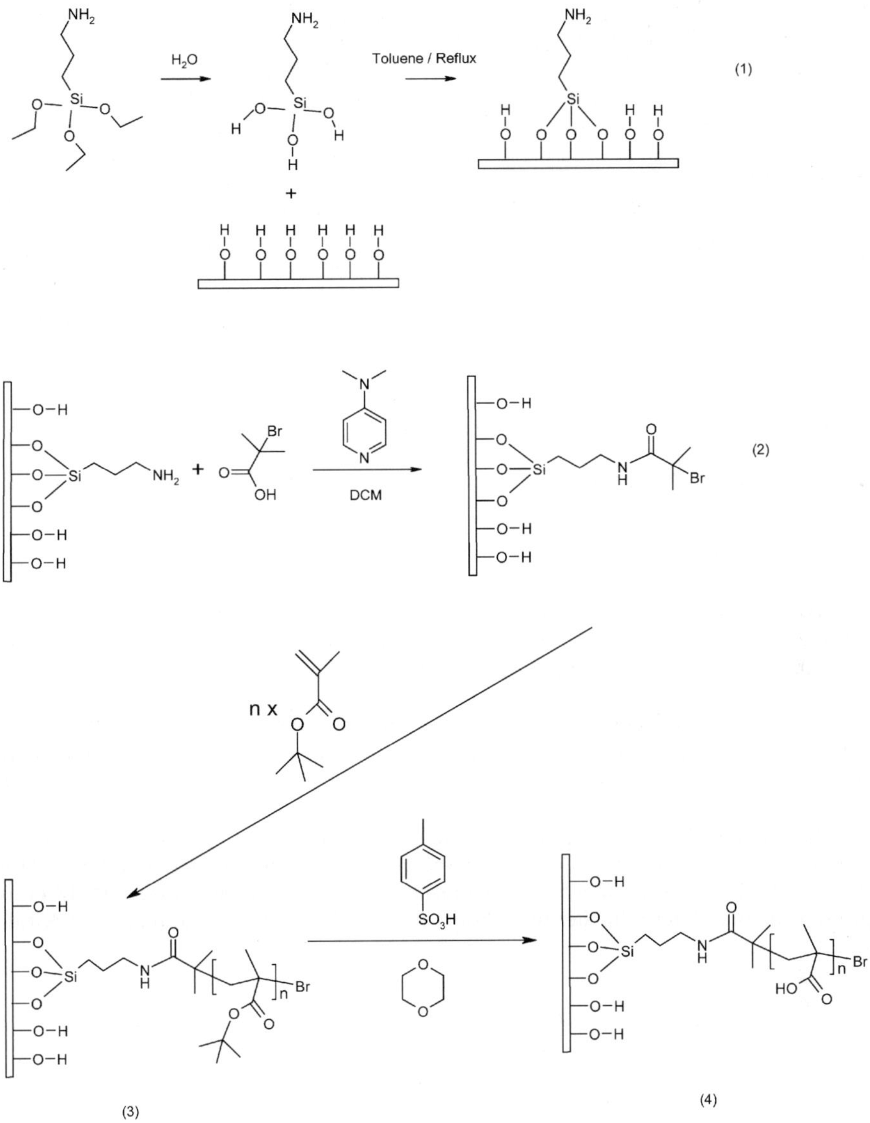

Scheme 1 Formation of polymethacrylic acid brushes.

All procedures were carried out under high vacuum conditions. The required amount of DDPE was weighed into a vessel equipped with a Young's tap, and then placed onto a high vacuum line and evacuated overnight to ensure dryness. Then approximately 10 ml of dry THF were distilled

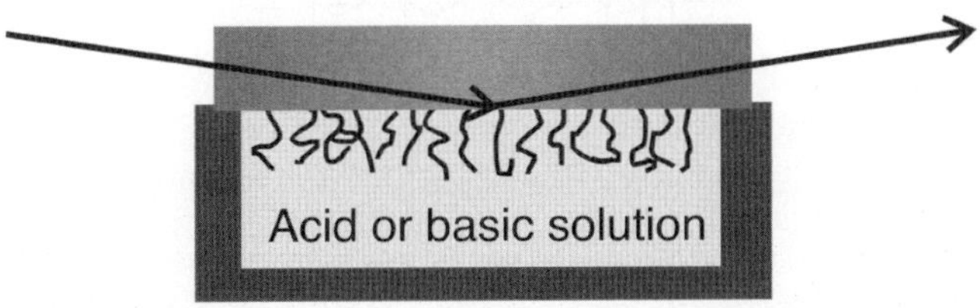

Fig. 1 A schematic representation of the neutron reflectivity geometry.

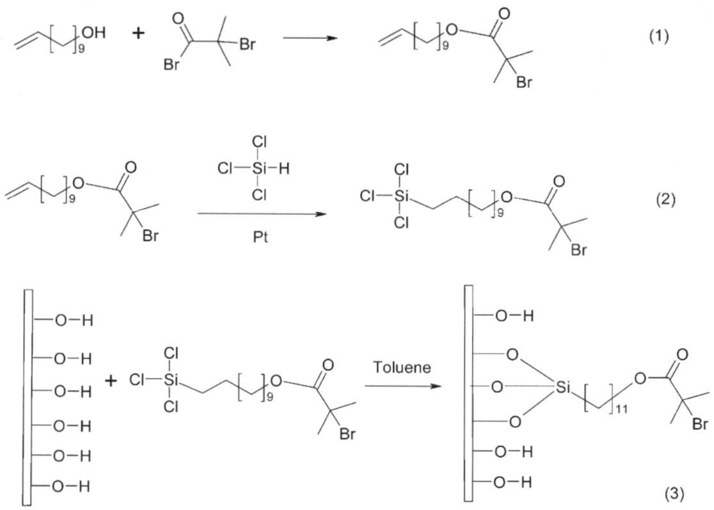

Scheme 2 Formation of the initiator surface for polybase brushes.

over. Into two similar vessels were distilled methyl methacrylate (MMA) and tBuMA. The three vessels and the required ampoule of *sec*-butyl lithium were then affixed to a four-armed vessel with a tap, containing a stainless steel breaker. The vessel was charged with LiCl (a five time molar excess with respect to *sec*-butyl lithium) and then evacuated. It was heated in an oil bath to 130 °C, and the combination of heat and vacuum used to dry the LiCl overnight. 100 ml of dry THF were then distilled into the vessel. The polymerisation was started by opening the initiator ampoule using the metal breaker, and allowing the contents into the vessel. The DDPE solution was then added very slowly, resulting in the development of a blood red colour. The reaction mixture was then cooled to −70 °C using a methanol/solid CO_2 bath and the tBuMA added. The red colour disappeared, and the solution became increasingly viscous over the course of approximately one hour. After three hours, the polymerisation was judged to have finished, and the second monomer (MMA) was added with vigorous mixing. After a further three hours the vessel was frozen and methanol distilled in to terminate the reaction.

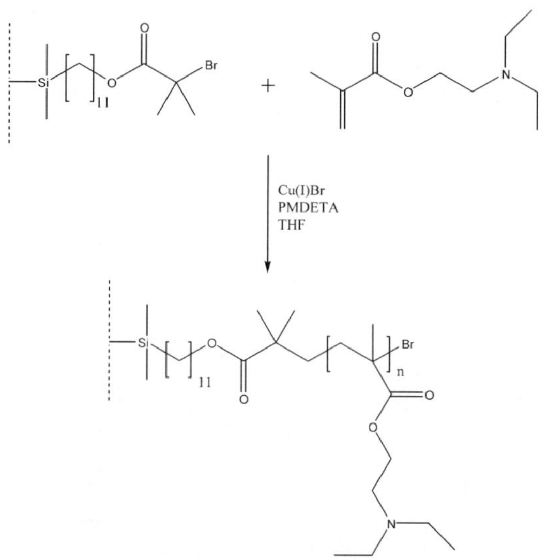

Scheme 3 Preparation of polybase brushes.

Table 1 Molecular characterisation

Code	M_n/kg mol^{-1}	Polydispersity index	Molar ratio$_{PMMA}$
TB56	90	1.24	0.13
TB64	56	1.17	0.13
TB58	155	1.01	0.13

The polymer was purified by precipitation into a methanol/water mixture (90/10 vol). Molecular weights and polydispersities of the polymers were determined using size exclusion chromatography against polystyrene standards, and the molar ratio of the comonomers determined using NMR spectroscopy and are reported in Table 1. The PMMA–tBuMA block copolymers were dissolved in dichloromethane and the tertiary butyl ester hydrolysed by stirring with a 5-fold molar excess of trifluoroacetic acid. The hydrolysed polymer was then precipitated into petroleum ether. NMR analysis showed the hydrolysis to be ~99% complete. The gels were then formulated by dissolving the hydrolysed polymer in 85 : 15 methanol acetone mixture (33% polymer w/v). The thick polymer solution was then sheared between two glass slides to orient the microdomains, and then allowed to dry in a methanol saturated atmosphere for 8 h.

The Landolt oscillating reaction

The reaction mixture used was that reported previously[13,14] and is a development of a range of pH oscillators described by Rabai, Orban and Epstein.[19] Potassium bromate (Aldrich, 99+%), sodium sulfite (Aldrich, 98+%), potassium ferrocyanide (Aldrich, 99+%), and concentrated sulfuric acid (Aldrich) were all used without further purification. Solutions were made using distilled, deionised water and filtered prior to use. The reactions were conducted in an open continuously stirred tank reactor (CSTR), with a peristaltic pump supplying the feed solutions of potassium bromate, sodium sulfite, potassium ferrocyanide and sulfuric acid; and also pumping out bulk solution to keep the volume in the CSTR constant. Measurements of pH were made using a standard gel-filled probe, and the output from the meter monitored by a PC. The optimum flow rate for reliable oscillations was found to be at approximately 0.029 ml s^{-1} with concentrations: $[SO_3^{2-}] = 0.075$ M, $[BrO_3^-] = 0.065$ M, $[Fe(CN)_6^{4-}] = 0.02$ M, $[H_2SO_4] = 0.01$ M.

Scanning force microscopy

Liquid-SFM measurements of the sample were made using a Digital Instruments Multimode Nanoscope IIIa fitted with a J-scan tube (125 μm × 125 μm). Oxide-sharpened silicon nitride (SN-20, Digital Instruments) probes were used. Thin armed 100 μm long cantilevers were driven in the acoustic mode using the standard multimode liquid cell. Resonant frequencies were determined to be 12–13 kHz and Q-enhanced using an Active Resonance controller (Infinitesima). The sample was mounted on a steel puck with a thin film of PTFE attached to the upper surface. The scan tube was protected by placing clingfilm on the head of the tube. Analysis was conducted using Digital Instrument software and the three-dimensional surface plots were generated using WSxM software (Nanotec). Measurements were taken over the same area of the sample with the liquid being introduced through one of the filling holes of the liquid-cell.

Single molecule force experiments

Measurements were made using a molecular force probe (MFP) instrument (Asylum Research Inc.) isolated from vibrations using a Halcyonics active vibration table. The experimental setup uses an open cell arrangement with a wetted SPM tip brought into contact with the wetted sample surface. Commercially available cantilevers (Veeco) were used which had ultra sharp, gold-coated (increased reflectivity) silicon nitride tips calibrated using the thermal method built into the MFP. The tips had a radius of curvature of less than 20 nm (MSCT-AUNM Veeco SPM). A new tip was used after each pH run to reduce tip contamination effects. The brush samples were cleaned with copious amounts of de-ionised water before and after each set of measurements and allowed to equilibrate with the solvent for 5–10 min. This timescale proved adequate as no noticeable change in the force curve data was observed. The pulling speed of the cantilever tip was kept constant for all data at 800 nm s^{-1} and the tip reversed and approached with no dwell time on the surface.

The MFP measures force as a function of cantilever deflection along the z-axis (perpendicular to the sample surface). The observed deflection is made up of the distance between the tip and the sample and the cantilever deflection, therefore to get a true value of the tip–sample distance D, and produce meaningful force–distance curves, the cantilever deflection must be subtracted from the z-piezo movement to give values for D. For a very hard surface, zero separation is defined as the region in the force curve in which the cantilever deflection is coupled 1 : 1 with the sample movement, this appears in the force curve as a straight line of unit slope. A complete force curve includes the measured force profile as the tip probe approaches the sample and is retracted to its initial starting position.

Neutron reflectometry

Neutron reflectometry experiments were performed using the time-of-flight EROS (Orphée reactor of the Laboratoire Léon Brillouin) neutron reflectometer. Although neutron reflectometry is an ideal tool for studying the collapse and swelling of polyelectrolyte brushes, the large sample area necessary for reflectometry experiments has in the past limited studies to polyelectrolytes prepared by the "grafting to" method,[20] although neutron reflectometry has recently been performed on PNIPAM brushes grafted from a silicon surface.[21] Deuterium labelling of the water used to swell the brush provides the necessary contrast with the hydrogenous polybase brush which was grafted from a silicon substrate as described above, and placed in a sample cell, where they could be swollen with a mixture of D_2O and NaOH or HCl as appropriate. The neutron beam passed through the silicon block and reflected from the brush and heavy water interface.

Such a sample geometry is desirable because it provides a highly reflective interface for neutrons, increasing the efficiency by which we can obtain information on the brush profile. In these experiments, we started from a high pH, in which the brush was in a collapsed state, and further experiments were performed with smaller values of pH. After each measurement of the reflectivity for a given swelling (*i.e.* pH value), the silicon wafer with the brush was removed from the sample cell and washed before the next solution was added. In some runs, the brush needed about 30 min to equilibrate (the reflectivity profile changed during the run). The reflectivity data were fitted to model profiles using a Monte Carlo fitting algorithm, as well as a steepest descent fitting routine.

Small angle X-ray scattering

Time resolved SAXS experiments were performed at the Dubble BM26B beam line of the European Synchrotron Radiation Facility (ESRF), Grenoble, France. The ESRF Beam line optics and construction are detailed elsewhere.[22] For comparison, similar SAXS/WAXS experiments were performed on station 16.1 at the Daresbury SRS, UK and the station specifications have been detailed elsewhere previously.[23] The scattering vector axis, Q, of the SAXS patterns were calibrated using an oriented sample of rat-tail collagen, while for the spatial calibration of the WAXS patterns use was made of HDPE and NBS (National Bureau of Standards) silicon.

The experiment set-up for studying the triblock gel with SAXS comprised of a Nylon cell with an internal liquid volume of 100 ml. Two hollow nylon bolts with mica windows situated on their ends were screwed in the walls of the cell into the liquid volume to leave a gap of approximately 3 mm. Perpendicular to this were placed two transparent polystyrene windows that allowed illumination and image capture of the gel's macroscopic behaviour. The gel sample was held in place using tweezers. The addition of the solutions of the oscillating chemical reaction was *via* a peristaltic pump and pH was recorded using a combined microelectrode (Mettler Tolledo) connected to a PC situated outside the experimental area.

SAXS images were collected at 60 seconds per frame for a total time of 120 frames (2 hours) whilst situated within the oscillating chemical reaction. The scattering pattern obtained showed a sharp ring indicating a single length-scale within the phase-separated gel. The 2 dimensional data was sector integrated to produce 1D data.

Results

Gels in oscillating reactions

Our general approach uses pH as the stimuli for the responsive polymer and an oscillating stimulus can be given by the the bromate/sulfite/ferrocyanide pH oscillator.[19] The experimental benefits of this system are a reliable oscillation at room temperature, a large pH range (four orders of magnitude in $[H^+]$), and, most importantly an oscillation that remains at the pH extremes for approximately equal times to allow the mass transport (of water and ions) associated with the change in polymer conformation.

The oscillating reaction mechanism of the process is very complicated, consisting of many mechanistic steps, but may be viewed at a simplified level as consisting of two subsystems; the oxidation of sulfite by bromate (1) and the oxidation of ferrocyanide by bromate (2). The first process serves as the proton production mechanism (positive feedback) with an appropriate "induction time", while the second acts as the negative feedback reaction consuming protons.

Careful tuning of both the reactant concentrations and the feed rate to the reactor are required to obtain reliable oscillations and there is obviously an interaction between the feedback process of reaction and the changing degree of dissociation of the polyelectrolyte. The size dependence of the swelling rate of polymer gels is well established,[24] the data in Fig. 2 were obtained for a piece of gel of characteristic dimension 20 μm that could attain equilibrium swelling in a time scale shorter than the period of the oscillating reaction.

Six oscillations are shown here, these were taken from a reaction that had been running for 170 min and continued to run for a further hour. The reliability of the pH change and the response of the gel is readily apparent from the graphs. Whilst the large dimension attained is stable the minimum dimensions seem to be affected by hysteresis, the most likely reason being the difference in the characteristic times for swelling and collapse. In materials which are glassy (or of high polymer concentration) in the collapsed state, collapse takes longer than swelling because the difference in the geometry—a particle that is swelling has a low density corona and a dense core whereas a collapsing particle has a dense shell surrounding a swollen core—which is, therefore, a barrier to solvent diffusion on collapse and mass transport is restricted compared to expansion. In this case, however, the polyelectrolyte is still substantially plasticized by water in the collapsed state and the hydrodynamics are dominated by ion transport and mechanical equilibration, the collapse transition has a characteristic time approximately half that of expansion.[14] The local pH change is

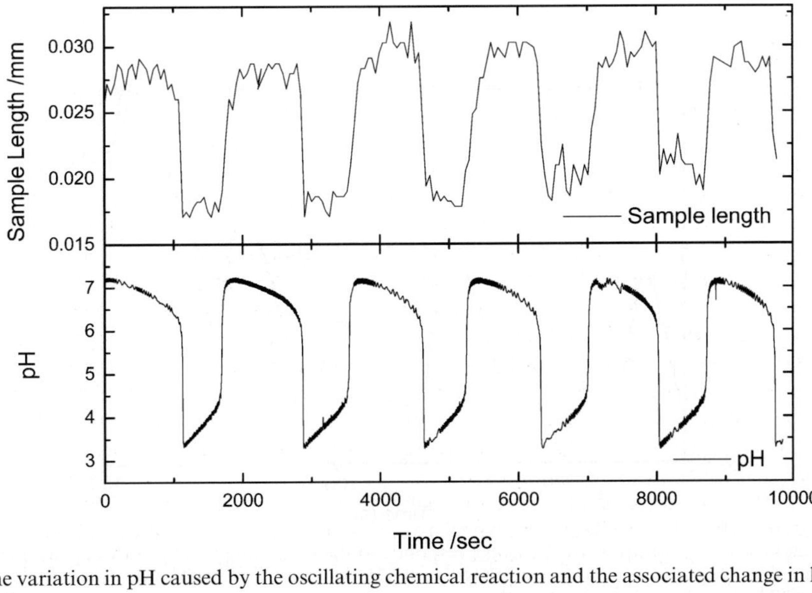

Fig. 2 The variation in pH caused by the oscillating chemical reaction and the associated change in length in the polymer gel.

$$BrO_3^- + 3HSO_3^- + H^+ \longrightarrow Br^- + 3SO_4^{2-} + 4H^+ \quad (1)$$

$$BrO_3^- + 6Fe(CN)_6^{4-} + 6H^+ \longrightarrow Br^- + 6Fe(CN)_6^{3-} + 3H_2O \quad (2)$$

Scheme 4 The fundamental reactions of the Landolt pH oscillator.

effectively instantaneous as protons diffuse rapidly. The sample can collapse completely within one pH cycle but cannot completely expand, this causes a gradual build-up in the ionic strength within the gel as the water diffuses faster than the solvated ions and the hysteresis due to the effective screening of charge.

Measurements were made of the force generated by such a gel in a modified JKR apparatus.[25] The sample is placed on a stage inside a beaker, and then a *Z*-translational stage, controlled by a worm microdrive, is lowered down onto the sample holding it in place. The solutions for the oscillating reaction are then pumped into the beaker with a peristaltic pump until the optimum level is reached, at which point waste is pumped out at an equal rate to maintain a constant volume. The reacting solution is then stirred using a paddle mounted on a 12 V motor, at as slow a rate as possible whilst maintaining good mixing, in order to minimize noise in the readings. The concentrations of the reactants used were the same as used for the swelling experiments. As the gel swells and deswells in the oscillating solution, the translational stage is set in position when the pH is low and the gel contracted, so that it is applying a force in the order of a gram. Changes in this force were then monitored *via* the balance over the course of numerous oscillations (Fig. 3). The pH data are presented as measured. The force data have been smoothed using a fast Fourier transform smoothing function to remove only the stirring noise, as this has a constant and definite frequency.

The oscillation appears to have a longer period, and is less regular, than that seen in Fig. 2. It is not clear whether this is due to a buffering effect of the gel or an effect of the size scale of the reaction vessel. Forces were measured in a reaction taking place in a 100 ml flask whereas the size was measured in a Petri dish with a volume of 25 ml. Although an attempt was made to keep the same residence time in both cases, some kinetic mismatch may occur due to the use of different apparatus volumes, and this could contribute to the change in oscillation period. Over the four hour time period, the force produced can be seen to follow the period of the oscillating reaction indicating that the mass transport through the gel is commensurate with the timescale of the oscillations. The maximum force produced by the gel is 0.012 N. The slight upward drift in the data is due to an increase in mass caused by a small mismatch between the flow rates of the pumps used.

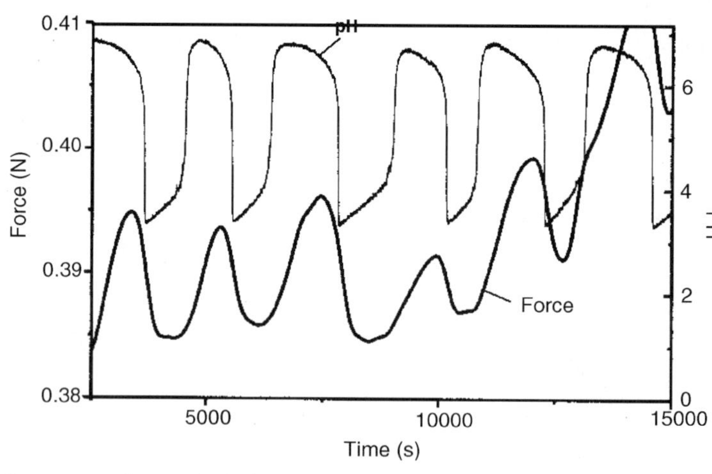

Fig. 3 pH and force *versus* time as the hydrogel responds to the oscillating reaction. Residence time 1000 s, at room temperature. Input concentrations: $[SO_3^{2-}] = 0.075$ M, $[BrO_3^-] = 0.065$ M, $[Fe(CN)_6^{4-}] = 0.02$ M, $[H_2SO_4] = 0.01$ M.

In summary a pH oscillator has been developed to cover the range 3–7 with a time constant of approximately 20 min. The reaction has been coupled with a responsive gel to generate a model system with spatio-temporal oscillations. We have observed macroscopic dimensional changes by optical microscopy and mechanically harnessed the gel to generate force. Of course, the measured force is only a small proportion of the potentially available force, as it is only measured in one dimension and is heavily damped by mass-transport considerations, but it displays that these systems can be used to directly convert chemical potential to mechanical work.

Responsive brushes

The polymer brush used in this study was generated by the "grafted-from" approach to generate a brush in the mid- to dense- brush regime. The grafted-from technique requires the anchoring of an initiator to a surface followed by exposure to a polymer generating medium/environment. The ATRP method used for the polymerisation is a living process where a halide is reversibly transferred from the initiator (or growing chain) to a metal halide to leave a radical. This radical will then propagate polymerisation for a short period of time before the halide is transferred back and the growing chain becomes dormant. This results in polymer chains that grow at the same rate coupled with a suppression of side reactions, as the concentration of radicals is always very small at any given time. Initial studies concentrated on the responsive nature of dense PMAA brushes which were grown on Si wafers with the maximum areal density of initiator.

To demonstrate the responsive nature of the dense "grafted-from" brushes the brush height was measured as a function of the pH (an ionic strength) of the solvent using SFM. From the images of a PMAA brush in Fig. 4 we can clearly see a marked difference in brush height and brush topography. This we attribute to the brush being in the collapsed (pH 3) and extended (pH 10) state, respectively. The height of the brush was measured by taking a histogram of heights from the upper right-hand corner of both images.

The height is quoted with reference to the bare silicon surface represented by the data at height = 0 nm in Fig. 5. A collapsed height of 86.4 nm with a standard deviation of 16.6 nm was

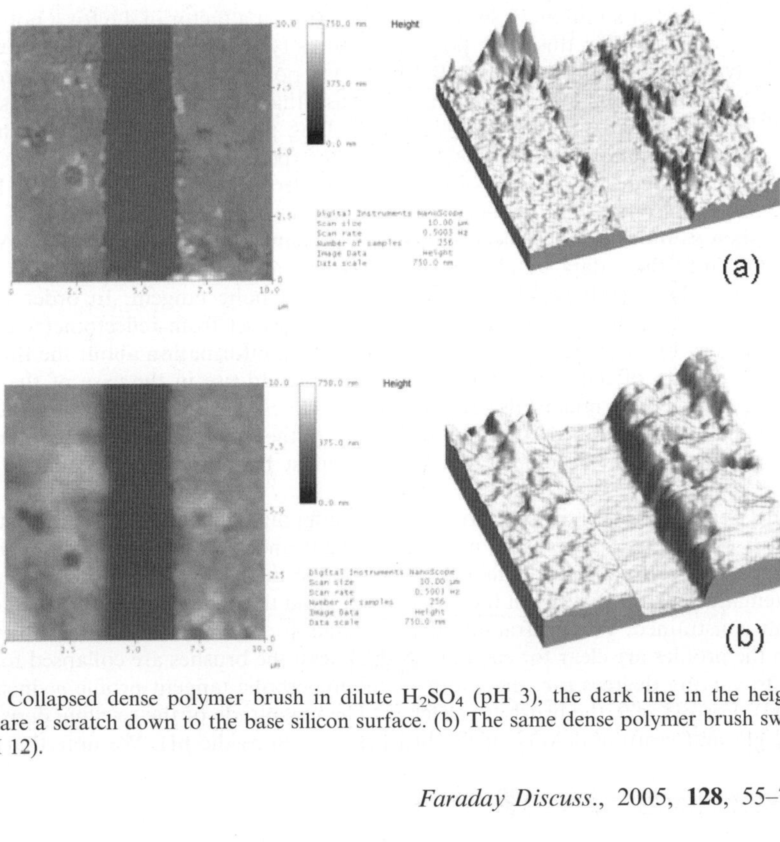

Fig. 4 (a) Collapsed dense polymer brush in dilute H_2SO_4 (pH 3), the dark line in the height image and the trough are a scratch down to the base silicon surface. (b) The same dense polymer brush swollen in dilute NaOH (pH 12).

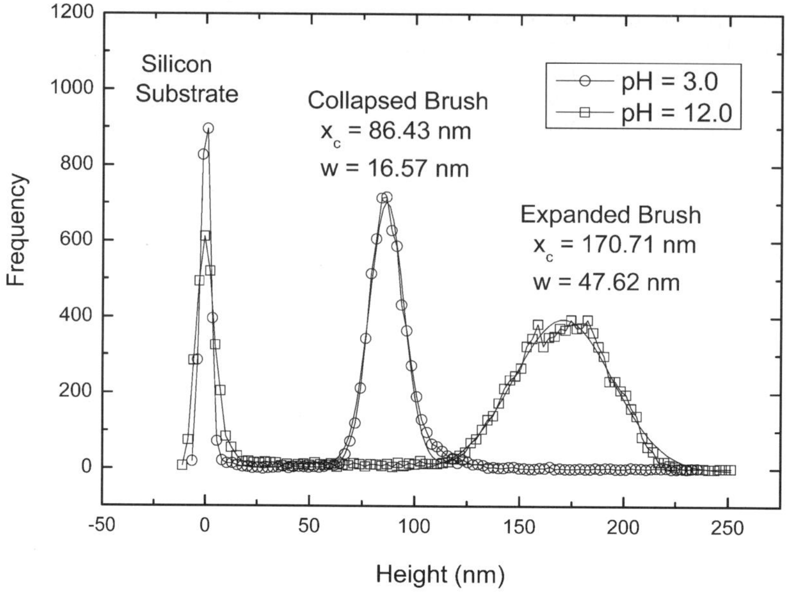

Fig. 5 Histogram of heights determined from the SFM images in Fig. 4. The data are reported as fits to the power spectra of position x_c and distribution w.

measured for acidic conditions. For basic conditions a height of 170.7 nm and a width of 47.6 nm was measured. This represents a swelling by a factor of $\times 1.97$ and is very low compared to that expected from the simple brush collapse model, discussed above. A conservative estimate of the degree of polymerisation of these chains is 300 (taking the dry-brush thickness to be the contour length). This would predict a minimum swelling ratio of at least a factor of 4.6 for a polymer chain.

The low degree of swelling in this thick polymer brush is typical of a polymer brush with chains that were densely packed during synthesis. Because each polymer chain is already stretched (for steric reasons) in the collapsed state the change in brush thickness with pH is rather small. That dense brushes like these had very low degrees of swelling prompted us to develop methods of controlling the surface initiator density as described in the experimental section. The brushes used in the subsequent neutron reflectivity and molecular force experiments had a significantly lower areal chain density to minimise the steric reduction in thickness change.

In Fig. 6 we show sample neutron reflectometry data for a medium areal density PDEAMA brush swollen in D_2O. To fit these data we chose a simple model accounting for single layers of silicon oxide and initiator. The brush profile was fitted to a hyperbolic tangent. In order to limit the number of parameters we used the scattering length densities set from reflectometry experiments performed on the dry brush in air. This enabled us to obtain information about the thickness and scattering length density of the initiator layer, which we could use in the case of the brushes in solution. The dry brush experiments also revealed the thickness of the brush layer in the dry state, which fixes the amount of PDEAMA that we could fit to in the swollen case (*i.e.* the integrated area under the hyperbolic tangent volume fraction profile must be constant and equal to the brush thickness).

Experiments were performed on two brushes with different thicknesses in the dry state (brush thickness increases with both increasing molecular weight and increasing grafting density, which forces the chains to stretch). Typical data and fits for one of the brushes are shown in Fig. 6. There are some systematic deviations between the model curves and the data. This reflects the fact that the fits were highly constrained. Volume fraction-depth profiles for the brush layers are shown in Fig. 7. The trends in the profiles are clear for each brush thickness; the brushes are collapsed for values of $pH > 4$, but to varying degrees (*i.e.* the width of the hyperbolic tangent profile σ, increases with decreasing pH). For $pH < 5$, the brush is stretched. The magnitude of the swelling depends on the molecular weight and grafting density of the brushes, as well as the pH. We note that the 130 Å

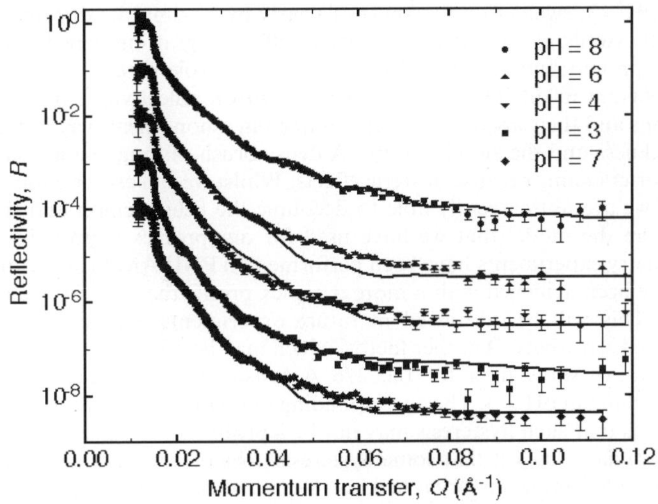

Fig. 6 Neutron reflectivity spectra for a PDEAMA brush in D_2O as a function of pH. For clarity of viewing the data are shifted down vertically by a factor of 10 from each other.

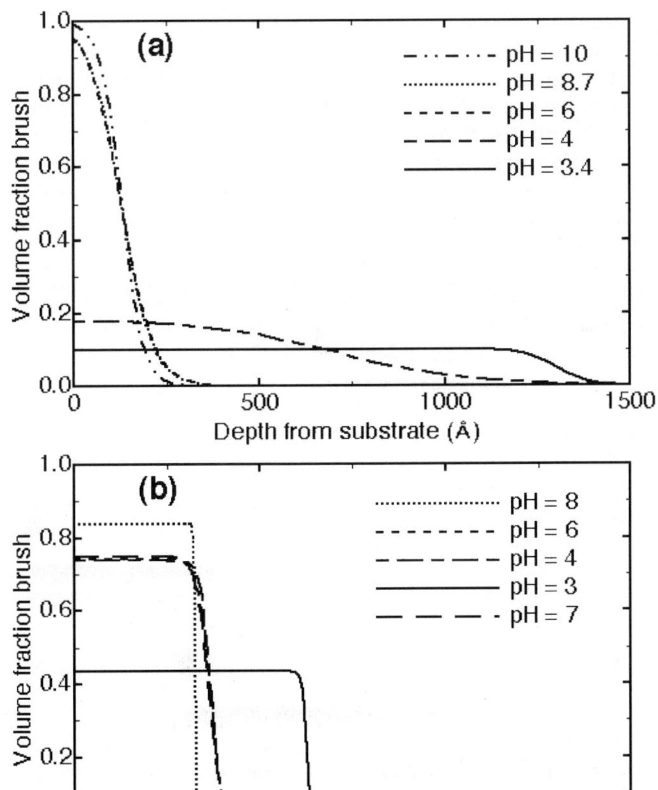

Fig. 7 Volume fraction–depth profiles for a brush of thickness 130 Å (a) and 276 Å (b) in the dry state. The profiles in part (b) correspond to the fits presented in Fig. 6.

brush swollen in pH = 3, swells to a thickness of nearly by 10 times its dry thickness, but that the 276 Å brush hardly swells at all, which suggests an effect of grafting density (a molecular weight effect may also be present, but is expected to play a smaller role here).

The problem inherent in the interpretation of the neutron reflectivity data is that the molecular weight of the chains and their areal density are coupled in a non-linear way in their effects on both the dry brush thickness and the swelling ratio. A dense brush of long chains may swell less than a dilute brush of shorter chains because of steric effects. Whilst we can exercise some degree of control over the initiator we are not currently able to decouple the brush length and the areal density of chains. Furthermore the model that we have used for our profiles is probably rather simplistic. Previous reflectometry experiments have been performed on PNIPAM brushes and the best fits that were achieved have been achieved with a more complex profile than the simple hyperbolic tangent presented here.[21] This will be addressed in future experiments with deuterated polyelectrolyte brushes for increased sensitivity. Another factor, which may be important is that of hysteresis in the swelling. In Fig. 7, we show data for the 276 Å brush swollen in pH = 7. This swelling was performed after swelling in pH = 3. The corresponding fit is consistent with the other data, shown in Fig. 7b, and suggests that such hysteresis may not be a problem, for these samples. However, other (as yet unpublished) data suggest that some hysteresis does indeed exist, and a full study of brush swelling as a function of sample history is lacking. Despite these caveats, neutron reflectometry does show that the PDEAMA brush swells as a function of pH, and that at low pH, the thinner brush indeed swells quite substantially compared to its dry state.

Single molecule force spectroscopy experiments have been previously reported for both natural[26] and synthetic[27] polyelectrolytes. In particular Ortiz and Hadziiannou[28] have studied the force–distance curve of polymethacrylic acid in water, using a *freely jointed chain* (FJC) model to interpret the single molecule behaviour. One of our concerns is that we can assess the "efficiency" of any device we might build in terms of the force developed by a single chain and we have studied the response of PDEAMA in the expanded and contracted states. A Si_3N_4 probe tip was used to pick up the chains *via* non-specific physisorption interactions.

A typical force–distance curve for a collapse (hydrophobic) chain is given in Fig. 8. Far away from the surface the force is zero and there is no interaction between the tip and the sample [region (a) in the figure]. As the cantilever approaches the surface, [region (b)], a nonlinear repulsive force is

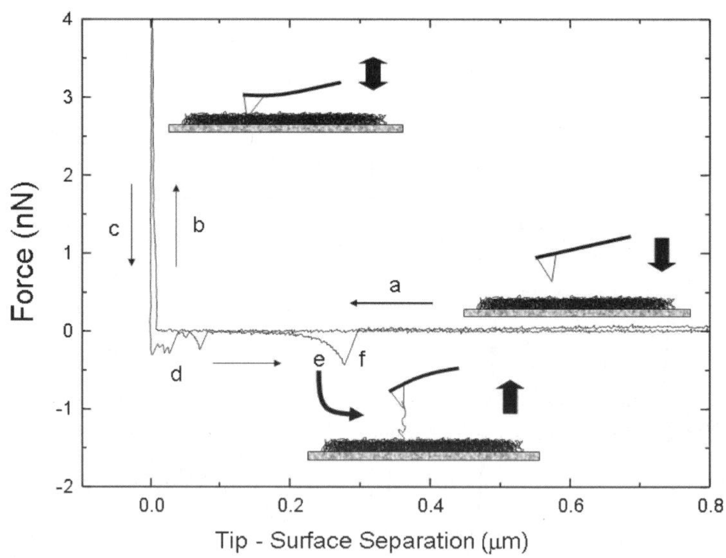

Fig. 8 The anatomy of a typical force–distance curve for a polyacid in the collapsed state capable of adsorption on the SFM tip. (a) Tip approaches surface—no interaction. (b) Tip compression of brush–"hard-sphere" type interaction. (c) Retraction—force acting on the tip decrease as the sample is moved away. (d) Tip pulled down—indicating adhesion/multiple-chain pulling events. (e) Stretching of a single molecule (force curve). (f) Polymer detaches from the tip.

observed corresponding to compression of the brush and this initial reaction can be seen more clearly in Fig. 9a. A detailed discussion of the transition from the squeezing of the brush to the hard-sphere repulsion on contact of the tip and the silicon wafer is beyond the scope of this discussion but the near vertical force–distance curve in region (b) denotes the contact point and compression of the Si wafer. Decompression of the Si is reversible, initially following the same force–distance curve [region (c)] as compression, and there is some hysteresis due to the tip–substrate adhesion but the majority of the attractive force and its instability in region (d) is due to the detachment of many chains. In contrast to previous reports[28] on dilute brushes there is a high likelihood of multiple chain attachment in these dense brushes. As the retraction proceeds a single attractive peak is observed with a nonlinear increase of force with distance [region (e)]. This attractive peak is attributed to the stretching of an individual polymer chain which had become physically adsorbed to the tip and bridges to the surfaces. That this part of the force–distance curve applies to a single chain is borne out by the statistical nature of the data fitted to many such curves. The chain conformation on the surface of the tip; that is a combination of loops, trains and tails; is not known. At sufficiently high extensions the elastic restoring force of the chains becomes greater than the adsorption force tethering the chain segments to the tip and the chain suddenly desorbs and the cantilever returns to its undeflected position corresponding to zero force (f).

Ortiz and Hadziiannou[28] tested a series of models for fitting the force–distance curves to the entropic restoring force due to uncoiling of individual chains. We have fitted our data to the FJC using both the contour length and the statistical segment length of the chain as fitting parameters as the length of chain that is physisorbed to the probe tip is variable so the effective contour length is also a variable even if the chains are all of a unique length. The FJC is expressed as

$$F_{\text{chain}} = {}^{k_{\text{B}}T}/_{a} L * (r/L_{\text{c}})$$

where F_{chain} (nN) is the restoring force produced by the stretched elastic polymer chain, r (nm) is the absorbed chain end to end separation (*i.e.* the tip to surface distance), k_{B} (J K^{-1}) is Boltzmann's constant, T (K) is temperature, a is the statistical segment length (nm) and L^* is the inverse Langevin function. L_{c} (nm) is the fully extended length of the chain (na) where n is the number of chain segments. L_{c} is a fitting parameter since the adsorption length is random even if all the chains have the same number of monomers.

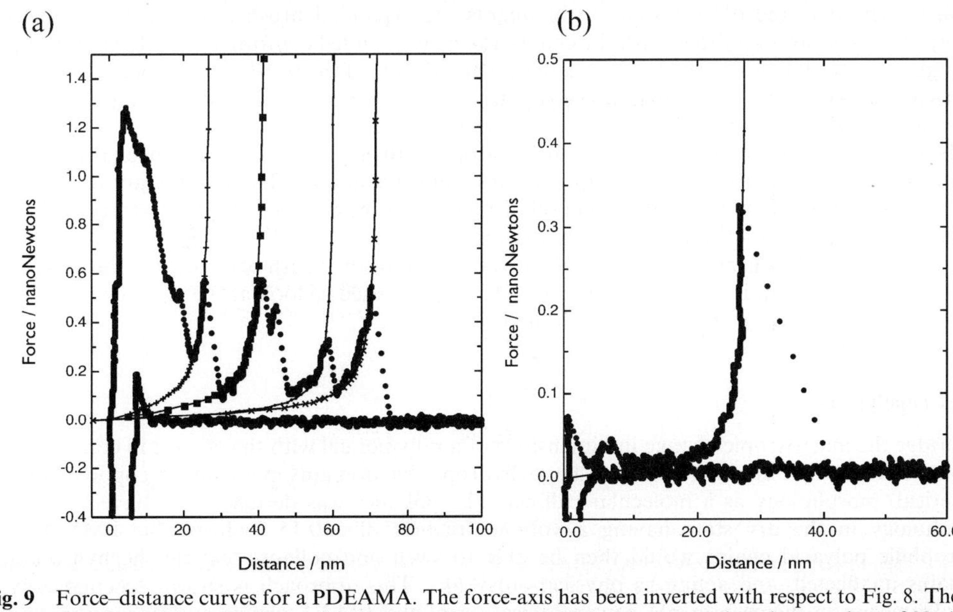

Fig. 9 Force–distance curves for a PDEAMA. The force-axis has been inverted with respect to Fig. 8. The line is a fit of the freely jointed chain model. (a) Pull-off curves at pH 4 with contour length, L_{c}, values of 28, 42, 63 and 73 nm and statistical segments lengths, a, of 0.09, 0.12, 0.18 and 0.23 nm respectively and (b) a pull-off curve in pH 10 buffer with contour length, L_{c}, of 48 nm and statistical segment length of 0.48 nm.

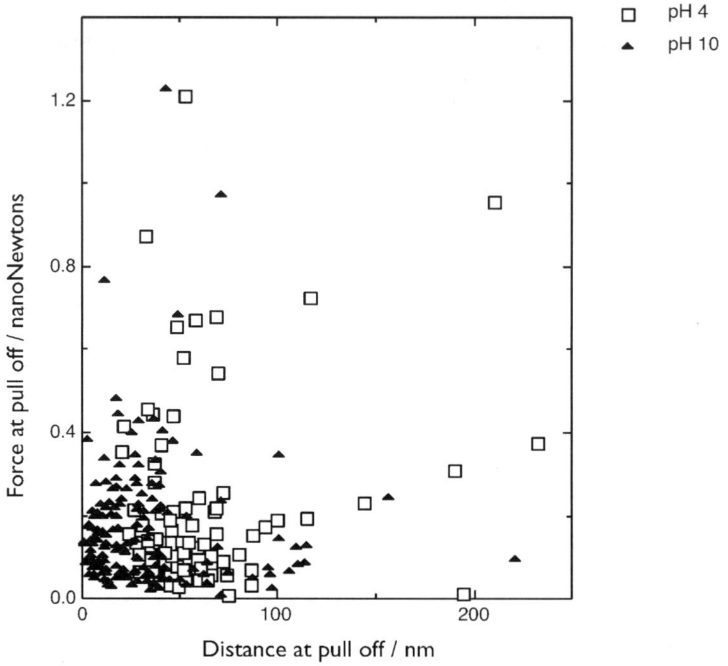

Fig. 10 Pull off values for the force and distance of the collapsed (filled triangles) and swollen (open squares) state of brush sample 12.

The (pH = 4) expanded brush state has a lower statistical segment length corresponding to a more flexible chain, a (pH = 10) = 0.37 nm ± 0.19 nm compared with a (pH = 4) = 0.19 nm ± 0.12 nm. The two sets of curves shown in Fig. 9 are representative of the data. The force–distance curve for the brush immersed in the pH = 4 solution highlights the expanded brush having a large adhesion energy. Near the surface of the brush the chains are in an extended conformation allowing the tip to entangle and attach to more chains than in the collapsed state of the brush. In the collapsed state the chains and the resulting force curves show very little adhesion near the surface, as they are less able to form attachments to the tip due to their collapsed nature.

The data presented in Fig. 10 are the final (pull-off) values of force and distance (L_{PO}) in each force distance curve as the cantilever tip detaches from the surface. The preliminary data for the swollen brush (pH 4) exhibits a larger pull off distance than the contracted (pH 10) force pulls. The average values are 62 ± 41 nm and 32 ± 31 nm respectively.

Further work will concern a more detailed investigation of the brush behaviour in the absence of added salt. Alongside this a variable pH mechanism will be used to look at the switching behaviour of the brush as contact is maintained with the tip.

Block copolymers

To bridge the macroscopic change in dimensions of a polymer gel with the change in conformation of the polyelectrolyte chains we use the glassy hydrophobic domains in a triblock copolymer cubic (spherical) morphology as a molecular calliper. The polymer was designed[29] to have a spherical morphology in the dry state having a volume fraction of ~0.15 hydrophobic PMMA. The hydrophilic polyacid chains would then be able to swell and collapse leaving the hydrophobic domains unaffected, and acting as physical crosslinks. This approach is shown schematically in Fig. 11 where an increase in pH causes ionisation of the PMAA chains and expansion on the nanoscale that propagates through to the unit-cell mesoscale and a serial addition of unit cell volume changes contribute to the macroscopic change in shape of the gel.

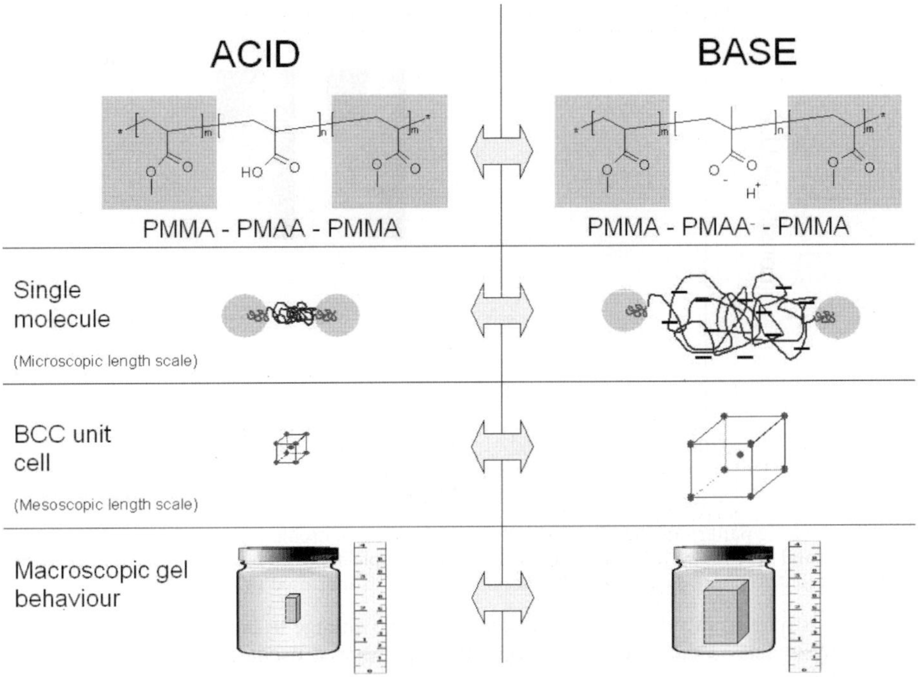

Fig. 11 The cascade of length scales probed in the SAXS study of responsive block copolymer gels.

The triblock copolymer used here was a PMMA–PtBuMA–PMMA (7k–141k–7k) triblock with a polydispersity of 1.01 that was subsequently hydrolysed to yield a mid-block of methacrylic acid and PMMA–PMAA–PMMA (7k–85k–7k). The preparation of a suitable sample for scattering proved difficult as annealing of the melt is not an option due to the glass transition of PMAA being greater than 250 °C. To form a gel approaching equilibrium a solvent casting method was developed using a mixed solvent system—2 miscible solvents that are selective solvents for each block. Acetone and methanol, solvents for PMMA and PMAA respectively, were used at 15 : 85 v/v ratio. (the same ratio as polymer volume fractions.) to make a concentrated solution at 33 wt.% polymer. High concentrations were required to prevent excessive looping of the chains so that the hydrophobes were not both in the same domain—*i.e.* to prevent flower micelle formation and encourage bridging of domains.[30] Acetone has a higher vapour pressure than MeOH and so this evaporates first to generate glassy PMMA domains in a MeOH swollen PMAA matrix. To generate a suitable sample, the concentrated solution was sheared between two glass slides held apart by 300 μm spacers. The slides were then run off each other and placed in a saturated methanol atmosphere for 8 h. To handle the gel, which is glassy in the absence of any solvent, the solvent annealed sample was submerged in water. The polymer gel floated off the glass surface and was cut into a suitable sized piece using a scalpel.

The response of the polymer was first studied in the quiescent state using a trapped gel that was exposed to both acidic and basic solutions. There was an appreciable shift in the peak position, q^*, with pH and the transition time was less than ten minutes. Monitoring the gel-size, by video microscopy, showed the shift in domain spacing to be commensurate with the swelling. The dynamic behaviour of the gels was then studied in the oscillating chemical reaction and the SAXS patterns and images of the gels are shown for both the collapsed and the expanded state in Fig. 12.

As expected the peak in the scattering pattern moved towards lower Q as the pH of the oscillating chemical reaction increased. As the pH suddenly dropped the peak moved towards higher Q. The position of the maximum (q^*) is plotted against the pH of the solution in Fig. 13. From this we see that the gel takes approximately 1 complete cycle of expansion–collapse to establish equilibrium with its surroundings. Also the moment at which the collapse and expansion begins is the point at

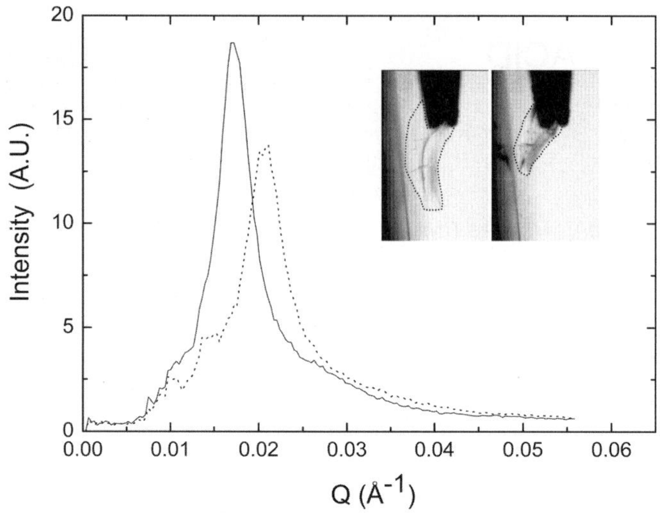

Fig. 12 1D scattering pattern of PMMA–PMAA–PMMA gel in an oscillating chemical reaction at $t = 5$ mins (pH = 6.19) (—) and $t = 15$ mins (pH = 3.66) (···). The macroscopic behavior is shown inset as the expanded and collapsed forms (left, right, respectively). The outline of the gel is indicated to aid the eye.

which the pH is seen to rapidly increase or decrease. The change in length scale shows no signs of damping or hysteresis as the reaction proceeds through several oscillations.

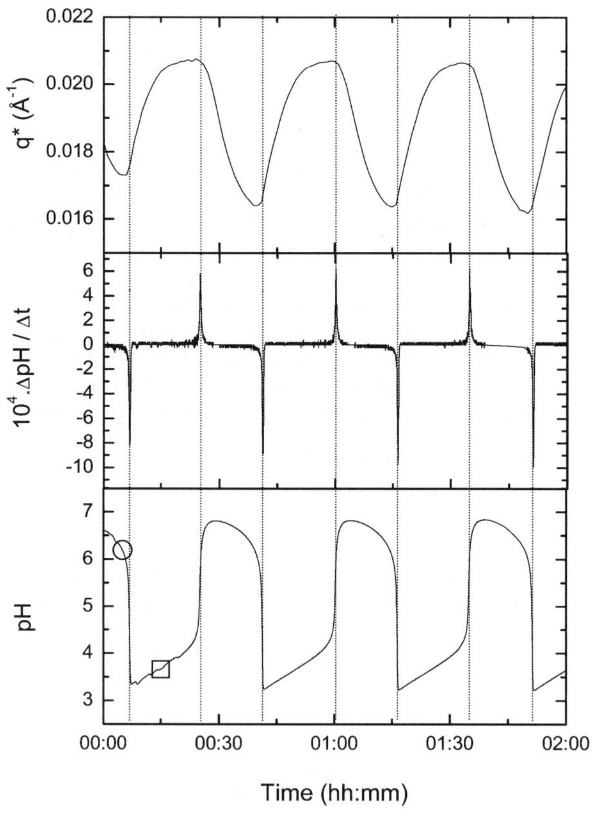

Fig. 13 1D SAXS peak position (q^*) showing the correlation between peak position and change in pH. The data shown in Fig. 12 is marked in the lower graph (○ and □).

Summary

This paper is essentially a report on a work in progress demonstrating that polyelectrolyte chains and oscillating chemical reactions can be coupled to give a system that shows responsiveness that scales additively from the molecular to the macroscopic. As active components we have utilised weak polyacids and polybases, which respond to changes in pH. These have been assembled into covalently linked networks, polymer brushes that are essentially responsive monolayers and triblock copolymers that combine the characteristics of self-assembly and responsiveness. At the macroscopic level, these systems, when coupled with a chemical reaction that cause spontaneous pH oscillations, allow us to make a free-running chemical motor. At the single molecule level, we have characterised the response of these brushes to changes in pH by using scanning force microscopy (SFM) and neutron reflectometry following the collapse transition *in situ.* Furthermore the mechanical properties of a single chain change in response to pH changes have been studied by single molecule force measurements in a molecular force probe, giving us the information to analyse a motor cycle in our target single molecule synthetic motor. In addition, we have quantified the degree to which our polyelectrolyte brushes act as surfaces with switchable adhesion. The macroscopic behaviour of responsive systems is linked to the molecular response by studying the response of triblock copolymers to pH oscillations. The active chains are the mid-blocks and are effectively crosslinked by the domains of the hydrophobic end block which act as molecular callipers. The expansion and contraction of the grafted polyelectrolyte chains causes the distance between hydrophobic domains to change, this separation has been monitored by SAXS allowing the molecular and macroscopic effects correlated for the first time.

There is much current activity in developing systems whose properties are responsive to environmental changes. The laws governing self-assembly processes operate at supra-molecular length scales and are universal, and the specific properties of any objects built this way are defined by their chemical structure. Templating with block copolymers, polypeptides and polyelectrolytes can create structures and build devices for a wide range of applications based on a limited set of physical principles. Polymer science and supra-molecular chemistry have developed to the stage that makes possible the construction of useful devices based on responsiveness and self-assembly. The major theme of this research is the identification of the underpinning technology required for a range of generic passive and active systems and the application of established methods in physical science to the production of *soft-nanotechnology* devices that are inspired by cell-biology. Complex, highly engineered structures are held together by weak bonds (hydrogen bonds, hydrophobic/hydrophilic interactions, screened Coulomb interactions) formed at room temperature and pressure from molecules that have information written into them. The macromolecules are used both as templates and the components of molecular machines. Devices made by these principles are self-assembling and responsive to their environment, they operate in a solvent (preferably water) at a constant, relatively high temperature. Early results from this research might find uses in drug delivery and opto-electronic devices. Longer-term efforts may be directed toward the making of self-propelled vehicles that recognise targets, creating a version of nanotechnology that looks like cell biology.

Acknowledgements

We would like to thank the European Commission for the provision of beamtime on EROS under the "Large Research Infrastructure" contract HPRI-CT-2002-00170, the ESRF for beamtime on DUBBLE under the long-term-proposal SC777 and CCLRC for beamtime on 16.1. LRP, CJC, PT, AJP, SJM and AC were supported by EPSRC grants (GR/R77544, GR/R74383, GR/R55573, GR/N02542, GR/M72449) and JRH was supported by ICI PLC.

References

1 M. Shibayama and T. Tanaka, *Adv. Polym. Sci.*, 1993, **109**, 1–62.
2 *Biorelated Polymers and Gels*, ed. T. Okano, Academic Press, San Diego, CA, 1998.
3 D. J. Beebe, J. S. Moore, J. M. Bauer, Q. Yu, R. H. Liu, C. Devadoss and B.-H. Jo, *Nature*, 2000, **404**, 588–590.

4 A. Karim, S. K. Satija, J. F. Douglas, J. F. Ankner and L. J. Fetters, *Phys. Rev. Lett.*, 1994, **73**, 3407–3410; M. Sferrazza, R. A. L. Jones and D. G. Bucknall, *Phys. Rev. E*, 1999, **59**, 4434–4440.
5 H. Iwata, I. Hirata and Y. Ikada, *Macromolecules*, 1998, **31**, 3671–3678.
6 D. L. Huber, R. P. Manginell, M. A. Samara, B. I. Kim and B. C. Bunker, *Science*, 2003, **301**, 352–354.
7 S. Leibler and D. A. Huse, *J. Cell Biol.*, 1993, **121**, 1357–68.
8 R. Yoshida, T. Takahashi, T. Yamaguchi and H. Ichijo, *J. Am. Chem. Soc.*, 1996, **118**, 5134–5135.
9 R. Yoshida, E. Kokufuta and T. Yamaguchi, *Chaos*, 1999, **9**, 260–266.
10 R. Yoshida, G. Otoshi, T. Yamaguchi and E. Kokufuta, *J. Phys. Chem. A*, 2001, **105**, 3667–3672.
11 R. Yoshida, T. Sakai, S. Ito and T. Yamaguchi, *J. Am. Chem. Soc.*, 2002, **124**, 8095–8098.
12 R. Yoshida, K. Takei and T. Yamaguchi, *Macromolecules*, 2003, **36**, 1759–1761.
13 C. J. Crook, A. Cadby, A. Smith, R. A. L. Jones and A. J. Ryan, *ACS Symp. Ser,*, 2003, **869**, 71–79.
14 C. J. Crook, A. Smith, R. A. L. Jones and A. J. Ryan, *Phys. Chem. Chem. Phys.*, 2002, **4**, 1367–1369.
15 R. Iscoff, *Semicond. Int.*, 1993, **16**, 58–63.
16 K. Matyjaszewski, P. J. Miller, N. Shukla, B. Immaraporn, A. Gelman, B. B. Luokala, T. M. Siclovan, G. Kickelbick, T. Vallant, H. Hoffmann and T. Pakula, *Macromolecules*, 1999, **32**, 8716–8724.
17 K. L. Beers, S. G. Gaynor, K. Matyjaszewski, S. S. Sheiko and M. Moller, *Macromolecules*, 1998, **31**, 9413–9415.
18 A. Ikker and M. Moller, *New Polym. Mater.*, 1993, **4**, 35–51.
19 G. Rabai, M. Orban and I. R. Epstein, *Acc. Chem. Res.*, 1990, **23**, 258–63.
20 Y. Tran, P. Auroy and L. T. Lee, *Macromolecules*, 1999, **32**, 8952–8964.
21 H. Yim, M. S. Kent, D. L. Huber, S. Satija, J. Majewski and G. S. Smith, *Macromolecules*, 2003, **36**, 5244–5251; H. Yim, M. S. Kent, S. Mendez, S. S. Balamurugan, S. Balamurugan, G. P. Lopez and S. Satija, *Macromolecules*, 2004, **37**, 1994–1997.
22 W. Bras, I. P. Dolbnya, D. Detollenaere, R. van Tol, M. Malfois, G. N. Greaves, A. J. Ryan and E. Heeley, *J. Appl. Crystallogr.*, 2003, **36**, 791–794.
23 J. P. A. Fairclough, C. L. O. Salou, A. J. Ryan, I. W. Hamley, C. Daniel, W. I. Helsby, C. Hall, R. A. Lewis, A. J. Gleeson, G. P. Diakun and G. R. Mant, *Polymer*, 1999, **41**, 2577–2582.
24 E. S. Matsuo and T. Tanaka, *J. Chem. Phys.*, 1988, **89**, 1695–703.
25 K. R. Shull, *Mater. Sci. Eng. R*, 2002, **36**, 1–45.
26 D. A. Smith and S. E. Radford, *Curr. Biol.: CB*, 2000, **10**, R662–4.
27 T. Hugel, M. Grosholz, H. Clausen-Schaumann, A. Pfau, H. Gaub and M. Seitz, *Macromolecules*, 2001, **34**, 1039–1047; C. Wang, W. Shi, W. Zhang, X. Zhang, Y. Katsumoto and Y. Ozaki, *Nano Lett.*, 2002, **2**, 1169–1172.
28 C. Ortiz and G. Hadziioannou, *Macromolecules*, 1999, **32**, 780–787.
29 A. K. Khandpur, S. Foerster, F. S. Bates, I. W. Hamley, A. J. Ryan, W. Bras, K. Almdal and K. Mortensen, *Macromolecules*, 1995, **28**, 8796–806.
30 J. Yao, P. Ravi, K. C. Tam and L. H. Gan, *Polymer*, 2004, **45**, 2781–2791.

Molecular dynamics modeling of polymer crystallization; from simple polymers to helical ones

Takashi Yamamoto,* Naohiko Orimi, Naohito Urakami and Kaoru Sawada

Department of Physics, Biology and Informatics Faculty of Science, Yamaguchi University, Yamaguchi 753-8512, Japan

Received 22nd April 2004, Accepted 14th May 2004
First published as an Advance Article on the web 13th August 2004

Crystallization of helical polymers is a very big challenge for molecular simulation. It involves many significant issues, such as folding in biomolecules and molecular recognition during crystal growth. Though direct molecular simulations of the process still involve very difficult problems, we here report our recent efforts toward better understanding of the crystallization in helical polymers. We begin with a brief review of our former studies on simple polyethylene-like polymers, and then we introduce several helical polymer models which are systematically made more complicated. We have already reported that a simple polyethylene-like polymer crystallizes very fast into chain folded lamellae from the melt. A slight modification of this simple polymer model by introducing proper bond angle and dihedral angle potentials gives one of the present models of the helical polymer. This helical polymer model is devised to be relatively rigid but mobile, to show easy helix-reversals, and to have a definite preference for *gauche* bonds. We find that this highly mobile helical polymer shows quick chain folded crystallization and forms approximate 4/1 helical structure. The intra- and the intermolecular order grow quite simultaneously suggesting highly cooperative nature of the phenomena. Further elaboration of the helical model, giving pendant side groups and higher energy barrier to the helix reversals, leads us to a realistic united atom model of iPP. The conventional and the multi-canonical Monte Carlo simulations are applied to find probable modes of chain folding and the ground state conformations. Though a very short chain readily forms a regular 3/1 helix of alternating *trans* and *gauche* bonds, much longer chains of 30- and 50-propylene units are not found to have energetic ground states in the regularly folded conformations.

The molecular mechanism of polymer crystallization has long been a very important but controversial issue in the history of polymer science. Innumerable studies on polymer crystallization have been hitherto reported, and interested readers will find many textbooks and proceedings on this topic.[1–6] Lauritzen–Hoffman (LH) theory[6] has succeeded in giving simple and elegant explanations for various aspects of polymer crystallization, and has dominated more than half a century. The theory only takes the string nature of polymer chains into account and involves no specific geometrical parameters for the molecular structures. The great success of the theory suggests that the polymer crystallization follows a universal path irrespective of the molecular details of the polymers. Despite the great success of the theory, the molecular level understanding of polymer crystallization is still considered far from satisfactory. The elementary molecular processes assumed in the LH theory were greatly simplified in order to make analytical treatments tractable, but there is no direct experimental evidence to verify the assumed molecular mechanisms. Indeed, various different molecular scenarios have been put forward which are more detailed from the molecular dynamics point of view.[7–11] However these molecular scenarios give more or less similar conclusions to the LH theory as to the macroscopic aspects of crystallization; it is still difficult to make decisive experiments to show which scenario is the most realistic.

Direct molecular simulation studies of polymer crystallization are recently appearing to break through the difficulties. In fact, many pioneering works have been reported on the primary nucleation[12–17] and on the steady state growth of chain folded lamellae[18–29] of simple polyethylene-like polymers. While a simple polyethylene-like polymer, hereafter we refer to it simply as *linear polymer* as opposed to a *helical polymer*, though this is not a conventional use of the word, is very suitable for study by molecular simulation methods, helical structures are quite common in polymers, for example the slightly twisted structure of polytetrafluoroethylene (PTFE), the approximate 9/5 helical structure of polyoxymethylene (POM), and the regular 3/1 helix of isotatctic polypropylene (iPP), to cite just a few. Molecular mechanisms of chain ordering or crystallization in helical polymers are however expected to be much more complex than that of a simple linear polymer. Let us consider a helical polymer which takes both the right-handed (R-handed) and the left-handed (L-handed) helices with equal probabilities. On the growth surface of ordered lamellar crystal, the molecule must strictly select either the R- or the L-handedness according to the structural symmetry of the growing crystal. Each molecule being taken into the crystal lattice must go through the selection process either by reorganizing itself or by selective deposition on the surface; it is definitely a sort of molecular recognition.[30]

Despite such a fascinating feature of the crystallization in helical polymers, the significance of the process seems to have attracted only scarce attention.[31,32] Especially, the molecular modeling of crystallization in helical polymers has not been attempted except very few studies,[33–35] but mostly without success. Molecular processes involved have been considered too complicated and slow to be investigated by molecular simulation methods. This paper reports on our recent efforts for better understanding of crystallization in helical polymers. The underlying difficulties in modeling crystallization of helical polymers are the central theme of the discussion.

Models and methods of simulations

MD simulations of linear polymers

The crystallization of a polymer is a very complex molecular process that has long defied approaches *via* direct molecular simulations. However, we have recently succeeded in simulating the steady state growth of chain folded lamellae.[20–23] We have there adopted a simplified model for a polyethylene-like linear chain, where the zigzag structure of the polymethylene chain is replaced by a linear sequence of beads, thereby facilitating chain diffusions along the chain and transverse to it. The polymer chains we considered were composed of many (from 100 to 1000) beads of mass 14 u that correspond to the methylene group. The beads were connected by harmonic springs, which had a force constant and an equilibrium bond length similar to that of the C–C bond.[20–23] Each pair of beads is assumed to interact *via* van der Waals forces of usual Lennard-Jones type of well-depth $\varepsilon = 598.64\ \mathrm{J\ mol^{-1}}$ and separation $\sigma = 0.3923$ nm; the interactions were cut off at $r = 2.5\sigma$. The molecular rigidity was adjusted through a bond angle bending potential in order to reproduce the characteristic ratio of a real polyethylene chain. We simulated the growth of lamellar crystals from the dense melt phase which was sandwiched between two growth substrates (Fig. 1); the substrate crystal was made of the same bead–spring chains closely packed. A conventional molecular dynamics (MD) simulation by use of the leapfrog method was applied, with the temperature being controlled by simple velocity scaling.

MC and MD simulations of helical polymers

(a) Monte Carlo (MC) simulations of realistic helical polymers. The simple linear polymer described above is a mere string of nearly uniform thickness and smooth contour. The molecule has low intermolecular friction, which undoubtedly gives high mobility and high crystallization rate. The most straightforward approach to the crystallization in helical polymers would be the study of

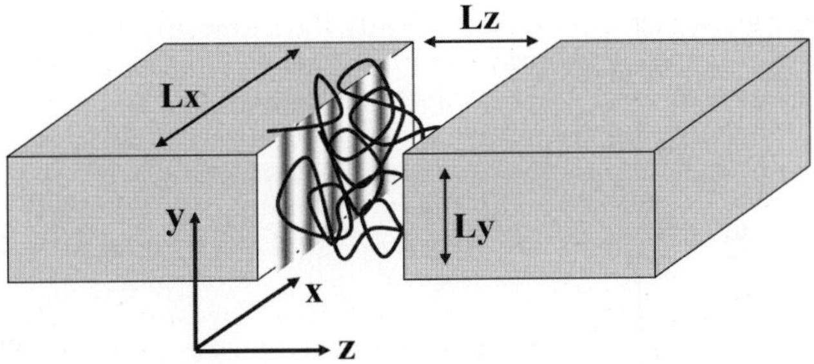

Fig. 1 Schematic picture of the polymer melt put between parallel substrates that correspond to the growth surfaces of two lamellae; the substrate lamellae are also made of the same model polymer chains. The *z*-axis is taken normal to the substrate, while the *y*-axis is along the chain direction of the substrate crystals.

realistic helical polymers like iPP having pendant side groups. We first present our efforts in this direction. A rather realistic model of single iPP chain was adopted where backbone CH, CH_2 groups and the pendant CH_3 groups were all treated as single united atoms with all the bonds connecting the united atoms having the same fixed lengths 0.154 nm (Fig. 2). Intermolecular and intramolecular potentials were obtained from the transferable potentials for phase equilibria (TraPPE) by Siepmann *et al.*[36] We here used both the usual Metropolis MC and the multi-canonical MC (MCMC),[37] where usual reptation motions and several local motions, rotation about axes connecting neighboring atoms, were considered.[38]

(b) MD simulations of simplified helical polymers. In the studies of realistic iPP, we will find it very difficult to reproduce crystal-like ordered arrangements of the helical chains. We consider the difficulty is partly due to the presence of helix reversal defects, the boundaries between the R-handed and the L-handed segments. In many cases where both the R- and the L-handed helices have the same intra-molecular energy, the crystallizing chain consists of fragmented helices of both hands and the consequent helix reversal defects. Since the helix reversal defects should be removed from the crystalline stems, there will be no route to crystallization if the helix reversal defects are very difficult to move. Thus the main origin of very slow crystallization in realistic helical polymers will lie in the difficulties in removing the helix reversal defects as well as in the general slow dynamics of chains with bulky groups.

We therefore examined a simple helical polymer which has no pendant side groups but that is nevertheless forced to adopt a helical conformation through the imposition of a torsional potential which has definite minima around *gauche* positions. The chain we here consider takes the approximate 4/1 helical conformation, which is similar to POM if we disregard the difference between oxygen and carbon atoms. The force field parameters used are similar to those of the conventional Rigby–Roe method for polyethylene (PE)[39] except for the dihedral angle potential, which is devised to give approximate 4/1 helix having a potential minimum around

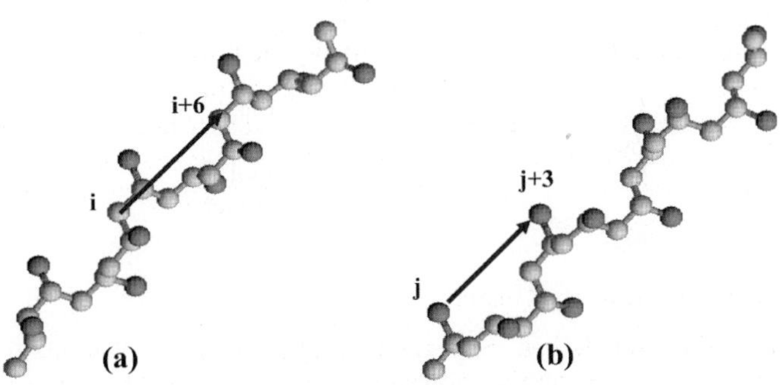

Fig. 2 Typical ordered conformations of realistic iPP chains of (a) the left-handed and (b) the right-handed helices.

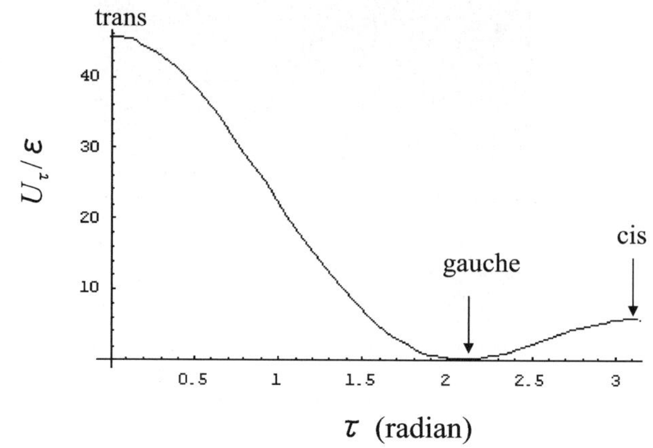

Fig. 3 A torsional potential for the simplified helical polymers with minimums at *gauche* positions and the low barrier to helix reversal at *cis* position.

gauche $\tau = 120°$(Fig. 3, Table 1);

$$U_\tau = k_\tau \sum_{n=0}^{3} a_n \cos^n \tau \tag{1}$$

Molecular dynamics simulations are performed there using a coarse graining MD-tool, COGNAC, in the recently developed simulation platform OCTA,[40] and both a single long chain made of 500 or 1000 atoms, and many chains of short oligomers are considered. The temperature is controlled with the conventional Andersen–Nosé–Hoover method, and the van der Waals interactions are truncated at 2.5σ.

Results of molecular simulations

Brief review of crystallization of linear polymer

We have already reported that the polymer chain whose molecular structure is very simple and straight shows rather fast crystallization into chain-folded lamellae.[18–23] In order to contrast it with later discussions on helical polymers, we here briefly report our previous studies on crystallization of

Table 1 Parameter values used

Parameters	Values
Mass (CH_2)	1.4×10^{-2} kg mol^{-1}
r_0	0.152 nm
k_b	3.46×10^7 J nm^{-2} mol^{-1}
θ_0	70.5°
k_θ	5.0×10^5 J mol^{-1}
k_τ	2.629×10^3 J mol^{-1}
a_0	1.0
a_1	3.983
a_2	3.868
a_3	−0.195
ε	500 J mol^{-1}
σ	0.38 nm
Cutoff	2.50σ

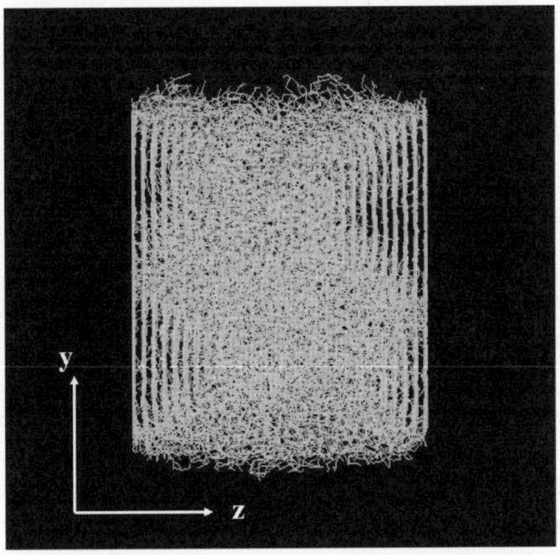

Fig. 4 A representative snapshot of the system made of 640 chains of C100, obtained at 12.8 ns after quenching to 350 K. We readily notice crystalline domains growing toward each other from both substrates.

the linear polymer. We there adopted a simplified molecular model for polymethylene-like chains; the chain was made of CH_2 beads connected by harmonic springs, and the lowest energy conformation was a linear stretched sequence of the beads with slight bending stiffness being imposed. The system of many molecules, typically 64 000 atoms made of 640 molecules of C_{100}, was placed between two parallel substrates that represent the growth surfaces of the lamellae growing toward each other, and the initial melt state at a sufficiently high temperatures was rapidly cooled down to various crystallization temperatures.

Due to its high molecular mobility, this linear polymer showed remarkably fast lamellar growth from the substrate, a crystal nucleator. Fig. 4 shows a typical snapshot of the growing lamellae from both crystal growth surfaces. Average growth rate of the lamellae was found to be nearly constant unless they collide, and depend very markedly on undercooling $\Delta T = T_m - T$ from the melting temperature T_m.[22,23] The molecular processes of crystallization at the growth front was studied *in situ*, and we found that chains were making frequent adsorption–desorption at the growth

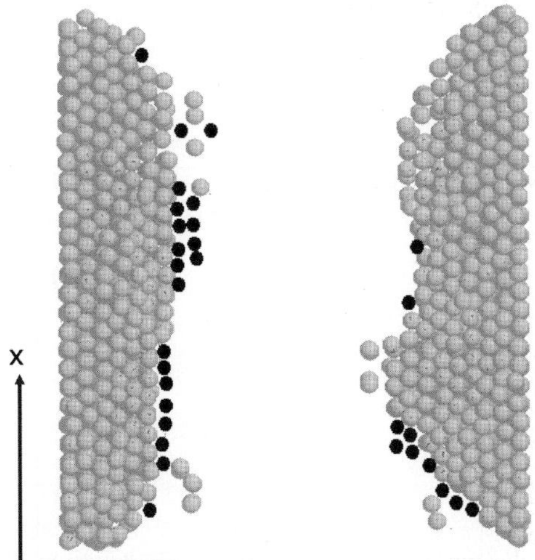

Fig. 5 Growing lamellae viewed along the chain axis at 330 K after 6.4 ns. Hexagonal chain packing and rather flat {100} surfaces are marked. Also shown are newly adsorbed crystalline stems (black circles) within the subsequent period of 0.128 ns.

surfaces but still leading, on the average, to the growth of crystalline patches on the growth surfaces (Fig. 5). The molecular mechanism showed a rather good correspondence to the usual Kossel mechanism of crystal growth. We have also investigated the structure of the undercooled melt, but we found no specific order in the chain orientation.[20–23]

Ordering in realistic helical polymer iPP

Straightforward MD modeling of crystallization in realistic helical polymers iPP was reported to be difficult, maybe taking an extremely long simulation time.[33,34] Therefore we began with the study of a short iPP chain of 10 propylene units (20 backbone carbons) by use of the MC and MCMC methods.[37]

Through the usual MC and MCMC simulations, we found that the ordering of relatively short chains free from folds was rather fast, giving well-defined 3/1 helices of iPP (Fig. 2). Conformational changes with temperature were investigated through MCMC simulations. Fig. 6 shows the changes in the radius of gyration R_g and the conformational order parameter P_{order}, where the latter parameter was defined as

$$P_{order} = \langle (3\cos^2\theta - 1)/2 \rangle; \quad (2)$$

θ denotes the angle between the vectors spanning the repeat units of the 3/1 helix, $\boldsymbol{r}_i$ and $\boldsymbol{r}_{i+6}$ of the backbone atoms, and those connecting pendant groups $\boldsymbol{r}_j$ and $\boldsymbol{r}_{j+3}$ (Fig. 2). The short iPP chain readily stretches from a random coil to highly ordered conformations, either the R-handed or the L-handed, around 400 K.

When we consider longer iPP with chain folding or multiple chains of short iPP, the ordering becomes much more sluggish. We here report on the similar MCMC simulations of a single iPP molecule, of 30- and 50-propylene units. The molecules of iPP in the usual α-form crystals are known to take one of four conformations of different helical senses, R or L, and methyl group directions, up (u) or down (d): R-u, R-d, L-u, L-d. In case of a 30-unit chain, we could pick up, out

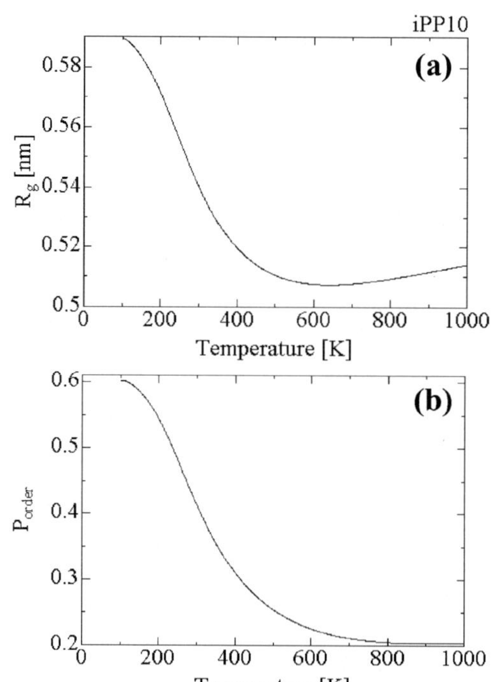

Fig. 6 Temperature dependence of (a) the radius of gyration R_g and (b) the order parameter P_{order} in the short iPP chain.

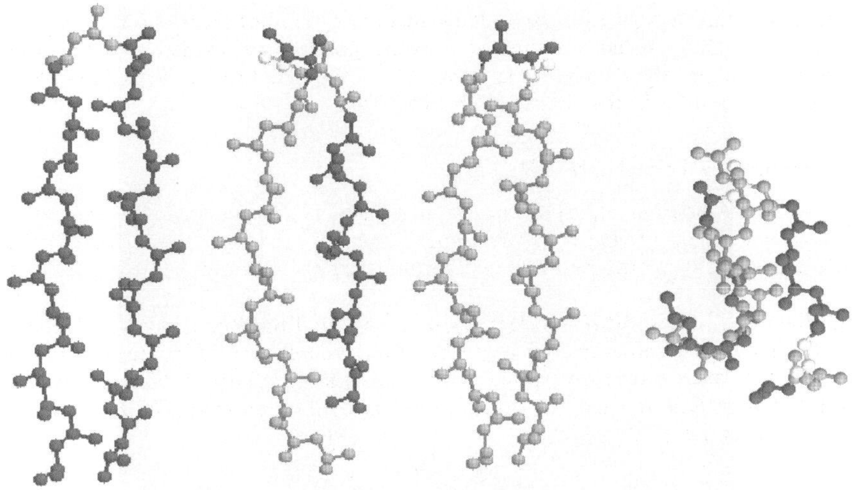

Fig. 7 Fold conformations of the iPP chain of 30-propylene units obtained by the multicanonical Monte Carlo simulations: (a) Ld-Lu, (b) Rd-Ld, and (c) Rd-Ru fold conformations, and (d) a ground state conformation. The right- and left-handed helical conformations are colored light grey and dark grey, respectively.

of many relatively ordered states, three types of one-fold conformations, Ld-Lu, Rd-Ld, and Rd-Ru (Fig. 7), but all these still had considerable conformational disorders. In fact, no conformation obtained was any of the most stable states. Surprising was the fact that the lowest energy conformation obtained through MCMC simulation was a globule-like state (Fig. 7). We also performed similar simulations for a 50-unit chain, but again we could not obtain a stable fold conformation. The lowest energy conformation obtained involved many helix reversal defects which seem to be nearly frozen by the competition between the R- and the L-handed helical fragments.

Ordering in simplified helical polymers

As has been described above, one of the factors dominating crystallization of helical polymers is the presence of helix reversal defects. In general, these kinds of defects greatly distort the linearity of the helical chains thereby seriously hinder the parallel packing of chain stems. In order to develop the crystalline order, the helix reversal defects must be annihilated or segregated into fold regions.

Here we consider a rather compact helical polymer without side groups, which is still forced to take a helical conformation. As a simple example we considered POM-like chain where each torsion angle has definite potential energy minimums around *gauche* states: G+ and G−. The simplest case we mainly discuss here is a 4/1 helical polymer whose energy barrier, at the *cis* position, to the helix reversal is as small as thermal energy (Fig. 3). In this case, the helix reversal defects are expected to drift very easily along the chain.

To gain basic understanding of this model helical polymer, we first studied the crystallization of short oligomer chains made of 20 carbons. Within a MD-cell under periodic boundary conditions, we placed a hundred of chains and held at 540 K where the molecules were in a gas phase. With decreasing temperature the oligomer gas turned into liquid, in which most of the chains took either the R- or the L-handed helix. By further cooling down to 240 K and subsequent annealing near the estimated melting point of 400 K for 20 ns, we could observe the development of lamellae of extended helical chains, though some of the chains were still floating between lamellae (Fig. 8). In the crystalline state, the chains were packed in an ordered hexagonal lattice of $a = 0.497$ nm slightly larger than that of the real POM crystal at 0.447 nm (Fig. 9), with a slight preference for chiral domains of the same helical sense being suspected. The chain conformation was found to be slightly rewound from the regular 4/1 helix, the torsion angle of which is 120°, toward the POM-like

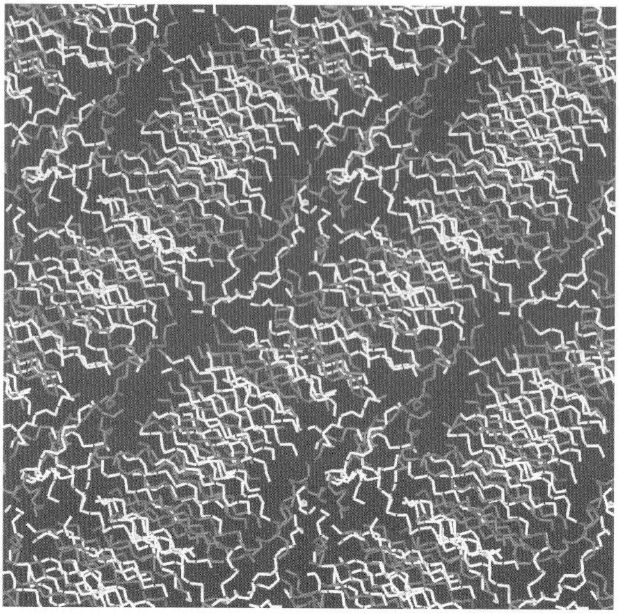

Fig. 8 Well-developed lamellae of the simplified helical oligomers, obtained after 20 ns of annealing at 360 K. The R- and the L-handed molecules are depicted in white and grey, respectively, against black background.

18/5 helix; the average torsion angle is about 108° which accidentally conforms to the 105° of the 9/5 helix of POM if the difference between C and O is disregarded.

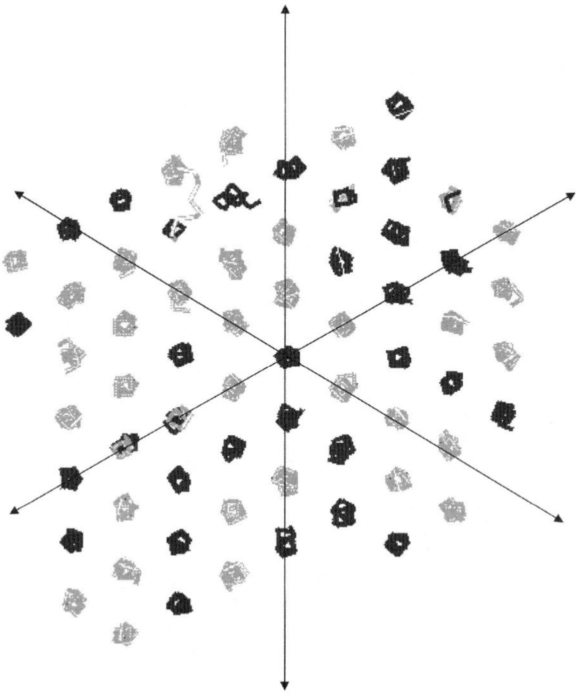

Fig. 9 Hexagonal packing of the oligomer chains after 2 ns simulation at 180 K. The presence of small chiral domains is suggested; black and grey chains are the R-handed and the L-handed helices, respectively.

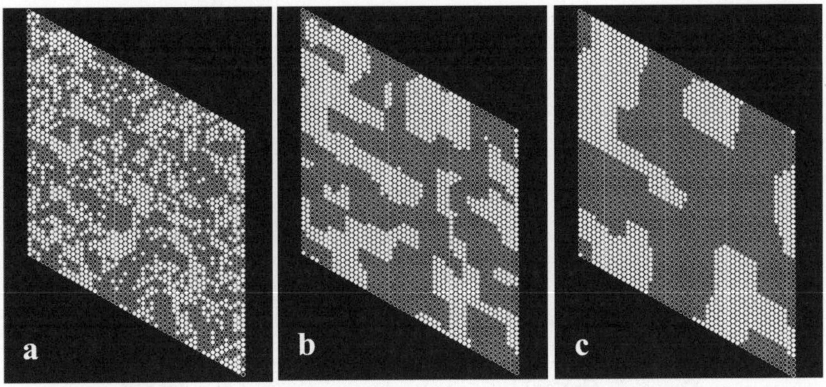

Fig. 10 Development of the chiral domains calculated with the lattice MC method in search of the most stable chain packing of the simplified helical chains. The R- and the L-handed chains viewed along the chain axis are depicted in white and black circles, respectively: (a) relatively random state after 10 MCS, (b) after 100 MCS, (c) after 1000 MCS.

It is very essential in crystallization studies whether or not the present model molecules tend to form chiral domains in the hexagonal lattice. We examined this by use of a 2D lattice MC simulation of stable chain packing; the chains were assumed perfectly regular helix of torsion angle 108° and were placed in the hexagonal lattice of $a = 0.497$ nm. The method of simulation was just the same as that was used in our previous studies of chain packing in crystals of PTFE, n-alkanes, iPP, *etc.*[41–43] We do not detail the simulation method here, but Fig. 10 clearly indicates that the present model helices prefer to pack into domains of the same handedness. The fact that the only imperfect chiral domains are observed in the present MD simulation (Fig. 9) thus suggests the presence of kinetic barrier to the helix reversals in the crystals.

Now we are ready to study crystallization in much longer helical polymers. Hereafter, we consider a single helical chain, made of larger number of atoms either 500 or 1000 backbone atoms, which has the same force field parameters described above. Quite in contrast to the realistic helical models of iPP described before, this model helical polymer showed very quick ordering into chain folded structures. This is maybe due to the absence of bulky side groups and to the easy helix reversals. Fig. 11 shows the changes in chain dimension, the radius gyration, during slow cooling. The molecule shows steady compaction down to around 400 K, then it begins to elongate along the direction along which helical stems orient.

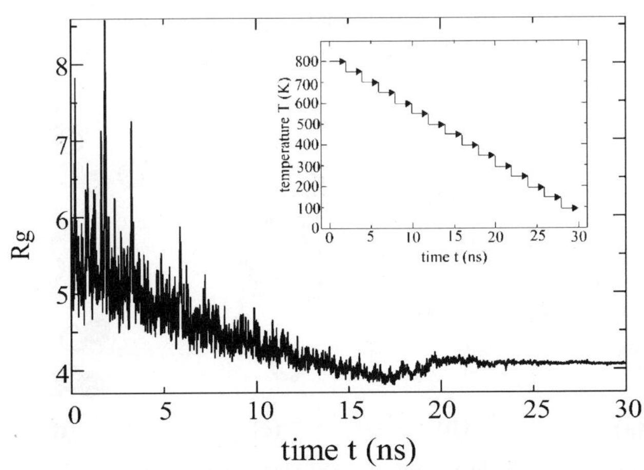

Fig. 11 Molecular process of chain compaction and ordering, represented by the radius of gyration, during stepwise cooling from 800 K down to 100 K. The inset is a temperature diagram during the cooling process.

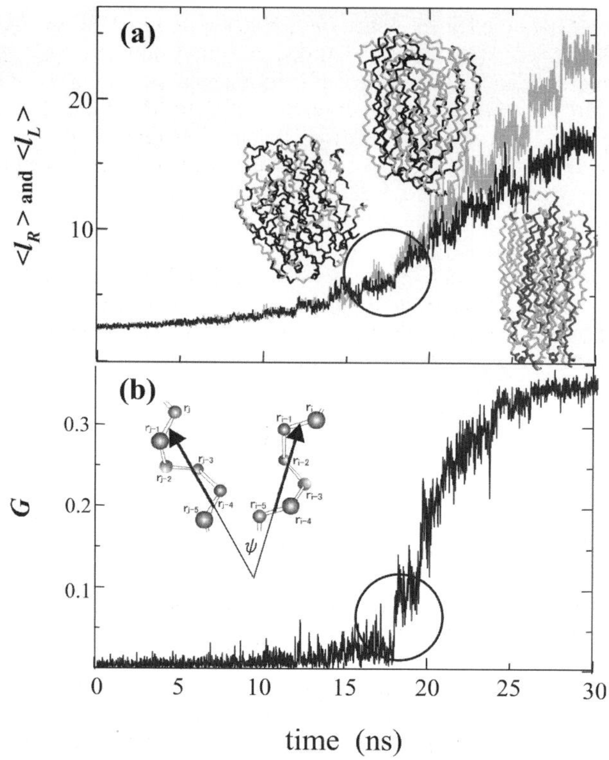

Fig. 12 Molecular processes of the intramolecular and the intermolecular ordering in the simplified helical polymer. (a) Increases in the average helical sequence length of the R-handed and the L-handed helices. The solid and the dotted lines correspond to the R-handed and the left-handed helices, respectively. Notice the change around 400 K. (b) Helical cord vectors to define global orientational order of the polymer, and the growth of interchain order parameter *G*. We can see the onset of ordering around 400 K very clearly.

Ordering or crystallization of long helical polymers necessitates removal or segregation of the helix reversal defects into regions of abundant chain defects, usually amorphous regions between lamellae. Good chain packing requires the removal of the defects from the crystalline stems, while the straight stems without the defects accelerate the chain packing into crystalline order; the intra- and the inter-molecular orders may be interrelated or cooperative. In order to investigate the cooperative actions during crystallization, we considered two kinds of order parameters. One is the

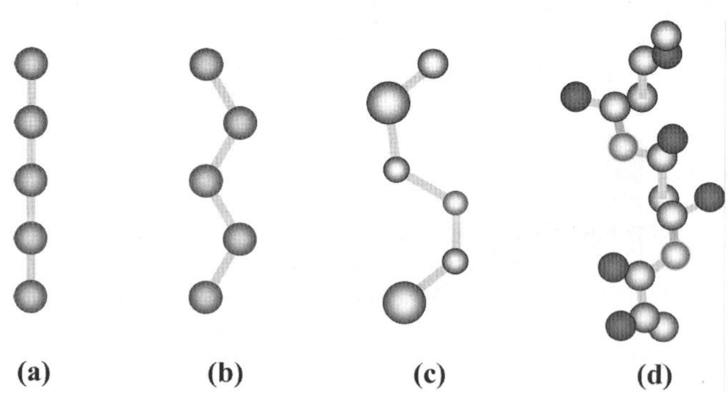

Fig. 13 Model step-up from (a) the linear polymer, (b) planar-zigzag polymer, that is the usual united atom model of polyethylene, (c) simplified helical polymer having approximate 4/1 helix, (d) realistic united atom model of iPP having pendant methyl units.

intra-molecular order of the helical polymer, which is represented by the average length of the consecutive helical sequences either the R-handed or the L-handed, $\langle l_R \rangle$ or $\langle l_L \rangle$, respectively; the increase in this length indicates the motion of helix reversal positions along the chain (Fig. 12a). The other is the global intermolecular order, which is defined taking the repeat period of the approximate 4/1-helix into consideration through helical cord vectors connecting every two center of bonds separated by three bonds (Fig. 12b); the intermolecular order parameter is here defined by $G = \langle (3\cos^2\psi - 1)/2 \rangle$; the parameter indicates the degree of parallel packing of the straight stems.

Fig. 12 shows changes in the average sequence lengths of the R- and the L-handed helices with time/temperature, and the global order parameter during slow stepwise cooling from 800 K down to 100 K at a cooling rate of 25 K ns^{-1} (see inset of Fig. 11). Though with little imbalance of the R-handed and the L-handed helix lengths, both helical sequences lengthen considerably around 17 ns (around 400 K) from a few bond lengths to more than 20-bond lengths. On the other hand, the global order shows a rather steep increase from that of a completely random state ($G \simeq 0$) to that of partially ordered state ($G \simeq 0.35$) (Fig. 12b) also around 400 K. It is thus strongly suggested that the intra- and the inter-molecular ordering cooperatively lead to crystallization in helical polymers.

Conclusions

The simplest linear chain (Fig. 13a) showed very fast crystallization. We could even observe the large scale crystallization, the steady state growth of the chain folded lamellae from the melt. In our previous papers on crystallization of n-alkanes, we showed that only the slight modification of the chain geometry by imposing the bond angle $\theta = 112°$ to the linear chain (Fig. 13b), which gives the usual united atom model of polyethylene, remarkably slows down the molecular processes of crystallization.[44,45]

In this paper, we further modified the chain geometry to have the simple helical polymer (Fig. 13c). We adopted somewhat artificial model in which the conventional united atom model for polyethylene was modified with respect to the torsion angle potential only; we considered the torsion potential that have the deep minimum at the gauche positions. We expected the ordering in this simple helix to be considerably slow, but we found that this helical polymer showed unexpectedly quick ordering into chain folded conformations within several tens of nanoseconds. The competition between the R-handed and the L-handed helices was found to be easily resolved if we took the torsional energy barrier against helix reversal to be sufficiently small.

By far the more difficult task was to observe the chain ordering in a realistic helical polymer model of iPP (Fig. 13d) which have the pendant methyl groups and the high energy barrier for helix reversals. The polymer showed only a collapsed state made of short helical fragments which did not easily lengthen by reversing their helical senses.

In every case studied, except the lattice MC simulation, the ordering in the helical senses of the crystalline stems seemed too much slower to be reproduced by the molecular level MD or MC simulations. The energy imbalance of the R–R, L–L, and R–L pairs is not large enough to give crystalline domains of definite chirality. In order to overcome the local energy barrier against helix reversals and to complete the reversal of the whole crystalline stem, sufficient thermal energy is needed for the helix reversal defects to be excited and propagate through the stem. This may be only possible at temperatures close to the melting point, but then the free energy drive to the ordered chiral structures can be smaller and can serve to work the other way.

Acknowledgements

T. Y. is very grateful to the Ministry of Education, Science, and Culture, Japan for the financial support through the Grant-in-Aid of Scientific Research on Priority area, "Mechanism of Polymer Crystallization" (No.12127206). Thanks are also to the members of the Doi-project for free use of the simulation platform OCTA and for their generous support.

References

1 B. Wunderlich, *Macromolecular Physics*, Academic Press, New York, 1976, vol. **1–3**.
2 F. C. Frank, *Faraday Discuss. Chem. Soc.*, 1979, **68**, 7.

3 K. Armistead and K. G. Goldbeck-Wood, *Adv. Polym. Sci.*, 1992, **100**, 219.
4 L. Mandelkern, *Crystallization of Polymers*, Cambridge University Press, Edinburgh, 2002.
5 *J. Macromol. Sci. B*, 2003, **vol. 42**, pp. 403–928, special issue on the International Symposium on Polymer Crystallization.
6 J. D. Hoffman and R. L. Miller, *Polymer*, 1997, **38**, 3135.
7 D. M. Sadler and G. M. Gilmer, *Polymer*, 1984, **25**, 1446.
8 J. P. K. Doye and D. Frenkel, *Phys. Rev. Lett.*, 1998, **81**, 2160.
9 J. J. Point, *Macromolecules*, 1979, **12**, 770.
10 M. Hikosaka, *Polymer*, 1990, **31**, 458.
11 G. Strobl, *Eur. Phys. J. E*, 2000, **3**, 165.
12 T. A. Kavassali and P. R. Sundararajan, *Macromolecules*, 1993, **26**, 4144.
13 C. Liu C and M. Muthukumar, *J. Chem. Phys.*, 1998, **109**, 2536.
14 P. Welch and M. Muthukumar, *Phys. Rev. Lett.*, 2001, **87**, 218302.
15 H. Meyer and F. J. Mueller-Plathe, *J. Chem. Phys.*, 2001, **115**, 7807.
16 H. Meyer and F. J. Mueller-Plathe, *Macromolecules*, 2002, **35**, 1241.
17 T. Miura, R. Kishi, M. Mikami and Y. Tanabe, *Phys. Rev. E*, 2001, **63**, 61 807.
18 T. Yamamoto, *J. Chem. Phys.*, 1997, **107**, 2653.
19 T. Yamamoto, *J. Chem. Phys.*, 1998, **89**, 2356.
20 T. Yamamoto, *J. Chem. Phys.*, 2001, **115**, 8675.
21 T. Yamamoto, *J. Macromol. Sci. B*, 2003, **42**, 629.
22 T. Yamamoto, *Polymer*, 2004, **45**, 1357.
23 T. Yamamoto, in *Interphases and Mesophases in Polymer Crystallization*, Adv. Polym. Sci., Springer, in print.
24 A. Koyama, T. Yamamoto, K. Fukao and Y. Miyamoto, *Phys. Rev. E*, 2002, **65**, 50 801.
25 C. M. Chen and P. G. Higgs, *J. Chem. Phys.*, 1998, **108**, 4305.
26 J. P. Doye and D. Frenkel, *J. Chem. Phys.*, 1998, **109**, 10 033.
27 J. P. Doye and D. Frenkel, *J. Chem. Phys.*, 1999, **110**, 2692.
28 N. Waheed, M. S. Lavine and G. Rutledge, *J. Chem. Phys.*, 2002, **116**, 2301.
29 W. Hu, D. Frenkel and V. B. F. Mathot, *Macromolecules*, 2003, **36**, 549.
30 I. Weissbuch, L. Addadi, M. Lahav and L. Leiserowitz, *Science*, 1991, **253**, 637.
31 L. Cartier, T. Okihara, Y. Ikada, H. Tsuji, J. Puiggali and B. Lotz, *Polymer*, 2000, **41**, 8909.
32 C. Mathieu, W. Stocker, A. Thierry, J. C. Wittman and B. Lotz, *Polymer*, 2001, **42**, 7033.
33 P. Choi, H. P. Blom, T. A. Kavassalis and A. Rudin, *Maromolecules*, 1995, **28**, 8247.
34 K. Nagarajan and A. S. Myerson, *Crystal Growth Des.*, 2001, **1**, 131.
35 D. C. Rapaport, *Phys. Rev. E*, 2002, **66**, 11 906.
36 M. G. Martin and J. I. Siepmann, *J. Phys. Chem. B*, 1999, **103**, 4508.
37 B. A. Berg and T. Neuhaus, *Phys. Lett. B*, 1991, **267**, 249.
38 A. Baumgartner, *J. Chem. Phys.*, 1980, **72**, 871.
39 D. Rigby and R. Roe, *J. Chem. Phys.*, 1987, **87**, 7285.
40 *Integrated simulation system for polymeric material design (OCTA)*, Nagoya University Doi Project, Japan, http://octa.jp/index.html.
41 T. Yamamoto, *J. Polym. Sci. Phys.*, 1985, **23**, 771.
42 T. Yamamoto, *J. Chem. Phys.*, 1988, **89**, 2356.
43 M. Hirose, T. Yamamoto and M. Naiki, *Comput. Theor. Polym. Sci.*, 2000, **10**, 345.
44 T. Shimizu and T. Yamamoto, *J. Chem. Phys.*, 2000, **113**, 3351.
45 H. Z. Li and T. Yamamoto, *J. Chem. Phys.*, 2001, **114**, 5774.

General Discussion

Professor Ryan opened the discussion of Professor Lodge's paper: (1) I suggest some names for your micelle cartoons based on current US politics. The Kerry micelle is fuzzy on the outside and two-faced on the inside whereas the George W. Bush micelle is fuzzy on the outside but confused on the inside.

(2) For a spherical micelle with a radius of 20 nm the molecular weights mean structure must be fully stretched (*i.e.* contour length). This must present enormous packing constraints.

Professor Lodge replied: (1) It is clear that inviting the audience to suggest names for any particular micellar structures that we might report is a high risk strategy! All I can say is that in the end, we expect our combination of analytical techniques—dynamic light scattering, small angle X-ray and neutron scattering, and especially cryoTEM—to provide a much clearer picture of micellar characteristics than one may expect to extract from the presidential candidates as viewed through the imperfect lens of the media.

(2) Indeed, the 20 nm radius of the OSB spherical micellar core does correspond to a nearly fully stretched conformation for some of the blocks. We have recently analyzed these data, and those of the OSF core/shell/corona flattened disks,[1] in terms of the criteria for accessing the proposed superstrong segregation regime.[2] Remarkably, perhaps, our analysis suggests indeed that OSB is on the brink of superstrong segregation, consistent with the nearly fully stretched core blocks, while OSF is well into this new regime.

1 T. P. Lodge, M. A. Hillmyer, Z. Zhou and Y. Talmon, *Macromolecules*, 2004, **37**, 6680.
2 A. N. Semenov, I. A. Nyrkova and A. R Khokhlov, *Macromolecules*, 1993, **26**, 3601.

Professor Gido asked: (1) How does the cryo-TEM contrast arise in the EOK materials? In the OSF linear triblocks the fluorinated blocks also contain I which gives good atomic number contrast in TEM. I do not think that there is a similar heavier atomic species in the Krytox block of the EOK.

(2) Why don't the beads in the EOK (2-7-2) structure merge to form the striped cylinders as in EOK (2-10-2)? It would seem energetically favourable for them to do so.

Professor Lodge responded: (1) It turns out that the Krytox materials have a mass density of approximately 1.8 g mL^{-1}, according to the manufacturer's specifications, and this is clearly the source of the significant natural contrast.

(2) This question raises an important point. Indeed, given the relatively shorter hydrophilic block in μ-EOK(2-7-2), one would expect that the driving force to aggregate as segmented cylinders would only be greater. There is not a simple answer, but it almost certainly reflects the non-equilibrium nature of the assembly process. The fact that we observe a variety of different micellar morphologies even for a single solution suggests that these micelles are non-ergodic, as previously proposed for other aqueous copolymer systems.[1] We suspect that in all cases the critical micelle concentration is so low that there is effectively no chance for micelle-to-micelle exchange of chains, and therefore no chance to find the global free energy minimum. At the same time, within a given micelle the chains are free to do the best they can; both hydrophobic blocks are liquid at this temperature. The prevalent packing motif is one in which the hydrocarbon and fluorocarbon blocks are nearly fully extended, again consistent with superstrong segregation, and the nanodomain interfaces are almost flat. In all the cryoTEM images shown the apparent concentration is much greater than that in the bulk solution (*ca.* 1% by volume) from which they were prepared. This implies a significant local

DOI: 10.1039/b413661f

enhancement in concentration during the blotting step. It is possible that individual μ-EOK(2-7-2) micelles were squeezed together at this time.

1 S. Jain and F. S. Bates, *Macromolecules*, 2004, **37**, 1511.

Professor Fraaije commented: In AB polymer solutions or melts, one has only a few basic structures (micelles, cylinders, lamellae *etc.*), and kinetic trapping is less of an issue, except close to phase boundaries. Here, in ABC systems, the structure space is much richer, and therefore it is more likely structures may be trapped. This offers a challenge for designing specified non-equilibrium structures, by employing external fields.

Professor Lodge responded: We certainly agree that the added complexity of ABC copolymers, compared to AB diblocks, is likely to lead to kinetic trapping and possibly to interesting metastable structures. Indeed, the non-ergodicity alluded to above indicates that this is the case for our system. However, possibly of more importance to metastability than the number of different chemical ingredients *per se* is the strength of the interactions. As our system is so strongly segregated, some local minima on the global free energy surface are likely to be quite deep. It is perhaps worth pointing out that long-lived metastable states are far from rare even in diblock systems. Two prevalent examples are the perforated layer phase in diblock melts, which is often a kinetic trap on the way to forming a gyroid, and the case of simple spherical micelles when the core block is glassy, such as polystyrene. Finally, your suggestion of employing external fields to guide structure formation in ABC systems is a very good one; I am not aware of extensive studies along these lines as yet.

Professor Iwuoha asked: What is the effect of temperature on the CryoTEM data? Does the flat sheet geometry being proposed agree with X-ray modelling?

Professor Lodge replied: (1) We have not varied temperature in the solutions immediately prior to their being used for cryoTEM, although of course the solution is both vitrified and imaged under cryogenic conditions. As all three pairwise interactions are so strong, and the hydrocarbon and fluorocarbon blocks so very hydrophobic, we do not anticipate much sensitivity to temperature in the solution structures, and the selected dynamic light scattering measurements that we have made at elevated temperatures support this.

(2) In the case of OSF disks, the SAXS and SANS data are completely consistent with the proposed structure.[1] For the μ-EOK miktoarm stars we have not yet modeled the SAXS data successfully. The distribution of micellar shapes and sizes evident in cryoTEM and DLS accounts for the rather featureless SAXS curves, and illustrates beautifully the importance of direct imaging techniques when the sample adopts a mixture of self-assembled structures.

1 Z. Zhou, Z. Li, Y. Ren, M. A. Hillmyer and T. P. Lodge, *J. Am. Chem. Soc.*, 2003, **125**, 10182.

Dr Gosal and **Professor Ross-Murphy** opened the discussion of Professor Donald's paper: It is interesting to note that over the last 50 or so years, globular protein gels have been discussed at least twice at Faraday Discussion meetings.[1,2] Many of these systems form "simple particulate gels", but under certain other conditions, *e.g.* lowered pH, they instead produce fibrillar networks, as described in Professor Donald's paper . Recently, some of these latter systems have received renewed interest due to the parallels with amyloidosis, and prion-related disorders. In the human body, the manifestation of amyloid can take many forms (*e.g.* spherulites, surface-associated, and complex gel-like networks *etc.*). In recent work, in collaboration with Prof. A. H Clark, then at Unilever Research, Colworth, we explored fibrils also formed by β-lactoglobulin, using both thermal and alcohol-induced methods. Using rheological measurements,[3] we were able to show that above certain concentrations, these systems behave like gels. Differences in fibril morphology between the two methods of denaturation (as revealed by microscopy) were used to assign distinct features in their respective rheological cure-curves. For example we observed a maximum in G'', the shear-loss modulus, in one of the alcohol induced fibrillar systems, probably due to the relaxation of pendant chains, and implying a more flexible network. However, and perhaps unexpectedly, the concentration dependence of both equilibrium and kinetic rheological parameters for the two different systems, were very similar. Both sets of data could be described adequately by a modified branching model of gelation, whereby fibrils are delivered into these self-assembled systems through nucleation and growth.[4]

The branching model for gelation implicitly requires a critical concentration, which is observed experimentally. Fibrils are still formed below these concentrations, as observed by microscopy, and it is in this regime that spherulites develop. The shift in mechanism from spherulitic growth to gel-formation is most intriguing, and needs to be explored further.

1 E. Barbu and M. Joly, *Faraday Discuss. Chem. Soc.*, 1953, **13**, 77–93.
2 M. P. Tombs, *Faraday Discuss. Chem. Soc.*, 1974, 158–164.
3 W. S. Gosal, A. H. Clark, P. D. A. Pudney and S. B. Ross-Murphy., *Langmuir*, 2002, **18**, 7174–7181.
4 W. S. Gosal, A. H. Clark and S. B. Ross-Murphy, *Biomacromolecules*, 2004, **5**(6), DOI:10.1021/bm049659d; W. S. Gosal, A. H. Clark and S. B. Ross-Murphy, *Biomacromolecules*, 2004, **5**(6), DOI:10.1021/bm049660c; W. S. Gosal, A. H. Clark and S. B. Ross-Murphy, *Biomacromolecules*, 2004, **5**(6), DOI:10.1021/bm0496615.

Professor Donald, **Dr Bromley** and **Dr Krebs** responded: Rheology is a powerful technique for the study of gelation, but is only a very indirect probe of fibril formation. The main advantage of the work described in our paper for determining the mechanism of fibril formation is that through the use of the Thioflavin T assay we may measure the fibrillar content directly. We were very encouraged to find values of the critical nucleus size which matched well with those found from the rheology studies.

Looking at the spherulitic structures that can form from the fibrils, our data suggest there is no change from one regime to another, *i.e.* between forming fibrils or spherulites. Because the protein is partially folded, several pathways are open and explored concurrently, so that fibrils and spherulites can—and do—coexist. Small changes in the conditions can, however, favour one over the other, shifting the balance of populations. Rheological methods are unlikely to tell us much about the formation of spherulites, since the fibrils in them are removed from the solution and cannot form gels. This strongly suggests the need for extreme care when studying BLG by rheology! The spherulites are better studied using other methods, such as ThT, electron/optical microscopy *etc.*

Dr Fahmi asked: How do you adjust for the rheological measurements of temperature, strain and frequency in the linear or non-linear regime?

Dr Gosal replied: The rheological measurements on these systems were made using a Peltier heating system to control the temperature, and operating under 'pseudo-strain' control at fixed frequency. Our initial control experiments showed only very slight frequency dependence of the cure-curves over two orders of magnitude, and no strain dependence, providing that this was sufficiently low as not to perturb the system.

Professor Fraaije asked: Is there a relation between the biological significance of the molecular structure of the de-naturated protein molecules and the fibril formation? If not, one expects a very broad distribution of denaturated molecule structures, and therefore an aspecific fibril formation mechanism.

Professor Donald, **Dr Bromley** and **Dr Krebs** replied: It is important to note that natively folded proteins form amyloid fibrils when partially unfolded; if the protein is 'too denatured' random aggregation is more likely to occur. Therefore, the number of accessible states is in reality confined and highly sensitive to the exact conditions chosen. Thus, although the range of states occupied by the partially denatured protein (which is the amyloid forming species in this work) is larger than the range of states occupied in the native state, the energy landscape is still far from being flat and a relatively small number of states are expected to be available. Indeed, different conditions lead to amyloid fibrils with different properties for a given protein: the coarse grained structure of amyloid fibrils will be conserved across different proteins and different preparation conditions for the same protein, but the fine structures may differ. Conversely, the same set of conditions may lead to different structures of amyloid fibrils for different proteins. Once a fibril nucleus forms, this is replicated very faithfully, since the arrangement in the β-sheet is almost crystalline. This specificity is also, for example, at least partly responsible for observations like the 'species barrier' in the transmissible spongiform encephalopathies. Finally it should be emphasised that, although extremes of pH and temperature do not occur in the body, genetic mutations can result in copies of proteins that are substantially destabilised and will be only partially folded even under native conditions. Misfolding of native proteins can also occur when the system of chaperones controlling folding breaks down, as may occur in the diseases of old age.

Professor Kabanov said: I appreciate all the phenomenological data presented. My concern relates only to the kinetic treatment. The issue is that homogeneous kinetics (in particular eqn. (2)) can hardly be applied to heterogeneous aggregation. The kinetic description of such processes should take into account the shape (growing surface) of propagating aggregates. For example, Avrami's theory gives

$$F = 1 - \exp(-kt^n)$$

where n is determined either by nucleation (homogeneous or heterogeneous), or the growth dimensionality (one, two, three dimensional). Therefore, n values obtained from log log$(1 - F)$ − log t plots either in Avrami or in any other realistic kinetic treatment would have nothing to do with the number of original species in critical nuclei.

Professor Donald, **Dr Bromley** and **Dr Krebs** responded: Fig. 13 is indeed a log log$(1 - F)$ − log t plot, and thus formally resembles an Avrami analysis. However, interpreting the slopes within the Avrami approach would yield values for the Avrami exponent n of 2.4, 5.5 and 5.8 for the different samples. Even allowing for a spherical geometry of growth (rather than the more applicable rod geometry one would expect for fibrils) this implies the nucleation phase must be breaking the Avrami assumption of a constant nucleation rate. Therefore we propose a model which allows an increasing probability of nucleation as time progresses, and which is strongly dependent on the critical nucleus size. In other words the underlying assumptions in the Avrami theory are not applicable to this case and cannot be applied.

Professor Lodge opened the discussion of Professor Klok's paper: Do you have any small-angle scattering patterns that indicate long-range order or a particular symmetry?

Professor Klok and **Professor Hamley** responded: The SAXS data presented in our paper were recorded from dilute aqueous solutions and no higher order reflections indicative of long range order were observed. We have, however, carried out SAXS experiments on related β-strand peptide –poly(ethylene glycol) block copolymers in the solid state.[1] Also in this case, we did not observe any higher order reflections, which would be needed to unambiguously determine the nature of the superstructure.

1 A. Rösler, H.-A. Klok, I. W. Hamley, V. Castelletto and O. O. Mykhaylyk, *Biomacromolecules*, 2003, **4**, 859–863.

Professor Ryan responded: The data presented here show a layer spacing (if the system is lamellar) of ∼100 Å. What is the volume fraction and would you expect lamellae?

Professor Klok and **Professor Hamley** responded: Please see our above response to the question by Professor Lodge.

Professor Donald asked: To what extent does the packing in solution carry over into the solid state when solvent molecules are not present? Do correlations between coiled coils carry over into the solid state?

Professor Klok responded: The information that we have on peptide secondary structure and packing in the solid state is mainly based on FTIR and WAXS experiments. These studies indicate that the coiled coil based peptides retain their α-helical secondary structure in the solid state and also indicate that the β-strand peptide sequences form antiparallel β-sheet structures in the solid state.

Dr Davidson asked: Your synthetic chemistry allows you to use small peptide segments that may have not only a structural role but also a biological function. Have you even thought or tried to use a synthetic peptide that would comprise a biologically meaningful sequence, triggering a signalling function for example?

Professor Klok responded: So far our efforts have concentrated on investigating the structural properties of synthetic–peptide hybrid block copolymers and on the question of whether biological structural peptide motifs can be used as auxiliaries to drive the self-assembly of such hybrid block copolymers. The ultimate goal is of course to go beyond structure formation and utilize functional peptide sequences or conjugate functional sequences with the systems that have already been prepared and try to design systems that would be capable of *e.g.* recognizing and binding proteins, cells or other biologically relevant targets.

Professor Gido commented: Fig. 1 of your paper suggests using the amphiphilic nature of a peptide sequence coupled with a hydrophilic–hydrophobic liquid–liquid interface to select a particular peptide chain conformation. The simple hexapeptide repeating sequence of *Bombyx Mori* silk-worm silk {$GAGAGS_n$} undergoes such a conformation switching at interfaces. It forms a β-sheet structure (silk II) in the bulk, but at air–water or aqueous–organic interfaces it takes a 3_1 helical structure (silk III). Given its simplicity and well defined conformational behaviour this would be an interesting sequence for incorporation in your block copolymers.

Dr Al-Hussein asked: How does the architecture of the polypeptide block influence the copolymer morphology? Have you seen all morphologies of amorphous–amorphous diblock copolymers.

Professor Klok replied: We have not yet systematically investigated the solid state morphology of the samples presented in this contribution. One may, however, expect that structure formation in the solid state of synthetic polypeptide hybrid block copolymers is not only governed by phase separation but is also influenced by the tendency of the peptide segments to aggregate and/or form superstructures. These competing processes may lead to unconventional morphologies not observed for conventional amorphous–amorphous diblock copolymers.

We have carried out, however, structural investigations on two different series of much simpler polystyrene–homopolypeptide diblock copolymers, which were prepared *via* the ring-opening polymerization of α-amino acid *N*-carboxyanhydrides.[1–3] In the course of these studies, we indeed found novel morphologies (*e.g.* double hexagonal structures), suggesting that the polypeptide block can affect the solid state morphology.

1 H.-A. Klok, J. F. Langenwalter and S. Lecommandoux, *Macromolecules*, 2000, **33**, 7819–7826.
2 H.-A. Klok and S. Lecommandoux, *Adv. Mater.*, 2001, **13**, 1217–1229.
3 S. Lecommandoux, M.-F. Achard, J. F. Langenwalter and H.-A. Klok, *Macromolecules*, 2001, **34**, 9100–9111.

Professor Ryan commented: The wrapping of the coiled coil peptide by the PEO, as illustrated in your cartoon, is difficult to conceive given that the R_g of PEO_{40} is ~17 Å and the R_g of PEO_{15} is ~8 Å whereas the length of the coiled coil is ~30 Å.

Professor Klok replied: There is certainly a discrepancy between the length of the peptide helix and the dimensions of the poly(ethylene glycol) (PEG) chains if one assumes a random coil conformation for the latter. Electron paramagnetic resonance (EPR) investigations on spin-labelled samples revealed, however, that in aqueous solutions the PEG chains do not assume a random coil conformation but instead are wrapped rather closely around the coiled coil peptide core.[1] In this case, the degree of polymerization of the PEG chains may be sufficient to wrap around the peptide core as illustrated in the cartoon. Indeed, continuous wave (CW) EPR studies, which were performed to study the dynamics of the PEGylated coiled coil nano-objects, suggested interactions between the PEG chains and the spin labels located at the peptide's C-terminus.[1] This is also in support of the encapsulation of the peptide core in a PEG shell.

1 G. W. M. Vandermeulen, D. Hinderberger, H. Xu, S. S. Sheiko, G. Jeschke and H.-A. Klok, *Chem. Phys. Chem.*, 2004, **5**, 488 494.

Professor Iwuoha commented: In the Results and Discussion section of the paper AFM images for compounds **8** and **9** were explained as evidence for the formation of discrete nano-sized supramolecular structures by the PEGylated block copolymers. However, the AFM data do not rule out the possible nanostructured self-organisation of the free peptides molecules stabilised by unreacted PEG.

Professor Klok responded: The paper only shows an AFM image of block copolymer **9** (Fig. 7). An AFM image of sample **8**, which reveals a similar structure, was published in an earlier paper.[1] Compounds **8** and **9** were purified *via* dialysis and neither contain free peptide nor unreacted PEG as confirmed *via* reversed phase HPLC.[2] The features observed in the AFM images therefore cannot be due to stabilization of free peptide by unreacted PEG.

1 G. W. M. Vandermeulen, D. Hinderberger, H. Xu, S. S. Sheiko, G. Jeschke and H.-A. Klok, *Chem. Phys. Chem.*, 2004, **5**, 488.
2 G. W. M. Vandermeulen, C. Tziatzios and H.-A. Klok, *Macromolecules*, 2003, **36**, 4107.

Professor Hillmyer commented: A claim is made that the AFM image in Fig. 7 is in support of the model shown in Fig. 6. In fact, the AFM image in question appears to show larger aggregates and the 1×10 nm dimensions quoted in the Results and Discussion section are not obvious.

Professor Klok responded: The 1×10 nm dimensions do not refer to the large structures which are visible in the AFM image but to the grainy fine structure, which does not appear very well in the printed paper.

Dr Semenov opened the discussion of Professor Rowan's paper: (1) Is the supramolecular gel reversible? (2) What is its structural relaxation time? (3) What are the energies of metal-ion/ligand bonds? (4) Why is gelation suppressed if transition metal ions are added first: this must not be so if the structural relaxation time is less than experimental time? (5) What is the volume fraction of ligands, and what is the fraction of ligands participating in macrocycles?

Professor Rowan responded: (1) Yes the gel shows reversible thermal and shear-responsive behaviour. Furthermore, while addition of formic acid results in a collapse of the gel if the formic acid (and acetonitrile) is removed in a vacuum oven the material can be reswollen with acetonitrile to once again yield the gel. (2) If you are asking about the stress relaxation time we have not done the rheological experiments on the gel yet to determine this but this is something we definitely have plans to do. If you are asking how long it takes for a sheared gel to fully reset that takes around 18 min for the Zn/La gel (11 wt.% solid swollen in acetonitrile[1]). (3) A rough estimation of free energy of the metal-ion–ligand bond based on the association constants is about 40 kJ mol^{-1} for the zinc(II) ions. (4) This is a complex process which we are still trying to elucidate. The first thing to say here is that we believe the response mechanism proposed in the paper only holds for the thermal and chemo responses of the gel (as mentioned in the paper). For example, there is no evidence of significant decomplexation of the Eu^{3+} ions from the BIP ligands upon addition of shear, based on photoluminescence data. We believe a different mechanism, which involves the disruption of higher ordered structures, is responsible for the shear sensitivity of the gel but we are still examining this. This is what we know so far about the system you asked about: Upon addition of the Zn^{2+} ions (97% based on the amount of ligand) to the monomer in acetonitrile some solid is formed, suggesting that some of the Zn^{2+}: monomer complexes have precipitated from solution. This has the effect of changing the stoichiometry of the components and thus results in the lack of gel formation. However, simply annealing the system at 80 °C for an hour does result in the formation of a gel upon cooling. (5) Based on estimations of the solvent-excluded volume of our monomer using molecular modelling, we find that the ligands (BIP units) account for just over 70% of the total volume of the molecule, roughly 35% of the volume at each end of the monomer. As for the fraction of monomers which are present in macrocycles, we do not have a good method of measuring this at the moment.

1 See Y. Zhao, J. B. Beck, S. J. Rowan and A. M. Jamieson, *Macromolecules*, 2004, **37**, 3529.

Professor Ross-Murphy said: I am very interested by the metal–ligand systems of Rowan and Beck, which seems to me to have some special features. In particular, many complex rheological systems portrayed as gels, including for example "shower gels" are more accurately described as structured liquids or fluids as opposed to "true gels" which are, in fact, soft solids. Initially I thought the systems in this paper were another example of this class of materials, but the apparent lack of creep seems to suggest not. I am not aware of any systems that can be sheared extensively, but then can reform, and still show true gel properties as do these. In the presentation it was mentioned that one of the disrupted gels can be reformed by heating 4 times with a hot air blower, and that after one heating cycle what is formed is a small amount of gel, and this increases on further cycles. It may be that this is simply an increase in the gel fraction, but I suspect not, in which case it seems inconsistent with the expected statistics/molecular weight distribution of the suggested step–growth type mechanism. Would the authors care to comment on this?

I understand some rheological measurements have now been published on this system. Have the authors considered using the approach of Stratton and Butcher[1] to characterise the reformation of structure? This is the method suggested in the standard work by Ferry[2] and we have used it in the past.[3,4] Other approaches which might be fruitful would be creep measurements on the gel state, and also steady shear and low strain dynamic oscillatory shear measurements to examine the validity (or, more likely, failure) of the so-called Cox–Merz superposition.[5]

1 R. A. Stratton and A. F. Butcher, *J. Polym. Sci.: Part B: Polym. Phys.*, 1973, **11**, 1747–1758.
2 J. D. Ferry, *Viscoelastic properties of polymers*, Wiley Interscience, New York, NY, USA, 1980.
3 R. K. Richardson and S. B. Ross-Murphy, *Int. J. Biol. Macromol.*, 1987, **9**, 250–256.
4 R. K. Richardson and S. B. Ross-Murphy, *Int. J. Biol. Macromol.*, 1987, **9**, 257–264.
5 S. B. Ross-Murphy and K. P. Shatwell, *Biorheology*, 1993, **30**, 217–227.

Professor Rowan replied: The experiment discussed was not on disrupted gels but on a system in which the Zn^{2+} ions were added to the monomer before the Eu^{3+} ions. If the metal ion addition is carried out in this order no gel is observed. The question is related to the previous question by Professor Semenov. The annealing experiments mentioned in the talk were very preliminary and initially all we wanted to do was to see if we could convert this sample into a gel. We were able to do this by repeated heating until everything was in solution and cooling. What we have shown subsequently is that you can convert this sample into a gel in one step by heating the sample to *ca.* 80 °C and keeping it at this temperature for *ca.* 30 min to an hour. Upon cooling of this annealed sample a gel is formed.

Indeed the rheological measurements we have published[1] utilized a version of the stop–start experiments of Stratton and Butcher, *viz.* the gel was sheared into the sol state, and then rheologically interrogated after increasing rest or recovery times. Thus we were able to record the apparent yield stress, yield strain, and storage and loss moduli at periods during reconstitution of the gel network.

Creep measurements have not been performed, but are planned. Low strain dynamic oscillatory shear measurements have been performed (0.2% strain) and indicate the storage modulus G' is frequency independent at low deformation frequencies ($G' \sim 1.6 \times 10^4$ Pa, complex viscosity $\eta^* \sim 1.6 \times 10^5$ Pa s at deformation frequency $\omega = 0.1$ rad s^{-1}). The material exhibits a catastrophic failure of the Cox–Merx superposition rule since the gel yields to form a low viscosity Newtonian sol at a very small strain ($\gamma \sim 1\%$), $\eta = 0.01$ Pa s.

1 Y. Zhao, J. B. Beck, S. J. Rowan and A. M. Jamieson, *Macromolecules*, 2004, **37**, 3529.

Professor Hillmyer asked: Do you have direct evidence for the ring structures? Do you worry about thermal degradation of PEG segment? Why is there no gel formation when you add 3% La to the ring/3% unbound ligand?

Professor Rowan responded: We do not have direct evidence of ring structures being formed. These should behave as step–growth polymers and with the monomer unit being flexible it would be expected that rings can and probably do form. We do observe low solution viscosities of the 1:1 mixture of Zn^{2+}: monomer which is consistent with the formation of a significant amount of rings.

As for the degradation of the PEG we are concerned with this but as of yet we have seen no evidence for degradation occurring in the samples we have studied. The final question is similar to the one asked by Professor Semenov and as I said to him we do think that the observed precipitation alters the stoichiometry of the system thus preventing formation of the gel but we have to look into this in more detail.

Dr Al-Hussein said: Have you synthesized multi-block copolymers using this route? If so, are they diblocks or triblocks?

Professor Rowan responded: As of yet we have not done this experiment but the nice thing about these self-assembled polymers is that this is easy to do and we are heading in that direction. We have now prepared a series of ditopic monomers which have a variety of different core units between the ligands. For example, we have made systems which have flexible polymeric cores and systems which have rigid conjugated cores. It will be interesting to see if such self-assembled rod/coil systems form controlled morphologies in the solid state.

Professor Ryan asked: Do these systems follow the well established theory for gelation if you consider them as classical $RA_2 + RB_2 + RB_3$ reactions, *e.g.*

$$(P_A P_B)_{gel} = \frac{1}{f_w - 1} \quad \text{at} \quad r = 1 = \frac{[A]}{[B]}$$

and the "normal" features of stoichiometry (an excess of A) inhibiting gelation.

Professor Rowan responded: If these supramolecular metallo-polymers are formed through a step–growth mechanism then it should of course be sensitive to the stoichiometry of the different components. What we generally do for the preparation of the gels is to add just enough metal ions to ensure that all the ligands are bound, assuming that zinc ions will bind two ligands while the lanthanoid ions bind a total of three ligands. To examine the effect that altering the component stoichiometry has on the system we have added an excess of either monomer (*e.g.* RA_2), Zn^{2+} ions (*e.g.* RB_2), or Eu^{3+} ions (*e.g.* RB_3) to a pre-made gel. The procedure we used to do this was addition of a solution containing 10% molar excess (based on the amount of monomer) of one of the components to the gel. This was followed by removal of the solvent and subsequent addition of enough acetonitrile to achieve the same concentration (based on total solid weight) that was present in the original gel. The result was that in all three samples, which contained excess monomer, Zn^{2+} ions or Eu^{3+} ions, the gel failed to reform.

Professor Iwuoha commented: The mechanism postulated for both the thermal and chemical responses of the supramolecular polylectrolyte gels is the decomplexation of the gel into transition metal ion bound oligomers and lanthanoid ions, which in the chemical response case results in the " 'switching off' of the aromatic ligands antenna effect". What is the role of the transition between the high and low states of the lanthanoid ions in the decomplexation and 'switching off' phenomena?

Professor Rowan responded: As of yet we have not looked at the nature of the spin states of the lanthanoid ions nor if there are any transitions between high and low states in these systems. However, the possibility of such a transition should not be ignored, and is certainly a viable avenue for future analysis.

Fiona Case commented: If I remember my ruthenium chemistry correctly, it is possible to design a complex (for example with polypyridyl ligands) which will break down when illuminated by light of a specific wavelength, and reform when the light is removed. Have you studied polymer systems linked by these types of Ru-ligand complexes? They might create an interesting, and potentially useful, material with light dependent rheology.

Professor Rowan responded: That would be nice and is definitely a direction we would like to go. One of the nice things about these materials is that there is a wealth of knowledge in the organometallic community that we can tap into. As for your question on the photolabilization of a ligand bound to a ruthenium complex, most examples of such photolabile ligands in the literature are either hindered bidentate ligands or those ligands with poor σ-donor character. To my knowledge I do not know of any tridentate ligands similar to ones that we use that undergo this photo-induced process but we might be able to tailor the chemistry of these ligands to achieve this effect or alternatively attach appropriate bidentate ligands to the ends of our core units.

Professor Fraaije asked: Does the toxicity of the metals limit the potential applications, or can it be used to advantage?

Professor Rowan responded: The answer to this is yes and no. I think the advantage of these systems is the ability to combine the functionality of the metal ions with the processing properties of a polymer, *e.g.*, processable light-emitting materials for display devices, rather than using them simply for their mechanical properties. However, there are a number of materials where metal ions have been added into polymers in order to utilize metals specific toxicity. For example, a range of metals have been put into paints to make anti-fouling coatings.

Professor De Jeu opened the discussion of Professor Ryan's paper: Concerning the interesting experiments measuring force–distance curves with the polymer attached to an AFM-tip, I understand that many experiments have been done and that only in some of these proper absorption of the polymer to the tip is accomplished. I wonder whether in these cases the absorption can be considered as fixed until pulling off occurs, or whether intermediate states of slip–stick sliding of the polymer take place at the tip. If this were the case possibly the force–distance curves could be seriously corrupted by situations in which the distance increases without an effective increase of the force.

Professor Ryan responded: Stick–slip certainly should not be a problem for multiple pull-off events. If one considers data such as that included in Fig. 9, we are searching for pull-off events with a given length scale. If the polymer chain were to slip–stick then you would expect to see a "saw tooth" pattern in the force–distance curve which could also be interpreted as a multiple chain pulling. If stick–slip were an issue, then there would be no reproducible length scale.

This sort of stick–slip should lead to a very characteristic shape of the force curve, as discussed by Haupt *et al.*[1] Although such curves have been observed by other workers in other systems we haven't seen strong evidence for them in ours.

1 B. J. Haupt, J. Ennis and E. M. Sevick, *Langmuir*, 1999, **15** 3886.

Professor Gido commented: I wish to understand the rationale for using single chain AFM measurements to understand the force response of an entire polymer brush. It seems to me that the force response of the brush is necessarily cooperative in nature involving many chains and is influenced by factors such as crowding of chains in the brush which are circumvented in the single chain AFM measurements.

Secondly, although the thickness change of the more densely grafted brush with pH (Fig. 7b in the paper) is smaller than in the more sparsely grafted brush (Fig. 7a) might the more dense brush be capable of exerting greater force?

Professor Ryan responded: The measurement of the single chain response is the underpinning feature of all of this work. By measuring this response we shall be able to determine cooperative and uncooperative processes occurring when we measure collections of single chains (dense brushes/gels *etc.*).

Obviously a denser brush could exert a greater force than a less dense brush, however, there would be some trade off with diffusion. The denser brushes don't expand so much so the work done might be less as work = force × distance. Also in the limit of a completely dense brush the molecules will be fully stretched whatever the pH and will not be able to respond.

Of course, the physics of uniformly compressing a polymer brush differs from the physics of extracting a single chain from that brush. The former situation is relevant for understanding the role of polymer brushes in conferring steric stabilisation, and has been experimentally addressed using the surface forces apparatus. We are interested in the second situation because of our ultimate goal of using a single polymer chain to convert chemical energy to mechanical energy.

My colleague Dr Geoghegan may also like to comment.

Dr Geoghegan commented: Of course the force response of a single isolated chain will provide information on that chain. An isolated brush will not be subject to entanglements, which will complicate the force–distance curve. The force distance curves should reveal some information on the nature of such a co-operative response. If the grafting density were to be decreased, different regimes would be observed in the force–distance curves, for example it is more likely that entanglements are an issue in less densely grafted chains (the "mushroom" regime) than in the "brush" regime. Single chain AFM experiments should be able to probe these regimes.

Professor Hamley said: Could you comment on the evidence for a BCC symmetry of the gel? If the gel has a cubic structure, why is the macroscopic shape change apparently anisotropic (see Fig. 12 of the paper (inset))?

In the paper, in the Results, Block Copolymers section, it is mentioned that the gel was shear aligned prior to SAXS/video microscopy—why? Could this break the isotropic symmetry?

Professor Ryan responded: We only have a single peak so cannot state that the structure has BCC symmetry. The sample has a volume fraction of hydrophobe block of $\sim$0.15 in the dry state. It is definitely ordered, since it is a gel and has domains of hydrophobe acting as physical crosslinks, so one would expect a spherical structure. The sample has a complex temperature and solvent history and is unlikely to be at equilibrium, this is despite our best attempts to maintain mobility during preparation and to anneal it in the presence of solvent. The most likely structure is a liquid-like arrangement of glassy domains which, as you point out in your book, is often seen in triblock copolymers. We use the BCC motif in Fig. 11 of our paper because it is the ground-state for these compositions and easy to draw. Like most schematic diagrams (or cartoons) it can lead to misunderstandings—but the link between the nanoscale response and the macroscopic behaviour would be difficult to draw for liquid-like spheres.

The macroscopic shape change in Fig. 12 is anisotropic because the sample has an irregular shape; the swelling is, however, isotropic.

The sample was sheared during preparation in an attempt to shear induce long-range order, and define the structure. The shearing also helps create a network structure with single chains spanning individual glassy domains rather than condensed flower micelles.

Professor Donald asked: Can you explain how you worked out the energy/work done in a single molecule motor stroke? The figure shown would appear to be fairly arbitrary in the way the area representing work done was evaluated, so what sort of precision do you think you can claim?

Professor Ryan responded: The energy referred to is presented in Fig. 1. It is a very crude attempt to get an order of magnitude estimate of the work that might be done in such a cycle.

You are quite correct that the area chosen was arbitrary! It is in the most responsive part of the curve where the chains are stating to strain harden, but at lower forces than the pull-off.

It is only an approximation of some work that could be achieved and has not actually been measured as such. The amount of work available depends upon the route the cycle takes *i.e.* do you jump from one curve to the other at either constant force or constant expansion? It was only ever meant as an order of magnitude guide to the energy that could be extracted should you get such a system to do work.

An alternative representation could be the force curves for acid and base but divided through by L (the contour length). This might give a measure of the energy that could be attained from a single monomer unit.

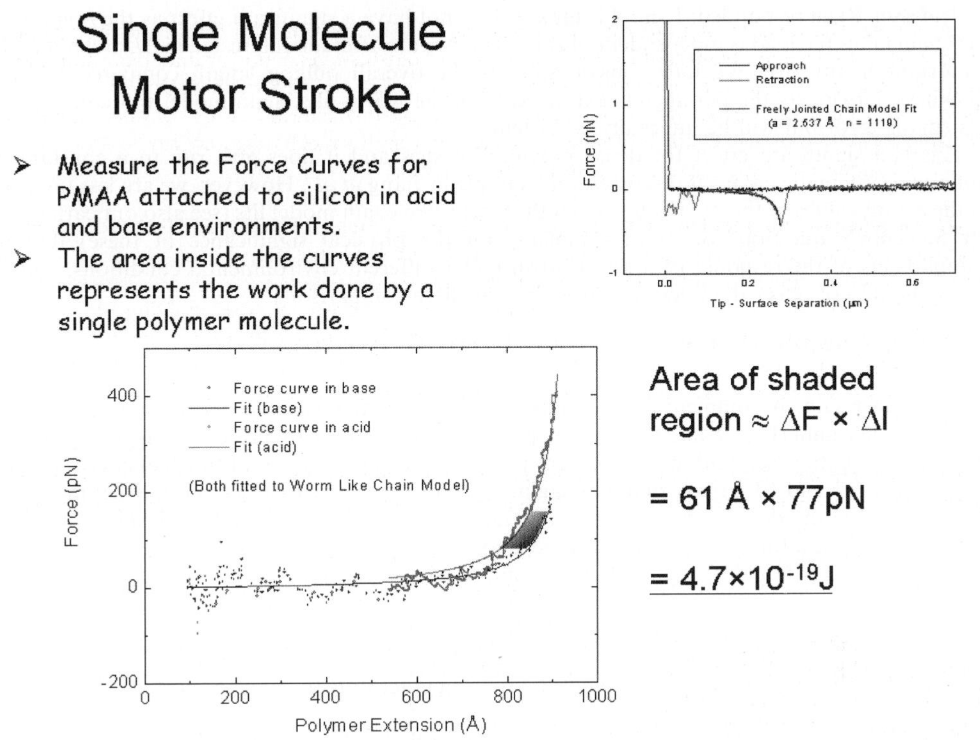

Fig. 1 Work done in a single molecule motor stroke.

Dr Semenov asked: (1) What is the level of efficiency of artificial molecular motors: what is the ratio of useful mechanical work produced to the chemical energy consumed?

(2) The statistical segment lengths you report for poly base brushes range from 0.09 nm to 0.23 nm depending on the chain length (at the same pH = 4). How would you explain such a dependence?

Professor Ryan responded: (1) Other than "very low", it is impossible to be more specific on the efficiency of this system. The polymer sits in a massive bath of continuously stirred tank reactor and reactants (energy) are pumped in. The way to make a more efficient artificial molecular motor is to put the catalyst for the oscillating reaction on the polymer backbone, this has been done for BZ reactions (ref. 8–12 in the paper) but no estimates of efficiency have been made.

(2) We should first observe that we, like most other workers in this field, use the worm-like chain model to fit our data in regimes where the conditions for its validity are not strictly met. Thus the model needs to be regarded as a convenient parameterisation of the force curves rather than a proper theory of the force response to extension of a single (possibly charged) chain in a polymer brush. Such a proper theory, of course, would need to include factors such as the interactions of the chain that is being pulled with the other brush chains. In the light of this one can see why the apparent persistence length might depend on the chain length; a chain that is picked up close to the surface will have more interactions with the other brush chains than one picked up close to its end, that is stretched to a distance much greater than the unperturbed brush height. Ideally, we would fit our data to a theory that properly accounts for all the relevant interchain and intrachain interactions, and we look forward to such a theory being developed.

Professor Biggs asked: Can you comment on the lack of certainty in the separation distance which must arise as a result of an unknown amount of trapped polymer between the tip and surface at the "apparent" zero (constant compliance region of the force curve).

What are the effects of this on your calculations of things such as the persistence length?

Professor Ryan responded: I should think it does not have a significant effect as the point where the chain takes the load is independent of this offset (we are after all, interested in Δl, and not l). The important feature to take into consideration is the overall pulling length compared to brush thickness. Obviously the parameters extracted from the worm-like chain model will contain some systematic error but will be internally consistent.

This is a significant effect for dense polymer brushes and it does give rise to a constant but unknown systematic error, as discussed for example in Jiang *et al.*[1] However, we are less interested in the actual values of the parameters from the worm-like chain model fits (see also our earlier reply Dr Semenov's question for more comments on the physical significance of these) than the comparisons of the response of a given brush under different environmental conditions, and this effect should not affect the internal consistency of these comparisons.

1 X. Jiang, C. Ortiz and P. T. Hammond, *Langmuir*, 2002, **18**, 1131067

Dr Webber said: A comment on the unknown nature of the exact zero-separation point in the force–distance data (Figs. 8 and 9 in your paper), as raised initially by Professor Biggs. Fig. 4 shows AFM images with a 'scratch' to define a zero point for the measurement of the thickness of polymer brushes at pH 3 and 12. I think that by imaging such a layer (*i.e.* with a 'scratch'), either in tapping-mode or contact-mode, at various values of imaging force one should be able to develop a profile of layer thickness as a function of applied force, and thus determine to thickness of an incompressible layer of polymer, should one exist. If an incompressible layer does exist, this must be taken into account when determining the point of zero-separation in the force–distance data. Indeed, in the discussion of Fig. 9 the authors state that the layer at pH 4 is highly extended, though at pH 10 the chains are collapsed. It is reasonable to expect, therefore, that at the high pH it will be more difficult for the AFM tip to penetrate the polymer layer to interact directly with the surface.

Professor Ryan responded: It is reasonable to expect that at high pH the tip cannot penetrate the polymer layer to interact directly with the surface, at least in the limit of very densely grafted chains. We will be assessing the zero point in calculating work done and pull-off.

Professor Biggs asked: Again, a lack of knowledge about zero distance in the pull-off force data for the 2 pH values must affect your calculated energy from the area in the plots. Is this not true? Can this have a significant effect?

Professor Ryan responded: Well, this basically goes back to the response to the previous questions. I should think it does not have a significant effect as the point where the chain takes the load is independent of this offset. It would make sense to use the strain hardening part of the curve in any working device.

Professor Iwuoha said: I don't understand the $\Delta pH/\Delta T$ used in one of your slides given in your presentation. Should it be $\Delta pH/\Delta t$?

Professor Ryan responded: Yes, it is a typographical mistake on the slide and I apologise for any confusion—$\Delta pH/\Delta t$ is correct and that is the way it is presented in the paper.

Mr Koga asked: Is there any damage to the substrate surface in basic conditions?

Professor Ryan responded: The change in pH can cause some problems in cleaving the bond between the polymer and the substrate, in our previous work we experienced the loss of brush layers. Judicious choice of linker chemistry means that the grafting chemistry is stable with respect to pH changes and we can check this by repeated neutron reflectivity measurements as the pH is cycled. Reproducibility in the reflectivity profiles is a very good indication that the brushes are stable to acid and base conditions.

Professor Kabanov asked: What is the principle difference, if it exists, between your constructed system and Kuhn's polyacrylic acid, which elongates at high and shrinks at low pH values?

Professor Ryan responded: We have been inspired by previous work on polyelectrolyte synthetic muscles and the zero-order response to your question is that there is no difference in principle. There are however some subtle differences. We have coupled the polyelectrolyte to an oscillating reaction and generated a cyclic force. We have recognised the limits imposed by mass diffusion and realised that polyelectrolyte based machines work best at small scale. This inspired us to measure the response of a polyelectrolyte brush and single polyelectrolyte chains. We have demonstrated, using the molecular callipers provided by the triblock copolymer structure, that the response is essentially affine—the molecular conformation changes are added to give the macroscopic shape change. We have built on Kuhn's original work.

Professor Vamvakaki communicated: How is the swelling behaviour of the polyelectrolyte gel influenced by the relation between the time constant of the oscillator reaction and the characteristic time of the swelling of the gel?

Professor Ryan communicated in response: In the case of Figs. 2 and 13 in the paper the response time of the gel was smaller than the cycle time of the oscillating reaction so the gel could fully expand and collapse inside each cycle. In the geometry used to collect the data for Fig. 3 the area available for diffusion was restricted, thus the gel could not fully expand and collapse inside each cycle and the oscillations in pH are damped and the force curve is much smoother. The competition between these two time scales is recognised and can even be used to advantage—creating a sample which oscillates in thickness along its length for example.

Professor Ryan opened the discussion of Professor Yamamoto's paper: iPP has a particular helix hand packing of chains. Do your results of tight chain folds bias a change of helix hand?

Professor Yamamoto responded: iPP has very complex and even perplexing geometry in the chain packing in crystals. In the most stable α-form, the nearest neighbour chains have definite bias for pairing in the opposite hand, while in the β-form all the chains tend to have same handed helixes. Our multicanonical Monte Carlo simulation, though it is still preliminary, showed both the same-handed and opposite-handed chain folds (see Fig. 7 of our paper). However, the opposite-handed pair has a higher order than those of same handed ones; in the former case both helical stems are nearly straight and parallel showing good chain packing (see Fig. 2(a) here), while in the latter case the two helixes are slightly twisted and apparently winding around each other (see Fig. 2(b)). These

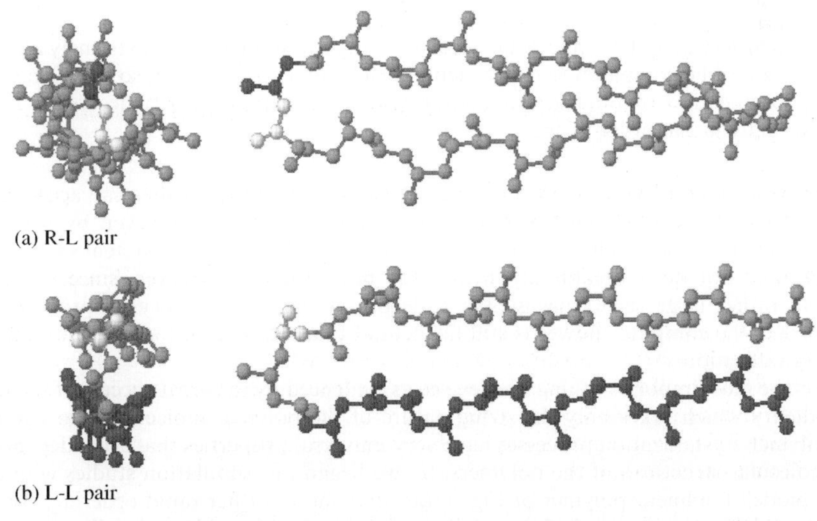

Fig. 2 Low energy conformations of iPP for (a) R–L and (b) L–L paired helices, viewed along the chain axis (left hand images) and perpendicular to the chain axis (right hand images).

results seem to suggest that the helix pair of opposite handed chains is the most favourable one from the energetic point of view.

Dr Al-Hussein asked: (1) Can your molecular simulations tell us something about the crystallization mechanism? (2) Does the crystallization process proceed by a sorting-out process of helices with different handedness?

Professor Yamamoto responded: (1) Our molecular simulations tell us much about the molecular mechanism of polymer crystallization. The mechanism inferred from the simplest polymer model (Fig. 13(a) of the paper) is fully discussed in a forthcoming review article.[1] For example we succeeded in observing *in situ* the chain folding processes at the growth surface. The conventional Kossel mechanism, which has been the very foundation of the Lauritzen–Hoffman theory of polymer crystallization, was found to be operative also in polymers. However, detailed inspections of the molecular processes showed that the molecular mechanism deviates considerably from the Lauritzen–Hoffman scenario, especially when the chain mobility at the growth front is very high.

(2) Crystallization of helical polymers is a very fascinating subject of research, and the present work started from the very motivation to observe the molecular recognition processes at the growth surfaces. As far as our present simulations are concerned, however, we have not yet succeeded in observing such a sorting process of helix hands in polymer crystallization. But various experiments for helical polymers, that do not have disordered crystalline phases before melting, definitely suggest that such mysterious molecular selection processes must be executed, and very efficiently.

1 T. Yamamoto, *Adv. Polym. Sci.*, 2004, in press.

Dr Perez Mendez asked: Is the simulation method able to explain the mechanism of stereospecific separation between helical enantiomers?

Professor Yamamoto responded: The stereospecific separation *in crystals* must come from the free-energy difference between the chain-packings into achiral domains and chiral domains. If the local R-R and R-L chain pairs have sufficient energy difference, the simulation gives the separation of the two enantiomers. One of our data presented, the POM-like crystals, showed such separation. Simulations for much more complex polymers, such as iPP in the β-form, also demonstrated clear separation into the chiral domains (see ref. 43 of our paper). However, the model and the method of simulation in ref. 43 were quite different from those adopted here for the studies of polymer crystallization.

Professor Schick asked: The time scale accessible in your simulation is extremely short (ns and below). This cannot be compared to experimental results on crystallization so far. Can you give an estimate of what period of time the simulation will reach experimentally accessible time scales, which are actually in the order of ms?

Professor Yamamoto responded: At first, the present-day MD simulations can access time scales of μs for sufficiently large systems of hundreds of thousands of atoms, even by using personal computers or relatively small cluster computers. If we make the size of the system smaller, even ms-simulations may not be a dream, say within the next decade. However, since we are mainly interested in rather large scale, mesoscopic scale phenomena such as crystal growth into chain folded lamellae, the computer power is still far behind that required in the realistic simulations of polymer crystallization.

However, the most important thing I believe is, as evidenced by the great success of the Lauritzen–Hoffman theory which takes only the string nature of the polymer molecules into consideration, that the polymer crystallization processes have very universal properties that are independent of the detailed molecular structures of the polymers. So we began our simulation studies with devising a molecular model, the linear polymer of Fig. 13(a), that shows rather rapid ordering. By using this model, we could observe the detailed molecular processes of chain folded crystallization *in situ*. The main criticism we have had is that the molecular model is artificial. But I believe that even this artificial model polymer should follow the same rule that governs the crystallization of real polymers, the rule by which we can find out the essentials of polymer crystallization.

The questioner may want to ask when we will be able to observe directly the crystallization processes of real polymers, such as iPP. When I reflect on the past ten years, I think rather optimistically that the rapid development of computer technology in the next decade must lead us to a considerably different perspective.

Mr Follestad said: My question concerns ordering/crystallisation of isotactic polypropylene, and whether computer limitations or something else is the problem. Do you think the present rules you have created and are using for the atoms and the bonds are satisfactory, so that with unlimited computer power you would have got a satisfactory solution on ordering/crystallisation?

Professor Yamamoto responded: This question is of a similar nature to the previous one. Our final goal is definite; the reproduction of the molecular processes of very realistic polymers. However, there can be various ways to reach our goal. One way might be to build up a complete quantum chemically satisfactory model first, and then try to speed up the calculation or just wait until the computers come to have unlimited power.

We did not follow this route, since we believed there must be very universal rules in polymer crystallization that are essentially independent of the molecular architectures. In our approach, both the molecular model adopted, the method of computation, and the development of computer technologies, proceed together to the final goal. So I must say that, at present, our model is not satisfactory, nor is the power of the computers.

The effect of molecular architecture on the grain growth kinetics of A_nB_n star block copolymers

Xiaochuan Hu,[a] Yuqing Zhu,†[a] Samuel P. Gido,*[a] Thomas P. Russell,*[a] Hermis Iatrou,[b] Nikos Hadjichristidis,[b] Ferass M. Abuzaina[c] and Bruce A. Garetz[c]

[a] *Polymer Science and Engineering Department, University of Massachusetts, Amherst, Massachusetts, 01003, USA*
[b] *Department of Chemistry, University of Athens, Panepistimiopolis Zografou, 15771, Athens, Greece*
[c] *Department of Chemical and Biological Sciences and Engineering, Polytechnic University, Brooklyn, New York, 11201, USA*

Received 12th March 2004, Accepted 14th May 2004
First published as an Advance Article on the web 30th September 2004

To investigate the effect of molecular architecture on the grain growth kinetics of star block copolymers, a series of A_nB_n miktoarm star block copolymers with different numbers of arms (n = 1, 2, 4 and 16) was studied. Across this entire series of materials, all the A arms are polystyrene (PS) blocks from the same anionically synthesized batch, and thus all the A arms are identical. Likewise, all the B arms are polyisoprene (PI) blocks from the same anionically synthesized batch, and thus all the B arms are identical. All the stars employed in this study are therefore composed of the same A and B arms liked together in symmetric numbers. The coarsening kinetics of grain growth was monitored in real space by transmission electron microscopy (TEM), followed by subsequent micrograph image analysis. It was found that the molecular architecture influenced the grain growth kinetics of these A_nB_n star copolymers dramatically. The grain coarsening kinetics was found to follow a scaling law as $V \sim t^{\beta}$, where V is the characteristic grain volume and t is time. The exponent, β, was found to be about 0.2 for the diblock copolymer (n = 1) and 0.4 for all three of the star block copolymers (n = 2, 4 and 16) in the series. It is postulated that the difference in grain growth rate between the diblock and the various stars is due to a reduction in molecular entanglements resulting from chain stretching near the junction points in the stars.

Introduction

Block copolymers can microphase separate into ordered structures composed of lamellae, cylinders, and spheres, as well as a variety of other structures at temperatures below the order-to-disorder transition temperature.[1] When these materials are quiescently quenched from the disordered state to the ordered state, coherently ordered regions or grains form, often with different orientations. Grain size and structure have been shown to play an important role in determining the mechanical and

† Present address: Eastman Chemical Co., Kingsport, TN, USA.

transport properties of block copolymers.[2–5] Recent efforts to use block copolymers in microelectronic and photonic applications also requires precise control of the grain size and/or structure.[6,7]

Coherent grain orientation can be achieved by application of an external field, such as electric field,[8–13] surface,[14–16] shear,[17–19] temperature gradient,[20] *etc.* The kinetics of grain growth of diblock copolymers, both in the bulk and in thin films, has been investigated with a variety of techniques such as small angle X-ray scattering (SAXS),[21–23] transmission electron microscopy (TEM),[24,25] depolarized light scattering (DLS),[24,26–28] and atomic force microscopy (AFM).[29] In the microphase separated state, coherent order is interrupted by the presence of topological defects that can be zero- (point), one- (line), or two- (grain boundaries) dimensional. Also, since the block copolymers are relatively soft mesophase materials one also observes more continuous change in the local microstructural orientation akin to focal conic smectic liquid crystalline textures. Grain growth occurs by the motion and annihilation of these defects.[30] Many variables have been found that influence the ordering of block copolymers such as temperature (quench depth),[20,31] molecular weight, and molecular relaxation time.[32]

Prior research on block copolymer grain ordering kinetics has been focused on diblock copolymers. Little is known about the grain coarsening kinetics of star copolymers, especially the effect of molecular architecture on the coarsening kinetics. However, it has been well established that star copolymers behave differently from conventional diblock copolymers due to the molecular architecture effect in microphase separation,[33,34] domain spacing,[35–37] rheology and dynamics.[38–41] Meanwhile, advances in anionic living polymerization and arm coupling chemistries have made it possible to synthesize well-defined star copolymers.[42–44] Building upon these prior developments, the objective of the current work is to explore the effects of molecular architecture on the ordering kinetics of *mikto*arm star block copolymers. To achieve this end, a series of $(PS)_n(PI)_n$ miktoarm star copolymers ($n = 1$, 2, 4, and 16) are prepared using anionic living polymerization and chlorosilane coupling chemistry.[35] Most importantly, all the PS arms used in this series come from the same polymerization batch. So they are all characteristically the same. Likewise, all the PI arms are also the same. This enables us to exclude any variables from the study that would arise due to differences in the composition of constituting arms.

Experimental section

The A_nB_n star copolymers used in this study are prepared by living anionic polymerization of styrene and isoprene respectively, followed by controlled chlorosilane coupling chemistry and purification. Table 1 shows their molecular characteristics. Based on the composition of these A_nB_n stars, lamellar morphology is expected and confirmed experimentally for all these stars. The detailed synthesis, molecular and basic morphological characterization of these star copolymers has been published previously.[35]

Sample preparation

Five weight percent solutions of each of the four star copolymers were prepared in toluene, a neutral solvent for the PS and PI blocks. Bulk films about 1 mm thick were obtained by evaporating the

Table 1 Molecular characteristics of the PS_nPI_n star block copolymers

Sample	$M_n{}^a$/g mol^{-1}	$M_w{}^b$/g mol^{-1}	$M_w/M_n{}^c$	PS (wt.%)^d
PS arm	19 000	—	—	100
PI arm	15 000	—	—	0
PS_1PI_1	36 200	36 300	1.05	55 ± 2
PS_2PI_2	64 000	66 000	1.04	56 ± 2
PS_4PI_4	121 000	127 500	1.04	54 ± 2
$PS_{16}PI_{16}$	^{e}	533 000	1.07	59 ± 2

[a] Membrane osmometry (MO) in toluene at 35 °C. [b] Low angle laser light scattering (LALLS) in THF at 25 °C. [c] Size exclusion chromatography in THF at 25 °C (DRI detector). [d] Measured by ^{1}H-NMR. [e] Too high to be measured by MO.

solvent from these solutions. Typically, studies of PS-PI block copolymer morphology involves covering the samples during solvent casting in order to slow solvent evaporation which takes place over one or two weeks.[35,45–48] In the present study, an initial microphase separated but not highly ordered state with small grains is desired. The size increase of these grains with thermal annealing will then be monitored. In order to produce an initial small grain size, the samples were cast without covering. The films thus produced were essentially solid after one day in air at room temperature, and were completely dry after two days. These dried films constituted the $t = 0$ initial point of the grain growth studies. Different pieces of these films were then annealed at 120 °C under vacuum for different periods of time to promote grain growth. To allow for quantitative comparison of grain size, all samples in this study are cast and annealed together to ensure that they experience identical external conditions. Different pieces of film for each of the samples were removed from the annealing oven at various times, from 30 min up to 60 days, and the grain size was characterized.

TEM characterization

All samples for transmission electron microscopy (TEM) were microtomed using a Leica Ultracut UCT cryoultramicrotome. Thin sections 50–100 nm in thickness are cut with a Diatome diamond knife at a temperature of −100 °C. In order to reduce sampling bias in the collection of TEM micrographs, several different sections of the sample were microtomed. These sections were then collected on 400 mesh copper TEM grids and stained in OsO_4 vapor for 6 hours. TEM was performed on either a JEOL 100CX or a JEOL 2000 FX instrument, operated at accelerating voltages of 100 kV or 200 kV respectively. TEM magnifications were calibrated with an etched silica size standard. Relatively low magnifications were used for all the micrographs to ensure that each micrograph encompasses as many entire grains as possible for the subsequent statistical analysis. A magnification of 8000× was used for samples during the early stages of grain growth, while a magnification of 6000× was used for the later stages of grain growth when the grains were larger. Forty to 50 TEM micrographs per sample per annealing time were used for the quantitative image analysis of grain size.

Quantitative image analysis

Grain size was obtained using the procedures described in detail by Garetz, Balsara and co-workers.[24,25,49] TEM micrographs were digitized and stored in TIFF files. Each digitized micrograph was divided into a total of 385 × 275 overlapping 32 pixel × 32 pixel squares with centers separated by 10 pixels, and the image within each square was stored as a 32 × 32 element array of 8-bit integers. The azimuthal angles, $\phi(\boldsymbol{r})$, which describe the local orientations of the lamellar normals at various points (given by the position vector $\boldsymbol{r}$) in the images relative to a fixed reference coordinate system, were computed by a local, two-dimensional Fourier transform (LFT). This LFT procedure produces a matrix of local azimuthal angles, $\phi(\boldsymbol{r})$, that describe the orientation of the projection of the lamellar normal in the image plane. This matrix is used to calculate the lamellar orientation correlation function, $C(\boldsymbol{r})$:[24]

$$C(|\boldsymbol{r}' - \boldsymbol{r}''|) = (15/8)\langle \sin^2\theta(\boldsymbol{r}')\sin^2\theta(\boldsymbol{r}'')\rangle\langle\cos\{2[\phi(\boldsymbol{r}') - \phi(\boldsymbol{r}'')]\}\rangle \approx \langle\cos\{2[\phi(\boldsymbol{r}') - \phi(\boldsymbol{r}'')]\}\rangle$$

The brackets $\langle\ldots\rangle$ represent an average over all pairs of points $\boldsymbol{r}'$ and $\boldsymbol{r}''$ in the ordered regions of a TEM micrograph. Since it is well established in the literature that lamellar microdomains are highly anisotropic in shape,[24,30,50,51] two different correlation functions, $C(\boldsymbol{r})_{\parallel}$ and $C(\boldsymbol{r})_{\perp}$, corresponding to the correlations in the direction of lamellar grain normal and perpendicular to this lamellar normal, have been evaluated. The parallel correlation function $C(\boldsymbol{r})_{\parallel}$ is calculated by averaging at points $\boldsymbol{r}'$ and $\boldsymbol{r}''$ where the vector $\boldsymbol{r}' - \boldsymbol{r}''$ is parallel to the lamellar grain normal within a prescribed tolerance of ±5°. Likewise, the perpendicular correlation function $C(\boldsymbol{r})_{\perp}$ is obtained by averaging at points $\boldsymbol{r}'$ and $\boldsymbol{r}''$ where the vector $\boldsymbol{r}' - \boldsymbol{r}''$ is perpendicular to the lamellar grain normal within a prescribed tolerance of ±5°. Both correlation functions were found to decay approximately exponentially. Thus, each correlation function curve was fit with an exponential form of $\exp(-r/\kappa)$, where κ is the correlation length. These correlation lengths are convenient parameters with which to characterize the grain size.

The procedure used in this study to obtain correlation lengths for TEM data was based on recently published methods[24] with the following exceptions: First, the correlation functions were obtained by systematically evaluating and summing the contributions from all pairs of 32 × 32 pixel boxes in each image. The analysis started with the first ordered box in the micrograph and

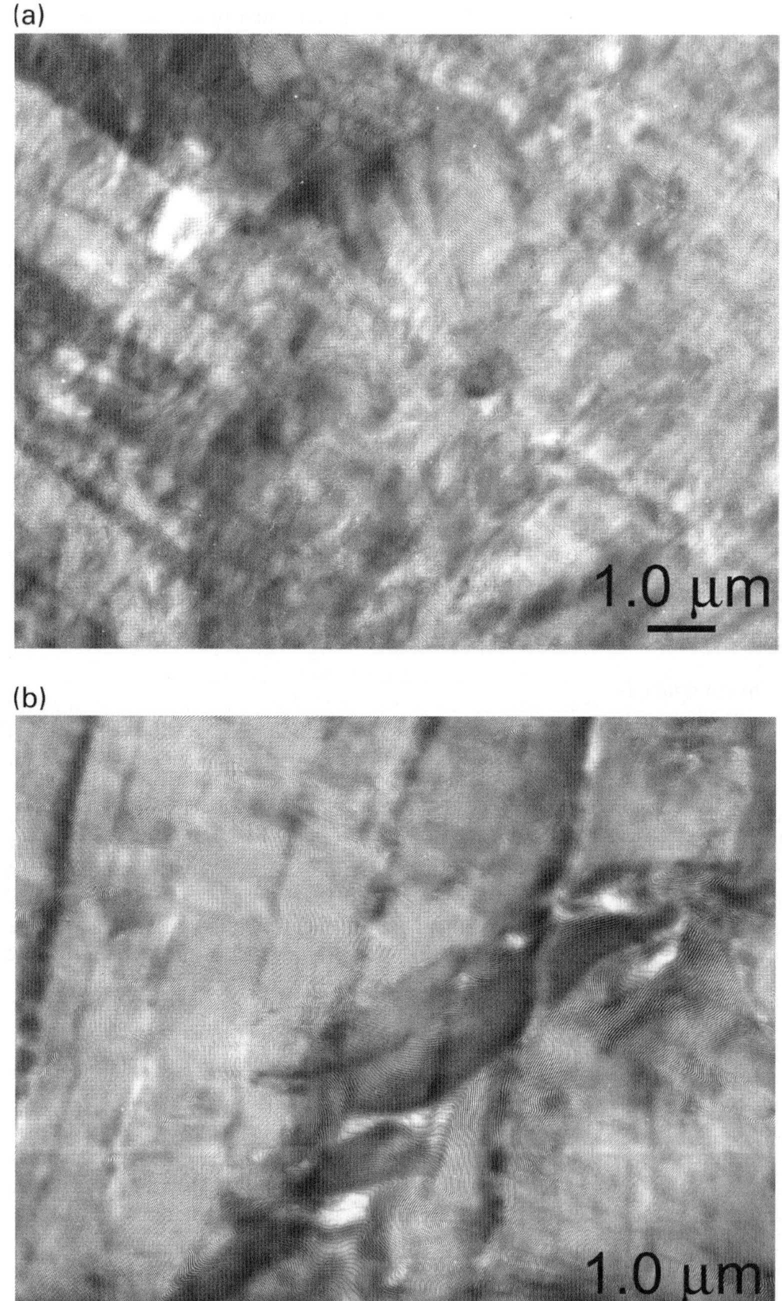

Fig. 1 Representative TEM images of PS_4PI_4 star block copolymer: (a) immediately after solution casting, *i.e.*, $t = 0$; and (b) after annealing for 60 days at 120 °C.

compared it with all the other ordered boxes in the same micrograph. Then the second box was compared to all other boxes, and so on until all the pairs of boxes in the micrograph have been compared. This contrasts with statistical sampling procedures used in a previously published study.[25] Also, since in this study the entire region in a micrograph is ordered with very few unclear regions and no disordered regions, the filtering process used in the previous studies[24] was not needed.

Results and discussion

Fig. 1a shows a TEM micrograph of the unannealed A_4B_4 sample, *i.e.* $t = 0$. Clearly the sample microphase separated in a lamellar morphology as a result of the solution casting, but the grain size is quite small. Fig. 1b shows a TEM image of the same A_4B_4 sample after annealing for 60 days, producing a significant increase in grain size.

As shown in Fig. 2a for the A_4B_4 material, both correlation functions, $C(r)_{\|}$ and $C(r)_{\perp}$, were found to approximately conform to the form of an exponential decay. Each curve is fit to an exponential function of the form $\exp(-r/\kappa)$, as shown in Fig. 2b, where κ is the correlation length, a characteristic dimension of the grain structure. The grains are generally anisotropic in shape, being larger parallel to the lamellar normal and smaller perpendicular to the lamellar normal. This has been reported previously in the literature for diblock copolymers[24,30,50,51] and is evident in our TEM of A_nB_n grain structure. Therefore, two characteristic correlation lengths are used to characterize each individual grain. The correlation length in the direction of lamellar normal is denoted as $\kappa_{\|}$, and the correlation length perpendicular to the lamellar normal is denoted as $\kappa_{\perp}$.

Both Table 2 and Fig. 3 show the time evolution of the characteristic grain size for the A_nB_n series of materials when annealed at 120 °C. The anisotropic grain shape is reflected in the fact that the $\kappa_{\|}$ values are consistently larger than the $\kappa_{\perp}$ values for all star functionalities and across the full range of annealing times, except for $A_{16}B_{16}$ at longer annealing times where $\kappa_{\|} \approx \kappa_{\perp}$. Least squares fits of power law functions of the form $\kappa = At^{\nu}$ are shown as lines on the graphs in Fig. 3. The A and ν parameters determined by these fits are given in Table 3. To within the error of the data, the $\kappa_{\|}$ and $\kappa_{\perp}$ fits, on the log-log scale in Fig. 3, are roughly parallel for each material. This indicates that the grain anisotropy does not change significantly during grain structure coarsening. The grain anisotropy for each sample can be approximated by the ratio of the constant, pre-exponential coefficients for the parallel and perpendicular correction lengths: $A_{\|}/A_{\perp}$. These values are 1.26, 1.52, 1.19, and 1.18 for A_1B_1, A_2B_2, A_4B_4, and $A_{16}B_{16}$, respectively. All of these grain anisotropy values are considerably lower than the previously reported value[50] of 2.37 for pre-impingement grain growth of lamellar forming diblock copolymers. The fact that the current studies focus on later stage grain size coarsening, after the initial growth and impingement process is complete, may unlie this discrepancy. A drive to reduce the overall amount of grain boundary surface area in the samples during grain coarsening will lead to lower grain anisotropy since surface to volume ratio is minimized for more compact, isotropic grain shapes.

In Fig. 4a all the parallel correlation lengths ($\kappa_{\|}$) for the A_nB_n series are plotted *vs.* time, and in Fig. 4b all the perpendicular correlation lengths ($\kappa_{\perp}$) are plotted *vs.* time. When the data is

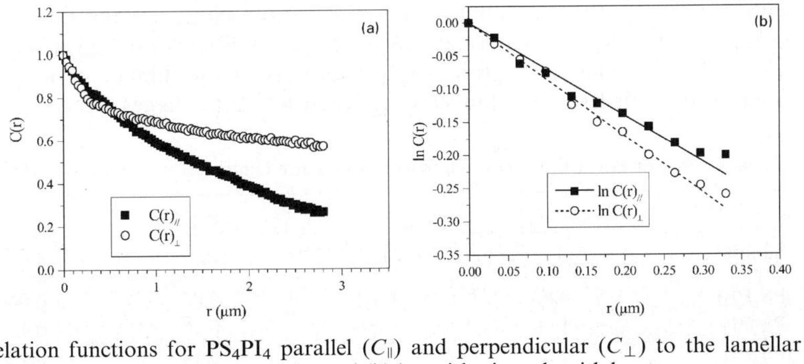

Fig. 2 Correlation functions for PS_4PI_4 parallel ($C_{\|}$) and perpendicular ($C_{\perp}$) to the lamellar normal after annealing at 120 °C for 60 days. (a) linear scale, and (b) logarithmic scale with least-squares exponential fit of the first 10 points of each correlation function.

Table 2 Correlation lengths determined by TEM image analysis. $\kappa_\parallel$ and $\kappa_\perp$ are correlation lengths parallel and perpendicular to the lamellar normal. $\sigma_\parallel$ and $\sigma_\perp$ are the standard deviations of $\kappa_\parallel$ and $\kappa_\perp$ respectively

	PS_1PI_1				PS_2PI_2			
Time/h	$\kappa_\parallel$/μm	$\sigma_\parallel$/μm	$\kappa_\perp$/μm	$\sigma_\perp$/μm	$\kappa_\parallel$/μm	$\sigma_\parallel$/μm	$\kappa_\perp$/μm	$\sigma_\perp$/μm
0	0.36	0.06	0.23	0.03	0.65	0.13	0.46	0.07
0.5	0.44	0.02	0.40	0.03	0.86	0.15	0.60	0.09
2	0.89	0.19	0.68	0.14	1.02	0.18	0.71	0.09
6	1.02	0.16	0.78	0.13	1.23	0.11	0.83	0.12
24	1.11	0.13	0.87	0.13	1.66	0.25	1.07	0.15
96	1.09	0.05	0.89	0.04	1.69	0.14	1.10	0.06
384	1.42	0.08	1.01	0.12	1.78	0.20	1.54	0.17
1440	1.45	0.16	1.04	0.04	1.97	0.14	1.71	0.07
	PS_4PI_4				$PS_{16}PI_{16}$			
0	0.42	0.04	0.43	0.04	0.37	0.03	0.32	0.03
0.5	0.47	0.10	0.45	0.11	0.45	0.03	0.33	0.03
2	0.68	0.10	0.55	0.07	0.43	0.03	0.38	0.03
6	0.82	0.15	0.55	0.09	0.49	0.05	0.48	0.04
24	0.99	0.12	0.84	0.16	0.62	0.07	0.61	0.07
96	1.02	0.03	0.87	0.04	0.63	0.05	0.62	0.05
384	1.39	0.04	1.15	0.08	0.85	0.09	0.84	0.11
1440	1.44	0.07	1.18	0.04	0.87	0.04	0.87	0.02

presented in this way, the relative grain size for different star functionalities can be seen. Fig. 5 shows plots of grain volume, estimated as $\kappa_\parallel\kappa_\perp\kappa_\perp$ *vs.* time for all the samples. Based on previous work by Balsara and co-workers[30] it is expected that the time evolution of average grain volume (V) obeys a power law: $V \sim t^\beta$. They reported that β for both pre- and post-impingement grain growth in a cylinder forming diblock copolymer of low molecular weight. The post-impingement β value of 1.20 is more relevant for comparison to our current results. Harrison *et al.*[29] has investigated the ordering kinetics in cylindrical diblock copolymer thin films and found that the correlation length increased with annealing time as $\sim t^{1/4}$.

The data in both Figs. 4 and 5 indicate that at any give time during the grain growth process the order of grain size among the samples from largest to smallest is generally A_2B_2, A_1B_1, A_4B_4, $A_{16}B_{16}$, except for the very longest times for which the order is: A_2B_2, A_4B_4, A_1B_1, $A_{16}B_{16}$. The volume *vs.* time plots are fit with power law functions of the form $V = At^\beta$, and the A and β parameters of this fit are given in Table 4. It is clear that the power law exponents (slopes in Fig. 5) for all the star architectures (A_2B_2, A_4B_4, and $A_{16}B_{16}$) are similar, in the range of 0.35 to 0.39. However, the power law exponent for the diblock, (A_1B_1), is very different, about 0.20. The grains of materials with star architectures are all growing at about the same rate and they are all growing faster than the grains of the diblock material. The grains of the diblock are bigger than those of the A_4B_4 and $A_{16}B_{16}$ stars, but the A_4B_4 grain size passes that of the diblock at long times. If the experiment could be extended to even longer times then the data suggests the $A_{16}B_{16}$ grain size

Table 3 Power law fitting parameters for correlation length growth data: $\kappa = At^\nu$

	A		ν	
Sample	$A_\parallel$	$A_\perp$	$\nu_\parallel$	$\nu_\perp$
PS_1PI_1	0.86	0.68	0.07	0.06
PS_2PI_2	1.00	0.66	0.10	0.13
PS_4PI_4	0.59	0.49	0.14	0.13
$PS_{16}PI_{16}$	0.44	0.37	0.10	0.13

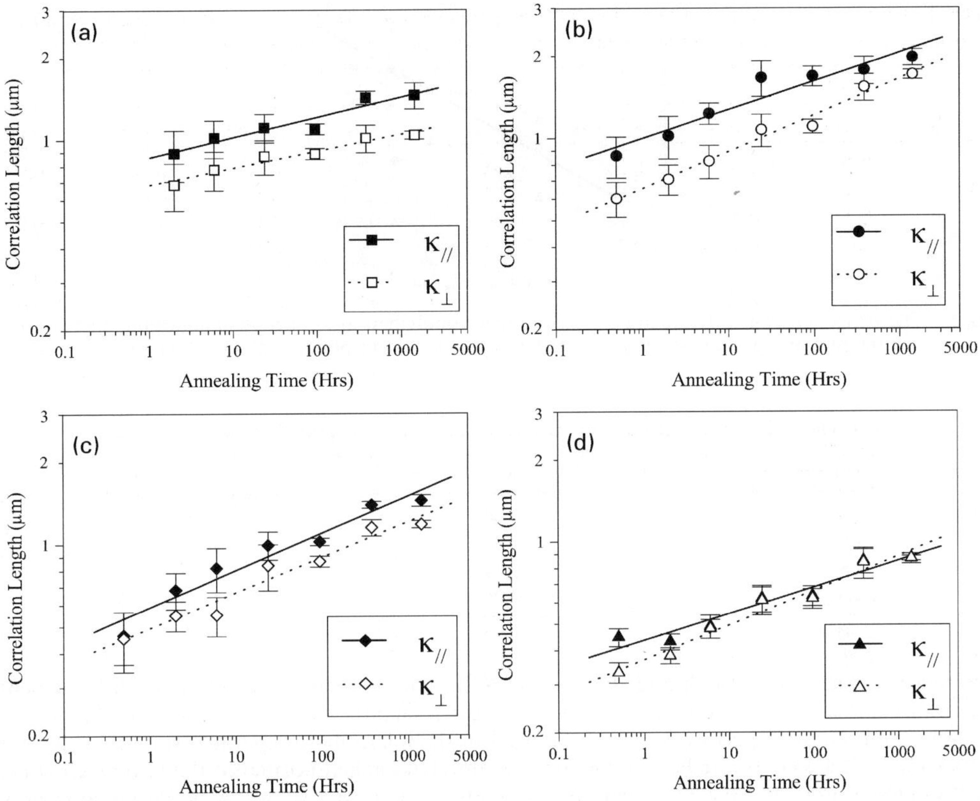

Fig. 3 The growth of correlation lengths in PS_nPI_n star block copolymers as a function of annealing time at 120 °C: (a) PS_1PI_1, (b) PS_2PI_2, (c) PS_4PI_4, and (d) $PS_{16}PI_{16}$. The lines in each plot are the least-squares, power law fits to the corresponding correlation length.

would eventually surpass that of the diblock. Grain growth experiments beyond the two-month maximum used in this study become difficult due to the limits of thermal and oxidative stability of the materials.

This data suggests a fundamental difference in the mechanisms of grain growth in the diblock verses all the A_nB_n stars for $n = 2, 4$, and 16. Dynamics in the case of the stars occurs by a faster type of mechanism than that of the diblocks. The increasing molecular weight as n increases from 2 to 16 in the A_nB_n series results in slower versions of the same mechanism. We hypothesize that the

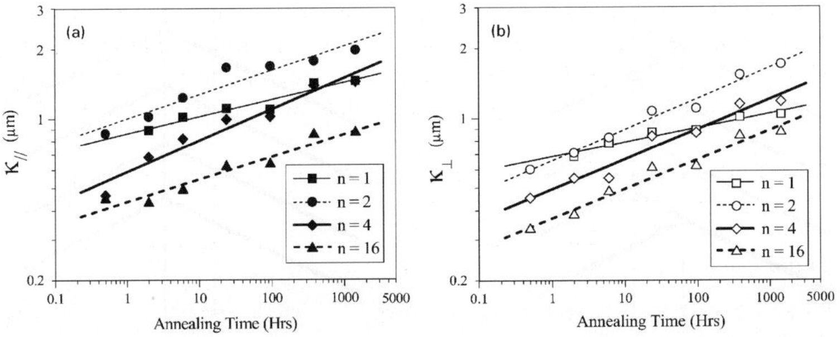

Fig. 4 The effect of arm number, n, on the growth of correlation lengths in PS_nPI_n star block copolymers as a function of annealing time at 120 °C: (a) $\kappa_{\parallel}$ values, (b) $\kappa_{\perp}$ values. The lines in each plot are the least-squares, power law fits to the corresponding correlation length.

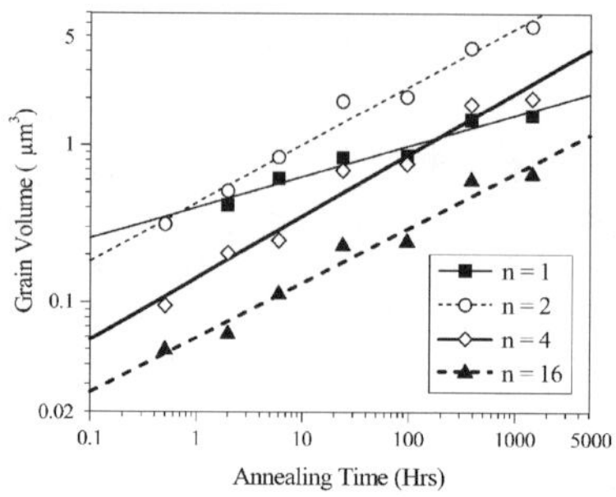

Fig. 5 The growth of grain volumes in PS_nPI_n star block copolymers as a function of annealing time at 120 °C: The lines in the plot are the least-squares, power law fits to the corresponding grain volume data.

Table 4 Power law fitting parameters for grain volume growth data: $V = A_v t^{\beta}$

Sample	A_v	β
PS_1PI_1	0.40	0.20 ± 0.02
PS_2PI_2	0.43	0.37 ± 0.02
PS_4PI_4	0.14	0.39 ± 0.03
$PS_{16}PI_{16}$	0.06	0.35 ± 0.03

difference between the diblock and the various A_nB_n stars is related to molecular entanglements. The molecular weight of the diblock suggests that it is slightly entangled. However, the previous work by Zhu, Gido and co-workers[35] shows that in the microphase separated state that the A_nB_n stars for $n = 2$, 4, and 16 are between 5 and 12% more stretched normal to the interface than the corresponding diblocks, due to the effect of arm crowding near the 2n functional junction point. This additional chain stretching normal to the interface may reduce entanglements with laterally neighboring chains thereby greatly increasing the rate at which the A_nB_n stars can diffuse within the lamellar layers relative to the corresponding rate for the entangled diblocks.[52–58] The diblock grains start out ($t = 0$) larger than the grains in the A_4B_4 and $A_{16}B_{16}$ samples due to the solvent casting process used to prepare the initial samples. The grains of the initial states for all samples are formed during solvent casting, where the diluting effect of the solvent will reduce the entanglements of the diblock during much if not all of the initial grain growth process. Once the solvent is fully removed, however, all the A_nB_n stars have faster growing grains, and even the A_4B_4 and $A_{16}B_{16}$ grain sizes should pass that of the diblock given enough time.

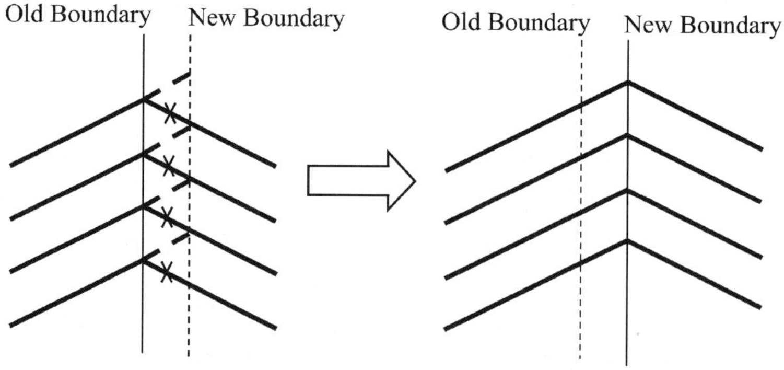

Fig. 6 Schematic of motion of a chevron tilt grain boundary. If the lamellar orientation on either side of the boundary is to be preserved, then the motion must involve the breaking and reforming of interfaces as indicated by ×'s and dashed lines in the diagram.

Conclusions

The scaling relationships we find for grain size growth as a function of time bears no simply discernible relationship to the kinetics of either entangled or unentangled block copolymer diffusion within, or normal to, lamellar layers as studied by Lodge and co-workers.[54–58] This is not necessarily surprising since the relationship between grain growth in three dimensions and diffusion of chains along two dimensional interfaces is likely complex and has not been elucidated. Simple geometric arguments, illustrated in Fig. 6, suggest that tilt grain boundaries (both symmetric and asymmetric) cannot move by molecular motions along the interface alone. Some interfacial breaking and reforming (*i.e.* diffusion of chains normal to the lamellar layers) is necessary to move this type of grain boundary. It may be possible for twist grain boundaries with Scherk saddle surface interfaces[59,60] to move through the sample *via* a mechanism in which the interfaces remain intact, *i.e.* only diffusion along existing interfaces is required. If this is true, then twist grain boundaries may move faster than tilt grain boundaries. However, any polygrain sample contains both twist and tilt boundaries as well as compound grain boundaries containing both twist and tilt character. Thus the situation is very complex.

Acknowledgements

S.P.Gido acknowledges funding from the U.S. Army Research Laboratory. The authors acknowledge Prof. Nitash P. Balsara for helpful discussion on ordering kinetics. The authors also acknowledge the use of central facility and the W. M. Keck Electron Microscopy in the Material Research Science and Engineering Center (MRSEC) at the University of Massachusetts–Amherst.

References

1 L. Leibler, *Macromolecules*, 1980, **13**, 1602.
2 K. Kawasaki and A. Onuki, *Phys. Rev. A*, 1990, **42**, 3664.
3 J. Csernica, R. F. Baddour and R. E. Cohen, *Macromolecules*, 1989, **22**, 1493.
4 D. H. Rein, J. Csernica, R. F. Baddour and R. E. Cohen, *Macromolecules*, 1990, **23**, 4456.
5 J. Csernica, R. F. Baddour and R. E. Cohen, *Macromolecules*, 1990, **23**, 1429.
6 M. Park, C. Harrison, P. M. Chaikin, R. A. Register and D. H. Adamson, *Science*, 1997, **276**, 1401.
7 R. R. Li, P. D. Dapkus, M. E. Thompson, W. G. Jeong, C. Harrison, P. M. Chaikin, R. A. Register and D. H. Adamson, *Appl. Phys. Lett.*, 2000, **76**, 1689.
8 K. Amundson, E. Helfand, X. Quan and S. D. Smith, *Macromolecules*, 1993, **26**, 2698.
9 K. Amundson, E. Helfand, X. N. Quan, S. D. Hudson and S. D. Smith, *Macromolecules*, 1994, **27**, 6559.
10 B. Ashok, M. Muthukumar and T. P. Russell, *J. Chem. Phys.*, 2001, **115**, 1559.
11 T. Xu, C. J. Hawker and T. P. Russell, *Macromolecules*, 2003, **36**, 6178.
12 Y. Tsori, F. Tournilhac and L. Leibler, *Macromolecules*, 2003, **36**, 5873.
13 G. G. Pereira and D. R. M. Williams, *Macromolecules*, 1999, **32**, 8115.
14 E. Huang, T. P. Russell, C. Harrison, P. M. Chaikin, R. A. Register, C. J. Hawker and J. Mays, *Macromolecules*, 1998, **31**, 7641.
15 E. Huang, S. Pruzinsky, T. P. Russell, J. Mays and C. J. Hawker, *Macromolecules*, 1999, **32**, 5299.
16 P. Mansky, T. P. Russell, C. J. Hawker, M. Pitsikalis and J. Mays, *Macromolecules*, 1997, **30**, 6810.
17 N. P. Balsara and B. Hammouda, *Phys. Rev. Lett.*, 1994, **72**, 360.
18 B. J. Dair, A. Avgeropoulos, N. Hadjichristidis, M. Capel and E. L. Thomas, *Polymer*, 2000, **41**, 6231.
19 T. J. Hermel, L. F. Wu, S. F. Hahn, T. P. Lodge and F. S. Bates, *Macromolecules*, 2002, **35**, 4685.
20 J. Bodycomb, Y. Funaki, K. Kimishima and T. Hashimoto, *Macromolecules*, 1999, **32**, 2075.
21 C. R. Harkless, M. A. Singh, S. E. Nagler, G. B. Stephenson and J. L. Jordansweet, *Phys. Rev. Lett.*, 1990, **64**, 2285.
22 M. A. Singh, C. R. Harkless, S. E. Nagler, R. F. Shannon and S. S. Ghosh, *Phys. Rev. B*, 1993, **47**, 8425.
23 R. T. Myers, R. E. Cohen and A. Bellare, *Macromolecules*, 1999, **32**, 2706.
24 M. Y. Chang, F. M. Abuzaina, W. G. Kim, J. P. Gupton, B. A. Garetz, M. C. Newstein, N. P. Balsara, L. Yang, S. P. Gido, R. E. Cohen, Y. Boontongkong and A. Bellare, *Macromolecules*, 2002, **35**, 4437.
25 F. L. Beyer, S. P. Gido, C. Buschl, H. Iatrou, D. Uhrig, J. W. Mays, M. Y. Chang, B. A. Garetz, N. P. Balsara, N. B. Tan and N. Hadjichristidis, *Macromolecules*, 2000, **33**, 2039.
26 H. Wang, M. C. Newstein, M. Y. Chang, N. P. Balsara and B. A. Garetz, *Macromolecules*, 2000, **33**, 3719.
27 M. C. Newstein, B. A. Garetz, N. P. Balsara, M. Y. Chang and H. J. Dai, *Macromolecules*, 1998, **31**, 64.
28 W. G. Kim, B. A. Garetz, M. C. Newstein and N. P. Balsara, *J. Polym. Sci. Part B*, 2001, **39**, 2231.

29 C. Harrison, D. H. Adamson, Z. D. Cheng, J. M. Sebastian, S. Sethuraman, D. A. Huse, R. A. Register and P. M. Chaikin, *Science*, 2000, **290**, 1558.
30 H. J. Dai, N. P. Balsara, B. A. Garetz and M. C. Newstein, *Phys. Rev. Lett.*, 1996, **77**, 3677.
31 W. G. Kim, M. Y. Chang, B. A. Garetz, M. C. Newstein, N. P. Balsara, J. H. Lee, H. Hahn and S. S. Patel, *J. Chem. Phys.*, 2001, **114**, 10 196.
32 N. P. Balsara, B. A. Garetz, M. Y. Chang, H. J. Dal and M. C. Newstein, *Macromolecules*, 1998, **31**, 5309.
33 S. T. Milner, *Macromolecules*, 1994, **27**, 2333.
34 D. M. A. Buzza, I. W. Hamley, A. H. Fzea, M. Moniruzzaman, J. B. Allgaier, R. N. Young, P. D. Olmsted and T. C. B. McLeish, *Macromolecules*, 1999, **32**, 7483.
35 Y. Q. Zhu, S. P. Gido, M. Moshakou, H. Iatrou, N. Hadjichristidis, S. Park and T. Chang, *Macromolecules*, 2003, **36**, 5719.
36 V. Grayer, E. E. Dormidontova, G. Hadziioannou and C. Tsitsilianis, *Macromolecules*, 2000, **33**, 6330.
37 M. W. Matsen and J. M. Gardiner, *J. Chem. Phys.*, 2000, **113**, 1673.
38 D. M. A. Buzza, A. H. Fzea, J. B. Allgaier, R. N. Young, R. J. Hawkins, I. W. Hamley, T. C. B. McLeish and T. P. Lodge, *Macromolecules*, 2000, **33**, 8399.
39 A. Jabbarzadeh, J. D. Atkinson and R. I. Tanner, *Macromolecules*, 2003, **36**, 5020.
40 K. Chrissopoulou, Y. Tselikas, S. H. Anastasiadis, G. Fytas, A. N. Semenov, G. Fleischer, N. Hadjichristidis and E. L. Thomas, *Macromolecules*, 1999, **32**, 5115.
41 K. Chrissopoulou, S. Harville, S. H. Anastasiadis, G. Fytas, J. W. Mays and N. Hadjichristidis, *J. Polym. Sci. Part B*, 1999, **37**, 3385.
42 J. Allgaier, R. N. Young, V. Efstratiadis and N. Hadjichristidis, *Macromolecules*, 1996, **29**, 1794.
43 H. Iatrou and N. Hadjichristidis, *Macromolecules*, 1993, **26**, 2479.
44 R. P. Quirk, T. Yoo and B. J. Lee, *J. Macromol. Sci., Pure Appl. Chem.*, 1994, **31**, 911.
45 Y. Q. Zhu, R. Weidisch, S. P. Gido, G. Velis and N. Hadjichristidis, *Macromolecules*, 2002, **35**, 5903.
46 F. L. Beyer, S. P. Gido, D. Uhrig, J. W. Mays, N. B. Tan and S. F. Trevino, *Journal of Polymer Science Part B-Polymer Physics*, 1999, **37**, 3392.
47 F. L. Beyer, S. P. Gido, Y. Poulos, A. Avgeropoulos and N. Hadjichristidis, *Macromolecules*, 1997, **30**, 2373.
48 F. L. Beyer, S. P. Gido, G. Velis, N. Hadjichristidis and N. B. Tan, *Macromolecules*, 1999, **32**, 6604.
49 B. A. Garetz, N. P. Balsara, H. J. Dai, Z. Wang, M. C. Newstein and B. Majumdar, *Macromolecules*, 1996, **29**, 4675.
50 N. P. Balsara, C. M. Marques, B. A. Garetz, M. C. Newstein and S. P. Gido, *Phys. Rev. E*, 2002, **66**.
51 T. Hashimoto and N. Sakamoto, *Macromolecules*, 1995, **28**, 4779.
52 D. Ehlich, M. Takenaka, S. Okamoto and T. Hashimoto, *Macromolecules*, 1993, **26**, 189.
53 D. Ehlich, M. Takenaka and T. Hashimoto, *Macromolecules*, 1993, **26**, 492.
54 M. C. Dalvi and T. P. Lodge, *Macromolecules*, 1993, **26**, 859.
55 M. C. Dalvi, C. E. Eastman and T. P. Lodge, *Phys. Rev. Lett.*, 1993, **71**, 2591.
56 T. P. Lodge and M. C. Dalvi, *Phys. Rev. Lett.*, 1995, **75**, 657.
57 T. P. Lodge, M. W. Hamersky, J. M. Milhaupt, R. M. Kannan, M. C. Dalvi and C. E. Eastman, *Macromol. Symp.*, 1997, **121**, 219.
58 M. W. Hamersky, M. Tirrell and T. P. Lodge, *Langmuir*, 1998, **14**, 6974.
59 S. P. Gido, J. Gunther, E. L. Thomas and D. Hoffman, *Macromolecules*, 1993, **26**, 4506.
60 S. P. Gido and E. L. Thomas, *Macromolecules*, 1994, **27**, 849.

On the theory of aggregation and micellization: PEO–PVP copolymer in water

Irina A. Nyrkova[a] **and Alexander N. Semenov**[b]

[a] *Physics Department, Moscow State University, Moscow 119992, Russia*
[b] *Institut Charles Sadron, 6 rue Boussingault, 67083 Strasbourg Cedex, France*

Received 5th March 2004, Accepted 13th May 2004
First published as an Advance Article on the web 13th September 2004

We develop a theoretical approach to micellization of the PEO–PVP block-copolymer in water. This copolymer is a weak polyelectrolyte due to protonation of VP blocks. The theory accounts for non-linear ion screening, and predicts strong position dependence of both ion concentration and the effective Debye length. We consider both the case when the local Debye length is small compared to the core radius and the case when it is large. We found that the effective (local) pH is not uniform even inside one micellar core, hence non-uniform protonation of the core with higher charge density near the surface. In many cases the core charge is concentrated in a relatively thin surface layer. Considering statistical weights of non-equilibrium micelles and their continuous evolution we show that kinetics of both formation and dissociation of typical block-copolymer or surfactant micelles can be extremely slow. Thus micelle formation at the genuine (equilibrium) critical micelle concentration (c.m.c.) is totally suppressed (involves astronomical time scales) if the micelles are big enough. An 'apparent' critical micelle concentration (c.m.c.*) is introduced to account for this effect. The apparent c.m.c.* could be much higher than the genuine equilibrium c.m.c., *i.e.* a significant hysteresis is inherent in these systems. We also determine the ranges of meta-stability of micelles depending on the experimental time-scales.

I. Introduction

Micelle formation is inherent in amphiphilic (surfactant) systems, in particular, in block-copolymers in selective solvents. In spite of numerous studies of the micellization, a number of important general issues are still either not clarified or are open to discussion:

(1) The aggregation is still routinely described in terms of the 'open association' and 'closed association' models which both are rather crude approximations to the real situations that are typically between these two extremes.

(2) A number of thermodynamic issues are still not clear, in particular: (i) whether (and to what extent) the typical size of the micelles, their polydispersity, *etc.* are concentration dependent at equilibrium; (ii) what the critical micelle concentration (c.m.c.) is and whether its value depends on the method of experimental determination.

(3) Finally, kinetic aspects of micellization have not yet been studied in sufficient detail.

The present paper addresses some of the above questions for micellization of block-copolymers in selective solvents. We focus on micelles formed by poly(ethylene oxide)–poly(vinylpyridine)

(PEO–PVP) diblock copolymers in water. In this case the inner (core) PVP block is a weak polyelectrolyte. This system is both of theoretical interest and of practical importance, hence the motivation for the study is several-fold:

(1) PEO–PVP micelles[1] are useful, *e.g.*, as nanoreactors for formation of metal nanoparticles[2] that can be employed as colloidal catalysts.[3]

(2) PVP blocks may be charged depending on pH due to protonation of VP units, *i.e.* PEO–PVP is a polyelectrolyte. Hence the micellar structure can be conveniently controlled by changing pH or ionic strength of the aqueous solution. These features are promising, *e.g.* in the area of drug delivery.[1,4,5] Yet micelles with charged core have not been considered theoretically so far.

(3) It turns out that the linear Debye–Hückel theory is often inapplicable to polyelectrolyte micelles, as the electrostatic potentials are high enough. Hence we develop an *analytical* theory of nonlinear electrostatic screening in polyelectrolyte solutions showing that the effective pH and Debye screening length vary significantly inside the micelle.

(4) We also consider kinetic aspects of micellization and show that micelle formation is typically an activation (nucleation) process.[1] This feature was missing in many dynamical models of micellization.[6–9]

II. Thermodynamics of PEO–PVP micelles in water

Let us consider an aqueous solution of PEO–PVP diblock copolymers with molecular weight $M = M_1 + M_2$, where subscripts 1 and 2 correspond to PEO and PVP blocks, respectively. Water is a selective solvent for PEO–PVP block-copolymer: Unprotonated PVP blocks are strongly insoluble (hydrophobic), so they tend to aggregate forming micellar cores. PEO is partially soluble in water, hence PEO blocks tend to form micellar shells. However, for low pH the pyridine rings in PVP become strongly protonated, leading to electrostatic destabilization of the micelles.

We assume that the critical micelle concentration, c.m.c., is low (which is typical when the PVP molecular weight M_2 is high enough), *i.e.* the dilute regime. Hence we consider an ideal gas of micelles and unaggregated (free) copolymer chains. Let $F(m)$ be the (internal) free energy of an aggregate (micelle) involving exactly m copolymer chains. The concentration of these aggregates is (the thermal energy k_BT is considered as the energy unit here and below)

$$c(m) = \frac{1}{V_0} e^{\mu m - F(m)} \tag{1}$$

where μ is the chemical potential of one chain (μ can be considered as an external parameter), $V_0 \equiv N_1 v_1 + N_2 v_2$ is the intrinsic volume of one copolymer chain, N_1, v_1 and N_2, v_2 are the polymerization degrees and monomer volumes of PEO and PVP blocks respectively.

A. The structure and the free energy of micelles

We anticipate that (at least at equilibrium) micelles are large ($m \gg 1$).[1,2] Let us consider a spherical micelle consisting of a PVP core (radius R_2), PEO corona (radius R_1) and two interfaces at $r = R_1$ and R_2 (see Fig. 1).

The micellar internal free energy can be thus written as

$$F = F_1 + F_2 + F_{10} + F_{21} + F_{ch} \tag{2}$$

where $F_1 = F_{osm1} + F_{el,1}$ is the corona free energy, $F_2 = F_{osm2} + F_{el,2}$ is the core free energy, and F_{10}, F_{21} are the energies of PEO/water and PVP/PEO interfaces. Here F_{osm} are the osmotic contributions (due to interactions between monomers) and $F_{el,}$ are the elastic free energies of (possible) stretching of blocks in the corona (1) or in the core (2). The last term F_{ch} accounts for the electrostatic interactions between charged PVP blocks and low-molecular ions; it is considered in the next section.

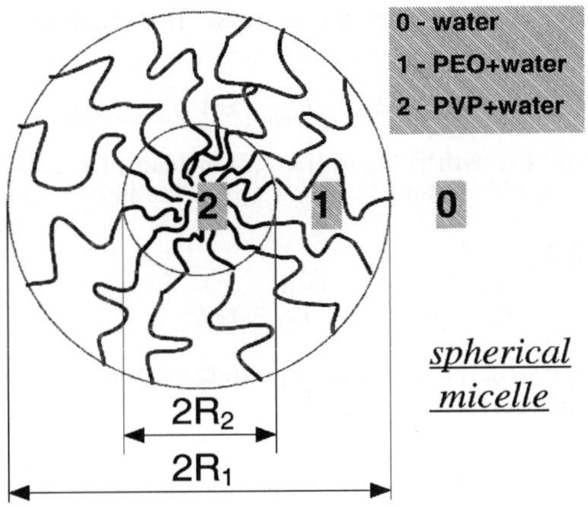

Fig. 1 Spherical PEO–PVP micelle in water (both the core and the corona are partially swollen by water; ions are not shown).

B. Non-electrostatic inputs to the micelle free energy

We adopt the Flory–Huggins approximation for the excess osmotic free energy density of monomer interactions in water:

$$f_{\text{osm}}(\phi) = \frac{1}{v}\left\{\frac{u}{2}\phi^2 + w\left[(1-\phi)\ln(1-\phi) + \phi - \phi^2/2\right]\right\} \tag{3}$$

where ϕ is volume fraction of copolymer (PEO in the corona or PVP in the core), v is the monomer volume, and u and w are the monomer interaction parameters (uv and wv^2 are the second and the third virial coefficients respectively). The parameters v and w can be estimated on geometrical grounds assuming that a monomer unit is a cylinder and that the density of undiluted polymer is known, see Table 1. As for the parameter u, a lot of experimental data[10–15] seem to indicate that PEO with molecular mass $M \gtrsim M_c \sim 10\,000$ tends to aggregate in water at room temperatures, which within the framework of the model, eqn. 3, implies that $u_{\text{PEO}} \approx -0.5$. u_{PVP} is considered as a fitting parameter.

As $u < 0$ for both polymers, long homopolymers of both types (PEO and PVP) tend to precipitate forming a phase with uniform density ϕ^{∞}, the latter is determined by the condition that the polymer osmotic pressure is zero (see Table 1). Although the monomer concentrations in the outer or inner parts of the micelles (ϕ_1, ϕ_2) do not necessary coincide with the corresponding ϕ^{∞}, the polymer

Table 1 Molecular parameters for PEO and PVP blocks. For water: $\varepsilon_s = 78$, $l_B = 7$ Å

	PEO	PVP	Comments
Monomer molar mass μ_{mol}/g	44	105	Chemical formula
Density ρ/g cm^{-3}	1.18	0.92	refs. 23 and 25
Monomer volume v/Å^3	62	190	$v = \mu_{\text{mol}}/\rho N_A$
Monomer length l/Å	4.5	2.5	Atomic model
Monomer diameter d/Å	4.2	9.8	$d = (4v/\pi l)^{0.5}$
Second virial parameter, u	−0.5	−2.85	
Third virial parameter, w	14.1	3.26	$w = (2dl)^3/v^2$
Bulk volume fraction Φ^{∞}	0.05	0.64	$\frac{u}{w} = 1 + \frac{2}{\phi^2}(\phi + \ln(1-\phi))$
Statistical segment b/Å	5.5	6.7	Refs. 23 and 25
Dielectric constant ε	—	12	Refs. 23 and 25
Bjerrum length l_B/Å	—	45	$l_B = e^2/\varepsilon k_B T$

densities should be nearly uniform and the osmotic free energies in eqn. (2) can be well approximated as

$$F_{\mathrm{osm1}} \approx V_1 f_{\mathrm{osm1}}(\bar{\phi}_1), \quad F_{\mathrm{osm2}} \approx V_2 f_{\mathrm{osm2}}(\bar{\phi}_2), \tag{4}$$

where $f_{\mathrm{osm},j}$ is defined in eqn. (3) with v_j, u_j, w_j (j = 1,2) instead of v, u, w, $\bar{\phi}_j = mN_jv_j/V_j$ is the mean polymer volume fractions and $V_1 = 4\pi(R_1^3 - R_2^3)/3$ and $V_2 = 4\pi R_2^3/3$ are the corona and the core volumes, respectively.

The interface free energies are

$$F_{10} = 4\pi R_1^2\gamma_{10}, \quad F_{21} = 4\pi R_2^2\gamma_{21} \tag{5}$$

We assume a smooth surface structure defined by excluded volume monomer interactions, eqn. (3). Then the bulk polymer/water surface tension is

$$\gamma = \min_{\phi(x)} \tilde{F}, \quad \tilde{F} = \int \frac{b^2}{24v}\frac{(\phi')^2}{\phi}\mathrm{d}x + \int f_{\mathrm{osm}}(\phi)\mathrm{d}x - \int \frac{\phi(x)}{\phi^\infty} f_{\mathrm{osm}}(\phi^\infty)\mathrm{d}x \tag{6}$$

where b is the statistical segment. Thus we get:

$$\gamma = \frac{b}{\sqrt{6}}\frac{1}{\sqrt{v}}\int_0^{\phi^\infty}\sqrt{\frac{f_{\mathrm{osm}}(\phi)}{\phi} - \frac{f_{\mathrm{osm}}(\phi^\infty)}{\phi^\infty}}\mathrm{d}\phi = \frac{b\sqrt{w}}{v}\tilde{\gamma}\left(-\frac{u}{w}\right), \quad \tilde{\gamma}(x) \approx \frac{3x^2}{16(1 + 1.41x + 0.24x^2)} \tag{7}$$

and find PEO/water and PVP/water coefficients γ_{10} and γ_{20}. We note that concentration of PEO is rather low ($\phi_{\mathrm{PEO}}^\infty \approx 5\%$), besides it is reasonable to assume that the interaction of PVP monomers with PEO is similar to their interaction with water molecules (both PEO monomers and water molecules are polar), hence γ_{21} must be close to the tension at the PVP/water interface: $\gamma_{21} \approx \gamma_{20}$.

In order to find the elastic free energy of PEO chain stretching, one can employ the parabolic molecular field approximation[16,17] valid if the blocks are far from being fully stretched:

$$F_{\mathrm{el},1} \approx \frac{\bar{\phi}_1}{v_1}\int U_1(z)\mathrm{d}^3r, \quad U_1(z) = \frac{3\pi^2}{8N_1^2b_1^2}z^2 \tag{8}$$

where $z = |r - R_2|$ is the distance to the PEO/PVP interface at $r = R_2$ and the integral is taken over the corona region. The result is:

$$F_{\mathrm{el},1} \approx \frac{\pi^2}{8}\frac{m(R_1 - R_2)^2}{N_1b_1^2}\frac{1 + 1.5\beta + 0.6\beta^2}{1 + \beta + \beta^2/3} \tag{9}$$

where $\beta = R_1/R_2 - 1$.

An additional complication arises for the elastic energy of the PVP core, as PVP blocks may be strongly stretched, exhibiting non-Gaussian elasticity. We assume the freely-jointed chain model for PVP (the Langevin-type elasticity) and use the generalized self-consistent molecular field $\tilde{U}_2(z)$ defined by the appropriate integral equation.[18] The result of integration of this $\tilde{U}_2(z)$ over the micelle core is (compare with eqn. (8))

$$F_{\mathrm{el},2} \simeq \frac{3\pi^2 m}{80N_2b_2^2}R_2^2\tilde{K}\left(\frac{R_2}{L_2}\right), \tag{10}$$

$\tilde{K}(x) \approx 1 + x^2(0.4093 + (x^2 - 1)(0.2598 + (x^2 - 0.4)(0.1593 + (x^2 - 0.2)(0.2385 + 0.7503(x^2 - 0.7)))))$
where L_2 is the PVP block contour length: $L_2 = N_2l_2$, l_2 is the contour length of a VP monomer.

III. Electrostatic contribution to the micelle free energy

A. Polyelectrolyte effects: the general equations

PVP blocks can be charged due to protonation of pyridine units in water: $\mathrm{VP} + \mathrm{H}^+ \rightleftharpoons \mathrm{VP}^+$. The fraction of protonated units, $\alpha = [\mathrm{VP}^+]/([\mathrm{VP}] + [\mathrm{VP}^+])$, is defined by the classic Henderson–Hasselbalch equation

$$\frac{\alpha}{1 - \alpha} = kc_{\mathrm{H}+} \quad \text{or} \quad \alpha = \frac{kc_{\mathrm{H}+}}{1 + kc_{\mathrm{H}+}}, \tag{11}$$

where $k = 10^{pK}\ M^{-1}$ is the reaction constant and c_{H^+} is the local concentration of protons (H^+). The latter is determined *via local* pH of the medium:

$$c_{H^+} = 10^{-pH}\ M = 10^{-pH_\infty} e^{-\Phi}\ M, \tag{12}$$

$$pH = pH_\infty + \Phi \ln 10, \quad \Phi = \frac{e\varphi}{k_B T} \tag{13}$$

Here $M = mol\ l^{-1}$, pH_∞ is pH of the medium at the reference point ∞ (far from the micelles), and φ is the local electric potential. At the reference point ∞: $\varphi(\infty) = 0$, $c_{H^+}(\infty) = 10^{-pH_\infty}$ M.

Usually the pK for polyelectrolytes is determined in the limit $\alpha \to 0$ where $pH = pH_\infty$. No reliable pK value for P2VP and P4VP in aqueous solutions is however available, as the polymer precipitates from the solution if $\alpha < 0.4$.[19–21] Some authors believe that a linear extrapolation to $\alpha = 0$ is good for pK, obtaining p$K \approx 5.1$ for P2VP and p$K \approx 5.0$ for P4VP.[21] Another option is to use the monomeric values of pK for the polymers:[1,22] p$K = 5.9$ for ethylpyridine.[23] We are tending to support the latter view, as the p$K(\alpha)$ dependencies for PVP used for linear extrapolation in ref. 21 are curved and lie far out of the range of interest (*e.g.* for P2VP in water: p$K \simeq 3.8$–2.3 for $\alpha \simeq 0.4$–0.8, see Fig. 1b in ref. 21†). Hence we assume:

$$pK \simeq 5.9. \tag{14}$$

If all low-molecular ions in the solution are monovalent then the density of charge is

$$\rho = e(\alpha c_{VP} + (c_+ - c_-)\phi_s) \tag{15}$$

where $c_{VP} = \phi_2/v_2$ is concentration of VP units, $c_+ = c_{H^+} + c'_+$ and $c_- = c_{OH^-} + c'_-$ are *reduced* concentrations of all positive and negative ions respectively and ϕ_s is the local volume fraction of water, $\phi_s \equiv 1 - (\phi_1 + \phi_2)$. The reduced concentrations c_+, c_- are defined as the number of ions *per unit volume of water*, hence the factor ϕ_s.‡Here

$$c_{OH^-} = 10^{pH-14}\ M \equiv 10^{pH_\infty - 14} e^{\Phi}\ M, \tag{16}$$

is the concentration of (OH^-) (*cf.* eqns. (12) and (13)), and c'_+, c'_- are concentrations of all other ions due to added salt/acid/base. We assume that at the reference point (∞): $\phi_s = 1$, $c_{VP} = 0$ and $\rho = 0$ (electroneutrality), *i.e.* $c_+ = c_- \equiv I_\infty$, where I_∞ is the ionic strength of the solution (see eqn. (15)). Thus by virtue of Boltzmann statistics

$$c_+ = I_\infty e^{-\Phi}, \ c_- = I_\infty e^{\Phi} \tag{17}$$

and

$$\frac{\rho}{e} = c_{VP} \frac{\alpha_\infty e^{-\Phi}}{1 + \alpha_\infty e^{-\Phi}} + I_\infty \phi_s \left(e^{-\Phi} - e^{\Phi}\right), \tag{18}$$

$$\alpha_\infty \equiv 10^{pK - pH_\infty} \tag{19}$$

(eqns. (11), (12), (15)). Note that α_∞ is a formal parameter which may exceed unity (there is no PVP at ∞).§

The non-dimensional potential Φ, eqn. (13), must satisfy the following Poisson equation:

$$\nabla \frac{\varepsilon}{\varepsilon_s} \nabla \Phi = -4\pi l_B \frac{\rho}{e}, \quad l_B \equiv \frac{e^2}{\varepsilon_s k_B T} \tag{20}$$

† Our own linear extrapolation (based on the data from Fig. 1b in the range $\alpha \simeq 0.4$–0.68 where p$K(\alpha)$ is nearly linear, p$K \simeq 3.8$–2.45) yields p$K_0 \simeq 5.73$.

‡ We assume totally repulsive excluded volume (steric) interactions between polymer chains and all low-molecular ions (except protons), *i.e.* no polymer solvatation by the ions except protonation of pyridine units.

§ If the VP protonation degree is small, $\alpha \ll 1$, eqn. (11) can be approximated as $\alpha \simeq \alpha_\infty e^{-\Phi}$ and hence the first term in the square brackets of eqn. (18) will be $c_{VP}\alpha_\infty e^{-\Phi}$.

where $\varepsilon = \varepsilon(r)$ is the local dielectric constant of the medium, ε_s is the dielectric constant of the solvent (water) and l_B is the Bjerrum length in the solvent. We approximate the function $\varepsilon(r)$ as:

$$\varepsilon = \begin{cases} \varepsilon_{core} \simeq \varepsilon_s(1-\phi_2) + \varepsilon_{PVP}\phi_2, & \text{inside the core}(r < R_2); \\ \varepsilon_{out} = \varepsilon_s, & \text{outside the core}(r > R_2), \end{cases} \tag{21}$$

where ε_{PVP} is the dielectric constant of the bulk PVP (note that the volume fraction of PEO in the corona is small: $1 - \phi_1 \simeq 1$), see Table 1.

Eqns. (18) and (20) for unknown functions $\{\Phi,\rho\}$ allow to determine the potential Φ, and hence the protonation degree α, eqn. (11), and the distributions of small ions, c_i, eqns. (12), (16) and (17), for known density profiles $\{\phi_1, \phi_2\}$ of copolymer units.

B. Free energy of ions, F_{ch}

The total free energy of ions is

$$F_{ion} = F_{Coul} + F_{id} + F_p \tag{22}$$

where

$$F_{Coul} = \frac{1}{2e}\int \Phi\rho\, d^3r \tag{23}$$

is the Coulomb energy considered as a functional of the charged density $\rho(\mathbf{r})$ defined in eqn. (15) (the Poisson eqn. (20) is used here),

$$F_{id} = \int \phi_s \left[c_{H+} \ln\frac{c_{H+}}{e} + c_{OH-} \ln\frac{c_{OH-}}{e} + \cdots \right] d^3r \tag{24}$$

is the ideal-gas free energy of unbound (free) protons and other ions (see the note about the factor ϕ_s after eqn. (15)), and

$$F_p = \int c_{VP} \left[\alpha \ln\frac{\alpha}{e} + (1-\alpha)\ln\frac{(1-\alpha)}{e} - \alpha E \right] d^3r \tag{25}$$

is the free energy of protonation including the relevant combinatorial contribution (the first two terms in square brackets) and the pyridine/proton interaction term ($E = \ln k = \mathrm{p}K \ln 10 - \ln \mathrm{M}$ is the energy of VP protonation).

The total electrostatic/ion/protonation contribution to the free energy is then defined as

$$F_{ch} = F_{ion} - F_{ion}^{(0)} \tag{26}$$

where $F_{ion}^{(0)}$ is the reference ideal-gas free energy of all ions in the limit $e \to 0$ (no charge effects).¶ Alternatively we can consider the thermodynamic potential Ω depending on chemical potentials of all types of ions, μ_{H^+}, $\mu_{OH^-}, \ldots$: $\Omega_{ion} = F_{ion} - \int[(1-\phi_s)(\mu_{H^+}c_{H^+} + \mu_{OH^-}c_{OH^-} + \cdots) + \mu_{H^+}c_{VP}\alpha]d^3r$ and define $F_{ch} = \Omega_{ion} - \Omega_{ion}^{(0)}$. Minimizing F_{ch} with respect to ion distributions and α we get for $\alpha \ll 1$:

$$F_{ch} = I_\infty \int (1-\phi)(2 - e^{\Phi} - e^{-\Phi}) d^3r - \alpha_\infty \int c_{VP} e^{-\Phi} d^3r - \frac{1}{2e}\int \Phi\rho\, d^3r \tag{27}$$

where ρ is defined in eqn. (18) and ϕ is local volume fraction of polymer. Note that eqns. (18) and (19) can be obtained as a result of minimization of eqn. (26).

The non-dimensional potential Φ must satisfy the Poisson eqn. (20):

$$\nabla \frac{\varepsilon}{\varepsilon_s} \nabla\Phi = \frac{1}{2}\chi^2 \left\{ (1-\phi)e^{\Phi} - \left[1 - \phi + c_{VP}\frac{\alpha_\infty}{I_\infty} \right] e^{-\Phi} \right\} \tag{28}$$

where we used eqn. (18), $\chi^2 = 8\pi l_B I_\infty$ ($r_D \equiv 1/\chi$ is the Debye length in the solvent at ∞).

¶ The numbers of ions in the reference system must coincide with those in the charged system. In particular, the total number of protons H^+ in the reference system must be equal to $\int[c_{H^+}(1-\phi_s) + \alpha c_{VP}]d^3r$, *i.e.* bound protons must be also taken into account.

Thus the free energy contribution due to protonation, F_{ch}, is defined in eqns. (18) and (27) including the non-dimensional potential Φ determined by eqn. (28). These equations represent a kind of Poisson–Boltzmann mean-field theory.

C. The electrostatic free energy: strong screening

Now we are in a position to solve eqn. (28) for the potential Φ around an isolated micelle. This is a strongly non-linear equation, and the optimal way to solve it depends on the value of the ratio of the micelle size to the electrical double layer width. We start with the case of strong screening.

1. Electrostatic potential near a micelle. We first consider the core region of the micelle ($r < R_2$, see Fig. 1). The dielectric constant there is approximated by the first line of eqn. (21) where ϕ_2 is volume concentration of PVP in the core. Substituting $\Phi = \Phi_0 - \delta\Phi$ in eqn. (28), and demanding that the rhs of this equation vanish at $\Phi = \Phi_0$ we get

$$\nabla^2 \delta\Phi = \kappa^2 \sinh \delta\Phi \tag{29}$$

with

$$\Phi_0 = \frac{1}{2}\ln\left[1 + \frac{\phi_2}{v_2(1-\phi_2)}\frac{\alpha_\infty}{I_\infty}\right], \quad \kappa^2 = \chi^2(1-\phi_2)\frac{\varepsilon_s}{\varepsilon_{core}}e^{\Phi_0} \tag{30}$$

where $\phi_2/v_2 = c_{VP}$. Note that the effective screening length $\overset{*}{r}_D \equiv 1/\kappa \sim r_D e^{-\Phi_0/2}$ can be much smaller than r_D since e^{Φ_0} is large: for example if no salt is added, pH < 7 and $\phi_2 = 0.65$, then $e^{\Phi_0} \approx 4 \times 10^3$ (*i.e.* $\Phi_0 \approx 8.25$) and $\overset{*}{r}_D \approx 0.02 r_D$.

Here we consider the case of strong electrostatic screening, $R_2 \gg \overset{*}{r}_D$, *i.e.* $\kappa R_2 \gg 1$ (note that the Debye length far from the micelle, r_D, may still be longer than R_2). In this regime any spherically symmetric solution of eqn. (29) (with no singularity at $r = 0$), $\delta\Phi$, must decrease exponentially as r is decreased. Eqn. (29) can be then approximated as

$$\frac{d^2}{dr^2}\delta\Phi = \kappa^2 \sinh \delta\Phi$$

with boundary condition $\delta\Phi \to 0$ as (formally) $r \to -\infty$. The general solution is

$$\frac{e^{\delta\Phi/2} + 1}{e^{\delta\Phi/2} - 1} = Ce^{\kappa(R_2 - r)} \tag{31}$$

where C = constant. The reduced electric field $E_{s-} = -d\Phi/dr$ at $r = R_2$ is

$$E_{s-} = 2\kappa \sinh \frac{\Phi_0 - \Phi_s}{2} \tag{32}$$

where $\Phi_s = \Phi(R_2)$ is the surface potential ($\Phi_s = \Phi_0 - \delta\Phi_s = \Phi_0 - 2\ln[(C+1)/(C-1)]$).

Let us turn to the region outside the PVP core, $r > R_2$, where $c_{VP} \approx 0$. We adopt the following approximations in this region: $\varepsilon \approx \varepsilon_s$, $1 - \phi \approx 1$ as the volume fraction of PEO in the corona is small. Hence eqn. (28) can be approximately written as $\nabla^2\Phi = \chi^2 \sinh \Phi$ with an obvious boundary condition: $\Phi \to 0$ as $r \to \infty$. In the case of strong screening (when Φ decreases appreciably in a relatively thin layer outside the core, $r - R_2 \ll R_2$) the Laplacian $\nabla^2\Phi$ can be approximated by $d^2\Phi/dr^2$. The solution for Φ is then indirectly defined by

$$\frac{e^{\Phi/2} + 1}{e^{\Phi/2} - 1} = \text{const} \times e^{\chi r} \tag{33}$$

The reduced electric field at the surface $r = R_2$ is

$$E_{s+} = 2\chi \sinh \frac{\Phi_s}{2} \tag{34}$$

where $\Phi_s = \Phi(R_2)$. Applying the electrostatic boundary conditions at $r = R_2$ (both Φ and εE are continuous, *i.e.* $\varepsilon_{core}E_{s-} = \varepsilon_s E_{s+}$) we get

$$\delta\Phi_s = \ln\frac{1+\omega}{1+\omega e^{-\Phi_0}}, \quad \Phi_s = \ln\frac{e^{\Phi_0}+\omega}{1+\omega} \tag{35}$$

where

$$\omega = \left[\frac{\varepsilon_s}{\varepsilon_{core}(1-\phi_2)}\right]^{1/2} \tag{36}$$

With typical core density $\phi_2 = 0.65$, $\varepsilon_s = 78$ (water) and $\varepsilon_{PVP} \approx 12$ we get $\omega \approx 2.5$ and $\delta\Phi_s \approx \ln(1+\omega) \approx 1.25$. Using the estimate on Φ_0 obtained after eqn. (29) we get $\Phi_s = \Phi_0 - \delta\Phi_s \approx 7$.

Assuming that $e^{\Phi/2} \gg 1$ and using eqn. (33) we may approximate the charge density ρ in the corona region just outside the core as

$$\rho \approx \rho_s e^{\Phi-\Phi_s} \simeq \frac{\rho_s}{\left[1+(r-R_2)/r_D^{**}\right]^2}$$

where $r_D^{**} = 2r_D e^{-\Phi_s/2}$ is the effective screening length near the core/corona interface. So $r_D^{**} \approx 0.06 r_D$ if $\Phi_s \approx 7$. Note that r_D^{**} was assumed to be small: $r_D^{**} \ll R_2$; typically $r_D^{**} \sim r_D^{*}$.

Using eqn. 31 we show that charge density in the core is exponentially decreasing with the distance from its surface: $\rho \propto \sinh\delta\Phi \propto e^{-\kappa(R_2-r)}$if $\kappa(R_2-r) \gg 1$ Hence the core (positive) charge is mostly concentrated near the core/corona interface, in a thin layer of thickness $\sim r_D^{*}$. The opposite (negative) charge of the corona region is also concentrated near the interface, in a layer of thickness r_D^{**}.

2. Free energy F_{ch} for the case of strong screening and $\alpha \ll 1$. The electrostatic/ion free energy is defined in eqn. (27). Using the electrostatic potentials in the core and corona regions obtained just above, we get after some algebra:

$$F_{ch} = V_2 f_{ch} + 4\pi R_2^2 \gamma_{ch}, \tag{37}$$

where V_2 is the core volume,

$$f_{ch} = -2I_\infty(1-\phi_2)(e^{\Phi_0}-1) \tag{38}$$

$$\gamma_{ch} = -\frac{\chi}{\pi l_B}\left[\left(\frac{1}{\omega}e^{\Phi_0/2}+e^{-\Phi_0/2}\right)\left(\frac{1+\omega}{1+\omega e^{-\Phi_0}}\right)^{1/2} - \left(\frac{1}{\omega}e^{\Phi_0/2}+1\right)\right] \tag{39}$$

where Φ_0 and ω are defined in eqns. (30) and (36).

Note that F_{ch} is always negative. This is expected since the driving force for the charge effects is protonation of PVP which is due to *negative* energy of proton/PVP attraction. We identify two contributions to F_{ch}: the first one is proportional to the core volume ($-f_{ch}$ is nearly twice the concentration of protonated units that is inversely proportional to $(r_D^{*})^2$); the second term is proportional to the surface area. Thus γ_{ch} can be considered as an electrostatic correction to the interfacial tension. This correction is negative and it is inversely proportional to r_D^{*}.‖

Let us briefly discuss the conditions of validity of eqns. (37)–(39). First of all, we have assumed that the effective screening length (the thickness of the electrostatic double layer near the core/corona interface) is small enough: $r_D^{*} \sim r_D^{**} \sim r_D e^{-\Phi_s/2} \ll R_2$. This condition can be written as (we assume that $\alpha_\infty/I_\infty v_2 > 1$ and that $\omega \sim 1$):

$$8\pi l_B R_2^2\left(\frac{I_\infty \alpha_\infty}{v_2}\right)^{1/2} \gg 1$$

‖ The surface term can be interpreted as an additional entropy of counterions (to protonated VP units) due to an effective increase of the available core volume by the volume of the electrostatic corona layer of thickness $\sim r_D^{*}$. Hence γ_{ch} is negative and it is proportional to the number of ions in the electrostatic layer per unit area.

Next, note that we assumed that $\alpha \ll 1$ before eqn. (27) (the more general case $\alpha \sim 1$ is considered in the next section). The highest $\alpha = \alpha_\infty e^{-\Phi_s}$ is attained at the surface. Hence the condition

$$\alpha_\infty I_\infty v_2 \ll 1 \tag{40}$$

If no salt is added, then this condition is satisfied for pH > 3.

Finally, we recall that we developed a mean-field Poisson–Boltzmann theory. The mean-field approximation is valid if the number of ions in the volume of screening is large, $c_-(r_D^*)^3 \gg 1$ (note that $c_+ < c_-$ as the electric potential Φ is positive). The latter condition is nearly equivalent to $c_- l_B^3 \ll 1$, *i.e.*

$$\alpha_\infty I_\infty \frac{l_B^6}{v_2} \ll 1 \tag{41}$$

(we take into account that the highest c_- is attained in the centre of the core: $(c_-)_{max} \approx (1-\phi_2) I_\infty e^{\Phi_0}$). Condition (41) is nearly equivalent to the previous condition (40) since $l_B^3 \sim v_2$.**

3. F_{ch} for strong screening and high protonation degree ($\alpha \sim 1$). So far we assumed that $\alpha \ll 1$. At low pH or at high ionic strength this condition may be violated. Hence we turn to the regime $\alpha \sim 1$ which is defined by two conditions:

$$\alpha_\infty \gtrsim 1 \text{ and } I_\infty \alpha_\infty v_2 \frac{1-\phi_2}{\phi_2} \gtrsim 1$$

Typically, these conditions are satisfied for pH < p$K \approx 5.9$. The above conditions also ensure that the strong screening approximation is valid if $R_2 \gtrsim 2$ nm and $\phi_2 \gtrsim 0.02$ (the latter inequalities always hold for copolymers with $M \gtrsim 10\,000$).

Below we calculate the dominant *volume contribution* to the protonation energy: $F_{ch} \approx V_2 f_{ch}$. Minimizing

$$\begin{aligned} F_{ch} &= \Omega_{ion} - \Omega_{ion}^{(0)} \\ &= F_{Coul} + \int (1-\phi) \sum_i \left(c_i \ln \frac{c_i}{e} - \mu_i c_i + c_{i\infty} \right) d^3r + \int c_{VP} \left[\alpha \ln \frac{\alpha}{\alpha_\infty} + (1-\alpha)\ln(1-\alpha) \right] d^3r \end{aligned}$$

(see eqn. 26) with respect to c_i and α (c_i are concentrations of all sorts of ions, α_∞ - $e^{E+\mu_{H+}} = 10^{pK - pH_\infty}$), we get

$$\alpha = \frac{\alpha_\infty e^{-\Phi}}{1 + \alpha_\infty e^{-\Phi}}$$

$$F_{ch} = -\frac{1}{2} \int \Phi \frac{\rho}{e} d^3r + \int (1-\phi) \sum_i (c_{i\infty} - c_i) d^3r + \int c_{VP} \ln(1-\alpha) d^3r$$

The last two terms can be rewritten as $\int(1-\phi) I_\infty (2 - e^{\Phi} - e^{-\Phi}) d^3r - \int c_{VP} \ln(1+\alpha_\infty e^{-\Phi}) d^3r$. Demanding approximate electroneutrality:

$$\rho/e = (1-\phi) I_\infty (e^{-\Phi} - e^{\Phi}) + c_{VP} \frac{\alpha_\infty e^{-\Phi}}{1+\alpha_\infty e^{-\Phi}} \approx 0$$

we get

$$F_{ch} \approx -V_2 f_{ch}, \quad f_{ch} = -\min_t \left\{ (1-\phi_2) I_\infty (t + 1/t - 2) + \frac{\phi_2}{v_2} \ln(1 + \alpha_\infty t) \right\} \tag{42}$$

** The mean-field theory is valid if $I_\infty e^{\Phi_0} l_B^3 \ll 1$. If $\alpha_0 \ll 1$, then this condition may be rewritten as $\alpha_0 l_B^3 \phi_2 / (v_2[1-\phi_2]) \ll 1$, where $\alpha_0 \simeq [I_\infty \alpha_\infty v_2 (1-\phi_2)/\phi_2]^{1/2}$ is the protonation degree at the core centre. (We assume that the factor e^{Φ_0} is large, otherwise the standard condition $I_\infty l_B^3 \ll 1$ is applicable everywhere.) If, on the contrary, α_0 is close to 1, then the mean-field condition is $l_B^3 \phi_2 / (v_2[1-\phi_2]) \ll 1$. This condition is satisfied if $\phi_2 \ll 1$.

where $t = e^{-\Phi}$. Thus after the minimization we get f_{ch}, the reduced potential $\Phi_0 = -\ln t$ and the protonation degree $\alpha_0 = \alpha_\infty t/(1 + \alpha_\infty t)$ in the core centre.

Note the two asymptotic regimes in eqn. (42):

(1) if $\alpha \ll 1$, *i.e.* either $\alpha_\infty \ll 1$ or $\alpha_\infty \nu_{2\infty} I_\infty (1 - \phi_2)/\phi_2 \ll 1$, then

$$f_{ch} \simeq -2I_\infty(1-\phi_2)\left[\sqrt{1+\frac{\alpha_\infty\phi_2}{\nu_2(1-\phi_2)I_\infty}}-1\right], \quad t \simeq \left[1+\frac{\alpha_\infty\phi_2}{\nu_2(1-\phi_2)I_\infty}\right]^{-1/2}$$

in agreement with eqns. (30) and (38); (2) if $\alpha \to 1$, *i.e.* $\alpha_\infty \gg 1$ and $\alpha_\infty \nu_{2\infty} I_\infty(1-\phi_2)/\phi_2 \gg 1$, then

$$f_{ch} \simeq -\frac{\phi_2}{\nu_2}\left[\ln\frac{\alpha_\infty \nu_2(1-\phi_2)I_\infty}{\phi_2}+1\right], \quad t \simeq \frac{\nu_2(1-\phi_2)I_\infty}{\phi_2}$$

D. The electrostatic free energy: weak screening

In section 3.3 we considered the electrostatic energy of a micelle in the case of strong screening when the *effective* (local) screening length $\overset{*}{r}_D$ is smaller than the core size, $\overset{*}{r}_D \ll R_2$. Now we turn to the opposite regime, $\overset{*}{r}_D \gg R_2$ (*i.e.* $r_D \gg R_2$ since r_D, the screening length in the bulk of the solvent is longer than $\overset{*}{r}_D$), when screening due to ions can be neglected. Hence the charge density (in the core) is $\rho = ec_{VP}\alpha = ec_{VP}\alpha_\infty e^{-\Phi}$. This equation can be obtained by formally setting $I_\infty = 0$ in eqn. (18). A similar reduction of eqns. (27) and (28) yields

$$\nabla\frac{\varepsilon}{\varepsilon_s}\nabla\Phi = -4\pi l_B c_{VP}\alpha_\infty e^{-\Phi} \tag{43}$$

$$F_{ch} = -\alpha_\infty c_{VP}\int e^{-\Phi}\left[1+\frac{\Phi}{2}\right]d^3r \tag{44}$$

where the integral is taken over the core region. Inside the core eqn. (43) can be written as

$$R_2^2\nabla(\nabla f/f) - qf = 0, \tag{45}$$

where $f \equiv e^{-\Phi}$ and

$$q \equiv 4\pi l_B \nu c_{VP}\alpha_\infty R_2^2 \tag{46}$$

is a non-dimensional parameter ($\nu \equiv \varepsilon_s/\varepsilon_{core}$). The parameter q is typically large (for equilibrium micelles) if pH < 10. So let us consider the case $q \gg 1$. An analysis shows that in this case $f(r)$ is concentrated in a relatively thin layer near the surface $r = R_2$. Hence eqn. (45) can be approximated by the one-dimensional analog: $R_2^2(f'/f)' - qf = 0$, where $f' = df/dr$, with the boundary condition $f \to 0$ for $r \to -\infty$. The solution is

$$f = \frac{2}{q}\frac{1}{(1+\beta-r/R_2)^2} \tag{47}$$

Here β is a constant to be determined by another boundary condition: The reduced electric potential just outside the core is

$$\Phi_s = \frac{eQ}{\varepsilon_s R_2}$$

where $Q = \int \rho\, d^3r = ec_{VP}\alpha_\infty \int f d^3r$ is the total core charge. Hence the condition

$$-\ln f(R_2) = \frac{l_B c_{VP}\alpha_\infty}{R_2}\int f\, d^3r \tag{48}$$

Integrating eqn. (45) and using eqn. (48) we get

$$R_2\frac{d}{dr}\ln f(r) + \nu \ln f(r) = 0 \text{ at} r = R_2 \tag{49}$$

thus obtaining the following equation defining β:

$$\ln \frac{q\beta^2}{2} \simeq \frac{2}{\beta v}$$

Using eqn. (44) we then get

$$F_{\rm ch} = -\frac{R_2}{l_{\rm B}} \frac{2}{\beta v}\left(2 + \frac{1}{\beta v}\right) \tag{50}$$

In the general case (arbitrary q) we use the following approximation

$$f \approx \frac{f_0}{\left(1 - pr^3/R_2^3\right)^2} \tag{51}$$

where p and f_0 are parameters that can be determined using eqns. (48) and (49). Thus we get

$$f_0 = \frac{18p}{q}$$

where p is defined by

$$\frac{6}{v}\frac{p}{1-p} + \ln\frac{18p}{q} - 2\ln(1-p) = 0 \tag{52}$$

The protonation free energy is now defined by eqns. (44) and (51):

$$F_{\rm ch} = -\frac{6R_2}{l_{\rm B}v}\left\{\ln(1-p) + \frac{2p}{1-p} + \frac{3}{v}\frac{p^2}{(1-p)^2}\right\} \tag{53}$$

Note that eqn. (53) for $q \gg 1$ is in agreement with eqn. (50). The surface potential $\Phi_{\rm s}$ (at $r = R_2$) is

$$\Phi_{\rm s} = -\ln f(R_2) = 2\ln(1-p) + \ln\frac{q}{18p} \tag{54}$$

IV. Micelle formation

A. Equilibrium c.m.c.

Eqn. (1) can be rewritten as:

$$c(m) = c(1){\rm e}^{-F^*(m)} \tag{55}$$

where

$$F^*(m) = F(m) - mF(1) - (m-1)\ln\phi(1) \tag{56}$$

is the effective free energy of a micelle, $\phi(1) = V_0c(1)$ is volume concentration of free copolymer chains, and $F(m)$ is the internal micelle free energy considered in the previous section.

The aggregation number of the most stable micelles is $m = m^*$ corresponding to the minimum of $F^*(m)$; $F^*(m^*) \gg 1$ at low concentrations well below c.m.c., $F^*(m^*) \sim 1$ ($k_{\rm B}T$) at the c.m.c., and $F^*(m^*) < 0$ well above c.m.c.. The typical plots of $F^*(m)$ are shown in Fig. 2. Obviously they correspond to the closed association scenario of micellization. Well above the c.m.c. most copolymer chains are aggregated in large micelles. In this regime concentration of free chains $\phi(1)$ is nearly constant: $\phi(1) \approx \phi_{\rm c.m.c.}$ if m^* is large (a ten-fold increase of the total polymer concentration would result in just 1% increase of $\phi(1)$ if $m^* = 250$).

B. Apparent c.m.c.

Micelle formation at the c.m.c. may be hindered by the high potential barrier $U_{\rm a} = \max_{(1<m<m^*)}\{F^*(m)\}$ (see Fig. 2, curve 1). The characteristic time of micelle formation can

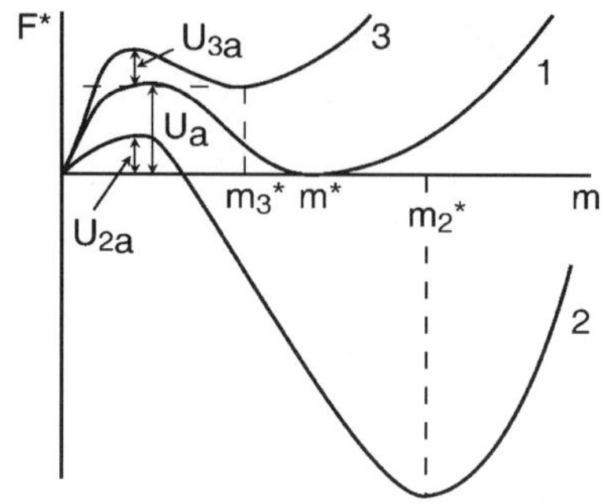

Fig. 2 The dependence of the effective free energy F^* on the micelle aggregation number m in aqueous solutions of PEO(15400)-PVP(14100) block-copolymers: $F^*(m)$ for the equilibrium critical micelle concentration (c.m.c., curve 1) and for the apparent c.m.c.'s (c.m.c.*, curve 2, and c.m.c.**, curve 3). Micelle formation and dissociation at the c.m.c. (curve 1) is hindered by the high energy barrier U_a, whereas their formation at c.m.c.* (curve 2) and dissociation at c.m.c.** (curve 3) are controlled by much lower barriers U_{2a} and U_{3a} respectively.

be estimated as

$$\tau_f \sim \frac{m_a^2}{\Gamma_0} e^{U_a} \tag{57}$$

where $\Gamma_0 = c(1)k_BT/\eta_s$ is (approximately) the rate of collisions of a Brownian particle with surrounding particles at concentration $c(1) = \phi(1)/V_0$, m_a is the aggregation number corresponding to the maximum of $F^*(m)$, $U_a = F^*(m_a)$.

Using eqn. (57) we typically get an extremely long τ_f at the c.m.c. (for example, $\tau_f = 10^{10}$–10^{60} years depending on pH for copolymer with molecular mass $M = 30\,000$; even longer formation times are predicted for higher M). Therefore micelle formation may be (and usually is) totally suppressed at the c.m.c. As $\phi \approx \phi(1)$ increases, the activation energy U_a and the time τ_f decrease. Therefore we define an *apparent* c.m.c. (c.m.c.*) corresponding to a realistic micelle formation time of, say, $\tau_f = 1$ h (Fig. 2, curve 2). The most favorable aggregation number m^* still corresponds to the minimum of $F^*(m)$. Note that micelles at c.m.c.* are larger than equilibrium micelles: m^* at c.m.c.* is larger than the aggregation number $m_{c.m.c.}$ at the equilibrium c.m.c.. The apparent c.m.c.* may be a few orders of magnitude higher than the equilibrium c.m.c. (see Fig. 3).

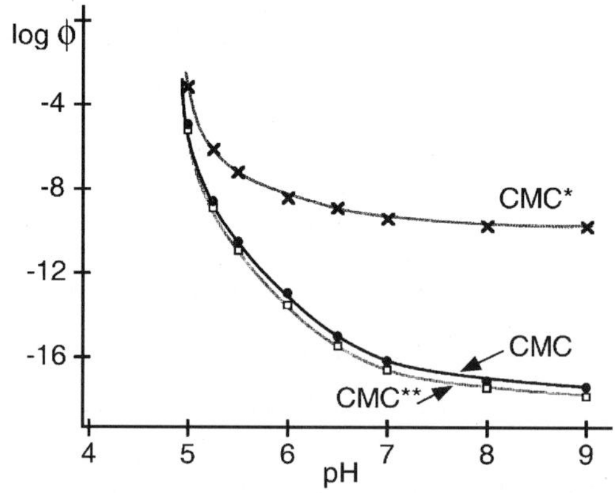

Fig. 3 pH dependencies of c.m.c., c.m.c.* and c.m.c.** for the same copolymer system as in Fig. 2.

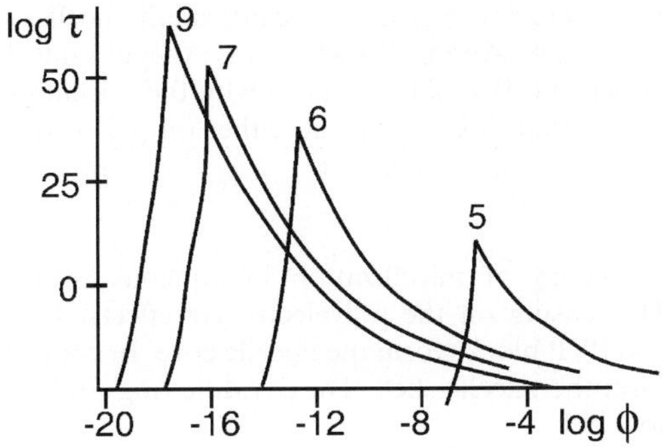

Fig. 4 The concentration ranges of micelle stability at various pH: life-time τ (in years) *vs.* polymer volume fraction $\phi(1)$. pH values are indicated above the corresponding lines; the number of micelles is nearly constant below these lines.

C. Hysteresis

Dissociation of micelles can be also suppressed (like micelle formation) at the genuine equilibrium c.m.c. To induce a dissociation during an experimental time-scale, the concentration must be decreased below c.m.c., down to c.m.c.**, the second apparent c.m.c. (see Fig. 2, curve 3). Hence, ones formed, the micelles are practically stable in a wide concentration range c.m.c.** $< \phi(1) <$ c.m.c.* The most favourable aggregation number m^* for $\phi(1) <$ c.m.c. is lower than $m_{\text{c.m.c.}}$.

Thus, for any given pH and unimer concentration $\phi(1)$, the number of micelles in solution is nearly not varying at time-scales shorter than the characteristic life-time τ (see Fig. 4). At time-scales shorter than τ micelles can only capture/release free (unassociated) unimers. As a result of this exchange the micelles become nearly monodisperse, with practically the same aggregation number $m^* \gg 1$ corresponding to the local minimum of $F^*(m)$ for the given $\phi(1)$. At longer time-scales (exceeding the life-time τ) either new micelles are formed (if $\phi(1) >$ c.m.c.), or some micelles dissociate (if $\phi(1) <$ c.m.c.). As a result the concentration of free unimers $\phi(1)$ becomes closer to c.m.c., and the life-time τ increases (note that τ is maximal at the c.m.c., see Fig. 4). If pH is changing during the experiment, the regions of micelle stability moves accordingly (see Fig. 4). That is, if micelles are formed from a unimer solution as a result of pH increase (as in ref. 1), then micelle formation starts when the right shoulder of the stability window (see Fig. 4) decreases down to the experimental time-scale $t \sim 1$–10 s for the given unimer concentration $\phi(1)$. For the typical $\phi(1) \sim 10^{-4}$ this corresponds to the threshold pH ≈ 4.8. Within a narrow pH range (pH ≈ 4.8–4.9) the residual unimer concentration $\phi(1)$ drops by a few orders of magnitude, hence for pH > 4.9 practically all copolymer chains are incorporated in micelles. The size R of the micelles which are formed at this point can be found in Fig. 5 (right shoulder of the stability regions): $R \approx 19$–21 nm. As was explained above, after the primary formation the micelles can exchange unimers but they cannot dissociate for time-scales shorter than their life-time, hence their number and their average size are virtually frozen. As can be seen from Fig. 5, even for very high pH ~ 9 (where the micelle

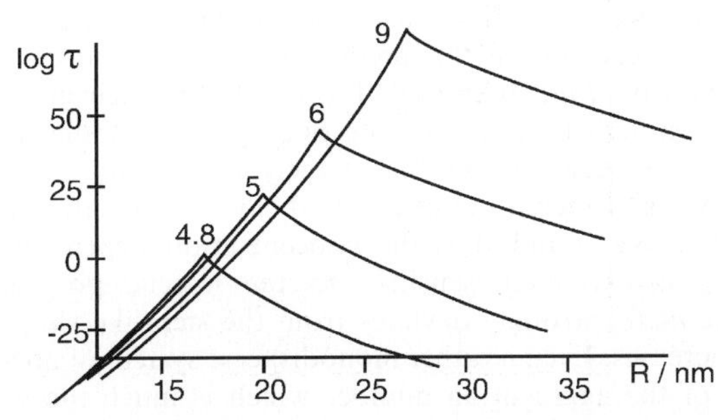

Fig. 5 The ranges of sizes (micelle radii R) of stable micelles at various pH: life-time τ (in years) *vs.* micelle radius R. pH values are indicated above the corresponding lines.

size at the equilibrium c.m.c. is much larger than 20 nm), small micelles with $R \approx 20$ nm are stable against dissociation for many years. And precisely because they do not dissociate, there is no extra material for a growth of other micelles. This explains why the micelles observed by Martin *et al.*[1] were all of nearly identical size that did not depend either on pH or on formation rate.

V. Conclusions

(i) We have developed a theory of micellization in aqueous solutions of PEO–PVP block-copolymers at various pH focusing on the polyelectrolyte effects. Unprotonated VP monomers are strongly hydrophobic, so PVP blocks form the micelle core. In contrast, PEO is partially soluble in water, so PEO blocks form the micelle shell. The pyridine rings in PVP are partially protonated; their charge increases at low pH.

The proposed general theory involves Poisson–Boltzmann (rather than Debye–Hückel) treatment accounting for the non-linear charge screening by ions and strong position dependence of ion concentration, local pH and effective Debye length. The screening can be very important for micellar structure even when the bulk Debye length r_D is (much) longer than the micellar size. The theory yields a variable effective Debye length r^*_D that can be much shorter near a micellar core than in the bulk; r^*_D is normally shorter than the core size for pH > 5. The much stronger screening in the core results in a lower electrostatic free energy and a higher degree of protonation of the micelles. Thus the theory accounts both for the case when the local Debye length is small compared to the core radius and for the case when it is large.

We found that the effective (local) pH is not uniform inside the core, hence non-uniform protonation of the core with higher charge density near the surface is predicted. In some cases (when micelles are large enough, *i.e.* at high copolymer concentration and 5 < pH < 9) the core charge is concentrated in a relatively thin surface layer.

(ii) The proposed theoretical approach enabled us to find the micelle geometrical parameters and to predict characteristic time scales of their formation/transformation.

(a) Micelles are formed if the degree of protonation of VP blocks is low enough, *i.e.* at high enough pH: pH $\gtrsim$ 4 for 4000 < M_{PVP}. In the regime of micelles their aggregation number is very high (up to *ca.* 4000), so that the micellar core is rather dense (VP volume fraction in the core may be as high as $f_2 = 0.65$) and the VP blocks are rather strongly stretched. The so-called critical micelle concentration (c.m.c.) is approximately defined by the condition $\min_m F^*(m) = F(m^*) = 0$ (eqn. (56)), where $m^* > 1$ is the aggregation number of the critical micelle. We predict that c.m.c. is strongly pH dependent.

(b) The height of the energy 'barrier' separating the minimums of $F^*(m)$ at $m = 1$ and $m = m^*$ determines the characteristic time of micelle formation. We show that the micelle formation time t_f is experimentally inaccessible at the c.m.c., being in the range 10^{10}–10^{60} years for copolymers with $M_{PVP} \gtrsim 15\,000$. Hence we define an 'apparent' c.m.c., c.m.c.*, for which the micelle formation takes place during an hour. The ratio c.m.c.*/c.m.c. is very high ranging from 10^2 for pH $\approx$ 5.5 to 10^8 at higher pH. The micelle parameters at c.m.c.* and c.m.c. differ significantly: the c.m.c.* micelles are much larger and denser.

(c) Once the micelles are formed, they are very stable and do not dissolve on dilution down to the 'apparent' threshold of stability c.m.c.** ~ c.m.c./2, *i.e.* c.m.c.** is much lower than c.m.c.* We showed that the regions of micelle stability c.m.c.** < c < c.m.c.* are pH and time dependent, and demonstrated that certain features of the predicted hysteresis effects are in agreement with experimental results of Martin *et al.*[1] on micelle formation and their growth.

(iii) We showed that multimerization in solutions of amphiphilic molecules must be characterized by the complete micelle size distribution that largely defines the equilibrium and dynamical properties of micelles.[24] We considered the generalized closed association model allowing for some polydispersity of micelles. We found that the concentration dependence of the mean micelle aggregation number m_w is extremely sensitive to the micelle polydispersity parameter $\delta = (m_w/m_n)_{mic} - 1$: the curve $m_w(c)$ strongly deviates from the step-like shape typical of the classical closed association as δ increases. Even a rather monodisperse system of micelles with $\delta = 0.02$ shows concentration behavior of the aggregation number which is much more similar to the behavior inherent to the classical open association than to that for the classical closed model. Therefore the characterization of the association mechanism by the qualitative shape of the $1/m_w$ *vs.* c curve is not enough: a more quantitative analysis of the experimental data is required. To this end we suggest a simple method of deriving the micelle size distribution $c(m)$ (for any total copolymer concentration) from the dependence $m_w(c)$ that could be obtained experimentally. Micellar sizes and polydispersity are found to be concentration dependent; a stronger dependence of the micellar size on concentration is predicted for more polydisperse micelles.[24]

It is remarkable that even above the apparent c.m.c. the system of micelles and unimers (plus their small aggregates) is typically not at the complete equilibrium; the latter can be achieved (for reasonable time-scales) only at extremely high (thus inaccessible) concentrations. Thus a surfactant (copolymer) solution is typically never at a complete equilibrium above c.m.c. unless it is prepared in a special way.

We analysed the ideal-gas (combinatorial) contribution to the micelle free energy, which was either neglected or not taken into account properly in the previous studies. The ideal-gas term is proportional to the aggregation number m, and therefore it is sometimes very important. For example it leads to a very high activation energy barrier prohibiting 'simultaneous' aggregation of many unimers involving a 'loose' micelle as an intermediate state.[24]

Acknowledgements

This work was partially supported by the NATO SfP Grant 974173. This work was also supported in part by the Alexander von Humboldt Foundation, Program for Investment in the Future (ZIP, W.Paul Award Project 'Sequence Design of Functional Copolymers').

References

1 T. J. Martin, K. Prochazka, P. Munk and S. E. Webber, *Macromolecules*, 1996, **29**, 6071.
2 L. M. Bronstein, S. N. Sidorov and P. M. Valetsky, *Langmuir*, 1999, **15**, 6256.
3 E. Sulman *et al.*, *Appl. Catal. A*, 1999, **176**, 75.
4 C. Ropert, M. Lavignon and C. Dubernet *et al.*, *Biochem. Biophys. Res. Commun.*, 1992, **183**, 879.
5 K. Seiki and D. Tirrel, *Macromolecules*, 1984, **17**, 1692.
6 J. Billingham and P. V. Coveney, *J. Chem. Soc. Faraday Trans.*, 1994, **90**, 1953.
7 E. E. Dormidontova, *Macromolecules*, 1999, **32**, 7630.
8 R. Ball and A. D. J. Haymat, *Phys. Chem. Chem. Phys.*, 2001, **3**, 4753.
9 P. A. Bachmann, P. L. Luisi and J. Lang, *Nature*, 1992, **357**, 57.
10 E. Bortel and A. Kochanowski, *Makromol. Chem., Rapid Commun.*, 1980, **1**, 205.
11 B. Nyström, S. Boileau, P. Hemery and J. Roots, *Eur. Polym. J.*, 1980, **17**, 249.
12 Y. Kambe and C. Honda, *Polym. Commun.*, 1984, **25**, 154.
13 W. F. Polik and W. Burchard, *Macromolecules*, 1983, **16**, 978.
14 H. Daoust and D. St-Cyr, *Macromolecules*, 1984, **17**, 596.
15 M. Polverari and T. G. M. van de Ven, *J. Phys. Chem.*, 1996, **100**, 13 687.
16 A. E. Likhtman and A. N. Semenov, *Macromolecules*, 1994, **27**, 3103.
17 A. E. Likhtman and A. N. Semenov, *Macromolecules*, 1997, **30**, 7273.
18 V. M. Amoskov and V. A. Pryamitsyn, *J. Chem. Soc. Faraday Trans.*, 1994, **90**, 889.
19 R. M. Fuoss and U. P. Strauss, *J. Polym. Sci.*, 1948, **3**, 246.
20 Yu. E. Kirsh, O. P. Komarova and G. M. Lukovkin, *Eur. Polym. J.*, 1973, **9**, 1405.
21 M. Satoh, E. Yoda, T. Hayashi and J. Komiyama, *Macromolecules*, 1989, **22**, 1808.
22 J. F. Gohy, S. K. Varshney, S. Antoun and R. Jerome, *Macromolecules*, 2000, **33**, 9298.
23 *CRC Handbook of Chemistry,Physics*, 66th edn., ed. R. C. Weast, CRC Press Inc., Boca Raton, FL, 1986.
24 I. A. Nyrkova and A. N. Semenov, *Macromolecules*, submitted.
25 *Polymeric Materials Encyclopedia*, ed. J. C. Salamone, CRC Press, 1996, vol. **1**, p. 817.

through a more quantitative analysis of the experimental data is required. To this end we suggest a [illegible]

Micellization in pH-sensitive amphiphilic block copolymers in aqueous media and the formation of metal nanoparticles

Maria Vamvakaki,[ab] Lampros Papoutsakis,[ac] Vasilios Katsamanis,[ac] Theodora Afchoudia,[ac] Panagiota G. Fragouli,[d] Hermis Iatrou,[d] Nikos Hadjichristidis,[d] Steve P. Armes,[e] Stanislav Sidorov,[f] Denis Zhirov,[f] Vasilii Zhirov,[f] Maxim Kostylev,[g] Lyudmila M. Bronstein[fg] and Spiros H. Anastasiadis*[ac]

[a] *Institute of Electronic Structure and Laser, Foundation for Research and Technology-Hellas, P.O. Box 1527, 71110, Heraklion Crete, Greece. E-mail: spiros@iesl.forth.gr*
[b] *Department of Materials Science and Technology, University of Crete, 710 03, Heraklion Crete, Greece*
[c] *Department of Physics, University of Crete, 710 03, Heraklion Crete, Greece*
[d] *Department of Chemistry, University of Athens, Panepistimiopolis Zografou, 15771, Zografou Athens, Greece*
[e] *School of Chemistry Physics and Environmental Science, University of Sussex, Falmer, Brighton, BN1 9QJ, UK*
[f] *Nesmeyanov Institute of Organoelement Compounds, Russian Academy of Sciences, 117813, Moscow, Russia*
[g] *Chemistry Department, Indiana University, Bloomington, IN, 47405, USA*

Received 4th March 2004, Accepted 11th March 2004
First published as an Advance Article on the web 17th September 2004

Dynamic light scattering, potentiometric titration, transmission electron microscopy and atomic force microscopy have been used to investigate the micellar behaviour and metal-nanoparticle formation in poly(ethylene oxide)-*block*-poly(2-vinylpyridine), PEO-*b*-P2VP, poly(hexa(ethylene glycol) methacrylate)-*block*-poly(2-(diethylamino)ethyl methacrylate), PHEGMA-*b*-PDEAEMA, and PEO-*b*-PDEAEMA amphiphilic diblock copolymers in water. The hydrophobic block of these copolymers (P2VP or PDEAEMA) is pH-sensitive: at low pH it can be protonated and becomes partially or completely hydrophilic leading to molecular solubility whereas at higher pH micelles are formed. These micelles consist of a P2VP or PDEAEMA core and a PEO or PHEGMA corona, respectively, where the core forming amine units can incorporate metal compounds due to coordination. The metal compounds (*e.g.*, H_2PtCl_6, K_2PtCl_6) can either be introduced in a micellar solution, where they are incorporated within the micelle core *via* coordination with functional groups, or can be added to a unimer solution at low pH, where they lead to a metal-induced micellization. In these micellar nanoreactors, metal nanoparticles nucleate and grow upon reduction with sizes in the range of a few nanometers as observed by TEM. The effect of the metal incorporation method on the characteristics of the micelles and of the synthesized nanoparticles is investigated.

I. Introduction

Block copolymers associate to form microphase-separated micellar structures in dilute solution in solvents that are selective for one of the blocks.[1–5] Polymeric micelles comprise a dense core formed by aggregation of the insoluble block(s) and a micellar shell formed by the soluble block(s) that extend in the solution and confer stabilization to the micelle structure. They have attracted considerable attention due to their wide range of applications as *e.g.* templates, or separation and delivery systems. One of the main features of block copolymer micelles is their ability to stabilize small molecules that are incorporated selectively into only one of the microphases.[6]

Remarkable progress has been made during the last decade in the development and investigation of micelles formed by supramolecular organization of amphiphilic block copolymers in aqueous systems.[7–13] The use of water as the solvent medium offers many advantages for potential applications in targeted drug delivery, isolation of metabolites, entrapment of environmental pollutants and catalysis.[14–17] In most cases spherical micelles are formed in water while different morphologies have also been reported ranging from cylindrical and platelike micelles, vesicles and complex micellar aggregates when varying the copolymer structure, molecular weight and composition.[18–21]

However, classical amphiphilic block copolymers suffer incomplete dissolution in water and usually require a co-solvent such as methanol, tetrahydrofuran or dimethylformamide to fully dissolve the polymer and form stable well-defined micelles. In this case, the co-solvent type and concentration have been found to influence the micellar morphologies.[22] Furthermore, kinetically frozen micelles, which are not in the equilibrium state, are formed in pure water especially when the cores are glassy.[23,24]

Progress in polymer synthesis has made it possible to prepare external-stimuli responsive block copolymers known as double-hydrophilic block copolymers. These materials respond to changes in their environment such as pH, temperature and salt concentration and undergo micellization in aqueous media.[25–32] The elimination of a cosolvent for micelle formation and the fully reversible micellization process controlled by proper adjustment of the external stimuli are considerable advantages of these systems.

Micellization induced by controlling the solution pH is advantageous for certain applications in particular when fully hydrophobic cores are required.[26] Such systems comprise either a pH-sensitive basic or acidic block that forms the micelle core surrounded by a corona formed by a neutral hydrophilic block,[25,26,33] or copolymers containing two ionisable weak base (acid) blocks of different pK_a s.[29,30,32,34] Of particular interest is also the pH-dependent micellization of block copolymers containing two weak ionisable blocks of opposite sign (weak acid and weak base blocks), which exhibit a complex association behaviour in water, from micelles to a soluble (or insoluble) polyampholytic system and, finally, to a reverse micellar solution.[35–38] Several studies have focused on the micellization process of pH sensitive diblock copolymers and examined the effect of polymer structure, solution pH and ionic strength on the micelle characteristics.[27,30] The micelle radius of gyration, hydrodynamic size and aggregation number have been studied in detail as a function of solution pH or degree of ionization of the two blocks, as well as a function of salt concentration, whereas the data were compared with a number of theoretical models providing additional insight into the micelle structure.

The spherical morphology of block copolymer micelles from a selective solvent was observed in adsorption experiments on various solid surfaces. It has been shown that this adsorption process leads to highly ordered two-dimensional structures for micelles adsorbed from an organic solvent or for water-based neutral micelles.[39–41] However, the arrangement of micelles with a polyelectrolyte or pH-sensitive corona is more complicated with their 2-D spatial distribution being affected by both the steric and the electrostatic corona–corona interactions.[42,43]

An interesting feature of block copolymer micelles is their potential use as nanoscopic reaction vessels to grow inorganic metal nanocrystals resulting in hybrid nanocomposite materials.[44–48] The properties of these colloidal metal nanoparticles, which are determined by their particle size, shape and internal structure, are greatly different from those of their bulk counterparts and make them particularly attractive for applications in catalysis, optical devices, nanoelectronics and superparamagnetic or ferromagnetic materials.[49–52] The synthesis of nanoparticles of small size, narrow size distribution and long term stability requires the use of a stabilizing system such as small molecule surfactants, polymers, zeolite cells, *etc.*[53–57] The advantages of block copolymer stabilizers with respect to thin film formation, control of the reagent diffusion in the polymer matrix, particle size and stability, and nanoparticle activity and selectivity as catalytic systems have been reported.[58,59]

The synthesis of metal nanoparticles within the block copolymer micelles requires one block, which normally forms the micelle core and is able to coordinate with the metal compounds, while the second block in the corona provides good stability in the solvent medium.[44,46,48] Double-hydrophilic block copolymer micelles allow adaptation of the nanoreactor approach of nanoparticle synthesis to aqueous systems since the functional core forming block can also interact with metal salts.[60–62] This can have certain advantages with respect to environmental considerations and the use of metal particles as catalysts for the synthesis of molecules of biological interest.[63]

In the present paper, the aqueous solution properties of double-hydrophilic block copolymers comprising a neutral hydrophilic block and a pH-sensitive tertiary amine block, which is able to coordinate metal compounds, are investigated. Poly(ethylene oxide)-*block*-poly(2-vinylpyridine), PEO-*b*-P2VP, poly(hexa(ethylene glycol) methacrylate)-*block*-poly(2-(diethylamino) ethyl methacrylate), PHEGMA-*b*-PDEAEMA, and PEO-*b*-PDEAEMA are utilized and their micellization behaviour and micelle characteristics are studied as a function of the solution pH. Platinum nanoparticles are grown within the cores of the micelles, formed by the hydrophobic tertiary amine block, by incorporation of metal compounds followed by reduction. The effect of micelle stability, the type and loading of the metal compound, and the nanoparticle synthetic procedure on the characteristics of the colloidal metal nanoparticles are also examined.

The remainder of this article is arranged as follows: Following the experimental section II, the results of the experimental investigations are presented and discussed in section III. The concluding remarks constitute section IV.

II. Experimental section

Materials

Poly(ethylene oxide)-*block*-poly(2-vinylpyridine), PEO-*b*-P2VP, diblock copolymers were synthesized by living anionic polymerization in tetrahydrofuran (THF) using high-vacuum techniques in glass reactors provided with break-seals for the addition of reagents and constrictions for the removal of intermediate aliquots. Benzyl potassium was used as the initiator whereas the 2-vinylpyridine monomer was polymerized first; the details of the synthesis have been reported elsewhere.[64] Diblock copolymers comprising hydrophilic hexa(ethylene glycol) methacrylate (HEGMA) and hydrophobic 2-(diethylamino)ethyl methacrylate (DEAEMA) units, PHEGMA-*b*-PDEAEMA, were prepared by group transfer polymerization (GTP) chemistry in THF using glass reactors under dry nitrogen and standard schlenk techniques in THF. Tetrabutyl ammonium bibenzoate was used as a catalyst and 1-methoxy-1-trimethylsiloxy-2-methyl-1-propene as the initiator whereas HEGMA was polymerized first.[65] In both cases, the diblock copolymers were synthesized by sequential monomer addition using rigorously purified monomers and solvents. A PEO-*b*-PDEAEMA diblock copolymer was synthesized by controlled atom transfer radical polymerization (ATRP) using a PEO macroinitiator for the polymerization of DEAEMA.[66]

The PEO-*b*-P2VP diblocks were characterized by size exclusion chromatography (SEC) with both refractive index and UV detectors, membrane osmometry, low-angle laser light scattering (LALLS) and ^{1}H NMR in order to provide the molecular characteristics of the materials.[64] The molecular weights and molecular weight distributions of the PHEGMA-*b*-PDEAEMA and the PEO-*b*-PDEAEMA copolymers were determined by SEC using poly(methyl methacrylate) standards, while ^{1}H NMR spectroscopy was used to assess their composition. SEC analysis of the diblock copolymers confirmed their narrow polydispersities ($M_w/M_n = 1.05–1.2$) and the good molecular weight control achieved. The copolymer compositions were in good agreement with the targeted values. The molecular characteristics of all the copolymers are shown in Table 1 together with those for two homopolymers PEO and P2VP.

Table 1 Molecular characteristics of the diblock copolymers

Sample	Polymer[b]	M_w	M_w/M_n[d]	Composition (mol%)[e]	$M_{w,\ amine\ block}$
PEO30	PEO_{681}	31,000[c]	1.03	—	—
P2VP30	$P2VP_{286}$	30,000[c]	1.07	—	—
D 25/25[a]	PEO_{420}-$P2VP_{56}$	24,300[c]	1.05	88–12	6,100
D 30/35[a]	PEO_{421}-$P2VP_{95}$	28,500[c]	1.07	81–19	9,800
D 30/50[a]	PEO_{339}-$P2VP_{142}$	29,800[c]	1.05	70–30	15,100
H_{13}–D_7	$PHEGMA_{13}$-$PDEAEMA_7$	6,600[d]	1.13	65–35	1,300
H_{10}–D_{10}	$PHEGMA_{10}$-$PDEAEMA_{10}$	6,100[d]	1.10	50–50	1,850
H_{17}–D_{16}	$PHEGMA_{17}$-$PDEAEMA_{16}$	10,400[d]	1.13	51–49	2,960
H_{13}–D_{27}	$PHEGMA_{13}$-$PDEAEMA_{27}$	10,600[d]	1.10	33–67	5,000
H_{50}–D_{50}	$PHEGMA_{50}$-$PDEAEMA_{50}$	13,500[d]	1.14	49–51	9,260
PEO_{45}–D_{37}	PEO_{45}-$PDEAEMA_{37}$	9,900[d]	1.21	55–45	6,850

[a] The first number indicates the approximate total molecular weight and the second the approximate P2VP weight content (ref. 64). [b] The numbers denote degrees of polymerization. [c] Determined by LALLS in CH_3OH. [d] Determined by SEC. [e] Determined by ^1H NMR.

Sample preparation

Dilute solutions of the polymers were prepared in Milli-Q water titrated to pH 1 or 2 using 12 M HCl. The appropriate amount of polymer was added in the low pH water and the samples were left stirring overnight to ensure complete polymer dissolution. For the solutions at an initial pH 1, the pH was increased to pH 2 by the addition of 10 M NaOH. Finally, all samples were adjusted to the required pH by the addition of 0.1 M NaOH and were allowed to stir for another 12 h.

Samples for the dynamic light scattering and the atomic force microscopy measurements were prepared in Milli-Q water, which was prefiltered through a 0.2 μm filter, as described above. After stirring for 12 h, the polymer solutions were filtered carefully through a 0.45 μm pore size filter to eliminate any dust and were left to equilibrate for about 1 h before being measured.

Potentiometric titrations

Polymer solutions at pH 2 were prepared as described above. Titration curves for all polymers were obtained by monitoring the increase of the pH after addition of 0.1 M NaOH (increments of 20 to 200 μL) in a 0.25 or 0.1 wt.% copolymer solution. The pH measurements were performed with the Crison (GLP 21) pH-meter and titration curves in the pH range from 2 to 12 were obtained.

Dynamic light scattering (DLS)-photon correlation spectroscopy (PCS)

The autocorrelation function of the polarized light scattering intensity $G_{VV}(q,t) = \langle I(q,t)I(q,0)\rangle / \langle I(q,0)\rangle^2$ was measured at different scattering angles, θ, using an ALV spectrophotometer and an ALV-5000 full digital correlator over the time range 10^{-7}–10^3 s; $I(q,0)$ is the mean scattered intensity. Generally both the incident and the scattered beam were polarized perpendicular to the scattering plane (VV geometry). An Adlas diode-pumped Nd-YAG laser was used as the light source with wavelength $\lambda = 532$ nm and single mode intensity 100 mW. The magnitude of the scattering wavevector is $q = (4\pi n/\lambda)\sin(\theta/2)$, where n is the refractive index of the medium.

Under homodyne conditions, $G_{VV}(q,t)$ is related to the desired scattered field autocorrelation function $C(q,t)$ by $C(q,t) = \{[G_{VV}(q,t) - 1]/f^*\}^{1/2}$, where f^* is an experimental factor calculated by means of a standard. The experimental correlation functions $C(q,t)$ are analyzed by performing the inverse Laplace transform (ILT) using the routine CONTIN assuming a superposition of exponentials for the distribution of relaxation times $L(\ln \tau)$, *i.e.* $C(q,t) = \int_{-\infty}^{\infty} L(\ln\tau)\exp[-t/\tau]d(\ln\tau)$. The rate Γ of each process is calculated as the inverse of its relaxation time $1/\tau$, whereas their dynamic intensities are calculated from the integrals under the peaks of $L(\ln \tau)$ multiplied by $I(q,0)$. In the case of a diffusive process, its diffusion coefficient D is obtained from the slope of Γ *versus* q^2 by $\Gamma = Dq^2$. The latter is related to the hydrodynamic radius R_h of the diffusing moiety by the Stokes–Einstein equation $R_h = k_BT/(6\pi\eta\ D)$, where, η is the viscosity of the solvent, k_B is the Boltzmann constant and T is the temperature of the sample (it is assumed that at these low concentrations, D corresponds to its limit for concentration $c \rightarrow 0$). All measurements were performed at 20 °C.

Atomic force microscopy

Atomic force microscopy (AFM) was used to study the micelle formation in a polymer solution as a function of solution pH. AFM measurements were performed on a Digital Instruments Multimode Scanning Probe Microscope using the Nanoscope IIIa, Version 4.23r3 software. All experiments were carried out in the tapping mode. Solutions at a polymer concentration of 0.5 wt.% and at different pH values were prepared as described above. Freshly cleaved mica sheets were used as substrates. For the adsorption experiments, the substrate was dipped into the polymer solution for about 30 s and then the surface was dried under a stream of nitrogen gas for a few minutes at room temperature.

Metal incorporation and metal nanoparticle formation

The incorporation of Pt species using $H_2PtCl_6 \times H_2O$ or K_2PtCl_6 was carried out by the method described previously.[61] Three different procedures were used for the introduction of the metal compounds within the polymeric matrices and their effects on both the micelle characteristics and the size of the metal nanoparticles was evaluated.

In the first method the appropriate amount of diblock copolymer to obtain a 0.33 wt.% polymer solution was dissolved in water at pH 2.0. The sample was stirred overnight to ensure complete polymer dissolution and then the pH was raised to 7 or 10 using 0.1 M NaOH to form the polymeric micelles. After stirring overnight at room temperature the metal precursor (H_2PtCl_6 or K_2PtCl_6) was added to the micelle solution at a 3 : 1 N : Pt molar ratio. More than 24 h were allowed for metal complexation, before the excess ionic species were removed by ultrafiltration using an Amicon® Stirred Ultrafiltration Cell by Millipore with regenerated cellulose ultrafiltration membranes, NMWL:3000. Reduction of the samples was carried out under vigorous stirring using a 3-fold excess of sodium borohydride ($NaBH_4$) in deionized water (12% of solution volume), and stirred overnight. Upon addition of the reducing agent, the solutions instantaneously turned black and hydrogen gas rapidly evolved.

The second method involved the molecular dissolution of the copolymer under stirring overnight in water at pH 2. Next the metal precursor was added to the molecular polymer solution. The solution was stirred for 24 h to allow for equilibration in metal incorporation, followed by ultrafiltration. Metal reduction was carried out similarly to the procedure followed for the first method.

In the third method the copolymer was added in de-ionized water without any previous adjustment of the solution pH. After stirring overnight at room temperature the metal precursor was added to the sample followed by ultrafiltration and metal reduction as described above.

Transmission electron miscroscopy (TEM)

The metal containing polymer micelles were characterized by transmission electron microscopy using a JEOL JEM1010 instrument at accelerating voltage of 60 kV. The polymer solutions were first diluted to 0.75 mg ml^{-1} with deionized water. A drop of the diluted sample was placed on a carbon-coated copper grid situated on a paper towel and allowed to dry in the open air for two days before the analysis.

The characteristics of the polymer micelles and the metal nanoparticles before and after metal reduction, respectively, were determined using Scion Image software in conjunction with Adobe Photoshop 6.0.

III. Results and discussion

Potentiometric titration curves

Since the hydrophobic blocks of the diblock copolymers are pH-sensitive, the properties of the polymer solutions were investigated as a function of the solution pH. Initially the effect of the

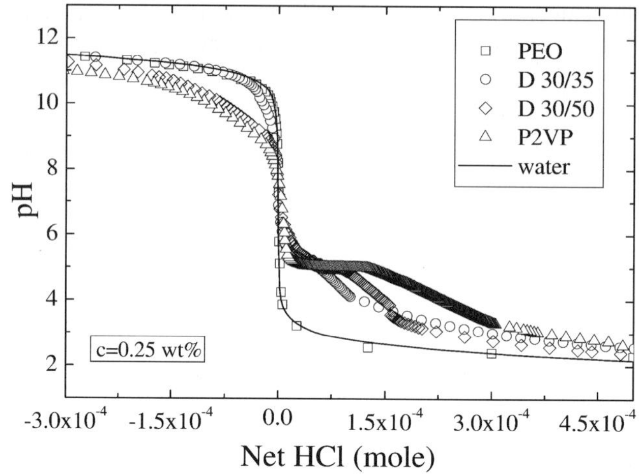

Fig. 1 Potentiometric titration curves for 0.25 wt.% solutions of the PEO30 (□) and P2VP30 (△) homopolymers and two PEO-*b*-P2VP (D 30/35 (○), D 30/50 (◊)) copolymers *versus* the net moles of HCl added. Also shown is the titration curve for water taken to pH 2 with the addition of HCl.

addition of acid or base in a dilute polymer solution on the pH was monitored. Fig. 1 shows the titration curves for two PEO-*b*-P2VP diblock copolymers, of similar overall molecular weight and different compositions, and of the respective homopolymers, PEO and P2VP, all at concentration 0.25 wt.%. The titration curves are shown *versus* the net moles of HCl added, *i.e.* moles of HCl minus moles of NaOH, and the titration occurs sequentially from right to left. Since PEO is a neutral polymer and is not affected by changes in the solution pH its titration curve is very similar to that obtained for water brought to pH 2 with the addition of HCl, also shown in Fig. 1. However, P2VP comprises tertiary amine groups of a weak base character and thus participates in an acid–base equilibrium. The pyridine units are protonated and hydrophilic at low pH, while, when the pH increases above ~5 (effective pK_a of the protonated P2VP), they become deprotonated and hydrophobic. This protonation/deprotonation process is indicated by the existence of a plateau region at pH 5 in the titration curve signifying the buffering capacity of the polymer. At pH values higher than 5, P2VP is hydrophobic and precipitates out of solution. A similar behaviour is found for the PEO-*b*-P2VP diblock copolymers, which also possess a weak base character due to the P2VP block and, thus, exhibit a plateau at pH 5 in their titration curves. Around the second inflection point of the titration curves of the copolymers at pH ~8, the solution becomes turbid indicating the aggregation of the micelles, which are formed above pH 5 as will be shown below by DLS. Furthermore, it is noted that the width of the plateau in the titration curves of the copolymer solutions increases with the size of the pyridine block, which, for the case of similar total molecular weight copolymers, suggests that the buffering capacity depends on the copolymer composition.

Similar results were obtained for the PDEAEMA containing polymers. The titration curves for the PHEGMA-*b*-PDEAEMA and the PEO-*b*-PDEAEMA diblock copolymers *versus* the net moles of HCl added are shown in Fig. 2. Similarly to P2VP, the PDEAEMA block in the copolymers behaves as a weak base and buffers the solution at a pH around 7, as shown by the plateau region at pH 7 of the titration curves. The appearance of the plateau at a higher pH value as compared to the PEO-*b*-P2VP copolymers is due to the higher effective pK_a of the protonated PDEAEMA, which is a weaker acid than the protonated P2VP. It is also evident that the width of the plateau increases with the DEAEMA content of the copolymer in good agreement with the discussion above for PEO-*b*-P2VP.

Fig. 2b shows the data of Fig. 2a plotted as pH *versus* the ratio α_{eff} = [HCl]/[DEAEMA monomer units] in the range $0 < \alpha_{eff} < 1$. α_{eff} approximates the degree of protonation, or the fraction of amine groups of the copolymer chains, which are protonated if one assumes that all H^+ from the added HCl protonate the amino groups.[27,30] Describing the net amount of acid added in terms of α_{eff} and pH provides two views of what is happening in the sample: the term α_{eff} describes the degree of

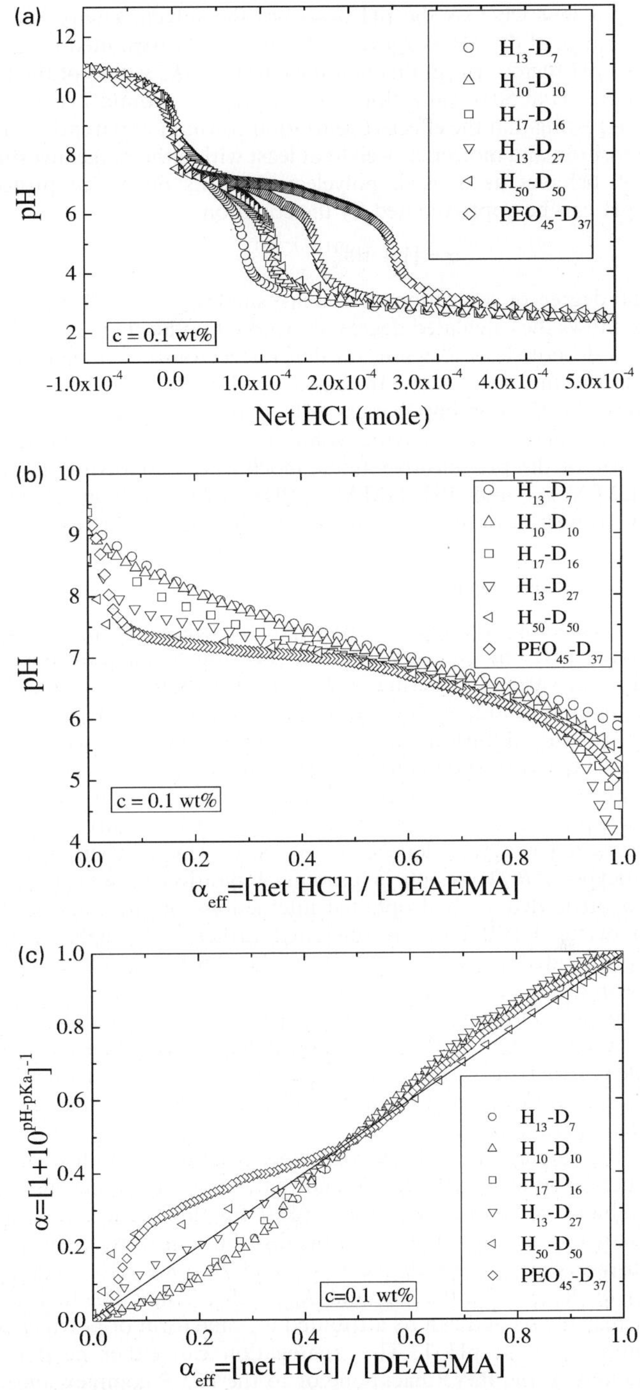

Fig. 2 (a) Potentiometric titration curves for 0.1 wt.% solutions of five PHEGMA-*b*-PDEAEMA copolymers (H_{13}-D_7 (○), H_{10}-D_{10} (△), H_{17}-D_{16} (□), H_{13}-D_{27} (▽), and H_{50}-D_{50} (◁)) and one PEO-*b*-PDEAEMA (PEO_{45}-D_{37} (◊)) copolymer *versus* the net moles of HCl added. (b) The data of Fig. 2a plotted as pH *versus* the ratio α_{eff} = [HCl]/[DEAEMA monomer units] in the range $0 < \alpha_{eff} < 1$. (c) The degree of ionization of a weak electrolyte calculated from the solution pH plotted *versus* α_{eff} for the data of Fig. 2b.

protonation of the polymer whereas the pH describes the solvent environment surrounding the polymer. The small change of the pH as α_{eff} varies from 0 to 1 is attributed to the buffering capacity of the polymer at this pH region. From this plot the effective $\text{p}K_\text{a}$ values of the diblock copolymers can be calculated as the pH at 50% ionization. Similar $\text{p}K_\text{a}$ s of around 7 (6.9 to 7.2) were obtained for all copolymers suggesting that the effective ionization constant apparently remains unaffected by the copolymer composition and molecular weight at least within the range investigated in this study.

Since PDEAEMA behaves as a weak polyelectrolyte, its degree of protonation *versus* the measured solution pH can be approximated by the equation

$$\alpha = [1 + 10^{\text{pH}-\text{p}K_\text{a}}]^{-1} \tag{1}$$

which describes the degree of ionization of a weak small-molecule electrolyte as a function of solution pH. Fig. 2c shows the calculated degree of protonation plotted *versus* α_{eff} as defined before. The data for high α_{eff} do not show a significant dependence on the copolymer characteristics and apparently lie close to the line of slope one. In contrast, at low degrees of ionization there is a strong differentiation among the various copolymers, which is attributed to the structure formation (micellization) that takes place in the polymer solution as will be discussed below. The deviations are stronger as the size of the pH-sensitive amine block of the copolymer increases, *e.g.* towards PHEGMA_{13}-*b*-PDEAEMA_{27} and PHEGMA_{50}-*b*-PDEAEMA_{50}. Note that PEO_{45}-*b*-PDEAEMA_{37} exhibits even stronger deviations due to its low ethylene oxide (hydrophilic) content.

Dynamic light scattering measurements

Photon correlation spectroscopy was used to investigate the behaviour of the polymer solutions as a function of the solution pH. Fig. 3a shows the intensity autocorrelation functions of a 0.1 wt.% P2VP solution at pH 1 for various scattering angles. The distributions of relaxation times, shown in the inset, exhibit a dominant mode with constant intensity and a q^2-dependent rate (not shown), which corresponds to the diffusion of single chains (unimers) with diffusion coefficient $D = 3.5 \times 10^{-7}$ cm^2 s^{-1} that corresponds to a hydrodynamic radius $R_\text{h} = 6.4 \pm 0.2$ nm. As the pH increases to pH 3, besides the diffusion of the free chains ($D_1 = 3.4 \times 10^{-7}$ cm^2 s^{-1}, $R_{\text{h},1} = 6.5 \pm 0.2$ nm), a slower diffusive process ($D_2 = 3.5 \times 10^{-8}$ cm^2/s, $R_{\text{h},2} = 64 \pm 2$ nm) is also evident (Fig. 3b).[67] This process possesses a strongly q-dependent intensity, which, when fitted with the Guinier expression $I(q) = I(0)\exp(-q^2R_g^2/3)$, results in a radius of gyration $R_\text{g} \approx 68 \pm 4$ nm; It signifies some kind of aggregation attributed to hydrophobic interactions of the polymer backbone or to a complex ionization of the P2VP units as suggested earlier.[29] At higher pH values the P2VP homopolymer is deprotonated and becomes hydrophobic. DLS shows large aggregates suggesting the macroscopic precipitation of the polymer, which prohibits the analysis at pH values above 5.

Fig. 4 shows the intensity autocorrelation functions for solutions of a linear PEO homopolymer of similar molecular weight to that of P2VP for three different pH values 2.0, 6.6 and 9.0. PEO is a neutral water-soluble polymer and, thus, only one diffusive process is consistently observed over the whole pH range with a diffusion coefficient $D = 3.7 \times 10^{-7}$ cm^2 s^{-1} that corresponds to single polymer chains of $R_\text{h} = 6.0 \pm 0.2$ nm.

The behaviour of the amphiphilic diblock copolymer solutions is studied next by dynamic light scattering as a function of solution pH. Fig. 5a shows the intensity autocorrelation functions for a 0.1 wt.% D 30/35 diblock copolymer solution at pH 1 for three different scattering angles. Two relaxation processes are consistently observed in the distributions of relaxation times, shown in the inset. The two diffusivities ($D_1 = 3.9 \times 10^{-7}$ cm^2 s^{-1} and $D_2 = 8.0 \times 10^{-8}$ cm^2 s^{-1}) correspond to hydrodynamic radii of $R_{\text{h},1} = 5.7 \pm 0.2$ nm and $R_{\text{h},2} = 28 \pm 2$ nm: the first corresponds to free polymer chains, whereas the second can be attributed to some form of polymer aggregates that are present in the solution even at pH 1. This aggregation can either be due to the remaining hydrophobic interactions of the P2VP backbone or to the P2VP complex ionization, similarly to the discussion above for the P2VP homopolymer, or, alternatively, to hydrogen bonding interactions, which may occur between the protonated tertiary amine groups and the ethylene oxide units of the PEO block as reported previously for PEO-*block*-poly(acrylic acid) copolymers at low pH.[68]

By increasing the solution pH, partial deprotonation of the P2VP units takes place and further polymer aggregation is induced. The correlation functions of a 0.1 wt.% D 30/35 solution at pH 4, which is one pH unit below the effective $\text{p}K_\text{a}$ value of the protonated P2VP, are shown in Fig. 5b for

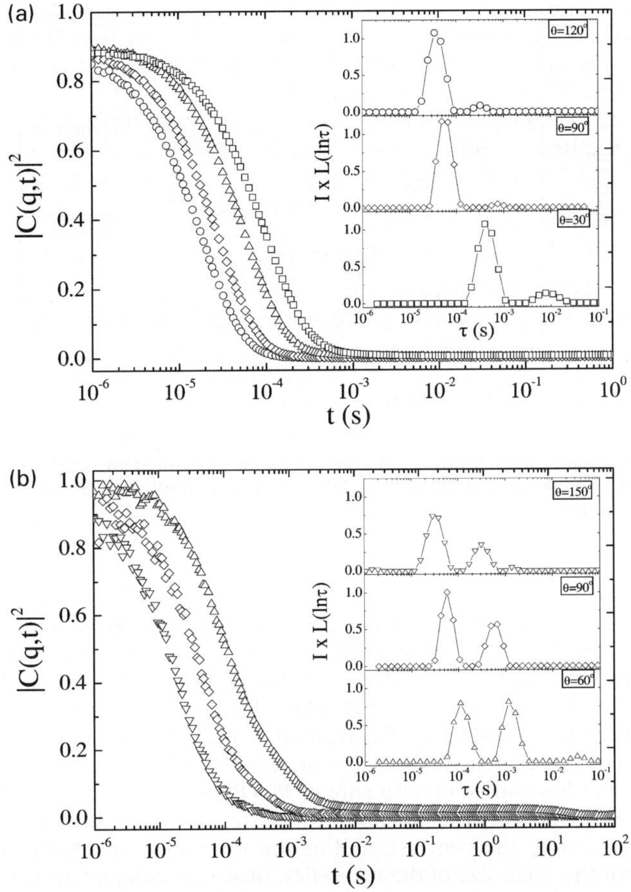

Fig. 3 Intensity autocorrelation functions of a 0.1 wt.% P2VP30 solution (a) at pH 1 for scattering angles 30° (□), 60° (△), 90° (◊) and 120° (○) and (b) at pH 3 for scattering angles 60° (△), 90° (◊) and 150° (▽). Insets show the distributions of relaxation times multiplied by the total scattering intensity (normalized to that of toluene) for the respective scattering angles.

various scattering angles. In the inset, the q-dependence of the intensities and the effective diffusion coefficients, $D = \Gamma/q^2$, for the two processes are shown. The two diffusion coefficients ($D_1 = 3.5 \times 10^{-7}$ cm^2 s^{-1} and $D_2 = 3.2 \times 10^{-8}$ cm^2 s^{-1}) correspond to hydrodynamic radii of $R_{h,1} = 5.8 \pm 0.2$ nm and $R_{h,2} = 60 \pm 2$ nm. The first is related to free copolymer chains whereas the second is related to polymer aggregates, which, however, are larger than those at pH 1 and possess a strongly q-dependent intensity (second inset of Fig. 5b). Fitting the intensities of the slower process with the Guinier expression results in a $R_g = 86 \pm 5$ nm. The larger aggregates at pH 4 are apparently due to the increased hydrophobic interactions as the P2VP units are now becoming deprotonated, which also allows for a larger number of chains in the aggregates as the electrostatic repulsions decrease. This behaviour is similar to that observed for PHEGMA-*b*-PDEAEMA copolymer solutions as the pH approached 7, and will be discussed below.

The pH-dependent behaviour of the copolymer solutions is depicted in Fig. 6, which shows the intensity autocorrelation functions at 90° scattering angle of 0.1 wt.% D 30/35 copolymer solutions at pH varying from 1 to 11. The distributions of relaxation times multiplied by the total scattering intensities are shown in the insets and can be directly compared. At low pH the two processes discussed above (single polymer chains and polymeric aggregates) are observed with low intensities. The low intensities of the aggregates signify a small number of aggregates and/or low scattering

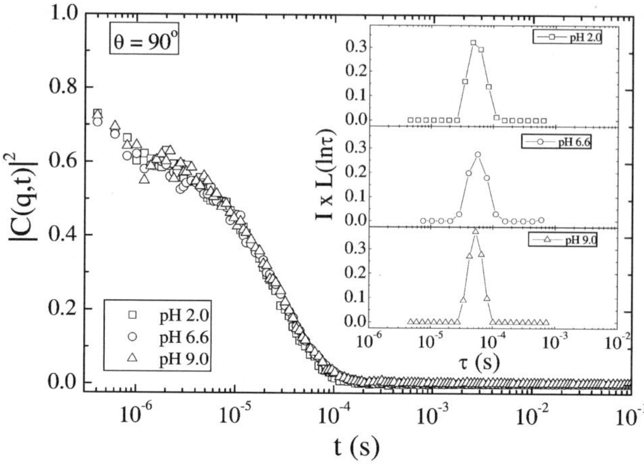

Fig. 4 Intensity autocorrelation functions of a 0.1 wt.% PEO30 solution at 90° scattering angle for pH 2.0 (□), pH 6.6 (○) and pH 9.0 (△). Inset: Distribution of relaxation times multiplied by the total scattering intensity (normalized to that of toluene).

contrast. As the pH increases, the slow process becomes even slower, thus corresponding to a larger hydrodynamic radius, with only a modest increase of the accompanying intensity.

For pH higher than 5, one process dominates the correlation functions with very strong intensities and dynamics that correspond to smaller hydrodynamic radii. This is due to the formation of polymer micelles when the P2VP block becomes deprotonated and insoluble in the aqueous medium. The process above pH 5 corresponds to a hydrodynamic radius $R_h = 33 \pm 1$ nm, which apparently does not change as the pH is increased further from pH 5 to pH 11 signifying stable, well defined micelles consisting of a core of P2VP and a corona of PEO chains. The sharp increase in the scattering intensity is consistent with the formation of the hydrophobic cores, which expel water, thus increasing the contrast, and/or the formation of a large number of micelles. Moreover, because of the small size of these micelles, their scattering intensities are independent of the wavevector q. It is noted that the same behaviour is observed even for copolymer concentrations as low as 0.033 wt.%.

One should comment on the fact that aggregates are observed even at pH < 5, when the P2VP units are protonated and, thus, in principle hydrophilic. Three possible explanations can be envisaged for the observed behaviour. First, the aggregation can be attributed to the inherent hydrophobicity of the P2VP backbone. A second explanation is related to the possibility of hydrogen bonding between the charged P2VP and the PEO units,[68] whereas a third can be the recently reported complex ionization of the P2VP chains.[29] Based on the latter, two neighbouring pyridine units can share one proton and, thus, the hydrophilic/hydrophobic character of the P2VP chain is not directly related to the complete protonation/deprotonation of the pyridine units. Thus, at lower than 100% degrees of deprotonation, interactions take place between the deprotonated and protonated pyridine units resulting in aggregation and the formation of "hydrophobic cores" surrounded by charged monomer units. Such an association can be both inter- and intra-molecular leading to the formation of aggregates known as quasi-aggregates because they consist of an ill-defined highly-hydrated core containing both charged and uncharged pyridine units. The hydrophobicity of the polymer backbone and the complex ionization of P2VP should be responsible for the behaviour of the P2VP homopolymer solutions at low pH discussed above in relation to Fig. 3.

The explanation should be more complicated for the aggregation phenomena at pH below 5 for the PEO-*b*-P2VP diblock copolymers. It is believed that at very low pH, where the P2VP units are fully protonated, it is the remaining hydrophobic interactions of the PV2P backbone and the hydrogen bonding between the charged P2VP units and PEO monomers, which are responsible for the aggregation and result in low density aggregates with a highly hydrated "core". The small hydrodynamic radius of these aggregates is due to the high ionization of the pyridine units, which results in strong electrostatic repulsion between the polymer chains and, thus, in structures of low

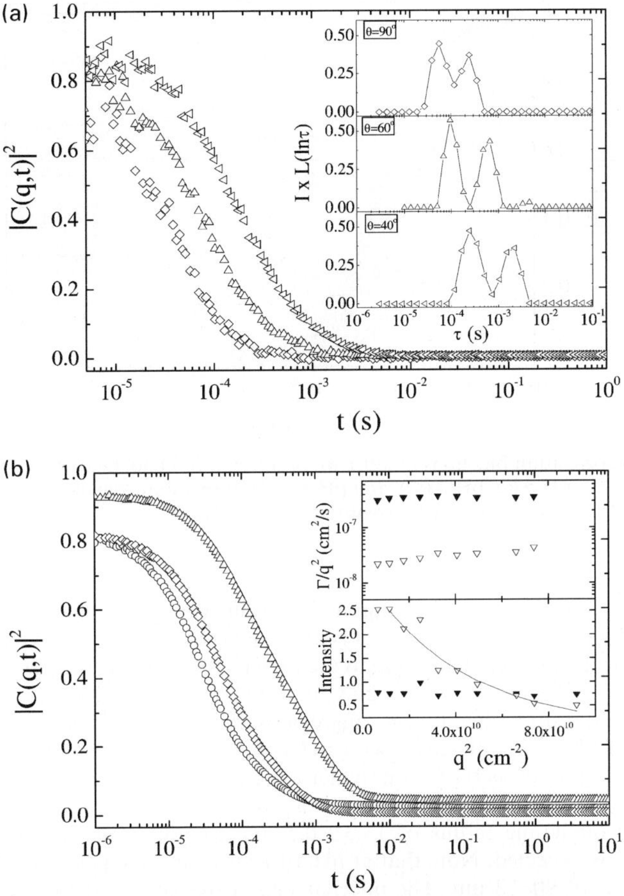

Fig. 5 (a) Intensity autocorrelation functions of a 0.1 wt.% D 30/35 solution at pH 1 for 40° (◁), 60° (△) and 90° (◇) scattering angles. Inset: distributions of relaxation times multiplied by the total scattering intensities (normalized to that of toluene) for the respective scattering angles. (b) Intensity autocorrelation functions of a 0.1 wt.% D 30/35 solution at pH 4 for 60° (△), 90° (◇) and 120° (○) scattering angles. Inset: wavevector dependence of the effective diffusion coefficients, $D = \Gamma/q^2$, and of the dynamic intensities (normalized to that of toluene) of the fast (▼) and slow (▽) processes. The line is the Guinier fit for the intensity of the slow process.

aggregation numbers. This is also consistent with the low scattering intensities obtained at this pH. As the pH is increased, but still below pH 5, the pyridine units are gradually deprotonated and, thus, the complex ionization mechanism of P2VP takes over. Moreover, the electrostatic repulsions are reduced and more polymer chains may participate in the aggregation process leading to larger moieties. However, the contrast and, thus, the scattering intensity of the solution remains relatively low and only modestly increases since the aggregate "core" is still sufficiently hydrated. Further increase of the pH above 5 leads to the deprotonation of the pyridine units and the formation of a fully dehydrated dense core containing the hydrophobic P2VP blocks. This is accompanied by a sharp decrease of the hydrodynamic radius of the micelles and a significant increase of the scattering intensity of the solution.

The effect of pH on the solution properties of the PHEGMA-*b*-PDEAEMA diblock copolymers is investigated as well. Fig. 7 shows the autocorrelation functions of 0.1 wt% $PHEGMA_{50}$-*b*-$PDEAEMA_{50}$ solutions for various pH values at scattering angle $\theta = 90°$. At pH 2.2, 4.0 and 6.0 the existence of both polymer chains and polymer aggregates are observed with hydrodynamic radii of about $R_{h,1} = 4.5 \pm 0.2$ nm and $R_{h,2} = 64 \pm 2$ nm, respectively. The polymer aggregates are attributed to possible clustering of the linear polymer chains discussed above for the PEO-*b*-P2VP

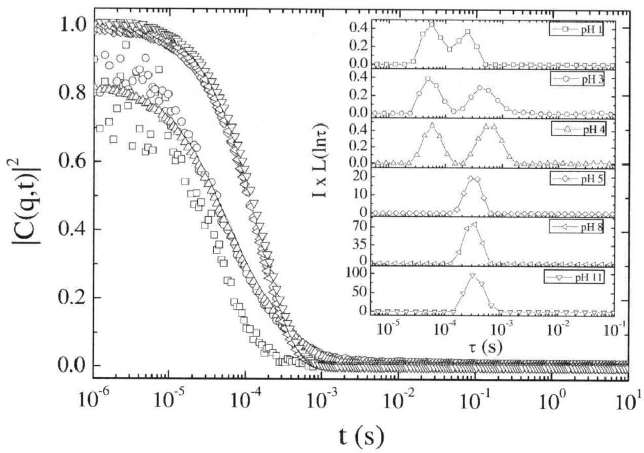

Fig. 6 Intensity autocorrelation functions of a 0.1 wt.% solution of D 30/35 at 90° scattering angle and pH 1 (□), pH 3 (○), pH 4 (△), pH 5 (◊), pH 8 (◁), and pH 11 (▽). Insets show the distributions of relaxation times multiplied by the total scattering intensity (normalized to that of toluene).

diblock copolymers due to the remaining hydrophobicity of the chain backbone and/or the hydrogen bonding interactions. Once more, it is observed that both processes exhibit low intensities, which is expected for the unimer chains due to their small size whereas for the polymer aggregates it suggests that they are either very few or highly hydrated, as discussed above. It is noted, however, that at pH 7 the intensities due to the aggregates increase. This is because the fraction of deprotonated DEAEMA units increases with pH and so do the hydrophobic interactions of the polymer chains leading to an increase in polymer aggregation. However, the size of the aggregates ($R_{h,2} = 63 \pm 2$ nm) was similar to that obtained at lower values of pH. At pH 10.1 micelles are formed with a hydrodynamic radius of 13.3 ± 0.5 nm accompanied by a sharp increase in the scattering intensity, as expected. Note that at pH 10.1 micellar aggregates are also observed with a hydrodynamic radius of 90 ± 3 nm. The micellar aggregates observed for the PDEAEMA-based diblock copolymers are attributed to the less effective steric stabilization conferred by the shorter hydrophilic PHEGMA blocks compared to the longer PEO stabilizing blocks of the PEO-*b*-P2VP copolymers.

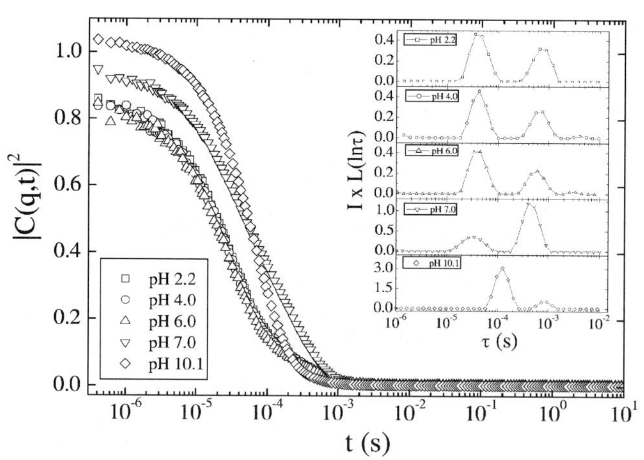

Fig. 7 Intensity autocorrelation functions of a 0.1 wt.% $PHEGMA_{50}$-*b*-$PDEAEMA_{50}$ solution at 90° scattering angle for various pH: pH 2.2 (□), pH 4.0 (○) and pH 6.0 (△), pH 7.0 (▽), and pH 10.1 (◊). Inset: Distributions of relaxation times multiplied by the total scattering intensity (normalized to that of toluene).

AFM studies

The formation of polymer micelles above a certain pH was investigated further by atomic force microscopy for both the PEO-*b*-P2VP and the PHEGMA-*b*-PDEAEMA systems. Images were obtained as a function of solution pH from pH 2 to 10. Fig. 8 shows the AFM images of PHEGMA$_{50}$-*b*-PDEAEMA$_{50}$ adsorbed onto a freshly cleaved mica surface from a 0.5 wt.% polymer solution at two different pH values 2 and 9. At pH 2 (Fig. 8a) the polymer adsorbs as a film of low roughness (0.4 nm rms roughness) and no structure formation is observed. However, above pH 7 polymer micelles were found, as for example at pH 9 (Fig. 8b), where both micelles and micellar aggregates are observed on the mica surface. The height of the micelles obtained by AFM is around 25 nm, which is in good agreement with the micelle characteristics discussed above and

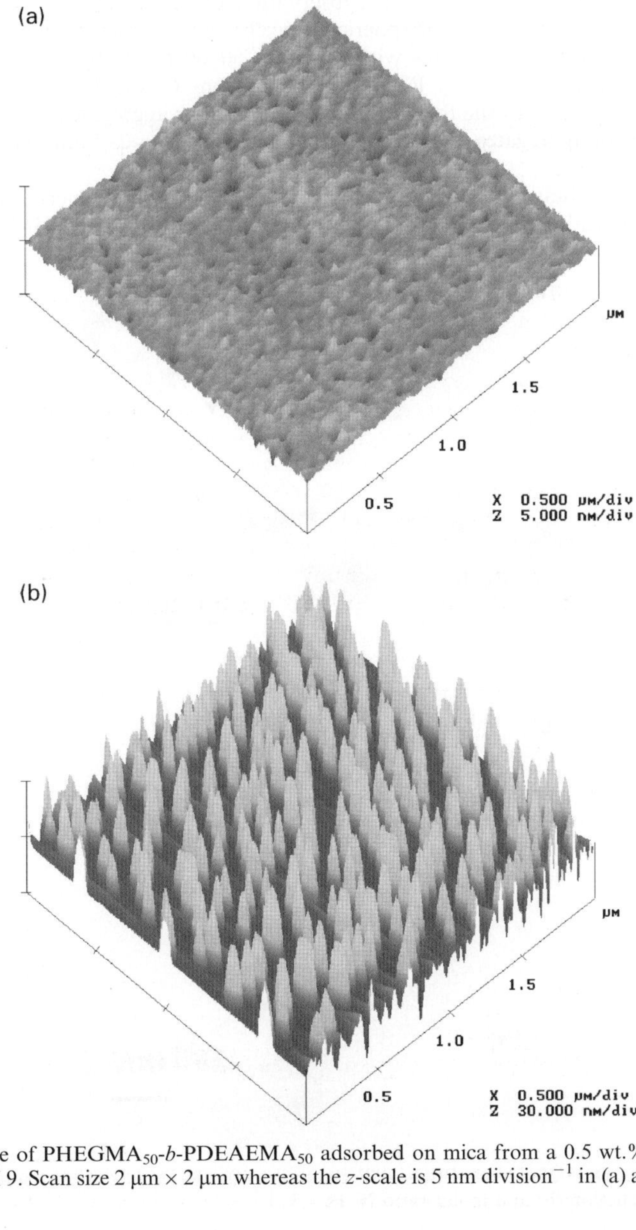

Fig. 8 AFM image of PHEGMA$_{50}$-*b*-PDEAEMA$_{50}$ adsorbed on mica from a 0.5 wt.% aqueous solution at (a) pH 2 and (b) pH 9. Scan size 2 μm × 2 μm whereas the *z*-scale is 5 nm division^{-1} in (a) and 30 nm per division in (b).

supports the DLS results. Similar results were obtained for all other copolymers with micelles observed above pH 5 for the PEO-*b*-P2VP copolymers and above pH 7 for the PHEGMA-*b*-PDEAEMA diblock copolymers, as expected.

Metal containing micelles

The incorporation of platinum within the cores of the copolymer micelles was carried out using either H_2PtCl_6 or K_2PtCl_6 compounds followed by metal reduction to form the metal nanoparticles. The micelle characteristics during the metallation process as well as the metal nanoparticle formation were studied by TEM. When hexachloroplatinic acid ($H_2PtCl_6 \times H_2O$) is added to a 0.25 wt.% solution of the D 25/25 diblock copolymer (initial pH 6.4) at a molar ratio N : Pt = 3 : 1 (final pH 2.6), it results in the incorporation of the metal species within the polymeric micelle core, which provides the contrast that allows the observation of the micelles by TEM without extra staining. Fig. 9a reveals the existence of spherical micelles with diameter about 20 nm (small grey circles), of spherical micellar aggregates with diameter 50–60 nm (large grey circles) and multi-vesicular structures of a similar size. Reduction by sodium borohydride ($NaBH_4$) leads to an increase of the pH to 9.3 and to the formation of large micellar aggregates of irregular shape. The latter consist of smaller aggregates filled with Pt nanoparticles of ~2–3 nm in diameter, as seen in Fig. 9b.

Similar results were obtained for PDEAEMA-based diblock copolymers. Fig. 10a shows the metal-containing spherical micelles observed by TEM when a 0.33 wt.% $PHEGMA_{17}$-*b*-$DEAEMA_{16}$

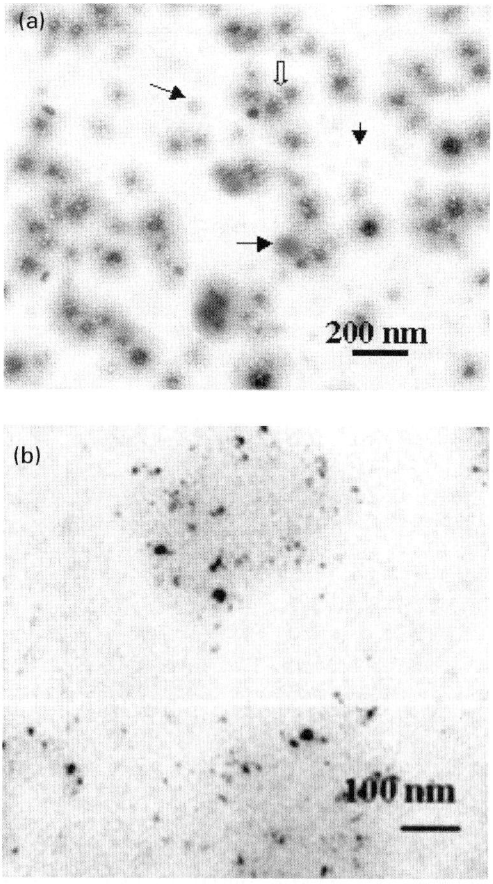

Fig. 9 TEM images of D 25/25 diblock PEO-*b*-P2VP micelles after $H_2PtCl_6 \times 6H_2O$ loading (a) and of the Pt nanoparticles after reduction (b) at a molar ratio N : Pt = 3 : 1.

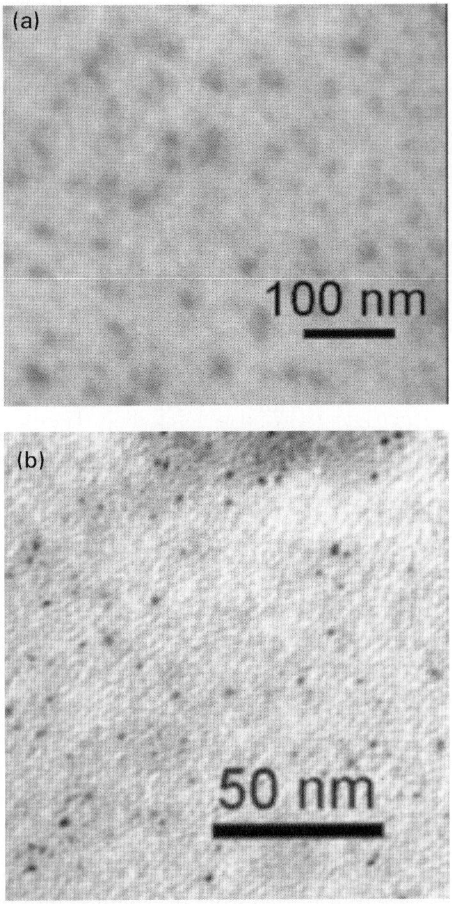

Fig. 10 TEM images of PHEGMA$_{17}$-*b*-PDEAEMA$_{16}$ micelles after $H_2PtCl_6 \times 6H_2O$ loading (a) and of the Pt nanoparticles reduction (b) at a molar ratio N : Pt = 3 : 1.

diblock copolymer solution is loaded with $H_2PtCl_6 \times H_2O$ at molar ratio N : Pt = 3 : 1. Fig. 10b shows the 1.4 nm in mean diameter Pt nanoparticles formed within the micelle cores following metal reduction with $NaBH_4$. The nanoparticle sizes and distributions are not affected by the overall copolymer length and block length ratio of the present diblock copolymers. This will be further investigated utilizing copolymers of higher molecular weights.

The metallation of PEO$_{45}$-*b*-PDEAEMA$_{37}$ with $H_2PtCl_6 \times H_2O$ was investigated as a function of the solution pH using two different methods for metal incorporation. As seen in Figs. 11a and b, the addition of the metal compound in already preformed micelles, which were obtained by a pH increase to 7 of a molecularly dissolved block copolymer solution at pH 2, followed by the $NaBH_4$ reduction results in small and narrowly distributed Pt nanoparticles (diameter 1.4 ± 0.3 nm). In contrast, when the metal compound is introduced to a molecularly dissolved block copolymer (at pH 2.0), metal-induced micellization takes place. This apparently yields swollen micelles and larger and more broadly distributed nanoparticles (diameter 1.8 ± 0.7 nm) as shown in Figs. 11c and d. This approach allows tuning of the particle characteristics by varying the metal nanoparticle preparation conditions.

The platinic acid loading was varied to find the best conditions for controlling the micelle size and shape and the size of the platinum nanoparticles. The local concentration of the metal precursor inside the micelle core was found to affect directly the particle formation during nucleation and the resultant particle size. Elemental analysis of the metallated PDEAEMA-based diblock copolymers

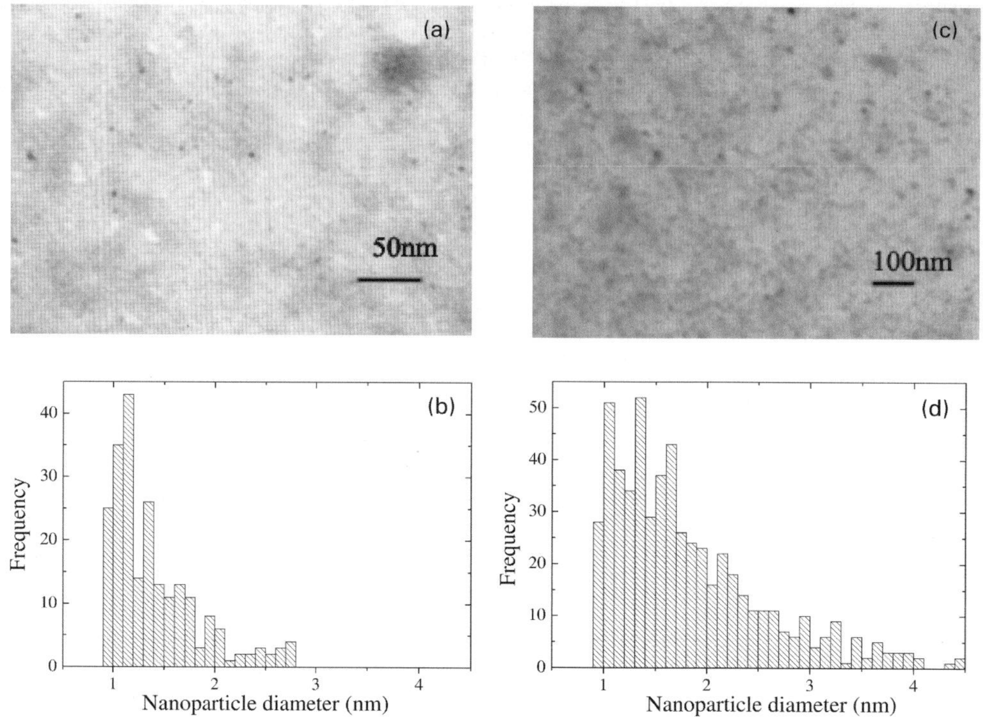

Fig. 11 TEM images (a, c) and histograms (b, d) of Pt nanoparticles formed within PEO_{45}-*b*-$PDEAEMA_{37}$ when metallation (molar ratio N : Pt = 3 : 1) takes place in a preformed micellar solution at neutral pH (a, b) and in a molecular solution at pH 2.0 (c, d).

obtained after incorporation of $H_2PtCl_6 \times H_2O$ at a molar ratio N : Pt of 3 : 1 reveals that nearly all of the precursor ions (95–98%) are transported from the solution into the micelle core. A number of syntheses with molar ratios N : Pt of 6 : 1 and 3 : 2 in addition to the original ratio of 3 : 1 were also carried out. The results do not show significant differences in the mean diameter of the nanoparticles synthesized with the three different ratios. Nonetheless, particles synthesized in solution with a molar ratio N : Pt of 3 : 2 tend to be larger (1.6 to 1.8 nm) and have stronger tendency to agglomerate than those synthesized with 3 : 1 and 6 : 1 ratios, whereas particles synthesized with a molar ratio N : Pt of 6 : 1 tend to have a slightly smaller diameter (1.3 to 1.5 nm) than those synthesized with a ratio of 3 : 1.

Protonation of the tertiary amine groups also provides some control over the local precursor concentration inside the micelle core. When $H_2PtCl_6 \times H_2O$ is used as a precursor for the $PtCl_6^{2-}$ ions, the amino groups become protonated and, thus, increase the transport rate of $PtCl_6^{2-}$ into the micelle core *via* ionic interactions. If a neutral salt, such as K_2PtCl_6, is used as a precursor rather than the acid, the solution pH influences the micelle and metal-nanoparticles characteristics at the same Pt loading (N : Pt 3 : 1). At low pH values the micelle structures and Pt nanoparticle sizes resemble those for $H_2PtCl_6 \times H_2O$. At high pH 5–8, the addition of the platinic salt has no effect on the solution pH and the incorporation of the metal compound was found to be less efficient as the tertiary amino groups are not protonated and the transport rate of $PtCl_6^{2-}$ into the micelle core is greatly reduced since no electrostatic attraction facilitates the interaction. Early experiments at high pH with the use of K_2PtCl_6, under the same conditions as those used with $H_2PtCl_6 \times H_2O$, produced no detectable particles after reduction. As with platinic acid, the solutions were subjected to ultrafiltration one day after the addition of the precursor. The yellow colour of the filtrate indicated that most of the $PtCl_6^{2-}$ ions were present in the solution outside the micelles. On the other hand, when the solutions were allowed to stir for five days after addition of the precursor, the subsequent results were the same as those obtained with H_2PtCl_6. This slow rate of precursor ion transport, which is observed for the deprotonated amine groups, provides a potential control mechanism for the nanoparticle size. The local precursor concentration inside the micelle core can be directly controlled by reduction of K_2PtCl_6-filled micellar polymer solutions at various time intervals after the addition of the salt. However, detailed time-controlled studies are required to establish further evidence and guidelines for this technique.

IV. Concluding remarks

The aqueous solution properties of pH-responsive diblock copolymers (utilized as nanoreactors for the formation of metal nanoparticles) were evaluated by potentiometric titration, dynamic light scattering and atomic force microscopy. At low pH, unimers were found in the polymer solution together with a few loose aggregates; these should be attributed to a synergy of the inherent hydrophobicity of the polymer backbone, of the possibility of hydrogen bonding between ethylene oxide units and the charged amino groups and of the complex ionization of the amine blocks. At high pH, micelles are formed above the effective pK_a of the tertiary amine block, where it becomes deprotonated and, thus, hydrophobic. AFM images support the light scattering results with micelles observed at high pH values. Metal compounds were incorporated within the micelle cores to form metal nanoparticles upon reduction, investigated by TEM. Different synthetic procedures for the preparation of Pt nanoparticles inside the polymer micelles were investigated. The present study indicates that the optimum conditions for nanoparticle synthesis include the incorporation of the metal precursor in the preformed block copolymer micelles and ultrafiltration of the micellar solutions after addition of the metal precursor. Platinum nanoparticles obtained under these conditions are well defined and have a mean diameter between 1.4 and 1.6 nm.

Potential Pt nanoparticle size control mechanisms may include variation of precursor concentration and time-controlled synthesis in a polymer micellar solution. An increase of the N : Pt molar ratio from 6 : 1 to 3 : 2 results in a slight increase of the particle size and eventually leads to particle agglomeration. The use of K_2PtCl_6 instead of $H_2PtCl_6 \times H_2O$ at high pH significantly reduces the precursor ion transport rate into the micelle core. It is presumed that in this case the local precursor concentration is directly proportional to the time after the addition of the precursor. Thus control over the resultant particle size may prove to be possible by metal reduction at various time intervals after the addition of the precursor in the polymer solution.

The noble metal nanoparticles synthesized in the present investigation exhibit significantly enhanced catalytic properties for hydrogenation and oxidation reactions. For example, Pt nanocatalysts were shown to be very active and selective in the direct oxidation of L-sorbose to 2-keto-L-gulonic acid, which is an intermediate step in vitamin C production.[69]

Acknowledgements

Part of this research was sponsored by NATO's Scientific Affairs Division in the framework of the Science for Peace Programme (grant SfP-974173) and by the Greek General Secretariat of Research and Technology. We thank Ms K. Tsagaraki for her collaboration with the AFM measurements and Dr K. Chrissopoulou for her critical reading of the manuscript.

References

1 I. W. Hamley, *The Physics of Block Copolymers*, Oxford University Press, Oxford, UK, 1998.
2 N. Hadjichristidis, S. Pispas and G. A. Floudas, *Block Copolymers: Synthetic Strategies, Physical Properties, and Applications*, John Wiley & Sons, Inc., Hoboken, NJ, 2003.
3 S. Foerster and M. Antonietti, *Adv. Mater.*, 1998, **10**, 195.
4 S. E. Webber, *J. Phys. Chem. B.*, 1998, **102**, 2618.
5 G. Riess, *Prog. Polym. Sci.*, 2003, **28**, 1107.
6 R. Nagarajan, *Polym. Adv. Technol.*, 2001, **12**, 23.
7 K. Procházka, T. J. Martin, P. Munk and S. E. Webber, *Macromolecules*, 1996, **29**, 6518.
8 L. Yu, H. Wang, H. H. Wang, V. S. Urban, K. C. Littrell and P. Thiyagarajan, *J. Am. Chem. Soc.*, 2000, **122**, 6855.
9 H. Schuch, J. Klingler, P. Rossmanith, T. Frechen, M. Gerst, J. Feldthusen and A. H. E. Müller, *Macromolecules*, 2000, **33**, 1734.

10 B. S. Lele and J.-C. Leroux, *Macromolecules*, 2002, **35**, 6714.
11 Y. Chang, S. C. Lee, K. T. Kim, C. Kim, S. D. Reeves and H. R. Allcock, *Macromolecules*, 2001, **34**, 269.
12 P. Matejicek, F. Uhlik, Z. Limpouchova, K. Prochazka, Z. Tuzar and S. E. Webber, *Macromolecules*, 2002, **35**, 9487.
13 Z. Zhou, Z. Li, Y. Ren, M. A. Hillmyer and T. P. Lodge, *J. Am. Chem. Soc.*, 2003, **125**, 10 182.
14 A. Gupte, R. Nagarajan and A. Kilara, *Ind. Eng. Chem. Res.*, 1995, **34**, 2910.
15 A. Gupte, R. Nagarajan and A. Kilara, *Biotechnol. Prog.*, 1991, **7**, 348.
16 V. Alakhov, E. Klinski, P. Lemieux, G. Pietrzynski and A. Kabanov, *Expert Opin. Biol. Ther.*, 2001, **1**, 583.
17 A. Rosler, G. W. M. Vandermeulen and H. A. Klok, *Adv. Drug Deliv. Rev.*, 2001, **53**, 95.
18 L. F. Zhang and A. Eisenberg, *Science*, 1995, **268**, 1728.
19 L. F. Zhang, K. Yu and A. Eisenberg, *Science*, 1996, **272**, 1777.
20 H. Shen, L. Zhang and A. Eisenberg, *J. Am. Chem. Soc.*, 1999, **121**, 2728.
21 I. C. Riegel, A. Eisenberg, C. L. Petzhold and D. Samios, *Langmuir*, 2002, **18**, 3358.
22 P. L. Soo and A. Eisenberg, *J. Polym. Sci.: Part B: Polym. Phys.*, 2004, **42**, 923.
23 L. Selb and Y. Gallot, in *Developments in Block Copolymers-2*, ed. I. Goodman, Elsevier Applied Science Publishers, New York, 1985, p. 27.
24 L. F. Zhang, H. Shen and A. Eisenberg, *Macromolecules*, 1997, **30**, 1001; L. Zhang and A. Eisenberg, *Macromolecules*, 1999, **32**, 2239.
25 T. J. Martin, K. Procházka, P. Munk and S. E. Webber, *Macromolecules*, 1996, **29**, 6071.
26 M. Vamvakaki, N. C. Billingham and S. P. Armes, *Macromolecules*, 1999, **32**, 2088.
27 A. S. Lee, A. P. Gast, V. Bütün and S. P. Armes, *Macromolecules*, 1999, **32**, 4302.
28 J.-F. Gohy, S. K. Varshney and R. Jérôme, *Macromolecules*, 2001, **34**, 3361.
29 J.-F. Gohy, S. Antoun and R. Jérôme, *Macromolecules*, 2001, **34**, 7435.
30 A. S. Lee, V. Bütün, M. Vamvakaki, S. P. Armes, J. A. Pople and A. P. Gast, *Macromolecules*, 2002, **35**, 8540.
31 J. Virtanen, M. Arotçaréna, B. Heise, S. Ishaya, A. Laschewsky and H. Tenhu, *Langmuir*, 2002, **18**, 5360.
32 S. Yusa, Y. Shimada, Y. Mitsukami, T. Yamamoto and Y. Morishima, *Macromolecules*, 2003, **36**, 4208.
33 J.-F. Gohy, S. K. Varshney, S. Antoun and R. Jérôme, *Macromolecules*, 2000, **33**, 9298.
34 L. I. Gabaston, S. A. Furlong, R. A. Jackson and S. P. Armes, *Polymer*, 1999, **40**, 4505.
35 V. Bütün, N. C. Billingham and S. P. Armes, *J. Am. Chem. Soc.*, 1998, **120**, 11 818.
36 J.-F. Gohy, S. Creutz, M. Garcia, B. Mahltig, M. Stamm and R. Jérôme, *Macromolecules*, 2000, **33**, 6378.
37 V. Bütün, S. P. Armes, N. C. Billingham, Z. Tuzar, A. Rankin, J. Eastoe and R. K. Heenan, *Macromolecules*, 2001, **34**, 1503.
38 S. Liu and S. P. Armes, *Langmuir*, 2003, **19**, 4432.
39 J. C. Meiners, A. Ritzi, M. H. Rafailovich, J. Sokolov, J. Mlynek and G. Krausch, *Appl. Phys. A*, 1995, **61**, 519.
40 J. C. Meiners, A. Quintel-Ritzi, J. Mlynek, H. Elbs and G. Krausch, *Macromolecules*, 1997, **30**, 4945.
41 S. D. Connell, S. Collins, J. Fundin, Z. Yang and I. W. Hamley, *Langmuir*, 2003, **19**, 10 449.
42 M. R. Talingting, Y. Ma, C. Simmons and S. E. Webber, *Langmuir*, 2000, **16**, 862.
43 G. B. Webber, E. J. Wanless, S. P. Armes, F. L. Baines and S. Biggs, *Langmuir*, 2001, **17**, 5551.
44 M. Moffitt, L. McMahon, V. Pessel and A. Eisenberg, *Chem. Mater.*, 1995, **7**, 1185.
45 J. P. Spatz, A. Roescher and M. Möller, *Adv. Mater.*, 1996, **8**, 337.
46 M. V. Seregina, L. M. Bronstein, O. A. Platonova, D. M. Chernyshov, P. M. Valetsky, J. Hartmann, E. Wenz and M. Antonietti, *Chem. Mater.*, 1997, **9**, 923.
47 L. M. Bronstein, E. Krämer, B. Berton, C. Burger, S. Föster and M. Antonietti, *Chem. Mater.*, 1999, **11**, 1402.
48 S. Mössmer, J. P. Spatz, M. Möller, T. Aberle, J. Schmidt and W. Burchard, *Macromolecules*, 2000, **33**, 4791.
49 A. Henglein, *Chem. Rev.*, 1989, **89**, 1861.
50 F. Palacio, C. Castro, J. Reyes, G. Sturgeon and F. J. Lázaro, in *Physics and Chemistry in Finite Systems: From Clusters to Crystals*, ed. P. Jena, Kluver Academic Publishers, Amsterdam, 1992, vol. **I**, p. 793.
51 F. Palacio, C. Castro, F. J. Lázaro and J. Reyes, *J. Magn. Magn. Mater.*, 1992, **104**, 2101.
52 L. N. Lewis, *Chem. Rev.*, 1993, **93**, 2693.
53 Y. N. C. Chan, G. S. W. Craig, R. R. Schrock and R. E. Cohen, *Chem. Mater.*, 1992, **4**, 885.
54 I. M. Papisov, Yu. S. Yablokov and A. I. Prokof'ev, *Polymer Science, Ser. A*, 1994, **36**, 291.
55 G. Schmidt, *Clusters and Colloids*, Wenheim, VCH, 1994, p. 300.
56 M. Antonietti, E. Wenz, L. M. Bronstein and M. V. Seregina, *Adv. Mater.*, 1995, **7**, 1000.
57 M. P. Pileni, *Langmuir*, 1997, **13**, 3266.
58 E. Sulman, Yu. Bodrova, V. Matveeva, N. Semagina, L. Cerveny, V. Kurtc, L. Bronstein, O. Platonova and P. Valetsky, *Appl. Catal. A-Gen.*, 1999, **176**, 75.
59 J. H. Youk, M.-K. Park, J. Locklin, R. Advincula, J. Yang and J. Mays, *Langmuir*, 2002, **18**, 2455.
60 L. M. Bronstein, S. N. Sidorov, A. Y. Gourkova, P. M. Valetsky, J. Hartmann, M. Breulmann, H. Colfen and M. Antonietti, *Inorg. Chim. Acta*, 1998, **280**, 348.
61 L. M. Bronstein, S. N. Sidorov, P. M. Valetsky, J. Hartmann, H. Colfen and M. Antonietti, *Langmuir*, 1999, **15**, 6256.

62 S. N. Sidorov, L. M. Bronstein, P. M. Valetsky, J. Hartmann, H. Colfen, H. Schnablegger and M. Antonietti, *J. Colloid Interface Sci.*, 1999, **212**, 197.
63 N. V. Semagina, A. V. Bykov, E. M. Sulman, V. G. Matveeva, S. N. Sidorov, L. V. Dubrovina, P. M. Valetsky, O. I. Kiselyova, A. R. Khokhlov, B. Stein and L. M. Bronstein, *J. Mol. Catal. A: Chem.*, 2004, **208**, 273.
64 P. G. Fragouli, H. Iatrou and N. Hadjichristidis, *Polymer*, 2002, **43**, 7141.
65 L. Bailey, M. Vamvakaki, N. C. Billingham and S. P. Armes, *Polym. Prepr.*, 1999, **40**, 263.
66 J. V. M. Weaver, S. P. Armes and S. Liu, *Macromolecules*, 2003, **36**, 9994.
67 A very weak slower process could also be observed at pH 1 in Fig. 3a, which is much more pronounced at pH 3 in Fig. 3b.
68 K. L. Smith, A. E. Winslow and D. E. Petersen, *Ind. Eng. Chem.*, 1959, **51**, 1361; T. Ikawa, K. Abe, K. Honda and E. Tsuchida, *J. Polym. Sci., Polym. Chem. Ed.*, 1975, **13**, 1505.
69 E. M. Sulman, N. Semagina, V. Matveeva, L. M. Bronstein, S. Sidorov, P. G. Fragouli, H. Iatrou, N. Hadjichristidis, M. Vamvakaki and S. H. Anastasiadis, in preparation, 2004.

Control of pore hydrophilicity in ordered nanoporous polystyrene using an AB/AC block copolymer blending strategy

Huiming Mao, Pedro L. Arrechea, Travis S. Bailey, Bret J. S. Johnson† and Marc A. Hillmyer*

Department of Chemistry, University of Minnesota, 207 Pleasant St. SE, Minneapolis, Minnesota, 55455-0431, USA. E-mail: hillmyer@chem.umn.edu

Received 2nd March 2004, Accepted 13th May 2004
First published as an Advance Article on the web 19th August 2004

Ordered nanoporous plastics with hydrophilic pore surfaces were prepared by the degradative removal of polylactide from a self-organised, multi-component composite containing two block copolymers: polystyrene-polylactide and polystyrene-polyethylene oxide. The solid-state characterization of blends containing up to 12 wt.% polyethylene oxide was consistent with nanoscopic cylinders of mixed polyethylene oxide and polylactide hexagonally packed in a polystyrene matrix. Orientation of these materials through simple channel die processing resulted in good cylinder alignment. Subsequent methanolysis/hydrolysis of the polylactide component gave nanoporous polystyrene with polyethylene oxide coated pores. The resulting nanoporous materials were able to imbibe water, in contrast to nanoporous polystyrene with no polyethylene oxide component.

Introduction

Block copolymers are a class of self-organising polymers that have utility in a wide variety of applications: from asphalt modifiers[1] to nanolithographic templates for magnetic storage media applications.[2] As demonstrated in the latter application, the evolution of nanotechnology has embraced soft material self-assembly[3] as a legitimate fabrication tool,[4,5] and block copolymers certainly represent a class of materials that exhibit unparalleled versatility in this regard.[6] Nanoporous materials derived from self-assembled block copolymers[7] are a specifc class of materials with many possible nanotechnological applications, for example, nanoscopic templates[8,9] or nanostructured separation media.[10] One can also envision these nanoporous materials as high surface area supports for catalysis applications[11] or as materials that confine chemical and physical processes on a nanometer length scale.[12] Since the first report of a nanoporous material from an ordered block copolymer precursor by Nakahama in 1988,[13] many groups have focused on the design, preparation, and applications of this class of versatile nanoporous materials.[8,10,14–26]

The need for nanoporous materials with predictable and reproducible pore size, alignment, distribution, and geometry is based on the critical nature of these morphological features to the efficacy of any application that involves their use (*e.g.*, separations,[27] photonic materials,[28] templates[29,30]). Furthermore, the precise pore wall functionality can be important in applications that require specific interactions between some substrate (or analyte) and the pore wall. Therefore

† Present address: Department of Chemistry, The College of St. Scholastica, Duluth, Minnesota, USA.

the controlled introduction of a specific chemical functionality within the pore space can be equally as critical to a material's designed purpose. While a great deal of attention has been directed at chemical functionalization of inorganic nanoporous silica phases, the introduction of functional groups must often be accomplished through modification of residual silanol groups following calcination, or through complex co-condensation reactions complicated by solution phase removal of the templating phase; a process often destructive to the nanoporous structure itself.[31] In contrast, the formation of polymeric nanoporous materials, based on block copolymer self-organisation in the melt state, can be designed to avoid chemically and thermally aggressive processing. As a result, the opportunities for functionalization in these systems can span the entire fabrication process, from the initial synthesis of the block copolymer to modification of the final nanoporous structure.

Nanoporous polymeric materials from ordered block copolymers are typically prepared by selective removal of a minority block from one of the traditional non-lamellar diblock copolymer phases, namely the sphere, gyroid, and cylinder morphologies. Through deliberate block selection (defining the Flory–Huggins interaction parameter, χ) and careful variation of the overall degree of polymerization (N) and composition (f_A, f_B, *etc.*), remarkable control over pore size, distribution, and geometry are afforded.[32] Furthermore, block copolymer morphologies have proven conveniently susceptible to macroscopic alignment both in monolithic and thin film environments, giving rise to highly ordered materials at length scales far beyond typical pore dimensions. With the recent evolution of "living" and controlled polymerization techniques, the number of monomer types that can be incorporated into low polydispersity block copolymers has been increased considerably. Combined with the flexibility of these systems to tolerate controlled chemical modification before or after formation of the porous solid, block copolymers systems provide an extremely facile and tunable route to functional ordered nanoporous materials.

Several example strategies for the functionalization of polymeric nanoporous materials are depicted in Fig. 1. The most basic strategy exploits the incorporation of a specific chemical functionality at the covalent junction between the matrix and sacrificial blocks. In this arrangement, removal or degradation of the sacrificial block conveniently exposes the chemical functionality at the pore wall. This strategy has been demonstrated by Zalusky *et al.*[33,34] and Wolf and Hillmyer[35] in which the degradation of a sacrificial polylactide (PLA) block generates hydroxyl functional groups at the pore surfaces in polystyrene (PS) and polycyclohexylethylene (PCHE) monoliths, respectively. This strategy is limited by the intrinsic maximum areal density of functional groups defined as the interfacial area per chain. Furthermore, the functional groups need to be accessible (*i.e.*, not buried as the result of surface reconstruction within the pores). In principle, moderate reduction of this intrinsic functional group density can be achieved through blending of non-functional matrix homopolymer, thus increasing the interfacial area associated with each functional group lining the pore wall.[36,37]

One method of circumventing this natural limitation in functional group density is to introduce a third block intermediate between the matrix and templating blocks, as depicted in Fig. 1(b). By incorporating blocks with pre-existing functionality, (*e.g.*, polyisoprene (PI) or polyvinylpyridine (PVP)) the brush-like arrangement of chains left following template removal generates a high density halo of functional groups along the pore's circumference. Control of the length of this middle block gives access to a wide range of functional group densities, given the volume fraction of this block is consistent with the desired morphology of the nanoporous material. Review of several references detailing the phase behavior of ABC triblock copolymer systems shows cylindrical and network morphologies consistent with desirable nanoporous structures, exist over a wide range of mid-block compositions.[38–40] We are currently exploring this approach using core-shell cylinder forming PS-PI-PLA, PI-PS-PLA and PS-PVP-polycaprolactone triblock copolymer systems,[41] and Liu *et al.* have described a related methodology.[42]

A final strategy, and the subject of this manuscript, is the ability to add functionality within the pore structure through the blending of AB and AC diblock copolymers. In this scenario, the A blocks common to both molecules form the matrix, the B and C blocks are miscible and form a composite templating phase, and only one of the two composite phase blocks (*e.g.*, B) is susceptible to the degradation process (Fig. 1(c)). The resultant material thus contains a disperse brush of the non-degradable C block confined within the porous structure. The brush density and overall C block composition in the pore space is defined by the degree of polymerization of the C block and

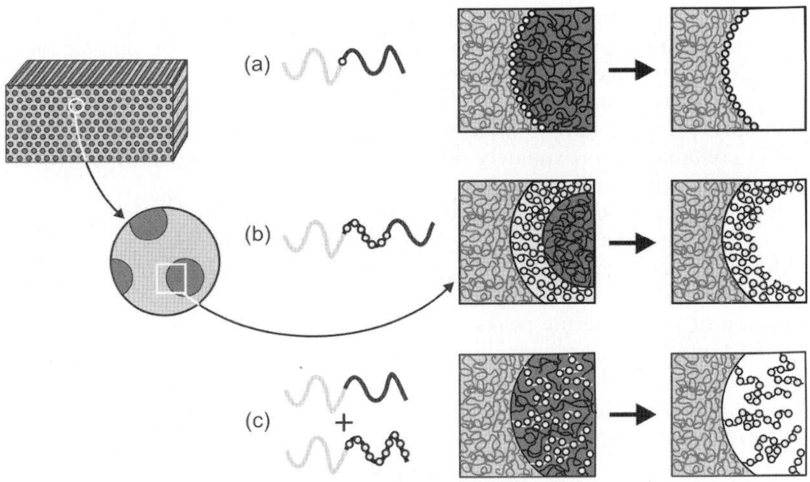

Fig. 1 Example routes to nanoporous materials with controlled pore-wall functionality. (a) A functional group is incorporated into the junction between matrix and sacrificial blocks, and is exposed upon template degradation. (b) A functionalized mid-block is inserted between the matrix and sacrificial end block, producing a functional polymer brush at the pore wall upon removal of the template. (c) An AB/AC diblock copolymer blend is formed in which the common A block serves as the matrix, the B and C blocks are miscible, and only one of the two blocks is susceptible to degradative removal. In this manner a functionalized non-degradable block can be introduced as a diffuse brush along the pore interior.

the overall composition of the blend. The overall functional group density in the pore can then be defined through control of the intra-chain density of functional groups in the brush as discussed in the second approach above. Provided block copolymers containing appropriate miscible B and C blocks can be prepared, this strategy is highly desirable, enabling systematic tuning of the above features.

As we will show in this report, the incorporation of a simple polyether into the pore space employing this method dramatically impacts the wettability of the resulting nanoporous material. Using this method, we can in principle generate monoliths capable of supporting catalysis, ion transport, and molecular separations in aqueous and biological media. Exploiting the partial miscibility of polyethylene oxide (PEO) and PLA,[43–53] we employed a blend of PS-PLA[33] and PS-PEO[54] diblock copolymers to generate a single-phase cylindrical morphology in which the PEO and PLA blocks form composite cylinders within a PS matrix. While the phase behavior of a few AB/AC diblock copolymer blends have been reported,[55–58] we are aware of only one report describing the formation of a single ordered morphology in which two of the components (B and C) completely mix to form a single domain (spherical or lamellar in this case).[59]

The parent blends and resultant nanoporous monoliths described here were examined using a collection of experimental techniques, revealing ordered materials with flow field induced pore anisotropy, narrow pore size distribution, and hydrophilic character completely absent in the unblended materials. The pore size and hydrophilic properties of these materials were easily adjusted through selection of the two component diblock copolymers and variation of their blending ratio. The hydrophilic nature of the pore walls imparted by this blending technique was demonstrated by absorption of water and observable density change and also by the diffusion of water-soluble dye and associated color change. Hydrophilicity was found to be exclusive to the PEO containing nanoporous monoliths.

Results and discussion

Preparation and characterization of parent PS-PLA and PS-PEO block copolymers

Four block copolymers were prepared for this study using established methods. Two PS-PLA copolymers and two PS-PEO copolymers were synthesized by anionic polymerization of styrene followed by end capping with ethylene oxide. The macroinitiators were used for the subsequent controlled "coordination-insertion" polymerization of D,L-lactide[33] or anionic polymerization of ethylene oxide,[60,61] to give model materials. The PLA component in the PS-PLA materials was atactic and completely amorphous. Molecular characterization of the four samples is given in Table 1. All samples exhibited narrow molecular weight distributions. The two PS-PLA copolymers (**SL1** and **SL2**) contained approximately 30 vol% PLA and the two PS-PEO (**SO1** and **SO2**) materials were nearly symmetric.

Small-angle X-ray scattering (SAXS) was used to determine the morphology of the parent materials. After annealing the samples at high temperature (120 °C for 20 min and then 60 °C for 20 min), SAXS data were acquired at 20 °C (see the Experimental section for details). The one-dimensional data (intensity *vs.* scattering wave vector q) for the block copolymers is shown in Fig. 2. From the position of the scattering peaks and consideration of the PS volume fractions we have assigned both **SL1** and **SL2** as hexagonally packed cylinders of PLA in a PS matrix and both **SO1** and **SO2** as alternating lamellae of PS and PEO. These assignments are consistent with the phase behavior of these block copolymers reported in the literature.[33,54] The principal domain spacing at 20 °C for the parent block copolymers is also given in Table 1.

Characterization of PS-PLA/PS-PEO binary blends

Block copolymers **SL1** and **SL2** were solution blended with **SO1** and **SO2**, respectively, to form a dozen optically transparent composites (see the Experimental section). The blend compositions were determined gravimetrically and agreed with ^{1}H NMR spectroscopic analysis. In all of the blends the volume fraction of PS was between 0.66 and 0.72. Provided the PLA and PEO homogenously mixed, all blends were likely to be composed of cylinders of mixed PLA/PEO in a matrix of PS. In addition to the optical clarity, SAXS analysis of the blends revealed a pronounced principal scattering peak and often higher order reflections consistent with microphase separation in these composites (see below). A complete compositional profile is given in Table 2.

One-dimensional SAXS patterns for a representative set of **SL1/SO1** blends are shown in Fig. 3 along with the parent material **SL1**. The principal spacing in these materials remained essentially constant at 28 nm in going from 0% **SO1** (*i.e.*, pure **SL1**) to 20.5% **SO1** (blend **4**). The complete set of data is given in Table 3. In each of the blends, scattering consistent with a hexagonally packed cylindrical morphology was observed. The other set of blends (**SL2/SO2**, blends **5–12**) gave similar results, although higher order reflections were more difficult to observe. To establish the cylindrical morphology in these blends, scanning electron microscopy (SEM) images were acquired on thin film samples. A thin film of blend **11** was cast from toluene on a Si substrate and stained with RuO_4 to provide domain contrast. The resultant SEM image is shown in Fig. 4. The dark worm-like features are likely the mixed PEO/PLA cylinders packed in a PS matrix. The average center-to-center distance for the cylinders was measured to be 49.2 nm from the SEM image, consistent with the corresponding value from SAXS of 51.8 nm.

To further understand the solid-state morphology of the blends, we obtained the wide-angle X-ray scattering (WAXS) and differential scanning calorimetry data (DSC) on the samples. While the WAXS data for **SO2** showed clear evidence of PEO crystallinity,[54,62] no PEO crystallinity was observed in the following blends: **1**, **3**, and **12** (samples tested). Even in the blend containing the highest fraction of PEO (**12**), no sharp peaks were observed by WAXS compared to the parent **SO2** (Fig. 5). The lack of a WAXS signature for PEO crystalline domains in the blends is consistent with suppression of crystallinity due to the miscibility of PEO and PLA.[43–53]

The lack of any significant melting endotherms in the DSC analysis of the blends **1**, **3**, **11**, and **12** also supports the absence of PEO crystallinity, again consistent with complete mixing of the PEO and PLA in the microdomains of these composites. A representative DSC trace for a **SL2/SO2** blend (blend **11**) is shown in Fig. 6. The glass transition temperatures (T_gs) for PEO and PS in the parent **SO2** were found to be −43 and 100 °C, respectively. Likewise, the T_gs for PLA and PS in the parent **SL2** were 52 and 105 °C, respectively. In blend **11**, two T_gs were observed at approximately 12 and 100 °C. These transitions are attributed to the T_g of the miscible PEO/PLA phase and the PS phase (containing PS from both the **SL** and the **SO** block copolymers), respectively. Using the weight fractions of PEO and PLA in blend **11**, the two parent T_gs, and the Fox equation for a miscible blend[63] we calculated a T_g of 20 °C for the mixed PEO/PLA phase, in rough agreement

Table 1 Characterization of the parent diblock copolymers

Samples	M_n/kg mol^{-1} PS[a]	M_n/kg mol^{-1} PLA (or PEO)[b]	M_n/kg mol^{-1} total	M_n/kg mol^{-1} total[c]	M_w/M_n[c]	f_{PLA} (or f_{PEO})[d]	D/nm[e]
SL1	21.7	11.1	32.8	33.7	1.09	0.300	27.8
SL2	41.1	19.4	60.5	61.4	1.11	0.284	43.9
SO1	15.5	14.0	29.5	30.4	1.08	0.452	24.6
SO2	29.8	25.7	55.5	57.1	1.09	0.440	40.7

[a] Determined by SEC *vs.* PS standards. These values are in agreement with the ^{1}H NMR end group analysis. [b] Determined by ^{1}H NMR spectroscopy. [c] Determined by SEC. [d] Calculated using the mass fraction of the minority component and the following densities at 140 °C PS = 0.969 g cm^{-3}, PLA = 1.154 g cm^{-3}, PEO = 1.064 g cm^{-3}. [e] Principal spacing from SAXS analysis at 20 °C.

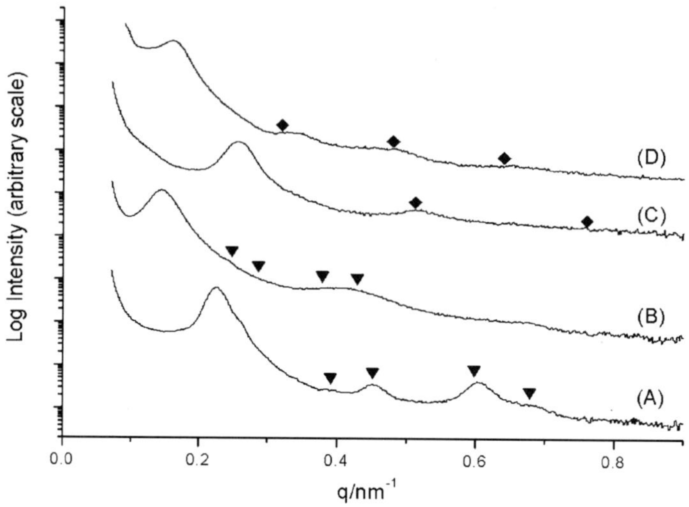

Fig. 2 1D SAXS profiles of the parent diblock copolymers: PS-PLA and PS-PEO (see Table 1 and the Experimental section). (A) **SL1** (B) **SL2** (C) **SO1** (D) **SO2**. The expected reflections for a cylindrical microstructure are marked by triangles ($\sqrt{3}:\sqrt{4}:\sqrt{7}:\sqrt{9}$) and the expected reflections for a lamellar microstructure are marked by diamonds ($\sqrt{4}:\sqrt{9}:\sqrt{16}$).

with the observed transition. All of the characterization data on the **SL**/**SO** blends strongly suggest that the samples consist of mixed and amorphous PLA/PEO cylinders embedded in an amorphous PS matrix.

Solid-state processing of the PS-PLA/PS-PEO blends

All of the blends were pressed in a channel die at 140 °C to orient the cylinders in the flow direction. This processing procedure was employed for PS-PLA materials and generally gave a high degree of cylinder orientation.[33] Two-dimensional SAXS data of the blends after channel die processing

Table 2 Characterization of the PS-PLA/PS-PEO blends

Blend[a]	f_{PS}[b]	SOx (wt.%)[c]	PEO (wt.%)[d]
1(SL1/SO1)	0.696	2.5	1.2
2(SL1/SO1)	0.689	7.6	3.6
3(SL1/SO1)	0.685	9.9	4.7
4(SL1/SO1)	0.669	20.5	9.8
5(SL2/SO2)	0.715	1.0	0.5
6(SL2/SO2)	0.712	2.5	1.2
7(SL2/SO2)	0.704	7.4	3.4
8(SL2/SO2)	0.697	12.3	5.7
9(SL2/SO2)	0.693	14.9	6.9
10(SL2/SO2)	0.689	17.3	8.0
11(SL2/SO2)	0.684	20.1	9.3
12(SL2/SO2)	0.676	25.3	11.7

[a] The block copolymers in parentheses are the parent materials used in the blend, see Table 1. [b] Volume fraction of PS was calculated using the following densities at 140 °C $\rho_{PS} = 0.969$ g cm^{-3}, $\rho_{PLA} = 1.154$ g cm^{-3}, and $\rho_{PEO} = 1.064$ g cm^{-3} (see the Experimental section). [c] Weight percent of **SO1** or **SO2** in the blend determined gravimetrically. [d] Weight percent of polyethylene oxide in the blend.

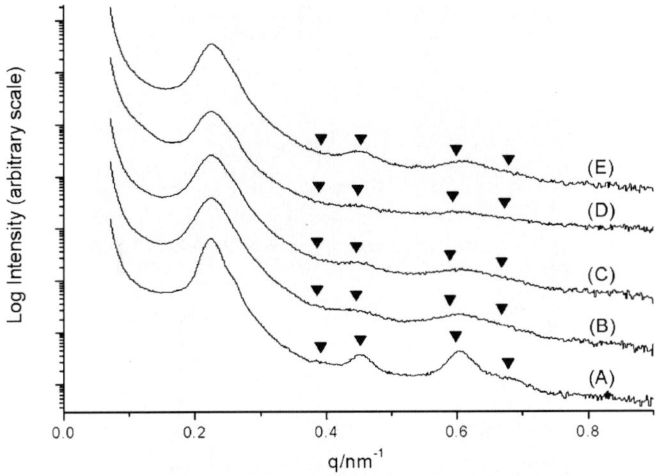

Fig. 3 1D SAXS profiles of the **SL1**/**SO1** blends (see Table 2 and the Experimental section). (A) **SL1** (B) blend **1** (C) blend **2** (D) blend **3** (E) blend **4**. The expected reflections for a cylindrical microstructure are marked by triangles ($\sqrt{3}:\sqrt{4}:\sqrt{7}:\sqrt{9}$).

showed two-spot patterns consistent with cylinder orientation in the flow direction. The degree of alignment in all of the blends was characterized by the second order orientation factor F_2.[64] The F_2 values were generally greater than 0.70 indicating a high degree of orientation. Small pieces of the oriented blends were subjected to a water/methanol sodium hydroxide solution to degrade the PLA for about one week at 65 °C. These monoliths were then washed with methanol and DI water, and dried overnight under reduced pressure at room temperature to afford polyether functionalized nanoporous polystyrene solids.

Table 3 Characterization of the nanoporous materials

Blend[a]	D_{pre}/nm[b]	D_{post}/nm[c]	d_{pore}/nm[d]	d_{pore}/nm[e]
1(SL1/SO1)	28.1	27.5	18.1	13.8 ± 0.9
2(SL1/SO1)	28.1	29.8	19.0	14.6 ± 2.2
3(SL1/SO1)	28.0	27.4	17.3	14.4 ± 1.7
4(SL1/SO1)	27.7	28.4	16.8	15.3 ± 3.0
5(SL2/SO2)	39.7	41.0	26.4	18.6 ± 4.0
6(SL2/SO2)	44.5	47.2	30.1	18.3 ± 3.8
7(SL2/SO2)	41.6	41.9	26.0	19.9 ± 5.1
8(SL2/SO2)	40.0	42.7	25.9	19.4 ± 3.8
9(SL2/SO2)	43.0	43.0	25.6	18.4 ± 3.7
10(SL2/SO2)	42.1	45.2	26.5	19.5 ± 4.7
11(SL2/SO2)	44.9	49.4	28.5	19.0 ± 4.2
12(SL2/SO2)	40.8	45.8	25.6	19.7 ± 4.4

[a] The block copolymers in parentheses are the parent materials used in the blend, see Table 1. [b] Principal domain spacing of the blends pre-degradation by SAXS. [c] Principal domain spacing of the blends post-degradation by SAXS. [d] Diameter of nanopores in degraded blends determined from the principle spacings of the degraded blends and the volume fraction of PS and PEO in the blends (see Table 1). [e] Nanopore diameters (± one standard deviation) as measured in the corresponding SEM images (see the Experimental section).

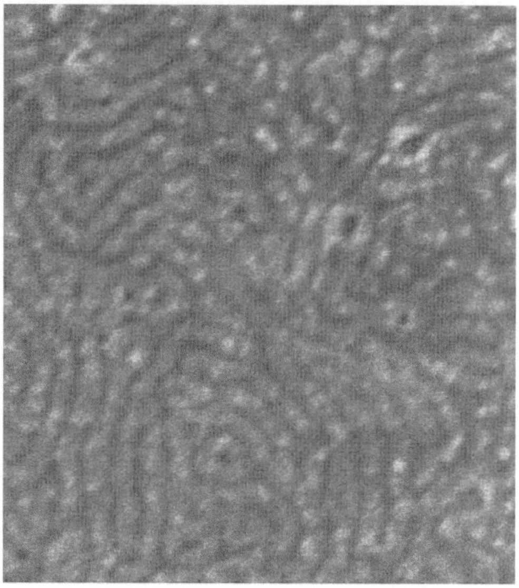

Fig. 4 SEM image of a thin film sample. Blend **11** was spin-cast from a 15 mg ml^{-1} toluene solution and then stained with RuO_4 for 5 min (the second largest PEO-containing blend). The SEM analysis was made on a high-resolution Hitachi S-900 FE-SEM using accelerating voltages of 1 keV.

Characterization of the degraded monoliths

The degraded monoliths could be readily dissolved in chloroform and analyzed by ^{1}H NMR spectroscopy. Fig. 7 shows a comparison of the ^{1}H NMR spectra for **SL1**, **SO1**, and blend **3** both before and after degradation. There are no resonances from the PLA component (methine protons at 5.2 ppm) in the degraded blend indicating complete removal of the PLA from the monolith. Also,

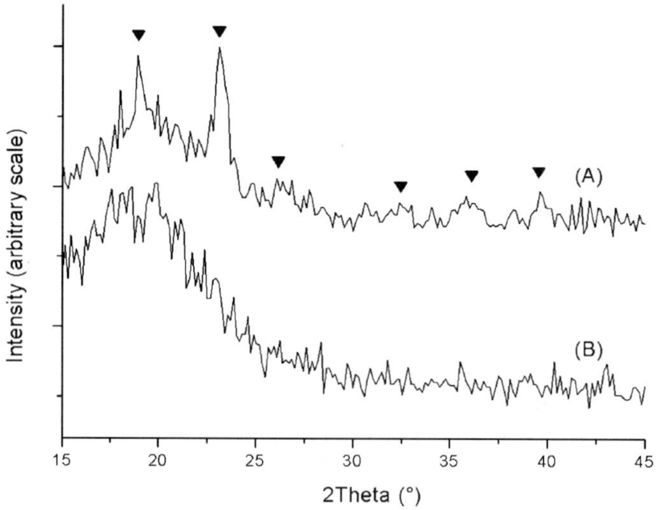

Fig. 5 WAXS data for (A) **SO2** and (B) blend **12**. Experiments were performed on a Siemens D5005 X-ray diffractometer with Cu Kα radiation ($\lambda = 1.542$ Å) and a 0.05° (2θ) scanning step.

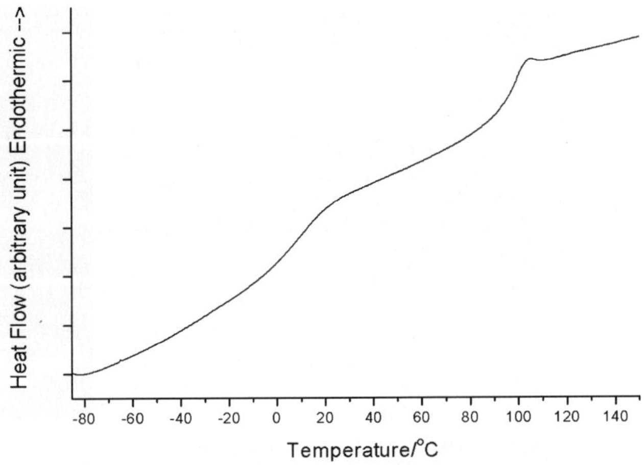

Fig. 6 DSC trace of blend **11** (10 °C min^{-1}, heating). The T_g for the mixed polyethylene oxide/polylactide phase is centered at 12 °C and the T_g for the polystyrene block is centered at 100 °C.

the degraded blend contains a distinct PEO component (methylene protons at 3.6 ppm). Integration of the PS resonances (between 6.3 and 7.2 ppm) and the PEO resonances, gave PEO compositions in the degraded blends that were generally consistent with the original blend composition data in Table 2.[65] Furthermore, IR spectroscopic analysis of the degraded monoliths confirmed both the absence of PLA and the presence of PEO in the degraded blends again corroborating the complete removal of only PLA from the monoliths.

SAXS analysis of the degraded blends showed a similar two spot pattern (in the shear gradient direction) as in the parent blends. A representative comparison of the one-dimensional and two-dimensional SAXS data between **SL1**, an **SL1**/**SO1** blend (blend **3**), and the corresponding degraded sample is shown in Fig. 8. The degraded sample and the parent blend have essentially the same principal spacing. While the higher order reflections are certainly less distinct in both the parent blend and the degraded sample compared to the parent **SL1**, the SAXS data support no

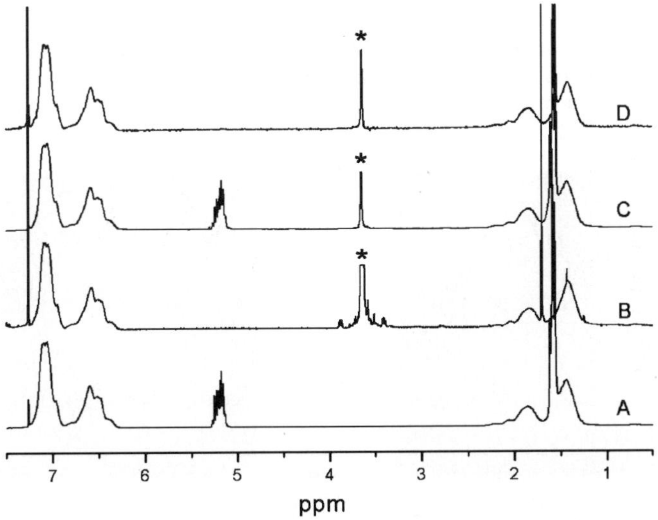

Fig. 7 ^{1}H NMR spectra for (A) **SL1**, (B) **SO1**, (C) blend **3**, and (D) degraded blend **3**. * PEO resonances were truncated for clarity.

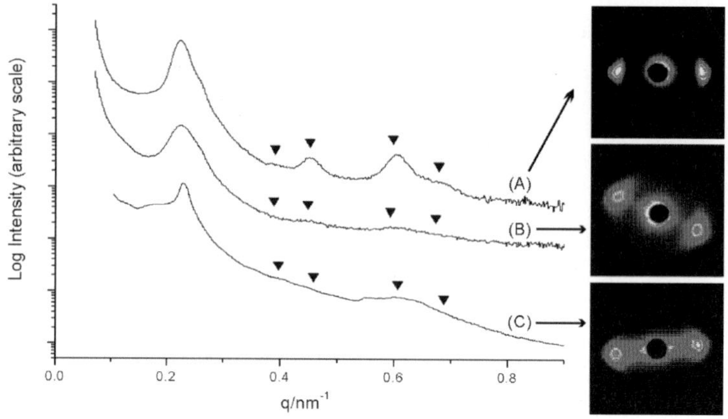

Fig. 8 1D SAXS profiles for (A) SL1, (B) blend **3**, and (C) degraded blend **3**. The expected reflections for a cylindrical microstructure are marked by triangles ($\sqrt{3}:\sqrt{4}:\sqrt{7}:\sqrt{9}$). The corresponding 2D SAXS patterns are also shown from the shear gradient direction (see the Experimental section).

significant change in ordered state symmetry, spacing, or microstructural alignment. In all of the degraded samples, the intensity of the principal scattering peak was significantly greater than the intensity pre-degradation, consistent with increased scattering contrast (see ref. 33).

The degraded monolithic samples were further characterized by SEM. Two representative images are shown in Fig. 9 (blends **3** and **11**). These images confirmed that the degraded monoliths consist of hexagonally-packed nanoscopic channels with a relatively narrow pore size distribution. The average pore sizes were measured using digital image analysis software, giving values in general agreement with those measured by SAXS experiments considering the estimated 2–3 nm coating of Pt on these samples (see the Experimental section). However, inspection clearly reveals defects present in the SEM images. This is reflected in the significant standard deviations in the average values given in Table 3.

DSC measurements were performed on selected degraded samples to confirm the presence of PEO in the nanoporous monoliths. Generally, samples with low PEO contents did not show any melting transitions, suggesting an amorphous PEO coating on the pore walls. In contrast, analysis of a degraded sample with high PEO loading (blend **12**) gave an endothermic peak centered around 50 °C. Upon integration of that peak a level of PEO crystallinity of approximately 14% was estimated.[39,66] Although the DSC data are consistent with PEO crystallinity in this degraded sample, no obvious diffraction peaks were identified by WAXS results, possibly due to the relatively small fraction of PEO in the composite. Taken together, the characterization data strongly support the successful preparation of ordered nanoporous PS monoliths with PEO-coated pore walls.

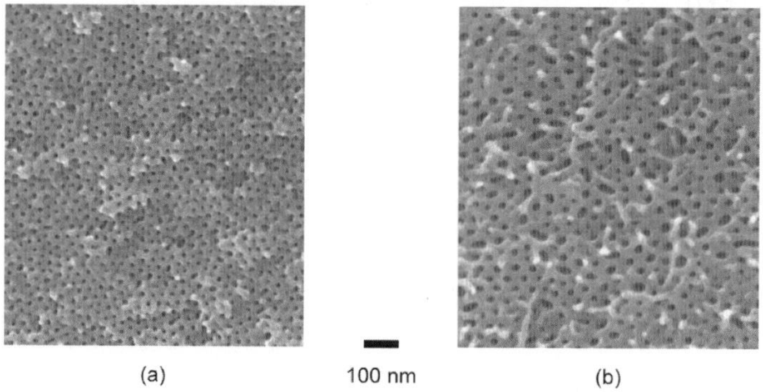

Fig. 9 SEM images of degraded blends (see the Experimental section). (a) A nanoporous monolith made from blend **3**, (b) A nanoporous monolith made from blend **11**.

Evaluation of pore-wall hydrophilicity

Amaranth, a red water-soluble dye, was used to demonstrate the significantly enhanced water compatibility of the PEO-coated nanoporous polystyrene. Standard degradation of the parent **SL2** gave a nanoporous material with no PEO incorporation. This material and a nanoporous monolith from the degradation of blend **12** were placed into an aqueous solution of amaranth (1.4 mg mL^{-1}). The nanoporous material from **SL2** (containing no PEO) floated on the surface of the water and imbibed none of the aqueous solution.[67] In contrast, the dye solution filled the voids in the degraded sample of blend **12** (containing less than 12 wt.% PEO) and the increased density of the material rendered it no longer buoyant.[68] After 5 days,[69] samples were taken out of the dye solution, washed rigorously with DI water, and then dried in a vacuum oven at room temperature overnight. The whole monolith and fractured samples containing PEO were dark red and the parent material was essentially colorless.

The water compatibility of the PEO lined nanoporous materials was also investigated as a function of PEO content in three degraded blends with essentially the same pore size (blends **1**, **3**, and **4**), through simple buoyancy experiments. The three PEO containing nanoporous materials and the parent nanoporous material from **SL1** were placed in four separate vials containing pure water. Consistent with all other experiments using non-PEO coated nanoporous materials, the nanoporous material derived from **SL1** floated on the surface of the water indefinitely. The degraded blends all sank to the bottom of the vessel after different times: **1** (62 h), **3** (14 h), and **4** (2.3 h). These times are correlated to the PEO content in each of the nanoporous materials; the time to sink (*i.e.*, when the density of the water-filled monolith is greater than 1.0 g $\cdot$ cm^{-3}) is inversely proportional to the PEO weight percent in these three blends. This phenomenon is currently under investigation.

Conclusions

We report a simple methodology for the preparation of ordered nanoporous PS using AB/AC diblock copolymer blends. The key components are a high T_g continuous matrix material (PS), a readily degradable minority component (PLA), and a material that is miscible with the sacrificial block, non-degradable, and water-compatible (PEO). We demonstrated that PS-PLA and PS-PEO meet these requirements, and corresponding binary blends of these two materials can be readily degraded to give a series of nanoporous materials with remarkable enhancements in water compatibility. The nanoporous plastics described in this manuscript will significantly expand applicability of ordered nanoporous materials derived from block copolymer precursors.

Experimental

Materials

All reagents and solvents were used as received unless otherwise noted. Two PS-PEO (**SO1**, **SO2**) and two PS-PLA (**SL1**, **SL2**) block copolymers were synthesized through two-step sequential polymerization methods previously reported.[33,60] These polymerization schemes employ the re-initiation of hydroxyl-terminated polystyrene for the subsequent growth of PEO (anionically *via* the potassium alkoxide) or PLA ("coordination–insertion" *via* triethyl aluminum catalyst) to form the final diblock copolymer.

Blend preparation

PS-PEO and PS-PLA blends were prepared through co-dissolution of the block copolymer samples in CH_2Cl_2 at room temperature, followed by evaporation of solvent at atmospheric pressure in air. Samples were subsequently dried *in vacuo* at 50 °C to ensure complete removal of solvent. All resulting films were colorless and transparent. Literature values for densities were used for PS,[70] PEO,[70] and PLA.[71]

General characterization methods

Size exclusion chromatography (SEC) data were collected using a Hewlett-Packard 1100 series liquid chromatograph equipped with a Hewlett-Packard 1047A refractive index detector and three

Jordi polydivinylbenzene columns (pore sizes of 10^4, 10^3, and 500 Å). Samples were run at 40 °C in HPLC grade THF at 1 mL min^{-1}. Polydispersity calculations were based on instrument calibration with polystyrene standards (Polymer Laboratories). ^1H NMR spectra were collected on a Varian Inova VI-500 spectrometer in $CDCl_3$ (Cambridge) at room temperature. Differential scanning calorimetry (DSC) measurements were performed on a Q1000 Autosampler (Thermal Analysis) calibrated with an Indium standard. Parent block copolymers and blends were equilibrated at 150 °C, above all thermal transitions. The degraded nanoporous monoliths were equilibrated at −100 °C, to maintain the integrity of the structure prior to the thermal scan. Samples were run at 10 °C min^{-1} under a nitrogen purge. IR spectra were acquired on a Nicolet Magna-IR spectrometer 550. Samples were crushed with dried KBr (1 : 10 by volume) using a mortar and pestle and then pressed into pellets with a screw-press.

X-ray scattering

Small-angle X-ray scattering (SAXS) measurements were performed at the University of Minnesota 2.3 m and 3.5 m SAXS lines. Cu Kα X-rays ($\lambda = 1.542$ Å) were generated by a Rigaku RU-200BVH rotating anode equipped with a 0.2 × 2 mm microfocus cathode and Franks mirror optics. Two-dimensional scattering data were recorded on a Siemens multi-wire area detector and corrected for detector response characteristics prior to analysis. Before degradation, samples were annealed 120 °C (20 min), 60 °C (20 min), and room temperature (10 min) *in situ* under vacuum, before a 10 min exposure at room temperature. The same procedure was used on the degraded monoliths, except they were not heated above 60 °C. Two-dimensional scattering images were azimuthally integrated to a one-dimensional plot of intensity *versus* scattering wavevector, q, where $q = (4\pi\lambda^{-1})\sin(\theta/2)$, and λ and θ are the radiation wavelength and scattering angle, respectively. Second-order orientation factors (F_2) were used to evaluate the long-range order in flow-field aligned samples.[64] ($F_2 = 1$ for perfectly aligned samples, $F_2 = 0$ for samples with no long-range order.) Wide-angle X-ray scattering (WAXS) was performed on a Siemens D5005 X-ray diffractometer with Cu Kα radiation ($\lambda = 1.542$ Å). Diffraction patterns were collected over 15–45° (2θ) at a scanning increment of 0.05° (2θ).

^1H NMR and IR spectroscopic characterization

The following ^1H NMR spectroscopy resonances are representative of polymeric samples reported in this study. All resonances are reported in ppm (δ) relative to $CHCl_3$ (s, 7.27 ppm), and are either broad (b) or contain multiple overlapping peaks (m). **PS-PLA**: 6.2–7.4 (m, aromatic protons), 5.2 (m, –CO–CH(CH_3)–O–), 1.7 (m, –CH(C_6H_5)–CH_2–), 1.6 (m, –CO–CH(CH_3)–O–). **PS-PEO**: 6.2–7.4 (m, aromatic protons), 3.6 (b, –O–CH_2–CH_2–O–), 1.7 (m, –CH(C_6H_5)–CH_2–). In the IR spectra the C=O stretch in PLA was observed near 1757 cm^{-1}. C–O and C–C stretching and CH_2 rocking modes present in PEO were observed near 1105 cm^{-1}.[72]

Microstructural alignment

Blends were aligned using a channel die (2 mm in width and 55 mm in length) as previously described for pure PS-PLA block copolymer samples.[33] Blends were pre-pressed into a stainless steel mold lined with two Teflon sheets at 140 °C. The resulting piece was placed vertically in the center of the channel die and pressed again at 140 °C. The PLA/PEO composite cylinders were aligned in the direction of polymer flow as confirmed by SAXS measurements and SEM observations.

Degradation of PLA in blend monoliths

Blend monoliths formed in the channel die were placed into a sodium hydroxide water/methanol (60 : 40 by volume) solution at 65 °C for about one week, the temperature deliberately held just above the T_g of PLA. The monoliths were then washed with methanol and DI water, and later dried under reduced pressure at room temperature overnight.

Scanning electron microscopy (SEM)

Visualization of the nanoporous structure in the nanoporous samples was achieved through SEM of samples both perpendicular and parallel to the original channel die flow field direction. Samples were cut with a razor blade at room temperature, mounted onto brass shims with colloidal graphite (Ted Pella), and sputter-coated with 2–3 nm of platinum (the thickness of platinum was estimated from a calculated deposition rate and experimental deposition time). Thin film samples of the undegraded blends were spin-cast from a solution (15–30 mg ml^{-1} toluene) onto a Si substrate at 2000 rpm using a Headway Research Inc. (Model 1-EC101D-R485) photo-resist spinner. Contrast in blends was achieved through selective PS staining with RuO_4 vapor (5 min exposure, 0.5% aqueous, Ted Pella) prior to SEM observation. All thin film samples were mounted onto brass shims with carbon tape and observed directly without metal coating. The SEM performed on a high-resolution Hitachi S-900 FE-SEM using accelerating voltages of 1 keV and 3 keV for thin film and degraded blends, respectively.

Acknowledgements

We thank the National Science Foundation (DMR-0094144) and the David and Lucile Packard Foundation for their support of this work. The work was also supported in part by the MRSEC program of the National Science Foundation (DMR-0212302). Dr. Joon-Hyung Lee is acknowledged for providing the two PS-PEO samples. We thank Roberto Olayo-Valles for helpful discussions and technical assistance.

References

1 A. Adedeji, T. Grünfelder, F. S. Bates and C. W. Macosko, *Polym. Eng. Sci.*, 1996, **36**, 1707.
2 K. Naito, H. Hieda, M. Sakurai, Y. Kamata and K. Asakawa, *IEEE Trans. Magn.*, 2002, **38**, 1949.
3 I. W. Hamley, *Angew. Chem., Int. Ed. Engl.*, 2003, **42**, 1692.
4 M. Lazzari and M. A. López-Quintela, *Adv. Mater.*, 2003, **15**, 1583.
5 I. W. Hamley, *Nanotechnology*, 2003, **14**, R39.
6 C. Park, J. Yoon and E. L. Thomas, *Polymer*, 2003, **44**, 6725.
7 M. R. Buchmeiser, *Angew. Chem., Int. Ed. Engl.*, 2001, **40**, 3795.
8 T. Thurn-Albrecht, J. Schotter, G. A. Kästle, N. Emley, T. Shibauchi, L. Krusin-Elbaum, K. Guarini, C. T. Black, M. T. Tuominen and T. P. Russell, *Science*, 2000, **290**, 2126.
9 T. Thurn-Albrecht, R. Steiner, J. DeRouchey, C. M. Stafford, E. Huang, M. Bal, M. Tuominen, C. J. Hawker and T. P. Russell, *Adv. Mater.*, 2000, **12**, 787.
10 G. Liu, J. Ding, T. Hashimoto, K. Kimishima, F. M. Winnik and S. Nigam, *Chem. Mater.*, 1999, **11**, 2233.
11 For an interesting related system see Y. Xu, W. Gu and D. L. Gin, *J. Am. Chem. Soc.*, 2004, **126**, 1616.
12 J.-M. Ha, J. H. Wolf, M. A. Hillmyer and M. D. Ward, *J. Am. Chem. Soc.*, 2004, **126**, 3382.
13 J.-S. Lee, A. Hirao and S. Nakahama, *Macromolecules*, 1988, **21**, 274.
14 P. Mansky, C. K. Harrison, P. M. Chaikin, R. A. Register and N. Yao, *Appl. Phys. Lett.*, 1996, **68**, 2586.
15 T. L. Morkved, M. Lu, A. M. Urbas, E. E. Ehrichs, H. M. Jaeger, P. Mansky and T. P. Russell, *Science*, 1996, **273**, 931.
16 T. Hashimoto, K. Tsutsumi and Y. Funaki, *Langmuir*, 1997, **13**, 6869.
17 V. Z.-H. Chan, J. Hoffman, V. Y. Lee, H. Latrou, A. Avgeropoulos, N. Hadjichristidis, R. D. Miller and E. L. Thomas, *Science*, 1999, **286**, 1716.
18 R. Mäki-Ontto, K. de Moel, W. de Odorico, J. Ruokolainen, M. Stamm, G. ten Brinke and O. Ikkala, *Adv. Mater.*, 2001, **13**, 117.
19 K. W. Guarini, C. T. Black and S. H. I. Yeung, *Adv. Mater.*, 2002, **14**, 1290.
20 O. Ikkala and G. ten Brinke, *Science*, 2002, **295**, 2407.
21 K. Shin, K. A. Leach, J. T. Goldbach, D. H. Kim, J. Y. Jho, M. Tuominen, C. J. Hawker and T. P. Russell, *Nano Lett.*, 2002, **2**, 933.
22 A. M. Urbas, M. Maldovan, P. DeRege and E. L. Thomas, *Adv. Mater.*, 2002, **14**, 1850.
23 U. Jeong, D. Y. Ryu, J. K. Kim, D. H. Kim, T. P. Russell and C. J. Hawker, *Adv. Mater.*, 2003, **15**, 1247.
24 A. Sidorenko, I. Tokarev, S. Minko and M. Stamm, *J. Am. Chem. Soc.*, 2003, **125**, 12211.
25 S. H. Kim, M. J. Misner, T. Xu, M. Kimura and T. P. Russell, *Adv. Mater.*, 2004, **16**, 226.
26 K. A. Cavicchi, A. S. Zalusky, M. A. Hillmyer and T. P. Lodge, *Macromol. Rapid Commun.*, 2004, **25**, 704.
27 S. B. Lee, D. T. Mitchell, L. Trofin, T. K. Nevanen, H. Söderlund and C. R. Martin, *Science*, 2002, **296**, 2198.
28 For a relevant example employing block copolymers see ref. 22.

29 J. Oh, Y. Tak and J. Lee, *Electrochem. Solid-State Lett.*, 2004, **7**, C27.
30 T. Ohgai, L. Gravier, X. Hoffer, M. Lindeberg, K. Hjort, R. Spohr and J.-P. Ansermet, *J. Phys. D: Appl. Phys.*, 2003, **36**, 3109.
31 K. Moller and T. Bein, *Chem. Mater.*, 1998, **10**, 2950.
32 I. W. Hamley, in *The Physics of Block Copolymers*, Oxford University Press, 1998, ch. 2.
33 A. S. Zalusky, R. Olayo-Valles, J. H. Wolf and M. A. Hillmyer, *J. Am. Chem. Soc.*, 2002, **124**, 12 761.
34 A. S. Zalusky, R. Olayo-Valles, C. J. Taylor and M. A. Hillmyer, *J. Am. Chem. Soc.*, 2001, **123**, 1519.
35 J. H. Wolf and M. A. Hillmyer, *Langmuir*, 2003, **19**, 6553.
36 M. W. Matsen, *Macromolecules*, 1995, **28**, 5765.
37 K. I. Winey, E. L. Thomas and L. J. Fetters, *Macromolecules*, 1991, **24**, 6182.
38 T. S. Bailey, H. D. Pham and F. S. Bates, *Macromolecules*, 2001, **34**, 6994.
39 T. S. Bailey, C. M. Hardy, T. H. Epps III and F. S. Bates, *Macromolecules*, 2002, **35**, 7007.
40 T. H. Epps III, E. W. Cochran, C. M. Hardy, T. S. Bailey, R. S. Waletzko and F. S. Bates, *Macromolecules*, accepted.
41 A. S. Zalusky, B. J. S. Johnson, S. Cho and M. A. Hillmyer, unpublished results.
42 G. Liu, J. Ding and S. Stewart, *Angew. Chem., Int. Ed. Engl.*, 1999, **38**, 835.
43 H. Younes and D. Cohn, *Eur. Polym. J.*, 1988, **24**, 765.
44 T. G. Park, S. Cohen and R. Langer, *Macromolecules*, 1992, **25**, 116.
45 A. J. Nijenhuis, E. Colstee, D. W. Grijpma and A. J. Pennings, *Polymer*, 1996, **37**, 5849.
46 M. Sheth, R. A. Kumar, V. Davé, R. A. Gross and S. P. Mccarthy, *J. Appl. Polym. Sci.*, 1997, **66**, 1495.
47 M. Baiardo, G. Frisoni, M. Scandola, M. Rimelen, D. Lips, K. Ruffieux and E. Wintermantel, *J. Appl. Polym. Sci.*, 2003, **90**, 1731.
48 Y. Hu, M. Rogunova, V. Topolkaraev, A. Hiltner and E. Baer, *Polymer*, 2003, **44**, 5701.
49 Y. Hu, Y. S. Hu, V. Topolkaraev, A. Hiltner and E. Baer, *Polymer*, 2003, **44**, 5711.
50 D. Kubies, F. Rypáček, J. Kovářová and F. Lednický, *Biomaterials*, 2000, **21**, 529.
51 W. Jiang and S. P. Schwendeman, *Pharm. Res.*, 2001, **18**, 878.
52 É. Kiss, I. Bertóti and E. I. Vargha-Butler, *J. Colloid Interface Sci.*, 2002, **245**, 91.
53 Y.-H. Na, Y. He, X. Shuai, Y. Kikkawa, Y. Doi and Y. Inoue, *Biomacromolecules*, 2002, **3**, 1179.
54 P. Huang, L. Zhu, S. Z. D. Cheng, Q. Ge, R. P. Quirk, E. L. Thomas, B. Lotz, B. S. Hsiao, L. Liu and F. Yeh, *Macromolecules*, 2001, **34**, 6649.
55 H. G. Jeon, S. D. Hudson, H. Ishida and S. D. Smith, *Macromolecules*, 1999, **32**, 1803.
56 K. Kimishima, H. Jinnai and T. Hashimoto, *Macromolecules*, 1999, **32**, 2585.
57 H. Frielinghaus, N. Hermsdorf, K. Almdal, K. Mortensen, L. Messé, L. Corvazier, J. P. A. Fairclough, A. J. Ryan, P. D. Olmsted and I. W. Hamley, *Europhys. Lett.*, 2001, **53**, 680.
58 H. Frielinghaus, N. Hermsdorf, R. Sigel, K. Almdal, K. Mortensen, I. W. Hamley, L. Messé, L. Corvazier, A. J. Ryan, D. van Dusschoten, M. Wilhelm, G. Floudas and G. Fytas, *Macromolecules*, 2001, **34**, 4907.
59 N. Y. Vaidya and C. D. Han, *Macromolecules*, 2000, **33**, 3009.
60 R. P. Quirk and J. J. Ma, *J. Polym. Sci. Part A: Polym. Chem.*, 1988, **26**, 2031.
61 H. Frielinghaus, W. B. Pedersen, P. S. Larsen, K. Almdal and K. Mortensen, *Macromolecules*, 2001, **34**, 1096.
62 L. Zhu, S. Z. D. Cheng, B. H. Calhoun, Q. Ge, R. P. Quirk, E. L. Thomas, B. S. Hsiao, F. Yeh and B. Lotz, *J. Am. Chem. Soc.*, 2000, **122**, 5957.
63 T. G. Fox, *Bull. Am. Phys. Soc.*, 1956, **1**, 123.
64 P. G. deGennes and J. Prost, *The Physics of Liquid Crystals*, Oxford University Press, New York, 1993.
65 A few of the **SL2**/**SO2** series blends showed some discrepancy between the PEO content in the original material and the degraded material. This is possibly due to inadvertent overheating during channel die alignment resulting in processing temperatures above the "normal" processing temperature of 140 °C described in the Experimental section. In separate experiments we determined that there was some minor decomposition (*i.e.*, chain cleavage) of the PEO in these blends after this errant heating. Furthermore, we did identify some PEO homopolymer contamination in the parent **SO2** material and this would be present in the blend as prepared, but not in the degraded blends. We were able to reproduce the data on these blends using the same materials, however we have since prepared new **SL** and **SO** materials for future experiments.
66 $\Delta H_f = 203$ J g^{-1}, see: B. Wunderlich, *Macromolecular Physics*, Academic Press, New York, 1980, vol. **3**, p. 67.
67 The estimated density of this nanoporous PS is approximately 0.6 g cm^{-3} (see ref. 32).
68 The density of pure PS at room temperature is about 1.05 g cm^{-3}.
69 The degraded sample of **SL2** remained buoyant for more than three months.
70 L. J. Fetters, D. J. Lohse, D. Richter, T. A. Witten and A. Zirkel, *Macromolecules*, 1994, **27**, 4639.
71 D. R. Witzke, R. Narayan and J. J. Kolstad, *Macromolecules*, 1997, **30**, 7075.
72 F. E. Bailey Jr. and J. V. Koleske, *Poly(ethylene oxide)*, Academic Press, Inc., New York, 1976, p. 115.

Vesicles made of PS-PI cyclic diblock copolymers: *In situ* freeze-drying cryo-TEM and dynamic light scattering experiments

Jean-Luc Putaux,[b] Edson Minatti,[c] Christelle Lefebvre,[a] Redouane Borsali,*[a] Michel Schappacher[a] and Alain Deffieux[a]

[a] *Laboratoire de Chimie des Polymères Organiques, LCPO-CNRS, UMR 5629, ENSCPB and Bordeaux University I, 16 Avenue Pey Berland, F-33600, Pessac, France. E-mail: borsali@enscpb.fr*
[b] *Centre de Recherches sur les Macromolécules Végétales, CNRS, affiliated with the Joseph Fourier University of Grenoble, BP 53, F-38041, Grenoble Cedex 9, France*
[c] *Departamento de Química da Universidade Federal de Santa Catarina, Florianópolis SC, 88040-900, Brazil*

Received 3rd March 2004, Accepted 9th March 2004
First published as an Advance Article on the web 11th August 2004

We have studied the morphology of self-assembled micelles made of linear and cyclic poly(styrene-*b*-isoprene) PS-*b*-PI block copolymers dispersed in selective solvents of the PI block (n-heptane, n-decane). Up to a copolymer concentration of 5 mg mL^{-1}, the micelles made from linear block copolymer chains adopt a spherical shape. Those arising from cyclic copolymer chains having exactly the same molar mass and volume fraction self-assemble into (i) planar sunflower-shaped particles at low concentration ($c < 0.1$ mg mL^{-1}), (ii) giant wormlike micelles at intermediate concentration (0.1 mg mL^{-1} $< c < 2$ mg mL^{-1}) and (iii) vesicles at higher concentration (2 mg mL^{-1} $< c$ < 5 mg mL^{-1}). Those results were obtained using dynamic light scattering and *in situ* freeze-drying cryo-transmission electron microscopy. In this contribution, we discuss the effects of concentration and temperature on the morphology of the self-assembled particles made from both linear and cyclic PS–PI copolymers, and highlight the surprising vesicle formation in cyclic block copolymer solutions.

1. Introduction

A substantial challenge in modern colloid chemistry is to find ways to control the size and morphology of colloids at the nanometric scale. In this context, the aggregates or self-assembled structures made from block copolymers have proven to be valuable tools. Micelles, microemulsions, vesicles and a huge variety of microphase textures may be used as structural templates for new materials. Diblock and triblock copolymers can form micelles when dissolved in a selective solvent (in which only one of the blocks is soluble). The increasing number of industrial applications for these systems attracted a large interest in both applied science and basic research.[1–8] For instance, copolymer micelles have been used as drug carriers in delivery systems,[9,10] stabilizers in organic reactions[11] as well as nanotemplates for nanotechnologies.[7,8]

Most of the published work deals with amphiphilic block copolymers, where the micelles are formed in water because one of the blocks is hydrophilic and the other hydrophobic. The size and thereby the aggregation number of those micelles are rather independent of the polymer concentration, but may change with the temperature or the copolymer composition (or volume fraction).[1,12] They usually have a spherical shape, with a very compact core and a more flexible corona. However, it was also shown in numerous studies that block copolymers can self-assemble into a large variety of objects such as rod-like[1] or wormlike[6] micelles, cylinders,[5] vesicles,[11–13] hollow spheres,[7] branched tubules,[9] *etc.* Commercial applications of those different nanostructures call for a high degree of control on the morphology of the resulting materials. As has been demonstrated by Eisenberg and coworkers,[12–16] as well as by Bates and coworkers[17–22] during the past decade, several parameters can be modified and controlled in order to tailor the micellar morphology: temperature, copolymer composition or concentration and sample preparation.

Micelles made from diblock copolymers can also be prepared in organic solvents.[23–29] In this case, the copolymer may have chemically distinct hydrophobic blocks. The most common block copolymers are constituted by an aromatic block attached to an aliphatic one. Copolymers consisting of polystyrene (PS) and polyisoprene (PI) blocks, as well as the corresponding self-assembled micelles are well documented in the literature.[2,23–26] Micelles made of highly asymmetric PS–PI copolymer chains are usually spherical.[2] The size and shape of the individual micelles do not change when the concentration of the copolymer within the micellar disordered regime increases. The core can be very compact and, depending on the solvent quality (or the insolubility of the inner block), these micelles can be 'frozen', and not exchange any unimer,[10] or mobile,[30] thus implying that the PS core is swollen.

The copolymer architecture plays a very important role in the micellar morphology.[31] The case of cyclic copolymers has recently been investigated.[29,31–36] In previous works,[34–36] we have shown that the micelles made from asymmetric linear PS–PI (at a PS volume fraction of $\Phi_{PS} = 0.78$) adopt a 'classical' spherical shape and keep the same morphology over the range of considered concentrations. In contrast, the micelles formed from asymmetric cyclic copolymer chains exhibit different sizes and shapes. They evolve from small individual 'sunflower' micelles at low concentration to giant wormlike micelles at higher concentration. In those studies, we used dynamic light scattering (DLS), small-angle X-ray scattering (SAXS), transmission electron microscopy (TEM) and atomic force microscopy (AFM) to characterize the micellar morphology of both linear and cyclic copolymers systems having the same degree of polymerization (DP), *i.e.* 290 for PS and 110 for PI (corresponding to a volume fraction $\Phi_{PS} = 0.78$) and dispersed in heptane.

In the present paper, we shall describe the effect of concentration in both systems using DLS in conjunction with *in situ* freeze-drying cryo-TEM in two selective solvents, namely n-heptane and n-decane. In particular, we document a new result: when dispersed in heptane at relatively high concentration (2–5 mg mL^{-1}), the cyclic PS–PI copolymer chains form vesicles whereas the analogous linear block copolymer chains self-assemble into spherical micelles.

2. Experimental section

2.1. Sample preparation

Linear diblock copolymers ($DP_{PS} = 290$ and $DP_{PI} = 110$) with a total molecular weight of 37×10^3 g mol^{-1} (determined by SEC-GPC analysis) were prepared using living anionic polymerization, as described elsewhere.[37] Cyclic poly(styrene-isoprene) was prepared by direct coupling of α-isopropylidene-1,1-dihydroxymethyl-ω-diethylacetal-heterodifunctional linear poly(styrene-*b*-isoprene). The fact that the cyclic copolymers are prepared by closure of the linear diblocks ensures that PS and PI blocks have exactly the same DP in linear and cyclic copolymers. Cyclic A–B copolymers have been also obtained by closing A–B–A triblock chains, as has been described by Hogen-Esch *et al.*[38,39] or, more recently, by Hadjichristidis *et al.*[29]

Copolymer solutions in heptane and decane were prepared according to two methods. The 'direct dissolution' method consisted of preparing solutions by directly dissolving proper amounts of copolymer to achieve predetermined concentrations. The solutions were continuously stirred for

72 h at 50 °C, under inert atmosphere. In the 'dilution' method, a mother solution was prepared at 5.0 mg mL^{-1} in the same conditions and subsequently diluted with pure solvent in order to achieve a range of lower concentrations. All solvents were HPLC grade. They were distilled and filtered through a 0.2 μm PTFE membrane prior use.

In the following, micelles made of the linear PS–PI block copolymer will be referred to as 'L-mic' whereas those made of cyclic copolymers will be called 'C-mic'.

2.2. Dynamic light scattering

The DLS measurements[40] were performed using an ALV Laser Goniometer, which consists of a cylindrical 22 mW HeNe linear polarized laser with a 632.8 nm wavelength and an ALV-5000/EPP Multiple Tau Digital Correlator with a 125 ns initial sampling time. The samples were kept at a constant temperature of 25.0 °C throughout the experiments. The accessible scattering angle of this equipment ranges from 12° up to 155°. All samples were systematically studied at 90° and some of them were studied at different scattering angles. The solutions were put in ordinary glass cells, 10 mm in diameter. The minimum sample volume required for an experiment was 1 mL. The data acquisition was done with the ALV-Correlator Control Software and the counting time varied for each sample from 300 s up to 15 min, depending on the sample, its concentration and the required time to build up fairly good correlation functions.

The autocorrelation functions of the scattered intensity were analyzed by means of the cumulant method to yield the effective diffusion coefficient as a function of the scattering angle. The analysis of the cumulant expansion of the correlation function is performed by fitting a polynomial up to third order to the function $\ln[g_2(t)-1]$. The polynomial coefficients are converted into the coefficients of the cumulant expansion of the field correlation function:

$$\ln(g_2(t)-1) = -\bar{\Gamma}t + \frac{\mu^2}{2!}t^2 - \frac{\mu^3}{3!}t^3 \tag{1}$$

The mean apparent hydrodynamic radius R_h is calculated from the time of 1st to 3rd order fitting according to the Stokes–Einstein equation.

$$R_h = \frac{kT}{6\pi\eta\bar{\Gamma}}q^2 \tag{2}$$

The polydispersity index I_p was calculated from the second moment μ_2 of 2nd and 3rd order fitting:

$$I_p = \frac{\mu^2}{\bar{\Gamma}^2} \tag{3}$$

In addition, the constrained regularization method CONTIN[41] was used to obtain the distribution $A(t)$ of decay times. For the latter, a statistical parameter 'probability to reject' is calculated for each solution, and the suggested one is that closest to 0.5.

$$\left(\frac{G^{(2)}(q,t)}{B}-1\right)^{1/2} = \int_0^{\infty} A(\tau)e^{-(t/\tau)}\mathrm{d}\tau \tag{4}$$

These methods are now routinely used to analyze the DLS data for polymer and colloidal systems and allow the determination of the relaxation modes which characterize the dynamics.

2.3. *In situ* freeze drying cryo-TEM

TEM observations were performed using the same solutions as those used for the DLS experiments. However, they generally had to be diluted with pure solvent to achieve concentrations compatible with cryo-TEM sample preparation. Typically, copolymer solutions prepared in heptane or in decane at concentrations of 0.1 mg mL^{-1} and 0.5 mg mL^{-1} were diluted 10× and those prepared at 2 mg mL^{-1} and 5 mg mL^{-1} were diluted 100×. In recent years, cryo-TEM has been successfully used to observe various polymer nanoparticles dispersed in water[42–45] and, in very few cases, in organic solvents.[46] According to the cryo-TEM procedure,[47] our specimens were prepared by quenching thin films of the suspensions formed on NetMesh (Pelco, USA) 'lacey' carbon films, using a guillotine-type device. However, as most organic solvents are known to dissolve in liquid ethane, fast-freezing was performed in liquid nitrogen. The specimens were then mounted onto a Gatan 626 specimen holder cooled with liquid nitrogen, transferred into the microscope and observed at −180 °C. As the freezing speed was not sufficiently high in liquid nitrogen, heptane appeared as crystalline lamellae and diffraction effects dramatically affected the contrast of the embedded objects. Therefore, the crystalline solvent was freeze-dried in the microscope by warming the sample to −115 °C. Around this temperature, frozen heptane completely sublimated and many embedded objects were revealed,[34–36] the smaller ones clinging to the edges of the supporting 'lacey' carbon membrane while the larger ones entangled across the holes in the film. Samples prepared from copolymers solutions in decane were warmed to about −70 °C. The heating was immediately stopped after evaporation and micrographs were recorded after a return to complete sample stability, at a temperature of −180 °C. All samples were observed under low dose conditions, using a Philips CM200 'Cryo' microscope operated at 80 kV. The micrographs were recorded on Kodak S0163 films.

3. Results and discussion

3.1. Dynamic light scattering

The DLS autocorrelation functions were measured for self-assembled L-mic and C-mic particles dispersed in PI-selective solvents (heptane or decane) at several concentrations, using the direct dissolution method. Typical results are illustrated in Figs. 1a and b. One observes that the autocorrelation functions measured for L-mic are mono-exponential (single relaxation time) and do not depend on the concentration from $c = 0.5$ mg mL^{-1} to 5.0 mg mL^{-1} (Fig. 1a). In contrast, those measured for C-mic present a multimodal distribution of relaxation times, as has been already reported.[34] In addition, the relaxation times deduced from the cumulant analysis increase with concentration. The variation of the corresponding diffusion coefficient D, calculated from $(\Gamma/q^2)_{q\to 0}$, as a function of the concentration is plotted in Fig. 2. In the case of linear block copolymer, the diffusion coefficient $D(c)$ and therefore the size are obviously independent of c while they rapidly decrease with concentration in the case of C-mic with a negative value of k_D in $D(c) = D_o(c)(1 + k_D c)$. This result indicates that the hydrodynamic size of C-mic particles is an increasing function of the concentration ($R_H \approx D^{-1}$) whereas that of L-mic is roughly constant.

Using the Stokes–Einstein relation (see eqn. (2)), R_H can be extracted from the relaxation time of frequency of the DLS autocorrelation functions. The hydrodynamic diameters $D_H = 2R_H$ of L-mic and C-mic are presented in Table 1 as a function of concentration. One observes that the size of C-mic strongly depends on the copolymer concentration, while this parameter is roughly constant for L-mic in the same conditions. Accordingly, L-mic have roughly a constant size of about 50 nm in both heptane and decane for all concentrations (0.01 up to 5 mg mL^{-1}) and temperature (see further in the discussion below). The dense core contains PS blocks while the more diffuse corona is made of PI. As one may expect, the excluded volume repulsion between the PI shells prevents an increase of the micelle size and at the same time promotes an increase in the number of micelles. The difference in size in heptane and in decane systems is directly related to the solubility of the inner PS block that decreases with the number of carbon atoms in the alkane chain used as a selective solvent. Indeed, as it has been shown by Quintana *et al.*[48,49] n-alkanes become worse solvents for polystyrene as the n-alkane carbon number increases. This behavior is remarkable for n-heptane and suggested less repulsive polystyrene/solvent interactions than higher n-alkanes. For that reason a slightly larger size of the micelles in heptane as compared to decane is systematically observed (see Table 1).

However, as seen in Table 1, the cyclization of the linear PS–PI leads to a different behavior: C-mic are larger in heptane than in decane and their size rapidly increases with the concentration. One may explain this result as follows: C-mic are not spherical but elongated objects where the PS core is more exposed or available for potential PS–PS interactions/attractions. Therefore, increasing the polymer concentration increases the probability that two particles interact by strong core–core (PS–PS) interactions. This explains the rapid increase of the C-mic size with the concentration: from 96.6 nm to 644.5 nm in heptane and from 38.2 nm to 180.0 nm in decane. The difference in size in

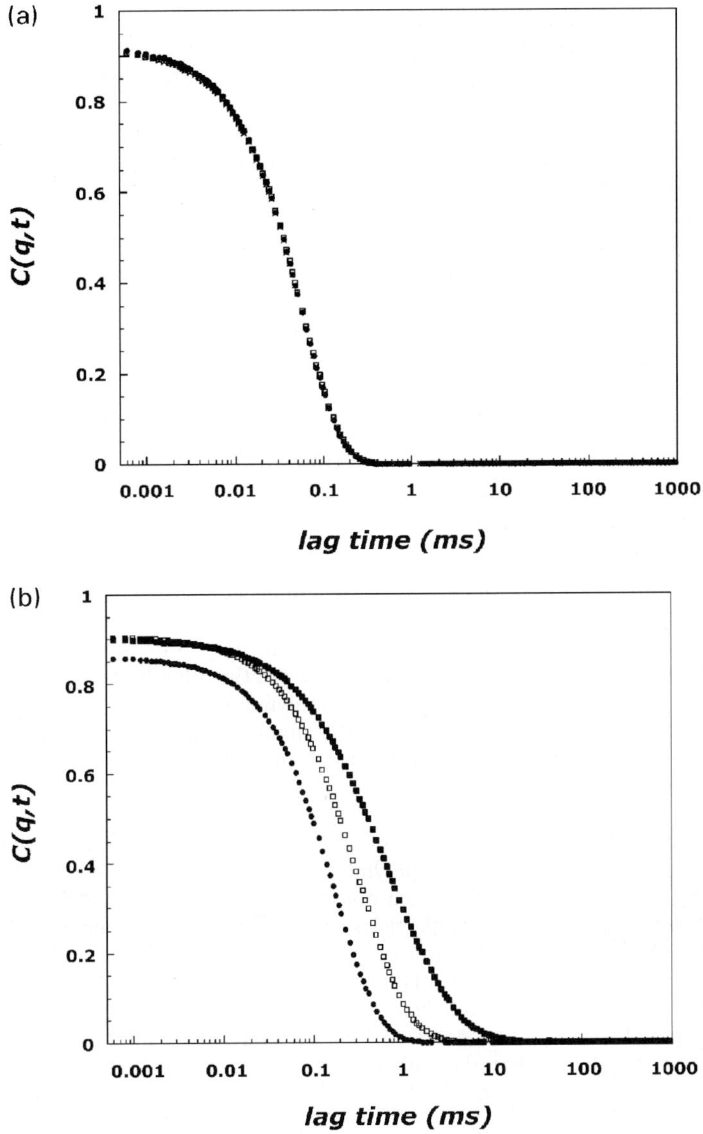

Fig. 1 DLS correlation functions measured at 25 °C: (a) micelles made of linear PS–PI at 0.5 mg mL^{-1} (open squares), 2 mg mL$_{-1}$ (filled circles) and 5 mg mL^{-1} (crosses) in heptane; (b) micelles made of cyclic PS–PI at 0.5 mg mL$_{-1}$ (filled circles), 2 mg mL^{-1} (open squares) and 5 mg mL^{-1} (filled squares) in heptane. The solutions were prepared using the direct dissolution method.

heptane and decane for cyclic systems could also be explained on the basis of the solubility of the inner PS block that decreases with the number of carbon atoms in the alkane chain used as a selective solvent. PS is less soluble in decane than in heptane leading to a more compact morphology in decane than in heptane and therefore a smaller size. For that reason, the dependence of C-mic size on copolymer concentration is less pronounced in decane than in heptane. At concentrations lower than 1 mg ml^{-1}, however, the size of C-mic is smaller than L-mic in decane. This result is still intriguing.

The concentration is not the only parameter that affects the size of C-mic. By decreasing the temperature to 10 °C, we observed an important increase in the size for C-mic, while that of L-mic did not depend on temperature. This is illustrated in Fig. 3, where we show the effect of

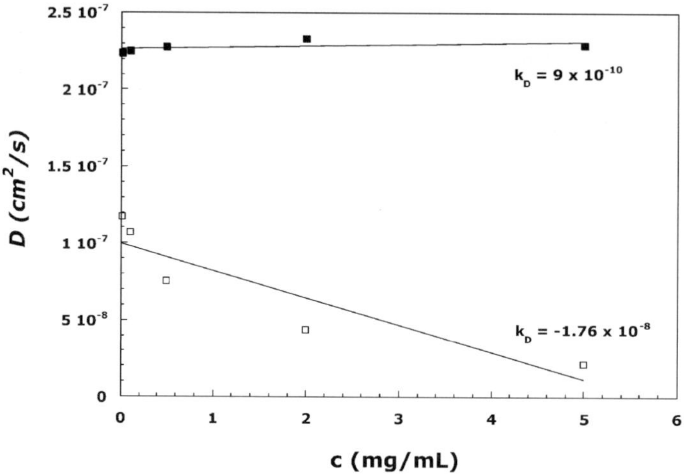

Fig. 2 Diffusion coefficient for cyclic PS–PI (open squares) and linear PS–PI (filled squares) as a function of copolymer concentration in heptane.

concentration on the R_H of L-mic and C-mic formed in decane at two different temperatures. It is important to note that the size of C-mic does not change with temperature at low copolymer concentration, but drastically increases at higher concentration. For instance, at $c = 5$ mg mL^{-1}, $R_H = 90$ nm and 173 nm at 25 °C and 10 °C respectively. On the other hand, Table 1 and Fig. 3 show that the L-mic size does not depend either on the copolymer concentration or on the temperature.

C-mic exhibit another distinct feature: in DLS, the angular dependency is more pronounced than in the case of L-mic. While the relaxation times for L-mic do not depend on the scattering angle, those measured for C-mic strongly depend on the scattering angle. This behavior is illustrated on Figs. 4a and b for linear and cyclic cases respectively where the distribution of the hydrodynamic radius is given using Contin analysis. This polydispersity in size and certainly in morphology becomes progressively important as the copolymer concentration increases. These results confirm that: (i) C-mic are not spherical but elongated objects—in this case, the angular dependency could be driven by the anisotropy of these nanoparticles coupled with their internal motion; (ii) C-mic are polydisperse since the relative contribution of smaller and bigger objects to the total scattering strongly depends on the scattering angle. This would also contribute to the origin of the angular dependency of the relaxation times and therefore to the deduced R_H values. The second result is supported by the polydispersity index deduced from the 2nd order cumulant fit of the autocorrelation functions. Fig. 5a shows that the polydispersity of C-mic increases as a function of the copolymer concentration, while it is constant for L-mic. In other words, while L-mic have a

Table 1 Hydrodynamic diameter calculated from DLS correlation functions 2nd order cumulant fit analysis of micelles made of linear and cyclic PS-PI

	Linear PS-PI L-mic hydrodynamic diameter/nm[a]		Cyclic PS-PI C-mic hydrodynamic diameter/nm[a]	
c/mg mL^{-1}	Decane	Heptane	Decane	Heptane
0.01	51.0	51.3	38.2	96.6
0.10	50.4	52.1	40.1	106.3
0.50	50.4	50.7	45.1	150.3
1.00	50.2	50.8	51.5	180.5
2.00	48.0	49.6	70.3	276.8
3.00	48.6	49.5	85.9	412.3
5.00	48.9	49.8	180.0	644.5

[a] Calculated from the diffusion coefficient at 25 °C.

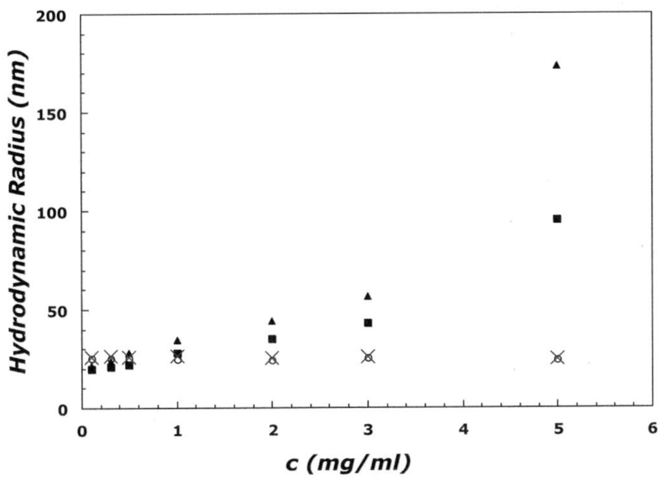

Fig. 3 Hydrodynamic radius as a function of copolymer concentration in decane for cyclic PS–PI at 10 °C (filled triangles), cyclic PS–PI at 25 °C (filled squares), linear PS–PI at 10 °C (empty circles) and linear PS–PI at 25 °C (crosses); scattering angle: 90°.

constant and monodisperse size in all the concentration ranges investigated, C-mic are growing in size, changing their morphology and becoming more polydisperse as the copolymer concentration increases. This is also illustrated in Fig. 5b.

Fig. 6 shows DLS autocorrelation functions for 3 sets of C-mic solutions prepared by both the direct dissolution and dilution methods (described in the Experimental Section). These samples, having the same final copolymer chain concentration, have distinct scattering properties. The micelles in samples prepared by dilution have the same size as in the mother solution, suggesting that the morphology of the micelles, once formed, is kept. As shown in the following section, these results were confirmed by TEM. Under such circumstances, it is worth noting that the only way to study the effect of concentration on C-mic is to prepare each sample by direct dissolution since the dilution would affect the number of particles and not their morphology. Interestingly, low concentration solutions obtained by the dilution method proved to be very useful in studying the shape of individual 'basic' micelles.

At higher concentrations (2–5 mg mL^{-1}) and particularly for C-mic in heptane, both elastic and quasi-elastic light scattering results indicated the presence of multimodal distributions in size and relaxation times. These features are typical of aggregate formation. Therefore, a series of direct observations using *in situ* freeze-drying cryo-TEM was carried out in order to shed light on the structure of the self-assembled aggregates. Some results obtained using AFM[35] and TEM[34,36] have already been reported.

3.2. *In situ* freeze-drying cryo-TEM

3.2.1. Solutions in heptane. Cryo-TEM observations were performed to visualize and compare the morphology of L-mic and C-mic. As described in the Experimental Section, solutions prepared by direct dissolution were subsequently diluted with pure solvent in order to achieve concentrations that were compatible with cryo-TEM. DLS experiments have shown that the morphology and size of the micelles were unaffected by the dilution (Fig. 6). As a consequence, the objects observed by cryo-TEM should be similar to those present in the mother solution prepared by direct dissolution. This was also supported by AFM imaging results.[35]

Fig. 7 shows micelles resulting from linear block copolymer chains dispersed in heptane. The micelles are seen spread on the surface of an extended area of carbon film after *in situ* freeze-drying of the embedding heptane. The observation does not provide information on the dimension parallel to the electron beam, *i.e.* on the thickness of the micelles spread on the carbon. However, the examination of many micelles clinging to the edges of the holes in the film, where there is no reason

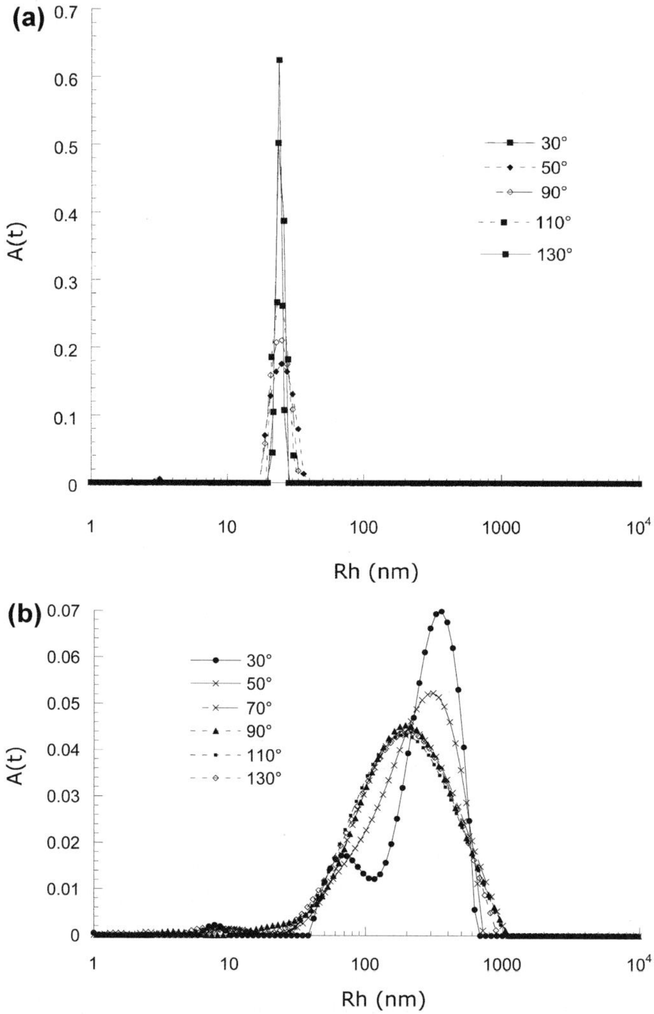

Fig. 4 CONTIN analysis from the DLS correlation functions of PS_{290}–PI_{110} at 5 mg mL^{-1} in heptane at different angles for (a) linear copolymer; (b) cyclic copolymer.

for the objects to adopt a preferential orientation, convinced us that the micelles were single and monodisperse spherical objects, with a diameter of about 35 nm. Their size and morphology are independent of the initial micellar concentration [(a) 0.5 mg mL^{-1}; (b) 5 mg mL^{-1}]. The slight difference in size observed between cryo-TEM and DLS data, as well as the reason why the PI corona could not be clearly resolved in TEM images, have been discussed in previous papers.[34–36] They were attributed to the low volume fraction of PI ($\Phi_{PI} = 0.22$) and probable shrinkage of the PI corona during the freeze-drying of the embedding solvent.

The morphologies developed by C-mic are readily illustrated in Fig. 8. They are clearly different from those observed for L-mic. In Figs. 8a and b, one can see a mixture of different objects, either spread on the surface of an extended carbon area (Fig. 8a) or clinging to the lacey carbon network (Figs. 8b): seemingly spherical particles about 20 nm in diameter, 70–100 nm-long rodlike and 250–300 nm-long wormlike objects, as well as toric micelles with a circumference of 200–300 nm. The last two classes of objects have a similar cross-sectional width, 20 nm, which is also the diameter of the individual particles. In these images, the darker parts on the micelles may correspond to thicker sections that point towards the viewer. During freeze-drying, the temperature did not exceed

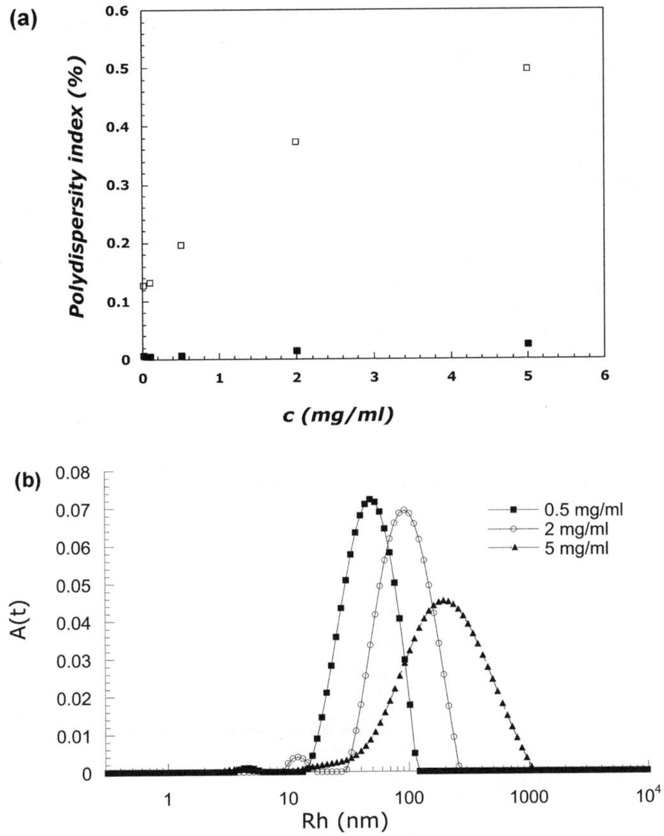

Fig. 5 (a) Polydispersity index calculated from the second moment μ_2 of 2nd order cumulant analysis according to relations (1)–(3) (see text) for cyclic PS–PI (open squares) and linear PS–PI (filled squares) solutions in heptane; (b) CONTIN analysis from the DLS correlation functions of cyclic PS_{290}–PI_{110} in heptane (angle 90°) at 0.5 mg mL^{-1} (filled squares), 2 mg mL^{-1} (open circles) and at 5 mg mL^{-1} (filled triangles).

−115 °C, which is much lower than the respective glass transition temperatures (T_g) of PS and PI. It is unlikely that the micelles could extensively deform on the supporting carbon membrane during freeze-drying. Thus, they certainly retained the 3D conformation they had in the thin liquid film of embedding heptane.

The circular 20 nm objects seen in Fig. 8a may correspond to individual sunflower micelles with a core formed by the PS half-loops and the corona by the PI half-loops. The smaller diameter of these micelles compared to that of the spherical L-mic is explained by the cyclic conformation of the copolymer chain, the looping of each block leading to shorter cores and coronas (see Scheme 1). Two opposing constraints control the shape of individual C-mic at low concentration in a PI-selective solvent: (i) the more diffuse PI corona 'protects' the dense PS core, favoring a spherical distribution of the molecules; (ii) the cyclic conformation of the PI loops extending out in the solvent promotes a lenticular sunflower-like arrangement. When the concentration increases, the sunflower-like C-mic pack unidirectionally to form more favorable cylindrical wormlike micelles.[34–36] The PS core is more efficiently protected, the effect of the opposing constraints being mostly felt at the ends of the wormlike micelles.

The wormlike C-mic formed in 0.5 mg mL^{-1} solutions appear to be tortuous, entangled and substantially longer (>1 μm) than those observed in 0.1 mg mL^{-1} solutions (Fig. 8c). However, they still have a similar and regular cross-sectional diameter of 20 nm. Looking carefully at different TEM images (Fig. 8c and d), so-called 'Y-junctions' can also be seen. The existence of such Y-junctions has been predicted[50,51] and recently observed in surfactant/oil/water[22] and PB–

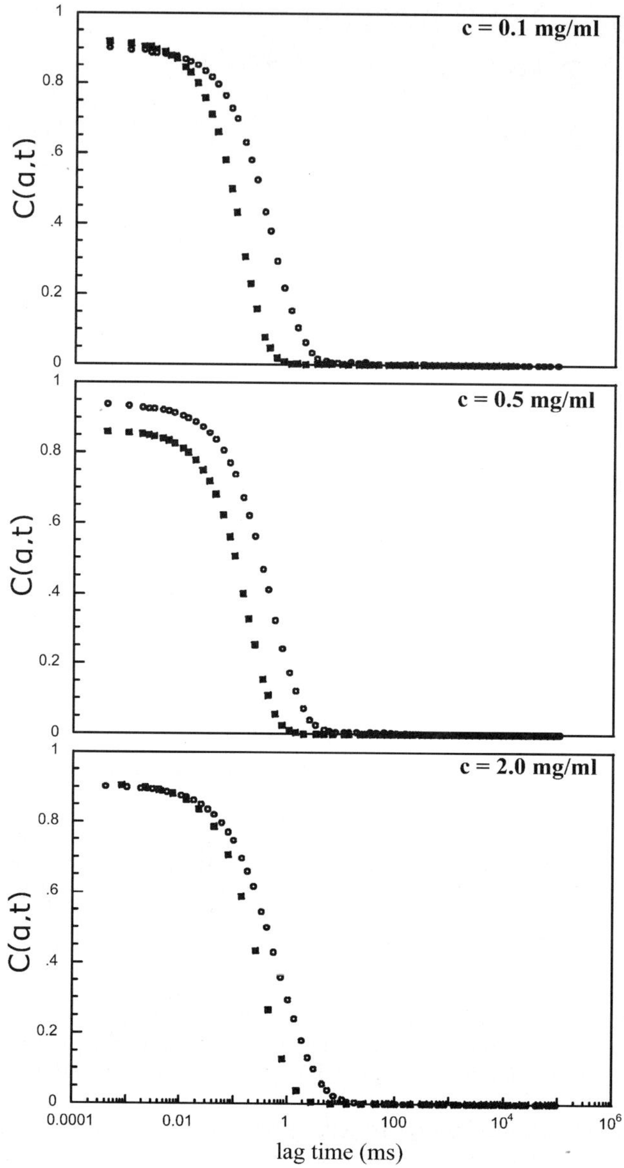

Fig. 6 DLS autocorrelation functions of micelles made from cyclic PS–PI in heptane at three different concentrations. We illustrate here the two methods of sample preparation: direct dissolution of the required amount of copolymer in heptane (filled squares) and dilution from a 5 mg mL^{-1} stock solution (open circles).

PEO/water[52,53] systems. It has been demonstrated that these Y-junctions were stable above a critical molecular weight. In the 2 mg mL^{-1} preparation, among a majority of endless wormlike micelles, new circular objects are observed (Fig. 8d) with a diameter of a few hundreds nm. Tilt experiments in the angular range of −30°/+30° (not shown) convinced us that these particles were spheroidal vesicles with a regular wall thickness of 20 nm. The walls appear to be darker because of the larger volume of material crossed by the electrons. In conventional TEM images recorded from similar suspensions air-dried on continuous carbon films, the vesicles appeared as deflated balloons (not shown). This means that while the rigid vesicles retained their spherical shape during the

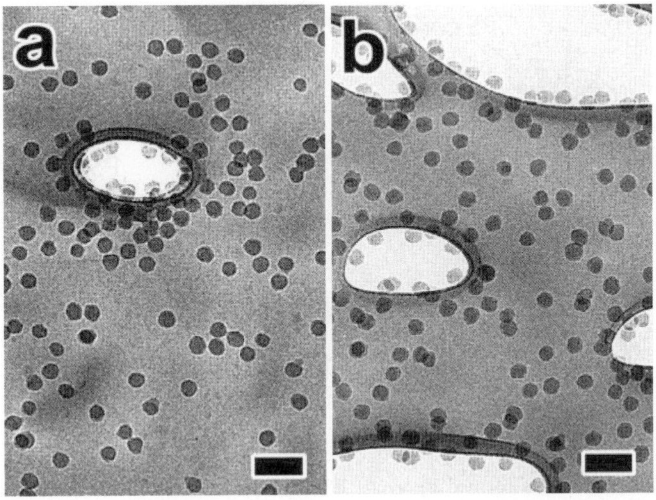

Fig. 7 Influence of the initial copolymer concentration on the morphology of micelles from linear PS–PI dispersed in heptane: (a) 0.5 mg mL^{-1}; (b) 5 mg mL^{-1}. The concentrations indicated are those at which the solutions were prepared. They were diluted 10× and 100× respectively before sample preparation. The embedding heptane was freeze-dried in the microscope. The gray areas correspond to extended surfaces of supporting carbon while the white areas are holes in the membranes (scale bars: 100 nm).

freeze-drying of heptane in vacuum (at −115 °C), they were softer at room temperature and did not withstand the surface tension forces during the air-drying of heptane.

Similar vesicles with a diameter ranging from 100 nm to 400 nm and wormlike micelles were observed in the samples that were prepared at 5 mg mL^{-1} although it was difficult to assess the

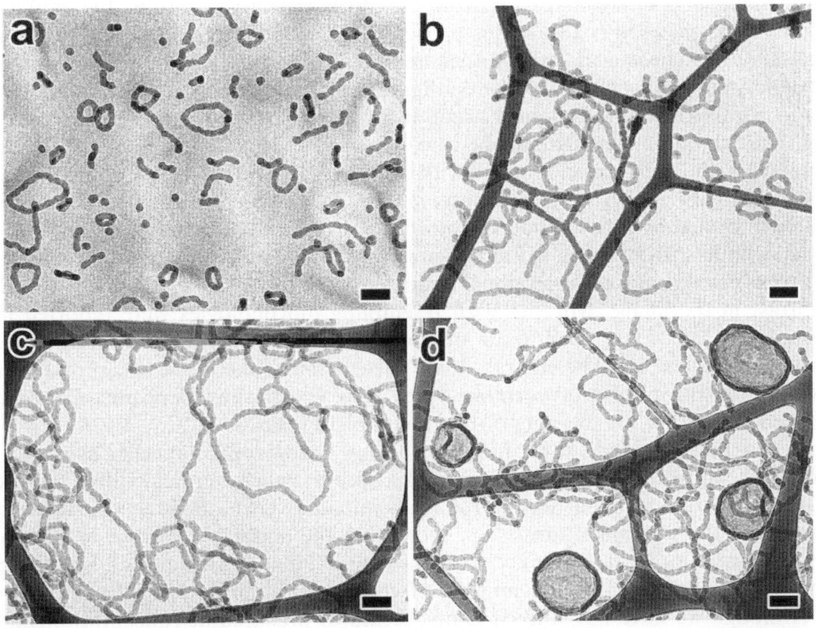

Fig. 8 Influence of the initial copolymer concentration on the morphology of micelles from cyclic PS–PI dispersed in heptane: (a), (b) 0.1 mg mL^{-1}; (c) 0.5 mg mL^{-1}; (d) 2 mg mL^{-1}. The concentrations indicated are those at which the solutions were prepared. They were diluted 10 × (a)–(c) and 100 × (d) before sample preparation. The embedding heptane was freeze-dried in the microscope. The gray area in (a) is an extended surface of supporting carbon while the dark areas correspond to the 'lacey' carbon network (scale bars: 100 nm).

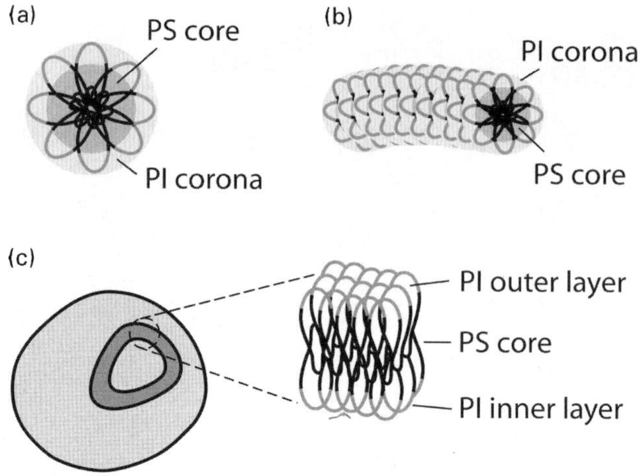

Scheme 1 (a) Sunflower micelles formed by cyclic PS–PI copolymer chains at low concentration. They are composed of a dense PS core and a more diffuse corona of PI half loops. (b) Formation of wormlike assemblies by cylindrical stacking of sunflower micelles. (c) Vesicular micelle whose envelope is a bilayer assembly with a PS core and PI outer and inner surfaces. In (b) and (c), for clarity, some molecules are schematized as opaque cycles.

change in vesicles/cylinders proportion by looking at the cryo-TEM images. Although the existence of bilayer vesicles has been reported in a number of diblock copolymer systems,[17,21,54] it is the first time that observations of vesicles made from cyclic copolymers are reported. Considering the principles that control the shape of spherical/sunflower and wormlike molecular assemblies, it is possible to describe how the cyclic molecules self-assemble to form surfaces. The wall of the vesicles is a bilayer formed by a core layer of densely packed PS half-loops while the inner and outer surfaces are made of PI half-loops (Scheme 1). With this organization, the initial sunflower organization of the micelles is totally lost. Local organizations of the molecules in the vesicle bilayer wall may be expected in order to accommodate the fact that the cyclic conformation of individual copolymer chains favors planar configurations. Sectors may form in which PI half-loops organize in a parallel fashion. This may account for the corrugated aspect of the vesicle surface, although a freeze-drying artifact cannot be ruled out.

In addition to well-defined vesicles mixed with cylinders (Fig. 8d), a number of hybrid assemblies, part-vesicular, part-wormlike, were also observed, as illustrated in Fig. 9. In Fig. 9a, a star-like object is shown formed by nine cylinders protruding out of a central 100 nm wide discoid bilayer. Some of the cylinders also clearly exhibit Y-junctions. We can assume that the thickness of the central bilayer is equal to the cross-sectional diameter of the arms, *i.e.* 20 nm, as they both exhibit a similar contrast in the image. Micelles with resembling morphologies were reported in different polymer systems.[14,15,54] Figs. 9b–d show octopus-looking assemblies that likely correspond to incompletely formed vesicles. The central 'dome' clearly looks like the skullcap of a vesicle out of which extends a varying number of wormlike 'arms'. Fig. 9e also shows a peanut-shaped vesicle that may have been formed by the merging of two incomplete vesicles.

The images in Fig. 9 provide some hints as to how the wormlike micelles assemble and form vesicles. When the local concentration of wormlike micelles reach a critical value, the fusion into a small lamellar bilayer such as that seen in Fig. 9a is initiated. Then, the junction zones between the small bilayer and the wormlike 'arms' zip, increasing the surface curvature and expanding the surface of the bilayer until a closed vesicle is formed (Figs. 9b and c; Scheme 2). The size of the final vesicle would thus be defined by the number of copolymer chains within the initially connected cylinders. Although we have not studied C-mic solutions prepared at concentrations above 5 mg mL^{-1}, a concentration range where only vesicles would be observed is likely to exist.

3.2.2. Solutions in decane. Self-assembled L-mic and C-mic were also prepared in n-decane. The evolution of their morphology with initial concentration is shown in Fig. 10. L-mic prepared at

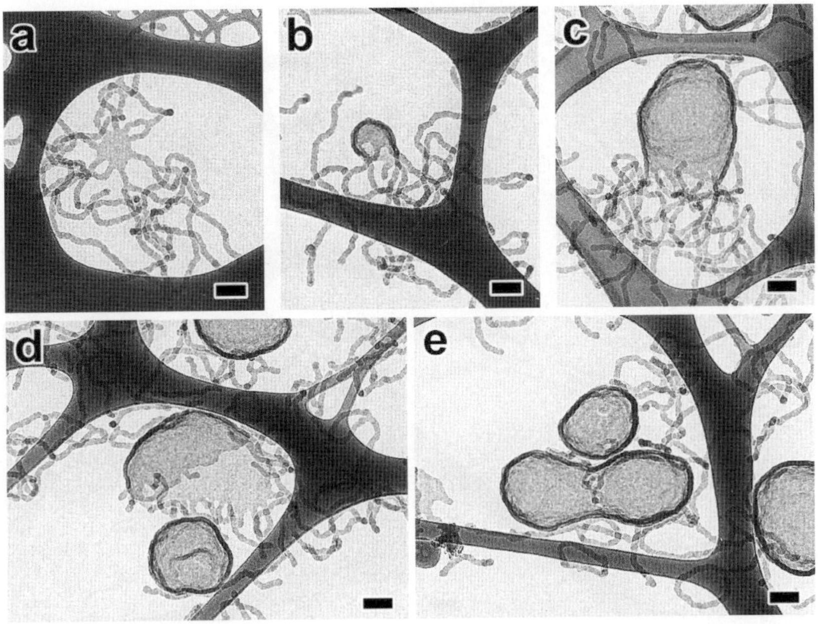

Fig. 9 Examples of incomplete vesicles observed in solutions of cyclic PS–PI copolymer in heptane: (a), (b) 2 mg mL^{-1}; (c)–(e) 5 mg mL^{-1}. The concentrations indicated are those at which the solutions were prepared. They were diluted 100× before sample preparation. The embedding heptane was freeze-dried in the microscope. The dark areas correspond to the 'lacey' carbon network (scale bars: 100 nm).

0.1 mg mL^{-1} (Fig. 10a) are similar in shape (spheres) and diameter (35 nm) to those made at the same concentration in heptane (Fig. 7a). However, among a majority of spheres, some extended wormlike micelles were observed in the sample prepared at 35 mg mL^{-1} and diluted 1000× for

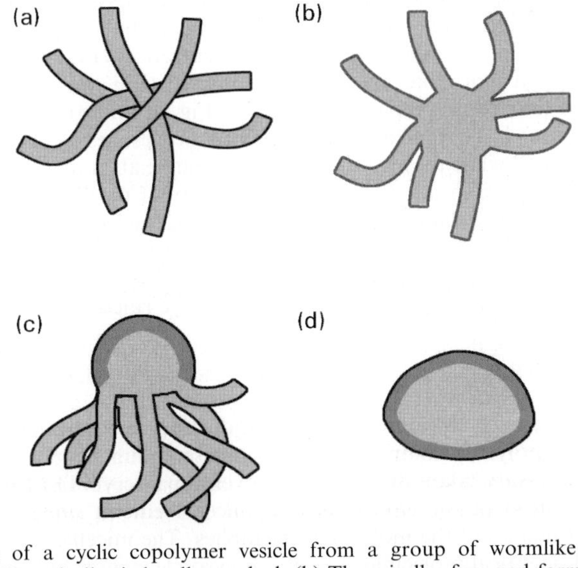

Scheme 2 Formation of a cyclic copolymer vesicle from a group of wormlike micelles: (a) A critical concentration of wormlike micelles is locally reached; (b) The micelles fuse and form a small bilayer; (c) The bilayer extends as the surface curvature increases by zipping the bilayer/cylinders junctions zones; (d) A closed vesicle is formed.

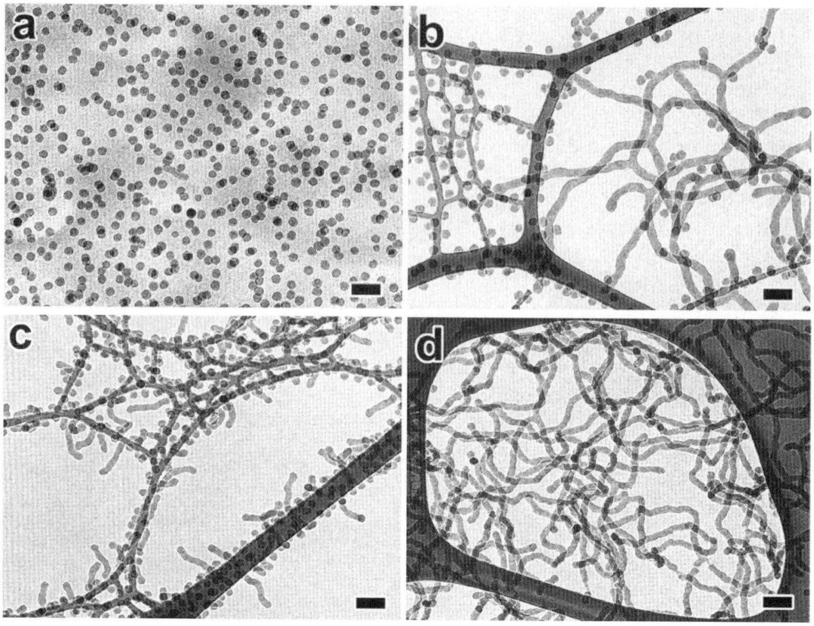

Fig. 10 Influence of the initial copolymer concentration on the morphology of PS–PI micelles dispersed in n-decane: (a) linear, 0.1 mg mL^{-1}; (b) linear, 35 mg mL^{-1}; (c) cyclic, 0.1 mg mL^{-1}; (d) cyclic, 8 mg mL^{-1}. The concentrations indicated are those at which the solutions were prepared. They were diluted 10 × (a,c), 100 × (d) and 1000 × (b) before sample preparation. The embedding heptane was freeze-dried in the microscope. The dark areas correspond to the 'lacey' carbon network (scale bars: 100 nm).

observation purpose (Fig. 10b). The width of the cylinders equals the diameter of the spheres: 35 nm. Y-junctions are also clearly observed along the wormlike micelles.

The C-mic solutions prepared in n-decane at 0.1 mg mL^{-1} contains a mixture of spheroidal particles that likely correspond to the individual sunflower micelles, and 100–150 nm-long rodlike cylinders. The diameter of the sunflower micelles is equal to the cross-sectional width of the rods, *i.e.* 20 nm (Fig. 10c). Qualitatively, the rod-like micelles appear to be less polydisperse than the short wormlike micelles formed in heptane at the same concentration (Fig. 8a). At 8 mg mL^{-1}, only 20 nm-wide tortuous giant wormlike micelles are observed (Fig. 10d). Again, they seem to be somewhat less tortuous than those prepared in heptane (Fig. 8c and d). The major difference when comparing C-mic solutions in heptane and decane is that no vesicles were observed in decane at a concentration of 8 mg mL^{-1} while they were detected in heptane at 2 mg mL^{-1}. This may illustrate a shift of the phase diagram due to the difference of solubility of the PS block in both solvents. As already explained from the DLS data presented in section 3.1, the formation of vesicles is promoted in heptane in the 2–5 mg mL^{-1} concentration range since PS is less soluble in this decane than in heptane. However, as we did not observe C-mic solutions in decane at concentrations higher than 8 mg mL^{-1}, the existence of vesicles in this solvent cannot be ruled out.

4. Conclusion

DLS and cryo-TEM have been efficiently used in conjunction to study the effect of cyclization of linear PS–PI diblock copolymer chains on their self-assembling properties in two PI-selective solvents. Despite the necessary adaptation of the 'conventional' cryo-TEM procedure to deal with the inevitable crystallization of the embedding organic solvent, *in situ* freeze-drying cryo-TEM provided high contrast images of the molecular assemblies. The micelles prepared from linear and cyclic copolymers with identical molecular weights showed clear morphological differences. L-mic kept the same aggregation number in the investigated concentration range while C-mic showed a large distribution in size (and therefore aggregation number) with a polydispersity that increased with concentration. Finally, a dilution of the solutions initially prepared at definite concentrations did not affect the shape and size of the micelles, which is consistent with previous results obtained from AFM images.[35]

Several conclusions can be drawn from the present results: (i) Within the investigated range of concentrations (0.001 mg mL^{-1} < c < 5 mg mL^{-1}), L-mic are spherical in heptane and decane and their sizes are slightly smaller in decane than in heptane and do not depend on the concentration or the temperature (10 °C < T < 25 °C). However, wormlike micelles were seen to co-exist with spherical micelles at a higher concentration in decane (35 mg mL^{-1}). (ii) At very low concentration in heptane, C-mic are more or less planar sunflower micelles which, as concentration increases, unidirectionally self-assemble into giant wormlike micelles by PS–PS core attraction. Above 2 mg mL^{-1}, vesicles build up from the fusion of giant wormlike micelles. (iii) Similar features are observed for C-mic in decane except that vesicles are not detected up to 8 mg mL^{-1}. We assume that vesicle formation is not promoted in decane in this concentration range since n-alkanes become worse solvents for polystyrene as the n-alkane carbon number increases. (iv) The morphologies of C-mic vary with temperature (10 °C < T < 25 °C) in heptane and decane.

The DLS and cryo-TEM results presented in this work clearly show that the morphology of self-assembled micelles made from cyclic PS–PI copolymer (sunflowers, giant wormlike micelles with or without Y-junctions, vesicles) and their behavior as a function of temperature and concentration are very different from those of spherical micelles made from linear PS–PI chains. Since both cyclic and linear copolymers have similar molecular mass and degree of polymerization, these differences are only induced by the cyclization of the copolymer chain.

Acknowledgements

R.B. acknowledges financial support from Centre National de la Recherche Scientifique (CNRS), Région Aquitaine and European Funds (FEDER) and Frank S. Bates for sending a preprint of ref. 55.

References

1 K. Mortensen, *Colloids Surf. A: Physicochem. Eng. Asp.*, 2001, **183–185**, 277.
2 J. H. Ahn, B. H. Sohn, W. C. Zin and S. T. Noh, *Macromolecules*, 2001, **34**, 4459.
3 A. Halperin, M. Tirrell and T. P. Lodge, *Adv. Polym. Sci.*, 1992, **100**, 31–71.
4 I. W. Hamley, J. S. Pedersen, C. Booth and V. M. Nace, *Langmuir*, 2001, **17**, 6386.
5 J. S. Pedersen, I. W. Hamley, C. Y. Ryu and T. P. Lodge, *Macromolecules*, 2000, **33**, 542.
6 Y. Y. Won, H. T. Davis and F. S. Bates, *Science*, 1999, **283**, 960.
7 S. A. Jenekhe and X. L. Chen, *Science*, 1999, **283**, 372.
8 M. Park, C. Harrison, P. M. Chaikin, R. A. Register and D. H. Adamson, *Science*, 1997, **276**, 1401.
9 V. P. J. Torchilin, *Controlled Release*, 2001, **73**, 137.
10 L. Yang and P. Alexandridis, *Curr. Opin. Colloid Interface Sci.*, 2000, **5**, 132.
11 I. Sapurina, J. Stejskal and Z. Tuzar, *Colloids Surf. A: Physicochem. Eng. Asp.*, 2001, **180**, 193.
12 Y. Yu, L. Zhang and A. Eisenberg, *Macromolecules*, 1998, **31**, 1144.
13 L. Zhang and A. Eisenberg, *J. Am. Chem. Soc.*, 1996, **118**, 3168.
14 L. Zhang and A. Eisenberg, *Science*, 1995, **268**, 1728.
15 K. Yu, L. Zhang and A. Eisenberg, *Langmuir*, 1996, **12**, 5980.
16 R. Savic, L. Luo, A. Eisenberg and D. Maysinger, *Science*, 2003, **300**, 615.
17 B. M. Discher, Y.-Y. Won, D. S. Ege, J. C.-M. Lee, F. S. Bates, D. E. Discher and D. A. Hammer, *Science*, 1999, **284**, 1143.
18 Y. Zheng, Y. Y. Won, F. S. Bates, H. T. Davis, L. E. Scriven and Y. Talmon, *J. Phys. Chem. B*, 1999, **103**, 10 331.
19 Y. Y. Won, H. T. Davis and F. S. Bates, *Science*, 1999, **283**, 960.
20 K. Krishnan, K. Almdal, W. R. Rurghardt, T. P. Lodge and F. S. Bates, *Phys. Rev. Lett.*, 2001, **87**, 98 301/1–98 301/4.
21 Y. Y. Won, A. K. Brannan, H. T. Davis and F. S. Bates, *J. Phys. Chem. B*, 2002, **106**, 3354.
22 S. Jain and F. S. Bates, *Science*, 2003, **300**, 460.
23 J. S. Pedersen, I. W. Hamley, C. R. Ryu and T. P. Lodge, *Macromolecules*, 2000, **33**, 542.
24 T. P. Lodge, X. Xu, C. Y. Ryu, I. W. Hamley, J. P. A. Fairclough, A. J. Ryan and J. S. Pedersen, *Macromolecules*, 1996, **29**, 5955.
25 M. Shibayama, T. Hashimoto and H. Kawai, *Macromolecules*, 1983, **16**, 1427.

26 C. Lai, W. B. Russel and R. A. Register, *Macromolecules*, 2002, **35**, 4044; C. Lai, W. B. Russel and R. A. Register, *Macromolecules*, 2002, **35**(3), 841.
27 K. Khougaz, X. F. Zhong and A. Eisenberg, *Macromolecules*, 1996, **29**, 3937.
28 J. Ding, J. Liu and M. Yang, *Polymer*, 1997, **38**, 5497.
29 H. Iatrou, N. Hadjichristidis, G. Meier, H. Frielinghaus and M. Monkenbusch, *Macromolecules*, 2002, **35**, 5426.
30 M. Pacovska, K. Prochazka, Z. Tuzar and P. Munk, *Polymer*, 1993, **34**, 4585.
31 C. Booth and D. Attwood, *Macromol. Rapid Commun.*, 2000, **21**, 501.
32 G.-E. Yu, Z. Yang, D. Attwood, C. Price and C. Booth, *Macromolecules*, 1996, **29**, 8479.
33 G.-E. Yu, C. A. Garrett, S. M. Mai, H. Altinok, D. Attwood, C. Price and C. Booth, *Langmuir*, 1998, **14**, 2278.
34 R. Borsali, E. Minatti, J.-L. Putaux, M. Schappacher, A. Deffieux, P. Viville, R. Lazzaroni and T. Narayanan, *Langmuir*, 2002, **19**, 6.
35 E. Minatti, P. Viville, R. Borsali, M. Schappacher, A. Deffieux and R. Lazzaroni, *Macromolecules*, 2003, **36**, 4125.
36 E. Minatti, R. Borsali, M. Schappacher, A. Deffieux, T. Narayanan and J.-L. Putaux, *Macromol. Rapid Commun.*, 2002, **23**, 978.
37 M. Schappacher and A. Deffieux, *Macromol. Chem. Phys.*, 2002, **203**, 2463.
38 R. Yin and T. E. Hogen-Esch, *Macromolecules*, 1993, **26**, 6952.
39 R. L. Lescanec, D. A. Hadjuk, G. Y. Kim, Y. D. Gan, R. Yin, S. M. Gruner and T. E. Hogen-Esch, *Macromolecules*, 1995, **28**, 3485.
40 B. J. Berne and R. Pecora, in *Dynamic Light Scattering with Applications to Chemistry, Biology and Physics*, 2nd edn., Dover Publications, New York, 2000.
41 S. W. Provencher, *Comput. Phys. Commun.*, 1982, **27**, 213.
42 C. L. Jackson, H. Chanzy, F. Booy, B. J. Drake, D. A. Tomalia, B. J. Bauer and E. J. Amis, *Macromolecules*, 1998, **31**, 6259.
43 J.-L. Putaux, A. Buléon, R. Borsali and H. Chanzy, *Int. J. Biol. Macromol.*, 1999, **26**, 145.
44 S. Chalaye, E. Bourgeat-Lami, J.-L. Putaux and J. Lang, *Macromol. Symp.*, 2001, **169**, 89.
45 M. Schappacher, A. Deffieux, J.-L. Putaux, P. Viville and R. Lazzaroni, *Macromolecules*, 2003, **36**, 5776.
46 G. T. Oostergetel, F. J. Esselink and G. Hadziioannou, *Langmuir*, 1995, **11**, 3721.
47 J. Dubochet, J. Lepault, M. Adrian, A. McDowall and J. C. A. Homo, *Quart. Rev. Biophys.*, 1998, **128**, 219.
48 J. R. Quintana, M. Villacampa, M. Munoz, A. Andrio and I. A. Katime, *Macromolecules*, 1992, **25**, 3125.
49 J. R. Quintana, M. Villacampa, A. Andrio, M. Munoz and I. A. Katime, *Macromolecules*, 1992, **25**, 3129.
50 T. Tlusty, S. A. Safran and R. Strey, *Phys. Rev. Lett.*, 2000, **84**, 1244.
51 J. T. Kindt, *J. Phys. Chem. B*, 2002, **106**, 8223.
52 A. Bernheim-Groswasser, T. Tlusty, S. A. Safran and Y. Talmon, *Langmuir*, 1999, **15**, 5448.
53 S. Y. Kwon and M. W. Kim, *Phys. Rev. Lett.*, 2002, **89**, 258 302-1, 258 302-4.
54 L. Zhang and A. Eisenberg, *Macromolecules*, 1999, **32**, 2239.
55 S. Jain and F. S. Bates, *Consequences of Nonergodicity in Aqueous Binary PEO-PB Micellar Dispersions*, (Preprint).

Self-assembled nanostructures from peptide–synthetic hybrid block copolymers: Complex, stimuli-responsive rod–coil architectures

Jérôme Babin,[a] Juan Rodriguez-Hernandez,[a] Sébastien Lecommandoux,*[a] Harm-Anton Klok[b] and Marie-France Achard[c]

[a] *Laboratoire de Chimie des Polymères Organiques, CNRS, ENSCPB-Université Bordeaux 1, 16 Avenue Pey Berland, 33607, PESSAC, France. E-mail: s.lecommandoux@enscpb.fr*

[b] *Ecole Polytechnique Fédérale de Lausanne (EPFL), Institut des Matériaux, Laboratoire des Polymères, CH 1015, Lausanne, Switzerland*

[c] *Centre de Recherche Paul Pascal (CNRS), Université Bordeaux 1, Avenue Albert Schweitzer, 33607, Pessac, France*

Received 2nd March 2004, Accepted 12th March 2004
First published as an Advance Article on the web 16th September 2004

The synthesis, solution and solid state self-assembly properties of a series of polyisoprene-*b*-poly(ε-benzyloxycarbonyl-L-lysine) PI-*b*-PZLys and polyisoprene-*b*-poly(L-lysine) PI-*b*-PLys block copolymers have been examined. The formation of stimuli-responsive micelles in water has been studied as a function of pH and ionic strength using static and dynamic light scattering, UV-circular dichroism and transmission electron microscopy. The observed change in the micelles dimensions has been directly attributed to the conformational transition in the secondary structure of polypeptide chains. In bulk, these rod–coil copolymers form self-assembled structures that have been characterized using DSC, DMA, WAXS and SAXS techniques. Hexagonal in lamellar (HL) and more interestingly hexagonal in hexagonal (HH) morphologies have been evidenced as a function of the chemical composition.

1. Introduction

Block copolymers are a well-known and developed class of macromolecules that have been synthesized in the laboratory during the last fifty years. Nevertheless, they still represent a subject in great expansion due to an increase in the potential industrial applications that can be obtained. Moreover, the recent progresses in the synthetic strategies that can be applied to vary the chemical composition and architecture of block copolymers also open new areas of development. One main reason for much of the current interest in these systems is related to the wide variety of micro and nano-structures that can be obtained. It is now well established, both from a theoretical[1] and an experimental[2] point of view, that linear AB diblock copolymers composed of immiscible blocks undergo microphase separation. They can form micellar aggregates when dissolved in a selective solvent for one of the blocks,[3] and phase separate into a variety of organized structures in the bulk state.[2,4] The corresponding morphologies (shape and size) depend on the total degree of polymerization N, which is proportional to the molecular weight, and on the chemical composition of the copolymer f_A, corresponding to the volume fraction of the block A ($f_B = 1 - f_A$). The thermodynamic interaction between the two blocks is given by the dimensionless Flory–Huggins interaction parameter χ_{AB} (or χ), which is inversely proportional to the temperature. Based on these three parameters, phase diagrams have been constructed theoretically.[1] In addition to simple linear block copolymers, a variety of architecturally more complex systems have been developed during the last few years, which include amongst others, star-shaped,[5] rod–dendron,[6] rod–coil[4,7] and cyclic[8] block copolymers. Among these, rod–coils present a particularly interesting class of diblock copolymers. It has been demonstrated, both theoretically[9] using the random phase approximation (RPA), and experimentally[10] that the phase separation in rod–coils can occur in the nano size range as compared to classical coil–coil systems for which phase separation typically occurs on larger characteristic length scales. It was evidenced that the transition from the disordered to an ordered phase occurs at (χN, qR_g) = (8.5, 2.2) for a symmetric rod–coil diblock copolymer (with q the wave-vector and R_g the radius of gyration) as compared to the classical result for linear coil–coil diblocks for which (χN, qR_g) = (10.5, 1.95).[9] Interesting examples of unique and non-predicted structures include the mushrooms of Stupp,[11] the zig-zag morphology of Thomas,[12] the microporous morphology of Jenekhe[13] or the double hexagonal assembly which has been found for styrene-b-(α-benzyl-L-glutamate) oligomers.[14] In spite of these findings, not much effort has been devoted so far to establishing theoretical or experimental phase diagrams of rod–coil block copolymer systems. The main theoretical work done up to now[15] describes that these copolymers should develop nematic, smectic C or smectic A structures, far from the experimentally observed ones. As far as the solution properties are concerned, micellization of rod–coils molecules seems to follow more regular rules in the sense that spherical micelles[16] or vesicles[17,18] have been observed in concentration and composition range comparable to those observed for coil–coil amphiphiles.[19]

This paper focuses on the use of complex "hybrid" rod-coil block copolymers based on a synthetic coil block and a polypeptide block with an α-helical conformation (rod). In contrast to the rod-segment of most other rod-coil molecules, the conformation of the α-helical peptide segment can be sensitive to environmental parameters like temperature, pH or ionic strength. We already proved this concept of stimuli-response of polypeptides to temperature in bulk on polystyrene-*b*-poly(γ-benzyl-L-glutamate)[14] (PS-*b*-PBLG) and to pH of water solutions on polybutadiene-*b*-poly(glutamic acid)[17] (PB-*b*-PGA). In this communication, these concepts are extended to another class of materials based on poly(L-Lysine) polypeptides. This paper will discuss the solution and solid state self-assembly properties of a series of polyisoprene-*b*-poly(ε-benzyloxycarbonyl-L-lysine) PI-*b*-PZLys and polyisoprene-*b*-poly(L-lysine) PI-*b*-PLys block copolymers.

2. Experimental

2.1. Materials

Isoprene purification proceeds as follows: first, the commercial isoprene was dried under CaH_2 and distilled. A second distillation was performed by addition of dibutylmagnesium. THF was also distilled in two steps: first from CaH_2 followed by sodium benzophenone. Cyclohexane was cryodistilled after removal of side reactive products with a polystyryllithium. All solvents used for anionic polymerization were stored in graduated burettes under vacuum until its utilization. *N*,*N*′-dimethylformamide (DMF) was distilled from CaH_2 under reduced pressure and stored over molecular sieves (4 Å). Ethylacetate was dried over molecular sieves (4 Å). All other solvents were used as received. ε-Benzyloxycarbonyl-L-lysine N-carboxyanhydride (Z-Lys NCA) was synthesized according to the method described by Poché *et al.*[20]

2.2. Analytical methods

^1H NMR spectra were recorded at room temperature on a Bruker Avance 250 spectrometer using the residual proton resonance of the deuterated solvent as the internal standard. GPC analysis was performed at 60 °C with a setup consisting of a Waters 510 pump and a series of 500, 10^5 and 10^6 Å, respectively. A 0.1 M solution of LiBr in DMF was used as the mobile phase, and the elution of the samples was monitored using simultaneous UV and RI detection. Elution times were converted to molecular weights using a calibration curve that was constructed with narrow-polydispersity poly(styrene) standards. FTIR spectra were recorded on a Nicolet 730 FTIR-spectrometer under nitrogen atmosphere. Samples were prepared by drop-casting a thin film from a diluted THF solution onto a piece of silicon wafer.

2.3. Physical characterization

DSC experiments were carried out on a Perkin-Elmer DSC-7 with heating- and cooling-rates of 10 °C min^{-1}. DMA experiments were performed on a Perkin-Elmer DMA-7 running in tensile mode at an oscillation frequency of 1 Hz and with heating rates of 10 °C min^{-1}. The samples were tested in compression mode between two parallel plates geometry. Static light scattering (SLS) and dynamic light scattering (DLS) experiments were performed on an ALV5000 goniometer equipped with an ALV5000/E Multiple Tau digital Realtime correlator that calculates the photon intensity auto-correlation function $g^2(t)$ and a helium–neon laser (wavelength $\lambda = 632.8$ nm). The R_H and PDI values of the aggregates were obtained by a cumulant and CONTIN analysis. The determination of the R_G and R_H was performed respectively by SLS and DLS measurements containing 1 M of NaCl. Distilled water was thoroughly filtered with 0.1 μm filters and directly employed for the preparation of the solution. All the experiments were performed at 298 K. Calculation of the dn/dc value proceeded as follows: four solutions were prepared with concentrations 8, 4, 2, and 1 g l^{-1} with a constant concentration of salt 1 M (NaCl) and pH ~ 2. The representation n–n_0 *versus* concentration, where n is the refractive index of the solution and n_0 is the refraction index of the solvent is according to $(1/K)(\mathrm{d}n/\mathrm{d}c)$ (K: apparatus constant). The resulting increment of the refractive index dn/dc obtained (slope of the plot) was 0.26. Circular dichroism spectra were measured on a Jobin Yvon CD6 spectropolarimeter (184–900 nm). The pH dependence curve was obtained by changing the pH with a solution 0.1 M of HCl and 0.1 M of NaOH to a polymer solution of 0.5 mg.ml^{-1}. TEM pictures were recorded on a JEOL JEM100S microscope working at 100 kV. Samples were prepared by spraying a 1 g l^{-1} solution of the block copolymer onto a TEM grid. Wide angle X-ray diffraction experiments were carried out using an 18 kW rotating anode X-ray source (Rigaku-200). A Germanium (111) crystal was used as monochromator to select the Cu Kα radiation. The scattered radiation was collected on a 2D imaging plate system (Mar Research). The reciprocal space FWHM (full-width at half-max) was 5×10^{-3} $Å^{-1}$. The sample to detector distance was 20 cm. The accessible wave-vector (q) range was 0.1 $Å^{-1} < q < 1.6$ $Å^{-1}$. The small angle X-ray scattering (SAXS) experiments were performed using Cu K radiation (1.54 Å wavelength) using a Nanostar from Bruker AXS (operating at 40 kV, 35 mA). A CCD detector (Siemens Hi-Star), at a sample to detector distance of 106.2 cm, was used to record scattering patterns. The accessible wave-vector (q) range was 0.01 $Å^{-1} < q < 0.1$ $Å^{-1}$.

3. Results and discussion

3.1. Synthesis

The preparation of the block copolymers takes place in a two step reaction. The first one involves anionic polymerization of isoprene in a THF/cyclohexane solvent mixture using *sec*-butyllithium as initiator.[21] The chain length was controlled *via* the molar ratio between monomer and initiator (the targeted degree of polymerisation was 50). The reaction was terminated with an agent that contains a chloro-function able to react with the carbanion of the end-chain and a protected-amino group. The tetramethyldisilaazacyclopentane endfunctionalized polyisoprene oligomers were deprotected under acidic conditions to obtain the corresponding ω-amino functionalized polymer (Scheme 1).[22] In a second synthetic step, the peptide block was grafted onto the polyisoprene *via* ring-opening polymerization of ε-benzyloxycarbonyl-L-lysine N-carboxyanhydride.[23] The polymerization of Z-lys NCA was performed in a THF/DMF solvent mixture and was allowed to proceed during 5 days under argon atmosphere. After this time the polymer was precipitated in diethylether and dried under vacuum (Scheme 2). Finally, the formation of amphiphilic block copolymers has been obtained after deprotection of the benzyloxycarbonyl groups in a mixture of HBr in acetic acid/TFA/dioxane (1/1/4 v/v) over 30 min at room temperature. A deprotected polyisoprene-*b*-poly(L-lysine) PI-*b*-PLys block copolymer has been obtained with a partial bromination of the isoprene double bonds (with the applied conditions, more than 50% remains intact).

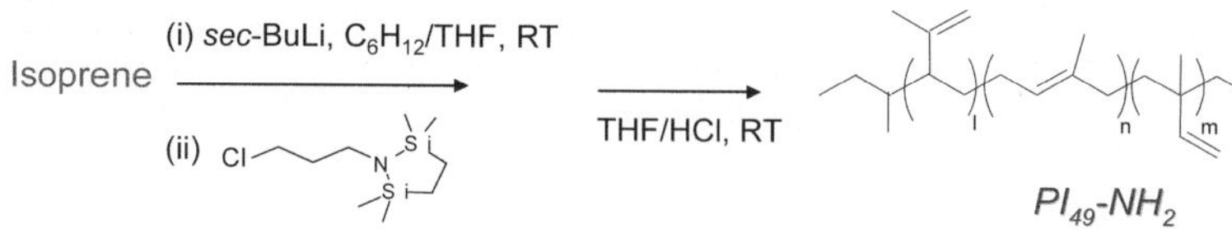

Scheme 1

In Table 1 are included the molecular characteristics of the prepared block copolymers obtained by ^{1}H NMR and GPC.

3.2. Solution self-assembly: stimuli-responsive micelles

The PI_{49}-*b*-$PLys_{123}$ sample could be directly dissolved in pure water. The pH of the solution was then adjusted by addition of HCl or NaOH. According to the time dependence of the hydrodynamic radius (R_H) as determined by dynamic light scattering (DLS), perfect equilibrium is established after several days of vigorous stirring. To further elucidate the supramolecular organization of PI_{49}-*b*-$PLys_{123}$, DLS experiments were performed on aqueous solutions at different concentrations (from 0.25 g l^{-1} to 2 g l^{-1}), at various angles (from 50° to 150°), and with or without added salt (1 M NaCl). The analysis of the normalized intensity autocorrelation functions $C(q,t)$, where q is the scattering vector ($q = [4\pi n_0 \sin(\theta/2)]/\lambda_0$) and t the time, was carried out following the method of cumulants, fitted with monoexponential decay functions. Once the first cumulant Γ is determined, the concentration and angular dependence can be expressed as $\Gamma/q^2 = D_Z(1 + k_D c + \ldots)$ where D_Z is the *z*-average diffusion coefficient, and k_D is the effective interaction parameter. From the diffusion coefficient, the *z*-average hydrodynamic radius R_H can be calculated from the Stokes–Einstein equation: $R_H = kT/(6\pi\eta D_Z)$ where η is the solvent viscosity. Dynamic light scattering data were also analyzed using the CONTIN method to determine the hydrodynamic size distribution. As a representative example, the angular dependence of the autocorrelation functions $C(q,t)$ are shown

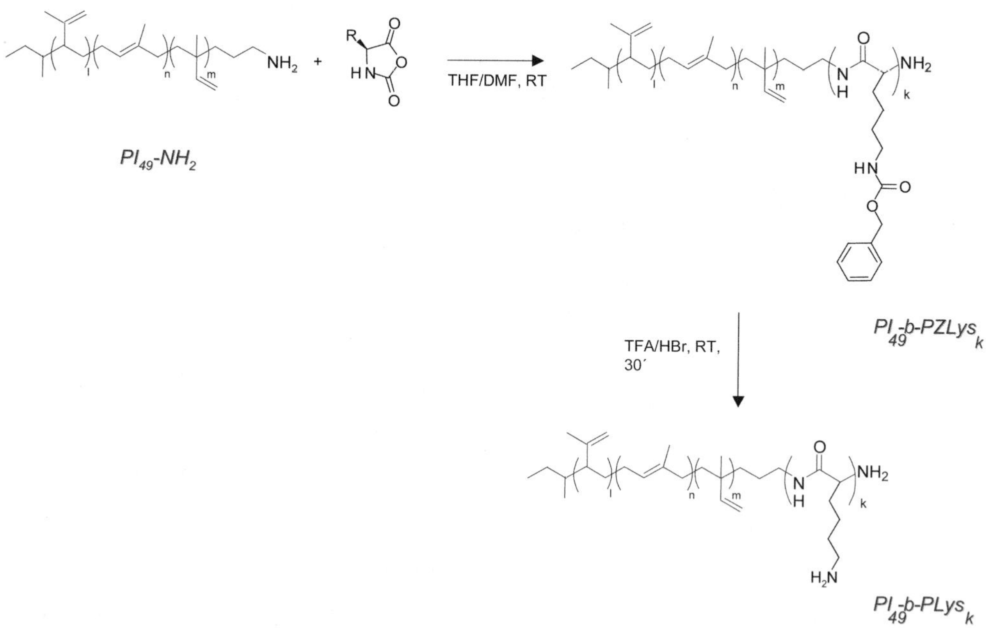

Scheme 2

Table 1 Molecular characterization of the polyisoprene-*b*-poly(ε-benzyloxycarbonyl-L-lysine) diblock copolymers

Sample	Yield[a](%)	$M_w{}^b$/g mol^{-1}	$M_n{}^b$/g mol^{-1}	$M_w/M_n{}^b$	$n_{NMR}{}^c$
PI_{49}-*b*-$PZLys_{35}$	69	8070	5230	1.54	35
PI_{49}-*b*-$PZLys_{61}$	66	9180	5680	1.61	61
PI_{49}-*b*-$PZLys_{92}$	79	11 960	7500	1.59	92
PI_{49}-*b*-$PZLys_{123}$	75	13 560	10 130	1.34	123
PI_{49}-*b*-$PZLys_{178}$	72	16 140	13 604	1.18	178

[a] Isolated yield, after precipitation and drying. [b] From GPC experiments, performed in DMF. [c] Number-average degree of polymerization of the peptide segment, determined from ^{1}H-NMR.

in Fig. 1 for PI_{49}-*b*-$PLys_{123}$ at pH = 1.5 with added salt. This graph shows a unimodal and narrow distribution, independent of the scattering angle. The hydrodynamic radii of the aggregates were determined at different scattering angles and analyzed using the cumulant method by extrapolation to 0° scattering angle. Fig. 2 presents the evolution of Γ *versus* q^2 for a PI_{49}-*b*-$PLys_{123}$ solution of 1 g l^{-1}. Fitting of these data gives a linear correlation with $r^2 = 0.99$, indicating that only translational diffusion occurs. From the Γ/q^2 *versus* q^2 plot, the intercept gives an apparent *z*-average diffusion coefficient D_Z at this concentration. The results show that the size of the obtained aggregates remains constant for the whole investigated range of concentration ($R_H \sim 40$ nm). The corresponding hydrodynamic radius in water was found to be $R_H = 41$ nm for solutions at pH = 1.5 without salt (by extrapolating $R_H(c)$ to $c \to 0$).

In a next step, the effect of pH-induced conformational changes on the size of the PI_{49}-*b*-$PLys_{123}$ aggregates was investigated by combination of UV-circular dichroism (CD) with DLS measurements. It is well-known that poly(L-lysine) can adopt different secondary structures (α-helical, β-sheet or coil) in water solutions depending on pH and temperature.[24] CD spectra were recorded from the same solutions as for light scattering analysis (with 1 M NaCl), but with varying the pH of the solution. Typical CD spectra obtained for PI_{49}-*b*-$PLys_{123}$ micelles at two extreme pH (pH = 10.8

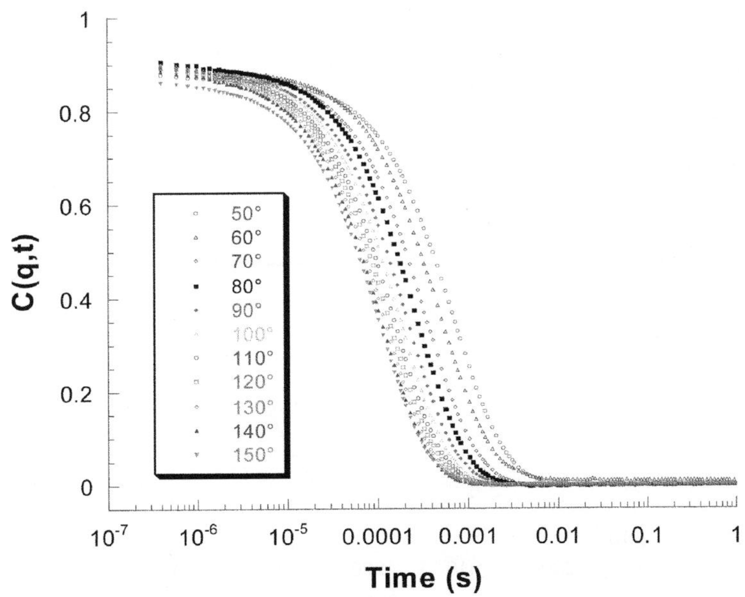

Fig. 1 Auto-correlation functions of PI_{49}-*b*-$PLys_{123}$ at pH = 1.5 with added salt ($c = 1$ g l^{-1}) at different scattering angles.

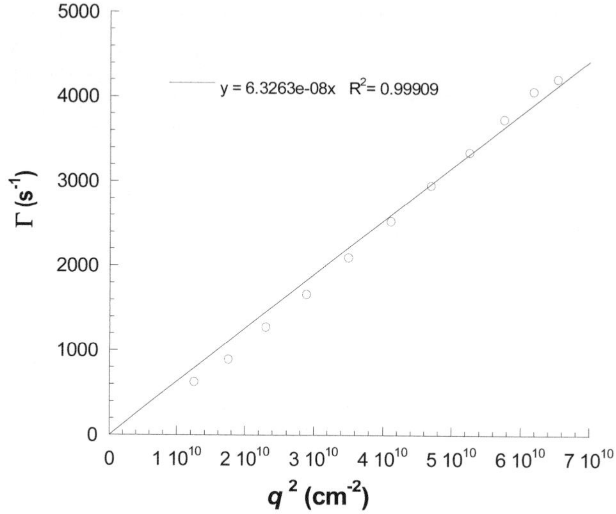

Fig. 2 Dependence of the first cumulant frequency (Γ) with the square magnitude of the scattering vector (q^2) for PI_{49}-*b*-$PLys_{123}$ at pH = 1.5 with added salt ($c = 1$ g l^{-1}).

and 2.1) values are shown in Fig. 3. This figure clearly evidences a change from a coil conformation of the polypeptide chains within the micelles in acidic conditions to an α-helical conformation in basic conditions, when the chains are no longer charged and thus able to form the intra-molecular hydrogen-bonding necessary to the formation of the α-helical structure.

Fig. 4 shows the pH-dependence of the hydrodynamic radius R_H measured by DLS (90°) on PI_{49}-*b*-$PLys_{123}$ ($c = 2$ g l^{-1}) with or without added salt (1 M NaCl). Going from acidic to basic

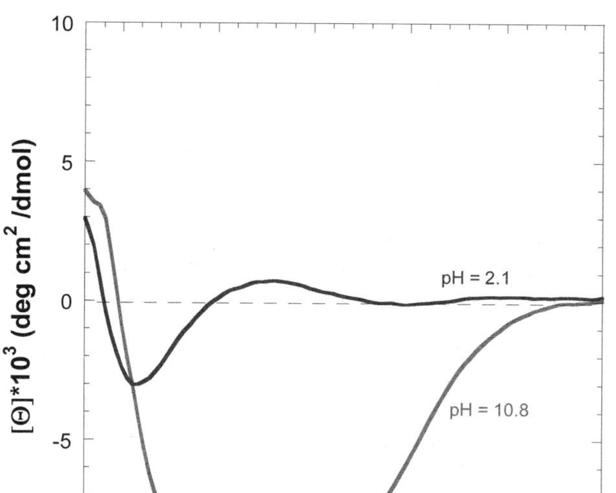

Fig. 3 CD spectra of PI_{49}-*b*-$PLys_{123}$ aggregates in aqueous solution (1 M NaCl) at different pH values.

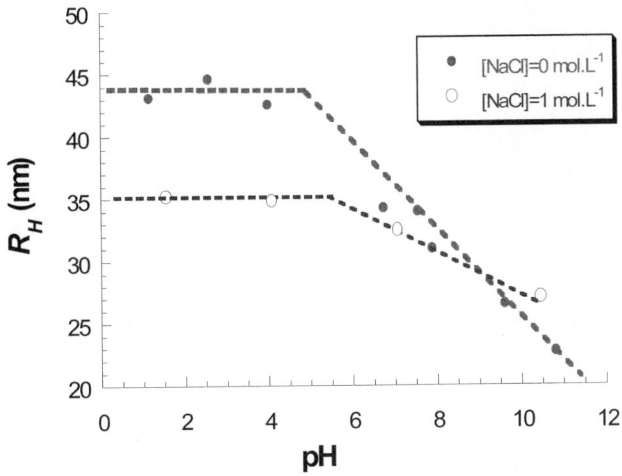

Fig. 4 Hydrodynamic radius (R_H) of PI_{49}-*b*-$PLys_{123}$ ($c = 2$ g l^{-1}) aggregates in water measured by DLS (90°) with and without added salt conditions as a function of pH.

pH values (and in the absence of added salt), first a plateau in the R_H values is observed up to about pH = 6 ($R_H \approx 44$ nm), followed by a strong decrease between pH = 6 and pH = 11 ($R_H \approx 23$ nm). This critical pH value around pH = 6, when the size of the micelle is changing, corresponds to the pH value where the peptide undergoes a transition from an α-helix to a random coil conformation, confirming that this change in size is related to conformational transition of the polypeptide segment. This phenomenon is still observed when 1 M NaCl is added to the solution, with a change in the micelle radius of about 10 nm.

In addition to DLS, static light scattering experiments (SLS) were performed in water under the same conditions. The analysis of the scattering intensities using a Zimm plot yields a mass average molar mass, a *z*-average radius of gyration and a second virial coefficient. From the apparent molecular weight obtained $M_w^{app} = 1.1 \times 10^7$ g mol^{-1}, and assuming a spherical micellar structure of the aggregates, one can calculate an apparent aggregation number of $N_{agg}^{app} = 31$.

From the aggregation number, the radius of the core of the micelles can be estimated using the following equation:

$$R_{core} = \left(\frac{3M_{PI}}{4\pi\rho^{PI}N_a}N_{agg}^{app}\right)^{1/3}$$

where M_{PI} is the molecular weight of the polyisoprene block ($M_{PI} = 3332$ g mol^{-1}), ρ_{PI} is the density of the polyisoprene block ($\rho_{PI} = 0.905$ g cm^{-3}), and N_a is Avogadro's number. We thus obtained a core radius $R_{core} = 3.5$ nm. When polypeptides are in their α-helical conformation, their length in nm can be calculated following the simple relation $L_{basic} = 0.15 \times$ DP (with DP the degree of polymerization of the peptide block). From this relation, we can calculate the length of the poly(L-lysine) in basic conditions $L_{basic} = 18.4$ nm. We can thus estimate the theoretical radius R_{basic}^{th} of the micelle in basic conditions $R_{basic}^{th} = L_{basic} + R_{core} = 219$ nm, which is close to the value determined from the scattering experiments ($R_{basic}^{exp} = 228$ nm) and which supports the micellar structure of the aggregated formed by PI_{49}-*b*-$PLys_{123}$. Considering now the situation at acid pH, the polypeptide is in a coil conformation. If no salt is added, the polyelectrolyte chains should be totally stretched, with a length that could be estimated $L_{acid} = 0.37 \times DP = 46.1$ nm leading to a theoretical radius $R_{acid}^{th} = L_{acid} + R_{core} = 496$ nm. This value is not surprisingly overestimated compared to the experimentally observed one ($R_{acid}^{exp} = 44$ nm), but nevertheless in good agreement.

The light scattering measurements were confirmed by TEM analysis as shown in Fig. 5. This figure clearly shows spherical micelles with low size distribution.

Finally, Fig. 6 presents a model for the solution self-assembly of PI_{49}-*b*-$PLys_{123}$ based on the data obtained from SLD, DLS and CD experiments. The core of the micelles is formed by the

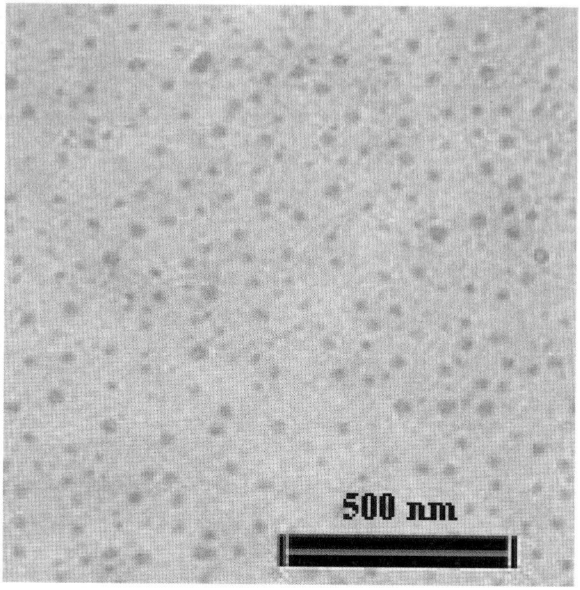

Fig. 5 TEM micrograph of PI_{49}-*b*-$PLys_{123}$ aggregates obtained from a spray of a solution ($c = 1$ g l^{-1}) onto a formvar coated copper grid.

hydrophobic polyisoprene (yellow) and the corona is based on the poly(L-lysine) block. In acidic conditions, this poly(L-lysine) segment is charged and therefore in its coil conformation, leading to large micellar size. On the contrary, when pH increases, the polypeptide undergoes transition to an α-helical conformation and the aggregate size decreases.

3.3. Bulk self-assembly

3.3.1. Thermal and mechanical behavior. The diblock copolymers were investigated by optical microscopy, differential scanning calorimetry (DSC) and dynamical mechanical analysis (DMA). The glass transition temperature (T_g) of the polyisoprene precursor (PI_{49}-NH_2) measured by DSC gives a value of about −5 °C. No other transitions were observed by DSC for the ε-benzyloxycarbonyl-L-lysine (Z-Lys) based molecules, even if the liquid crystalline textures observed by polarized optical microscopy remained unchanged until decomposition of the material

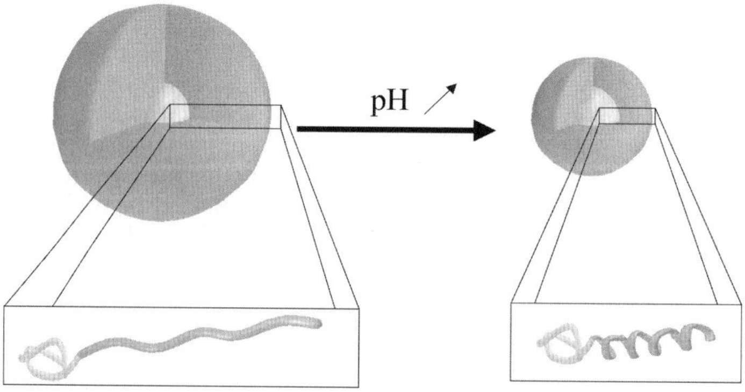

Fig. 6 Schematic representation of the spherical micelles obtained in acid and basic conditions.

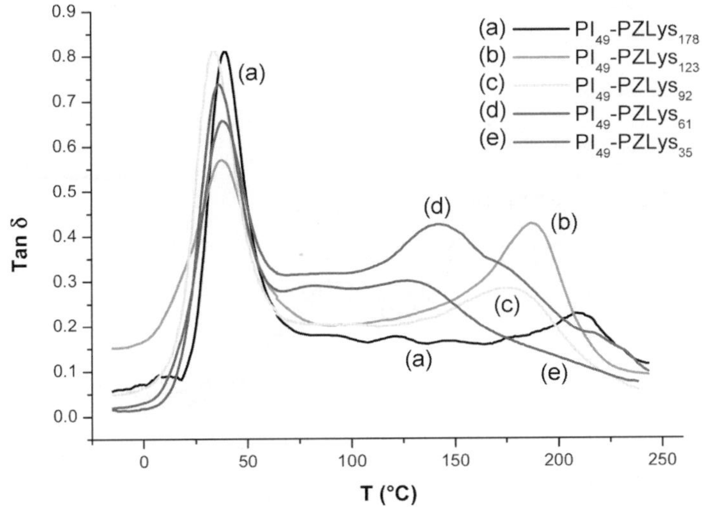

Fig. 7 DMA traces of polyisoprene-*b*-poly(ε-benzyloxycarbonyl-L-lysine) PI-*b*-PZLys block copolymers.

around 220 °C. This is in agreement with earlier investigations on related systems.[25] Fig. 7 plots the temperature dependence of tan δ obtained from DMA experiments.

The presence of the polyisoprene coil in the diblock copolymers is reflected in a glass transition between 30 and 50 °C. For all the block copolymers under study, the T_g is observed in the same temperature range. The observed increase compared to the pure PI_{49}–NH_2 is due to the confinement of the coil blocks in the phase segregated domains, as usually observed in block copolymers. This observed shift in the measured transition is also related to the fact that we are comparing DMA and DSC values (actually, the T_g of PI_{49}–NH_2 was impossible to be determined from DMA because the material was too sticky!). In addition, a second transition was observed for all the block copolymers at higher temperatures. This transition was found to shift to higher temperature with increasing length of the polypeptide block. Fig. 8 represents the evolution of this transition temperature as a function of the degree of polymerization of the PZLys block.

Using WAXS, the transition temperature plotted in Fig. 8 could be assigned to an α-helix to β-sheet conformational transition of the peptide block. An illustrative example of variable

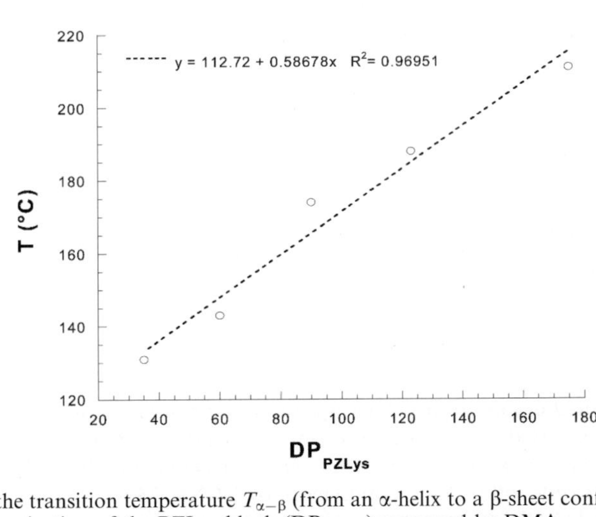

Fig. 8 Evolution of the transition temperature $T_{\alpha-\beta}$ (from an α-helix to a β-sheet conformation) as a function of the degree of polymerization of the PZLys block (DP_{PZLys}) measured by DMA.

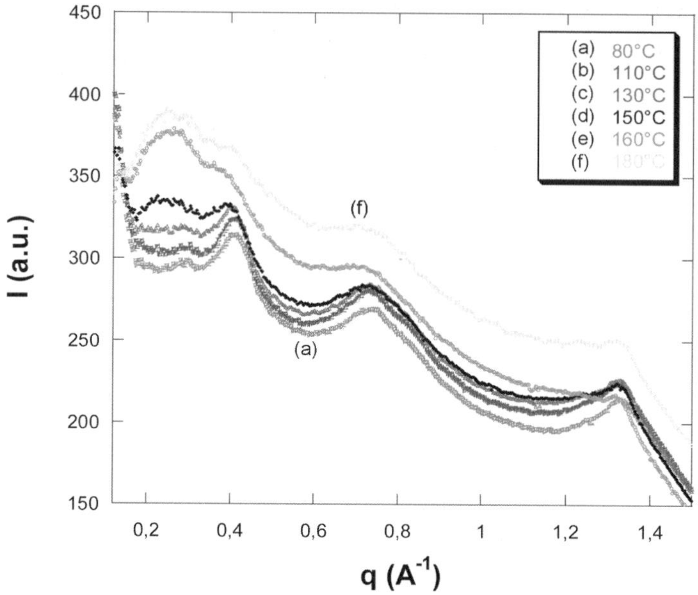

Fig. 9 WAXS patterns of the PI_{49}-*b*-$PZLys_{60}$ block copolymer as a function of temperature.

temperature WAXS spectra obtained for PI_{49}-*b*-$PZLys_{61}$ is represented in Fig. 9. One can see in Fig. 9 that for temperatures up to 130 °C, the SAXS patterns remain identical with characteristic peaks at positions around 0.41 $Å^{-1}$, 0.71 $Å^{-1}$, 0.82 $Å^{-1}$ (ratio $1:3^{1/2}:2$) corresponding to the hexagonal structure of the PZLys chains in their α-helical conformation. At higher temperatures (160 °C and 180 °C), the signal changes and the WAXS patterns display two major Bragg peaks at $q \sim 0.23$ $Å^{-1}$ and $q \sim 0.70$ $Å^{-1}$, respectively corresponding to the d-spacing of the lamellar phase formed by the assembly of the PZLys chains in their β-sheet conformation, and to the distance between two neighboring Z-Lys side-chains.[25] Finally, at an intermediate temperature (150 °C), coexistence of the two structures can be observed, reflecting the gradual conversion of α-helical secondary structures into β-sheets. These observations, identical but shifted in temperature for the other compositions, clearly support the fact that the measured transitions by DMA (Fig. 8) correspond to the α-helix to β-sheet conformational transition of the PZLys blocks. Moreover, the DMA curves show an interesting behavior that is intermediate between that of a thermoplastic elastomer and a semi-crystalline thermoplastic, in the sense that the PI rubber is reinforced by copolymerization with PZLys that creates liquid crystalline interactions. As a result, after the T_g of PI, the rubber plateau which was around 10^2 Pa for the pure polyisoprene increased to values varying from 2×10^6 to 5×10^6 Pa (as a function of the degree of polymerization of the polypeptide block, see Table 2). Moreover, the temperature range of this plateau is enlarged up to the transition from the α-helical to β-sheet conformation of the polypeptide (Fig. 8). At this transition, the storage modulus (E') is increasing up to the thermal degradation of the PZLys chains. This increase in E' is certainly due to the formation of intermolecular hydrogen-bonding during the formation of the β-sheet structure of the polypeptide chains. These relatively high modulus values observed for those low molecular weight compounds, associated with the thermally-induced conformational changes (modifying the local interactions and in the meantime the value of E'), and good film forming properties, make these block copolymers attractive for mechanical purposes.

3.3.2. Nanostructured organizations. The supramolecular organization of the PI-*b*-PZLys block copolymers was investigated by SAXS at room temperature on films prepared from solvent casting. Almost thirty years ago, Gallot *et al.*[26] investigated the solid state behavior of polypeptide–polyvinyl rod–coil diblock copolymers by SAXS. More recently, the self-assembly of similar compounds (PS-*b*-PBLG and PS-*b*-PZLys) in the oligomer regime[14,25] was studied in our

Table 2 SAXS characterization of the polyisoprene-*b*-poly(ε-benzyloxycarbonyl-L-lysine) diblock copolymers

Sample	F_{PZLys}[a]	E'^{b}/Pa	Solvent	q_{max}^{c}/Å^{-1}	Morphology[d]/d-spacing[e]	γ
PI_{49}-*b*-$PZLys_{35}$	0.70	2×10^6	THF/DMF	0.0236	HL (d = 266 Å)	0.9
			Dioxane	—	—	—
PI_{49}-*b*-$PZLys_{61}$	0.80	2.5×10^6	THF/DMF	0.0263	HL (d = 236 Å)	1
			Dioxane	0.034	HL (d = 183 Å)	1.2
PI_{49}-*b*-$PZLys_{92}$	0.86	3.5×10^6	THF/DMF	0.026	HL (d = 243 Å)	1.3
			Dioxane	0.044	HL (d = 141 Å)	2.2
PI_{49}-*b*-$PZLys_{123}$	0.89	5×10^6	THF/DMF	0.0277	HH (d = 261 Å)	—
			Dioxane	0.0319	HH (d = 227 Å)	—
PI_{49}-*b*-$PZLys_{178}$	0.92	4.5×10^6	THF/DMF	0.0285	HH (d = 254 Å)	—
			Dioxane	0.030	HH (d = 241 Å)	—

[a] Calculated using the following densities: polyisoprene, $D = 0.905$ g cm^{-3}, poly(ε-benzyloxycarbonyl-L-lysine) (α-helix), $d = 1.208$ g cm^{-3}. [b] Storage modulus determined by DMA analysis in the "elastomeric" domain. [c] q_{max} represents the position of the first Bragg reflection. [d] HL signifies "hexagonal in lamellar" and HH "hexagonal in hexagonal" morphology. [e] d-spacing (d) is a function of the morphology: for lamellae $d = 2\pi/q_{max}$, for cylinders $d = \sqrt{4/3} \times 2\pi/q_{max}$.

group. Furthermore, Schlaad *et al.*[27] compared the behavior of linear and bottle brushes PS–PZLys copolymers. For all these studies, the different authors only observed, regardless of the volume fraction of the two components, a lamellar morphology of alternating polyvinyl and polypeptide layers. This organization has been called lamellar in hexagonal (LH) morphology referring to the local hexagonal packing of the α-helical polypeptide chains and the phase-segregation into a lamellar structure. Only in one system based on oligomers[14] (PS_{10}-*b*-$PBLG_{20}$), we have evidenced the possibility for such molecules to phase separate into an hexagonal lattice, which was referred to as an hexagonal in hexagonal (HH) morphology. All the SAXS results (position of Bragg reflections and corresponding morphologies) on the PI-*b*-PZLys copolymers under study obtained on films cast from THF/DMF (50/50 v/v) and dioxane are summarized in Table 2. We used these two kinds of non-selective solvents in order to study the casting conditions and in particular the influence of the evaporation speed. The X-ray patterns (results not presented in Table 2) at high scattering vectors ($q > 0.1$ Å^{-1}) show for all the systems obtained from the two solvents the same set of three peaks with Bragg spacing in the ratio $1:\sqrt{3}:2$ which arise from the hexagonal packing of the PZLys in their α-helical conformation. The intercolumnar distance remains constant and is around $d_\alpha \approx 17.7$ Å. Table 2 summarizes the position of the first Bragg reflection (q_{max}), the corresponding morphology and its d-spacing. The data in Table 2 reveal that the hexagonal in lamellar (HL) morphology is present in the copolymers with the lowest rod volume fraction (F_{PZLys}). Indeed, generally one broad scattering peak (Fig. 10) was observed which is typical of this lamellar organization as already observed by Schlaad *et al.*[27] on PS-*b*-PZLys block copolymers. The presence of this single peak, even if the sample has a long-range order, has been attributed to an undulated or zigzag morphology. These fluctuations are supposed to be the result of an hexagonal packing of polydisperse rod-like polypeptides within the layers.

One could gain at this stage information on the conformation of the rod part within the lamellae, as largely done by Gallot.[26] Indeed, dividing the molecular volume by the average lateral extension of a chain (meaning the interface area per chain) and the cross-section of a helix, yields to a parameter γ that is a function of the polypeptide arrangement within the layers (Table 2). From these data, we can conclude that depending on the block length ratio and also on the solvent, the PZLys helices can be stacked ($\gamma = 1$), once-folded ($\gamma = 2$) or partially folded and/or tilted ($1 < \gamma < 2$). Even if dioxane and DMF/THF mixture are both non-selective and good solvents, these results clearly evidence that the solvent interactions and/or the speed of evaporation are major factors in the resulting structures. More interestingly, the data obtained with the highest rod-volume fraction clearly evidence a different morphology with peak ratio $1:\sqrt{3}:2:\sqrt{7}$ as shown in Fig. 10 for instance for PI_{49}-*b*-$PZLys_{178}$ film cast from dioxane. This morphology is observed independent of the solvent and for rod volume fractions $F_{PZLys} \geq 0.89$.

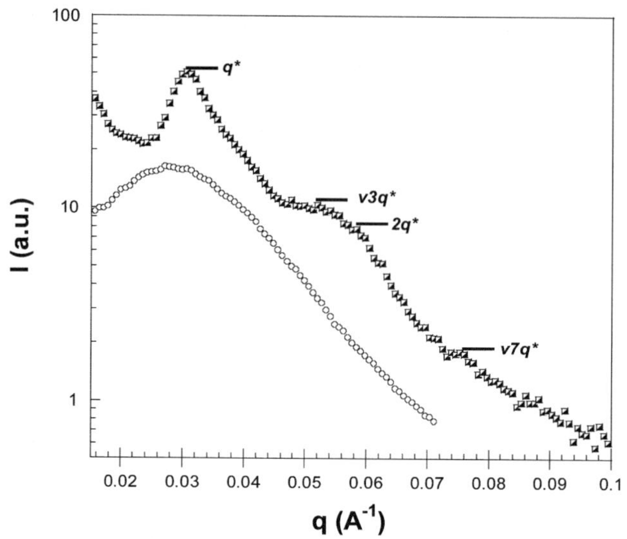

Fig. 10 SAXS patterns of the PI_{49}-*b*-$PZLys_{178}$ (upper curve) and PI_{49}-*b*-$PZLys_{62}$ (lower curve).

3.3.3. Model of bulk organization. A schematic representation of the HL and HH morphologies is given in Fig. 11 and 12, respectively. Fig. 11 describes the hexagonal in lamellar (HL) morphology of PI_{49}-*b*-$PZLys_{92}$ copolymer cast from dioxane solution, taking into account the parameters determined by SAXS (Table 2).

For this system, the geometric factor obtained is close to $\gamma = 2.2$, indicating that the α-helical segments are interdigitated. All the HL structures are based on this basic structure, consisting of alternating layers of PI and PZLys. In addition, all the polyvinyl–polypeptide systems studied so far,[26,27] regardless of the rod (or peptide) volume fraction (in the range 0.15–0.75) and its chemical composition (poly(γ-benzyl-L-glutamate), poly(ε-benzyloxycarbonyl-L-lysine), poly(N-hydroxypropylglutamine), poly(N-trifluoroacetyl-L-lysine)), were forming this HL structure. The data obtained in this work show that when the polypeptide volume fraction is more than 0.89 (or equal), the system undergoes transition to another phase separated structure, the hexagonal in hexagonal (HH) morphology described in Fig. 12.

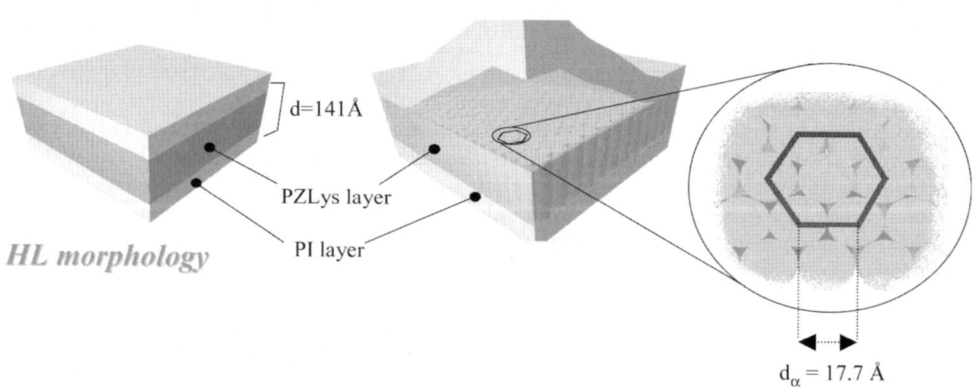

Fig. 11 Schematic model representation of the hexagonal in lamellar (HL) morphology obtained for PI_{49}-*b*-$PZLys_{92}$ copolymer cast from dioxane solution.

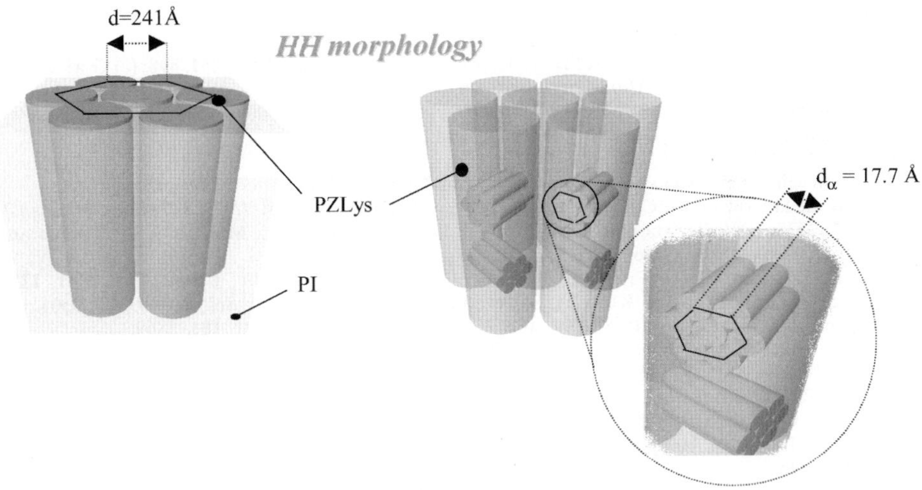

Fig. 12 Schematic model representation of the hexagonal in hexagonal (HH) morphology obtained for PI_{49}-*b*-$PZLys_{178}$ copolymer cast from dioxane solution.

4. Conclusions and outlook

This paper focuses on the use of complex "hybrid" rod–coil block copolymers based on a synthetic coil block and a polypeptide block with an α-helical conformation (rod). We developed a synthetic approach to prepare well defined diblock copolymers combining anionic polymerization and NCA-ring opening polymerization methods. Then, the synthesis, solution and solid state self-assembly properties of a series of polyisoprene-*b*-poly(ε-benzyloxycarbonyl-L-lysine) PI-*b*-PZLys and poly-isoprene-*b*-poly(L-lysine) PI-*b*-PLys block copolymers have been examined, the purpose being to bring an overview of the potentiality of such systems in nanotechnologic applications.

The formation of stimuli-responsive micelles in water has been studied as a function of pH and ionic strength using static and dynamic light scattering, UV-circular dichroism and transmission electron microscopy. The observed change in the micelles dimensions has been directly attributed to conformational transition in the secondary structure of the polypeptide chains. This example illustrates the potentiality of polypeptide-based copolymers in the design of "smart nanocapsules" with a high level of control over the dimensions and shapes through the formation of local intra- and/or inter-molecular interactions.

In bulk, these rod–coil copolymers form self-assembled structures at the nano-scale that have been characterized using DSC, DMA, WAXS and SAXS techniques. Hierarchically ordered structures that we called "hexagonal in lamellar" (HL) and more interestingly "hexagonal in hexagonal" (HH) have been evidenced as a function of the chemical composition. Joining all the efforts done in literature on similar systems, we concluded that the most of them generally formed the HL morphology, excepted when the composition of the rod part becomes very asymmetric (rod volume fraction ≥ 0.89): the system then undergoes transition to a HH morphology. Future work in this project will focus on performing long-range ordering of the structures, associated with a detailed characterization using in particular TEM analyses on microtomed films. Of course, the influence of the modification of the secondary structure of the polypeptides induced by changes of the amino-acids used, their sequence or the solvent used, on the resulting structures would be of interest.

Acknowledgements

S.L. acknowledges financial support from the CNRS, the Ministère des Affaires Etrangères (PROCOPE), la "Région Aquitaine" and FEDER. H.A.K. thanks support by the Deutsche Forschungsgemeinschaft (KL 1049/3-1 and/3-2) and the Deutscher Akademischer Austausch Dienst (PROCOPE, D/0031015).

References

1 (*a*) L. Leibler, *Macromolecules*, 1980, **13**, 1602; (*b*) F. S. Bates, *Science*, 1991, **251**, 898; (*c*) G. H. Fredrickson and F. S. Bates, *Annu. Rev. Mater. Sci.*, 1996, **26**, 501.

2 (*a*) A. Ciferri, *Supramolecular Polymers*, Marcel Dekker, New York, 2000; (*b*) I. W. Hamley, *The Physics of Block Copolymers*, Oxford University Press, Oxford, 1998.

3 G. Riess, *Prog. Polym. Sci.*, 2003, **28**, 1107.

4 Review on self-assembly:H.-A. Klok and S. Lecommandoux, *Adv. Mater.*, 2001, **13**, 1217.

5 (*a*) O. M. de la Cruz and I. C. Sanchez, *Macromolecules*, 1986, **19**, 2501; (*b*) S. T. Milner, *Macromolecules*, 1994, **27**, 2333; (*c*) E. L. Thomas, D. B. Alward, D. J. Kinning, D. C. Martin, D. L. Handlin and L. J. Fetters, *Macromolecules*, 1986, **19**, 2197.

6 (*a*) E. R. Zubarev, M. U. Pralle, E. D. Sone and S. I. Stupp, *J. Am. Chem. Soc.*, 2001, **123**, 4105; (*b*) S. Lecommandoux, H.-A. Klok, M. Sayar and S. I. Stupp, *J. Polym. Sci. A*, 2003, **41**, 3501.

7 Review on rod-coils:M. Lee, B.-K. Cho and W.-C. Zin, *Chem. Rev.*, 2001, **101**, 3869.

8 (*a*) J. F. Marko, *Macromolecules*, 1993, **26**, 1442; (*b*) A. N. Morozov and J. G. E. M. Fraaije, *Macromolecules*, 2001, **34**, 1526; (*c*) R. Borsali, E. Minatti, J.-L. Putaux, M. Schappacher, A. Deffieux, P. Viville, R. Lazzaroni and T. Narayanan, *Langmuir*, 2002, **19**, 6; (*d*) R. L. Lescanec, D. A. Hadjuk, G. Y. Kim, Y. D. Gan, R. Yin, S. M. Gruner, T. E. Hogen-Esch and E. L. Thomas, *Macromolecules*, 1995, **28**, 3485; (*e*) H. Iatrou, N. Hadjichristidis, G. Meier, H. Frielinghaus and M. Monkenbusch, *Macromolecules*, 2002, **35**, 5426.

9 R. Borsali, S. Lecommandoux, R. Pecora and H. Benoît, *Macromolecules*, 2001, **34**, 4229.

10 J. S. Crespo, S. Lecommandoux, R. Borsali, H.-A. Klok and V. Soldi., *Macromolecules*, 2003, **36**, 1253.

11 S. I. Stupp, V. LeBonheur, K. Walker, L. S. Li, K. E. Huggins, M. Keser and A. Amstutz, *Science*, 1997, **276**, 384.

12 J. T. Chen, E. L. Thomas, C. K. Ober and S. S. Hwang, *Macromolecules*, 1995, **28**, 1688.

13 S. A. Jenekhe and X. L. Chen, *Science*, 1999, **283**, 372.

14 H.-A. Klok, J. F. Langenwalter and S. Lecommandoux, *Macromolecules*, 2000, **33**, 7819.

15 (*a*) A. N. Semenov and A. V. Subbotin, *Sov. Phys. JETP A. V.*, 1992, **74**, 690; (*b*) R. R. Netz and M. Schick, *Phys. Rev. Lett.*, 1996, **77**, 302; (*c*) D. R. M. Williams and G. H. Fredrickson, *Macromolecules*, 1992, **25**, 3561.

16 S. A. Jenekhe and X. L. Chen, *Science*, 1998, **279**, 1903.

17 F. Chécot, S. Lecommandoux, Y. Gnanou and H.-A. Klok., *Angew. Chem. Int. Ed. Engl.*, 2002, **41**, 1339.

18 H. Kukula, H. Schlaad, M. Antonietti and S. Forster, *J. Am. Chem. Soc.*, 2002, **124**, 1658.

19 G. Riess, *Prog. Polym. Sci.*, 2003, **28**, 1107.

20 D. S. Poché, M. J. Moore and J. L. Bowles, *Synth. Commun.*, 1999, **29**, 843–854.

21 Anionic polymerization technique has been described elsewhere: H. L. Hsien, R. P. Quirk, in *Anionic polymerization: principles and practical application*, ed. M. Dekker, 1996, p. 212. The solvent mixture was optimized in order to obtain the maximum amount of 3–4 double bond insaturations for further crosslinking purposes (150 ml cyclohexane and 2.15 ml THF).

22 K. Ueda, A. Hirao and S. Nakahama, *Macromolecules*, 1990, **23**, 939.

23 J. Rodríguez Hernández and H.-A. Klok, *J. Polym. Sci. Part A: Polym. Chem.*, 2003, **41**, 1167.

24 (*a*) Y. P. Myer, *Macromolecules*, 1969, **2**, 624; (*b*) A. Harada, S. Cammas and K. Kataoka, *Macromolecules*, 1996, **29**, 6183.

25 S. Lecommandoux, M. F. Achard, J. F. Langenwalter and H.-A. Klok, *Macromolecules*, 2001, **34**, 9100.

26 (*a*) B. Perly, A. Douy and B. Gallot, *Makromol. Chem.*, 1976, **177**, 2569; (*b*) J.-P. Billot, A. Douy and B. Gallot, *Makromol. Chem.*, 1976, **177**, 1889; (*c*) J.-P. Billot, A. Douy and B. Gallot, *Makromol. Chem.*, 1977, **178**, 1641; (*d*) A. Douy and B. Gallot, *Polymer*, 1982, **23**, 1039.

27 H. Schlaad, H. Kukula, B. Smarslya, M. Antonietti and T. Pakula, *Polymer*, 2002, **43**, 5321.

Tunable diblock copolymer micelles–adapting behaviour *via* subtle chemical modifications

Grant B. Webber,*[a] Erica J. Wanless,[b] Steven P. Armes[c] and Simon Biggs[a]

[a] *School of Process, Environmental and Materials Engineering, University of Leeds, Clarendon Road, Leeds LS2 9JT, UK. E-mail: g.b.webber@leeds.ac.uk; Fax: +44 (0) 113 343 2377; Tel: +44 (0) 113 343 2380*
[b] *School of Environmental and Life Sciences, The University of Newcastle, N.S.W. 2308 Australia*
[c] *Department of Chemistry, School of Life Sciences, University of Sussex, Falmer, Brighton, BN1 9QJ, UK*

Received 1st March 2004, Accepted 12th May 2004
First published as an Advance Article on the web 16th September 2004

Selectively quaternising the PDMA block of poly(2-(dimethylamino)ethyl methacrylate)-*block*-poly(2-(diethylamino)ethyl methacrylate) (PDMA–PDEA) copolymers modifies both their solution and adsorption behaviour. These copolymers exist as free unimers in aqueous solution at low pH and form micelles, with PDMA coronas, at high pH. The critical micellisation pH, the hydrodynamic micelle diameter and electrophoretic behaviour are all affected by the degree of quaternisation of the PDMA block. Highly quaternised copolymers form smaller, more highly charged micelles at lower pH than weakly or non-quaternised copolymers. The adsorption of the copolymer micelles onto muscovite mica is studied by *in situ* atomic force microscopy. The adsorbed micelle monolayer becomes increasingly disordered as the degree of quaternisation increases. No micelle desorption occurs on removal of the bulk copolymer solution. Addition of acid to the overlying solution leads to different responses from the surface-adsorbed micelles. Unquaternised micelles undergo a reversible change in morphology due to the formation of localised polymer brushes, whereas lightly quaternised micelles are characterised by irreversible changes. Highly quaternised micelle monolayers are disrupted by the addition of acid. Such differences can be rationalised by simple electrostatic arguments. This behaviour has been confirmed by quartz crystal microbalance studies, which show that the adsorbed mass decreases with increasing degrees of quaternisation.

1. Introduction

The self-assembly of block copolymers into structures with highly-ordered morphologies is often exploited in the preparation of novel "nanomaterials".[1–4] The self-assembled polymeric structure can represent either the final material itself,[5–7] or merely serve as a scaffold for further assembly processes.[8–11] Bulk self-assembly of block copolymers is routinely used in the preparation of mesoporous SiO_2 and TiO_2 networks *via* the sol–gel process,[12] while block copolymer thin-films can be used to prepare nano-sized silica particles,[13] or metal sols based on gold,[11,14–16] iron,[16] platinum[15] and palladium.[15,17]

The ever-expanding range of nanomaterials, and the application-specific demands placed upon those materials, has led to a shift in focus from so called "top-down" manufacturing methods, to techniques described as "bottom-up".[18–24] Recent advances in lithographic techniques allow the fabrication of structures with length scales of a few tens of nanometres.[10,25] However such techniques are, in general, expensive, time-consuming and require multiple processing steps, thereby increasing the potential for the loss of sample integrity. Block copolymer templates offer real advantages for the efficient generation of nano-scaled surface structures. As these techniques become more refined it is important that they adapt to meet the needs of industry today, satisfying issues of function, relevance, cost and zero emissions. Consequently, attention is shifting toward simple, aqueous-based *in situ* techniques, which eliminate the requirement for expensive, environmentally-unfriendly organic solvents and remove time-consuming drying steps.

The self-assembly of (co)polymers in bulk solution and at interfaces has been studied extensively, both experimentally and theoretically, in recent years.[26–36] Various research groups have produced self-assembled block copolymer aggregates of different shapes by changing parameters such as copolymer molecular weight, block ratio or architecture, solvent quality, or copolymer concentration.[37–48] A significant limitation with many of the diblock copolymer systems studied to date is the non-equilibrium nature of the self-assembled aggregates. This is a particular problem if one block has a high glass transition temperature (T_g) or when micellisation is initiated by the addition of a solvent that is poor for one of the blocks. In such cases kinetically frozen aggregates are usually formed, so the self-assembly is irreversible, and the final aggregate structure is dependent on the method of their formation.[30,31,39]

Diblock copolymers have also been shown to self-assemble to form adsorbed layers of diverse, well-defined morphologies on a variety of substrates, either *via* the organisation of copolymer unimers during an annealing process, or by direct adsorption of solution aggregates.[49–56] However, many studies utilise *ex situ* techniques, such as transmission or scanning electron microscopy (TEM or SEM)[42,51,53–55] or atomic force microscopy (AFM)[49,50,52] in air to probe the morphology of copolymer adsorbed layer. Such methods require drying of the polymer layer, which in turn may influence the final observed morphology, and prevent *in situ* processing of the layer by changing the overlying solution or adsorbing further material. To date, there are few studies that utilise *in situ* techniques to monitor the development of diblock copolymer adsorbed layers.[57–60]

Herein we show that subtle chemical modification (selective quaternisation) of a PDMA–PDEA diblock copolymer significantly alters both its solution and adsorption properties. The degree of quaternisation affects the micelle dimensions, as well as the spatial arrangement of micelles in the adsorbed layer. Additionally, the transition in adsorbed layer morphology in reaction to pH alteration is dependent on the degree of quaternisation.

2. Experimental

Diblock copolymer synthesis

The diblock copolymer poly(2-(dimethylamino)ethyl methacrylate)-*block*-poly(2-(diethylamino)ethyl methacrylate) (PDMA-PDEA, Fig. 1) used in the present study was synthesised using group transfer polymerisation as described elsewhere by Bütün *et al.*[48] The molecular weight of this precursor diblock was 19 100 g mol^{-1} as measured by gel permeation chromatography (THF eluent, PMMA calibration standards), and the PDMA content was determined to be 79 mol% by nuclear magnetic resonance (NMR) spectroscopy. Based on these data, the degree of polymerisation was calculated to be 93 for the PDMA block and 25 for the PDEA block. The tertiary amine residues on the PDMA chain of the precursor diblock were selectively quaternised using a substoichiometric amount of methyl iodide (MeI), as described previously.[61] The degree of quaternisation of the PDMA block was assessed by NMR spectroscopy. Three quaternisations were performed, resulting in degrees of quaternisation of 10, 50 and 100 mol%, respectively. Hereafter, the diblock copolymer samples are referred to as XqPDMA–PDEA, where X refers to the level of quaternisation.

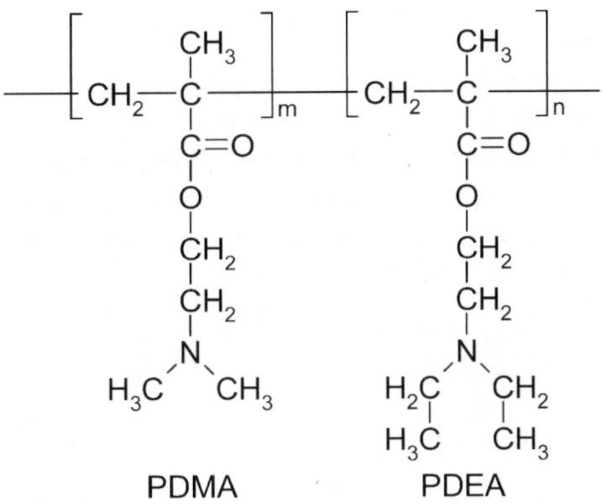

Fig. 1 Structure of the unquaternised PDMA–PDEA diblock copolymer, with $m = 93$ and $n = 25$. Quaternisation of the tertiary amine residues with methyl iodide is restricted to the PDMA block and is random along its length.

Potentiometric titrations

The aqueous solution properties of the partially quaternised copolymers were assessed by potentiometric titration, fluorescence spectroscopy, dynamic light scattering and zeta potential measurements. All aqueous solutions were prepared by molecularly dissolving the copolymer in dilute HNO_3 (pH 3; Milli-Q grade water), with constant background electrolyte of 10^{-2} M KNO_3 (Ajax Chemicals/Sigma-Aldrich, AR grade, used without further purification). Where required, pH was adjusted using small quantities of HNO_3 or KOH, allowing 24 h for solutions to equilibrate after pH adjustment, and solution pH was remeasured immediately prior to use. Potentiometric titrations were performed by addition of 0.05 M KOH to 10 mL aliquots of 500 ppm solutions of the copolymers. The initial pH of the copolymer solution was ~2.25 and, after titration, the final solution pH was 10.5.

Fluorescence spectroscopy

A Perkin-Elmer LS-5B Luminescence Spectrometer was used for the fluorescence studies undertaken in this work, with pyrene used as the chromophore. The ratios I_1/I_3, where I_1 and I_3 are the intensities of the first and third peak respectively in the emission spectrum ($\lambda_{ex} = 340$ nm), and I_{338}/I_{333}, where I_{338} and I_{333} are the intensities at 338 and 333 nm respectively in the excitation spectrum ($\lambda_{em} = 420$ nm), were monitored as a function of pH for aqueous copolymer solutions. These ratios are sensitive to the local pyrene environment. In particular, a higher I_1/I_3 ratio indicates a more polar pyrene environment, while the reverse behaviour is observed for the I_{338}/I_{333} ratio. Pyrene is particularly useful for detecting the presence of micellar aggregates in aqueous solution due to its extremely poor solubility in water; pyrene will preferentially partition into the non-polar, hydrophobic regions of the aggregates when present, which in turn results in a change in the intensity ratios.[62,63]

Dynamic light scattering

An ALV/DLS/SLS-5022F dynamic light scattering instrument (ALV, Germany) was used to assess changes in micelle radius as a function of quaternisation level. Scattering data was collected at a range of detection angles from 30° to 120°. The measured scattering data was fitted using number-weighted, non-linear single exponential analysis to determine the mean hydrodynamic radii (R_H). The raw scattering intensity, at constant incident intensity, was also recorded. It was found that the calculated hydrodynamic radius was independent of detection angle. Additionally, there was a narrow distribution of micelles sizes at each angle. Thus, for simplicity, only data collected at 90° is presented. All copolymer solutions analysed by DLS were passed through a syringe-mounted 0.2 µm filter (GHP Acrodisc, Pall Gelman Sciences, MI) as they were loaded into the measuring cell.

Zeta potential

A Malvern NanoSeries ZetaSizer (Malvern, UK) was used to measure the zeta potential of 500 ppm aqueous solutions of the diblock copolymers. Zeta potential measurements were recorded at a solution pH of 9.0 ± 0.1, with at least 500 individual measurements recorded and averaged for each sample.

Atomic force microscopy

In situ imaging of the adsorbed layer of the four copolymers on muscovite mica was performed with either a Nanoscope III or Nanoscope IV AFM (Veeco, USA). Cantilevers with an integral silicon nitride tip (NanoProbe, Veeco, USA) were used for all AFM experiments, and were cleaned using UV irradiation immediately prior to use. The muscovite mica substrates were freshly cleaved before AFM assembly. Buffer and copolymer solutions were passed through a syringe-mounted filter as they were injected into the AFM fluid cell. Images were collected using the soft-contact method,[64] which uses the minimum force necessary to obtain an image, thereby minimising scanning-induced deformations of the adsorbed layer. All images presented are deflection images, and have been zero-order flattened. At the beginning of each experiment, Milli-Q water, adjusted to the pH of the copolymer solution under investigation, was injected into the cell and left for approximately 30 min to allow the mica to reach charge equilibrium. This also allowed the cleanliness of the mica substrate to be assessed prior to the injection of the copolymer sample.

Quartz crystal microbalance

A commercial (Q-Sense, Göteburg, Sweden) quartz crystal microbalance with dissipation function (QCM-D) was used to measure adsorbed amounts. The features of this instrument are described in greater detail elsewhere.[65,66] The QCM technique for measuring adsorbed masses has attracted much interest in recent years, since its adaptation for use with liquid systems. In a typical QCM experiment, the resonant frequency of an AT-cut piezoelectric quartz crystal is monitored as it changes due to adsorption of materials onto the substrate. The adsorbed mass can then be calculated by applying the Sauerbrey equation:[67]

$$\Delta f = -\frac{f_0^2}{\rho_q v_q}\Delta m = -\frac{f_0}{\rho_q t_q}\Delta m = -C\Delta m \tag{1}$$

where Δf is the change in resonant frequency, f_0 is the resonant frequency of the 'clean' substrate, ρ_q and ν_q are the specific density and shear wave velocity of the quartz, t_q is the thickness of the quartz crystal, and Δm is the mass of material adsorbed. The Sauerbrey equation was developed for QCM instruments operating under vacuum conditions; its applicability in fluid environments lessens when the adsorbed material is viscoelastic in nature, since the crystal oscillations are damped by the adsorbed film. Furthermore, the mass measured using the QCM technique is the total mass of material coupled to the oscillation. Consequently, solvent molecules bound to the adsorbed layer will also contribute to the measured mass and adsorbed amounts determined using QCM exceed those measured using optical techniques such as ellipsometry or optical reflectometry.

An extra feature of the Q-Sense quartz crystal microbalance is the simultaneous measurement of dissipation. Here, the voltage driving the oscillation of the crystal is turned off, and the decay of the oscillation is monitored. For a rigid layer, the decay is relatively slow, whilst for a viscoelastic layer the decay is relatively fast, due to the dampening effect of the adsorbed layer. Thus, by comparing values of dissipation it is possible to assess the relative rheological properties of adsorbed layers. The QCM-D data presented here are those measured at the third overtone ($\sim$15 MHz), since this overtone offers better signal-to-noise than the fundamental frequency and is less affected by the cell mounting than the higher (5th and 7th) overtones. These data were also used in calculations involving the Sauerbrey equation.

A single sensor crystal, with a silica coating, was used for all QCM-D experiments. At the beginning of each experiment, Milli-Q water (adjusted to the pH of the copolymer solution to be studied) was passed into the cell and left for at least one hour. This enabled the silica surface charge to equilibrate to the solution pH, and also allowed the measurement cell to "settle" after assembly.

Experience has shown that the values of resonant frequency and dissipation, across all overtones, drift in the period immediately after assembly. This is most likely due to thermal fluctuations and relaxation of the rubber O-ring that seals the cell/sensor assembly.[68]

3. Results and discussion

The tertiary amine residues of PDMA–PDEA block copolymers become protonated at low pH, enabling their molecular dissolution as unimers in aqueous solution. As the solution pH is increased, both blocks become progressively deprotonated, until at some critical pH the difference in hydrophobicity between the two blocks leads to micellar self-assembly. Previous studies have shown that the PDEA blocks form the micelle cores, with the more hydrophilic PDMA blocks being located in the corona.[48] Thus, varying the degree of quaternisation of the PDMA block affects the cationic charge density of the coronal chains.

Potentiometric titrations were performed to assess effect of varying the degree of quaternisation on the solution deprotonation of the PDMA–PDEA copolymer, Fig. 2. The titration curves for each copolymer show a characteristic profile, with two inflection points indicating the onset (at low pH) and cessation (at high pH) of deprotonation of the tertiary amine residues. The lack of a clear break-point, and thus pK_a value, is characteristic of weak polyelectrolytes, since the removal (or addition) of each proton from the polymer chain affects the ease with which the remaining protons can be removed (or added).[69] Quaternisation of (some of) the DMA residues necessarily precludes (de)protonation, thus the number of titratable amine residues decreases as the degree of quaternisation is increased. However, the pH at which the inflections occur is relatively constant, indicating that the ease of (de)protonation is unaltered by the degree of quaternisation. Furthermore, there was no evidence for precipitation in any of the copolymer solutions across the pH range studied, signifying that the copolymers micelles, once formed, remained colloidally stable up to pH 10.5.

Fluorescence spectroscopy, using a pyrene chromophore, was employed to investigate the formation of micellar aggregates in response to changes in solution pH. The I_1/I_3 and I_{338}/I_{333} intensity ratios, measured from the emission and excitation spectra respectively, were monitored as a function of solution pH for a 500 ppm solution of 0qPDMA–PDEA, Fig. 3a. Clear transitions are observed within a narrow pH range of 7.2–8.2 for both ratios. This indicates that the 0qPDMA–PDEA copolymer is molecularly dissolved below this pH range, and forms micelles at higher pH. Cycling of the solution pH above and below this critical region confirmed that the micellar self-assembly (and dissociation) was reversible. Aqueous solutions of the selectively quaternised copolymers show similar reversible pH-dependent micellisation behaviour. A schematic representation of this reversible pH-induced micellisation behaviour is shown in Fig. 3b.

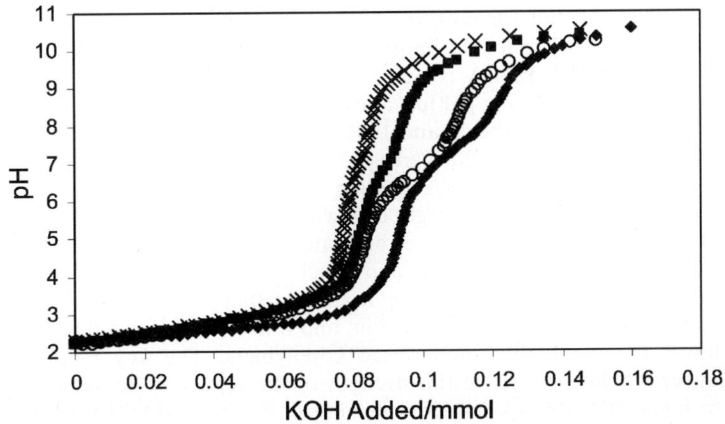

Fig. 2 Potentiometric titrations of 500 ppm, 10^{-2} M KNO_3 aqueous solutions of the $PDMA_{93}$–$PDEA_{25}$ based diblock copolymers; (♦) 0qPDMA–PDEA, (○) 10qPDMA–PDEA, (■) 50qPDMA–PDEA and (×) 100qPDMA–PDEA.

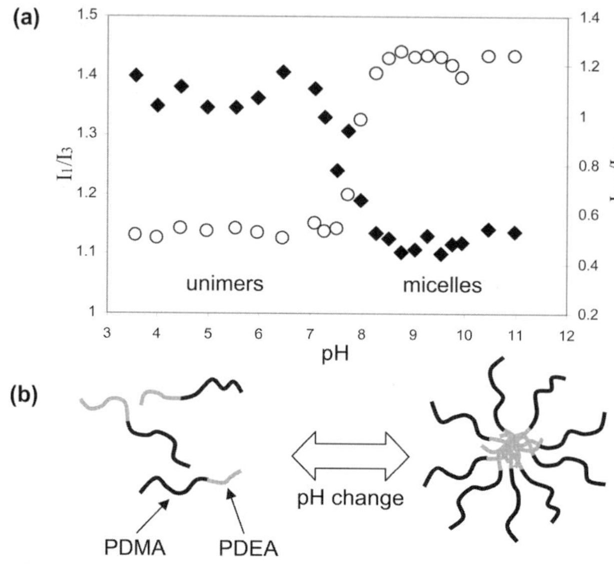

Fig. 3 (a) Fluorescence intensity ratios (♦) I_1/I_3 and (○) I_{338}/I_{333}, from emission and excitation spectra respectively, as a function of pH for 500 ppm, 10^{-2} M KNO_3 aqueous solution of 0qPDMA–PDEA. (b) Schematic representation of the reversible, pH-induced micellisation process observed in aqueous solutions of PDMA–PDEA.

The size of the micellar aggregates as a function of solution pH was studied by dynamic light scattering, Fig. 4. As with the fluorescence data, the light scattering intensity is strongly pH-dependent. Weak scattering is observed below pH 8, Fig. 4a. At higher pH, the scattered light intensity increases significantly; this effect is most noticeable for 0qPDMA–PDEA, though it is apparent for all the copolymers. The increase in light scattering is due to the formation of micelles in solution.[46,70] It is obvious that the critical micellisation pH (cm-pH) is affected by the degree of quaternisation of the PDMA block. The highly quaternised copolymers form micelles in aqueous solution above pH ~7.8, while the lightly quaternised copolymers and unquaternised copolymer form micelles above pH ~8.3.

The degree of quaternisation influences the size of the micelles formed by the PDMA–PDEA block copolymers, Fig. 4b. For the unquaternised diblock, the hydrodynamic radius, R_H, of the micelles is approximately 11 nm. This R_H value is essentially constant for all pH values greater than the cm-pH. For degrees of quaternisation of 50% and 100%, the micelle dimensions are relatively independent of solution pH, but for both these diblocks the micellar R_H is approximately 4 nm, which is substantially smaller than for the unquaternised diblock. Interestingly, the behaviour of the lightly quaternised 10qPDMA–PDEA is quite different, since the size of these micelles is pH-dependent. For example, at a solution pH just above the cm-pH, the measured R_H is ~4 nm, which is similar to the size of micelles of the 50 and 100% quaternised diblock copolymers. However, as the solution pH is increased further, R_H increases, reaching ~9 nm at pH 10. A summary of the solution data for the diblock copolymers, including micellar size and charge, is given in Table 1.

On the basis of the coronal chain charge, one would expect the hydrodynamic radius of the copolymer micelles to increase with the degree of quaternisation; highly quaternised PDMA chains will adopt a more extended conformation and repel each other more strongly. However, the opposite trend is observed. These changes in R_H imply that the aggregation of the copolymer micelles changes with the degree of quaternisation, and it may be inferred that micelle aggregation numbers are lower for the highly quaternised copolymers. The key factor determining the aggregation number would seem to be the strength of the repulsion between the coronal chains since the core-forming PDEA block remains unchanged. For the 0qPDMA–PDEA diblock above pH 8, the formation of larger micelles must be due to reduced repulsion between coronal chains compared to the 50q or 100q diblocks. The 10qPDMA–PDEA copolymer exhibits intermediate behaviour, although the reasons for this are as yet unclear.

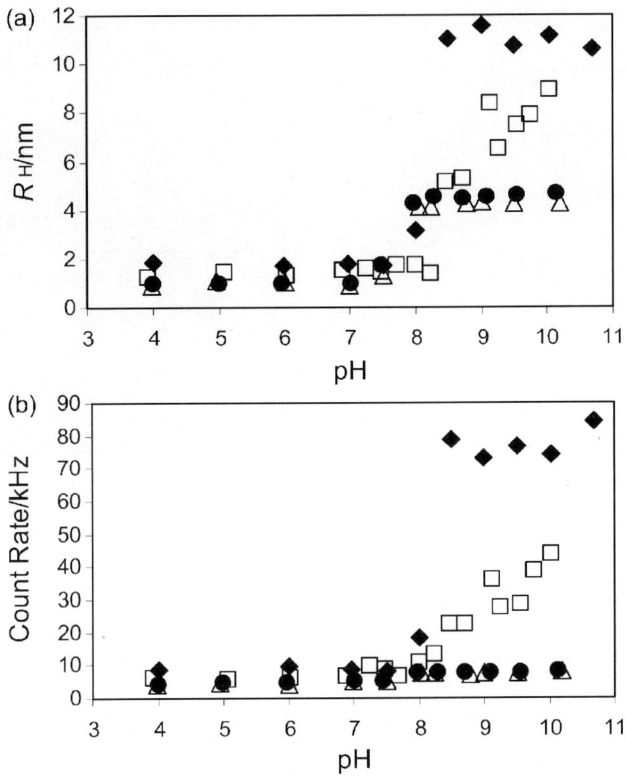

Fig. 4 (a) Calculated hydrodynamic radius and (b) measured scattering intensity as a function of pH for 500 ppm, 10^{-2} M KNO_3 aqueous solutions of the diblock copolymers (◆) 0qPDMA-PDEA, (□) 10qPDMA-PDEA, (●) 50qPDMA-PDEA and (△) 100qPDMA-PDEA.

The adsorption of the diblock copolymers from 500 ppm aqueous solutions onto muscovite mica substrates has been investigated using *in situ* atomic force microscopy. A soft-contact image of a bare mica surface and 0qPDMA–PDEA adsorbed from a pH 4 solution is shown in Fig. 5. The image of the bare mica, Fig. 5a, is representative of the freshly cleaved surfaces used in all AFM experiments; surfaces showing any sign of contamination were re-cleaved. The image of 0qPDMA–PDEA adsorbed from a pH solution, Fig. 5b, was recorded after the AFM fluid cell was flushed with copious amounts of pH 4 water. The image is laterally featureless, and shows no indication of any structures formed by association of the copolymer unimers at the mica surface. Images recorded throughout the experiment, both before and after replacing the bulk copolymer concentration with pure solvent, all show a similar lack of morphology, indicating that the adsorbed layer is resilient to the removal of the bulk copolymer concentration. Polymer adsorbed layers are well known to exhibit very slow rates of desorption, due to the large number of 'attachment points' per polymer chain.[71]

Table 1 Characteristics of PDMA–PDEA copolymers in 500 ppm, 10^{-2} M KNO_3 aqueous solution

PDMA quaternisation (%)	cm-pH[a]	Zeta potential[b]/mV	R_H[b]/nm
0	8.3 ± 0.1	+10.82	11.56
10	8.3 ± 0.1	+17.53	6.49
50	7.8 ± 0.1	+22.62	4.54
100	7.8 ± 0.1	+26.80	4.32

[a] Values of cm-pH taken from dynamic light scattering data. [b] Values for micelles in solutions at pH 9.

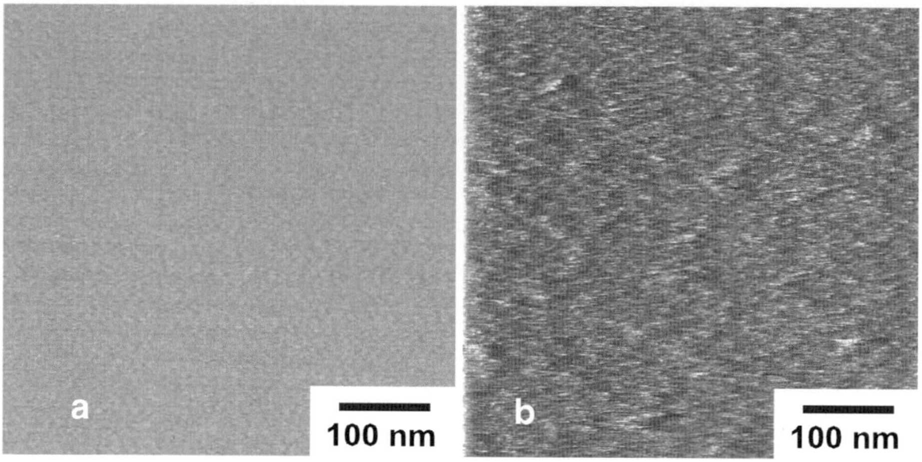

Fig. 5 Soft-contact *in situ* AFM images of (a) bare mica in pH 4, 10^{-2} M KNO_3 aqueous solution and (b) 0qPDMA–PDEA adsorbed from a 500 ppm, 10^{-2} M KNO_3, pH 4, aqueous solution.

Protonation of both the DMA and DEA residues occurs in acidic solution. Thus, independent of their degree of quaternisation, all the copolymers exist as highly cationic polyelectrolyte chains at low pH and similar adsorption behaviour was expected under these conditions. This has been verified by quartz crystal microbalance (QCM) measurements.

QCM data collected for the adsorption of 0qPDMA–PDEA from a 500 ppm solution onto a silica crystal are shown in Fig. 6. The quaternised copolymers behaved similarly, but these data are not shown for clarity. Values for Δf_3, the change in resonant frequency of the third overtone, dissipation, adsorbed mass and layer thickness for all the copolymers, measured 24 h after the onset of adsorption, are summarised in Table 2. The adsorbed mass calculated using the Sauerbrey equation is directly proportional to the change in resonant frequency, and thus follows the same trend, while the layer thickness is calculated using the Voight viscoelastic representation.

It is clear from the data in Fig. 6 that there are two stages in the adsorption of the PDMA–PDEA unimers. The electrostatic attraction between the cationic chains and the bare anionic silica surface leads to rapid initial adsorption, which accounts for around one-third of the overall adsorbed amount. This is followed by a much slower adsorption regime, where the adsorbed unimers undergo conformational changes to allow further adsorption of copolymer from the bulk. Such behaviour is common for the adsorption of strong polyelectrolytes onto charged surfaces.[72] The data in Table 2 show that the adsorption behaviour of the copolymers at pH 4 is insensitive to their degree of quaternisation. There is very little variance in the amount of copolymer adsorbed after 24 h and, apart from the 10qPDMA–PDEA copolymer, the calculated layer thickness values are also consistent.

Micelle adsorption from 500 ppm PDMA–PDEA copolymer solutions at pH 9 has also been studied by AFM and QCM-D. Unlike the unimers, the micelle adsorption behaviour is significantly affected by the degree of quaternisation. This is reasonable, since both the micelle dimensions and charge density strongly depend on the degree of quaternisation. In a typical AFM experiment, the fluid cell was flushed with pH 9 water at least four hours after the injection of the sample solution. The layer was then left immersed at pH 9 for at least six hours. For all copolymers this resulted in an increase in image clarity and definition, but did not cause any discernible change in adsorbed layer morphology. The increase in image resolution after flushing with water is due to the removal of weakly or non-adsorbed copolymer located at the adsorbed layer/solution interface, which interact with the AFM cantilever as it scans across the surface. It is also likely that adsorption of the copolymer will result in a polymer depletion layer close to the interface. The existence of a depletion layer will result in an attractive interaction between the cantilever and the interface and may decrease image resolution. Such a depletion layer is not present after the bulk copolymer solution is

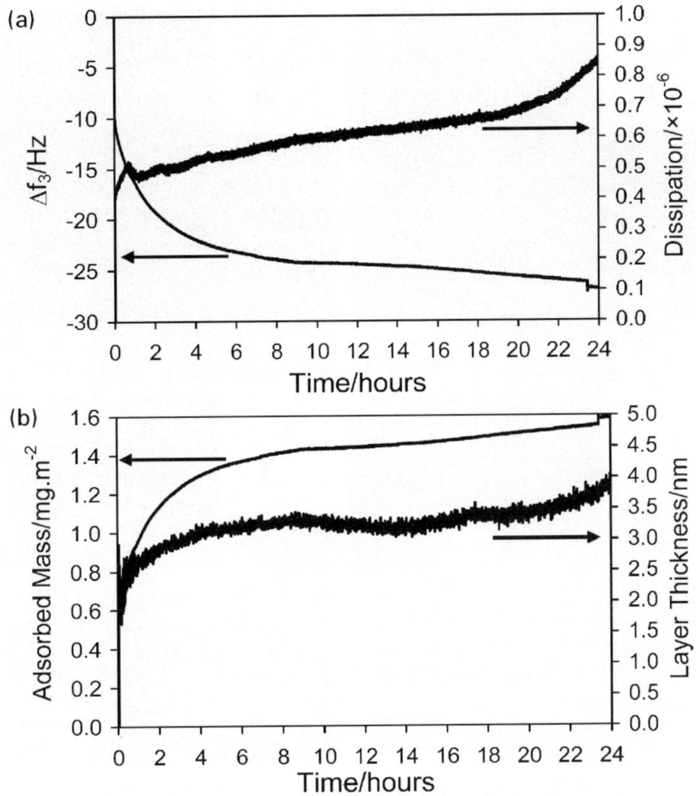

Fig. 6 QCM-D data for the adsorption of 0qPDMA–PDEA from a 500 ppm, 10^{-2} M KNO_3, pH 4 solution onto a silica sensor crystal; (a) change in the third overtone resonant frequency (Δf_3) and value of dissipation, and (b) the mass and thickness of the adsorbed layer as a function of time.

exchanged for solvent. All images of the adsorbed micelle monolayers presented here were recorded after this initial rinsing of the fluid cell.

Soft-contact *in situ* AFM images of the adsorption of 0qPDMA–PDEA micelles are shown in Fig. 7. The adsorbed micelles form close-packed monolayers, with localised hexagonal ordering apparent, Fig. 7a. Additionally, as evidenced by Fig. 7b, there is significant long-range structural order. From the images, the mean centre-to-centre adsorbed micelle spacing was determined to be 30 ± 2 nm. This value is larger than the hydrodynamic diameter of the micelles in solution, and indicates that the micelles adopt a flattened conformation at the interface. Each copolymer micelle adsorbs to the mica substrate *via* electrostatic interactions between the cationic coronal chains and anionic surface sites. The mobility of these chains is restricted by the points of initial adsorption. It is likely, therefore, that unadsorbed, highly mobile coronal chains on the sides and top of the micelle will relax down to the mica substrate and adsorb at surrounding exposed charged sites. This collapsed corona model of diblock copolymer micelle adsorption has been proposed previously.[58,73]

Table 2 Summary of QCM-D data for the adsorption of the PDMA–PDEA copolymers from 500 ppm, 10^{-2} M KNO_3, pH 4 aqueous solution. Data collected 24 h after sample injection.

Copolymer quaternisation (%)	Δf_3/Hz	Dissipation ($\times 10^{-6}$)	Adsorbed mass/mg m^{-2}	Layer thickness/nm
0	−44.87	0.85	2.65	3.9
10	−37.35	0.22	2.20	1.4
50	−38.85	1.15	2.29	2.9
100	−43.79	0.75	2.58	4.0

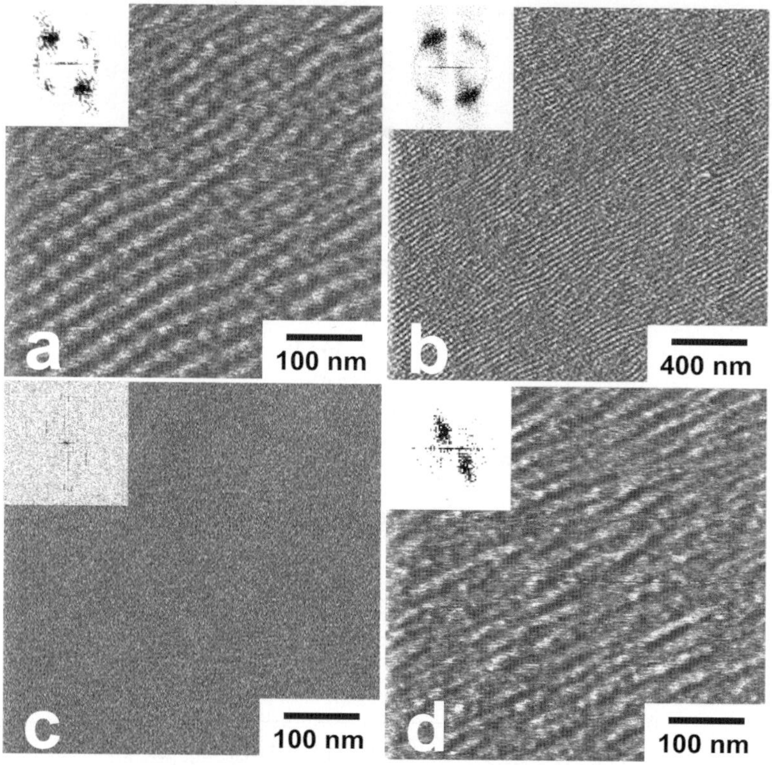

Fig. 7 Soft-contact *in situ* AFM images of the adsorbed layer of 0qPDMA–PDEA micelles adsorbed from a 500 ppm, 10^{-2} M KNO_3, pH 9 aqueous solution onto muscovite mica. Images were recorded (a, b) after the fluid cell was flushed with pH 9 water, (c) after flushing with pH 4 water and (d) after flushing with pH 9 water again. Insets show the respective Fast Fourier Transforms (FFTs).

Calculated values of the mean centre-to-centre inter-aggregate spacing in the adsorbed layer of micelles for all the copolymers are given in Table 3.

In bulk aqueous solution it is well known that the copolymer micelles dissociate rapidly to form unimers on addition of acid.[59,74] We wished to examine whether the surface influenced the dissociation behaviour of the micelle, hence the response of the *adsorbed* copolymer micelles to changes in the solution pH was investigated. The pH of the bulk solution above the adsorbed layer of 0qPDMA–PDEA micelles was adjusted from pH 9 to pH 4. AFM images revealed that the adsorbed micelles underwent a dramatic morphological transition, with a laterally featureless layer being observed at low pH, Fig. 7c. We believe that protonation of the adsorbed copolymer chains increases the electrostatic repulsions within the layer, causing the chains to extend into solution to form a localised polymer brush. Such polymer brushes are *always* observed when the aqueous

Table 3 Mean aggregate centre-centre spacing, calculated from AFM images, and summary of QCM-D data, for the adsorption of the PDMA–PDEA copolymers from 500 ppm, 10^{-2} M KNO_3, pH 9 aqueous solution. QCM-D data recorded 24 h after sample injection

Copolymer quaternisation (%)	Aggregate spacing/nm	Δf_3/Hz	Dissipation ($\times 10^{-6}$)	Adsorbed mass/mg m^{-2}	Layer thickness/nm
0	30 ± 2	−82.08	1.88	14.52	17.1
10	30 ± 3	−62.63	2.93	11.07	11.5
50	40 ± 8	−47.89	4.61	8.47	20.2
100	44 ± 4	−15.16	0.68	2.68	5.8

solution in the fluid cell is at pH 4. Furthermore, when the AFM fluid cell is readjusted to pH 9, the polymer brush layer collapses and, to a first approximation, the original micelle morphology is regained, Fig. 7d. While these re-formed micelles are not as well defined as the initial surface micelles observed prior to pH cycling, discrete aggregates are nevertheless clearly evident within the layer.

Images of the comparable AFM study with the lightly quaternised 10qPDMA–PDEA copolymer are shown in Fig. 8. The initial adsorbed layer, Figs. 8a and b, is similar to that observed for the unquaternised copolymer and, though less ordered, some regions of hexagonal close-packing are still evident. The centre-to-centre inter-aggregate spacing was calculated to be 30 ± 3 nm, which is more than twice the solution hydrodynamic diameter of ~ 13 nm. Thus, the inter-aggregate spacing is greater, relative to the micellar hydrodynamic size in solution, than for the unquaternised copolymer, indicating that the collapse of the coronal chains is more pronounced in this case. This is to be expected, given the increased charge density on the 10qPDMA–PDEA micelles at pH 9 compared to the 0qPDMA–PDEA micelles. Similar changes in structural order have been previously observed by Johnson and Lenhoff for the adsorption of latex particles, though in this case the electrostatic interactions were modified by the use of electrolyte.[75]

Interestingly, the layer of adsorbed 10qPDMA–PDEA micelles responds quite differently to the unquaternised micelles with respect to pH cycling. In this case, the layer retains its symmetry after the AFM cell is flushed with pH 4 water. However, the micelles appear to have swollen to some extent, presumably due to the increased intra-micelle electrostatic repulsions caused by protonation. We believe that the greater inter-micelle spacing (larger 'footprint') allows a greater proportion of the PDMA coronal chains to adsorb to the mica surface, and hence results in a greater number of attachment points per chain. Consequently, the ability of the copolymer chains to extend into

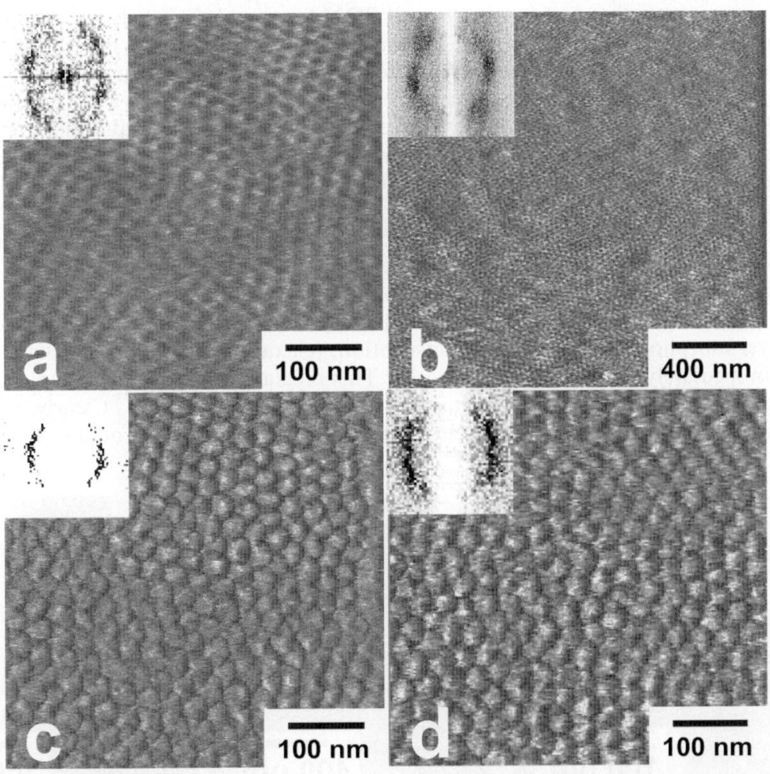

Fig. 8 Soft-contact *in situ* AFM images of the adsorbed layer of 10qPDMA–PDEA micelles adsorbed from a 500 ppm, 10^{-2} M KNO_3, pH 9 aqueous solution onto muscovite mica. Images were recorded (a, b) after the fluid cell was flushed with pH 9 water, (c) after flushing with pH 4 water and (d) after flushing with pH 9 water again. Insets show the respective FFTs.

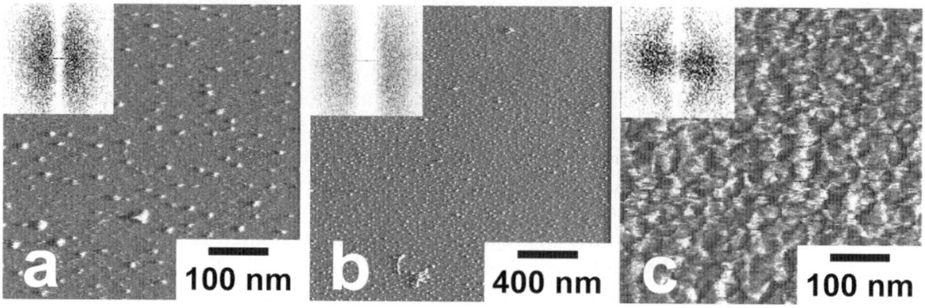

Fig. 9 Soft-contact *in situ* AFM images of the adsorbed layer of 50qPDMA–PDEA micelles adsorbed from a 500 ppm, 10^{-2} M KNO_3, pH 9 aqueous solution onto muscovite mica. Images were recorded (a, b) after the fluid cell was flushed with pH 9 water and (c) after flushing with pH 4 water. Insets show the respective FFTs.

solution at lower solution pH is hindered, leading instead to "swelling" of the micelles, rather than formation of an extended polymer brush. Furthermore, this swollen micelle morphology is *preserved* after pH cycling, rather than returning to the approximate original morphology as was the case for the unquaternised micelles. So, in the case of 10qPDMA–PDEA, the adsorbed micelle layer responds irreversibly to pH cycling, with the change in morphology limited to swelling of the adsorbed micelle layer.

AFM images reveal the adsorption behaviour of 50qPDMA-PDEA and 100qPDMA-PDEA micelles to be quite similar, Figs. 9 and 10 respectively. In these images the white structures are inferred to be the hydrophobic PDEA cores, positioned at the points of original micelle adsorption, located within a surrounding matrix of collapsed PDMA coronal chains, as previously described.[58] In both cases the mean inter-aggregate distance is more than four times greater than the hydrodynamic diameter of the respective micelles in solution. This indicates that, for both copolymers, the micelles are widely spaced within the adsorbed layer, as a result of dramatic coronal collapse during adsorption. There is no evidence for any structural order over either short or long length scales. Addition of acid to the overlying bulk solution has an irreversible, deleterious effect on the adsorbed layers formed by both copolymers, with complete loss of the original surface morphology.

A schematic representation of the adsorption of this PDMA–PDEA family of copolymer micelles, and the response of the respective layers towards pH cycling, is shown in Fig. 11. We believe that the mechanism of micelle adsorption is similar for all copolymers, though the differences in degree of quaternisation, and thus coronal charge, lead to markedly different adsorbed layer morphologies. After initial adsorption, relaxation of the coronal chains occurs over the surface immediately surrounding the micelle. This relaxation is driven by simple electrostatics; the cationic coronal chains interact with anionic surface sites on the mica.[73] Clearly, the strength of this interaction will increase with the level of quaternisation, and the coronal chains will interact

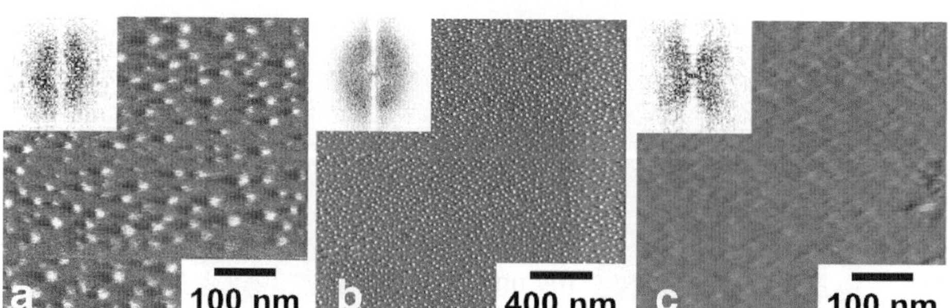

Fig. 10 Soft-contact *in situ* AFM images of the adsorbed layer of 100qPDMA–PDEA micelles adsorbed from a 500 ppm, 10^{-2} M KNO_3, pH 9 aqueous solution onto muscovite mica. Images were recorded (a, b) after the fluid cell was flushed with pH 9 water and (c) after flushing with pH 4 water. Insets show the respective FFTs.

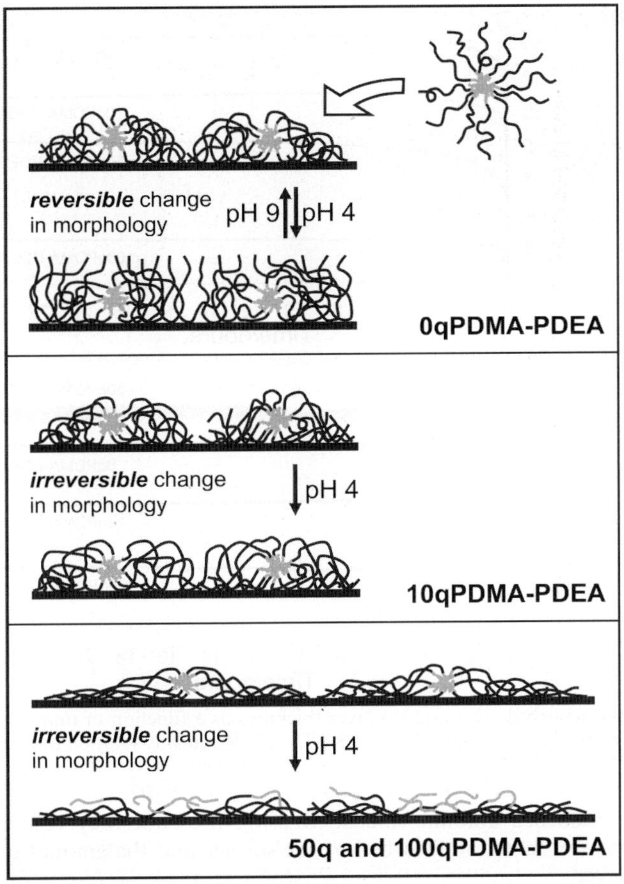

Fig. 11 Schematic representation of the adsorption of micelles of the PDMA–PDEA copolymers and the response of the adsorbed layers to changes in solution pH.

over a correspondingly larger surface area around the adsorbed micelle. The adsorption of coronal chains prevents micelle/surface interactions, and so hinders further micelle adsorption. Consequently, the development of structural order between adsorbed micelles requires a balance between the rate of micelle adsorption and the rate of coronal chain relaxation. As the degree of quaternisation and micellar charge density is increased, the relaxation of coronal chains is more rapid. Hence the inter-aggregate spacing increases and the location of micelles within the adsorbed layer becomes increasingly random. We have previously observed such behaviour with PDMA–PMMA micelles.[58] Furthermore, we believe it is the close-packed nature of the 0qPDMA–PDEA and 10qPDMA–PDEA micelle monolayers that prevents micelle dissociation at low pH.

Adsorption of the copolymer micelles from 500 ppm, pH 9 aqueous solution over 24 h has also been studied by QCM-D. Considering the distinct morphological differences of the adsorbed micelles depending on their degree of quaternisation, the QCM-D data was also expected to show significant variations within the family of copolymers. This is indeed the case. Fig. 12 shows the change in the adsorbed mass, and calculated layer thickness for each of the copolymers. It is clear that the adsorbed mass depends on the copolymer charge density, with the final plateau value decreasing as the degree of quaternisation is increased. Moreover, the slow nature of the adsorption is apparent, as plateau values of both adsorbed mass and layer thickness are only reached after approximately 10 h.

It is clear that the adsorbed amount decreases as the level of quaternisation is increased, Fig. 12a. This is due to the decrease in micelle density within the adsorbed layer as quaternisation increases, as observed in the AFM images, and supports the proposed model of adsorption. It is worth

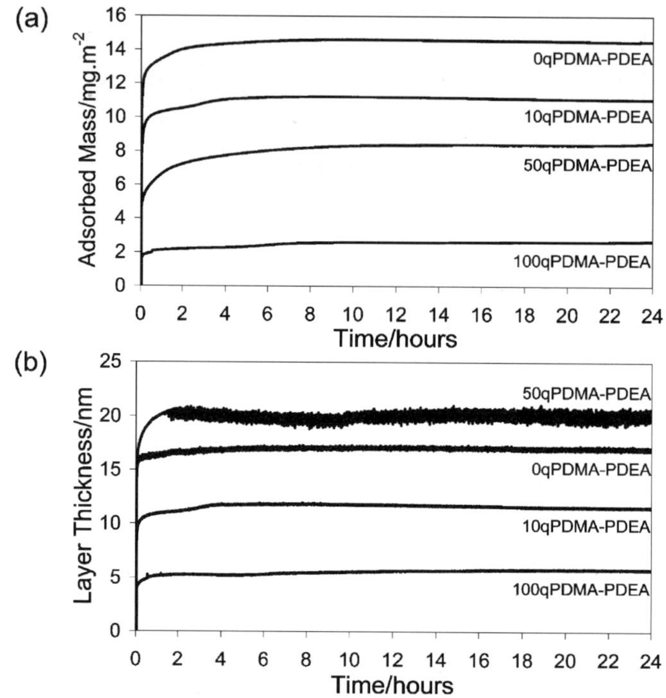

Fig. 12 Plots of (a) the adsorbed mass and (b) layer thickness as a function of time for the adsorption of the PDMA–PDEA copolymers from 500 ppm, 10^{-2} M KNO_3, pH 9 aqueous solution onto a silica QCM-D sensor crystal.

reiterating that the adsorbed amount calculated using the Sauerbrey equation in this method includes both the mass of material adsorbed to the surface and the amount of solvent associated with the adsorbed layer.

The layer thickness values do not correlate so well with the degree of quaternisation. However it is important to recall the nature of the dissipation measurement, from which the layer thickness is derived. The dissipation is, in essence, the extent to which the oscillatory motion of the sensor is damped by the adsorbed layer; a rigid layer has a small dissipation value, while a soft layer has a large value. Furthermore, the amount of water associated with the layer, and thus coupled to the oscillatory motion, influences the measured dissipation. A large volume of water associated with the micelle layer will increase the damping effect, and so increase the dissipation, while a smaller amount of associated water will have less impact. Since the charge density on the copolymer micelles increases with their degree of quaternisation, it is reasonable to expect a concomitant increase in the amount of water associated with the adsorbed layer, and thus the extent of dissipation. Except for 100qPDMA–PDEA, this is observed to be the case.

Examining the DLS and AFM data, it is clear that the ratio of inter-aggregate spacing (in the adsorbed layer) to hydrodynamic radius (in solution) is greatest for the 100qPDMA–PDEA micelles. One can infer from this that these micelles are the most collapsed at the solid/liquid interface, with the coronal chains from each individual micelle adopting a highly extended and flattened conformation. In the case of 50qPDMA-PDEA, the lateral extension of the coronal chains is less and so, given the identical coronal chain length for all copolymers, it is expected that the coronal PDMA chains will adopt conformations with more "loops and tails" normal to the interface. It is known that relatively few adsorbed polymer "tails" extending normal to the interface can have a significant impact on the hydrodynamic flow over a surface.[76] For QCM-D measurements, this affect may be very important since the apparent adsorbed mass includes "immobile" solvent.

The growth of the adsorbed mass over the first five minutes of the adsorption process for each of the copolymer micelles is shown in Fig. 13. Each micelle solution was only passed into the

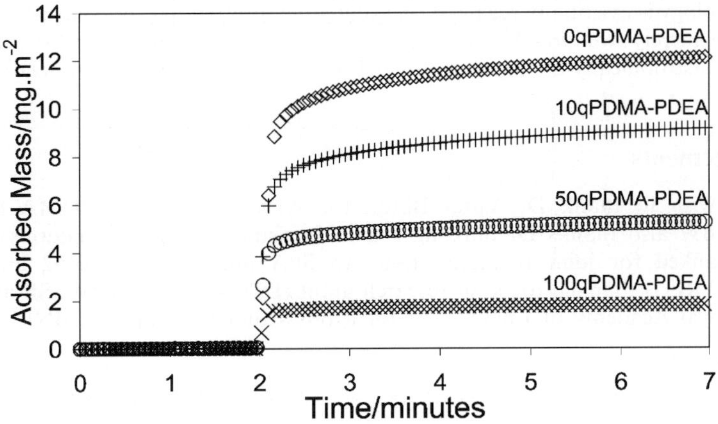

Fig. 13 Development of the adsorbed mass over the first five minutes of the adsorption of PDMA–PDEA micelles from 500 ppm, 10^{-2} M KNO_3, pH 9 aqueous solution onto a silica QCM-D sensor crystal.

measurement chamber after recording a stable baseline for two minutes. These data show that the rate of initial (rapid) adsorption is independent of the degree of quaternisation. However, the slower second-stage process is clearly influenced by the degree of quaternisation, since the ability of the adsorbed micelles to rearrange and hence accommodate further adsorption is greatest for those micelles with the lowest charge density. In other words, 100qPDMA–PDEA copolymer micelles adsorb with such a large number of attachment points, and over such a large surface area, that it is not possible for them to relax in order to allow the adsorption of further micelles from solution. Furthermore, electrostatic repulsions between the adsorbed micelles and those micelles remaining in solution will be greater at higher degrees of quaternisation, which will further hinder the build-up of micelle monolayers.

4. Conclusions

In this study we have investigated how subtle chemical modifications may be employed to alter the aqueous solution and adsorption properties of PDMA–PDEA diblock copolymers. These copolymers undergo reversible aggregation in aqueous solution due to the pH-responsive character of the core-forming PDEA blocks. By selectively quaternising the DMA residues, we have been able to impart varying degrees of permanent cationic charge to the corona-forming PDMA block. This chemical modification affects both the critical micellisation pH and the micelle dimensions. Increasing the degree of quaternisation led to a decrease in micelle diameter, presumably due to a change in the micelle aggregation number. Furthermore, the coronal charge density of the micelles formed in alkaline solution increased with the degree of quaternisation.

Quaternisation of the PDMA–PDEA copolymers also has a significant impact on their adsorption behaviour. At pH 4, all the copolymers behaved similarly, since at this pH the chains carry the same maximum cationic charge due to a combination of quaternisation and protonation. The adsorption of the copolymer micelles from alkaline solution is more interesting. *In situ* soft-contact atomic force microscope images show that the micelles of 0qPDMA–PDEA adsorbed in an ordered monolayer, with localised regions exhibiting hexagonal close-packing. Long-range structural order is also evident, and this is resilient to the replacement of the bulk copolymer concentration with pure water. The mean centre-to-centre micelle separation increases with the degree of quaternisation. While the 10qPDMA–PDEA micelle monolayer exhibits some degree of order, the highly quaternised copolymer micelles form much more random surface structures. In all cases these adsorbed layers are not affected by replacing the bulk copolymer solution with water at pH 9. QCM-D studies confirm that the adsorbed amount decreases with increasing degree of quaternisation. The response of the adsorbed micelles to the addition of acid to the overlying bulk solution is markedly dependent on the degree of quaternisation. The 0qPDMA–PDEA micelles form localised polymer brushes, with the cationic copolymer chains extending into aqueous solution; the original micelle morphology is essentially regained on cycling the solution pH. In contrast, the 10qPDMA–PDEA micelles merely swell *irreversibly*, while the adsorbed layers of 50q and 100qPDMA–PDEA micelles are catastrophically damaged at low pH, suggesting that micelle dissociation occurs.

Acknowledgements

The authors wish to thank Dr Vural Bütün for synthesising the 0qPDMA–PDEA diblock copolymer. GBW also thanks Dr Shiyong Liu for assistance with the quaternisation reactions. EPSRC is thanked for joint research grants to SPA and SB (GR/60402 and GR/S60419, respectively). ARC is thanked for joint research grant (DP0343783) to EJW, SB and SPA. GBW thanks the Royal Academy of Engineering for a one-year fellowship and BNFL for additional funding.

References

1 A. Fery, B. Schöler, T. Cassagneau and F. Caruso, *Langmuir*, 2001, **17**, 3779.
2 Y. Boontongkong and R. A. Cohen, *Macromolecules*, 2002, **35**, 3647.
3 T. Kowalewski, N. V. Tsarevsky and K. Matyjaszewski, *J. Am. Chem. Soc.*, 2002, **124**, 10 632.
4 Y. Ma, L. Qi, J. Ma and H. Cheng, *Langmuir*, 2003, **19**, 4040.
5 C. Allen, Y. Yu, D. Maysinger and A. Eisenberg, *Biconjug. Chem.*, 1998, **9**, 564.
6 A. Rösler, G. W. M. Vandermeulen and H.-A. Klok, *Adv. Drug Deliv. Rev.*, 2001, **53**, 95.
7 E. R. Gillies and J. M. J. Fréchet, *Chem. Commun.*, 2003, 1640.
8 L. Bronstein, E. Krämer, B. Berton, C. Burger, S. Förster and M. Antonietti, *Chem. Mater.*, 1999, **11**, 1402.
9 T. K. Bronich, M. Ouyang, V. A. Kabanov, A. Eisenberg, F. C. J. Szoka and A. V. Kabanov, *J. Am. Chem. Soc.*, 2002, **124**, 11 873.
10 M. Haupt, S. Miller, A. Ladenburger, R. Sauer, K. Thonke, J. P. Spatz, S. Riethmuller, M. Moller and F. J. Banhart, *J. Appl. Phys.*, 2002, **91**, 6057.
11 R. Glass, M. Arnold, J. Bluemmel, A. Kueller, M. Moeller and J. P. Spatz, *Adv. Funct. Mater.*, 2003, **13**, 569.
12 E. Krämer, S. Förster, C. Göltner and M. Antonietti, *Langmuir*, 1998, **14**, 2027.
13 K. Yu, A. J. Hurd, A. Eisenberg and C. J. Brinker, *Langmuir*, 2001, **17**, 7961.
14 S. Mössmer, J. P. Spatz, M. Möller, T. Aberle, J. Schmidt and W. Burchard, *Macromolecules*, 2000, **33**, 4791.
15 J. P. Spatz, S. Moessmer, C. Hartmann, M. Moeller, T. Herzog, M. Krieger, H.-G. Boyen, P. Ziemann and B. Kabius, *Langmuir*, 2000, **16**, 407.
16 B.-H. Sohn, J.-M. Choi, S. I. Yoo, S.-H. Yun, W.-C. Zin, J. C. Jung, M. Kanehara, T. Hirata and T. Teranishi, *J. Am. Chem. Soc.*, 2003, **125**, 6368.
17 S. Klingelhöfer, W. Heitz, A. Greiner, S. Oestreich, S. Förster and M. Antonietti, *J. Am. Chem. Soc.*, 1997, **119**, 10 116.
18 G. Decher, *Science*, 1997, **277**, 1232.
19 F. Caruso, *Adv. Mater.*, 2001, **13**, 11.
20 W. A. Lopes and H. M. Jaeger, *Nature*, 2001, **414**, 735.
21 G. M. Whitesides and B. Grzybowski, *Science (Washington, D.C.)*, 2002, **295**, 2418.
22 N. Nath and A. Chilkoti, *Adv. Mater.*, 2002, **14**, 1243.
23 T. P. Russell, *Science (Washington, D.C.)*, 2002, **297**, 964.
24 J. Lahann, S. Mitragotri, T.-N. Tran, H. Kaido, J. Sundaram, I. S. Choi, S. Hoffer, G. A. Somorjai and R. Langer, *Science (Washington, D.C.)*, 2003, **299**, 371.
25 D. S. Ginger, H. Zhang and C. A. Mirkin, *Angew. Chem. Int. Ed. Engl.*, 2004, **43**, 30.
26 L. Leibler, H. Orland and J. C. Wheeler, *J. Chem. Phys.*, 1983, **79**, 3550.
27 M. R. Munch and A. P. Gast, *Macromolecules*, 1990, **23**, 2313.
28 Y. J. Zhan and W. L. Mattice, *Macromolecules*, 1994, **27**, 683.
29 N. G. Hoogeveen, M A. Cohen Stuart and G. J. Fleer, *Faraday Discuss.*, 1994, **98**, 161.
30 A. P. Gast, *Curr. Opin. Colloid Interface Sci.*, 1997, **2**, 258.
31 S. E. Webber, *J. Phys. Chem. B*, 1998, **102**, 2618.
32 S. Förster, N. Hermsdorf, W. Leube, H. Schnablegger, M. Regenbrecht, S. Akari, P. Lindner and C. Böttcher, *J. Phys. Chem. B*, 1999, **103**, 6657.
33 E. Y. Kramarenko, I. I. Potemkin, A. R. Khokhlov, R. G. Winkler and P. Reineker, *Macromolecules*, 1999, **32**, 2495.
34 I. W. Hamley, C. Daniel, W. Mingvanish, S.-M. Mai, C. Booth, L. Messe and A. J. Ryan, *Langmuir*, 2000, **16**, 2508.
35 O. V. Borisov and E. B. Zhulina, *Macromolecules*, 2003, **36**, 10 029.

36 S. H. Kim, M. J. Misner, T. Xu, M. Kimura and T. P. Russell, *Adv. Mater.*, 2004, **16**, 226.
37 L. Zhang and A. Eisenberg, *Science*, 1995, **268**, 1728.
38 L. Zhang, K. Yu and A. Eisenberg, *Science*, 1996, **272**, 1777.
39 L. Zhang and A. Eisenberg, *J. Am. Chem. Soc.*, 1996, **118**, 3168.
40 L. Zhang and A. Eisenberg, *Polym. Adv. Technol.*, 1998, **9**, 677.
41 K. Yu and A. Eisenberg, *Macromolecules*, 1996, **29**, 6359.
42 Y. Yu, L. Zhang and A. Eisenberg, *Langmuir*, 1997, **13**, 2578.
43 H. Shen, L. Zhang and A. Eisenberg, *J. Am. Chem. Soc.*, 1999, **121**, 2728.
44 W. Groenewegen, S. U. Egelhaff, A. Lapp and J. R. C. van der Maarel, *Macromolecules*, 2000, **33**, 3283.
45 J.-F. Gohy, S. Creutz, M. Garcia, B. Mahltig, M. Stamm and R. Jerôme, *Macromolecules*, 2000, **33**, 6378.
46 J.-F. Gohy, S. Antoun and R. Jérôme, *Macromolecules*, 2001, **34**, 7435.
47 M. Štepánek, K. Podhájecká, E. Tesarová, K. Procházka, Z. Tuzar and W. Brown, *Langmuir*, 2001, **17**, 4240.
48 V. Bütün, S. P. Armes and N. C. Billingham, *Polymer*, 2001, **42**, 5993.
49 J. P. Spatz, S. Sheiko and M. Möller, *Macromolecules*, 1996, **29**, 3220.
50 M. Regenbrecht, S. Akari, S. Förster and H. Möhwald, *J. Phys. Chem. B*, 1999, **103**, 6669.
51 J. C. Meiners, A. Quintel-Ritzi, J. Mlynek, H. Elbs and G. Krausch, *Macromolecules*, 1997, **30**, 4945.
52 I. I. Potemkin, E. Y. Kramarenko, A. R. Khokhlov, R. G. Winkler, P. Reinecker, P. Eibeck, J. P. Spatz and M. Möller, *Langmuir*, 1999, **15**, 7290.
53 M. Breulmann, S. Förster and M. Antonietti, *Macromol. Chem. Phys.*, 2000, **201**, 204.
54 J. Hahn and S. E. Webber, *Langmuir*, 2003, **19**, 3098.
55 J. Hahn and S. E. Webber, *Langmuir*, 2004, **20**, 1489.
56 V. M. De Cupere, J. F. Gohy, R. Jerome and P. G. Rouxhet, *J. Colloid Interface Sci.*, 2004, **271**, 60.
57 M. Regenbrecht, S. Akari, S. Förster, R. R. Netz and H. Möhwald, *Nanotechnology*, 1999, **10**, 434.
58 G. B. Webber, E. J. Wanless, S. P. Armes, F. L. Baines and S. Biggs, *Langmuir*, 2001, **17**, 5551.
59 G. B. Webber, E. J. Wanless, V. Bütün, S. P. Armes and S. Biggs, *Nano Lett.*, 2002, **2**, 1307.
60 S. D. Connell, S. Collins, J. Fundin, Z. Yang and I. W. Hamley, *Langmuir*, 2003, **19**, 10 449.
61 V. Bütün, S. P. Armes and N. C. Billingham, *Macromolecules*, 2001, **34**, 1148.
62 I. Astafieva, X. F. Zhong and A. Eisenberg, *Macromolecules*, 1993, **26**, 7339.
63 I. Astafieva, K. Khougaz and A. Eisenberg, *Macromolecules*, 1995, **28**, 7127.
64 S. Manne, J. P. Cleveland, H. E. Gaub, G. D. Stucky and P. K. Hansma, *Langmuir*, 1994, **10**, 4409.
65 M. Rodahl, F. Höök, A. Krozer, P. Brzezinski and B. Kasemo, *Rev. Sci. Instrum.*, 1995, **66**, 3924.
66 M. Rodahl and B. Kasemo, *Rev. Sci. Instrum.*, 1996, **67**, 3238.
67 G. Sauerbrey, *Z. Phys.*, 1959, **155**, 206.
68 V. S. J. Craig and M. Plunkett, *J. Colloid Interface Sci.*, 2003, **262**, 126.
69 G. S. Manning and A. Holtzer, *J. Phys. Chem.*, 1973, **77**, 2206.
70 A. R. Eckert and S. E. Webber, *Macromolecules*, 1996, **29**, 560.
71 G. J. Fleer, M. A. Cohen Stuart, J. M. H. M. Scheutjens, T. Cosgrove and B. Vincent, *Polymers at Interfaces*, Chapman & Hall, London, 1st edn., 1993.
72 N. G. Hoogeveen, M. A. C. Stuart and G. J. Fleer, *J. Colloid Interface Sci.*, 1996, **182**, 133.
73 M. R. Talingting, Y. Ma, C. Simmons and S. E. Webber, *Langmuir*, 2000, **16**, 862.
74 A. S. Lee, A. P. Gast, V. Bütün and S. P. Armes, *Macromolecules*, 1999, **32**, 4302.
75 C. A. Johnson and A. M. Lenhoff, *J. Colloid InterfaceSci.*, 1996, **179**, 587.
76 M. A. Cohen Stuart, F. H. W. H. Waajen, T. Cosgrove, B. Vincent and T. L. Crowley, *Macromolecules*, 1984, **17**, 1825.

General Discussion

Professor Hamley opened the discussion of Professor Gido's paper: Are your samples fully dry initially? Is it possible that your growth kinetics are influenced by solvent loss? Why not start from a well-defined initial state, such as the disordered phase, and quench from there?

Professor Gido replied: The samples were dried first in air and then in a vacuum oven at room temperature before the growth kinetics studies were started. If solvent was present, it would lead to a foaming of the samples at the high temperature and vacuum used for annealing during the kinetic studies. We of course took steps to avoid this situation.

One can divide the grain growth in these materials into two regimes. The first is growth of grains starting from nucleation in the disordered to until initial impingement. The second is a coarsening of the space filling grain structure, in which some grains grow at the expense of others *via* the motion of grain boundaries. Our present study is focused on the second regime, *i.e.* post impingement grain growth.

We certainly are also interested in pre-impingement grain growth from the disordered state, however, the arm molecular weights of the A_nB_n materials that we had available for this study did not allow thermal access of the ODT without degrading the materials. We are currently pursuing experiments in collaboration with Nitash Balsara in which we are using DOP to lower the ODT, and then following initial grain growth from the disordered state using depolarized light scattering. We also hope to obtain model A_nB_n materials of lower molecular weight for future studies in which the ODT may be accessed thermally without the use of a diluent.

Dr Davidson asked: In the field of lamellar liquid crystalline phases, scientists usually consider the positional correlation function that determines the X-ray reflection lineshapes and gives information about the elastic constants of the phase. You use instead an orientational correlation function that shows an exponential decay. Are there any theoretical predictions for this orientational correlation function in a lamellar phase and is there any way of deriving from it more information about the compression and splay elastic constants of the phase?

Professor Gido responded: A detailed model for X-ray reflection lineshapes has been published by Shibayama and Hashimoto.[1] This model includes contributions to Bragg peak broadening resulting from grain size, interfacial thickness, and fluctuations in lamellar thickness.

As discussed by Gido and Thomas,[2] the long range ordering of lamellar block copolymers has a dual nature. TEM images of these materials clearly show grain structures, similar to that of a metal, in which regions of coherent lamellar orientation are separated from one another by grain boundaries. This grain structure is generally accepted to be the result of nucleation and growth of grains analogous to grain growth in metals or ceramics. However, block copolymer layers are also materials with the symmetry of smectic A liquid crystals. Their order is inherently soft and much more susceptible to structural distortions than inorganic grain forming materials. Thus one frequently observes gentle changes in lamellar direction even within what might be classified as a single grain. Also classical liquid crystalline defects, disclinations, are frequently observed in lamellar block copolymers, mixed in with the grain structure.

The analysis of X-ray line shape by Shibayama and Hashimoto,[2] relies on the grain picture of the lamellar morphology. It does not account for slight fluctuations in lamellar direction within a given grain and it does not include the possibility of the type of continuous lamellar orientation change associated with disclinations. (By contrast grain boundaries represent a discontinuous lamellar orientation change.)

DOI: 10.1039/b413662b *Faraday Discuss.*, 2005, **128**, 211–229

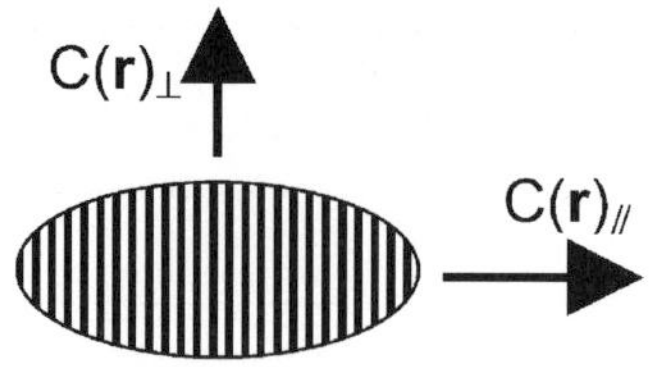

Fig. 1 Anisotropic grain with its long axis parallel to the lamellar normal. The directions in which correlation functions $C(\boldsymbol{r})_{\parallel}$ and $C(\boldsymbol{r})_{\perp}$, parallel and perpendicular to the lamella normal, are indicated.

Hashimoto's line shape calculations already employ a number of adjustable parameters (interfacial width, grain size, paracrystalline distortion). Combined with the imperfect nature of experimental data, this leads to a lack of uniqueness in the determination of these parameters by fitting data with the model. If one compounded the situation by introducing additional aspects to account for liquid-crystal-like changes in lamellar orientation, then this fitting process would only become less meaningful.

1 M. Shibayama and T. Hashimoto, *Macromolecules*, 1986, **19**, 740.
2 S. P. Gido and E. L. Thomas, *Macromolecules*, 1994, **27**, 849.

Dr Borsali said: Looking at Fig. 2 of your paper, it doesn't seem to me that the correlation function is described by a single exponential decay. Certainly not the perpendicular to the lamellar. Should you extract two relaxations and what could their physical meanings be?

Professsor Gido responded: Yes, there clearly is a kink in the correlation function perpendicular to the lamellar normal: $C(\boldsymbol{r})_{\perp}$. As shown in Fig. 1 here, the grains are generally anisotropic in shape, with their long axis parallel to the lamellar normal.

Thus $C(\boldsymbol{r})_{\perp}$ is measured perpendicular to the long axis of a grain. The initial part of the $C(\boldsymbol{r})_{\perp}$ curve results from intragrain correlation. It is this part of the curve that we have fit with an exponential to obtain a grain size. The part of the curve after the kink suggests a stronger intergrain orientational correlation perpendicular to the grain long axis than parallel to this axis. This observation agrees with our recent, unpublished, computer simulations of grain growth, which suggest that excluded volume interactions coupled with sporadic nucleation result in enhanced orientation correlation perpendicular to the long axis of the grains.

Professor Lodge commented: The mechanism of grain growth depends critically on whether $n = 1$ (diblock) or $n = 2, 4 \ldots$ (stars). The entanglement/disentanglement transition is not usually that sharp, so I wonder whether it is entanglement that makes the difference, or rather the thermodynamic penalty for pulling the junction point off the interface. In this case, $n = 2, 4, \ldots$ appears in the exponent of the enthalpic barrier to motion.

Professor Gido responded: But in this case, one would expect to see different growth exponents for the different star functionalities. However our data indicates that all stars with $n = 2$ or greater have the same growth exponent.

Additional results on grain growth during annealing in supercritical CO_2 which were obtained after the submission of our paper show that the diblock ($n = 1$) growth exponent increases dramatically to 0.82 upon swelling with supercritical CO_2. However, the stars ($n = 2$, 4, and 16) show the same growth exponents in supercritical CO_2 as in the present thermal annealing study. This suggests that in the presence of the supercritical CO_2 another mechanism of molecular motion is greatly enhanced for the diblock, but not for the stars. Perhaps this corresponds to motion of the junction point off the interface. Enhancement of this type of chain motion would be expected to greatly facilitate grain boundary motion that requires the breaking and reforming of lamellar layers as illustrated in Fig. 6 of our paper.

Professor Ryan asked: (1) Doesn't the chevron grain boundary just move like a ripple in a carpet? (2) Is CO_2 the best solvent to use? What about a high boiling neutral solvent like those used by Lodge *et al.* referred to in the Introductory Lecture.

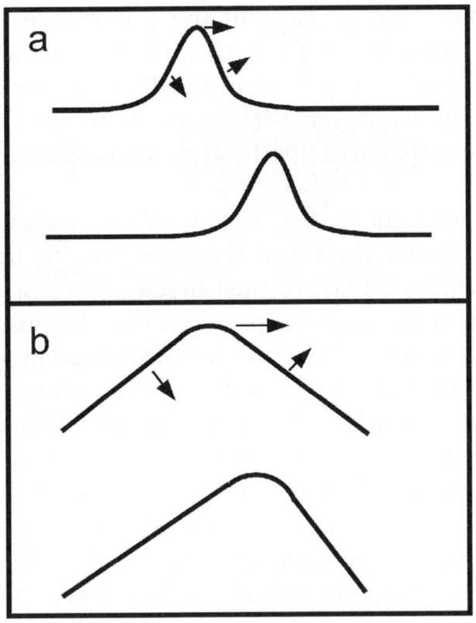

Fig. 2 Mechanisms of motion of a soliton (a) and a chevron grain boundary (b). In each case, the arrows indicate the directions of motion required of different parts of the structure to produce motion.

Professor Gido responded: (1) My initial thought was that a chevron grain boundary can move as a soliton, *i.e.* a ripple in a carpet. However, closer examination of the geometry involved convinced me that this was not the case. Fig. 2 here illustrates the situation.

Part a of the figure shows the motion of a ripple in a carpet. The arrows indicate the directions of motion required of different parts of the carpet to move the soliton along. In part b of the figure, the lines represent a single lamellar layer in a chevron grain boundary. Application of the same type of motion, as in part a, in this case necessitates the changing of the orientations of the lamellar layers on either side of the boundary. This wholesale reorientation of all the lamellae in the grains adjoining a moving chevron grain boundary is clearly unlikely.

(2) This question refers to additional results on grain growth during annealing in supercritical CO_2 which were obtained after the submission of our paper. Some of these additional results were discussed in response to Professor Lodge's comment.

Certainly a high boiling, neutral solvent such as dioctyl phthalate (DOP) could be useful in the study of grain growth kinetics. We are, in fact using DOP to lower the ODT of these materials in order to study the kinetics of grain growth from the disordered state. However, the use of such a solvent poses some difficulties because once introduced into the block copolymer material it is difficult to remove. Thus one cannot use TEM, which subjects the sample to high vacuum, to follow grain growth in the presence of such a solvent. It may be possible to use light scattering techniques instead. Also from the practical point of view, we are interested in controlling self-assembly kinetics to enhance the long range order in nanostructured materials. We typically want to end up with a solid material without entrapped solvent. CO_2 can be easily removed once the ordering process is complete, while DOP cannot.

Professor Hamley asked: Is supercritical CO_2 a neutral solvent for your copolymers?

Professor Gido responded: This question also refers to additional results on grain growth during annealing in supercritical CO_2 which were obtained after the submission of our paper. Some of these additional results were discussed in response to Professor Lodge's earlier comment.

Supercritical CO_2 is very slightly selective for polyisoprene, and the degree of selectivity may be tuned slightly by changing pressure. See Fig. 4 of ref. 1 below. The fact that we continue to observe the lamellar morphology in supercritical CO_2 suggests that solvent selectivity is not strong enough to alter the morphology.

1 B. D. Vogt, G. D. Brown, V. S. RamachandraRao and J. J. Watkins, *Macromolecules*, 1999, **32**, 7907.

Dr Semenov asked: Does the grain growth, in the general case, imply that A blocks have to penetrate through B layers (or *vice versa*)? If yes, then in the case of many arms, do all A-arms of a block-copolymer penetrate through an unfavourable B-environment simultaneously (this would imply a high energy barrier, perhaps, proportional to the number of arms) or one after another (this process may be hindered by entanglements with other arms of the same copolymer)?

Professor Gido responded: As indicated in Fig. 6 of our paper and further elaborated in my response to Professor Ryan's earlier question, the geometry of the chevron tilt grain boundary indicates that the microphase separated layers must break and reform for the boundary to move. As I speculated in the Conclusions section of our paper, it may be possible for twist grain boundaries composed of arrays of saddle surfaces (Scherk's First Surface) to move through the sample while the interfaces remain intact. To the extent that many grain boundaries in real materials contain a tilt character (either pure tilt or compound twist–tilt character) a breaking and reforming of the layer connections is required to move many, if not most, of the grain boundaries in a lamellar block copolymer. Thus the facility with which this can occur is a controlling factor for grain growth.

I do not know whether this rearrangement of the layer connectivity occurs on a molecule-by-molecule basis in the kind of diffusion process suggested by Prof. Semenov's question. It might also occur *via* a cooperative distortion of the interface in which layers pinch off and reconnect while the blocks remain segregated in their domains. The fact that swelling the samples in supercritical CO_2 enhances grain growth in the diblock suggests that facilitating a cross-domain diffusion process will facilitate faster grain growth.

All the A_nB_n stars ($n = 2$, 4, and 16) show grain growth according to the same power law. This experimental result suggests that if a cross-domain diffusion process of the arms is important to grain growth, it cannot be controlled by an energy barrier that is proportional to the arm number (n).

Professor Müller asked: Is the polyisoprene stable at 120 °C for 60 days or even longer? Is it possible that it degrades?

Professor Gido responded: The samples were annealed at 120 °C in a vacuum oven to reduce the chances of oxidation, the main mechanism by which the samples are degraded. However, some oxidation does take place. The two month duration of our annealing experiments represents a practical limit to the extent of annealing beyond which results start to become noticeably influenced by degradation. Longer annealing times may be possible with PS–PI materials if they are evacuated thoroughly on a high vacuum line and sealed into glass ampoules. One could also consider other polymer materials with greater thermal and/or oxidative stability.

Professor Lodge asked: Your grain growth regime corresponds to "late stage" kinetics, where the space is filled with grains that compete to grow. Presumably if you could examine the early stage growth of independent grains, which would follow a linear dependence on time, as in the Goveas–Milner theory?[1] This is what we observed in several PS–PI solution systems.

The late stage exponent of nearly 1/3 is reminiscent of polymer blends, where late stage coarsening follows $t^{1/3}$ for either evaporation/condensation or coalescence mechanisms. Is this similarity in exponent accidental, or is there a deeper correspondence?

1 J. L. Goveas and S. T. Milner, *Macromolecules*, 1997, **30**, 2605.

Professor Gido responded: My intuition is that late stage grain growth in microphase separated block copolymers must be a complex process in which fundamental modes of molecular mobility may not contribute directly to grain growth. But, rather molecular motion must contribute to grain boundary motion, and it is this motion that allows the grains to grow. The mechanisms of grain boundary motion are likely influenced by grain boundary morphology and topological constraints of domain connectivity as illustrated in Fig. 6 of our paper. Thus I suspect that the exponents observed in our study represent a convolution of multiple fundamental processes reflecting both chain motion and grain boundary motion. However, your observation suggests that perhaps one of these processes dominates, and that it is analogous to late stage coarsening of blend morphology. Considerable additional work would be necessary to understand the connection.

Professor Hamley opened the discussion of Dr Semenov's paper: This is a very interesting model, but I am not convinced of its applicability to PEO–PVP block copolymers, since the glass transition of PVP has not been considered. This could have a dramatic effect on the exchange kinetics of molecules between micelles and could lead to very slow exchange, as for example in the observations of Hadziioanou *et al.*[1] who reported extremely slow processes (up to many months). Thus other mechanisms may be responsible for very slow micellization processes. Please could you comment on this?

1 F. J. Esselink, E. E. Dormidontova and G. Hadziioannou, *Macromolecules*, 1998, **31**, 2925; F. J. Esselink, E. E. Dormidontova and G. Hadziioannou, *Macromolecules*, 1998, **31**, 4873.

Dr Semenov responded: The glass transition temperature of bulk (undiluted) PVP is around 100 °C (ref. 1) (the corresponding absolute temperature $T_{g0} \sim 370$ K). In the micellar core, however, PVP is mixed with water. Our calculations show that the water fraction in the core is at least 35% even at high pH, or more at low pH when PVP blocks are protonated (charged). Using Vogel–Fulcher–Tamann superposition law, and taking into account that water is not glass forming, we get an estimate of T_g in the core: $T_g = \Phi^* T_{g0} < 240$ K (here $\Phi < 0.65$ is the PVP volume fraction in the core), *i.e.* well below room temperature. In addition, finite size effects (the core size is typically only *ca.* 20 nm) may further suppress the glass transition. These conclusions are consistent with experimental observations[2] that the micelle aggregation number can change following a decrease of pH.

1 J. Wu *et al.*, *Polymer Preprints*, 2002, **43**, 105.
2 T. J. Martin *et al.*, *Macromolecules*, 1996, **29**, 607.

Professor Kabanov asked: (1) Am I correct that in the case of open association Flory distribution is established? (2) How did you control the charge and charge distribution within the partly protonated PVP core? I would expect that in the case of a glassy core it is hardly homogeneous. In other words at intermediate pH values a still glassy core interior may coexist with strongly ionized PVP shell. Is it taken into account in your kinetic treatment?

Dr Semenov responded: (1) Yes, in the case of pure, "classical" open association (living polymerization with constant bond energy). (2) The core is not likely to be glassy at intermediate pH since its strongly charged surface drives its significant expansion which necessarily implies that water penetrates in the core interior thus suppressing the glass transition (see also my response to Professor Hamley's question earlier), hence no additional complications for the kinetics. However, the charge distribution and protonation degree distribution are still highly inhomogeneous within the core as we showed by solving Poisson–Boltzmann equations.

Professor McLeish asked: Are you able to defend in a little more detail the use of the free-energy function $F(m)$ as the appropriate *kinetic* landscape? There are in reality many more parameters or degrees of freedom in the process of micelle formation than the single dimension of degree of micellisation. Have you also checked that coalescence of smaller micelles is not faster than the "transition pathway" you have assumed?

Dr Semenov responded: Other parameters (micelle size, shape and structure) are normally defined by m. We checked that this is true for the model of PEO–PVP block-copolymer micelles considered in the paper, hence m is the essential parameter. In a more general case other parameters may be involved in the calculation of $F(m)$. Our general analysis is then applicable once this function is known.

We did check that coalescence of smaller micelles is slower than their continuous growth. This is true since the second derivative of $F(m)$ is negative for m smaller than the critical (activated) micelle.

Fiona Case asked: Do you think that your model will hold for higher concentration block copolymer systems? In solutions of lower molecular weight surfactants (particularly surfactants with non-ionic head-groups) mesoscale simulations have suggested that micelle coalescence is an important mechanism for structure growth above the c.m.c. It is very interesting that block copolymers follow a different mechanism.

Dr Semenov responded: The model is valid if interactions between micelles are negligible which is normally true for low micelle volume fractions.

As for your simulations of non-ionic low-molecular-weight micelles, it is possible that kinetics of their formation is not controlled by activation (this is likely if aggregation number is less than *ca.* 50 and/or the concentration is really much larger than equilibrium c.m c.). In this case small micelles are formed because they are kinetically more accessible (*i.e.* their formation is faster), then smaller micelles coalesce into larger ones which are more favorable thermodynamically.

Dr Borsali opened the discussion of Professor Vamvakaki's paper: (1) What are the driving forces to move from a system having free chains ($\sim$6 nm) and "aggregates" (60 nm) at a pH = 4 to a micelle (single relaxation) corresponding to 33 nm at pH > 5. (2) Do you have any idea about the morphology of the micelles at pH > 5 or the system at different pH? Some imaging techniques would certainly be useful. (3) Is it reversible? Changing the pH and coming back to the same value (lets say 4), do you observe the same size of the aggregates?

Professor Vamvakaki responded: (1) Micellar structures are formed in dilute aqueous solutions of smart polymers by a simple change in the solution pH, temperature, additive concentration, *etc.* The diblock copolymers discussed in the manuscript comprise a neutral, hydrophilic PEO-based block and an ionisable, pH-sensitive, hydrophobic block. These copolymers can self-assemble into micellar structures in water by simply adjusting the solution pH. Thus, at low pH the copolymer is in its "unimer" state due to the hydrophilicity of the protonated tertiary amine units, while an increase in the solution pH results in the deprotonation of the amine residues, which become hydrophobic, and leads to the formation of micelles.

The large aggregates, formed at low pH, are attributed to hydrophobic interactions of the polymer backbone, hydrogen bonding and complex ionization phenomena as discussed in the manuscript. The rather low scattering intensity of these large aggregates suggests the formation of highly hydrated structures with only weak inter-chain interactions. As the pH increases, the amine units are gradually deprotonated, the electrostatic repulsions are reduced and more polymer chains participate in these aggregates. The increase in the scattering intensity of the aggregates at this pH range is attributed to the increase of the core density as water is expelled from the micelle core, which becomes progressively more hydrophobic. However, the contrast and the scattering intensity of the sample remain relatively low as the micelle core is still hydrated. Further increase of the pH above a critical value leads to the deprotonation of all amine units and the formation of a fully dehydrated dense core containing the hydrophobic blocks. This is accompanied by a sharp decrease of the hydrodynamic radius of the micelles and an increase of the scattering intensity of the solution.

(2) The most appropriate technique in order to assess the morphology of the micelles in an aqueous medium is cryo-TEM which however, has not been used in this study. Instead, *ex situ* AFM was employed to investigate the solution behavior of our copolymers following adsorption on a mica surface from both a low and a high pH aqueous solution. Furthermore, TEM was used to observe the polymer structure after incorporation of the metal compounds. The latter serve as staining agents for the micelle core and thus render them visible in TEM. If one assumes that there is no significant effect of the solid substrate on the micelle morphology, although some deformation is expected, both these techniques lead us to believe that the micelles formed at high pH are of spherical shape. At lower pH values and before micellization takes place AFM studies suggest that the polymer adsorbs as a film of low roughness and no indication of the larger aggregates, observed by DLS, is evident. On the other hand, TEM shows the formation of micellar structures even for the low pH values (below pH 5) due to the metal induced micellization discussed in the manuscript.

(3) Dynamic light scattering measurements have shown that the micellization process observed, upon increasing the pH in both PEO-*b*-P2VP and PHEGMA-*b*-PDEAEMA aqueous solution, is reversible. In all cases, micelles are formed upon increasing the pH of a dilute polymer solution, above a critical pH value determined by the effective pK_a of the protonated amine block, while free polymer chains and loose aggregates are obtained when decreasing the pH to values lower than the effective pK_a. Furthermore, the dimensions of both the polymer micelles and the loose aggregates, observed at low pH (*i.e.* pH 4), were found to be the same upon switching up and down the solution

pH. The latter suggests a fully reversible micellization process which is consistent with results published earlier.[1]

1 V. Butun, N. C. Billingham and S. P. Armes, *J. Am. Chem. Soc.*, 1998, **120**(46), 12135.

Professor Lodge commented: If you examine any polyelectrolyte in solution by dynamic light scattering, there is a high probability of observing a "slow mode", it does not require any particular "aggregate" structure. Is it possible your PEO–PVP polymers are just acting as common polyelectrolytes? One simple experimental test would be to add salt and see if the slow mode goes away.

Professor Vamvakaki responded: Indeed, certain features of the slower process observed at low pH in our systems resemble the slow mode documented for low ionic strength polyelectrolyte solutions. That mode has been thought to result from strong Coulombic interactions of the unscreened charges along the polymer chain, leading to the formation of large polyelectrolyte domains with dimensions of the order of 60–100 nm. These interactions could also contribute to the formation of the long-range fluctuations discussed in the present manuscript as loose aggregates. However, two observations led us to think that this is only a weak contribution. The first is the existence of the slower process even for solutions of the PHEGMA homopolymer, which is not a polyelectrolyte. The second is the fact that addition of excess salt eliminates the slow mode in polyelectrolyte solutions, whereas, in the present systems, the slower process remains unaffected when NaCl is added up to concentrations of 0.012 M (these data will be discussed in a subsequent publication).

The above led us to propose the three possible explanations (poor solubility of the polymer backbone, complex ionization and hydrogen bonding, as discussed in the text) as the main reasons for the formation of the loose aggregate structures that we observe.

Professor Ryan asked: Is the slow mode observed in PHEGMA (homopolymer) due to residual acid making it a polyelectrolyte?

Professor Vamvakaki responded: Dynamic light scattering studies have shown the existence of the slower process attributed in the manuscript to loose aggregates, even for solutions of the PHEGMA homopolymer (data to be discussed in a forthcoming publication). HEGMA monomer is synthesised by an esterification reaction between a mono hydroxyl terminated oligoethylene glycol with a methacrylic moiety which could in principle lead to the contamination of the final product with residual acid groups. The copolymerization of the residual methacrylic acid units with the HEGMA monomer would result in the statistical distribution of ionisable monomer units along the polymer chain and thus the formation of a polyelectrolyte. We believe that such residual acid is negligible for two main reasons. First, such traces of acid would have been removed when the impure monomer is passed twice through a basic alumina column before being polymerized. Second, the polymerization of HEGMA using a living technique (group transfer polymerization, GTP), leads to the formation of living polymer chains which allow the polymerization of a second monomer (DEAEMA) and the formation of diblock copolymers with narrow polydispersities ($M_w/M_n < 1.15$). Even the presence of traces of acidic impurities during the polymerization reaction would result in premature termination of the living chain ends, and thus, in broadening of the molecular weight distribution of the final polymer and could even hinder the formation of the block copolymers.

Dr Webber said: I have two questions for the authors. (1) Fig. 8 in the paper shows *ex situ* AFM images of adsorbed polymer layers, where the layers were dried prior to imaging. Have the authors attempted to image the adsorbed polymer layer using *in situ* AFM, which will eliminate the possibility of changes in layer structure during the drying process? (2) Additionally, have the authors attempted to use layers of adsorbed micelles, such as that shown in Fig. 8b, for the preparation of metal nanoparticles (rather than forming the nanoparticles in bulk solution as detailed in the paper)?

Professor Vamvakaki responded: (1) When using *ex situ* AFM it is likely to observe changes in the adsorbed polymer layer, which may occur during the drying process. In all our studies, which have been conducted using *ex situ* AFM, polymer micelle structures were observed only at high pH values as discussed in the manuscript. One cannot exclude the deformation of the micelles upon adsorption on the mica surface, however we believe that the micelles are pre-formed in solution before being adsorbed on the solid surface and micellization is not simply induced by the mica substrate. This is further supported by the agreement between our DLS and AFM results regarding the dimensions of the micellar structures.

(2) The formation of metal nanoparticles within ordered micelles adsorbed on a solid surface is a very interesting suggestion. At present, work is underway to adsorb an ordered layer of metal nanoparticle containing micelles on the solid substrate. However, alternatively one could follow the incorporation of metal compounds and the formation upon reduction of the metal nanoparticles in layers of adsorbed micelles using *in situ* imaging techniques such as AFM.

Dr Lecommandoux asked: You mentioned in your paper that these nanocatalysts are very active and selective for oxidation reactions. What is the evolution of selectivity and reactivity when pH is varied?

Professor Vamvakaki responded: The platinum nanocatalysts discussed in the manuscript have been used for the direct oxidation of L-sorbose to 2-keto-L-gulonic acid which is the intermediate in vitamin C production. Catalytic reactors were prepared by deposition of the metal nanoparticle containing micelles on the surface of alumina particles. These catalytic systems exhibit high selectivities of up to 99% at 97% conversion of L-sorbose. The reaction is conducted in alkali media (pH $\geq$ 8) by the addition of an equimolar quantity of $NaHCO_3$ to the initial L-sorbose concentration. It has been reported that under basic conditions the performance of platinum as a catalyst in the oxidation of L-sorbose improves significantly while, at lower pH values, over-oxidation and side-product formation are observed which result in lower selectivity and lower L-sorbose conversions.[1] For this reason the effect of the pH of the reaction medium on the performance of the polymer stabilized nanoparticles has not been examined.

1 T. Mallat, C. Bronnimann and A. Baiker, *Appl. Catal., A*, 1997, **149**(1), 103.

Professor Hillmyer opened the discussion of his paper by commenting: There has been much discussion on the use of cartoons at the meeting. In a recent manuscript in which I was a co-author was written "A cartoon illustrating these structural changes is presented in Fig. 4". The reviewer wrote "I have a special plea to the senior author regarding the last sentence of that paragraph. Please help us try to stop the insidious growth of the use of "cartoon" when what is meant is "schematic" or "diagram". A cartoon is a humorous or satirical drawing. So when people call their schematic a "cartoon" they are saying it is laughable!"

I propose the following phrase to replace cartoon: "conceptual representation of actual phenomena".

Professor McLeish responded: I cannot resist a comment defending your use of "cartoon" as an explanatory instrument. In fifteenth century Europe, the word had no satirical or laughable connotations at all: it was a cartoonists task to transfer an artists design to a full-scale diagram used in the production of tapestries. The cartoonist needed a deep understanding of both the artists ideas, and of the technical limitations of the wearer. The cartoon was therefore an indispensable bridge between theoretical conception and technological realisation!

Professor Hillmyer said: Thank you for your enlightening comments! I will remember this when using a "cartoon" in a future manuscript.

Professor Biggs asked: Are you able to deposit porous films as thin "monolayers"? Can these films be put onto particles following an idea developed by my colleague, Professor Hamley?

Professor Hillmyer responded: There has been much work on preparing nanoporous thin films from ordered block copolymer precursors. I can refer you to the exemplary work of Professor Tom Russell at the University of Massachusetts, Amherst. We have recently shown that nanoporous thin films can be prepared from polystyrene–polylactide block copolymers.[1] We used these films as nanolithographic masks for the preparation metal nanodot arrays. Most of the thin film work has been done on relatively flat surfaces. I am not aware of any work on the coating of particles with nanoporous thin films from ordered block copolymer precursors.

1 R. Olayo-Valles, M. S. Lund, C. Leighton and M. A. Hillmyer, *J. Mater. Chem.*, 2004, **14**, 2729.

Professor Müller said: In the paper (towards the end of the Results and Discussion section) it is stated that for the degraded samples, the PEO cannot crystallize unless its content is high (*i.e.* 11.7%). Did you cool the samples to very low temperatures? Is it possible that if all samples are cooled to −50 °C, the PEO may be able to crystallise since it is being confined in nanocylinders, in this case it should undergo homogeneous nucleation and thereby crystallise at approx. −40 °C to −45 °C as has been observed in for instance PCL–*b*–PEO diblock copolymers.

Professor Hillmyer responded: At this stage in our research, we have not tried to cool the degraded monoliths with "PEO-coated" walls to very low temperatures. I agree with you that the PEO may be able to crystallize at low enough temperatures. This would be quite interesting to explore as a function of the areal density of PEO chains on the pore walls.

Professor Hamley asked: You indicate miscibility of PEO and PLA within the cylinders. Are there conditions under which lateral phase separation between PEO and PLA can occur, either in the melt or crystal state? Lateral phase separation to produce "patches" of polymer in the nanopores could be of interest for applications.

Professor Hillmyer responded: There is no question that this would be an interesting phenomenon. In fact, Kimishima *et al.* observed phase separation of fully hydrogenated polyisoprene (PEP) and partially hydrogenated polyisoprene (HPI) in the rubber microdomains of a microphase separated blend of polystyrene–PEP/polystyrene–HPI (AB/AC) binary blend.[1] At this stage in our work with polysytrene–polylactide (PS–PLA)/polystyrene–polyethylene oxide (PS–PEO) blends we do not have direct evidence for "lateral" phase separation. However, we do observe a discontinuity in the principal spacing of the microphase separated melt structures when the PEO content is above about 50% in the binary blends. At this point, the origin of the sudden change in spacing is unclear, but lateral phase separation could certainly result in a volume change within the PEO/PLA microdomain and thus lead to a change in spacing. We are currently evaluating this hypothesis.

1 K. Kimishima, H. Jinnai and T. Hashimoto, *Macromolecules*, 1999, **32**, 2585.

Professor Gido asked: Have you thought about the possibility of lateral phase separation of the B and C blocks of the AB/AC blends in order to produce patchiness in the functionalization of your pore walls?

Professor Hillmyer responded: Yes. Please see my earlier response to Professor Hamley's question.

Professor Ryan asked: (1) The phase behaviour of crystallisable polymers shows some unusual features (see Ryan *et al.*[1]) with a different $\chi(T)$ behaviour below T_m. This could also provide a mechanism for unusual phase behaviour. (2) Could you generalise your work to provide a framework of rules for making AB/BC blends as finding systems that have ODTs before macrophase separation is limited?

1 A. J. Ryan, S-M. Mai, J. P. A. Fairclough, I. W. Hamley and C. Booth, *Phys. Chem. Chem. Phys.* 2001, **3**, 2961–2971.

Professor Hillmyer responded: I have read the excellent contribution on this topic co-authored by you in *PCCP*.[1] In Fig. 2 of that work the change in the temperature dependence of χ around the melting point of the polyethylene oxide (PEO) block is noted. However, I'm not sure what "unusual phase behavior" you are referring to in our work on polystyrene–polylactide (PS–PLA)/PS–PEO blends. We have not yet attempted to extract the temperature dependence of χ(s) for our system, but in the event we do, I will be sure to recall your interesting measurements of χ for polypropylene oxide–PEO diblock copolymers at temperatures near the crystallization point of PEO. To be clear about the generalization of our work with respect to the order–disorder transition (ODT) and macrophase separation, we saw neither in the PS–PLA/PS–PEO blends we examined. The materials we used were quite high in molecular weight (see Table 1 in our manuscript), and given that χ for these systems is quite large, we were unable to observe an ODT in any of the four parent materials or their blends (for the temperature dependence of χ for PS–PLA see Zalusky *et al.*[2]). To prepare AB/AC block copolymer blends that are miscible (*i.e.*, do not show macrophase separation of the two diblocks), one should simply choose B and C blocks that form miscible homopolymer blends. For example, PEO and PLA (our work) and 1,4-polyisoprene and 1,2-polybutadiene (see Vaidya and Han[3]).

1 A. J. Ryan, S-M. Mai, J. P. A. Fairclough, I. W. Hamley and C. Booth, *Phys. Chem. Chem. Phys.*, 2001, **3**, 2961–2971.
2 A. S. Zalusky, R. Olayo-Valles, J. H. Wolf and M. A. Hillmyer, *J. Am. Chem. Soc.*, 2002, **124**, 12761.
3 N. Y. Vaidya and C. D. Han, *Macromolecules*, 2000, **33**, 3009.

Professor Iwuoha asked: Is one week enough to degrade the PLA blend at 65 °C to form polyether functionalised nanoporous polystyrene solids? What is the ratio of the water/methanol sodium hydroxide solution? Will the concentration of sodium hydroxide have an effect on the degradation of PLA?

Professor Hillmyer responded: Yes, one week is enough to degrade all of the polylactide (PLA). I refer you to Fig. 7 in our paper. The degradation solution was prepared using 60 vol% water and 40 vol% methanol. Please see the Experimental section of our paper. For certain thin film applications we reduced the concentration from 1 or 0.5 M NaOH to 0.05 M NaOH. At the lower concentration, we are still able to degrade all of the PLA in the thin films. While we have not made a detailed study on the influence of NaOH concentration on the degradation rate of PLA in the polystyrene(PS)–PLA/PS–polyethylene oxide blends (or just PS–PLA alone), the basic hydrolysis of PLA has been well studied for PLA homopolymers.

Professor Gido opened the discussion of Dr Borsali's paper: Your data comparing the morphologies of linear block copolymer to corresponding cyclic block copolymers in micelle forming solutions seems consistent with the general trend that we have recently reported in a comparison of bulk morphology of ABA linear triblocks and their cyclic analogs.[1] In this study we found that when cyclic and linear block copolymer analogs form different morphologies, that the cyclics tend to favor a decrease in interfacial curvature.

1 Y. Zhu, S. P. Gido, H. Iatrou, N. Hadjicristidis and J. W. Mays, *Macromolecules*, 2003, **36**, 148.

Dr Borsali responded: I am aware of this very interesting work and the results in solution that I present in this discussion are in agreement with the reported bulk morphology which shows that the cyclics tend to favor a decrease in the interfacial curvature. To stress a general conclusion, we should study the whole phase diagram, *i.e.* at various volume fractions. This is under investigation in our group in solution and in bulk.

Professor de Jeu commented: In connection with the comparison cyclic–linear block copolymers the simplest zero-order model seems to me to consider the cyclic block copolymer with a hairpin at the midpoint of each block. In such a model a cyclic block copolymer would effectively be equivalent to two linear triblocks coupled at their end *via* the hairpin. This allows, in turn, a comparison with other information from triblock copolymers. Uniform lamellar triblock copolymer films have been contrasted with similar diblock films.[1] In this situation a triblock $A_nB_{2n}A_n$ with a single hairpin in the middle block gave similar sublayer spacings and structures as a A_nB_n diblock. Hence, it may be worth while to push the analogy between half a cyclic block copolymer and a triblock system further.

1 W. H. de Jeu, P. Lambooy, I. W. Hamley. D. Vaknin, J. K. Pedersen, K. Kjaer, R. Seyger, P. van Hutten and G. Hadziioannou, *J. Phys. II France*, 1993, **3**, 139.

Professor Gido commented: During the discussion, I was confused by Prof. de Jeu's use of the term "hairpin" to describe the looping over of a block in a cyclic block copolymer or of the middle block of a triblock. To me a "hairpin" turn is a very tight turn of fairly specific chain conformation, as for instance in a chain folded crystal. The materials studied in Dr Borsali's paper and in the reference given in Prof. de Jeu's question are amorphous block copolymers in which some blocks must form looped chain conformations. These loops are not necessarily tight and their chain conformations can fluctuate within the constraint that the blocks stay looped.

Professor Lodge asked: It is very unusual for micelles, as discrete structures, to change morphology with concentration. If preferred interfacial curvature is the dominant difference between linear and cycle, which seems reasonable, that should not change with concentration. Is it possible, therefore, that the vesicles are the equilibrium structure, and the spheres and worms are non-equilibrium structures formed due to "lack" of materials to form sheets?

Dr Borsali responded: It is certainly true that the interfacial curvature is the dominant difference between linear and cyclic block copolymers. While at relatively high concentrations the observed vesicles (as well as the Y-shaped particles) arising from cyclic block copolymer chains "seem" to be stable and in equilibrium, it is not the case for the "sun-flowers" and worm-like micelles formed in the dilute regime. Indeed, discrete "sun-flowers" exist but their cores (PS) are "available" and can easily stack together after collisions giving rise to worm-like particles whose diameter is exactly the size of the discrete "sun-flower" micelles. Consequently, in this dilute regime, the size increases, the morphology changes (from sun-flower to worm-like cylinder) with the time, and a quick experiment show that these self-assembled particles are non-equilibrium structures: the size decreases with the effect of ultrasound (ultimately only the sun-flower exists) and increases after cessation (co-existence of "sun-flowers" and worms), and after a long time, *i.e.* a couple of days, giant worm-like particles exist alone.

Dr Webber commented: During the course of the discussion, the presenter stated that he believed the vesicle structure is likely to be the equilibrium aggregate structure for the cyclic PS–PI copolymers. Could the authors comment on the trend in the data shown in Fig. 3 in the paper, that show an increased aggregate size at lower temperature for the cyclic copolymers? This suggests that the lower temperature drives the system to the equilibrium structure, which seems counter-intuitive.

Dr Borsali responded: Let me recall that the vesicles were only observed in *heptane* and *not in decane*. Decane becomes a worse solvent for PS at lower temperature, the trend in the data presented in Fig. 3 is coherent and the size increases for "sun-flower" or worm-like particles whose cores (extremities) are no longer or hardly protected by the PI. In other words, what is observed in relatively high concentrations in decane is somehow similar to lower concentrations in heptane.

Dr Semenov asked: Is it possible to get the same sequence of morphologies which you observed for cyclic PS–PI also for a linear PS–PI (with the same molecular weight) by appropriately adjusting its composition?

Dr Borsali responded: Generally for linear block copolymer chains, we could certainly observe a change in the morphology by simply changing the composition. In cyclic block copolymer chains, a variety of morphologies is observed also by increasing the concentration. So, playing with both the composition and the concentration gives rise to more complex and interesting morphologies in cyclic as compared to linear block copolymer chains. Moreover, it seems that the effect of solvent quality (decane and heptane are both bad solvents for PS in this case) and temperature (from 10° to 25 °C) also induces a change in the morphology in the cyclic case whereas it is less sensitive for linear block copolymer chains.

Mr Murray asked: (1) How efficient is the cyclisation reaction? (2) What behaviour would you expect from a blend of the linear and cyclic block copolymers?

Dr Borsali and **Miss Lefebvre** responded: (1) After careful and clean controlled reactions, we could get up to 90% of the linear block copolymer chains that end up linked. The 10% are removed after successive fractionations.

(2) We have actually done these experiments[1] and we can conclude that micelles formed either by the pure linear (spherical) or cyclic copolymers (worms) or by blending the two copolymers in various ratios (spheres or worms depending of the ratio) remain stable over a large temperature range. A change in the micellar morphology is observed, however, at higher temperatures.

1 N. Ouarti, P. Viville, R. Lazzaroni, M. Schappacher, A. Deffieux and R. Borsali, *Langmuir*, in press.

Mr Follestad asked: Do you see differences in morphology depending on the intensity of the stirring (implicating that the stirring intensity might play a role through grinding action)?

Dr Borsali responded: We did not check the effect of stirring on the size and morphology of the formed micelles. We, however, observed that the size decreases with the effect of ultra-sounds and increases after cessation—only on self-assembled particles obtained from cyclic block copolymer chains.

Professor Iwuoha asked: According to Table 1, linear poly(styrene-*b*-isoprene) exhibits similarity in its hydrodynamic diameter when dispersed in decane or heptane. This is not the case with cyclic poly(styrene-*b*-isoprene) which showed clear differences in hydrodynamic diameter on changing from decane to heptane. Is there an explanation for the observed solvent effect?

Dr Borsali responded: The fact that the linear diblock copolymer chains form spherical micelles that have about the same size in decane and heptane indicates that the two solvents lead to similar interfacial tensions. A difference exists for cyclic block copolymers and is probably due to the fact that PS is less soluble in decane than in heptane. While this difference in the solvent quality is not "seen" in spherical micelles since the core (PS) is protected by the corona (PI), it becomes observable for "sun-flower" or worm-like particles whose cores (PS: extremities of the cylinders) are no longer or hardly protected by the corona (PI: surface of the cylinders).

Professor Lodge commented: The fact that the linear diblock forms spherical micelles that have about the same size in decane and heptane indicates that the two solvents lead to similar interfacial tensions; the micelles would get bigger if the interfacial tension went up. Doesn't this imply that any differences in micelle size for cycles between heptane and decane must be due to something other than differences in interfacial tension?

Dr Borsali responded: While in linear block copolymer chains the size of the formed micelles is about the same in heptane and decane, this is not the case for cyclic block copolymer chains. Besides the interfacial tension, the n-alkanes become worse solvents for PS as the n-alkane carbon number increases, meaning that PS is less soluble in decane than in heptane. While this difference in the solvent quality is not "seen" in spherical micelles since the core (PS) is protected by the corona (PI), it apparently becomes observable and measurable for "sun-flower" or worm-like particles whose cores (PS located at the extremities) are no longer or hardly protected by the corona (PI forming the surface of the cylinders) as I stated previously..

Professor Vamvakaki communicated: In Fig. 3 in your paper, is the difference in the hydrodynamic size observed at 10 °C and 25 °C for the C-mic (at $c = 5$ mg ml^{-1}) accompanied by a change in the aggregation number of the micelles or is it due to the different extension of the polymer chains comprising the micelle? Have you observed these structures by TEM?

Dr Borsali communicated in response: The size, the morphology and the aggregation number are not sensitive to the temperature (from 10–25 °C) for micelles made from linear block copolymer chains in decane. These parameters are found to be temperature dependent (from 10° to 25 °C) for self-assembled particles made from cyclic copolymer chains. This is probably due to the ability of the corona (PI) to protect the core (PS). Indeed, besides the interfacial tension, decane becomes an even worse solvent at lower temperature. While this difference in the solvent quality is not "seen" in linear block copolymer chains (spherical micelles since the core (PS) is protected by the corona (PI), it becomes observable for "sun-flower" or worm-like particles whose core (located at extremities for the case of worms) are no longer or hardly protected by the PI (corona: surface of the cylinders).

Unfortunately, we did not do the experiments at 10 °C using TEM or Cryo-TEM. I agree, it would certainly be interesting to do it.

Professor Hillmyer opened the discussion of Dr Lecommandoux's paper: (1) Do you know the impact of HBr addition to PI on the glass transition of that block in the PI-PLys materials? Are you concerned that the $Tg > RT$ and the micellization is influenced by this glassy block?

(2) Do you have direct evidence for the HH morphology? Specifically, how do you know the PZLys forms the cylinders?

Dr Lecommandoux responded: (1) As we described in the paper, the HBr addition resulted in a partial bromination of the isoprene double bonds. Nevertheless, the DSC measurements did not show any significant increase in the *Tg* of the PI block. One can explain this data first by the fact that the most important contribution in the modification of the *Tg* of PI is related to its microstructure (1, 2 or 1, 4 or 3, 4). With the experimental conditions used for the synthesis of the polyisoprene (mixture cyclohexane/THF 98/2), the resulting polymer had 13% of microstructure 1,4 and 87% of microstructure 1, 2 and 3, 4. Under these conditions the *Tg* of the polyisoprene obtained is pretty high (around −5 °C) and the bromination does not have a high impact. In addition, the fact that we are dealing with oligomers (40 units) also tends to reduce the *Tg* of the PI block (even if bromination occurs). Of course we were aware of the fact that the micellization of block copolymers with a $Tg > RT$ can be more dependent on external parameters (sample preparation, co-solvent added for instance) because the thermodynamic equilibrium is difficult to reach. The nice work of Professor Eisenberg on PS–*b*–PAA block copolymers demonstrates the complexity that can be attained using this phenomenon. Nevertheless, that was not our goal and we really wanted to be sure of being at thermodynamic equilibrium (or at least as close as possible) and we never needed any additional trigger such as temperature or the use of a co-solvent to obtain micelles with a controlled size and a narrow size-distribution (which is itself a good argument to say that we should be very close to the equilibrium).

(2) We do not have direct evidence of the hexagonal in hexagonal morphology (HH). Our model is essentially based on Small Angle X-ray Scattering (SAXS) results. On this kind of systems, basically only the so-called hexagonal in lamellar morphology has been observed so far and this was some time ago.[1] This is also the only structure that has been theoretically predicted for rod–coil systems.[2] According to SAXS data, it appears to be clear that the phase separation observed for some of our systems is no longer lamellar, and the assignment of the scattering peaks perfectly fits with a hexagonal morphology. Nevertheless, and in order to have clear evidence of this original structure, a direct observation with a TEM analysis would be crucial, especially if we want to prove, or at least observe whether the PZLys rigid block really forms the cylinders as we believe and as we've shown in our schematic representation of the proposed model for this morphology.

1 (a) J.-P. Billot, A. Douy and B. Gallot, *Makromol. Chem.*, 1976, **177**, 1889; (b) J.-P. Billot, A. Douy and B. Gallot, *Makromol. Chem.*, 1977, **178**, 1641.

2 (a) A. N. Semenov and A.V. Subbotin, *Sov. Phys. JETP A. V.*, 1992, **74**, 690; (b) R. R. Netz and M. Schick, *Phys. Rev. Lett.*, 1996, **77**, 302; (c) D. R. M. Williams and G. H. Fredrickson, *Macromolecules*, 1992, **25**, 3561.

Professor Gido said: You have proposed, based on SAXS and WAXS data, a HH (hexagonal–hexagonal) morphology in which the rod-like PZL blocks are microphase separated and inside hexagonally packed cylinders in a PI matrix. The compositions of the block copolymers in Table 2 indicate that the PZL volume fractions that must be incorporated inside the microphase separated cylinder domains (0.86, 0.89, and 0.92) approach or even exceed the hard contact packing limit for cylinders which is 0.91. Additionally, it may be difficult to even approach the volume fraction of this limit for the relative amount of material that can be incorporated into hexagonally packed cylinders due to severe distortions in the chains of the matrix material in order to fill the corners of the Wigner –Seitz cells around each cylinder. The space filling problems imposed on the matrix block as a result of a large volume fraction of the cylinder forming block(s) can be appreciated from the arguments in ref. 1 here. More recently we[2] have found that these packing considerations put an effective upper limit of about 0.50 on the volume fraction that can be forced into the inside of cylindrical domains by increasing arm number asymmetry in miktoarm star block copolymers. Based on these considerations, it would be very helpful to add real space imaging in the form of TEM to your evidence for an HH morphology.

1 S. Gido, Schwark, Thomas and Goncalves, *Macromolecules*, 1993, **26**, 2636.
2 Yang, Hong, S. Gido, Velis and Hadjichristidis, *Macromolecules*, 2001, **34**, 9069.

Dr Lecommandoux replied: I completely agree with this comment and I also believe that it would be very helpful to have direct imaging analysis on this morphology in order to clearly evidence the structure. Actually these experiments are planned now and are the basis of a collaboration that we will develop together!!

The space filling problems that you mentioned and that are described in the two papers cited are, in the situation of flexible copolymers, (even for mikto-arm stars) a limiting factor on the volume fraction that can be forced into the inside of cylindrical domains (limit around 50% in volume). In our situation, the rod volume fraction is typically between 0.85 and 0.90 and I can reasonably think preliminarily that the rod part would form the matrix of the hexagonal phase separated morphology. However, this hypothesis would suggest that during the formation of this structure as the solvent evaporates, the phase separation occurs first and then the hexagonal structure due to the rod–rod packing appears later. Nevertheless, I do not believe this hypothesis to be realistic in the sense that rod–rod interactions of the polypeptides in their α-helical conformation are very strong and they should appear first. In addition, in all the studies based on rod–coil systems performed in the groups of Stupp,[1] Lee,[2] Ober[3] or us,[4] everyone agrees that the driving force for the self-assembly is the "liquid crystal type interactions" related to the rod blocks. Even more recently, Stupp and myself[5] evidenced that in rod–dendron systems, where the dendron itself has been shown to be able to self-assemble by Percec,[6] still the rod–rod interactions dominate the aggregation process. Based on these observations and taking this hypothesis into account, we resulted in the proposed model: when solvent evaporates, first the rod blocks are interacting together thus forming the inside of the structure (columnar hexagonal structure classically observed for PBLG or PZLys), and then the copolymer phase separates into a lamellar (HL) or a hexagonal (HH) structure depending on the composition of the copolymer, the low volume fraction of the diblock thus forming the matrix of the structure.

Finally, I would say that the two arguments or hypotheses can be realistic: either the rod–rod interactions or the incompatibility (phase separation) dominates the structure formation. Our feeling is that the liquid crystal like interactions are predominant but I completely agree that it would be very helpful to add real space imaging in the form of TEM to our evidence for an HH morphology.

1 See for example: S. I. Stupp, V. LeBonheur, K. Walker, L. S. Li, K. E. Huggins, M. Keser and A. Amstutz, *Science*, 1997, **276**, 384.
2 See for example: M. Lee, B.-K Cho and W.-C. Zin, *Chem. Rev.*, 2001, **101**, 3869 and references therein.
3 See for example: J. T. Chen, E. L. Thomas, C. K. Ober and G.-P. Mao, *Science*, 1996, **273**, 343.
4 See for example: (a) H.-A. Klok, J. F. Langenwalter and S. Lecommandoux, *Macromolecules*, 2000, **33**, 7819. (b) H.-A. Klok and S. Lecommandoux, *Adv. Mater.*, 2001, **13**, 1217.
5 S. Lecommandoux, H.-A. Klok, M. Sayar and S. I. Stupp, *J. Polym. Sci. A*, 2003, **41**, 3501.
6 See for example: (a) S. D. Hudson, H.-T. Jung, V. Percec, W.-D. Cho, G. Johansson, G. Ungar and V. S. K. Balagurusamy, *Science*, 1997, **278**, 449. (b) V. Percec, C.-H. Ahn, G. Ungar, D. P. J. Yeardley, M. Möller and S. S. Sheiko, *Nature*, 1998, **391**, 161.

Dr Davidson said: I have two questions about your hexagonal in lamellar morphology: (1) Did you even try to produce an aligned domain by applying an external field or a shear flow? (2) This morphology is formally comparable to the hexatic phase of liquid crystals where layers showing long-range hexagonal (bond-orientational) order are more or less coupled along the layer normal direction. Here, would you be able to tune your PI layer thickness and therefore adjust this coupling, which would possibly lead to crystal-B or hexatic B phases? The use of a single domain would be very helpful in this respect.

Dr Lecommandoux responded: (1) We never tried to align the system with an external field or a shear flow. In principle this would be possible using a magnetic field applied during the solvent casting process. This has already been done by Professor Tirrell on monodisperse homopolypeptide (PBLG) from chloroform/TFA solutions.[1]

(2) You are absolutely right that the hexagonal in lamellar (HL) morphology can be compared with the hexatic phase of liquid crystals, and the analogy is very interesting. In addition, the ability to tune the PI layer thickness is completely realistic and has been done either in our group[2] or even previously in Gallot's group.[3] This can be easily done and controlled over a large domain of composition, especially because, as we demonstrated, the HL morphology is stable over the most part of the phase diagram. Of course, the use of a single domain would be very helpful in this respect, but its formation is still challenging in the block copolymer world!!

1 S. M. Yu, V. P. Conticello, G. Zhang, C. Kayser, M. J. Fournier, T. L. Mason and D. A. Tirrell, *Science*, 1997, **389**, 167.
2 F. Chécot, PhD thesis, University Bordeaux 1, 2003.
3 (a) J.-P. Billot, A. Douy and B. Gallot, *Makromol. Chem.*, 1976, **177**, 1889; (b) J.-P. Billot, A. Douy and B. Gallot, *Makromol. Chem.*, 1977, **178**, 1641.

Dr Perez Mendez asked: Solvent must play an important role on the secondary structure of the polypeptide. Helical diameter of the coil must change depending on solvent and on concentration.

Dr Lecommandoux responded: Of course the solvent is playing an important role on the secondary structure of the polypeptide.[1] As a result, the helical diameter of the rod (and not the coil!) can change depending on the solvent used affording different helical structures. It is well known also that other solvents can disturb the intramolecular H-bonding such as trifluoro-acetic acid (TFA), so the helicoidal structure of the polypeptide can be disrupted.

1 See for example: (a) H. R. Kricheldorf, *α-Amino acid-N-carboxyanhydrides and related heterocycles*, Springer-Verlag, Berlin, 1987. (b) H. Block, *Poly(γ-benzyl-L-glutamate) and other glutamic acid containing polymers*, Gordon and Breach Science Publishers, New York, 1983.

Mr Papadopoulos asked: Why does the conformation start to change at pH $\sim$5? The pK_a of the poly(L-lysine) side group has a much higher value (around 9–10), so even at pH = 7 or 8 the side group would be completely protonated and the electrostatic energy high enough to destabilize the helix.

Dr Lecommandoux replied: There are several factors that can modify the pH of the transition between α-helix and random coil. First, we have prepared fairly polydisperse materials: polydispersity can modify the range where the transition takes place, increasing the window. Secondly, the first block which is hydrophobic can alter the stability of the α-helix modifying as a result the pH of the conversion of the coil into an α-helix. In fact, for instance the stabilization of peptide secondary structures by conjugation of PEG has been reported.[1] This phenomenon has been explained by the ability of PEG to form a microhydrophobic environment which interacts with the amphiphilic secondary structure of the peptide segment. One can expect that this behaviour can be emphasized by our real hydrophobic PI second block. One has also to detail that the doted line in Fig. 4 is just a help for eyes that is probably not very appropriate, and the real transition is probably more between pH $\sim$7–8 which is more realistic (no experimental data have been taken in this pH range between 5 and 7 because its stability is very hard to reach).

1. (a) M. Pechar, P. Kopeckova, L. Joss and J. Kopocek, *Macromol. Biosci.*, 2002, **2**, 199; (b) A. Harada, S. Cammas and K. Kataoka, *Macromolecules*, 1996, **29**, 6183.

Fiona Case commented: In your polyisoprene-*b*-poly(L-lysine) system the polypeptide coils with increasing pH. You have also studied polyisoprene-*b*-poly(L-glutamic acid) copolymers, which coil with decreasing pH. Have you ever tried to mix the two materials? The resulting blend system might show interesting pH dependant behaviour as the mixed micelles changed from positively charged to negatively charged and showed a size maximum at pH around 4 or 5.

Dr Lecommandoux responded: In addition to the polyisoprene-*b*-poly(L-lysine) system described in the present communication, we also studied polybutadiene-*b*-poly(L-glutamic acid) copolymers in solution and with varying pH and salt conditions.[1] These two systems were chosen effectively because their behaviour as a function of pH is completely opposite. However, we have never tried to mix the two kinds of diblocks, even if I agree that the resulting blend system might show interesting pH dependance and might also show a size maximum at pH around 4 or 5. We have to go in that direction and thus synthesize diblocks that can be compatible in size and composition.

1 (a) F. Chécot, S. Lecommandoux, Y. Gnanou and H.-A. Klok. *Angew. Chem. Int. Ed. Engl.*, 2002, **41**, 1339; (b) F. Chécot, S. Lecommandoux, H.-A. Klok and Y. Gnanou. *Eur. Phys. J. E*, 2003, **10**, 25.

Professor Lodge asked: In Fig. 4 in your paper, and the associated cartoon Fig. 6, the chain end-to-end distance shrinks when going from coil to helix. Isn't this surprising? Given that the end to end distance of a helix grows as N, whereas the coil as $\sqrt{N}$, the helix must be bigger than the coil at large N, whereas you infer the opposite result.

Dr Lecommandoux responded: I agree that the chain end-to-end distance scales as N for a rod and as $\sqrt{N}$ for a coil. No problem! . . . As we described and calculated in the paper, this distance L can be calculated (the result is given for $N = 123$) in basic conditions (α-helical conformation) as $L_{\text{basic}} = 0.15 \times N (= 18.45 \text{ Å})$ and in totally stretched conditions (acid with no added salt) as $L_{\text{acid}} = 0.37 \times N (= 46.1 \text{ Å})$. If we now consider a real coil conformation of the polypeptide, we should have $L_{\text{coil}} = 0.37 \times \sqrt{N} (= 4.1 \text{ Å})$, which is effectively much smaller that the distance expected for a rod conformation. However, the R_H values measured by DLS that are given in Fig. 4 do not correspond to homopolypeptides but to the hydrodynamic radii of the micelles. This clearly shows that the polypeptide conformation in the aggregates is modified, and that the chains are stretched. That is the reason why we only considered in our models and calculations the two extreme situations where we have a pure α-helix or a charged polyelectrolyte. The intermediate case where the electrostatic interactions are screened and where a coil conformation can be expected has been called in the paper a "stretched coil" due to the micelle formation. What we also should not forget is that these polypeptides are not "simple" polyelectrolytes (even if the behaviour of homopolyelectrolytes is still not fully understood!!): when pH or ionic strength are modified, not only the degree of charge or the electrostatic forces, but also the chain conformation are changing.

Professor Hamley commented: Referring to Fig. 4 in your paper, it is surprising that even when the charge on the polypeptide is screened in 1 M NaCl, the hydrodynamic radius of the PI–PLys diblock is larger for the coil at low pH than for the helix at high pH.

Dr Kostarelos commented: It would be interesting if you could provide us with some further information on the vesicle formation observation you mentioned using the peptide di-block copolymers, and whether your conclusion that micellisation depends on the pH holds for the vesicle formation as well. If this is the case, it will dramatically determine any biomedical applications of these vesicle systems. Also, can the bilayer thickness of these vesicles be altered?

Dr Lecommandoux responded: We have a couple years of experience in vesicle formation and size dependence as a function of pH and ionic strength on different kinds of polypeptide based copolymers (using glutamic acid or lysine).[1–3] For some of our systems, the micellisation depends on the pH hold for vesicle formation.[2] As we also recently determined by small angle neutron scattering (SANS) measurements, the membrane (bilayer) thickness of these vesicles can be modified by pH variations.[3] Based on these data, we effectively believe that these micelles and vesicles are potentially very interesting candidates for biomedical applications.[4]

1 (a) F. Chécot, S. Lecommandoux, Y. Gnanou and H.-A. Klok. *Angew. Chem. Int. Ed. Engl.* , 2002, **41**, 1339; (b) F. Chécot, S. Lecommandoux, H.-A. Klok and Y. Gnanou, *Eur. Phys. J. E*, 2003, **10**, 25.

2 (a) J. Rodriguez-Hernandez and S. Lecommandoux, in preparation.
3 F. Chécot, S. Lecommandoux, Y. Gnanou, H.-A. Klok and A. Brûlet, in preparation.
4 J. Rodriguez-Hernandez, F. Chécot, Y. Gnanou and S. Lecommandoux, *Prog. Polym. Sci.*, in press.

Professor Vamvakaki opened the discussion of Dr Webber's paper: In Fig. 11 in your paper, why is the response of the 0qPDMA–PDEA copolymer different to the 50q and 100qPDMA–PDEA copolymers at low pH? Do you not observe any micelle dissociation perhaps accompanied with the release of polymer chains at low pH for the 0qPDMA–PDEA sample as well? Furthermore, is the change in morphology for the 0qPDMA-PDEA copolymer fully reversible upon cycling the pH from 4 to 9?

Dr Webber responded: Lowering the pH of the overlying solvent causes a change in the morphology of the adsorbed layers of copolymer micelles. This is due to the protonation of the tertiary amine residues of both PDMA and PDEA blocks. In the case of 0qPDMA–PDEA, protonation leads to dissolution of the micellar cores. Upon pH increase the cores reform, and the layer is observed to return to its original morphology. Quartz crystal microbalance (QCM-D) data indicate the loss of some material in the initial 9–4–9 pH cycles, but further cycles do not result in substantial loss of material. We believe that the close-packed nature of the initial layer facilitates the reformation of the hydrophobic cores at high pH.

For the 50q and 100qPDMA-PDEA copolymers, the micelles are more widely spaced in the initial adsorbed layer. Hence, when the micellar cores breakdown at low pH the change is irreversible; since the PDEA chains are constrained by the adsorption they must remain near each other at low pH to reform hydrophobic domains upon return to high pH conditions.

Professor Penfold commented: Comparing the micelle dimensions (or $2 \times R_H$) from Table 1 in your paper and the thickness of the adsorbed layer (Table 3) as a function of degree of quaternisation shows a trend that implies that the surface and bulk structures are rather different:

Degree of quaternisation	$2 \times R_H$	d
0	23	17
50	9	20
100	9	6

Can this trend be fully rationalised between the bulk and surface structures?

Dr Webber responded: Indeed the surface and bulk structures are different. We believe that adsorption is *via* complete micelles, and that once at the surface relaxation of coronal chains occurs, leading to a flattening of the micelles at the interface. The spacing of the micelles in the adsorbed layer is thus determined by the balance between coronal chain relaxation and further micellar adsorption; highly charged micelles, where corona–surface interactions will be strong, form adsorbed layers with large inter-aggregate separation whereas micelles of low charge are able to pack more closely within the layer. Such behaviour has been previously observed in the adsorption of ionic copolymer micelles.[1,2]

The determination of adsorbed layer thickness is done using quartz crystal microbalance with dissipation monitoring (QCM-D). It is important to remember here that in QCM-D measurements all material coupled to the oscillation of the crystal is sensed. Thus the hydrodynamics of the interaction between the adsorbed material and the solvent is crucial, in particular for the measurement of dissipation. As discussed in the paper the value of thickness is derived from the dissipation measurement.

As the degree of quaternisation increases the micelles become smaller in bulk solution, while the inter-aggregate distance in the adsorbed layer and the strength of the interaction between the micelles and the surface will increase. Thus one would expect that the layers become 'thinner' with increasing quaternisation. Indeed, apart from the 50qPDMA–PDEA this is the case. As discussed in the paper we believe that the coronal chains of the 50qPDMA–PDEA adopt a 'loop and train' type conformation at the interface. It is known that polymer chains extending normal to an interface have a significant impact on the flow of fluid over a surface.[3] Such a conformation will increase the amount of bound solvent sensed by the crystal, which in turn increases the measured dissipation and calculated layer thickness.

1 M. R. Talingting, Y. Ma, C. Simmons and S. E. Webber, *Langmuir*, 2000, **16**, 862.
2 G. B. Webber, E. J. Wanless, S. P. Armes, F. L. Baines and S. Biggs, *Langmuir*, 2001, **17**, 5551.
3 M. A. Cohen Stuart, F. H. W. H. Waajen, T. Cosgrove, B. Vincent and T. L. Crowley, *Macromolecules*, 1984, **17**, 1825.

Professor Iwuoha said: I do not think that soft-contact *in situ* AFM experiments provided sufficient evidence for monolayer adsorption behaviour postulated for the micelles of PDMA–PDEA.

Dr Webber responded: We do not agree with this statement. Force–distance data recorded during AFM experiments indicate monolayer formation, see Fig. 3. These data, recorded during the adsorption of 0qPDMA–PDEA micelles, are representative of those recorded for all copolymers studied. There is a clear repulsive electrosteric interaction as the cantilever approaches the surface, which is used in the soft-contact imaging technique, before the cantilever jumps into contact with the substrate. In this case, the jump-in occurs at an apparent separation of approximately 7 nm. This short distance is indicative of pushing through a single layer of micelles. Whilst there will exist an unknown thickness of incompressible polymer adjacent to the mica substrate, an adsorbed morphology comprising of two or more layers of micelles will result in a larger jump-in distance. For higher degrees of quaternisation the jump-in distance is decreased.

Dr Borsali asked: How can you explain a jump in R_H while the scattering intensity seems to be constant as a function of pH? Is it a scale problem?

Dr Webber responded: The increase in RH for 50q and 100qPDMA–PDEA, with apparent constant scattering intensity, is merely a problem with scale. There is in fact an increase in scattering intensity upon formation of micelles for both these copolymers.

Dr Perez Mendez asked: As both N atoms protonate in PDMA and PDEA units at low pH, and at $\sim$pH $= 8$ micelle formation is observed, which N atom deprotonates first? Or do they lose both their H atoms at the same time? PDEA should be more hydrophobic.

Dr Webber responded: At low pH all tertiary amine residues of both the PDMA and PDEA blocks are protonated. As pH is increased protons are removed from these tertiary amines. The rate

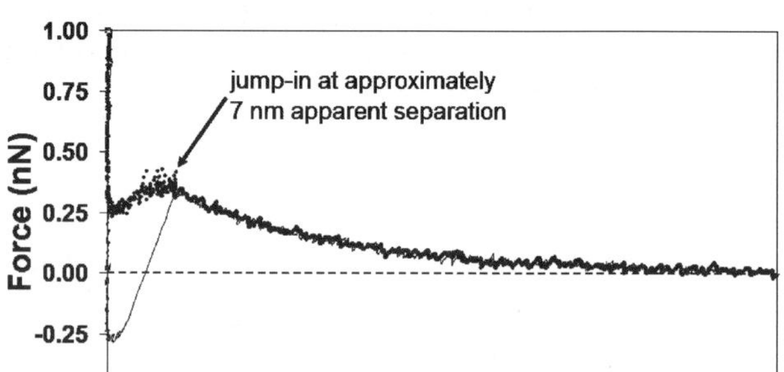

Fig. 3 Force–distance data recorded normal to the adsorbed layer of 0qPDMA–PDEA as the AFM cantilever approaches (dots) and retracts (line) from the surface. The dashed line represents the points of zero force.

of proton removal is essentially equal for both blocks, though crucially the two blocks exhibit slightly different hydrophobicities in the low-charge states. It is this difference, with the PDEA block being more hydrophobic, that drives the micellisation of the copolymers. With quaternisation, the copolymers are able to exist as stable unimers at higher pH, and so the cm-pH is increased.

Professor Hamley said: Could you comment on the relationship between the adsorption behaviour of your polyelectrolyte diblocks and neutral diblocks, especially in regard to reversibility of adsorptions?

Please note that your citation of our paper (ref. 60) mistakenly mentions cationic coronal chains. Actually we studied adsorption of neutral diblocks *via in situ* liquid cell AFM, see also our recent paper.[1]

1 I. W. Hamley, S. D. Connell and S. Collins, *Macromolecules*, 2004, **37**, 5337.

Dr Webber responded: Adsorption of polyelectrolyte diblock copolymer micelles differs from neutral diblock copolymers in distinct ways. By utilising charged micelles and electrostatic adsorption, it is possible to alter adsorbed layer morphology by controlling the charge, either permanent or pH-controlled, of the micelles. This is the focus of this paper and earlier work by our group.[1] Layer morphology may also be modified by judicious control of surface charge or salt concentration. Additionally, since adsorption is electrostatic in nature, and adsorption occurs at multiple points on each chain, the adsorbed films are generally very resilient, in particular toward solvent exchange. Such robustness is critical if an adsorbed micellar layer is to be utilised in further processes such as capture/release or nano-fabrication.

1 G. B. Webber, E. J. Wanless, S. P. Armes, F. L. Baines and S. Biggs, *Langmuir*, 2001, **17**, 5551.

Professor Biggs commented: Since these polymers protonate/deprotonate as a function of pH in the same way for both blocks, the transition from unimers to micelles is very sharp. This is an important advantage over many other copolymer systems, for layers of this type, in terms of stimulus responsive behaviour.

Professor Penfold asked: Will the potentially different pH dependence of the silica surface used in the QCM measurements compared to the mica surface used in the AFM measurements compromise the combination of the data from the two techniques?

Dr Webber responded: Comparing the adsorption of material at the mica/solution and silica/solution interface is always complicated by the different charge characteristics of the two surfaces, both with respect to their pH dependence and level of charge. However we believe that it is the solution properties of the copolymers that is critical in the observed phenomena; that is, their existence as highly charged unimers at low pH and micelles of differing charge at high pH. We have used AFM to study the morphology of 0qPDMA–PDEA micelles adsorbed from pH 9 solution to both mica and silica and observed no significant difference. Thus we are confident that comparison of adsorption at high pH, where both surfaces are near their maximum surface charge,[1] is justified. At pH 4 we are operating above the isoelectric point of both surfaces, which are thus still anionic though with lower charge.

1 P. G. Hartley, I. Larson and P. J. Scales, *Langmuir*, 1997, **13**, 2207.

Self-nucleation and crystallization kinetics of double crystalline poly(*p*-dioxanone)-*b*-poly(ε-caprolactone) diblock copolymers†

Alejandro J. Müller,*[a] Julio Albuerne,[a] Leni Marquez,[a] Jean-Marie Raquez,[b] Philippe Degée,[b] Philippe Dubois,[b] Jamie Hobbs[c] and Ian W. Hamley[d]

[a] *Grupo de Polímeros USB, Departamento de Ciencia de los Materiales, Universidad Simón Bolívar, Apartado 89000, Caracas 1080-A, Venezuela. E-mail: amuller@usb.ve*
[b] *Laboratory of Polymeric and Composite Materials (LPCM), University of Mons-Hainaut, Place du Parc 20, 7000, Mons, Belgium*
[c] *Department of Chemistry, University of Sheffield, Sheffield S3 7HF, UK*
[d] *Department of Chemistry, University of Leeds, Leeds LS2 9JT, UK*

Received 1st March 2004, Accepted 4th April 2004
First published as an Advance Article on the web 17th August 2004

The crystallization kinetics of each constituent of poly(*p*-dioxanone)-*b*-poly(ε-caprolactone) diblock copolymers (PPDX-*b*-PCL) has been determined in a wide composition range by differential scanning calorimetry and compared to that of the equivalent homopolymers. Spherulitic growth rates were also measured by polarized optical microscopy while atomic force microscopy was employed to reveal the morphology of one selected diblock copolymer. It was found that crystallization drives structure formation and both components form lamellae within mixed spherulitic superstructures. The overall isothermal crystallization kinetics of the PPDX block at high temperatures, where the PCL is molten, was determined by accelerating the kinetics through a previous self-nucleation procedure. The application of the Lauritzen and Hoffman theory to overall growth rate data yielded successful results for PPDX and the diblock copolymers. The theory was applied to isothermal overall crystallization of previously self-nucleated PPDX (where growth should be the dominant factor if self-nucleation was effective) and the energetic parameters obtained were perfectly matched with those obtained from spherulitic growth rate data of neat PPDX. A quantitative estimate of the increase in the energy barrier for crystallization of the PPDX block, caused by the covalently bonded molten PCL as compared to homo-PPDX, was thus determined. This energy increase can dramatically reduce the crystallization rate of the PPDX block as compared to homo-PPDX. In the case of the PCL block, both the crystallization kinetics and the self-nucleation results indicate that the PPDX is able to nucleate the PCL within the copolymers and heterogeneous nucleation is always present regardless of composition. Finally, preliminary results on hydrolytic degradation showed that the presence of relatively small amounts of PCL within PPDX-*b*-PCL copolymers substantially retards hydrolytic degradation of the material in comparison to homo-PPDX. This increased resistance to hydrolysis is a complex function

† Electronic supplementary information (ESI) available: Video showing the sequential isothermal crystallization of $D^{8}_{23}C^{27}_{77}$ according to the same temperature program described in Fig. 2. See http://www.rsc.org/suppdata/fd/b4/b403085k/

of composition and its knowledge may allow future prediction of the lifetime of the material for biomedical applications.

Introduction

Poly(*p*-dioxanone), PPDX, and poly(ε-caprolactone), PCL, are considered as biocompatible and biodegradable materials suitable for biomedical applications in view of their molecular structures; their respective repeat units are shown in Scheme 1.

In particular PPDX applications include a series of bioabsorbable materials like sutures, pins for fracture fixation, surgical clips and fasteners. These applications are based on the relatively fast hydrolytic degradation of PPDX which can in some cases be too fast for certain biomedical applications. The idea to copolymerize it with PCL could provide a way to tune the lifetime of the biomedical devices within the human body since PCL hydrolytic degradation is several orders of magnitude lower than that of PPDX.[1,2]

Crystallization within block copolymer microdomains (MD) is an issue which has attracted increasing interest in recent years. The structure development in semicrystalline block copolymers depends on two competing self organizing mechanisms: microphase separation and crystallization. Depending on the segregation strength in the melt, crystallization can be either confined within the copolymer MD for strongly segregated systems, or crystallization can drive structure formation for weakly segregated melts (overwriting any previous MD) or homogeneous systems.

The most commonly studied of the semi-crystalline block copolymer systems in the literature have been AB diblock copolymers or ABA triblock copolymers where one block is amorphous and the other semicrystalline. It is generally accepted that the changes of state as a function of temperature can determine the final morphology according to three key transition temperatures: the order–disorder transition temperature T_{ODT}, the crystallization temperature (T_c) of the crystallisable block, and the glass transition temperature T_g of the amorphous block. A generalization can be made to describe four cases that have been described in the literature: (1) Homogeneous melt, $T_{ODT} < T_c > T_g$. In diblock copolymers exhibiting a homogeneous melt, microphase separation is driven by crystallization if T_g of the amorphous block is lower than the T_c of the crystallisable block. This generally results in a lamellar morphology where crystalline lamellae are sandwiched by the amorphous block layers and spherulite formation can be observed depending on composition.[3–7] (2) Weakly segregated systems, $T_{ODT} > T_c > T_g$ with soft confinement. In this case, crystallization often occurs with little morphological constraint enabling a "breakout" from the ordered melt structure and the crystallization overwrites any previous melt structure usually forming lamellar structures and in many cases spherulites depending on composition.[5,8–15] (3) Strongly segregated systems, $T_{ODT} > T_c > T_g$ with soft confinement. If the segregation strength is sufficiently strong, the crystallization can be confined within spherical, cylindrical or lamellar MD in strongly segregated systems with a rubbery block.[7,12–14,16–24] (4) Strongly segregated systems, $T_{ODT} > T_g > T_c$ with hard confinement. A strictly confined crystallization within MD has been observed for strongly segregated diblock copolymers with a glassy amorphous block.[25–38]

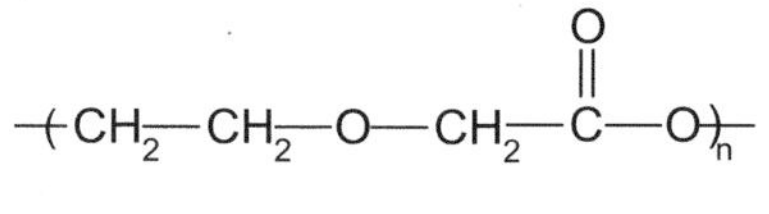

PPDX

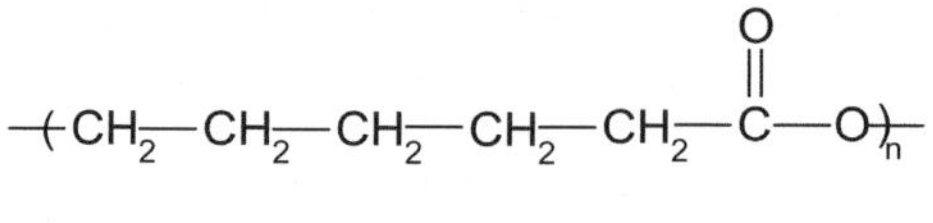

PCL

Scheme 1 Molecular structures of PPDX and PCL repeating units.

There have been relatively few reports dealing with double crystalline block copolymers.[39–45] The PPDX-*b*-PCL diblock copolymers have only been recently prepared[46] and we are unaware of other studies in similar materials except for the contribution of Lendlein and Langer[47] where multiblock copolymers of PPDX and PCL have been employed to prepare shape memory polymers for biomedical applications.

The objective of the present paper is to study the isothermal crystallization kinetics and self-nucleation of each one of the two constituents of the PPDX-*b*-PCL diblock copolymers as a function of composition. In this work, we have been able to follow the overall crystallization kinetics of the PPDX block by accelerating the overall crystallization rate through a self-nucleation procedure. The application of the Lauritzen and Hoffman theory to spherulitic growth rate data and to the overall crystallization data obtained after self-nucleation allowed us to calculate the energy barriers to crystal growth and to crystal nucleation. Using such a procedure, we were able to quantitatively estimate for the first time the effect of the molten PCL block on the energy barrier for crystallization of the PPDX block (melting points of PPDX and PCL are approximately 107 °C and 57 °C respectively). In the case of the PCL block, we first crystallize the PPDX block until saturation and then quench to suitable temperatures to determine overall isothermal crystallization kinetics. In this case, the nucleation effect of PPDX on PCL can be ascertained and the energy barriers for nucleation and growth determined as a function of composition. Additionally, atomic force microscopy experiments revealed the morphology at the lamellar scale of one selected PPDX-*b*-PCL diblock copolymer during the sequential isothermal crystallization of each block. Finally, preliminary results on hydrolytic degradation are reported.

Experimental

Materials

The controlled synthesis of the well defined PPDX-*b*-PCL diblocks has been described earlier.[44,46] Table 1 denotes the main characteristics of the copolymers such as composition and molecular weight. We employ subscripts to indicate the content of a particular component in weight% and superscripts to indicate the number average molecular weight in kg mol^{-1}.

Polarised optical microscopy (POM)

The polarised optical microscopy experiments were described previously and representative images have been reported.[42,44] The films were prepared by pressing the material between two glass cover slips at 130 °C for 3 min, and then quenched to the desired crystallisation temperature in a Linkam TP-91 hot stage. The samples were observed in a Zeiss MC-80 optical microscope, equipped with crossed polars and a digital camera. Typical errors in the measurement of spherulite growth rates are in the order of 5% or less.

Atomic force microscopy (AFM)

A Veeco Dimension D3100 was used in Tapping Mode using standard silicon cantilevers with a nominal spring constant of 50 Nm^{-1} and a resonant frequency of approximately 250 kHz. The temperature of the sample was controlled using a Linkam heater, using the same experimental set-up as described previously[48] but without the heat-shield as the temperatures required were relatively low. In the phase images shown, soft/sticky areas appear dark and hard materials appear light. Images were collected continuously, each image being collected in approximately 30 s. To allow these high scan rates to be obtained, an ActivResonance Controller from Infinitesima Ltd was used to electronically reduce the effective quality factor of the cantilever from its natural value of around 300 to around 100, giving it a shorter response time and thus allowing faster imaging. The imaging force was adjusted to the minimum that allowed the surface to be successfully tracked by the AFM probe.

Table 1 Molecular characteristics of the materials employed and calorimetric data determined by DSC

				PCL segment						PPDX segment					
				Cooling (10 °C min^{-1})		Heating (10 °C min^{-1})									
Sample	F_{PCL}	M_n^{PCL}/g mol^{-1} [a]	M_n^{PPDX} /g mol^{-1}	T_c/°C	ΔH_c/J g^{-1}	T_{mo}/°C	T_{mp}/°C	ΔH_m/J g^{-1}	χ_c (%)	T_c/°C	ΔH_c/J g^{-1}	T_{mo}/°C	T_{mp}/°C	ΔH_m/J g^{-1}	χ_c (%)
$D_{23}^{8}C_{77}^{27}$	0.77	7510	27 190 [b]	31.3	−51.2	54.3	58.2	54.5	50.8	75.4	−1.8	88.3	96.7	14.1	43.4
$D_{40}^{5}C_{60}^{7}$	0.40	7100	4750 [b]	30.5	−62.8	55.0	57.0	46.2	55.0	79.2	−1.3	93.7	99.2	34.2	61.0
$D_{65}^{26}C_{35}^{15}$	0.35	15 320	26 200 [b]	29.1	−58.3	54.1	56.4	27.0	55.2	77.7	−4.8	93.7	103.9	45.9	50.0
$D_{77}^{32}C_{23}^{10}$	0.23	10 280	32 280 [b]	26.3	−52.6	53.0	55.2	17.1	53.2	77.9	−5.8	84.8	104.5	54.1	49.8
PCL[11]	1.00	11 360	—	34.1	−73.1	54.7	56.7	71.9	52.0	—	—	—	—	—	—
PPDX[15]	0.00	—	15 190 [c]	52.1	−65.9	—	—	—	—	87.0	−12.1	101.4	107.4	81.0	57.4

[a] Molecular weight of PCL macroinitiator was characterised by GPC in THF at 35 °C.[45] [b] M_n in the copolymers were determined by ^{1}H NMR comparing the relative intensity of both CL and PDX repeating units and knowing M_n of the PCL block.[45] [c] Average molecular weight determined by capillary viscometry (M_ν) determined in phenol/1,1,2,2 tetrachloroethane (2 : 3 in vol.) at 25 °C.[2,51] T_{mo} and T_{mp} stand for onset and peak melting temperatures.

Differential scanning calorimetry (DSC)

Aluminium pans were used to encapsulate 5 mg of each sample. A Perkin-Elmer DSC-7 calibrated with indium and hexatricontane standards was employed; all measurements were performed with an ultra high purity nitrogen atmosphere.

Standard DSC experiments

The samples were first heated to 130 °C for 3 min in order to erase the thermal history. Then DSC cooling scans were recorded from 130 °C to −25 °C at 10 °C min^{-1}, followed immediately by a subsequent heating scan also at 10 °C min^{-1} up to 130 °C.

Self-nucleation experiments

The self-nucleation experiments performed are based on experimental protocol previously reported in detail.[25,49] The procedure is carried out as follows: (1) Erasure of thermal history by heating the samples to 130 °C. (2) Creation of a "standard" thermal history by cooling at 10 °C min^{-1} down to −25 °C. (3) Subsequent heating at 10 °C min^{-1} until a temperature is reached that we shall term T_s (or self-nucleation temperature). (4) Thermal conditioning at T_s for 3 min. Depending on the T_s value the sample will be in one of the three self-nucleation domains. If T_s is too high complete melting will occur and the sample will be in Domain I or "Complete Melting" Domain, at which the nucleation density will be constant. If T_s is high enough to melt almost completely all crystals but low enough to leave self-nuclei that increase the nucleation density of the polymer, then the sample is within Domain II or exclusive "Self-nucleation" Domain. Finally, if T_s is too low as to only partially melt the sample, then the sample will be in Domain III or "Self-nucleation and Annealing" Domain, since the not melted crystals will anneal during the 3 min at T_s. (5) Cooling from T_s at 10 °C min^{-1} during which the effect of self-nucleation will be revealed and (6) final subsequent heating scan, at 10 °C min^{-1}, where the presence or absence of annealing effects will be seen. The DSC scans during steps 5 and 6 are usually presented to examine the effects of the self-nucleation treatment.

Isothermal DSC experiments

Isothermal crystallisation of the PCL and PPDX homopolymers were performed after melting the samples for 3 min at 130 °C and then they were quenched (at 80 °C min^{-1}) to the desired isothermal crystallization temperature or T_c.

Isothermal crystallisation of PCL block

Samples were first melted at 130 °C for 3 min, then they were cooled down to −25 °C at 10 °C min^{-1} to promote the crystallisation of both PPDX and PCL blocks. After that, the samples were heated up to 62 °C, a temperature above the melting temperature of the PCL component. The samples were annealed at this temperature for 70 min to guarantee that PPDX had crystallized until saturation. Finally, the samples were quenched (80 °C min^{-1}) to T_c and the overall isothermal crystallization was monitored by the DSC.

Self-nucleation followed by isothermal crystallization

In some cases, in order to be able to register with the DSC the isothermal crystallization of the PPDX block which proceeded very slowly in some diblock copolymers, the sample was first self-nucleated employing a T_s temperature within Domain II and then it was quenched (80 °C min^{-1}) to T_c. Experiments were performed to check that the sample did not crystallize during the cooling to T_c and that a full crystallization exotherm was recorded at T_c.

Hydrolytic degradation experiments

The PPDX[15], $D^{26}_{65}C^{15}_{35}$ and $D^{32}_{77}C^{10}_{23}$ were hydrolytically degraded in a 0.2 M (pH 7.4) phosphate buffered saline solution at 37.2 ± 0.2 °C for 9 months. The buffers were replaced every 10 days, in order to maintain a constant pH during the whole degradation process. The samples were extracted monthly, dried at 25 °C under vacuum for 48 h, and then weighted. The compositions of the diblock copolymers were recalculated after hydrolysis by ^{1}H NMR spectroscopy knowing the relative intensities of methylene protons of the PDX repeat units at 4.38 ppm and methylene protons of CL repeat units at 2.32 ppm, using a Bruker AMX-300 apparatus in d^2-1,1,2,2-tetrachloroethane/TMS.[42,46]

Results and discussion

General thermal behaviour

Fig. 1 presents DSC cooling and subsequent heating scans for the PPDX-*b*-PCL diblock copolymers in a wide composition range, for comparable molecular weight homopolymers and for a 50/50 PPDX/PCL blend. The cooling scans indicate that regardless of the composition in the diblock copolymers only one crystallization exotherm is observed. In a previous work[42] this behaviour was examined by time resolved WAXS and isothermal crystallization kinetics and it was concluded that a coincident crystallization process takes place.

During cooling at 10 °C min^{-1} the PPDX block cannot crystallize at the same temperature as the equivalent homopolymer because its secondary nucleation is hindered by the covalently bonded molten PCL block, therefore it needs a larger undercooling to start its crystallization. New evidence that supports this explanation will also be presented in the present work. Once the crystallization of the PPDX block starts, it is quickly followed by the crystallization of the PCL block which can be nucleated by the PPDX block. As a result, the crystallization of both blocks occurs in the same temperature range and in a quick succession leading to an overlap of the DSC exothermic signal during cooling. This overlap can not be overcome by the use of slower cooling rates, at least down

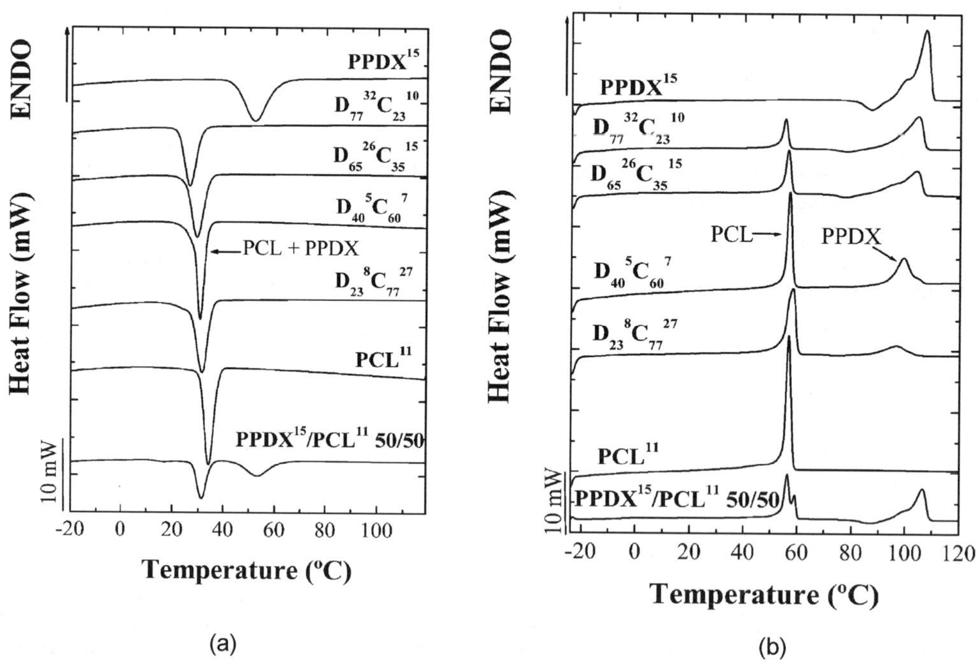

Fig. 1 (a) DSC cooling scans (10 °C min^{-1}) for PCL and PPDX homopolymers, diblock copolymers and a 50/50 blend. (b) Subsequent heating scans (10 °C min^{-1}).

to 1 °C min^{-1}. This may be related to the fact that the PCL only crystallizes after the PPDX block starts to crystallize, this was clearly demonstrated by previous WAXS results.[42]

In a related double crystalline polyethylene-*b*-poly(L-lactide) diblock copolymer (with an approximate 50/50 weight ratio) a similar behaviour has been observed.[50] When the diblock copolymer is cooled from the melt at 10 °C min^{-1} only one crystallization exotherm is observed indicating that the L-lactide block is also being affected by the molten polyethylene block that is covalently bonded to it.[50] However in this case, the crystallization exotherms can be separated by reducing the cooling rate to 2 °C min^{-1}.[41]

The data shown in Fig. 1(a) for a 50/50 blend supports the above explanation since the individual crystallization of each blend component can clearly be observed. It will also be demonstrated below that when self-nucleation is applied, the crystallization signals of each block can be separated.

The DSC heating scans in Fig. 1(b) show that each phase melts in a very different temperature range and that the melting temperatures are close to those exhibited by the equivalent homopolymers (see Table 1). The ability of PPDX to reorganize during heating has been reported previously[51] and a cold crystallization exotherm can be seen in some of the heating DSC scans in Fig. 1(b).

Table 1 indicates that the melting point of the PPDX block reduces as its content within the copolymer is decreased, while the melting point of the PCL block remains constant. Table 1 also reports the enthalpy of fusion of each constituent and the degree of crystallinity calculated with those values once they have been normalized by the composition, no clear trend is observed as a function of composition.

Morphology

The observation of the MD morphology in PPDX-*b*-PCL diblock copolymers has been challenging. PPDX and PCL are apparently not miscible in spite of their chemical similarities.[42] We have prepared a series of PPDX and PCL blends employing polymers in a wide range of molecular weights and none of the blends have shown signs of miscibility. The lower limit in M_n explored so far was a blend of 5000 g mol^{-1} PPDX with a PCL of just 600 g mol^{-1}. In this case, as in all others, pieces of evidence of immiscibility were obtained: (a) the blend exhibited two glass transition temperatures at values close to those corresponding to the homopolymers; (b) there was no significant melting point depression for the PPDX block (or any change in the melting point of the PCL block) regardless of the PCL content in the blend and (c) a two phase morphology was observed in scanning electron microscopy experiments.

In view of the above, and of rather similar results obtained with our diblock copolymers (similar trends regarding T_g and T_m values being independent of composition and similar to values exhibited by homopolymers of equivalent molecular weights), we have considered that the PPDX-*b*-PCL diblock copolymers should exhibit phase segregation in the melt. If they are phase segregated in the melt they must be in the weak segregation limit considering their molecular structures and solubility parameters.[42] However, we have not been able to obtain so far any direct evidence of the MD structure in the melt. The order–disorder transition has not been directly detected since decomposition of the PPDX at temperatures higher than 140 °C mask thermal events at such higher temperatures.[42] Since PPDX has a higher susceptibility to both hydrolytic degradation and thermal degradation than PCL,[1,2] we have not been able to anneal the samples in the melt for a long time in order to perfect the possible heterogeneous melt morphology. As a consequence of this, previous attempts to measure SAXS have resulted in broad scattering peaks with no higher order reflections indicating the absence of long range order,[42] this may also indicate that the melt is homogeneous. So far, staining attempts with several agents have failed to reveal the morphology by TEM, although work in still in progress in this respect.

If we refer to the introduction, we have previously[42] made the reasonable assumption that PPDX-*b*-PCL copolymers fall within case number 2, *i.e.*, a weakly segregated system with soft confinement for the PPDX block, since it crystallizes first, so that $T_{ODT} > T_c > T_g$. It should be noted that if crystallization proceeds from a homogeneous melt (case 1) or from a weakly segregated melt with soft confinement (case 2), the crystallization should occur in both cases with a lamellar morphology (which either comes from a homogeneous melt or from a "break-out" morphology that overwrites the MD morphology present in the melt[17]).

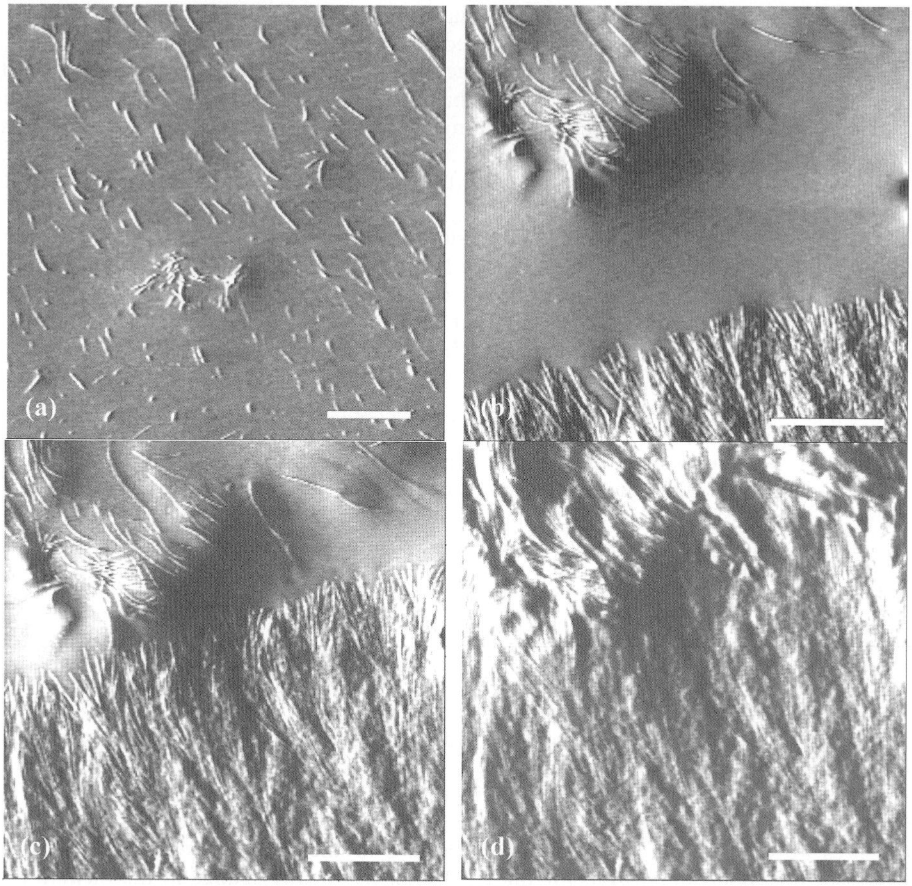

Fig. 2 A series of AFM images taken during sequential isothermal crystallization of $D^{8}_{23}C^{27}_{77}$. (a) The sample was isothermally crystallized at 60 °C, only PPDX can crystallize at this temperature. (b) Through (d), the sample was quenched to 40 °C and images are taken at increasing times: (b) $t_b = t$, (c) $t_c = t + 2$ min, (d) $t_d = t + 5.5$ min. Scale bars represent 1 μm.

Previous reports[42,44] dealing with PPDX-*b*-PCL diblock copolymers have shown by POM that superstructural aggregates with a predominant granular texture are obtained at temperatures around 50 °C or higher (at such high T_c values only the PPDX block is able to crystallize and in copolymers with a higher content of PPDX the Maltese cross extinction pattern was observed in a clearer spherulitic texture) while banded spherulites composed of both blocks are present at lower temperatures.

Recently, we have been able to perform AFM observations in $D^{8}_{23}C^{27}_{77}$. The observations in the melt indicated a grainy appearance and it was difficult to ascertain whether a MD melt morphology was present or not. For this sample, and based on the composition, a PCL matrix with PPDX spheres or cylinders should be expected, if the melt were heterogeneous.

Fig. 2(a) shows the morphology obtained after lowering the temperature to 60 °C, a temperature too high for the PCL block to crystallize (peak T_m value for the PCL block was 58 °C, see Table 1). PPDX lamellae can be seen in Fig. 2(a) in a molten PCL matrix. They seem to be approximately randomly distributed although some connections between lamellae can be observed as well as some lamellar aggregates. The fact that lamellae are present may be an indication that some form of "breakout"[17] has been generated during crystallization facilitated by the soft confinement or that crystallization from a homogeneous melt is in place and PPDX lamellae are being formed. In any case, it is clear that crystallization is driving structure formation.

The temperature was then lowered to 40 °C, a temperature at which the PCL block can now crystallize isothermally. Fig. 2b shows in the top part previously crystallized PPDX lamellae (at 60 °C) and in the bottom part newly formed PCL lamellae that belong to a growing spherulite (PCL amounts to 77% for this copolymer). In the sequence of micrographs shown in Figs. 2(b)–(d) the spherulite can be seen as it grows past the previously crystallized PPDX lamellae and these are engulfed by the growing superstructure. A mixed spherulite containing lamellae of both PPDX and PCL results and can be observed in Fig. 2(d). See also a video in the ESI showing the sequential isothermal crystallization of $D^{8}_{23}C^{27}_{77}$ according to the same temperature program described in Fig. 2.†

Crystallization kinetics from growth rate data

We have followed by POM the radial growth of spherulites at temperatures where only the PPDX can crystallize for a $D^{32}_{77}C^{10}_{23}$ diblock copolymer and PPDX[15] homopolymer. This diblock copolymer showed clearly spherulitic features with Maltese cross when crystallized at temperatures where the PCL was molten. Fig. 3(a) shows the experimentally determined spherulitic growth rates (G) as a function of crystallization temperature. The difference in growth rates is very pronounced between the homopolymer and the PPDX block within the diblock copolymer and amounts to an order of magnitude at certain crystallization temperatures. This difference demonstrates that the PPDX block is experiencing secondary nucleation problems because it is covalently bonded with the molten and highly flexible PCL. This indicates that the coincident crystallization phenomenon observed in Fig. 1 is caused not only by a reduction in nucleation that the PPDX block may be experiencing as compared to PPDX homopolymer.

We have applied the Lauritzen and Hoffman kinetic treatment to the growth rate data by considering that G can be expressed as:[52,53]

$$G(T) = G_0 \exp\left(-\frac{U*}{R(T_c - T_\alpha)}\right) \exp\left(-\frac{K_g^G}{T\Delta Tf}\right) \qquad (1)$$

where G_0 is a growth rate constant, U^* is the activation energy for the transport of the chains to the growing nuclei, a value of 1500 cal mol^{-1} is normally employed,[53] R is the gas constant and T_c the isothermal crystallization temperature. T_α is a temperature at which chain mobility ceases and it is usually taken as T_g−30 (K),[53] however, we found a better fit to our data with a value of T_g−15 (K). ΔT is the supercooling defined as $(T_m{}^\circ - T_c)$, where $T_m{}^\circ$ is the equilibrium melting point. The factor f is a temperature correction term equal to: $2T_c/(T_c + T_m{}^\circ)$.[52,53] Finally, K_g^G is an important

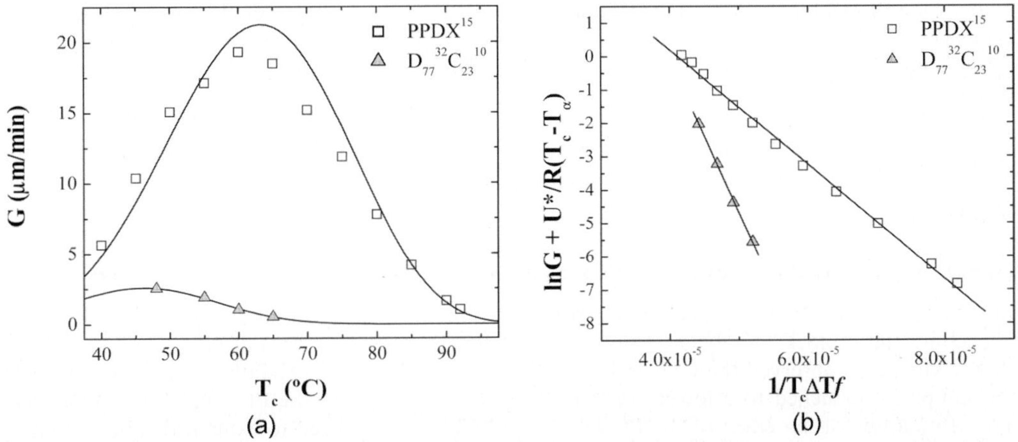

Fig. 3 (a) Spherulitic growth rates for PPDX and the PPDX block within $D^{32}_{77}C^{10}_{23}$ diblock copolymer. Solid lines are fits to the Lauritzen and Hoffman theory according to eqn. (1). (b) Lauritzen and Hoffman kinetic theory plot for PPDX and the PPDX block within $D^{32}_{77}C^{10}_{23}$ diblock copolymer.

parameter since it can be considered proportional to the energy barrier for secondary nucleation or growth and is given by:[52,53]

$$K_g^G = \frac{jb_0\sigma\sigma_e T_m^0}{k\Delta h_f} \tag{2}$$

where j is taken as 2 for Regime II, b_0 is the width of the chain, σ is the lateral surface free energy, σ_e is the fold surface free energy, k is the Boltzmann constant and Δh_f the heat of fusion of a perfect crystal.

Fig. 3(b) shows the usual way of fitting the experimental data to the theory by plotting $\ln G + U/R(T_c - T_\alpha)$ *versus* $1/(T_c\Delta Tf)$, a straight line should be obtained and K_g^G can be determined from the slope of the plot. We have assumed that crystal growth occurs under Regime II and the values of all parameters employed for the calculation are listed in Table 2. Once we determined K_g^G, we calculated the theoretical prediction from eqn. (1) and plotted the results over the experimental data points in Fig. 3a, the use of a linear scale highlights the differences between experiment and theory which are masked in Fig. 3b by the use of logarithms and reduced parameters.

A typical bell shape curve in the spherulitic growth kinetics of PPDX (Fig. 3a) has been obtained.[54] Previously, we had only been able to access (with different molecular weight PPDX samples) the nucleation controlled right hand side of the curve.[2,51] Other authors[55] have obtained similar bell shape curves from overall PPDX homopolymer crystallization rate data.

Table 2 shows the values obtained for K_g^G for PPDX[15] and the PPDX block within $D_{77}^{32}C_{23}^{10}$. It can be observed that the K_g^G value for the PPDX block within the copolymer is more than twice that of neat PPDX[15]. This gives an idea of the difficulty that this block encounters to crystallize with a covalent PCL chain attached to it as compared to neat PPDX.

From the values of K_g^G the product $\sigma\sigma_e$ is obtained according to eqn. (2). The following expressions allow the calculation of σ (and therefore σ_e) and q, the work done by the chain to form a fold:

$$\sigma = 0.1\Delta h_f\sqrt{a_0 b_0} \tag{3}$$

$$q = 2a_0 b_0\sigma_e \tag{4}$$

where a_0b_0 is the cross sectional area of the chain. As expected from the K_g^G values, both the fold surface free energy and the work for chain folding are much higher for the PPDX block within $D_{77}^{32}C_{23}^{10}$ than for the PPDX[15] homopolymer, as shown in Table 2. The range of values obtained for the fold surface free energies are consistent with previous literature reports for PPDX, PCL and polyesters in general.[2,51,53]

In this section we have shown that the energy barrier for growing spherulitic superstructures is higher for a PPDX block within $D_{77}^{32}C_{23}^{10}$ than for the PPDX[15] homopolymer. This supports previous evidence[42,44] suggesting that the molten PCL block covalently bonded to the crystallizing PPDX block is hindering the secondary nucleation process. More new evidence in this respect is presented below.

The dramatic depression of the growth rate in the PPDX block within the $D_{77}^{32}C_{23}^{10}$ diblock copolymer as compared to an equivalent molecular weight homopolymer is the determining factor in the coincident crystallization behaviour encountered upon cooling the copolymer from the melt in Fig. 1.

Overall crystallization kinetics for the PCL block within PPDX-*b*-PCL diblock copolymers

In a previous work,[42] it was demonstrated that if both blocks are simultaneously crystallized by quenching the diblock copolymer from the melt directly to a low crystallization temperature, a slower kinetics is observed than when the PPDX block is allowed to crystallize at high T_c and then the sample is quenched to a lower crystallization temperature and the PCL block is crystallized. Therefore a nucleating effect of the PPDX on the PCL was proposed to explain this behaviour.

The PPDX block was first crystallized until saturation and then the temperature was lowered to the crystallization temperature of the PCL block (see experimental). The quality of the data obtained can be appreciated in Fig. 4, where examples of isothermal DSC scans for the

Table 2 Values obtained by fitting the Lauritzen and Hoffman theory to experimental data and parameters employed in eqn. (1)–(4) and (6)

Sample	T_c range/°C	T_α/K	$K_g^G \times 10^{-4}$/K^2	$K_g^\tau \times 10^{-4}$/K^2	σ_e^G/erg cm^{-2}	σ_e^τ/erg cm^{-2}	$q^G \times 10^{-13}$/erg	$q^\tau \times 10^{-13}$/erg	σ/erg cm^{-2}
PCL[11]	40–50	192.7	7.0 ± 0.1	10.0	80.1	113.6	3.0	4.2	7.0
$D_{23}^{8}C_{77}^{27}$	40–50	192.7		9.6		108.8		4.0	7.0
$D_{40}^{5}C_{60}^{7}$	40–50	192.7		12.6		143.6		5.3	7.0
$D_{65}^{26}C_{35}^{15}$	40–44	192.7		12.7		144.6		5.3	7.0
$D_{77}^{32}C_{23}^{10}$	44–46	192.7		16.3		185.6		6.8	7.0
PPDX[15]	40–90	248.2	17.2 ± 0.4	31.0	168.1	302.6	6.2	11.2	8.1
$D_{77}^{32}C_{23}^{10}$	48–65	248.2	46 ± 2		448.0		16.5		8.1
PPDX[15] sn 117 °C	85–95	248.2		17.0		166.3		6.1	8.1
$D_{77}^{32}C_{23}^{10}$ sn 110 °C	60–70	248.2		31.0		303.4		11.2	8.1
$D_{65}^{26}C_{35}^{15}$ sn 110 °C	55–65	248.2		42.7		417.6		15.4	8.1

Values for eqn. (1) and (3) include: $a_o = 4.5$ Å, $b_o = 4.1$ Å and $U^* = 1500$ cal mol^{-1} which correspond to PCL and were employed for all materials.[51,53] For PCL and PCL blocks the value of $T_m°$ employed was 70.9 °C (literature values range from 70.3–74 °C[53]) and for PPDX and PPDX blocks $T_m°$ was taken as 127.1 °C.[51] Superscripts G and τ are used to indicate that a value was calculated from spherulitic growth rate data (G) or from DSC overall crystallization rate data ($1/\tau_{50\%}$) respectively. Values of ΔH_f for PPDX and PCL were taken as: 141.2 J g^{-1} and 139.5 J g^{-1} respectively.[2,51,63] The unit cell densities used were 1.175 g cm^{-13}, for the PCL and PCL block, and 1.332 g cm^{-13} for the PPDX and PPDX block, respectively.[63] Errors in K_g^τ are estimated to be less than 2.5%. sn: self-nucleated material at the indicated T_s temperature, see text.

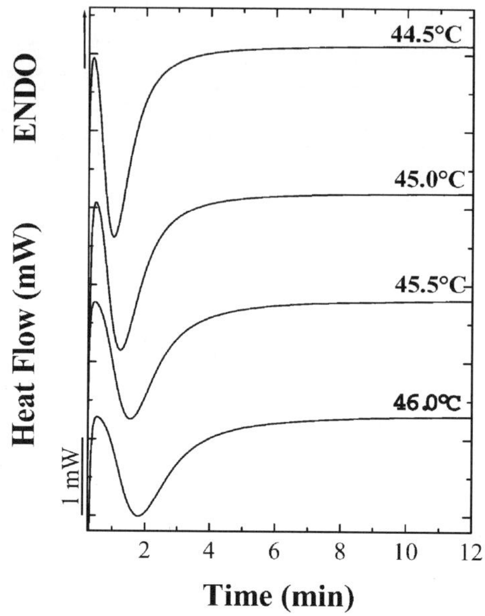

Fig. 4 Examples of DSC isothermal crystallization scans for the PCL block within $D^{5}_{40}C^{7}_{60}$ diblock copolymer. The measurements are performed after the PPDX block had been previously crystallized until saturation (see experimental).

crystallization of the PCL block are presented. The data obtained has been initially analysed with the Avrami equation:[53,54,56]

$$X(t) = 1 - \exp(-Kt^{n}) \tag{5}$$

where $X(t)$ is the relative crystalline volume fraction of the polymer as a function of time. The parameters K and n are dependent on the nucleation type and the crystal growth geometry, K can be considered an overall transformation rate constant and n is the Avrami index.

The overall crystallization rate, expressed as the inverse of the experimental crystallization half-time, for PCL[11] and for the PCL block of the diblock copolymers employed in this work are shown in Fig. 5, where the nucleation effect of the previously crystallized PPDX block on the PCL block is confirmed. The overall crystallization rate of the PCL block in the case of $D^{8}_{23}C^{27}_{77}$ and $D^{5}_{40}C^{7}_{60}$ is higher than that of PCL[11]. This effect was reported in a previous work when the PPDX content in the copolymer was lower than 53%.[44] In cases where the PPDX content in the diblock copolymer was equal or more than 53%, the nucleation effect of the PPDX block on the PCL seems to be overtaken by topological restrictions. Fig. 5 shows that the PCL blocks within $D^{32}_{77}C^{10}_{23}$ and $D^{26}_{65}C^{15}_{35}$ crystallize at a slower overall rate than neat PCL. In these cases, the PPDX matrix has already crystallized in spherulites and the 23% or 35% PCL must crystallize within them. Interestingly, we have found no fractionated crystallization effects during cooling from the melt in Fig. 1 for $D^{32}_{77}C^{10}_{23}$ or $D^{26}_{65}C^{15}_{35}$, which can be explained by the nucleating influence of the PPDX block. In fact, Fig. 5 demonstrates that even under isothermal conditions the PCL block in these copolymers can crystallize in a similar temperature range to the PCL homopolymer, a clear indication of a heterogeneous nucleation mechanism.[25]

Even though the nucleating effect of the PPDX on the PCL prevents any homogeneous nucleation, increasing amounts of PPDX within the PPDX-*b*-PCL diblock copolymers hinders crystallization of the PCL block as already indicated above by an overall reduction in kinetics (Fig. 5). Fig. 6 shows how the Avrami index depends on the PCL content in the diblock copolymers for a wide temperature range. In this plot we have collected all the data from previous works[42,44] as well as recent data. The Avrami index is found to decrease as the PCL content in the copolymer decreases. Such a decrease in Avrami index can be interpreted as a decrease in the dimensionality of growth of the PCL block superstructures as more previously crystallized PPDX is present in the sample.

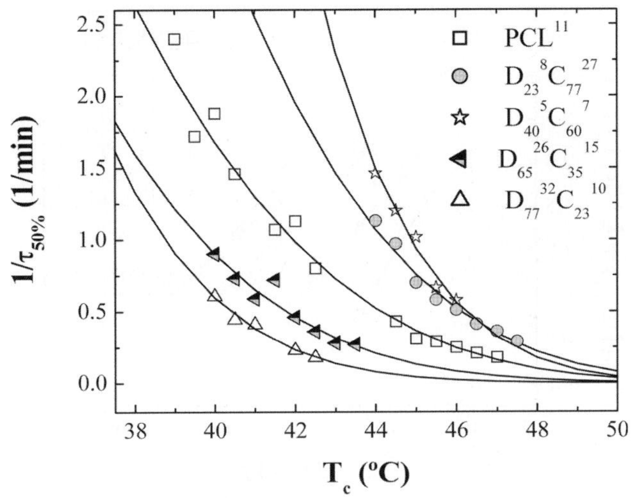

Fig. 5 Inverse of the crystallization half-time as a function of isothermal crystallization temperature for PCL[11] homopolymer and for the PCL block of the indicated copolymers. All experiments were performed after the PPDX block had been previously crystallized until saturation (see experimental). Solid lines are fits to the Lauritzen and Hoffman theory according to eqn. (6).

It is well known that when a polymer is confined into a great number of isolated MD (*e.g.*, spheres or cylinders) it crystallizes from homogeneous nuclei that form at the largest possible supercooling.[25] When such a situation arises, Loo *et al.*[38] have shown that the Avrami index is usually 1.

For the $D_{77}^{32}C_{23}^{10}$ case, where the highest possible confinement of the PCL block would be expected, the Avrami index was found to be larger than 1, a result which confirms the absence of homogeneous nucleation. The Avrami index obtained for this copolymer was of the order of 2 and could correspond to one-dimensional crystals grown from sporadic nuclei, or two-dimensional growth from instantaneous nuclei. If the PCL phase is not isolated in a large number of MD but connected throughout by percolation paths, this allows the spread of secondary nucleation and an

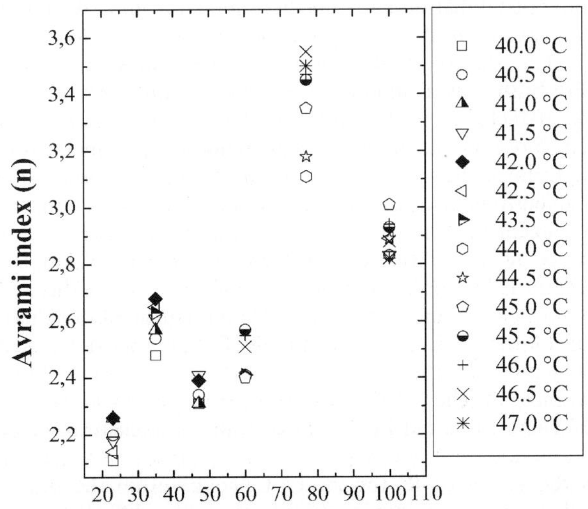

Fig. 6 Variation of the Avrami index with PCL content in PPDX-*b*-PCL diblock copolymers at the indicated crystallization temperatures.

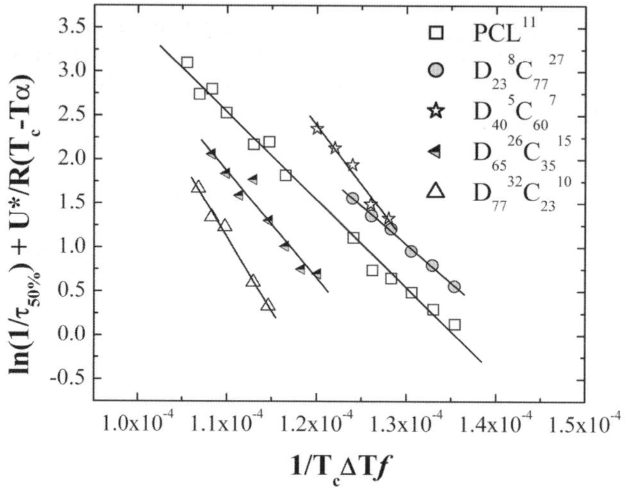

Fig. 7 Lauritzen and Hoffman kinetic theory plots for PCL[11] homopolymer and for the PCL block of the indicated copolymers. All experiments were performed after the PPDX block had been previously crystallized until saturation (see experimental).

Avrami index larger than 1 would also be obtained. Nevertheless, many features are consistent with a nucleation effect of the PPDX crystals on the PCL phase (and more will be given below, see Fig. 9) that this must be the dominant factor, although interconnection of MD could also be present.

As the PCL content in the copolymer increases the Avrami index increases to values that are similar to those of PCL homopolymer, *i.e.*, around 3, indicating spherulitic growth that has been confirmed by POM observations.[42]

The temperature dependence of the inverse of the half-crystallization time can be assumed to have a form similar to eqn. (1):[54,57,58]

$$\frac{1}{\tau_{50\%}}(T) = G_o^\tau \exp\left(-\frac{U^*}{R(T_c - T_\infty)}\right) \exp\left(-\frac{K_g^\tau}{T\Delta Tf}\right) \tag{6}$$

where $\tau_{50\%}$ is the overall crystallization half-time determined from DSC measurements. The inverse of $\tau_{50\%}$ represents a measure of the overall crystallization rate that takes into account both nucleation and growth. The pre-exponential factor and the term K_g in eqn. (6) have a τ superscript to indicate that they are both now a function of nucleation and growth.

Fig. 7 shows a plot of $\ln(1/\tau_{50\%}) + U^*/R(T_c - T_\infty)$ *versus* $1/(T_c\Delta Tf)$ according to eqn. (6) for PCL[11] and for the PCL block within four different diblock copolymers. It can be seen that in all cases good fits to straight lines allow the determination of K_g^τ. The values obtained are reported in Table 2 and were employed to fit the experimental data of Fig. 5 with eqn. (6), the quality of the fit can be appreciated in a linear scale in this case, all fits were excellent.

It is interesting to see how the value of K_g^G compares with K_g^τ in the case of the PCL[11] homopolymer reported in Table 2. K_g^τ is larger than K_g^G by 30% indicating that the process of nucleation and growth has a larger energetic barrier than the process of spherulite growth only. The fact that the K_g^τ value of the PCL block within $D_{23}^8C_{77}^{27}$ is almost the same as that of PCL[11] homopolymer may indicate that the nucleating effect of the PPDX block is already counter balanced by the presence of the PPDX crystals, since a value of the order of K_g^G (*i.e.*, $7.0 \pm 0.1 \times 10^4$ K^2, see Table 2) would be expected if 100% nucleation efficiency of the PPDX is assumed.

Table 2 shows that the values of K_g^τ for the PCL block within the copolymers increase as the content of PPDX in the copolymer increases, indicating the difficulty encountered by the PCL chains to crystallize as more crystalline PPDX material is present in the copolymer. As a result of the increase in K_g^τ the values for the fold surface free energy and the work to fold the PCL chains also increase as the content of PCL within the copolymers decreases.

We tried to follow the same procedure employed to determine the overall crystallization kinetics of the PCL block within the PPDX-*b*-PCL diblock copolymers in POM experiments but once the PPDX was crystallized the nucleation density was so high that the PCL spherulites were very small and their growth could not be followed by POM.

Self-nucleation studies

The technique of self-nucleation was applied initially to PPDX and PCL and later to the PPDX-*b*-PCL diblock copolymers. In the case of PPDX[15], Fig. 8(a) shows DSC cooling scans from selected self-nucleation temperatures, T_s, and Fig. 8(b) the corresponding subsequent heating scans after self-nucleation at T_s.

If the T_s temperature is equal or higher than 125 °C the melting is complete and no new nuclei are generated after 3 min at such temperatures. Therefore the sample is considered to be under Domain I or complete melting Domain and its crystallization temperature during cooling from T_s does not change since it reflects a constant nucleation density.

When the T_s temperature is lower than 125 °C, the crystallization temperature increases because self-nuclei have been generated by the thermal treatment at T_s, and the samples are in Domain II. The high temperature melting peak does not show any signs of annealing until the T_s temperature is lowered into Domain III. The low temperature shoulder does change with T_s temperature within Domain II. These changes are related with the increased crystallization ability of PPDX once it has been self-nucleated in Domain II and how this reflects in its reorganization capacity during the subsequent heating scan, the reader is referred to ref. 51 for more details.

Finally, if the T_s temperature is too low, partial melting and annealing occurs together with self-nucleation. The annealing can be seen in the subsequent heating runs, like that shown in Fig. 8(b) for $T_s = 105$ °C, indicated by an arrow. The self-nucleation behaviour of PPDX[15] is the standard behaviour of semi-crystalline homopolymers.[49] A similar standard behaviour was obtained for the PCL[11] homopolymer (results not shown).

It is interesting to examine the self-nucleation behavior of the diblock copolymers for two reasons. First, to try to separate by temperature the coincident crystallization shown in Fig. 1 by

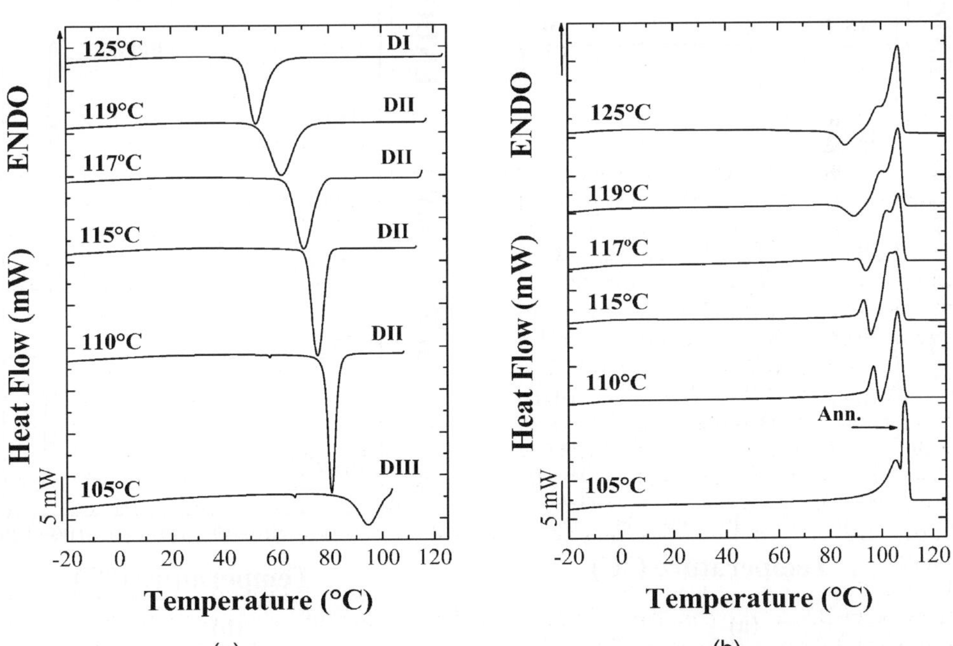

Fig. 8 Self-nucleation of PPDX[15]: (a) DSC cooling scans from the indicated T_s temperatures; (b) subsequent heating scans.

self-nucleating just the PPDX block, and second, because it has been shown in a wide range of crystallizable diblock and triblock copolymers that if a particular phase is isolated and crystallizes in a large number of individual microdomains (like spheres or cylinders for instance) Domain II usually disappears and a direct transition from Domain I to Domain III has been observed.[25,59–61]

Fig. 9(a) presents self-nucleation results for $D^{32}_{77}C^{10}_{23}$ in a T_s temperature range where self-nucleation affects only the PPDX block. At $T_s = 115\,°C$ the sample is in Domain I and the coincident crystallization of both blocks occurs with a single exotherm that peaks at around 26 °C. At $T_s = 110\,°C$ the sample is under Domain II and self-nucleation occurs as indicated by the higher peak crystallization temperature of the coincident main exotherm. Close observation of that cooling DSC scan reveals a downwards drift of the baseline upon cooling from approximately 60 °C that indicates that some PPDX crystals are crystallizing at higher temperatures, in Fig. 9(a) a close up of a section of this curve highlights this observation. For a $T_s = 108\,°C$ it is very clear that the PPDX has been self-nucleated and is now crystallizing at higher temperatures in a complex bimodal exotherm. A similar behaviour is seen for $T_s = 107\,°C$ except that the exotherms corresponding to the PPDX are slightly shifted to higher temperatures. For $T_s = 104\,°C$ the crystallization occurs immediately upon cooling, a typical sign that indicates partial melting and annealing.[25,49] In fact the annealing can be observed in the heating scan corresponding to $T_s = 104\,°C$, where a very sharp high temperature second melting peak is observed and indicated by an arrow (Fig. 9(b)).

Another interesting point from the self-nucleation results reported in Fig. 9a is the fact that at rather low T_s temperatures within Domain II, *e.g.*, at $T_s = 107\,°C$, or even within Domain III, *i.e.*, $T_s = 104\,°C$, most of the PPDX that was going to be self-nucleated by the treatment is crystallizing at much higher temperatures than the PCL block (an estimate of the amount of PPDX that has been self-nucleated, based on the differences between the crystallization and fusion enthalpies is given in Fig. 9a for two T_s temperatures). Therefore, the remaining exotherm located at around 30 °C must correspond mainly to the crystallization of the PCL block and its crystallization temperature is also shifted to higher temperatures as compared to the standard sample (*i.e.*, that with $T_s = 115\,°C$). Therefore, this result also indicates that the PPDX block can nucleate the PCL block and it seems

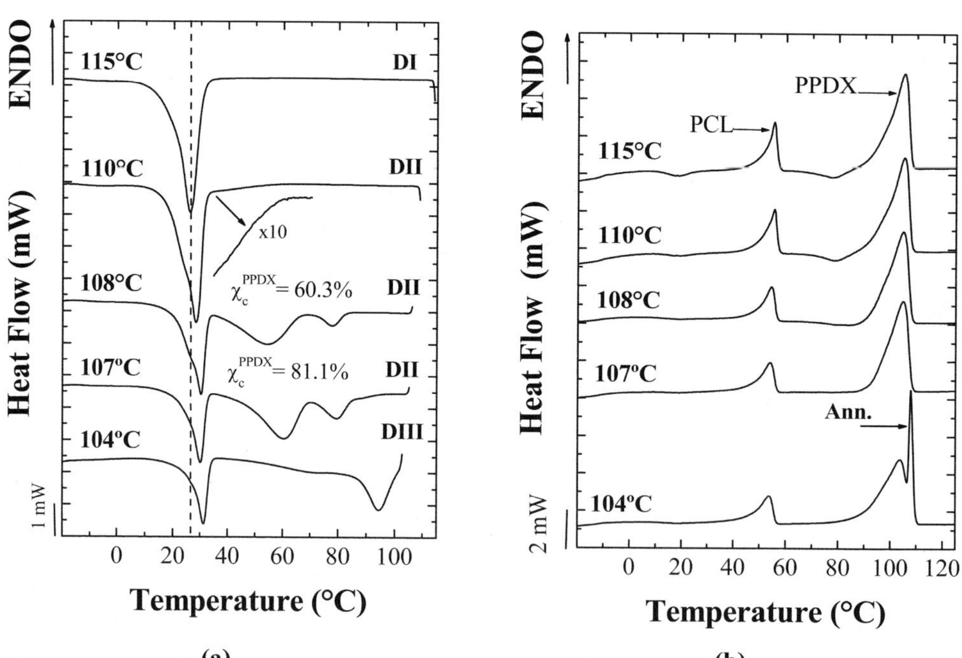

Fig. 9 Self-nucleation of the PPDX block within $D^{32}_{77}C^{10}_{23}$: (a) DSC cooling scans from the indicated T_s temperatures. χ_c^{PPDX} represents the amount of PPDX that has been self-nucleated; (b) Subsequent heating scans. The arrow points to the melting endotherm of annealed crystals during treatment at T_s.

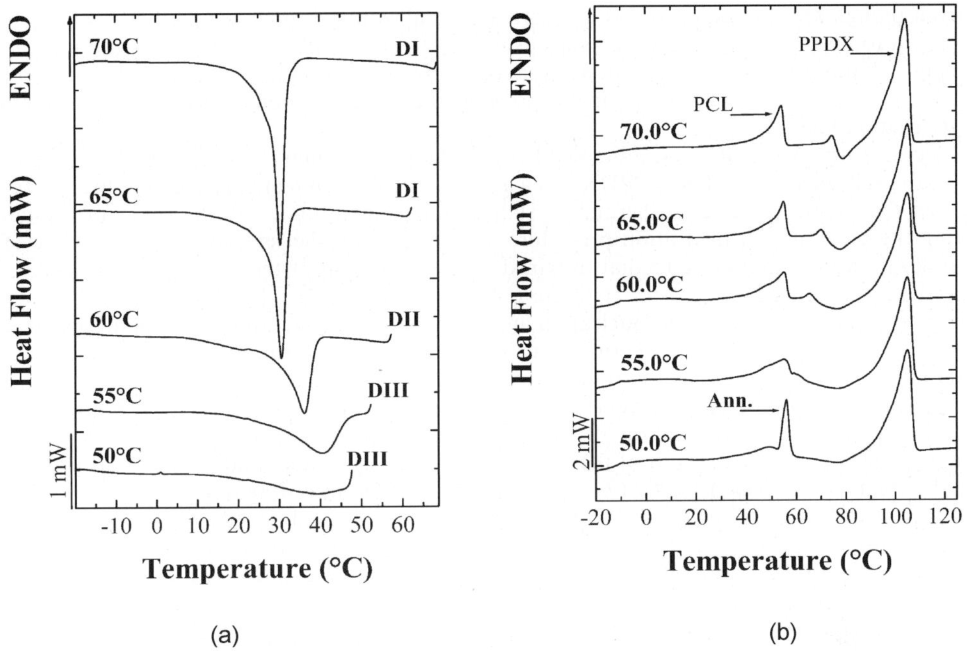

Fig. 10 Self-nucleation of the PCL block within $D_{77}^{32}C_{23}^{10}$: (a) DSC cooling scans from the indicated T_s temperatures; (b) subsequent heating scans. The arrow points to the melting endotherm of annealed crystals during treatment at T_s.

the effect is more pronounced when the PPDX nucleation density has been greatly increased by the self-nucleation treatment.

Fig. 10a shows cooling scans from T_s for $D_{77}^{32}C_{23}^{10}$ in a temperature range where the PCL block self-nucleation can be studied. Fig. 10(b) presents the subsequent heating scans. The analysis of Fig. 10 indicates that the PCL block can be self-nucleated in a standard way and all three self-nucleation Domains can be clearly identified, and have been labelled in Fig. 10(a). This result has an important implication since it corroborates that in spite of its small amount within the copolymer, the PCL block still shows a clear Domain II and therefore does not show signs of homogeneous nucleation.[25,59–61] Since we have presented evidence which indicates that the previously crystallized PPDX block can nucleate the PCL block, then the lack of homogeneous nucleation or fractionated crystallization can be attributed to a heterogeneous nucleation effect produced by the PPDX crystals on the PCL chains. On the other hand, the existence of a percolation path of the PCL crystals may also contribute to spread nucleation.

Isothermal crystallization after self-nucleation

We tried to measure the overall isothermal crystallization kinetics by DSC for the PPDX block within the PPDX-*b*-PCL diblock copolymers in a temperature range where only the PPDX can crystallize (*i.e.*, at T_c higher than 50 °C). Unfortunately, the overall crystallization rate was so slow that the DSC was not able to measure any significant exothermic signal in the isothermal mode. One way to circumvent this limitation is to self-nucleate the material first and then quench it (at a controlled cooling rate of 80 °C min^{-1}) to its isothermal crystallization temperature. In this way two interesting results could be obtained. Firstly, the acceleration of the overall kinetics provided by the self-nucleation treatment allows the DSC to measure the isothermal crystallization kinetics. Secondly, the measured kinetics should be closer to that obtained by POM because they should reflect mostly the growth of the crystals since the nucleation process has been performed previously during the self-nucleation step. This last point will depend on the efficiency of the self-nucleation treatment and therefore on the T_s employed.

Even though self-nucleation before isothermal crystallization is a classic method that has been employed to obtain solution single crystals as well as to study bulk crystallization and nucleation kinetics,[54,62] this is the first time, as far as the authors are aware, that the technique has been applied by DSC in the bulk as a tool to make possible the determination of subsequent isothermal crystallization kinetics.

Selected isothermal DSC scans obtained by quenching (80 °C min^{-1}) the samples to T_c immediately after self-nucleation of neat PPDX[15] at a T_s of 117 °C for 3 min are presented in Fig. 11. It can be seen that at the selected isothermal crystallization temperatures, PPDX[15] did not crystallize during the previous self-nucleation procedure as indicated by the levelling of the baseline after crystallization to a level close to that obtained at the beginning of the scans. The self-nucleation procedure was applied in Domain II where only self-seeds remain, therefore heating DSC scans performed immediately after self-nucleation and quenching to T_c, at equivalent conditions to $t=0$ min in Fig. 11, exhibited no endothermic signals. This type of check runs was performed in order to ensure that no prior crystallization took place before the isothermal measurements.

Fig. 12a shows a plot of the inverse crystallization half time *versus* T_c for PPDX and for self-nucleated PPDX. The acceleration of the overall crystallization kinetics for the self-nucleated sample is clearly seen by the shift of the curve to higher $1/\tau_{50\%}$ values and higher temperatures. In Fig. 12b plots of $\ln(1/\tau_{50\%}) + U^*/R(T_c - T_\infty)$ *versus* $1/(T_c\Delta Tf)$ according to eqn. (6) for PPDX and self-nucleated PPDX are presented. It can be seen that a good fit is obtained and the values of K_g^τ are presented in Table 2. The results are very satisfactory since the K_g^τ value obtained for neat PPDX is reduced from 31.0 to 17.0 ($\times 10^4$ K^2) once the self-nucleation treatment is performed before the isothermal crystallization. This indicates that the energy barrier for nucleation and growth has been reduced by performing the nucleation in a previous step. A value of $K_g^G = 17.2 \times 10^4$ K^2 is obtained from fitting the Lauritzen and Hoffman theory to spherulitic growth rate experiments in PPDX[15], this value is in remarkably close agreement to the K_g^τ value obtained after self-nucleation has been performed and therefore corresponding to mostly crystal growth. As expected, the values of the fold surface free energy and the work to fold the chains obtained for self-nucleated PPDX[15] are also in close agreement with those obtained for neat PPDX[15] from the application of the Lauritzen and Hoffman theory to spherulitic growth rate data. This agreement indicates that the efficiency of the previous self-nucleation of PPDX[15] at $T_s = 117$ °C within Domain II was very high.

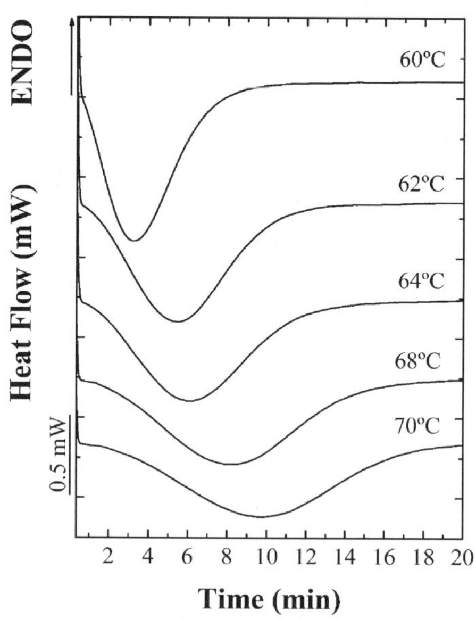

Fig. 11 Examples of DSC isothermal crystallization scans for the PPDX block within $D_{77}^{32}C_{23}^{10}$ diblock copolymer. The measurements are performed after the PPDX block had been previously self-nucleated at $T_s = 110$ °C and then quenched (80 °C min^{-1}) to the indicated T_c temperatures.

Fig. 12 also shows data for two block copolymers whose PPDX block was first self-nucleated and then isothermally crystallized in the DSC. In this case, the molten PCL block covalently attached to the PPDX block slows down the overall crystallization rate, this result being consistent with those presented above in the section of spherulitic growth rate. The higher the PCL content the slower the kinetics for the crystallization of the PPDX block when the two diblock copolymers employed are compared. The fittings to eqn. (6) are also very good in these cases and the values of K_g^τ for the PPDX block tend to increase as compared to self-nucleated PPDX[15] as shown in Table 2. As the PCL content in the self-nucleated copolymer increases the values of K_g^τ, σ_e and q for the corresponding PPDX block also increase as can be seen in Table 2. For copolymers with higher PCL contents the crystallization kinetics of the PPDX block within the copolymers was so slow that even after self-nucleation the measurements could not be performed.

Our results differ from those obtained previously by Bogdanov *et al.*[40] when they studied by DSC the crystallization kinetics of 80/20 poly(ε-caprolactone)-*b*-(polyethylene glycol) diblock copolymer (PCL-*b*-PEG). In their case, the PCL block crystallized first from a homogeneous melt and the Avrami parameters K and n were found to be similar to the kinetic parameters of the isothermal crystallization of a corresponding PCL homopolymer. A significant crystallization retardation was found for the PEG block that crystallized second. The retardation was attributed to the "mutual influence between the PEG constituent and the PCL crystal phase which fixes (hardened) the total copolymer structure".[40] In our case, conversely, when the PPDX block crystallizes it does so at a slower rate than a comparable PPDX homopolymer. The crystallization of the PCL block, on the other hand, strongly depends on the composition of the diblock copolymer as shown in Fig. 7. However, for $D_{77}^{32}C_{23}^{10}$, the overall kinetics are retarded, which can be regarded as similar to that experienced by the PEG block in the 80/20 PCL-*b*-PEG diblock copolymer studied by Bogdanov *et al.*,[40] the Avrami index in both cases was also of the order of 2.

Recently, Ueda *et al.*[22] have examined the crystallization kinetics of a flow oriented polyethylene-*b*-(atactic polypropylene), PE-*b*-aPP, with a polyethylene volume fraction 0.48. In this case, the crystallization of the PE block occurs with a molten aPP block covalently bonded to it. The components are reported to be in the strong segregation limit and therefore phase separated well before crystallization. They found that the crystallization kinetics (as determined by small angle light scattering) was substantially retarded as compared to an equivalent molecular weight homo-PE. They attributed the effect to a mobility reduction for the chains close to the interfacial region and to the presence of the non-crystallizable aPP chains close to the growth face which could obstruct the growth process. These results are similar to those obtained here for the crystallization of the PPDX block at higher temperatures, when the PCL component is molten. Similar results have been also obtained by Shiomi *et al.*[64] since they have detected a reduction in crystallization rate

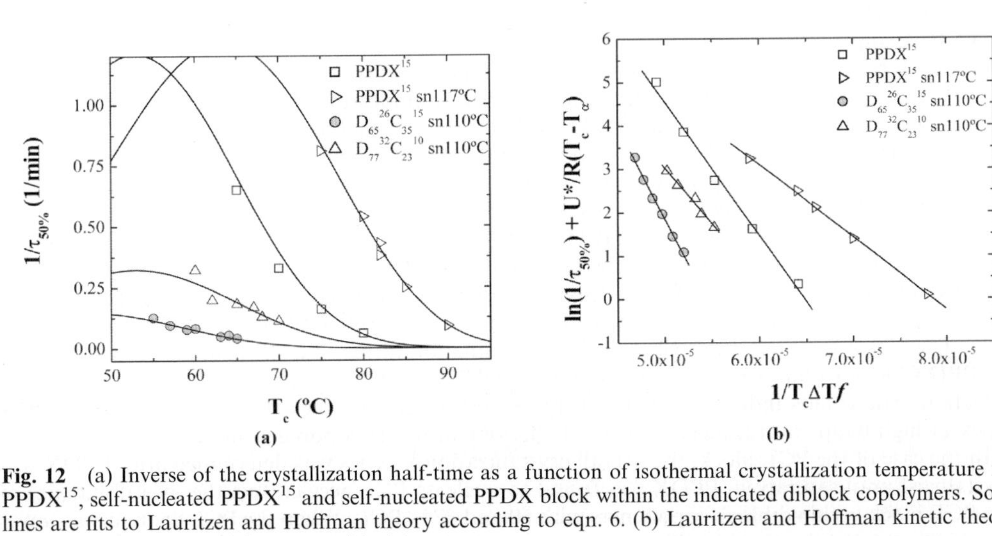

Fig. 12 (a) Inverse of the crystallization half-time as a function of isothermal crystallization temperature for PPDX[15], self-nucleated PPDX[15] and self-nucleated PPDX block within the indicated diblock copolymers. Solid lines are fits to Lauritzen and Hoffman theory according to eqn. 6. (b) Lauritzen and Hoffman kinetic theory plots for the same materials presented in (a).

when they compared a poly(ethylene glycol) homopolymer, PEG, with the PEG block within PEG-*b*-PB diblock copolymers. In this case, the crystallization of the PEG block is carried out with a covalently bonded rubbery block and similar energetic restrictions for its crystallization decrease its crystallization rate as compared to homo-PEG.

The results presented in this paper show for the first time a quantitative measure of the energetic restrictions imposed by the molten PCL block on the crystallization kinetics of the PPDX block within PPDX-*b*-PCL diblock copolymers. As a result of these energetic restrictions the PPDX-*b*-PCL diblock copolymers crystallize at a much slower rate than PPDX homopolymer in the temperature range where only the PPDX can crystallize. This reduction in crystallization rate is the responsible for the coincident crystallization phenomenon experienced by the materials when they are cooled from the melt as in Fig. 1.

The preceding sections have shown the complex influence exerted by one block on the other, such that the nucleation and crystallization kinetics of the material are affected. At temperatures where the PCL block is molten, the PPDX block within PPDX-*b*-PCL diblock copolymers encounters complex changes in its interfacial area as compared to a homo-PPDX that results in a higher energy barrier for growth. Once the PPDX block crystallizes, it nucleates the PCL block. Nevertheless, the PCL block also encounters energetic restrictions for secondary nucleation that are more pronounced as the PPDX content in the copolymer increases.

Finally, some preliminary results on the hydrolytic degradation of $PPDX^{15}$ and two PPDX-*b*-PCL diblock copolymers have been examined. $PPDX^{15}$ completely degrades within a 6 month period. However, when the diblock copolymer containing just 23% PPDX is examined, the degradation rate is substantially reduced as compared to $PPDX^{15}$. A further increase in the PCL content to 35% greatly increases stability. A comparison of weight retention values after 6 months of exposure to the hydrolysis medium reveals that while $PPDX^{15}$ has been completely degraded, $D^{26}_{65}C^{15}_{35}$ has only lost 10% of its weight. The complex dependence of hydrolysis rate on composition derives from the different attack rates on the different phases. Hydrolysis reactions are faster in the amorphous regions of PPDX where water penetration is also faster, and slower in the crystalline regions.[1,2] The access to such regions will also be hindered by the disposition of the PCL lamellae and amorphous regions within the mixed spherulites in the copolymer, since PCL will offer protection against hydrolysis to the material in view of its much greater resistance to it.

Conclusions

AFM observations performed during sequential isothermal crystallization of $D^{8}_{23}C^{27}{}_{77}$ indicate that the final morphology is composed of mixed spherulites where lamellae of both constituents are inter-dispersed. Considering both AFM and POM measurements we can conclude that in PPDX-*b*-PCL diblock copolymers crystallization drives structure formation.

We have applied the Lauritzen and Hoffman kinetic theory to spherulitic growth rate data for PPDX and the PPDX block within $D^{32}_{77}C^{10}_{23}$. The results show that at temperatures higher than 50 °C the spherulitic growth of the PPDX block is energetically hindered by the covalently bonded molten PCL chain, resulting in a much larger fold surface free energy and higher values of the work needed for the chains to fold as compared to homo-PPDX. These energetic restrictions imposed by the molten PCL block on the PPDX are responsible for the coincident crystallization observed when any PPDX-*b*-PCL copolymer is cooled from the melt.

The application of the Lauritzen and Hoffman theory to overall crystallization rate data was highly successful for PPDX and the diblock copolymers. This was corroborated by applying it to data from DSC measurements of isothermal crystallization of previously self-nucleated PPDX and obtaining a perfect match with the energetic parameters determined from spherulitic growth rate data. The use of self-nucleation allowed the determination of the overall crystallization kinetics of the PPDX block within two copolymers (with 65 and 77% PPDX) which was otherwise too slow for detection. The results indicate that the energetic restrictions for the crystallization of the PPDX block at high temperatures increase as the PCL content in the copolymer increases.

In the case of the PCL block, the overall crystallization kinetics was determined after PPDX was crystallized until saturation. PPDX was found to nucleate PCL accelerating the overall kinetics of the PCL block within diblock copolymers with 60 and 77% PCL. When the PCL content was lower than 50%, a reduction in crystallization rate as well as a reduction in the Avrami index indicated that the previously crystallized PPDX (which is the matrix in these cases) was imposing some topological restrictions on the PCL block. Nevertheless, the crystallization was always heterogeneous and in the same undercooling range regardless of composition. The self-nucleation behaviour of the diblock copolymers also indicated that the PPDX within the copolymers is capable of nucleating the PCL and that its nucleation was always heterogeneous, since all three self-nucleation Domains were present. Through self-nucleation a separation of the crystallization exotherms of PPDX and PCL during the cooling scan in the DSC can be achieved.

Acknowledgements

The Simón Bolívar University team acknowledges sponsorship by Fonacit, through grant S1-2001000742. Discussion with Drs. Arnal and Balsamo from Simón Bolívar University Polymer Group are gratefully acknowledged. LPCM members are very grateful to "Région Wallonne" and European Community (FEDER, FSE) for general support in the frame of "Objectif 1-Hainaut: Materia Nova". This work was partly supported by the Belgian Federal Government Office of Science Policy (SSTC-PAI 5/3). J.-M. Raquez is Scientific Research Collaborator of the Belgian F.N.R.S.

References

1 R. S. Bezwada, D. D. Jamiolkowski and K. Cooper, in *Handbook of Biodegradable Polymers*, eds. A. J. Domb, J. Kost and D. M. Wiseman, Harwood, Singapore, 1997, ch. 2, pp. 29–61.
2 M. A. Sabino, J. Albuerne, A. J. Müller, J. Brisson and R. E. Prud'homme, *Biomacromolecules*, 2004 **5**, 358.
3 P. Rangarajan, R. A. Register and L. J. Fetters, *Macromolecules*, 1993, **26**, 4640.
4 P. Rangarajan, R. A. Register, D. H. Adamson, L. J. Fetters, W. Bras, S. Naylor and A. J. Ryan, *Macromolecules*, 1995, **28**, 1422.
5 A. J. Ryan, I. W. Hamley, W. Bras and F. S. Bates, *Macromolecules*, 1995, **28**, 3860.
6 P. H. Richardson, R. W. Richards, D. J. Blundell, W. A. MacDonald and P. Mills, *Polymer*, 1995, **36**, 3059.
7 D. J. Quiram, R. A. Register, G. R. Marchand and A. J. Ryan, *Macromolecules*, 1997, **30**, 8338.
8 K. C. Douzinas, R. E. Cohen and A. F. Halasa, *Macromolecules*, 1991, **24**, 4457.
9 I. W. Hamley, J. Patrick, A. Fairclough, N. J. Terrill, A. J. Ryan, P. M. Lipic, F. S. Bates and E. Towns-Andrews, *Macromolecules*, 1996, **29**, 8835.
10 P. Rangarajan, R. A. Register, L. J. Fetters, W. Bras, S. Taylor and A. J. Ryan, *Macromolecules*, 1995, **28**, 4932.
11 S. Nojima, K. Kato, S. Yamamoto and T. Ashida, *Macromolecules*, 1992, **25**, 2237.
12 D. J. Quiram, R. A. Register, G. R. Marchand and A. J. Ryan, *Macromolecules*, 1997, **30**, 4551.
13 A. Rohadi, R. Endo, S. Tanimoto, S. Sasaki and S. Nojima, *Polym. Int.*, 2000, **32**, 602.
14 A. Rohadi, S. Tanimoto, S. Sasaki and S. Nojima, *Polym. J.*, 2000, **32**, 859.
15 A. J. Ryan, J. P. A. Fairclough, I. W. Hamley, S.-M. Mai and C. Booth, *Macromolecules*, 1997, **30**, 1723.
16 M. A. Hillmyer and F. S. Bates, *Macromol. Symp.*, 1997, **117**, 121.
17 Y.-L. Loo, R. A. Register and A. J. Ryan, *Macromolecules*, 2002, **35**, 2365.
18 S. Nojima, M. Toei, S. Hara, S. Tanimoto and S. Sasaki, *Polymer*, 2002, **43**, 4087.
19 G. Reiter, G. Castelein, J.-U. Sommer, A. Röttele and T. Thurn-Albrecht, *Phys. Rev. Lett.*, 2001, **87**, 226 101.
20 C. Robitaille and J. Prud'homme, *Macromolecules*, 1983, **16**, 665.
21 D. J. Quiram, R. A. Register, G. R. Marchand and D. H. Adamson, *Macromolecules*, 1998, **31**, 4891.
22 M. Ueda, K. Sakurai, S. Okamoto, D. Lohse, W. J. MacKnight, S. Shinkai, S. Sakurai and S. Nomura, *Polymer*, 2003, **44**, 6995.
23 G. Reiter, *J. Polym. Sci.: Part B: Polym. Phys.*, 2003, **41**, 1869.
24 *Polymer Crystallization, Observations, Concepts and Interpretations*, ed. J. U. Sommer and G. Reiter, Springer, Berlin, 2003.
25 A. J. Müller, V. Balsamo, M. L. Arnal, T. Jakob, H. Schmalz and V. Abetz, *Macromolecules*, 2002, **35**, 3048.
26 M. L. Arnal, V. Balsamo, F. López-Carrasquero, J. Contreras, M. Carrillo, H. Schmalz, V. Abetz, E. Laredo and A. J. Müller, *Macromolecules*, 2001, **34**, 7973.
27 G. Floudas and C. Tsitsilianis, *Macromolecules*, 1997, **30**, 4381.
28 M. Gervais and B. Gallot, *Polymer*, 1981, **22**, 1129.
29 B. Lotz and A. J. Kovacs, *Polym. Prepr. (Am. Chem. Soc., Div. Polym. Chem.)*, 1969, **10**, 820.
30 J. J. J. O'Malley, *Polym. Sci., Polym. Symp.*, 1977, **60**, 151.

31 L. Zhu, Y. Chen, A. Zhang, B. H. Calhoun, M. Chun, R. P. Quirk, S. Z. D. Cheng, B. S. Hsiao, F. Yeh and T. Hashimoto, *Phys. Rev. B*, 1999, **60**, 10 022.
32 L. Zhu, S. Z. D. Cheng, B. H. Calhoun, Q. Ge, R. P. Quirk, E. L. Thomas, B. S. Hsiao, F. Yeh and B. Lotz, *Polymer*, 2001, **42**, 5829.
33 L. Zhu, B. R. Mimnaugh, Q. Ge, R. P. Quirk, S. Z. D. Cheng, E. L. Thomas, B. Lotz, B. S. Hsiao, F. Yeh and L. Liu, *Polymer*, 2001, **42**, 9121.
34 L. Zhu, S. Z. D. Cheng, P. Huang, Q. Ge, R. P. Quirk, E. L. Thomas, B. Lotz, B. S. Hsiao, F. Yeh and L. Liu, *Adv. Mater.*, 2002, **14**, 31.
35 L. Zhu, P. Huang, W. Y. Chen, Q Ge, R. P. Quirk, S. Z. D. Cheng, E. L. Thomas, B. Lotz, B. S. Hsiao, F. Yeh and L Liu, *Macromolecules*, 2002, **35**, 3553.
36 J. T. Xu, J. J. Yuan and S. Y. Cheng, *Eur. Polym. J.*, 2003, **39**, 2091.
37 P. A. Weimann, D. A. Hajduk, C. Chu, K. A. Chaffin, J. C. Brodil and F. S. Bates, *J. Polym. Sci.: Part B: Polym. Phys.*, 1999, **37**, 2053.
38 Y. L. Loo, R. A. Register, A. J. Ryan and G. T. Dee, *Macromolecules*, 2001, **34**, 8968.
39 Y. K. Choi, Y. H. Bae and S. W. Kim, *Macromolecules*, 1998, **31**, 8766.
40 B. Bogdanov, A. Vidts and H. Berghmans, *Macromolecules*, 1999, **32**, 726.
41 Y. Wang and M. Hillmyer, *J. Polym. Sci.: Part A: Polym. Chem.*, 2001, **39**, 2755.
42 J. Albuerne, L. Márquez, A. J. Müller, J. M. Raquez, Ph. Degée, Ph. Dubois, V. Castelletto and I. W. Hamley, *Macromolecules*, 2003, **36**, 1633.
43 R. M. Ho, P. Y. Hsieh, W. H. Tseng, C. C. Lin, B. H. Huang and B. Lotz, *Macromolecules*, 2003, **36**, 9085.
44 A. J. Müller, J. Albuerne, L. M. Esteves, L. Márquez, J. M. Raquez, Ph. Degée, Ph. Dubois, S. Collins and I. W. Hamley, *Macromolecular Symp.*, 2004, in press.
45 N. Bhattarai, H. Y. Kim, D. I. Cha, D. R. Lee and D. I. Yoo, *Eur. Polym. J.*, 2003, **39**, 1365.
46 J.-M. Raquez, P. Degée, R. Narayan and P. Dubois, *Macromol. Rapid Commun.*, 2000, **21**, 1063.
47 A. Lendlein and L. Langer, *Science*, 2002, **296**, 1673.
48 J. K. Hobbs, A. D. L. Humphris and M. J. Miles, *Macromolecules*, 2001, **34**, 5508.
49 B. Fillon, J. C. Wittmann, B. Lotz and A. Thierry, *J. Polym. Sci., Part B: Polym. Phys.*, 1993, **31**, 1383.
50 V. Castillo, A. J. Müller and M. Hillmyer, 2004, unpublished results.
51 M. A. Sabino, J. L. Feijoo and A. J. Müller, *Macromol. Chem. Phys.*, 2000, **201**, 2687.
52 J. I. Lauritzen Jr. and J. D. Hoffman, *J. Appl. Phys.*, 1973, **44**, 4340.
53 *Physical Properties of Polymers Handbook*, ed. J. E. Mark, AIP Press, New York, 1996.
54 J. Schultz, *Polymer Crystallization*, American Chemical Society, Washington, 2001.
55 S. Andjelic, D. Jamiolkowski, J. Mcdivitt, J. Fischer, J. Zhou and R. Vetrecin, *J. Appl. Polym. Sci.*, 2001, **79**, 742.
56 M. Avrami, *J. Chem. Phys.*, 1939, **7**, 1103; M. Avrami, *J. Chem. Phys.*, 1940, **8**, 212; M. Avrami, *J. Chem. Phys.*, 1941, **9**, 177.
57 L. Mandelkern, in *The Crystalline State*, eds. J. E. Mark, A. Eisenberg, W. W. Graessley, L. Mandelkern and J. L. Koenig, *Physical Properties of Polymers*, 2nd edn., ACS, Washington, 1993.
58 W. S. Lambert and P. J. Phillips, *Macromolecules*, 1994, **27**, 3537.
59 V. Balsamo, Y. Paolini, G. Ronca and A. J. Müller, *Macromol. Chem. Phys.*, 2000, **201**, 2711.
60 H. Schmalz, A. J. Müller and V. Abetz, *Macromol. Chem. Phys.*, 2003, **204**, 111.
61 M. L. Arnal, F. López-Carrasquero, E. Laredo and A. J. Müller, *Eur. Polym. J.*, 2004, **40**, 1461.
62 B. Wunderlich, *Macromolecular Physics, Volume 2, Crystal Nucleation, Growth, Annealing*, Academia Press, New York, 1976, pp. 52–70.
63 *Polymer Handbook*, ed. J. Brandrup, E. H. Immergut, E. A. Grulke, A. Abe and D. R. Bloch, 4th edn., John Wiley and Sons, New York, 1999.
64 T. Shiomi, H. Takeshita, H. Kawaguchi, M. Nagai, K. Takenaka and M. Miya, *Macromolecules*, 2002 **35**, 8056.

Oriented primary crystal nucleation in lamellar diblock copolymer systems

Wenbing Hu and Daan Frenkel

FOM Institute for Atomic and Molecular Physics, Kruislaan 407, 1098 SJ Amsterdam, The Netherlands

Received 1st March 2004, Accepted 24th May 2004
First published as an Advance Article on the web 11th August 2004

We report a numerical study of the primary crystal nucleation of diblock copolymers in a lamellar phase. Only one of two polymer blocks is crystallizable; the other is maintained in a glassy state. We find that crystals in this lamellar geometry tend to nucleate with the chain axis perpendicular to the lamellar plane. However, if in the same lamellar structure, we break the junction between the crystallizable and non-crystallizable blocks of the polymers, we find that crystallites tend to align parallel to the lamellar plane. This observation clarifies the molecular origin of the competition between parallel and perpendicular crystallite orientations in real block-copolymer systems.

I Introduction

Crystallites that form in lamellar diblock copolymer systems often exhibit an appreciable degree of orientational order.[1,2] Depending on the circumstances, the crystallites may be oriented either parallel or perpendicular to the plane of the lamellae. The tendency of crystallites to orient perpendicular to the microphase lamellae has been attributed to the preferential crystallization of polymer chains that have been pre-stretched in the process of lamellar microphase separation.[3] In contrast, the formation of crystallites that are aligned parallel to the lamellae is thought to be caused by surface nucleation near the microphase interfaces.[4,5] The latter scenario is consistent with the observation of parallel crystallite orientations in high-molecular-weight block copolymers since, for long-chain polymers, the pre-stretching should be relatively unimportant.[6,7] Although both of the scenarios sketched above are likely to operate under the appropriate conditions, it is still unclear at what stage of crystallization the crystallite alignment takes place: alignment could either emerge during crystal growth or it could be imposed by the nucleation of aligned crystallites.

X-ray diffraction measurements indicate that the crystallites are already aligned by the stage that they are large enough to give clear Bragg-diffraction peaks.[8] However, with X-ray diffraction it is not possible to probe alignment during the initial nucleation stage, as the relevant crystal nuclei are too small to give a detectable diffraction signal. The aim of the present simulation is to gain insight into the processes that lead to crystallite alignment during the nucleation stage of crystallization of polymers in a lamellar confinement.

The advantage of a simulation approach is that it is possible to distinguish between two scenarios under otherwise identical conditions. To this end, we look at crystal nucleation in a lamellar diblock-copolymer system where the non-crystallizable component is in a glassy state. We can consider two cases: (1) crystallization in the original block copolymer system where the crystallizable chains are significantly stretched and (2) crystallization in the same system, but with the connection between two blocks broken: in that case only the aligning effect of the interfaces between the microphases is present.

In what follows, we first introduce the "sample preparation" of our model systems. Then, we describe the structural characteristics of two model systems, in particular the coil sizes, bond orientations and crystallite orientations. We also discuss the isothermal crystallization behavior. In our concluding remarks, we place the present results in a broader context.

II Simulation techniques and sample preparation

The first step in the simulations was the preparation of a system of symmetric diblock copolymers with a well-defined lamellar structure. From earlier studies it is known that microphase structures in the molecular simulations of symmetric diblock copolymers are very sensitive to both the chain length and the system size.[9] To minimize these effects, we chose short diblock copolymers (32 units per chain) and a large cubic lattice box (a linear size of 64). We put 7680 chains in the box so that the polymer volume fraction was as high as 93.75%, thus mimicking a bulk polymer system. To speed up the time-evolution of the system, a small number of athermal voids were introduced in the system (thus creating "free volume"). The chains "live" on the periodic lattice box. Their motion is modelled using a micro-relaxation model, which allows single-site jumping, with partial sliding diffusion if necessary. The algorithm avoids bond crossing and hard-core overlaps.[10] Bonds may be oriented either along the lattice axes or along the (body or plane) diagonals. Hence the coordination number of each lattice site is $6+12+8=26$. This implies that there are 13 distinct orientations possible for every bond. Initially, the chains were annealed at high temperatures to the coil state. We make one half of each chain (16 monomers) crystallizable and the other half non-crystallizable by switching on or off the interaction parameter E_p which measures the attraction of parallel bonds. A large value of E_p favors crystallization.[10,11] The acceptance of the micro-relaxation moves was determined using the Metropolis criterion, where the potential energy penalty for a trial move was given by $\Delta E=(bB/E_c-pE_p/E_c-c)\ E_c$. Here, b is the change in the number of "mixing contacts" between crystallizable and non-crystallizable monomers, p is the change in the number of pairs of parallel nearest-neighbor bonds between crystallizable chains, and c is the change in the number of collinear connections of consecutive bonds along the chains. B measures the mixing energy penalty for bringing two unlike monomers in contact, E_p is the energy penalty for disrupting the parallel alignment of two neighboring bonds (of the crystallizable chains) and E_c is the energy cost of making a "kink" in a chain. In practice, B/E_c is mainly responsible for microphase separation, and E_p/E_c determines the propensity for crystallization, while $E_c/(k_BT)$ can be considered as a measure for the (inverse) temperature of the system.[11] In what follows, we fixed E_p/E_c at the values $E_p/E_c=1$, in order to mimic semiflexible chains. The value of B/E_c was varied in order to adjust the tendency for micro-phase separation.[12] In order to mimic a cooling experiment, we performed a cooling run where we increased $E_c/(k_BT)$ in steps from zero (infinite temperature) with a step size of 0.002. A temperature step was taken every 300 MC cycles (one cycle is one trial move per chain unit). During every temperature step, the first 100 MC cycles were discarded for equilibration. The reported data were averaged over the remaining 200 MC cycles. During cooling, we could detect microphase separation through a steep increase in the demixing parameter, defined as the mean fraction of neighbors around each non-crystallizable monomer occupied by other non-crystallizable monomers. Simultaneously, we monitored the crystallization by following the fraction of crystallizable bonds that have more than five parallel crystallizable bonds as neighbors. We chose the threshold of the number of parallel neighbors at five, in order to account for the fact that the number of parallel bonds at the surfaces and, in particular at the edges and vertices of crystallites, is much less than the maximum value (24).

Fig. 1a shows the results of one such a cooling run. As can be seen from the figure, cooling down to $B/E_c=0.5$ is sufficient to induce microphase separation. Moreover, this transition occurs well before crystallization. Fig. 1b shows a snapshot of the sample at an inverse temperature $E_c/(k_BT)=0.25$. The figure shows that the micro-phase separated domains have formed a well-defined lamellar structure.

In what follows, we focus our attention on the structure shown in Fig. 1b. To make the non-crystallizable domains glassy, we simply eliminated all mobility of the non-crystallizable blocks

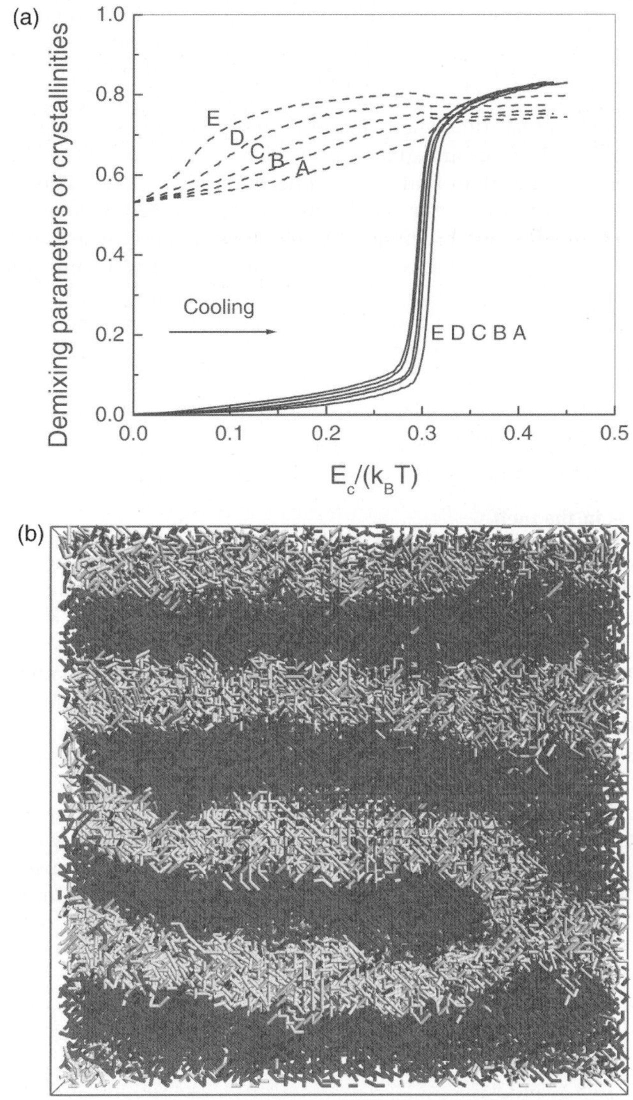

Fig. 1 Preparation of lamellar microphase domains for the confined crystallization of symmetric diblock copolymers. (a) Cooling curves of demixing parameter (dashed lines) and crystallinity (solid lines) of the sample with $E_p/E_c = 1$ and variable B/E_c as 0.1, 0.15, 0.2, 0.3, 0.5 from curve A to curve E. (b) Snapshot of the sample system with $E_p/E_c = 1$ and $B/E_c = 0.5$ at the temperature of $E_c/k_BT = 0.25$. All the bonds are drawn in cylinders. Non-crystallizable blocks are shown in black, and crystallizable blocks are in gray.

(*i.e.* we rejected all trial moves of non-crystallizable monomers). To prepare a comparable sample where the crystallizable and non-crystallizable chains are not connected, we simply removed the bonds joining two blocks of each chain. Starting from the structure of Fig. 1b, we then annealed the short chains for 1×10^4 MC cycles at the same temperature and glassy environment. This was long enough to remove all memory of the location of the initial block junctions.

III Results and discussion

A Coil sizes in the melt

As micro-phase separation separates two blocks of a copolymer chain on different sides of the interface, it is obvious that the formation of a lamellar phase leads to a "polarization" of the block-copolymers. It is interesting to consider to what extent the individual blocks are deformed. Earlier simulations found that the microphase separation has little effect on the size of the individual blocks in the copolymer.[13,14] In the present case, we also found that the mean squared end-to-end distance of the crystallizable block in the state of Fig. 1b is 25.203, very close to the result (24.900) at infinite temperature where microphase separation plays no role. However, looking in more detail, we find that the mean square end-to-end distance (12.491) of the eight monomers closest to the junction is larger than that of the other half end (9.841). This suggests that in each block, only the chain segments near the block junction are aligned as a result of the microphase separation. The direction of local stretching near the block junction is dictated by the microphase separation and should be perpendicular to the microphase interfaces. From a thermodynamic point of view, the pre-stretching of a chain decreases the entropy penalty for crystallization and hence increasing the thermodynamic driving force for freezing. Even when supercooling is insufficient to cause crystal nucleation in the isotropic bulk, the local stretching may facilitate primary crystallization nucleation of properly aligned polymers. In an earlier study, we showed that such effects can be quite pronounced, as even a single prealigned chain can induce the formation of a polymer crystallite with shish-kebab morphology.[15]

B Bond orientations in the melt

In our cubic lattice, the coordination number of each lattice site is 26 and therefore, there are totally 13 possible bond orientations. In an isotropic liquid, the probability of bond orientations should be constant within each subset (axes, face diagonals or body diagonals), see Fig. 2. This symmetry is broken in the lamellar phase. Fig. 2 also shows that the bond orientations of crystallizable blocks have a weak distribution peak at the *y*-axis which is perpendicular to the lamellar plane in Fig. 1b. This is consistent with the fact that the deformation of the coil due to microphase separation is localized in the vicinity of the block junction. Breaking the block junction will lead to the multiple peak at the *x*, *z*, *xz* and $-xz$ directions which are the parallel orientations located in the lamellar plane. Comparing to the athermal and thermal bulk liquids of 16-mers, the strong peaks of parallel orientations can be attributed to the introduction of microphase interfaces. This suggests that the parallel orientations of the bonds will be favored by the contacts of coil with the flat microphase interfaces, while the perpendicular orientations (along the *y* axis) are only possible in the bulk phase away from the interfaces. Indeed in Fig. 2, after breaking the block junction, the probability of the *y* orientation approaches that of bulk 16-mer liquids.

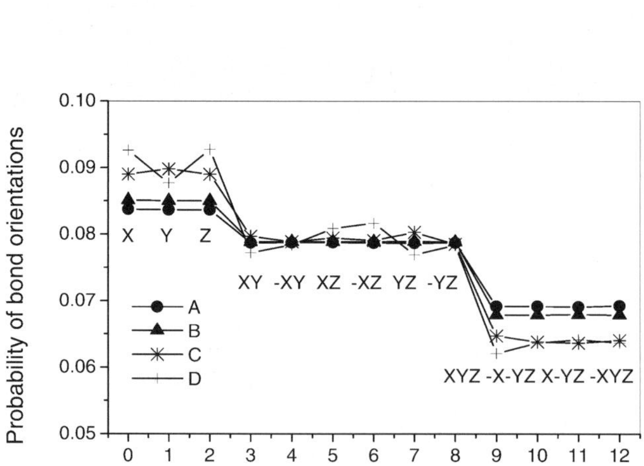

Fig. 2 Probability distribution of bond orientations of the variable systems on 13 directions as denoted in the figure. The segments are drawn to guide the eye. Curve A is for the athermal bulk 16-mers in the 64-sized cubic box, curve B is for the bulk 16-mers at $E_c/k_BT = 0.25$ with $E_p/E_c = 1$ and $B/E_c = 0.1$, curve C is for the crystallizable 16-mers in the state of Fig. 1b and with a glassy environment, and curve D is for the crystallizable 16-mers under the conditions similar to curve C but without the block junction.

C Crystallite orientations in the thermal fluctuations of the melt

In order for crystal nucleation to occur, a spontaneous fluctuation in crystallinity should result in the formation of a crystallite with a size that exceeds a well-defined threshold.[16] To monitor the orientation of nucleating crystals, we performed a large number of simulations that started in the supercooled state shown in Fig. 1b. During these simulations, we monitored the onset of crystallinity. We then computed the size of the largest crystallite in the system by applying a cluster analysis to the crystalline bonds in the system. Adjacent crystalline bonds are assumed to belong to the same cluster. Having determined the size of the largest crystallite, we then determine its overall orientation. In Fig. 3, we have separated the results into three parts according to whether the crystallite is oriented perpendicular to the interface, parallel or otherwise. Among the total 13 bond orientations, the perpendicular orientation accounts for one degree of freedom, the parallel corresponds to four, and the others to eight. Fig. 3a shows that in the case with the block junction, sufficiently large crystallites tend to have fewer parallel than perpendicular bonds even though the number of possible parallel bonds is four times larger. Other orientations are found to be even more strongly suppressed.

In the case of short chains with broken block junction, the computed size distribution of the largest crystallites is quite different from the previous case, as shown in Fig. 3b. Perpendicular orientations are now much less likely, while the enhanced probability to observe parallel bonds

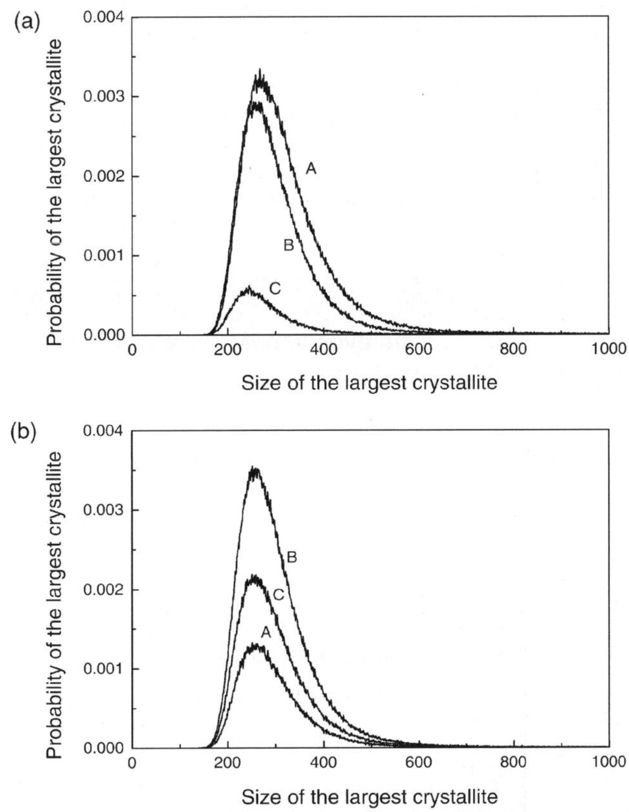

Fig. 3 Size distribution of the largest crystallite decomposed in the contributions from orientations perpendicular (curve A), parallel (curve B) or otherwise (curve C) to the lamellar microphase domains. (a) Same state point as shown in Fig. 1b. (b) Same conditions as in (a) but without block junctions. The crystallite size is defined as the total number of crystalline bonds (more than five parallel neighbors of other crystallizable bonds) that belong to a single cluster. The results cover over one million ensemble observations. If the size of the largest crystallite goes beyond 1000, a new array of observations with a 10 MC cycle interval will be switched on from the state of Fig. 1b both with and without the block junction.

remains. In other words: the initial tendency to form crystallites that are oriented perpendicular to the interface has disappeared and, instead, crystallites now tend to align parallel to the interface.

D Isothermal crystallization

A study of isothermal crystallization at a slightly lower temperature provides even clearer evidence for the preferred orientations of crystallites in the primary crystal nucleation process. We quenched the sample of Fig. 1b to a slightly lower temperature ($E_c/(k_BT) = 0.27$), and traced the time evolution of the crystallinity as well as the orientational order parameter of crystallites defined by

$$P = \frac{3\langle\cos^2\theta\rangle - 1}{2}, \tag{1}$$

where θ is the angle of crystalline-bond orientation to the y-axis and $\langle\ldots\rangle$ is an average over all crystalline bonds. In order to make the order parameter P more sensitive to changes in the crystal

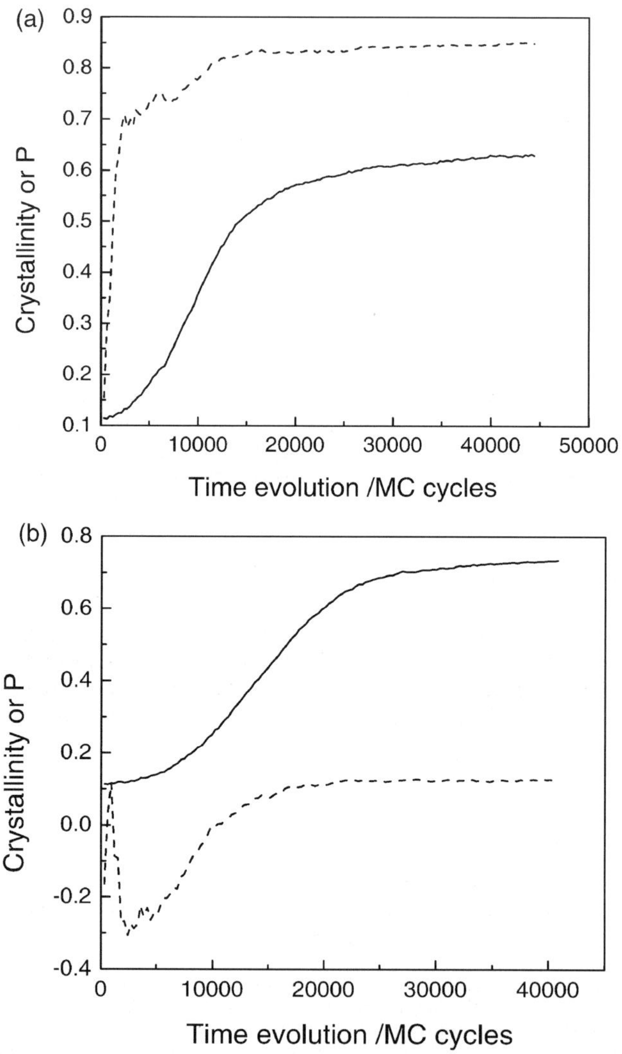

Fig. 4 Time evolution of crystallinity (solid line) and orientational order parameter (dashed line) quenched from the state of Fig. 1b to $E_c/k_BT = 0.27$ (keeping the non-crystallizable chains immobile): (a) diblock copolymers and (b) short chains without the block junction.

alignment, we used a slightly stronger definition of a "crystalline" bond, namely one that has ten (rather than five) parallel crystallizable bonds as neighbors. If all the crystalline bonds are normal to the microphase interfaces, P will be one; if their orientations are random, P will be close to zero; if they are all parallel along microphase interfaces, P goes to -0.5. Fig. 4a shows the time evolution of the crystallinity and the crystalline orientational order in a system of diblock copolymers. P rises even faster than the crystallinity order parameter itself. This indicates that the very first crystallites that form are already aligned perpendicular to the lamellar interface. By comparison, Fig. 4b shows the time-evolution of crystallinity and orientational order parameter in the corresponding case with the broken block junction. In this case, the orientational order initially becomes fairly strongly negative ($P \approx -0.3$). This indicates that the first crystallites tend to be aligned parallel to the interface. Subsequently, the orientational order parameter increases and it saturates at a small positive value. This suggests that the subsequent growth of crystallites with orientations that are not parallel to the interface becomes dominant, presumably because such crystallites can grow longer before impinging on the lamellar domain walls. These crystallites tend to increase the average value of P.

In experimental measurements of the total crystallization rate during the confined crystallization of diblock copolymers, a low value of the Avrami index has been observed.[17–19] For the homogeneous nucleation and two-dimensional growth, the ideal value of the Avrami index is 3.[20] In our cases of Fig. 4, we find a value of 1.4 for the Avrami index of diblock copolymers and a value of 1.6 for short chains. Such a low value of the Avrami index is often observed in strain-induced crystallization of polymers.[21,22] This supports our assumption that crystal nucleation is induced by a local pre-ordering of the crystallizable chains, either perpendicular or parallel to the lamellar planes.

IV Concluding remarks

Our simulations indicate that, although the limited dimension of the microphase domains makes polymer crystal nucleation more difficult (less possibility of occurring in a smaller space),[23] the local chain ordering near the block junction or, alternatively, close to the microphase interface, facilitates primary crystal nucleation and hence causes the emergence of perpendicular (or parallel) crystallite orientations. In this respect, our observations support the notions put forward in refs. 3–8. We point out that the tendency to align crystallites perpendicular or parallel to the lamellae is not an artefact of our lattice model, because no such preference exists in the bulk melt. Indeed, off-lattice molecular-dynamics simulations by Meyer[24] also show a similar preference of perpendicular alignment of crystallites in lamellar diblock-copolymer systems. Again, cutting the block junction results in a change of this preference to parallel.

Acknowledgements

W. H. acknowledges helpful discussions with Prof. S. Z. D. Cheng (University of Akron), Prof. L. Zhu (University of Connecticut), and Dr L. Li (Amolf). The work of FOM institute is part of the research program of the "Stichting voor Fundamenteel Onderzoek der Materie" (FOM), which is financially supported by the "Nederlandse organisatie voor Wetenschappelijk Onderzoek" (NWO).

References

1 I. W. Hamley, *Adv. Polym. Sci.*, 1999, **148**, 113.
2 L. Zhu, S. Z. D. Cheng, B. H. Calhoun, Q. Ge, R. P. Quirk, E. L. Thomas, B. S. Hsiao, F. Yeh and B. Lotz, *J. Am. Chem. Soc.*, 2000, **122**, 5957.
3 J. P. Fairclough, S.-M. Mai, M. W. Matsen, W. Bras, L. Messe, S. Turner, A. J. Gleeson, C. Booth, I. W. Hamley and A. J. Ryan, *J. Chem. Phys.*, 2001, **114**, 5425.
4 K. C. Douzinas and R. E. Cohen, *Macromolecules*, 1992, **25**, 5030.
5 R. E. Cohen, A. Bellare and M. A. Drzewinski, *Macromolecules*, 1994, **27**, 2321.
6 I. W. Hamley, J. P. A. Fairclough, N. J. Terrill, A. J. Ryan, P. M. Lipic, F. S. Bates and E. Towns-Andrews, *Macromolecules*, 1996, **29**, 8835.
7 I. W. Hamley, J. P. A. Fairclough, A. J. Ryan, F. S. Bates and E. Towns-Andrews, *Polymer*, 1996, **37**, 4425.
8 L. Zhu, B. H. Calhoun, Q. Ge, R. P. Quirk, S. Z. D. Cheng, E. L. Thomas, B. S. Hsiao, F. Yeh, L. Liu and B. Lotz, *Macromolecules*, 2001, **34**, 1244.
9 K. Binder, *Adv. Polym. Sci.*, 1994, **112**, 1.

10 W.-B. Hu, *J. Chem. Phys.*, 1998, **109**, 3686.
11 W.-B. Hu, *J. Chem. Phys.*, 2000, **113**, 3901.
12 W.-B. Hu, D. Frenkel and V. B. F. Mathot, *J. Chem. Phys.*, 2003, **118**, 10 343.
13 K. Binder, in *Monte Carlo and Molecular Dynamics Simulations in Polymer Science*, ed. K. Binder, Oxford University Press, New York, 1995, ch. 7, p. 419.
14 A. Hoffmann, J.-U. Sommer and A. Blumen, *J. Chem. Phys.*, 1997, **106**, 6709.
15 W.-B. Hu, D. Frenkel and V. B. F. Mathot, *Macromolecules*, 2002, **35**, 7172.
16 K. F. Kelton, in *Crystal Nucleation in Liquids and Glasses*, ed. H. Ehrenreich and D. Turnbull, Academic Press, Boston, 1991, vol. 45.
17 J. C. Mitchell and D. J. Meier, *J. Polym. Sci., Part A*, 1968, **6**, 1689.
18 D. J. Blundell, D. H. MacKerron, W. Fuller, A. Mahendrasingam, C. Martin, R. J. Oldman, R. J. Rule and C. Riekel, *Polymer*, 1996, **37**, 3303.
19 B. Lotz and A. J. Kovacs, *ACS Polym. Prepr.*, 1969, **10**, 820.
20 B. Wunderlich, *Macromolecular Physics Crystal Nucleation Growth Annealing*, Academic Press, New York, 1976, vol. **2**.
21 Y.-L. Loo, R. A. Register and A. J. Ryan, *Phys. Rev. Lett.*, 2000, **84**, 4120.
22 G. Reiter, G. Castelein, J. U. Sommer, A. Rottele and T. Thurn-Albrecht, *Phys. Rev. Lett.*, 2001, **87**, 226 101.
23 C. W. Frank, V. Rao, M. M. Despotopoulou, R. F. W. Pease, W. D. Hinsberg, R. D. Miller and J. F. Rabolt, *Science*, 1996, **273**, 912.
24 H. Meyer, Charles Sadron Institute, France, personal communication.

Isothermal reorganization of poly(ethylene terephthalate) revealed by fast calorimetry (1000 K s^{-1}; 5 ms)

Alexander A. Minakov,† Dmitry A. Mordvintsev and Christoph Schick*

University of Rostock, Dept. of Physics, Universitätsplatz 3, 18051, Rostock, Germany. E-mail: christoph.schick@physik.uni-rostock.de; Fax: +49-381-498 6882

Received 5th March 2004, Accepted 10th May 2004
First published as an Advance Article on the web 16th September 2004

Reorganization of semicrystalline polymers on heating is a fast process. For poly(ethylene terephthalate) (PET) heating rates of several thousand Kelvin per second are needed to prevent reorganization of unstable crystals. Utilizing a thin film vacuum gauge as a fast calorimeter we are able to extend the scanning rate range of commercial DSC's (μK s^{-1} to 10 K s^{-1}) to rates as high as 10 000 K s^{-1} on heating and cooling. Because of the fast equilibration time isothermal experiments can be performed after scanning at several thousand Kelvin per second. The dead time after such a quench is in the order of 10 ms and the time resolution is in the order of milliseconds. These ultra fast calorimeters allow us to study the kinetics of extremely fast processes in semicrystalline polymers like reorganization. For PET crystallized at 130 °C reorganization needs less than 40 ms between 150 °C and 200 °C. Isothermal reorganization at 223 °C is about two orders of magnitude faster than isothermal crystallization from the isotropic melt at the same temperature. The melt memory for the remaining structures needed for reorganization is removed 25 K above the equilibrium melting temperature of PET.

Introduction

Mesophase precursors to crystallization are considered important in the crystallization process of polymers[1–6] and other complex materials.[7] Only a few examples are known where the precursors can be detected and the development with time can be studied.[1,3,5] This is on one hand caused by the small contrast between the isotropic melt and the preordered state. On the other hand the preordered structures may exist as transient states in front of the growing lamellae only and lifetime may be very short.

One can address the question of precursors to crystallization not only by investigating the early stages of polymer crystallization but also by looking for remaining structures after partial or very short time melting of polymer crystals. This melt memory is well known for polymers[8,9] and causes significant differences in crystallization kinetics depending on the remaining order after partial melting. In order to study the influence of partial melting the semicrystalline sample is commonly heated at a certain temperature for a certain time and successively cooled to the recrystallization temperature to study kinetics of recrystallization. As long as some order remains the recrystallization

† On leave from the Natural Science Research Centre of A. M. Prokhorov General Physics Institute of Russian Academy of Science, Vavilov st. 38, 119991 Moscow, Russia.

DOI: 10.1039/b403441d

is much faster compared to crystallization from the isotropic melt.[10] Because recrystallization is in general a fast process it is difficult to prevent recrystallization on heating to the partial melting temperature for most polymers at common heating rates (K min^{-1}). Therefore the remaining structures are not originated by the originally formed crystalline structure but the most stable structures due to reorganization remain. The reorganization process on heating is in general very fast and an *in situ* study requires techniques with millisecond time resolution.

We have developed a thin film (chip) calorimeter allowing heating and cooling rates up to 10 000 K s^{-1},[11,12] which allows the prevention of recrystallization of poly(ethylene terephthalate) (PET)[13] on heating. Having a tool in hand for fast calorimetry enables us to study the kinetics of recrystallization as well as the influence of partial melting. We report a study of recrystallization kinetics and melt memory for PET on time scales ranging from 10 ms to hours.

Experimental

The PET sample was from KoSa GmbH & Co. KG, Germany, and had a molecular weight of 44 000 Da. From the amorphous granule a tiny piece of the order of a few hundred nanograms was cut. The sample was moved on top of the heater of the calorimetric sensor, see Fig. 1. To avoid damaging the sensor membrane (*ca.* 1 μm thick) the sample was moved by a soft copper wire (diameter 50 μm). A stereo microscope was used to control the movement. When the sample was in the right place an electrical current through the heater was switched on to melt the sample for the first time. This way the sample was fixed at a position just on top of the heater. Because of strong adhesive forces the sample-membrane thermal contact was good and very stable after a few heating–cooling cycles. Thus, a small and reproducible thermal contact resistance was created, which is important for calorimetric measurements.[14,15]

The thermal conductivity gauge TCG-3880[16] shown in Fig. 1 consists of 0.5 μm Si_3N_x membrane with a thin-film thermopile and a resistive film-heater placed at the center of the membrane. All electrical connections are covered by an additional 0.7 μm SiO_2 layer for electrical insulation and protection. The six thermopile hot junctions—the white spots around the central region in the

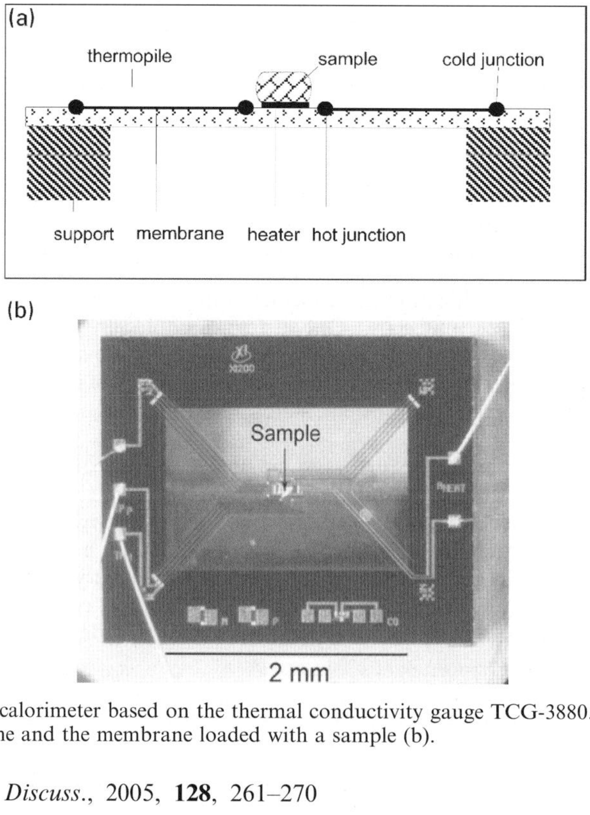

Fig. 1 Thin-film chip calorimeter based on the thermal conductivity gauge TCG-3880. Scheme (a) and micro photograph of the frame and the membrane loaded with a sample (b).

photograph—are placed around the central heated area, approximately $50 \times 100\ \mu m^2$. The cold junctions are placed at the silicon frame fixing the membrane $\approx$ 1 mm from the center. Thus the cold-junction temperature equals the temperature of the holder which is close to the temperature of the thermostat. An additional copper–constantan thermocouple was utilized for the measurement of the holder temperature, which was used as the reference temperature T_0.

To allow fast cooling the cell was operated in an ambient gas. The heat flow from the heated region to the environment can be described by Newton's law: $\Phi = \xi \Delta T$, where Φ is the heat flow from the heated region, ξ is the heat exchange parameter measured in W K^{-1}, and $\Delta T = (T - T_0)$, where T is the temperature of the heated region of the membrane, T_0 is the temperature of the environment, which is close to the temperature of the holder. The heat exchange parameter ξ is approximately 2×10^{-5} W K^{-1} in air at pressures in the range 10^2–10^5 Pa and room temperature.[11] The parameter ξ is 6 times larger, when the calorimeter is operated in helium gas. Because the membrane is very thin the heat transfer through the ambient gas is dominant. The heat transfer through the membrane in a radial direction is at least 10 times smaller than that through the gas in the perpendicular direction.[11] The measurements can be performed at atmospheric pressure, as well as at pressures in the range 10–10^5 Pa, where the gas thermal conductivity is still relatively large. In fact, the measurements were performed in air at 2×10^3 Pa.

The expected maximal cooling rate $(dT/dt)_{max}$ of a 500 ng sample, with heat capacity $C \approx 500$ nJ K^{-1}, can be estimated as follows: $(dT/dt)_{max} = \Delta T \xi/(C + C_0)$. The addenda heat capacity C_0 equals $\approx$ 150 nJ K^{-1} at room temperature.[11] Thus, $(dT/dt)_{max}$ is of the order of 10^4 K/s at $\Delta T = 500$ K, which was confirmed experimentally.[11,12] Fast linear cooling is only possible, when the scanning rate is several times smaller than $(dT/dt)_{max}$. On the other hand, sample thickness d must be small enough to avoid a large temperature gradient in the sample. The temperature difference δT across a plate-like sample can be estimated as follows: $\delta T = (dT/dt)(d^2 \rho c/\lambda)$, where ρ, c and λ are the density, the specific heat capacity and the thermal conductivity of the sample, respectively. Thus, δT equals 1–3 K at $d = 10\ \mu m$ and $dT/dt = 3 \times 10^3$ K s^{-1} for PET with $c = 1$–2 J g^{-1} K^{-1},[17] $\rho = 1.3$ g cm^{-3} and $\lambda \approx 0.3$ W K^{-1} m^{-1}.[18] As a temperature gradient exists in the radial direction along the membrane, the sample should be placed just on the heater. Otherwise, there will be a temperature gradient on the periphery of the sample outside the heated area.

The thermal contact between the heater and the thin PET sample is sufficiently good due to strong adhesive forces. Nevertheless, the measured temperature does not represent the temperature of the heater/sample interface, because the thermopile measures the temperature at the membrane around the heater. Thus the measured temperature was calibrated as described in ref. 11 The heat capacities and thermal resistances of the film-heater and of the thermopile are negligibly small. The addenda heat capacity is therefore the effective heat capacity of the heated part of the membrane. Thus, the sample-membrane system can be described by the following parameters: the effective heat capacity of the central part of the membrane C_0, the heat capacity of the sample C and the heat-exchange coefficient ξ. The resistive film-heater, $\approx 600\ \Omega$, provides the heat flow $\Phi_0(t)$, which is supplied to the membrane/sample interface and propagates through the sample, the membrane and the ambient gas. The heat balance equation is as follows:

$$(C + C_0)(dT/dt) = \Phi_0(t) - \xi(T(t) - T_0), \quad (1)$$

This equation is correct, provided the thermal thickness of the sample is small enough and the heat transfer from the cell to the thermostat can be described by Newton's law. The temperature measured by the thermopile is approximately 10 ms delayed with respect to the heater current.[11] The duration of a cooling or heating scan must be large enough with respect to the delay time $\tau_0 \sim 10$ ms, otherwise we have to consider thermal waves rather than a monotonic heating or cooling process. Thus, the scanning rate must be much smaller than $\Delta T/\tau_0 \sim 5 \times 10^4$ K s^{-1} at $\Delta T = 500$ K. This limitation is of the same order of magnitude as the restriction, which follows from the limited heat exchange through the gas. In the present work the scanning rate was less than or equal to 3×10^3 K s^{-1}, *i.e.* the rate was *ca.* 10 times smaller compared to both limits. The heat flow $\Phi_0(t)$ is determined by the electric current in the heater, $I_H(t)$, and its resistance $R_H(T)$. The resistance $R_H(T)$ was calibrated in advance. The electric current in the heater was monitored during its scanning simultaneously with the temperature difference $T(t) - T_0$, measured by the thermopile. The temperature dependence of the thermopile sensitivity, S *ca.* 1 mV K^{-1}, and the thermopile thermal lag were calibrated as described in ref. 11

Thus, there are three unknown parameters in eqn. (1)–ξ, C_0 and C. The parameter $C_0(T)$ was determined in advance for the empty cell. It increases monotonically from $\approx$ 100 to 200 nJ K^{-1} in the temperature range 100 to 600 K. The parameters ξ and C can be determined from heating–cooling scans as described in ref. 11. As the gas thermal conductivity depends on temperature, the heat exchange coefficient ξ is also a function of the surrounding gas temperature. The dependence $\xi(T)$, which can be approximated by a smooth polynomial function,[11] was determined as follows. First the heating–cooling cycle for an amorphous sample was investigated. Outside the glass transition range the heat capacity on cooling equals heat capacity on heating. This was used to determine $\xi(T)$. Therefore, the dependences $C(T)$ and $\xi(T)$ were determined simultaneously from cooling and heating curves for an amorphous sample. Next, the heat capacity of the crystallized sample was determined on heating assuming the same dependence $\xi(T)$ as in the previous experiment, provided the gas pressure was kept constant. In order to determine the specific heat capacity, $c = C/m$, the sample mass m has to be known. The mass was not measured independently, because it was too small (100 ng). It was determined from the measured heat capacity and the known specific heat capacity in the molten state at temperatures above 280 °C.[17]

The current in the heater and the signal from the thermopile were monitored in real time during fast scanning of the membrane temperature as described in ref. 11. To simplify the analysis of the time dependences $T(t)$ and dT/dt we tried to scan temperature almost linearly with time. A function generator, Stanford Research Systems DS 340, was utilized to realize an approximately constant scanning rate. The generator output voltage supplied to the heater was proportional to the square root of time. The data acquisition and control electronics are described in ref. 12. The temperature of the thermostat was stabilized at T_0 = 30 °C. The scanning range ΔT was $\approx$ 350 K.

DSC and Hyper DSC™ measurements were performed utilizing a PerkinElmer Pyris Diamond DSC equipped with an Intracooler II and nitrogen as purge gas. The instrument was calibrated by indium and zinc for temperature at the scanning rate of interest and by sapphire for heat flow. The sample was wrapped in aluminum foil of a few mg only to minimize thermal lag.[19]

For DSC measurements the samples were crystallized from the melt (melt crystallization). First the crystallization time t_C for each crystallization temperature T_C was chosen so that primary crystallization was completed. For the fast scanning experiments (chip calorimeter) the samples were quenched below glass transition before crystallization (cold crystallization) in order to allow a more simple scanning program. No difference between cold and melt crystallization was observed because of the high cooling and heating rates applied.

Results

The multiple melting peaks in isothermally crystallized PET are well known, see *e.g.* ref. 20. For PET it was shown that the multiple melting peaks are only due to melting–recrystallization–remelting sequences and not because of bimodal distributions of crystal stability.[13] Fig. 2 shows the melting curves for PET crystallized at 210 °C and 130 °C, respectively, at different heating rates. For slow heating multiple peaks are observed, witch disappear at higher heating rates. At the highest rates only one broad peak is seen indicating a broad distribution of lamellae thicknesses or stabilities. A similar result was obtained in real space AFM and TEM observations by Haubruge *et al.*[21]

For the PET crystallized at 210 °C a heating rate of 200 K min^{-1} is fast enough to prevent recrystallization. Much higher rates are needed for the sample crystallized at 130 °C. Even at 162 000 K min^{-1} a shoulder at the high temperature flank can be seen indicating some recrystallization on heating. Recrystallization needs only a few ms in the temperature range 160 °C to 220 °C.

To study the kinetics of reorganization we performed some isothermal experiments. To be sure that there is only negligible reorganization during the scan we first studied the melting and recrystallization behaviour of PET crystallized at different temperatures. The sample was prepared as follows: it was quenched from molten to glassy state—from 350 °C down to 30 °C—with the rate $\approx 10^4$ K s^{-1}. Then it was heated (with the same rate) up to the crystallization temperature T_C and crystallized during the crystallization time t_C. After the crystallization it was quenched again to 30 °C. Thus, the sample was ready for the experiment.

At crystallization temperatures of 114 °C and 133 °C some recrystallization occurs, see Fig. 3, curves 8 and 9, respectively. Only a slight shoulder due to reorganization during the scan is seen for

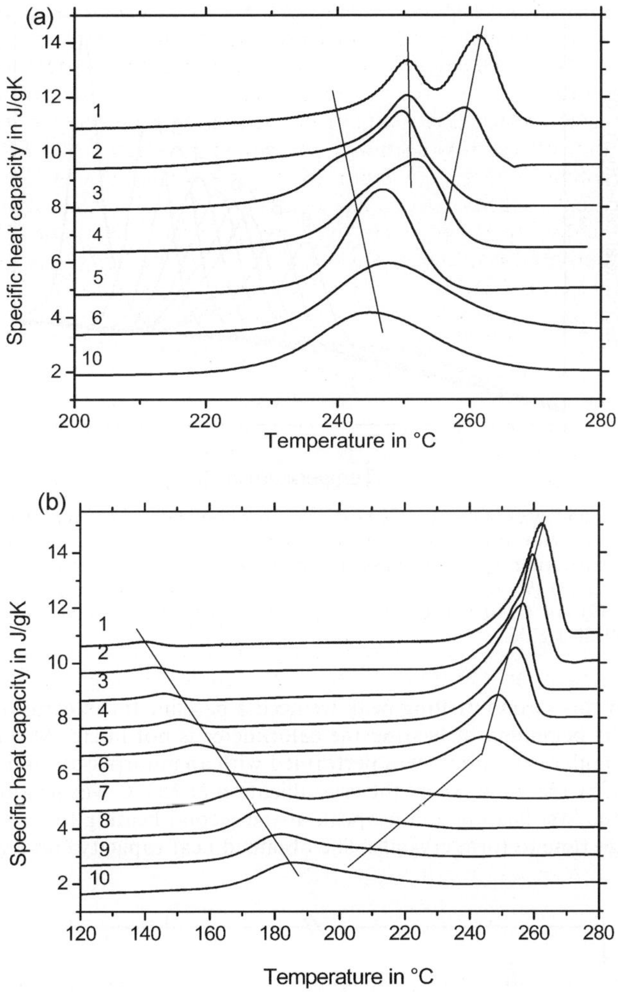

Fig. 2 Temperature dependences of the specific heat capacity of 0.0005–16 mg PET samples at the following heating rates: (1) 2 K min^{-1} (0.033 K s^{-1}) 16 mg, (2) 5 K min^{-1} (0.083 K s^{-1}) 16 mg, (3) 20 K min^{-1} (0.33 K s^{-1}) 6 mg, (4) 50 K min^{-1} (0.83 K s^{-1}) 6 mg, (5) 200 K min^{-1} (3.3 K s^{-1}) 0.5 mg, (6) 500 K min^{-1} (8.3 K s^{-1}) 0.5 mg, (7) 9000 K min^{-1} (150 K s^{-1}), 480 ng, (8) 33 000 K min^{-1} (550 K s^{-1}) 480 ng, (9) 66 000 K min^{-1} (1100 K s^{-1}) 480 ng, (10) 162 000 K min^{-1} (2700 K s^{-1}) 480 ng. The samples were crystallized at 210 °C (a) and 130 °C (b) for 1 h. The straight lines are guides to the eyes only.

the sample crystallized at 153 °C, curve 7. Therefore the crystallization temperature 153 °C, which yields a heat of fusion of 42 J g^{-1}, was chosen for further experiments. The temperature time profile we used for the study of recrystallization kinetics is shown in Fig. 4.

The sample was first cooled from the melt at $\approx$ 350 °C to the crystallization temperature of 153 °C and crystallized for 15 min. Then the sample was cooled at 10 000 K s^{-1} at 30 °C. The cold crystallized sample, T_C = 153 °C, t_C = 15 min, was heated at 2700 K s^{-1} to 223 °C. At that temperature most of the original formed crystals are molten, compare Fig. 3 curve 7 and the first peak in Fig. 4. To achieve a fast change from scan to isothermal conditions the power profile shown in the inset of Fig. 4 was used. The small power spike at the end of the heating scan results in an abrupt change from scanning to isothermal conditions. After the sample was partially melted, the peak area equals *ca.* 32 J g^{-1} compared to 42 J g^{-1} at continuous scan, it was allowed to recrystallize for a certain time at 223 °C. At further heating to the isotropic melt at about 350 °C at 2700 K s^{-1} a second melting peak due to recrystallization was observed, see Fig. 4 at about 2.95 s. To determine

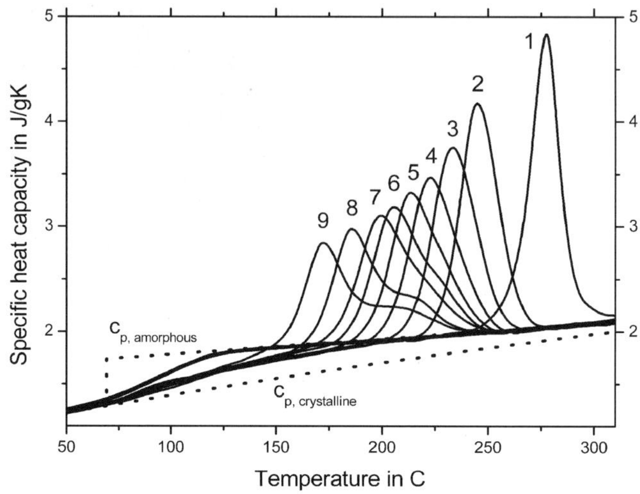

Fig. 3 Temperature dependences of the specific heat capacity of poly(ethylene terephthalate) sample, $\approx$ 480 ng, measured by fast scanning calorimetry at heating and cooling rates 2700 K s^{-1} after fast, $\approx 10^4$ K s^{-1}, cooling from crystallization temperature T_C below glass transition. The sample was crystallized at (1) $T_C = 230$ °C, $t_C = 10$ h; (2) 207 °C, 1 h; (3) 196 °C, 20 min; (4) 185 °C, 5 min; (5) 174 °C, 5 min; (6) 163 °C, 5 min; (7) 153 °C, 15 min; (8) 133 °C, 1 h; (9) 114 °C, 10 h. The heat of fusion, ΔH_f, calculated from the area under the melting peak of the curves equals 40–50 J g^{-1}.

the heat of fusion of this second melting peak we need a baseline for integration. Because the peak appears just after the beginning of heating the calorimeter is not in steady state yet. To obtain a reliable base line a similar experiment was performed with an amorphous sample. The temperature time profile was exactly the same except the crystallization at 153 °C was missing, see Fig. 4 dashed curve. No peak at the first heating and no peak at the second heating too were observed because there was not enough time to form crystals. The obtained heat capacity curve was then used as the

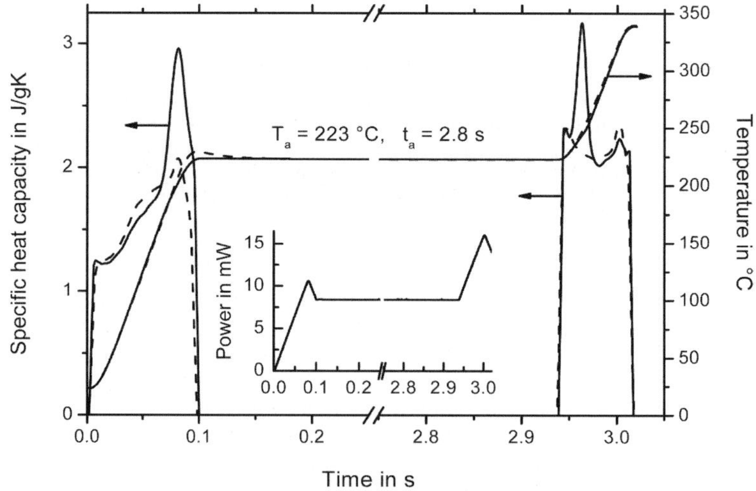

Fig. 4 Time dependences of the specific heat capacity $c(t)$ and the temperature $T(t)$ at scanning rate 2700 K s^{-1} of the amorphous sample—dashed lines and of the sample crystallized at $T_C = 153$ °C during $t_C = 15$ min—solid lines (the sample was the same as in Fig. 2). The amorphous and crystalline samples were exposed at 223 °C for 2.8 s. After the first melting peak of the crystallized sample and further recrystallization at 223 °C the next melting peak was observed. No peak was observed for the amorphous sample after annealing at 223 °C for 2.8 s. Time dependence of the heating power $P(t)$ is shown in the insert.

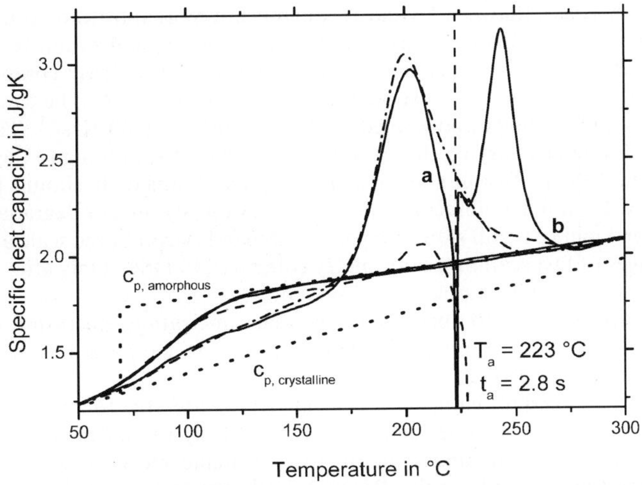

Fig. 5 Temperature dependences of the specific heat capacity $c(T)$ corresponding to the time dependences $c(t)$ in Fig. 4—dashed and solid lines, as well as $c(T)$ in the normal scan for the crystallized sample—dot-dashed line. The dashed line can be used as a base line for integration of the second peak.

base line for peak integration. To illustrate the behaviour the curves of Fig. 4 are shown in Fig. 5 as a function of temperature instead of function of time as in Fig. 4.

After 2.8 s at 223 °C the second melting peak has an area of *ca.* 17 J g^{-1} compared to 32 J g^{-1} of the first one. This indicates significant reorganization of the material within 2.8 s. Next, the time for reorganization at 223 °C was varied to study the kinetics of isothermal recrystallization. Similar experiments were performed for poly(butylene terephthalate) (PBT) by Robertson[10] utilizing a PerkinElmer DSC 4. The advantage of our method (thin-film calorimeter) is the much higher scanning rate preventing reorganization on heating to the recrystallization temperature. Robertson was therefore limited to high crystallization temperatures and only a very narrow time and temperature range for reorganization was accessible. In contrast, we were able to study recrystallization of PET at 223 °C in the time range from 10 ms to 2000 s. The heat of fusion for the second melting peak after recrystallization is shown in Fig. 6.

Recrystallization is indeed a fast process. Already after 15 ms a well developed peak is observed resulting in a heat of fusion of about 2 J g^{-1}. Until 1 s a slight increase of the heat of fusion due to

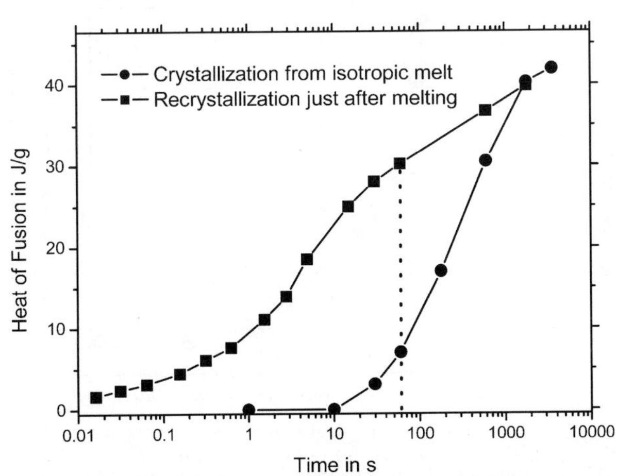

Fig. 6 Heat of fusion *vs.* time: circles–for normal crystallization from amorphous state at 223 °C and squares–for recrystallization at 223 °C just after partial melting of the sample crystallized at 153 °C during 15 min (the temperature program shown in Fig. 3).

recrystallization is observed. Between 1 s and 10 s significant reorganization occurs yielding a step in the curve. At longer times a slight increase due to secondary crystallization is seen. The observed curve shape is somehow similar to a crystallization isotherm. For a direct comparison points from isothermal crystallization from the isotropic melt are added to Fig. 6. The sample was heated to $\approx$ 350 °C for a few milliseconds and cooled at 223 °C both at 2700 K s^{-1}. Then the isothermal crystallization was studied by varying the crystallization time and scanning to the melt at 350 °C. At the heating scan the heat of fusion shown in Fig. 6 was obtained. It should be mentioned that heating up to 350 °C for a few milliseconds does not result in significant degradation of the sample because degradation kinetics is still slow compared to the millisecond time scale applied here. This is one of the advantages of fast scanning calorimetry compared to DSC. Only after about 100 scans to 350 °C degradation of the PET sample was observed.

Crystallization starts at about 10 s because of necessary nucleation. Recrystallization is about two orders of magnitude faster than crystallization at the same temperature. A similar result was obtained by Robertson[10] for PBT.

Utilizing the thin film vacuum gauge as a calorimeter we are able to study recrystallization and isothermal crystallization at the same temperature. Obviously partial melting, as used for recrystallization, results in some remaining structures which make recrystallization much faster than crystallization from the isotropic melt. Because both processes can be studied at the same temperature we can address the question: At what temperature will the remaining structures be destroyed? This is a question about melt memory. Because we can prevent recrystallization due to the high scanning rates we can address this question for semicrystalline structures formed at relatively large super cooling rather than for structures formed at low super cooling as is common in such studies.[8]

At a time of 60 s at 223 °C we can observe a small effect due to crystallization from the isotropic melt, as well as an appreciable effect due to recrystallization after partial melting. Therefore this time was chosen for the next experiment. The experiment is as follows. The sample is crystallized at 153 °C for 15 min. After cooling at 30 °C the sample is heated to a temperature T_{max} equal or higher than 223 °C. After exposing the sample for a few milliseconds to such a temperature it is cooled at 223 °C and recrystallized for 60 s, as shown in the insert of Fig. 7. Then, the sample is heated to the melt at 350 °C, and the melting peak due to reorganization is observed. In Fig. 7 the heat of fusion for the melting peak after recrystallization at 223 °C is shown as a function of T_{max}—the temperature to which the sample was exposed prior to recrystallization.

If T_{max} equals just the recrystallization temperature 223 °C the same value of ΔH_{f} as in Fig. 6 at 60 s is measured (compare the value for recrystallization at the dashed line). The question arises:

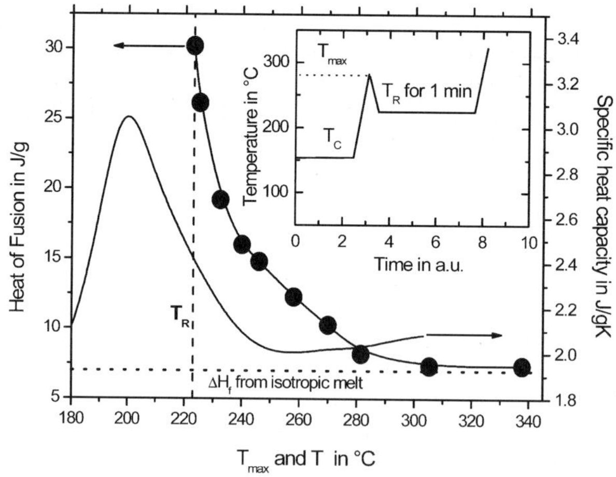

Fig. 7 Heat of fusion, ΔH_{f}, *vs.* the highest temperature, T_{max}, of the sample heating before recrystallization at 223 °C during t_{R} = 60 s. The sample of 480 ng was crystallized at 153 °C for 15 min. The scheme of the experiment is shown in the insert. The melting curve at continuous heating shows the position of the melting peak with respect to T_{max}.

what T_{max} is needed to destroy all remaining structures and to create an isotropic melt? At increasing T_{max} the melting peak and the heat of fusion decrease rapidly. But even at the end of the continuous melting peak, at about 250 °C, the observed heat of fusion is significantly higher compared to the value ΔH_f obtained from the isotropic melt, see horizontal line in Fig. 7. In fact, the end of melting may even be below 250 °C because of the fast scanning and the expected smearing of the melting curve. The value ΔH_f for crystallization from the isotropic melt, horizontal line, is reached for T_{max} larger than 305 °C, which is even higher than the published equilibrium melting temperature, 280 °C, of PET.[17]

Discussion

Recrystallization of PET is a very fast process. For samples crystallized at 110 °C it needs less than 10 ms to recrystallize part of the just molten material slightly above the crystallization temperature; see Fig. 3 and.[13] We have utilized a commercially available vacuum gauge as a fast calorimeter allowing us to prevent recrystallization of PET on heating and to study isothermal processes with millisecond time resolution.[11–13]

Our results confirm the observation by Robertson[10] that recrystallization of a partly molten polymer is about two orders of magnitude faster than crystallization from the isotropic melt, see Fig. 6. Because of the much wider dynamic range of our calorimeter, compared to the DSC used by Robertson,[10] we are able to follow isothermal crystallization as well as reorganization from a partly molten state at the same relatively low temperature. Our technique allows us to address the question of melt memory of less stable crystalline morphologies in a very direct way.

Most of the studies performed so far are limited in dynamic range, see *e.g.* the review by Rault.[8] Therefore they are limited to crystallization conditions where relatively stable morphologies are created (low super cooling). Under such conditions it is not unexpected that a melt memory also exists at relatively high temperatures. In our study we were able to study the melt memory for less stable morphologies created at much larger super cooling. PET isothermally crystallized at 153 °C melts in the temperature range between 160 °C and 250 °C. Interestingly, also these morphologies create a melt memory, which is destroyed above 305 °C only. Even 50 K above the end of the melting peak or, in other words, 15 K above the equilibrium melting temperature some order exists, which allows fast recrystallization.

Crystallization from the isotropic melt needs about 10 s for nucleation as seen from Fig. 6 (crystallization curve). The recrystallization curve is not simply shifted for these 10 s to shorter times but follows totally different, much faster, kinetics. There must be significantly more remaining order in the melt than just nuclei. We can conclude that the presence of nuclei at the beginning of recrystallization has a negligible influence on the overall position of the recrystallization curve compared to the crystallization curve. Suppose nuclei exist in the isotropic melt at the very beginning of crystallization. Then the crystallization curve should be shifted for 15 s to shorter times. Obviously, only at short times, until 100 s let's say, the crystallization curve will be shifted more than the symbol size after subtracting the 15 s needed for nucleation. There is no way to transform one curve to the other by shifting along the time axis only.

From Fig. 7 we see that there is a very smooth decrease in the observed heat of fusion due to recrystallization rather than a sharp transition to the isotropic melt. This means that the remaining structures must be localized. At increasing temperature T_{max} they are successively destroyed. In fact, at the high temperature flank of the melting peak crystalline structures are remaining, which lead to a heat of fusion observed at further heating. This is not the case at temperatures above 260 °C. The remaining structures do not show a detectable heat of fusion at higher temperatures, see the curve for continuous heating in Fig. 7. But for the recrystallization this makes only a quantitative not a qualitative difference. The structures remaining above 260 °C must be related to a small fraction of the material only but they must be very stable. Therefore these can not be small crystals or nuclei, which should be destroyed at much lower temperature. On the other hand, if these structures are large and stable crystals, they should somehow show a heat of fusion, which is also not observed. Coming back to the starting point these structures may be some mesophase precursors of polymer crystals, which are large enough to be stable at high temperature. Although this discussion is highly speculative and gives, if any, only a very weak hint to the existence of mesophase precursors, fast scanning calorimetry may be a way to obtain deeper insight into the melting recrystallization processes in polymers. This may help to better understand polymer crystallization and melting.

Conclusion

Fast scanning calorimetry utilizing a thin film vacuum gauge as a calorimeter in combination with conventional DSC covers a scanning rate range between 10^{-4} K s^{-1} and 10^4 K s^{-1}. These 8 orders of magnitude allow the study of the kinetics of different processes. Our results on melting at rates as high as 2700 K s^{-1} give clear evidence for the validity of a melting–recrystallization–remelting process for PET at low scanning rates (DSC). At isothermal melt crystallization PET forms crystals, which all melt within a few 10 K slightly above the isothermal crystallization temperature. There is no evidence for the formation of multimodal distributions of crystals with different stability (melting temperatures) at isothermal melt crystallization of PET. The single broad melting peaks at high heating rates show that broad monomodal distributions of lamellae thickness exist. This is in agreement with microscopic and X-ray scattering results by Haubruge *et al.*.[21]

Next, we will investigate the melting of polymers for which populations of lamellae with different stability are assumed as the reason for the multiple melting peaks, like isotactic polystyrene (iPS).[22]

Acknowledgements

This work was financially supported by the German Science Foundation (DFG) Grant Schi 331/7, 436 RUS 17/49/03 and PerkinElmer Inc. We acknowledge valuable discussions with M. Merzliakov, Lubbock, TX, USA, L. Allen, Urbana, IL, USA and V. Mathot, Geleen, Nl.

References

1 G. Strobl, *Eur. Phys. J. E*, 2000, **3**, 165.
2 S. Ran, Z. Wang, C. Burger, B. Chu and B. S. Hsiao, *Macromolecules*, 2002, **35**, 10 102.
3 L. Li and W. H. de Jeu, *Macromolecules*, 2003, **36**, 4862.
4 P. W. Agarwal, R. H. Somani, W. Weng, A. Mehta, L. Yang, S. Ran, L. Liu and B. S. Hsiao, *Macromolecules*, 2003, **36**, 5226.
5 A. Wurm, R. Soliman and C. Schick, *Polymer*, 2003, **44**, 7467.
6 T. A. Ezquerra, E. Lopezcabarcos, B. S. Hsiao and F. J. Baltacalleja, *Phys. Rev. E: Stat. Nonlinear Soft Matter Phys.*, 1996, **54**, 989.
7 M. Sutton, Y. S. Yang, J. Mainville, J. L. Jordansweet, K. F. Ludwig and G. B. Stephenson, *Phys. Rev. Lett.*, 1989, **62**, 288.
8 J. Rault, *CRC Crit. Rev. Solid State Mater. Sci.*, 1986, **13**, 57.
9 G. C. Alfonso and P. Scardigli, *Macromol. Symp.*, 1997, **118**, 323.
10 H. G. Kim and R. E. Robertson, *J. Polym. Sci. Part B: Polym. Phys.*, 1998, **36**, 133.
11 S. A. Adamovsky, A. A. Minakov and C. Schick, *Thermochim. Acta*, 2003, **403**, 55.
12 S. Adamovsky and C. Schick, *Thermochim. Acta*, 2004, **415**, 1.
13 A. A. Minakov, D. A. Mordvintsev and C. Schick, *Polymer*, 2004, **45**, 3755.
14 I. Hatta and A. A. Minakov, *Thermochim. Acta*, 1999, **330**, 39.
15 A. A. Minakov, *Thermochim. Acta*, 2000, **345**, 3.
16 http://www.xensor.nl/.
17 B. Wunderlich, *Pure Appl. Chem.*, 1995, **67**, 1019 and on the web: http://web.utk.edu./~athas/.
18 L. Olenka, É. N. da Silva, W. L. F. dos Santos, A. F. Rubira, E. C. Muniz, A. N. Medina, L. P. Cardoso, M. L. Baesso, L. C. M. Miranda and A. C. Bento, *Anal. Sci.*, 2001, **17**, 387.
19 M. F. J. Pijpers, V. B. F. Mathot, B. Goderis, R. Scherrenberg and E. van der Vegte, *Macromolecules*, 2002, **35**, 3601.
20 P. J. Holdsworth and A. Turner-Jones, *Polymer*, 1971, **12**, 195.
21 H. G. Haubruge, A. M. Jonas and R. Legras, *Macromolecules*, 2004, **37**, 126.
22 H. Xu, B. S. Ince and P. Cebe, *J. Polym. Sci. Part B*, 2003, **41**, 3026.

The complex phase behaviour of suspensions of goethite (α-FeOOH) nanorods in a magnetic field

Bruno J. Lemaire,[a] Patrick Davidson,*[a] Jacques Ferré,[a] Jean-Pierre Jamet,[a] Denis Petermann,[a] Pierre Panine,[b] Ivan Dozov,[c] Daniel Stoenescu[c] and Jean-Pierre Jolivet[d]

[a] *Laboratoire de Physique des Solides, UMR CNRS 8502, Bâtiment 510 Université Paris-Sud, 91405, Orsay, France*
[b] *European Synchrotron Radiation Facility, B.P.220, 38043, Grenoble, France*
[c] *Nemoptic, 1 rue Guynemer, 78114, Magny-les-Hameaux, France*
[d] *Laboratoire de Chimie de la Matière Condensée, UMR CNRS 7574, Université Paris 6, 4 Place Jussieu, 75252, Paris, France*

Received 1st March 2004, Accepted 18th May 2004
First published as an Advance Article on the web 27th September 2004

In 1902, Majorana reported the magneto-optical properties of aqueous colloidal suspensions of mixed iron oxides. Oddly enough, the magnetic-field induced birefringence displayed a non-monotonic dependence upon field intensity. This behaviour was later interpreted as due to the existence in these sols of at least two different chemical species. During the course of our studies of mineral liquid crystals, we have revisited this problem by examining aqueous suspensions of pure goethite (α-FeOOH) nanorods. Although they are comprised of a single chemical species, these suspensions show the same odd behaviour reported by Majorana. Moreover, we show that, as the volume fraction increases, the suspensions have an isotropic liquid/nematic/rectangular columnar phase sequence, with first-order transitions between these phases. The non-monotonic dependence of the field-induced birefringence can be explained by the existence of a remanent magnetic moment of the nanorods and the negative anisotropy of their magnetic susceptibility. Therefore, the nanorods align parallel to a weak field but realign perpendicular to the field beyond $B_c \approx 375$ mT. In addition, other interesting phenomena appear upon application of a magnetic field: the disordered (*i.e.* isotropic in zero-field) phase becomes highly anisotropic and difficult to distinguish from the nematic phase. Both phases then acquire not only quadrupolar order but also dipolar order. The rectangular columnar phase is strongly stabilised *versus* the nematic one. Our experimental observations raise new theoretical questions about the phase diagram of these suspensions with respect to volume fraction and magnetic field intensity.

Introduction

In a recent article, Scholten gave a historical review of the magneto-optical properties of colloidal suspensions.[1] The scientific interest in this subject dates back, at least, from Faraday who is well known for his discovery of the Faraday rotation.[2] Later, in 1902, Majorana studied the magnetic-field induced birefringence of aqueous colloidal suspensions, namely the "fer de Bravais" which was a pharmaceutical prescription at the time.[3] Strangely enough he observed in this dispersion, a non-monotonic dependence on field intensity of the birefringence that started positive, reached a maximum value, then decreased, became zero at some critical field and finally reached negative values (Fig. 1). At the beginning of the twentieth century, Cotton and Mouton reconsidered Majorana's observations using more refined equipment and slightly better control of the chemistry.[4] They realised that the fer de Bravais was a mixture of various iron oxide and hydroxide particles in suspension. They separated some of the constituents by slow sedimentation and interpreted Majorana's results as due to the superposition of a positive magnetic birefringence due to one type of particle and a negative one due to another type of particle, possibly goethite. They also properly understood that this effect is related to the orientation of anisometric particles by the field. It should be noted here that, since this time, the magnetic-field induced birefringence of colloidal suspensions has been called the "Majorana phenomenon" and that of molecular liquids the "Cotton–Mouton effect". Later, Zocher and Heller studied the properties of goethite suspensions in detail and compared them with those of common thermotropic liquid crystals.[5] During our search for new mineral liquid crystals of novel properties and mostly unaware of this exciting historical perspective, we recently undertook a detailed investigation of the influence of a magnetic field on the nematic and isotropic phases of goethite suspensions.[6] Besides, goethite (α-FeOOH) is one of the most widespread and stable iron oxides.[7,8]

We synthesised aqueous suspensions of pure goethite nanorods according to a well documented procedure.[9] A NaOH (1 M) solution is slowly added at room temperature under stirring to a $Fe(NO_3)_3$ (0.1 M) solution until pH 11 is reached. An ochre ferrihydrite precipitate forms that is left to age at room temperature for 10 days. The supernatant is then removed by centrifugation and the precipitate is washed twice with distilled water. Finally, the goethite nanorods formed are dispersed in a nitric acid solution (pH 3). Goethite particles bear hydroxyl surface groups that can be deprotonated, neutral or protonated according to pH. The point of zero charge of goethite being 9.2, the nanorods bear at pH 3 a large positive (0.2 C m^{-2}) surface charge, which gives rise to

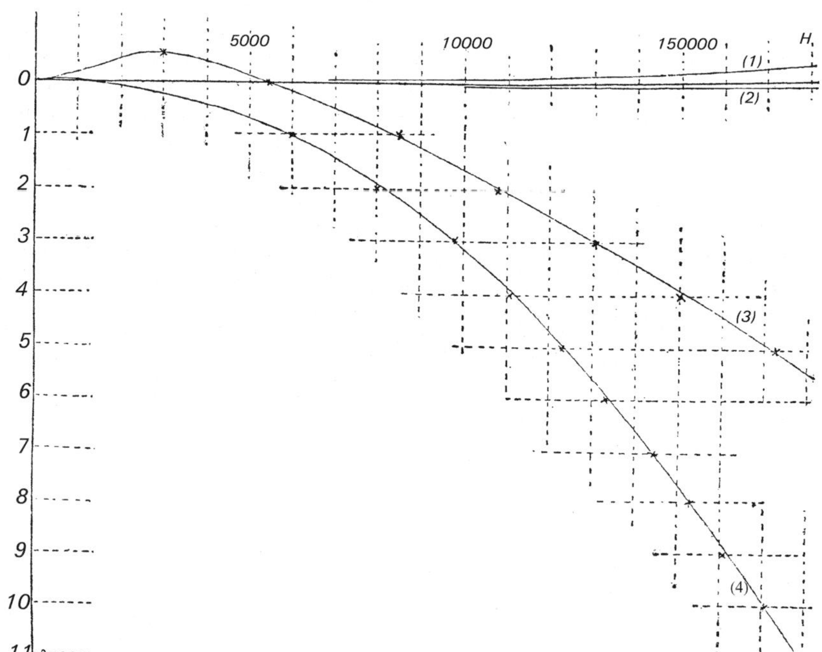

Fig. 1 Original figure published by Majorana (ref. 3b), reproduced with kind permission from the Accademia Nazionale dei Lincei. Evolution of the field-induced birefringence *versus* field intensity for 4 different types of colloidal suspensions. Note the similarity between type (3) behaviour and that of our goethite suspensions (Fig. 6a).

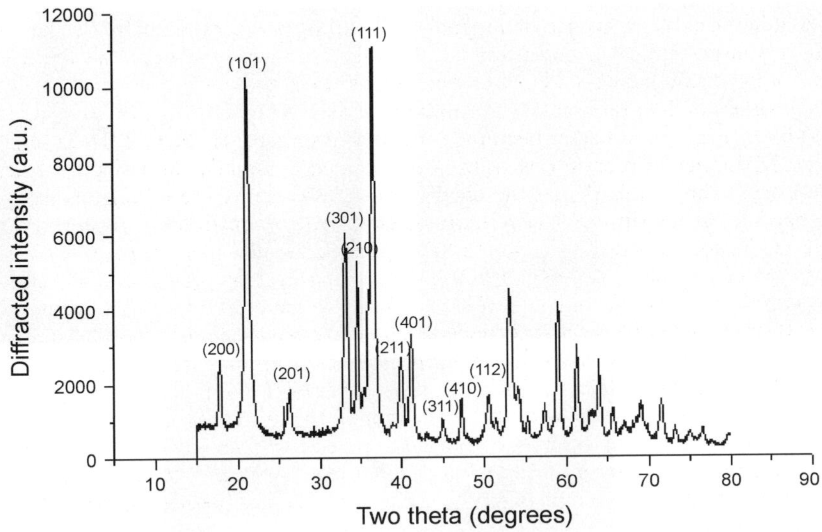

Fig. 2 Powder X-ray diffractogram of a dried goethite suspension. All diffraction lines are indexed according to the goethite space group.

solvation and electrostatic repulsions that ensure colloidal stability.[10] Moreover, goethite nanorods tend to dissolve in solutions of pH smaller than 3. Under these conditions, our goethite suspensions proved to be stable over years but their detailed properties slowly drifted over about a year. Powder X-ray diffractograms of dried suspensions only display the diffraction lines of goethite that crystallises in the *Pnma* space group with $a = 0.995$ nm, $b = 0.302$ nm and $c = 0.460$ nm. No trace of haematite was detected (Fig. 2) and the suspensions can be considered as chemically pure.

The goethite nanorod dimensions were determined by a combination of various X-ray scattering and electron microscopy techniques, as described elsewhere.[11,12] The nanorods have average length $L_b = 150$ nm, width $L_a = 25$ nm and thickness $L_c = 10$ nm. The size (length, width, thickness) polydispersities were found to be fairly large and could be well approximated by gaussian distributions of standard deviations $\Delta L_i/L_i = 0.4$. This polydispersity must be considered in order to describe quantitatively the magneto-optical properties of the suspensions. Electron microscopy also showed that goethite nanorods were mostly monocrystalline.

It is very important to note that Fe^{III} ions in the octaedric sites of goethite bear spins that are ferromagnetically coupled within chains running along the *b* crystallographic direction but these chains are antiferromagnetically coupled in the perpendicular plane. Consequently, bulk goethite is regarded as a typical antiferromagnetic material with a Néel temperature of 400 K.[13] Nevertheless, previous studies have shown that goethite nanorods bear a small remanent magnetic moment probably due to uncompensated surface spins[14] and that the Néel temperature is decreased to about 350 K.[15] Moreover, the magnetic anisotropy energy is very large so that the "spin–flop" transition does not take place even in a field as large as 5 T, well beyond those used in our experiments.

The isotropic/nematic phase transition in a magnetic field

In a suitable range of volume fractions, $0.055 < \phi < 0.085$, in zero-field, aqueous suspensions of goethite nanorods spontaneously demix into a birefringent fluid phase and an isotropic one (Fig. 3). Microscopic observations in polarised light of the birefringent phase reveals a threaded texture with topological defects of the director field called disclination lines and typical of nematic phases.[16] Small-angle X-ray scattering (SAXS) experiments were performed at the ID02 experimental station of the ESRF, Grenoble, France, with an already described setup.[17] The diffractograms of the birefringent phase only display diffuse rings, which demonstrates the absence of any positional long-range order of the nanorods[18] and therefore the nematic nature of the mesophase. The fact that goethite suspensions readily demix, within a day, into coexisting nematic (N) and isotropic (I)

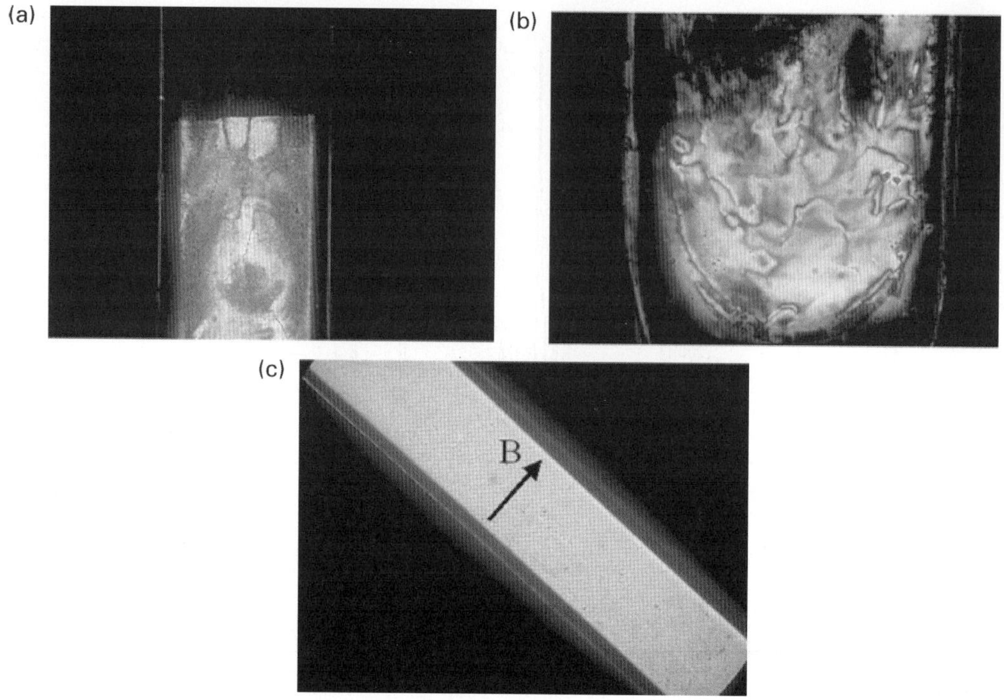

Fig. 3 Optical textures in polarised light microscopy of: (a) a biphasic sample of goethite suspension. The birefringent nematic phase (bottom) coexists with the dark isotropic phase (top). (b) The Schlieren texture of the nematic phase. (c) A nematic sample, uniformly aligned in a 100 mT magnetic field, oriented at 45° (maximum transmission). Note the disappearance of all topological defects. All samples are held in optical flat glass capillary tubes 0.5 mm wide and 50 μm thick. The polariser and analyser are parallel to the edges of the photographs.

phases proves that thermodynamic equilibrium is easily reached. In fact, this phase transition can be described, at least semi-quantitatively, in the frame of the Onsager model and related theories.[19,20]

At this stage, we need to recall the theoretical predictions of Khokhlov and Semenov about the effect of a magnetic field on the I/N phase transition of liquid-crystalline polymers.[21] Among several cases, these authors also considered rigid rod-like macromolecules. A distinction must be made, according to the sign of the coupling with the magnetic field, *i.e.* whether the particles tend to align parallel or perpendicular to the magnetic field. In a (volume fraction ϕ, magnetic field intensity B) phase diagram (Fig. 4), the isotropic/nematic coexistence is limited to a finite region that ends with a

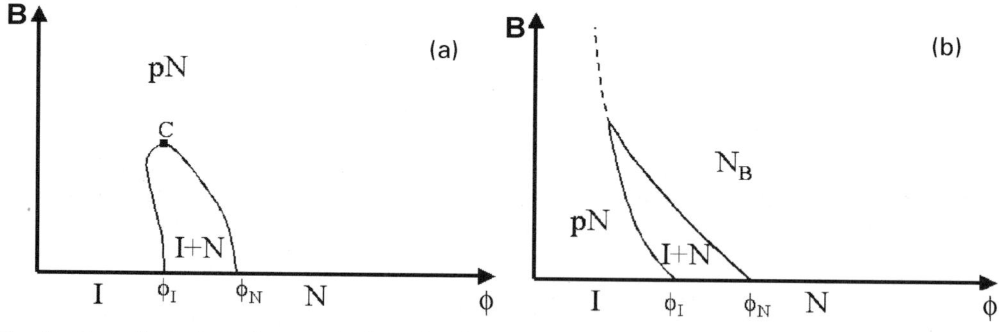

Fig. 4 Theoretical phase diagrams (volume fraction, magnetic-field intensity) predicted by Khokhlov and Semenov for hard-rod polymers that align (a) parallel to the field, (b) perpendicular to the field (adapted from ref. 21).

critical point in the case of parallel alignment. Indeed, as soon as a field is applied to the system, the disordered phase (isotropic in zero-field) becomes anisotropic and has the same uniaxial symmetry as the nematic one. It is then sometimes called a "paranematic" phase.[22] The situation is somewhat more complicated in the case of perpendicular alignment because a nematic single domain oriented perpendicular to the field loses its uniaxial symmetry and becomes biaxial. Since both phases now have different symmetries, then the isotropic/nematic coexistence region ends on a tricritical point from which a second-order transition line starts that still separates the paranematic and biaxial nematic regions. Note also that, in both cases, the coexistence region is tilted towards smaller volume fractions, which means that the ordered phase is favoured over the disordered one by application of the field, as intuitively expected. However, the magnitude of this stabilisation will depend on the microscopic details of the system. Recent experiments by van der Beek with suspensions of gibbsite plate-like particles can probably be discussed in this context.[23]

Let us now examine in detail the influence of a magnetic field on isotropic and nematic suspensions of goethite nanorods. The nematic phase is very easily aligned even by a weak magnetic field (Fig. 3c). Actually, the minimum magnetic field that can affect a single domain of the nematic phase, expressed by its Frederiks threshold[16] of 20 mT for a 20 μm thick sample, is extremely small. This is at least an order of magnitude smaller than required to align common thermotropic liquid crystals or other lyotropic nematic phases of rod-like particles such as the tobacco mosaic virus (TMV).[16,24] We also performed SAXS experiments with a sample of the nematic phase aligned in the field (Fig. 5a). The recorded SAXS pattern does not display any sharp reflection, which proves the absence of positional long-range order and the nematic character of the mesophase. Moreover, diffuse scattering spots are detected in a direction perpendicular to the magnetic field. These spots correspond, in reciprocal space, to the intersection with the Ewald sphere of a diffuse scattering torus. This torus, located in a plane perpendicular to the director, arises from the lateral interferences between nanorods and its position gives us the average distance between them (55 nm). The width of the diffuse scattering spots gives an estimate of the correlation length (250 nm) of the liquid-like positional short-range order, which is rather large for a lyotropic nematic phase of rod-like particles. From the orientation of the torus with respect to the magnetic field, we deduce that the nanorods are aligned parallel to the field at this weak field intensity. Using classical methods,[25,26] the value $S_2 = 0.95 \pm 0.05$ of the nematic order parameter (*i.e.* the second moment of the orientational distribution function, $S_2 = \frac{1}{2}\langle 3\cos^2\theta - 1\rangle$, where θ is the angle between a nanorod and the director) can be obtained from an azimuthal scan of the scattered intensity going

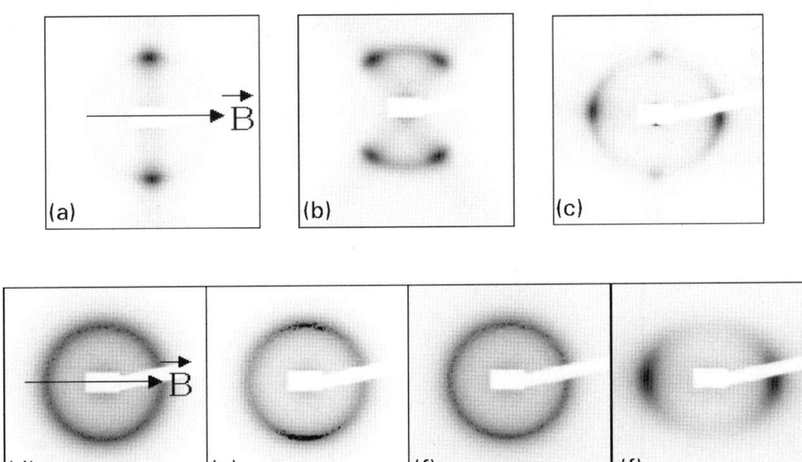

Fig. 5 Small angle X-ray scattering patterns of a nematic sample of volume fraction $\phi = 8.5\%$ submitted to a horizontal magnetic field of intensity (a) $B = 150$ mT, (b) $B = 275$ mT, (c) $B = 650$ mT. An isotropic sample of volume fraction $\phi = 5.5\%$ submitted to a horizontal magnetic field of intensity (d) $B = 0$, (e) $B = 250$ mT, (f) $B = 400$ mT, (g) $B = 1400$ mT. Note that the SAXS pattern recovers its isotropy around B_c and compare with Figs. 1 and 6a.

through the diffuse spot. This very large value of S_2 is far above the predictions of theoretical models and the values observed in other similar systems, which already suggests the existence of strong magnetic interactions with the field.

The isotropic phase shows a rather unusual behaviour when submitted to a magnetic field. When observed under a polarising microscope, it becomes strongly birefringent in a rather small magnetic field of order 0.1 T. In itself, this property is not unexpected for a paranematic phase but the magnitude of this effect is many orders of magnitude larger than in the few cases already reported in the literature. In fact, it is also detectable by SAXS, which is quite uncommon (Fig. 5d–g). The influence of the magnetic field on the isotropic phase was further examined with a previously described magneto-optical setup.[27,28] Strangely enough, as the magnetic-field intensity is increased, the birefringence of the disordered phase reaches a maximum around 250 mT, then it decreases, becomes zero around $B_c = 375$ mT (Fig. 5f) and finally becomes negative (Fig. 6a). This behaviour is precisely that described by Majorana and then by Cotton and Mouton, but we know now that our system is chemically pure and that goethite nanorods are the only colloidal species in suspension. The linear birefringence is directly proportional to S_2 that reaches positive values of about 0.05, which may not seem very large but still represents an effect at least five orders of magnitude larger than observed with TMV suspensions.[29] Moreover, SAXS patterns recorded at different field intensities prove that the nanorods simply reorient perpendicularly to the field beyond B_c (Fig. 5g). Note that, in this case, the phase could be called "antinematic" because the elongated particles tend to align *away* from a common direction. (The orientation distribution function of the phase can be represented as an oblate ellipsoid rather than the prolate one for usual nematics). Such an orientational ordering is characterised by a negative order parameter that takes values comprised between −0.5 and 0. Here, we measured values as low as −0.3, which corresponds to a very strong effect. The same reorientation transition is also observed with the nematic phase for which SAXS experiments show that a single domain, initially aligned along the magnetic field direction, tilts in the perpendicular plane at field intensities larger than 250 mT (Fig. 5c). However, the reorientation takes place in a more complicated way due to the influence of nematic anchoring on the vial surfaces

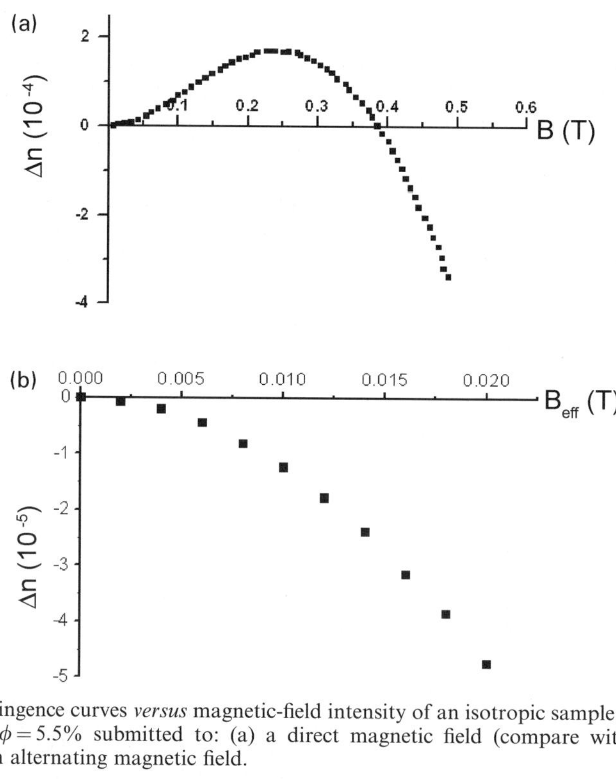

Fig. 6 Linear birefringence curves *versus* magnetic-field intensity of an isotropic sample of goethite suspension of volume fraction $\phi = 5.5\%$ submitted to: (a) a direct magnetic field (compare with curve (3) in Fig. 1 by Majorana), (b) an alternating magnetic field.

and to the slower dynamics. For instance, Fig. 5b demonstrates the occurrence of a transient zig-zag instability. In addition, experiments in an electric field also revealed a complex orientational behaviour, depending on field intensity and frequency.[11]

The fact that this reorientation transition occurs whatever the volume fraction (for $0.005 < \phi < 0.1$), even though it is best observed in the nematic phase due to collective effects, strongly suggests that it is due to the magnetic properties of individual nanorods. We have therefore performed detailed magnetisation measurements, with a SQUID device, of the isotropic and nematic phases.[11] These investigations have indeed shown that goethite nanorods bear a longitudinal remanent magnetic moment of roughly $\mu = 1000\ \mu_B$ (μ_B is the Bohr magneton) (Fig. 7). In addition, the anisotropy of magnetic susceptibility was measured: $\Delta\chi = \chi_{\parallel} - \chi_{\perp} = -3 \times 10^{-4}$. It is very important to note that $\Delta\chi$ is actually negative. Moreover, a non-zero magnetisation in the disordered phase could be measured in zero-field by freezing the suspensions below 0 °C under different field intensities. This magnetisation arises from the partial ordering of the remanent nanorod moments and is directly proportional to $S_1 = \langle\cos\theta\rangle$, the first moment of the orientational distribution function, which vanishes for symmetry reasons in the absence of field. In fact, very large values of S_1 have been measured, for instance $S_1 = 0.4$ at $B = 0.3$ T. Then, a simple model can explain qualitatively the strange observations of Majorana. Indeed, the magnetic energy of a goethite particle of volume V can be written as:

$$E_{\mathrm{m}}(\theta) = -\mu B\cos\theta - \frac{\Delta\chi V}{2\mu_0} B^2\cos^2\theta$$

The first term, linear with the magnetic field B, represents the contribution of the remanent magnetic moment; it drives the parallel orientation of the nanorods at low fields. In contrast, the second term, quadratic in B, represents the contribution of the induced magnetisation; it explains the perpendicular orientation of the nanorods at high fields. A reorientation transition will occur when the two terms are roughly equal, which predicts a reasonable value for the critical field. This interpretation is further supported by magneto-optical experiments in an alternating field. At field frequencies larger than about 100 Hz, the suspensions only display a negative field-induced birefringence, which demonstrates that they always align perpendicular to the field. This much simpler behaviour is due to the fact that the nanorod permanent moment cannot follow the field at such frequencies so that the induced magnetisation ensures perpendicular alignment at all field intensities (Fig. 6b). The magnetic energy for a single nanorod can further be incorporated in a statistical physics frame, and excluded volume effects can also be considered to describe concentrated suspensions. Provided that the polydispersity distribution is properly taken into account, this model explains quantitatively the complex behaviour of the magnetic-field induced birefringence.[11]

Nevertheless, the magnetic properties of goethite suspensions still raise many open theoretical questions. For instance, the reorientation transition largely complicates the thermodynamics

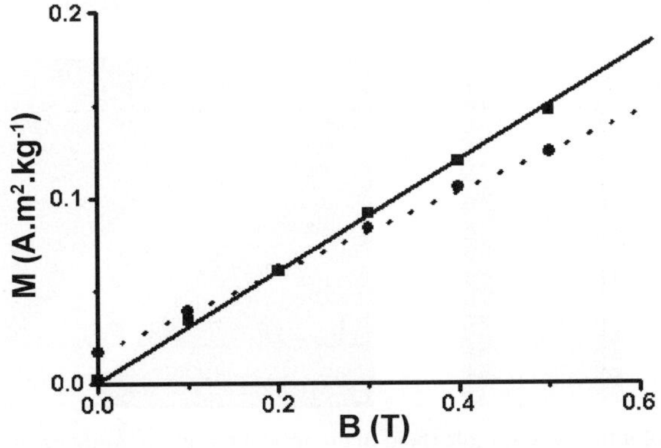

Fig. 7 SQUID measurements of the magnetisation (per kg of dried mass) *versus* field intensity of a nematic suspension ($\phi = 8.5\%$) frozen either in a 1 T field (solid squares) or in a 0.1 T field (solid circles). Straight lines are linear fits to the data whose slopes give the parallel and perpendicular magnetic susceptibilities.

because the phase diagram is now divided by a horizontal line at $B_c = 375$ mT into two regions where parallel and perpendicular orientations respectively prevail so that none of the phase diagrams shown in Fig. 4 strictly prevail. Moreover, from the tilt of the coexistence region with increasing magnetic field, one could expect that an isotropic suspension of volume fraction close enough to the transition should display the nucleation of nematic domains when submitted to a strong enough magnetic field, as was already observed for instance in similar cases of thermotropic liquid crystals and other lyotropic suspensions submitted to electric and magnetic fields.[30,31] Unfortunately, we never detected the nucleation under field of the nematic phase in the paranematic one. Nevertheless, let us now examine what happens when a magnetic field is applied to biphasic samples stored vertically (Fig. 8). In zero-field, the I/N phase coexistence is obvious in polarised light because the nematic phase is birefringent whether the isotropic one is not; moreover, the phase coexistence also shows up in natural light due to the absorption contrast between the isotropic phase and the nematic one that is denser, which results in a clear meniscus between the two. When a magnetic field is applied, as mentioned before, the isotropic phase acquires a birefringence comparable to that of the nematic phase. At strong enough fields, the meniscus between the two phases fades and finally vanishes on the time scale of a week. When examined in polarised light microscopy, the whole sample then appears uniformly birefringent and no absorption contrast can be detected any more in natural light (Fig. 8). This strongly suggests the disappearance of the coexistence region typical of the first-order character of the I/N transition beyond some value of the magnetic field, in good agreement with the theoretical predictions. Moreover, this phenomenon is fully reversible as a sharp meniscus separating the two phases reappears within hours when the magnetic field is removed. The difference in time scales between the disappearance and reappearance of the meniscus is most probably related to the influence of the magnetic field on the dynamics at a microscopic scale. Indeed, in zero-field, strong fluctuations are clearly visible in goethite

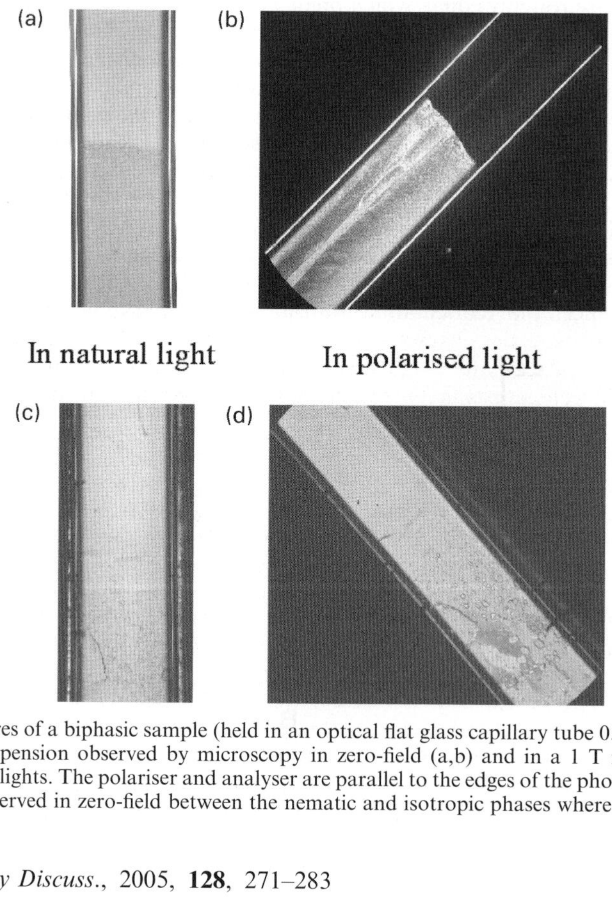

Fig. 8 Optical textures of a biphasic sample (held in an optical flat glass capillary tube 0.5 mm wide and 50 μm thick) of goethite suspension observed by microscopy in zero-field (a,b) and in a 1 T magnetic field (c,d) in natural and polarised lights. The polariser and analyser are parallel to the edges of the photographs (b,d). A clear meniscus is easily observed in zero-field between the nematic and isotropic phases whereas is has vanished in a 1 T magnetic field.

suspensions under the microscope whereas they are considerably weakened and slowed down in a strong magnetic field. (Preliminary recent X-ray photon correlation spectroscopy experiments,[32] performed at ESRF, seem to support this slowing down of the dynamics by the magnetic field.) Since the disappearance and establishment of the meniscus involve particle diffusion, it is then natural that different time-scales be involved in the presence or absence of the magnetic field.

The nematic/rectangular columnar phase transition in a magnetic field

So far, we have only discussed phases, namely the isotropic, nematic, "paranematic" and "antinematic" phases, that completely lack positional long-range order. However, numerical simulations have shown that cylindrical rod-like particles may form, at high volume fractions, a hexagonal phase in which the particles are organised on a 2-dimensional lattice.[33] The particles then form columns uncorrelated in the third dimension, which ensures the phase fluidity and distinguishes it from a regular 3-dimensional crystal. Experimental examples of lyotropic columnar phases comprised of rod-like particles are not very common. This is possibly due to the large volume fractions that need to be reached, so that the visco-elastic properties of the suspensions can hamper the formation of the columnar phase due to the occurrence of a gel phase for instance. This situation explains the somewhat conflicting results that have been reported in the literature and due care must be taken experimentally to check that samples have indeed reached thermodynamic equilibrium. It is now clear that a number of suspensions of biological rod-like macromolecules do form a hexagonal columnar phase, including DNA, and poly-L-benzyl-glutamate.[34,35] The nematic/hexagonal phase transition was found to be first order (there is evidence for a coexistence region) as intuitively expected.

It is known in the fields of common small thermotropic mesogens and of main-chain thermotropic polymers that when these molecules do not have a circular cross-section and when the density is large enough that they cannot rotate around their long axis, a columnar phase of symmetry lower than hexagonal may form.[36,37] Usually, a rectangular lattice is observed with two molecules per unit cell, which corresponds to a small distortion of a reference hexagonal lattice. The 2-dimensional rectangular lattice has two parameters a and b that often obey $b \approx a\sqrt{3}$. Since a true hexagonal lattice can also be described as a rectangular lattice with $b = a\sqrt{3}$, the quantity $(b/a - \sqrt{3})$ defines the rectangular distortion and the effective anisotropy of the particle cross-section. As mentioned in the introduction, goethite nanorods are lath-shaped with a width of 25 nm and a thickness of 10 nm, which implies that goethite suspensions are likely to form a rectangular columnar phase rather than a hexagonal one.

Indeed, in zero-field and at volume fractions larger than 15%, we observed by optical microscopy small facetted crystallites of another phase nucleating inside the nematic one (Fig. 9). The coexistence region extends up to 20% where the nematic phase completely disappears. This behaviour proves the first-order character of the transition. SAXS diffractograms of powder (*i.e.* unaligned) samples of large volume fractions, recorded at ESRF, showed very sharp, almost resolution-limited diffraction lines that prove the onset of long-range positional order (Fig. 10). The corresponding lattice spacings cannot be indexed on a hexagonal lattice and, at this stage, the production of a single domain should be very helpful in order to obtain a better insight into the structure of this phase.

Let us now examine the effect of a magnetic field on the nematic/columnar phase transition. To our great surprise and luck, applying a magnetic field of about 1 T upon a nematic sample of volume fraction 8.5% resulted in the appearance of very sharp, diffraction limited, spots in its SAXS pattern (Fig. 11). The sharpness of the spots proves the existence of long-range positional order. Moreover, the detection of diffraction spots rather than lines strongly suggests that a single domain has grown within the volume ($100 \times 100 \times 50$ μm^3) probed by the X-ray beam instead of a powder distribution of small domains. All the diffraction spots are clearly located on layer lines and could therefore be indexed with a single rectangular 2-dimensional lattice of parameters $a = 96$ nm and $b = 75$ nm. The lattice is centred and has two particles per unit cell, related by symmetry elements, which gives rise to systematic extinction conditions: $h + k = 2n + 1$ for (hk) reflections. This field-induced ordering is fully reversible as the sharp reflections completely disappear when the magnetic field is suppressed. Moreover, the production of a single domain could be repeated many times.

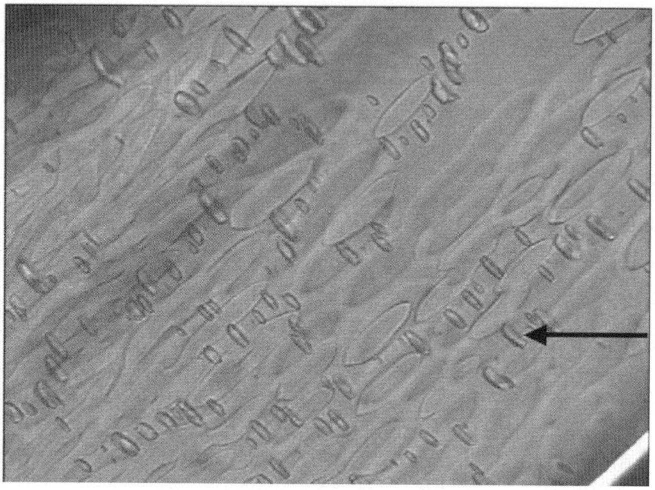

Fig. 9 Optical texture in polarised light microscopy and in zero-field of a sample of volume fraction larger than 15%. The sample is inhomogeneous with domains of the 2-dimensional rectangular columnar phase coexisting with the nematic phase. The arrow points to a columnar domain anchored at the surface of the capillary.

More precisely, when the magnetic field intensity is increased from zero, the sharp spots appear beyond some field value that depends on the sample volume fraction. The sharp spots are weak and they coexist with the diffuse ring typical of the nematic phase. As the field intensity is further increased, the sharp spots become stronger and the diffuse ones become weaker until they completely vanish beyond some field value. The superposition of the sharp and diffuse spots again illustrates the first-order character of the transition as the field intensity is varied. However, the kinetics involved in the phase transition clearly depends on the strength of the magnetic field applied. For instance, the sharp reflections are fully developed in less than a minute at $B = 1.5$ T whereas they take about an hour to appear at about 1 T. Very detailed studies of the slow kinetics could not be undertaken at ESRF where time constraints are stringent. By applying Scherrer's formula,[18] a typical domain size can be derived from the width of the reflections. Thus, it was found

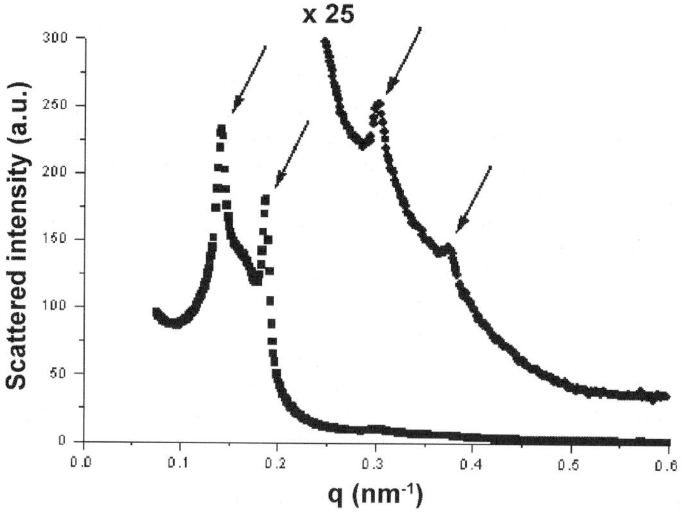

Fig. 10 SAXS powder pattern in zero-field of a sample of volume fraction 15.3% in the 2-dimensional rectangular columnar phase. The arrows point to the sharp reflections arising from the rectangular lattice.

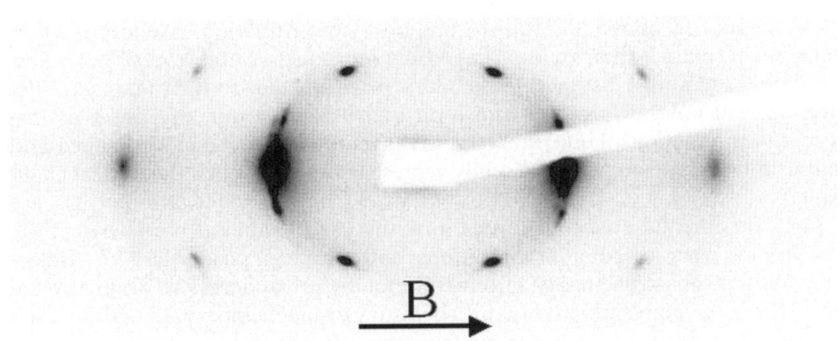

Fig. 11 SAXS pattern of a sample of goethite suspension of volume fraction 8.5% submitted to a magnetic field $B = 1.35$ T. Sharp peaks located in layer lines are due to a single domain of a rectangular columnar phase in the X-ray beam. B is the magnetic-field direction that was applied in the horizontal direction, perpendicular to the X-ray beam.

that the positional order of the 2-dimensional rectangular phase extends over about a micron. It should also be noted that the positions of the sharp reflections in the columnar phase are very close to that of the diffuse spots of the nematic phase, which means that the average distances of goethite nanorods in the plane perpendicular to their main axis are almost the same in the columnar and nematic phases. Therefore, the positional ordering is not due to flocculation under the influence of strong attractive forces.

We failed to detect any additional reflection in the third dimension, along the particle main axis direction, which proves the 2-dimensional character of the phase. Unfortunately, we did not observe either any diffuse scattering that would be the sign of a columnar periodicity of the nanorods. This is hardly surprising however because of the very large length polydispersity of the particles.

Discussion and conclusions

Let us first discuss the case of the disordered (*i.e.* isotropic in zero-field) phase. Due to the existence of the remanent magnetic moments of the nanorods, this phase acquires a very peculiar symmetry under the influence of an external magnetic field. Indeed, its symmetry is not simply quadrupolar, which is the case for usual nematics, because the mirror symmetry perpendicular to the director is lost. (Note that we use here a classical image where the magnetisation is represented as a dipolar object.) Then, the phase is characterised not only by the value of S_2 but also by that of $S_1 = \langle \cos \theta \rangle$. This results in a very original symmetry where both dipolar and quadrupolar orders are simultaneously present, with opposite signs. The dipolar contribution, new in this context, is similar to the reorientational contribution to the dielectric constant of a polar isotropic liquid. We have therefore developed a simple statistical physics model that is somewhat hybrid between those classically used for liquid crystals and ferrofluids.[11,38,39] One of the results of this model is that the nanorods can have a conical distribution and the reorientation will occur gradually as the cone angle goes from zero (parallel alignment) to $\pi/2$ (perpendicular alignment) when the field increases. There is a value of the field at which the cone angle will then be equal to the magic angle (the angle $\theta_m = 54.7°$ for which $\cos^2 \theta_m = 1/3$) and the birefringence will vanish, in agreement with experimental findings. In the nematic phase, the same combination of dipolar and quadrupolar orders prevails when the magnetic field is applied. Then, the suspensions also have features typical of both a nematic phase and a dipolar liquid, which results in complicated phenomena upon nanorod reorientation around B_c when there is a strong conflict between the dipolar and Onsager-type orders. Incorporating these kinds of considerations into the calculation of the phase diagram of these suspensions *versus* volume fraction and field intensity is a tantalising but challenging theoretical problem.

Another interesting feature of this system is the existence of the rectangular columnar phase. Even though such a phase could be expected from theoretical considerations in zero-field and in view of the lath-like shape of the nanorods, we still need to understand its microscopic organisation. Moreover, its spectacular stabilisation *versus* the nematic phase by applying a magnetic field is quite surprising. As mentioned above, the transformation of the nematic phase into a columnar phase single domain at fields larger than about 1 T is a fully reproducible and reversible phenomenon. The fact that the average nanorod distance, in the plane perpendicular to their main axis, does not vary much at the nematic/columnar phase transition seems to rule out any onset of magnetic-field induced attractive interactions. Unfortunately, we have not yet reached any detailed understanding of the stabilisation mechanism, which actually represents another open theoretical challenge raised by this system.

Finally, our experimental study illustrates how the use of mineral particles in soft condensed matter can provide very original organisations and physical properties.[40–42] Indeed, the low temperature, solution-phase chemistry of mineral moieties gives access to original building blocks of a kind out of reach of conventional organic chemistry or biochemistry. In addition, it has allowed us to revisit an old problem in the linear birefringence of colloidal suspensions.

Acknowledgements

We are deeply indebted to P.C. Scholten for his help in understanding the historical context of this subject and to L.Fruchter, H.N.W.Lekkerkerker, G.J.Vroege, P.Pincus, C.Santangelo for pleasant and stimulating discussions.

References

1 P. C. Scholten, *Indian J. Eng. Mater. Sci.*, in press.
2 M. Faraday, *Philos. Trans.*, 1846, **136**, 1.
3 (*a*) Q. Majorana, *Comptes Rendus Acad. Sci.*, 1902, **135**, 159; Q. Majorana, *Comptes Rendus Acad. Sci.*, 1902, **135**, 235; (*b*) Q. Majorana, *Rend. Accad. Lincei*, 1902, **11–1**, 374; Q. Majorana, *Rend. Accad. Lincei*, 1902, **11–1**, 463; Q. Majorana, *Rend. Accad. Lincei*, 1902, **11–1**, 531; (*c*) Q. Majorana, *Rend. Accad. Lincei*, 1902, **11–2**, 90; Q. Majorana, *Rend. Accad. Lincei*, 1902, **11–2**, 139.
4 (*a*) A. Cotton and H. Mouton, *Comptes Rendus Acad Sci*, 1905, **141**, 317; A. Cotton and H. Mouton, *Comptes Rendus Acad Sci*, 1905, **141**, 349; A. Cotton and H. Mouton, *Comptes Rendus Acad Sci*, 1906, **142**, 203; (*b*) A. Cotton and H. Mouton, *Comptes Rendus Acad Sci*, 1907, **145**, 229; (*c*) A. Cotton and H. Mouton, *Ann. Chim. Phys.*, 1907, **11**(8), 145; A. Cotton and H. Mouton, *Ann. Chim. Phys.*, 1907, **11**(8), 289.
5 W. Heller and H. Zocher, *Z. Phys. Chem.*, 1933, **164**, 365.
6 B. J. Lemaire, P. Davidson, J. Ferré, J. P. Jamet, P. Panine, I. Dozov and J. P. Jolivet, *Phys. Rev. Lett.*, 2002, **88**, 125 507.
7 R. M. Cornell and U. Schwertmann, *The Iron Oxides*, VCH, Weinheim, Germany, 1996.
8 J. P. Jolivet, *Metal oxide chemistry and synthesis. From solution to solid state*, Wiley, Chichester, UK, 2000.
9 J. Hernandez, PhD Thesis, University of Paris VI, 1998.
10 J. Israelachvili, *Intermolecular and Surface Forces*, 2nd edn., Academic Press, London, 1991.
11 B. J. Lemaire, P. Davidson, J. Ferré, J. P. Jamet, D. Petermann, P. Panine, I. Dozov and J. P. Jolivet, *Eur. Phys. J. E*, 2004, **13**, 291; B. J. Lemaire, P. Davidson, J. Ferré, J. P. Jamet, D. Petermann, P. Panine, I. Dozov and J. P. Jolivet, *Eur. Phys. J. E*, 2004, **13**, 308.
12 B. J. Lemaire, PhD Thesis, Université Paris-Sud, #6905, 2002.
13 J. M. D. Coey, A. Barry, J.-M. Brotto, H. Rakoto, S. Brennan, W. N. Mussel, A. Collomb and D. Fruchart, *J. Phys.: Condens. Matter*, 1995, **7**, 759.
14 L. Néel, *J. Phys. Soc. Jpn.*, 1962, **17**(suppl. B-I), 676.
15 S. Bocquet, R. J. Pollard and J. D. Cashion, *Phys. Rev. B*, 1992, **46**, 11 657.
16 P. G. de Gennes and J. Prost, *The Physics of Liquid Crystals*, 2nd edn., Clarendon, Oxford, UK, 1993.
17 T. Narayanan, O. Diat and P. Bösecke, *Nucl. Instr. Methods Phys. Res. A*, 2001, **467**, 1005.
18 A. Guinier, *X-Ray Diffraction in Crystals, Imperfect Crystals and Amorphous Bodies*, Dover, New York, USA, 1994.
19 L. Onsager, *Ann. N. Y. Acad. Sci.*, 1949, **51**, 627.
20 G. J. Vroege and H. N. W. Lekkerkerker, *Rep. Prog. Phys.*, 1992, **55**, 1241.
21 A. R. Khokhlov and A. N. Semenov, *Macromolecules*, 1982, **15**, 1272.
22 I. Lelidis and G. Durand, *Phys. Rev. E*, 1993, **48**, 3822.
23 D. Van der Beek, personal communication.
24 S. Fraden, A. J. Hurd, R. B. Meyer, M. Cahoon and D. L. D. Caspar, *J. Phys. Colloq.*, 1985, **46**, 85.
25 A. J. Leadbetter and E. K. Norris, *Mol. Phys.*, 1979, **38**, 669.
26 P. Davidson, D. Petermann and A. M. Levelut, *J. Phys. II*, 1995, **5**, 113.
27 J. Ferré and G. A. Gehring, *Rep. Prog. Phys.*, 1984, **47**, 513.
28 E. Hasmonay, E. Dubois, J.-C. Bacri, R. Perzynski, Y. L. Raikher and V. I. Stepanov, *Eur. Phys. J. B*, 1998, **5**, 859.

29 S. Fraden, G. Maret and D. L. D. Caspar, *Phys. Rev. E*, 1993, **48**, 2816.
30 (*a*) Y. Poggi and J. C. Filippini, *Phys. Rev. Lett.*, 1977, **39**, 150; (*b*) I. Lelidis, M. Nobili and G. Durand, *Phys. Rev. E*, 1993, **48**, 3818.
31 J. X. Tang and S. Fraden, *Phys. Rev. Lett.*, 1993, **71**, 3509.
32 See for instanceJ. Lal, D. Abernathy, L. Auvray, O. Diat and G. Grübel, *Eur. Phys. J. E*, 2001, **4**, 263.
33 A. Stroobants, H. N. W. Lekkerkerker and D. Frenkel, *Phys. Rev. A*, 1987, **36**, 2929.
34 F. Livolant and A. Leforestier, *Prog. Pol. Sci.*, 1996, **21**, 1115.
35 V. M. Ugaz, D. K. Cinader Jr and W. R. Burghardt, *J. Rheol.*, 1998, **42**, 379.
36 A. M. Levelut, *J. Chim. Phys. Phys.-Chim. Biol.*, 1983, **80**, 149.
37 W. R. Krigbaum, J. Watanabe and T. Ishikawa, *Macromolecules*, 1983, **16**, 1271.
38 J. P. Straley, *Mol. Cryst. Liq. Cryst.*, 1973, **22**, 333.
39 J.-C. Bacri, in *Proc. Conf. Magnetic Fluids and Powders, New Technological Materials, ECMF*, Minsk, URSS, 1991.
40 J. C. P. Gabriel and P. Davidson, *Adv. Mater.*, 2000, **12**, 9.
41 J. C. P. Gabriel and P. Davidson, *Top. Curr. Chem.*, 2003, **226**, 119.
42 J. C. P. Gabriel, F. Camerel, B. J. Lemaire, H. Desvaux, P. Davidson and P. Batail, *Nature*, 2001, **413**, 504.

Micro-convection, dissipative structure and pattern formation in polymer blend solutions under temperature gradients

Takeshi Nambu, Yuji Yamauchi, Takahiro Kushiro and Shinichi Sakurai*

Department of Polymer Science & Engineering, Kyoto Institute of Technology, Matsugasaki, Sakyo-ku, Kyoto 606-8585, Japan. E-mail: shin@ipc.kit.ac.jp; Fax: +81-75-724-7800; Tel: +81-75-724-7864

Received 1st March 2004, Accepted 19th April 2004
First published as an Advance Article on the web 19th August 2004

We report experimental results on time evolution of phase separation in a polymer blend solution under a temperature gradient. This study is aimed at making a self-organized structure by utilizing convection in a polymer blend solution comprising polystyrene (PS) and polybutadiene (PB) as solute with dioctylphthalate (DOP) as solvent. The polymer blend solution (PS/PB/DOP) with various compositions (PS/PB = 20/80, 50/50 or 80/20) was prepared with a total polymer concentration (PS + PB) at 3.0 wt.%. The solution was confined in a layer (0.1 mm thick), which was further subjected to a temperature gradient by controlling top and bottom temperatures (T_{top} and T_{bot}) independently with a 10 or 20 °C temperature difference ($T_{top} < T_{bot}$). For all compositions, micro-convection was generated due to the Soret effect, which is relevant to the fact that the temperature gradient induces a concentration gradient. An unusual phase separating process was observed as a result of dynamic coupling with convective flows. Especially, an interesting hierarchical structure was formed for a PS/PB = 50/50 solution. This structure is characterized by a doubly surrounding "sea-island". Namely, the "sea" contains large "islands" which involve tiny "islands". The large "islands" are spatially arranged and embedded in polygonal convection cells.

Introduction

Intensive studies have been made on the kinetics and dynamics of phase separation in polymer blends to understand the nature of phase demixing transition in a binary mixture of liquid[1–4] and further to reveal characters of polymers affecting the demixing transition.[5–7] Very recently, roles of the viscoelastic property of polymers have been studied intensively.[8] The studies of phase separation in polymer blends have also been conducted from the requirement of controlling and designing phase-separated structures.[6,7,9] Isothermal phase separation in a quiescent state is the simplest type of experiment and has been well studied. Then, experiments have been designed in the presence of an external field, such as a flow field.[9] Rich phenomena are found as a result of dynamic coupling of phase separation with flows. As for the temperature field, however, no explicit effort to incorporate it in the study of phase separation of polymer blends has been made so far, besides the fact that inhomogeneous temperature distribution is frequently encountered in the technological processing. Therefore, this study is aimed at making a self-organized structure by utilizing convection in a polymer blend solution comprising polystyrene (PS) and polybutadiene (PB) as solute with dioctylphthalate (DOP) as solvent.

When a temperature gradient is maintained in a fluid mixture, a concentration gradient is spontaneously induced. This is known as the Soret effect.[10–15] This induced concentration gradient generally results in a density gradient and further in convection if the concentration gradient is relevant to inversion of density along the gravitational direction.[16] Therefore, the driving force of the Soret convection is much more powerful compared to the case of the Rayleigh–Bénard convection (RBC),[17,18] to which buoyancy due to thermal expansion is relevant. In fact, it is difficult to generate the RBC pattern in a fluid layer several millimeters thick. Bearing in mind that the size of the convection pattern can be adjusted by changing the thickness of a fluid layer, it is required to reduce the layer thickness in order to generate a micro-convection pattern, of which the characteristic size of the mesh pattern is of the order 0.1 mm. However, it is not so straightforward because an unrealistic temperature difference ΔT ($=T_{bot} - T_{top}$; where T_{bot} and T_{top} stand for temperatures at the bottom and the top surfaces of the fluid, respectively) is required to induce micro-convection in such a thin layer, considering the relationship $\Delta T \sim d^{-3}$ where d stands for the layer thickness.[17,18] Thus, the RBC has necessarily a lower limit in the size of the pattern. When considering applications of convection as a tool of morphological control, the micro-convection is essential and therefore a driving force other than thermal expansion is needed. In this regard, the Soret-driven concentration (density) gradient is promising such that the micro-convection can be generated.

In our previous study,[19] we examined the Soret-driven micro-convection in a polymer solution using PS or PB as solute and DOP as solvent. The micro-convection was generated in a binary PS/DOP solution but not in a binary PB/DOP solution. For both a binary PS/DOP solution and a ternary polymer blend solution (PS/PB/DOP), it has been reported[19] that the magnitude of convective flow was oscillating with time until reaching a stationary state upon a sudden change of ΔT, so that the temporal change was expressed by a damped oscillatory function. Interestingly, it was found that the convective oscillation was more exaggerated in the blend solution than in the binary solution. Note however that the blend solution examined was dilute in polymer concentration so that no phase separation into PS-rich and PB-rich phases took place. Increasing the polymer concentration induces the phase separation and rich phenomena may be followed as a result of dynamic coupling of phase separation with the Soret-driven micro-convection. In this paper, we report experimental results of an unusual phase separation process in ternary polymer blend solutions (PS/PB/DOP) under an applied temperature gradient, comprehensively with various PS/PB compositions.

Experimental

PS and PB samples having similar molecular weights were used, which were purchased from Pressure Chemical and Scientific Polymer Product, respectively. The number average molecular weight, M_n, and index of the molecular weight distribution, M_w/M_n, were measured by size-exclusion chromatography (SEC), where M_w denotes the weight average molecular weight. Calibration of the SEC column was conducted using molecular weight standards of PS and PB. Therefore, the measured values can be considered as the absolute molecular weights. For the PS sample, $M_w = 2.00 \times 10^5$ and $M_w/M_n = 1.05$, and for the PB sample, $M_w = 2.06 \times 10^5$ and $M_w/M_n = 1.49$ were obtained. The microstructure of the PB sample was analyzed by NMR and 47.1% of *cis*-1,4-linkage, 44.5% of *trans*-1,4-linkage, and 8.4% of 1,2-linkage were found.

Predetermined amounts of the PS, PB and DOP were weighed and dissolved together in an excess amount of methylene chloride. After complete evaporation of the methylene chloride at room temperature, a PS/PB/DOP blend solution was obtained. However, it underwent phase separation at room temperature. Therefore, we paid special attention to homogenize the blend solution above its phase separation temperature before a moiety of the solution was picked up on the optical microscopic observation.

It is important to know concentration and blend composition dependences of some physical properties relevant to this work, which are density, refractive index, and solution viscosity. Density values are available from literature that are 1.05, 0.895, and 0.978 g cm^{-3} for PS, PB, and DOP, respectively. Nevertheless, we measured carefully the density of a PS/DOP solution using a

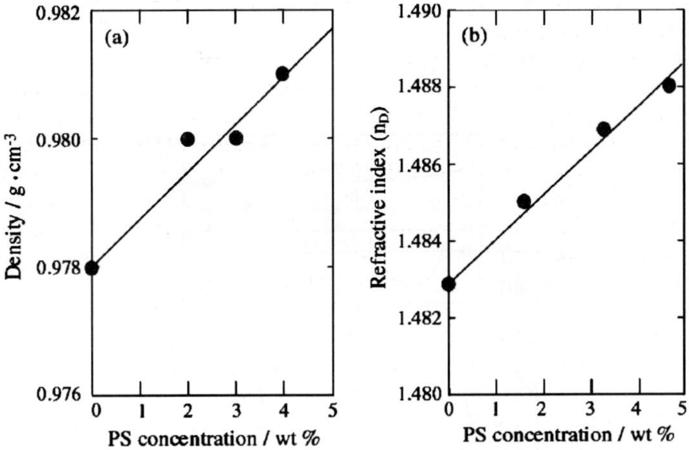

Fig. 1 Concentration dependences of (a) density and (b) refractive index for a PS/DOP solution, measured at 35 °C and 28.1 °C, respectively.

pycnometer at 35 °C, in order to know how the concentration gradient induced by the Soret effect in PS/DOP reflects density modulation (not necessarily for PB/DOP because of no appreciable Soret effect). Furthermore, the concentration dependence of refractive index for the PS/DOP solution is required to examine convection structures observed with a phase-contrast optical microscope (PCOM), which visualize structures according to modulation of the refractive index. Note here that under the PCOM, matter with lower refractive index appears brighter. An Abbe refractometer NAR-1T (Atago Co., Ltd.) was used for refractive index measurements (n_D) at 28.1 °C, although the literature values are available (PS: $n_D = 1.59$, PB: $n_D = 1.52$, DOP: $n_D = 1.48$). Fig. 1 shows the density and the refractive index. Both exhibit more likely linear dependences on the polymer concentration in a PS/DOP solution. The viscosity of the solution was measured at 50 °C using a compact rheometer MCR300 (Paar Physica). In Fig. 2, the result was plotted as a function of the PS weight fraction (w_{PS}) of the total polymer weight (PS + PB) in a PS/PB/DOP solution. The total

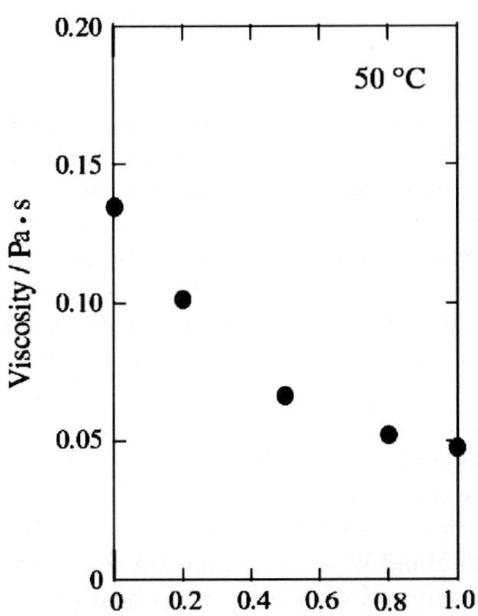

Fig. 2 Composition dependence of solution viscosity measured at 50 °C, showing the results as a function of the PS weight fraction (w_{PS}) of the total polymer weight (PS + PB) in a PS/PB/DOP solution. The total polymer concentration is 3.0 wt.%.

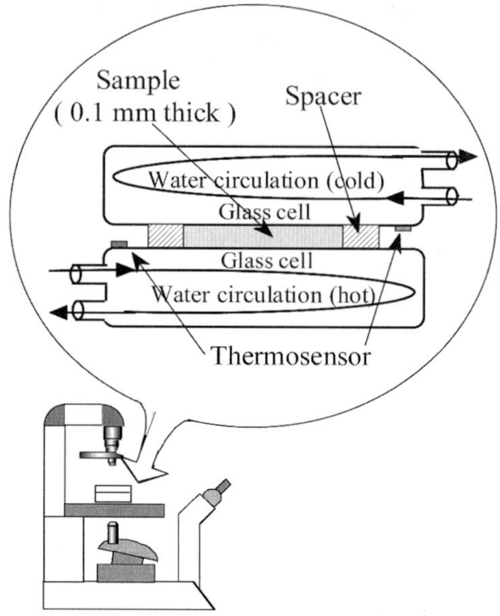

Fig. 3 Schematic view of an experimental set-up with a phase-contrast optical microscope (PCOM) for observation of a structural change in a polymer blend solution (PS/PB/DOP) under a temperature gradient. The system is composed of an inverted microscope providing a long working distance for a pair of glass cells between the sample stage and the condenser lens. The pair of glass cells sandwich the solution sample of which thickness was maintained at 0.1 mm using a spacer. In the glass cells, temperature-controlled water is circulating from individual reservoirs so as to perform individual temperature controls of top and bottom surfaces of the sample layer.

polymer concentration is 3.0 wt.%. For our particular sample, the viscosity monotonically decreases with an increase in the PS fraction.

Fig. 3 presents a schematic view of an experimental set-up with a PCOM for observation of a structural change in a polymer blend solution (PS/PB/DOP) under a temperature gradient. An inverted microscope (Nikon ECLIPSE TE200) provides a long working distance that is required to place a pair of glass cells between the sample stage and the condenser lens. The pair of glass cells sandwich the solution sample of which thickness was maintained at 0.1 mm using a spacer. In the glass cells, temperature-controlled water is circulating from individual reservoirs so as to perform individual temperature controls of top and bottom surfaces of the sample layer. This system enables ΔT as large as 30 °C within 0.3 °C tolerance for the 0.1 mm thick sample. A digital image of 1280×1000 pixel size was captured by frame grabber software (Fujix Photograb-300Z) through a CCD camera (Nikon-Fujix HC-300i). The brightness of the PCOM image corresponds to a distribution of refractive index, and the component having a larger value of the refractive index appears darker. In our particular case, the PS-rich phase appears darker. Furthermore, it has been reported in our previous work that a convection mesh appearing brighter comprises up-flows.[19]

Results and discussion

To discuss phase separation phenomena, a phase diagram is required. Since phase separation in a PS/PB/DOP blend solution was slow upon quenching shallowly a specimen from a homogeneous to an unstable state, special attention was paid to determine precisely the phase separation temperature. Instead of a conventional cloud point measurement where intensity of laser beam transmittance through a specimen is monitored with continuous heating or cooling, we conducted PCOM observations using the system shown in Fig. 3 to monitor whether phase-separated domains appear or do not appear by changing the temperature of the specimen stepwise with a 0.5 °C cooling interval. Starting from a homogeneously mixed state (after maintaining the sample at 80 °C for 2 h), a series of observations was continued until phase-separated domains appeared within 1 h at a fixed

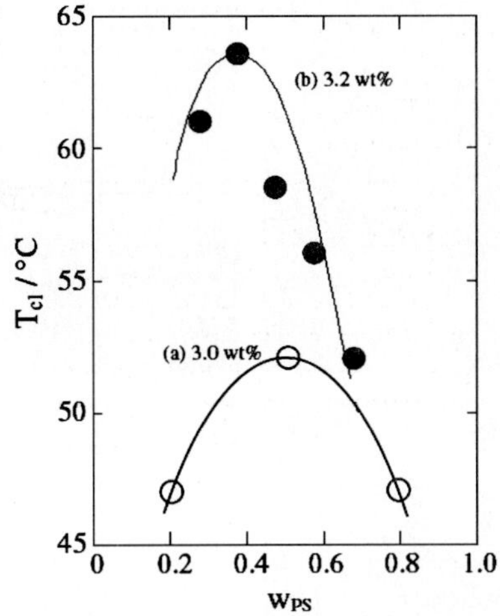

Fig. 4 Phase diagrams for PS/PB/DOP blend solutions with slightly different total polymer (PS + PB) concentrations (a) 3.0 wt.% and (b) 3.2 wt.%. The phase separation temperature (T_{cl}) was plotted as a function of the PS weight fraction (w_{PS}) of the total polymer weight (PS + PB) in a PS/PB/DOP solution.

temperature. The thus determined phase-separation temperature (T_{cl}) is presented in Fig. 4, where the total polymer (PS + PB) concentration was slightly changed in (a) 3.0 wt.% and (b) 3.2 wt.%. Although T_{cl} was sensitively affected by the trivial change in the polymer concentration, both the blend solutions exhibit the UCST (upper critical solution temperature) type phase diagrams with the UCST located around PS/PB = 50/50. Thus, we adjusted the total polymer concentration so as to locate the UCST around 50 °C for experimental convenience.

Figs. 5–7 display temporal changes in the PCOM images for the solution with 3.0 wt.% total polymer concentration (phase diagram shown in Fig. 4(a)) under $\Delta T = 10$ °C, where the mean

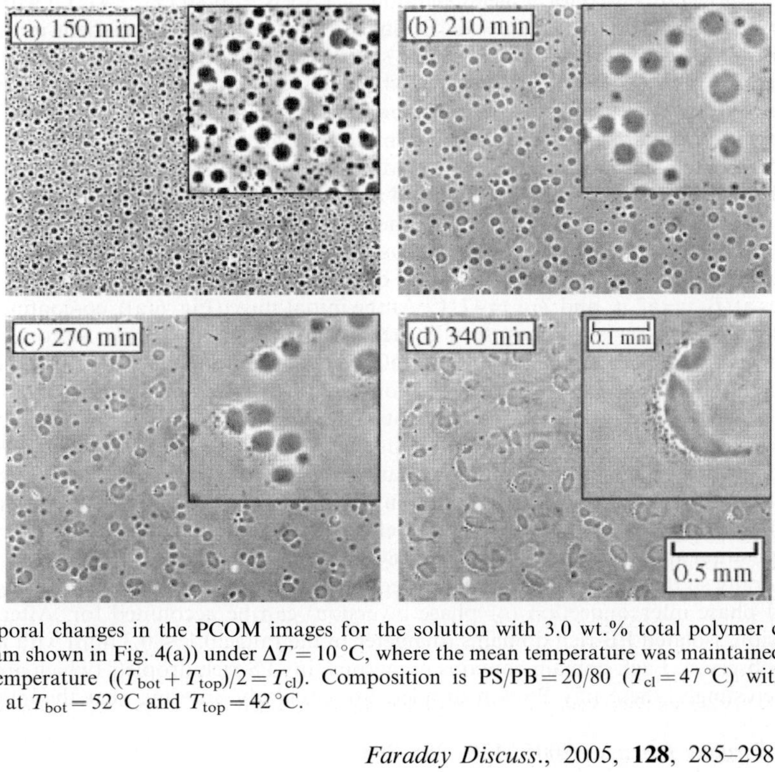

Fig. 5 Temporal changes in the PCOM images for the solution with 3.0 wt.% total polymer concentration (phase diagram shown in Fig. 4(a)) under $\Delta T = 10$ °C, where the mean temperature was maintained at the phase separation temperature ($(T_{bot} + T_{top})/2 = T_{cl}$). Composition is PS/PB = 20/80 ($T_{cl} = 47$ °C) with controlling temperatures at $T_{bot} = 52$ °C and $T_{top} = 42$ °C.

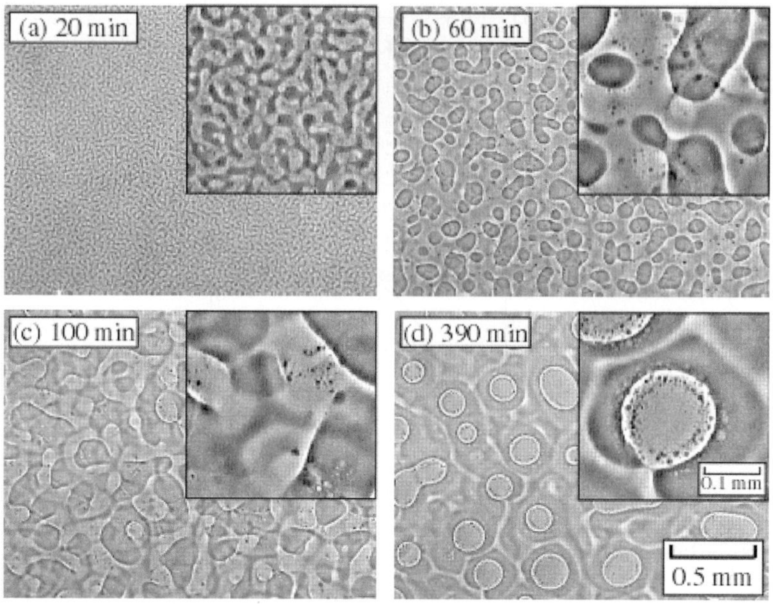

Fig. 6 Temporal changes in the PCOM images for the solution with 3.0 wt.% total polymer concentration (phase diagram shown in Fig. 4(a)) under $\Delta T = 10\,°C$, where the mean temperature was maintained at the phase separation temperature ($(T_{bot} + T_{top})/2 = T_{cl}$). Composition is PS/PB = 50/50 ($T_{cl} = 52\,°C$) with controlling temperatures at $T_{bot} = 57\,°C$ and $T_{top} = 47\,°C$.

temperature was maintained at the phase separation temperature ($(T_{bot} + T_{top})/2 = T_{cl}$). Again, note here that the PS-rich phase appears darker and that a convection mesh appearing brighter comprises up-flows.[19]

Results are shown in Fig. 5 for a blend solution of PS/PB = 20/80 ($T_{cl} = 47\,°C$) with controlling temperatures at $T_{bot} = 52\,°C$ and $T_{top} = 42\,°C$. Since PS is the minor component in the blend, the PS-rich (dark) droplets were formed in the initial stage. Although the droplets grew, their shape was not round at 270 min as shown in Fig. 5(c) where it seems that linear boundary between two merging droplets stays for a while without spontaneous annihilation. Furthermore, the PS-rich domains were not round droplets but exhibit complicated shape and more likely were being faded out in Fig. 5(d). These features in Fig. 5(c) and (d) are unusual for the isothermal phase separation process and might be ascribed to flows induced at the later stage of the phase separation in the presence of the temperature gradient. To check the effect of flows upon the phase separation process even at the initial stage, more quantitative analyses are conducted and presented later.

Fig. 6 shows results for a blend solution of PS/PB = 50/50 ($T_{cl} = 52\,°C$) with controlling temperatures at $T_{bot} = 57\,°C$ and $T_{top} = 47\,°C$. At the initial stage (Fig. 6(a)), co-continuous domain structure was observed, as is similarly seen in the isothermal phase separation process for a 50/50 blend (rigorously near the critical point). At 60 min, the continuous PS-rich phase undergoes transition to clusters, which is so-called percolation-to-cluster transition governed by mismatches in viscosity and interfacial tension of the components. Although the PS-rich domains become clusters, those clusters interconnect to recover a continuous shape at 100 min. Furthermore, it seems that the PB-rich domains undergo transition to clusters and then a phase inversion takes place. Bearing in mind that the Soret effect is most exaggerated in the PS/PB = 80/20 blend,[20] it can be considered that convections are being induced by the Soret effect in an isolated cluster of the PS-rich phase of which composition may be close to PS/PB = 80/20. The convective flows expand the PS-rich domains laterally so that two neighboring isolated domains have a chance to merge together. Thus the revisited phase interconnection (or phase inversion) can be accounted for. After a complete phase inversion, isolated PB-rich domains (diameter in the range 0.1 ~ 0.3 mm) with neatly round shape were observed. Each domain contains numerous tiny PS-rich droplets (which are as small as 10 μm). Interestingly, these tiny PS-rich droplets are circularly ordering near the interface of the

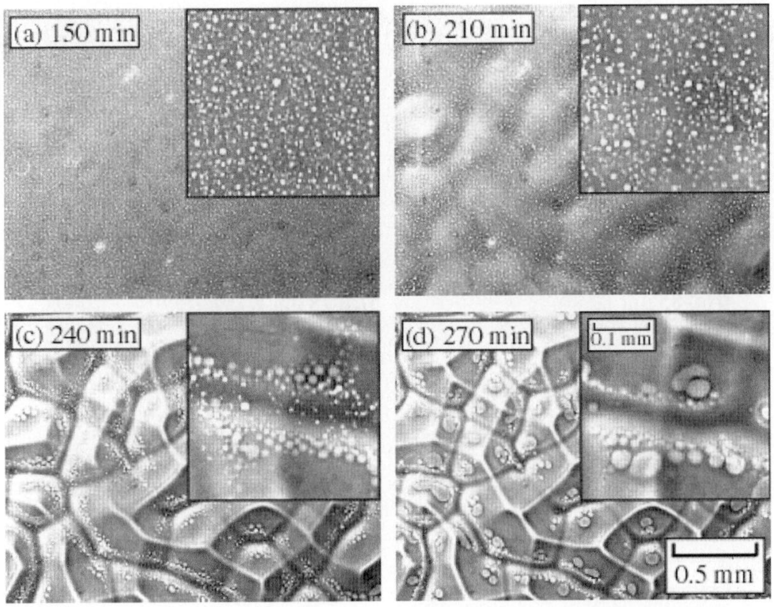

Fig. 7 Temporal changes in the PCOM images for the solution with 3.0 wt.% total polymer concentration (phase diagram shown in Fig. 4(a)) under $\Delta T = 10\ ^\circ C$, where the mean temperature was maintained at the phase separation temperature ($(T_{bot} + T_{top})/2 = T_{cl}$). Composition is PS/PB = 80/20 ($T_{cl} = 47\ ^\circ C$) with controlling temperatures at $T_{bot} = 52\ ^\circ C$ and $T_{top} = 42\ ^\circ C$.

large PB-rich domain. Moreover, numerous tiny PB-rich droplets with similar size ($\sim$10 μm) are found to be circularly ordering near the interface but outside of the large PB-rich domain. These features, which can be referred to as "doubly-surrounded sea-island structure", are quite unusual and never observed in the isothermal phase separation process. Considering that the matrix ("sea") is the PS-rich phase, the Soret-driven convections may play a crucial role for maintaining this peculiar structure. As a matter of fact, bright mesh strings can be seen in the matrix, some of which form polygonal cells surrounding the large PB-rich domains. Since these strings disappeared when the applied temperature difference was removed, these can be ascribed to convective flows, more rigorously upflows that should appear bright under the PCOM. This curious structure is long lasting as long as the temperature difference is maintained. Therefore, it is more likely a dissipative structure. However, in our experimental set up, it is impossible to obtain a side view of the structure. Three-dimensional details of the structure are to be examined using a confocal microscope.

Results are shown in Fig. 7 for a blend solution of PS/PB = 80/20 ($T_{cl} = 47\ ^\circ C$) with controlling temperatures at $T_{bot} = 52\ ^\circ C$ and $T_{top} = 42\ ^\circ C$. PB-rich droplets appearing bright were formed in Fig. 7(a) and coarsening of the droplets was confirmed in Fig. 7(b). Although this seems to be a usual coarsening process in the isothermal phase separation, unusual long-ranged fluctuations are also observed here, which become obvious as convective polygonal cells with both bright and dark meshes at 240 min. It is observed that convections organize distribution of the PB-rich droplets in Fig. 7(c) and (d). Interestingly, the PB-rich droplets are localized near the dark strings (or meshes). To examine the three-dimensional detail of the structure, the PCOM observations were performed with higher magnification. The thickness of the focal plane is thin enough so that three-dimensional reconstruction was possible from the PCOM images obtained by changing precisely the height of a focal plane. Examples are shown in Fig. 8 for the PS/PB = 80/20 solution at 360 min . Most of the PB-rich droplets were just-focused on the top plate (100 μm height), while the bright strings (or meshes) were defocused. Fig. 8(f) presents the 3d-structural model based on the images 8(a)–(e), where the lighter PB-rich droplets are floating near the top plate and they were accumulated by the convective flows near their sinking positions (downflows appearing dark in the PCOM images). Thus the PCOM observations with higher magnification were successful to present a reasonable 3d model. Again, note that the exaggerated convection patterns with both bright and dark meshes can

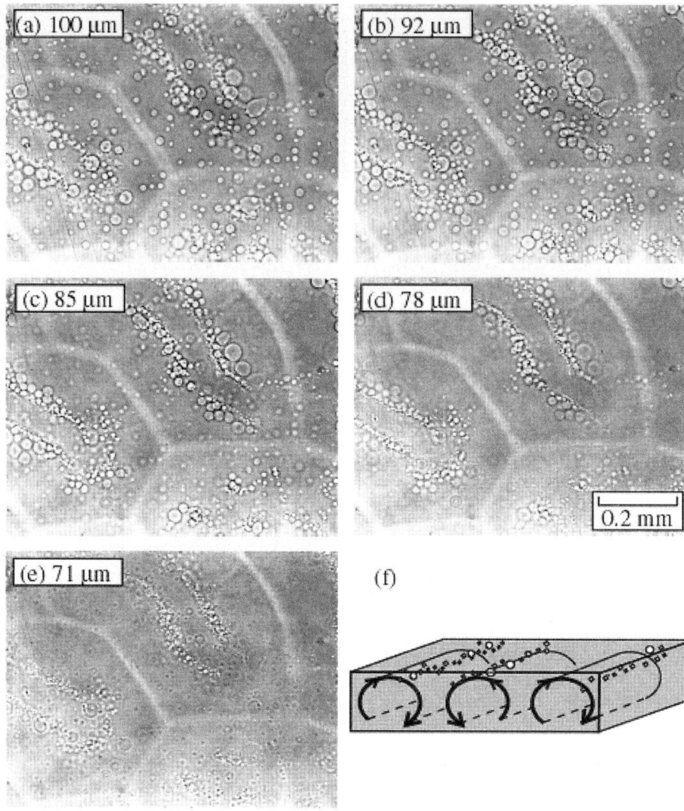

Fig. 8 (a)–(e) Changes in the PCOM images (higher magnification) by changing precisely the height of a focal plane for the PS/PB = 80/20 solution at 360 min . The focal positions were (a) 100 µm, (b) 92 µm, (c) 85 µm, (d) 78 µm, and (e) 71 µm from the bottom surface of the sample solution; (f) presents the 3d-structural model based on images (a)–(e), where the lighter PB-rich droplets are floating near the top plate and they were accumulated by the convective flows near their sinking positions (downflows appearing dark in the PCOM images).

be ascribed to the most powerful driving force of the convection due to the most evident Soret effect in the PS/PB = 80/20 blend solution among the three different compositions compared herewith.[19,20]

As discussed so far, in all blend solutions the structural developments were more or less affected by the applied temperature gradient through (convective) flows. To check the effect of flows upon the phase separation process even at the initial stage, the domain coarsening process is examined quantitatively by conducting the two-dimensional fast Fourier transformation (2d-FFT) technique. For this purpose, the software for image analyses "NIH image" was used. Examples are presented in Fig. 9 for the initial stage of the PS/PB = 20/80 solution. The insets are the resulting 2d-FFT pattern, displaying isotropic and circular ring, corresponding to the isotropic distribution of the PS-rich droplets. Circular averaging was further conducted to obtain one-dimensional FFT spectra, as shown in Fig. 10. A single broad peak is definitely observed. The peak shifts towards lower-wavenumber side, indicating the coarsening of the droplets. From the peak position, q^*, the characteristic length Λ ($=2\pi/q^*$) of the phase-separated structure was evaluated. Fig. 11 shows double logarithmic plots of Λ *vs.* time, as a function of the blend composition and the applied temperature difference (ΔT). Note here that results are plotted for the initial stage where the effect of flows did not explicitly appear in the PCOM image. Each of the data sets can be fitted with a straight line, indicating a power law growth of the domain structures, as expressed by $\Lambda \sim t^{-\alpha}$. The value of the exponent α can be rationalized with a mechanism of the domain growth.[1–4] As shown in Fig. 12, it is known that $\alpha \sim 1$ for the percolation (co-continuous domain) regime and $\alpha \sim 1/3$ for the cluster (droplet) regime. It is especially noted that hydrodynamically controlled growth, which is mediated

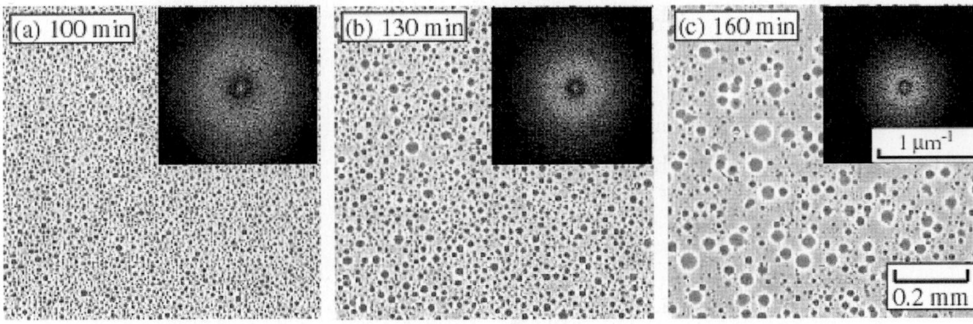

Fig. 9 Two-dimensional fast Fourier transformation (2d-FFT) of the corresponding PCOM images at the initial stage of the PS/PB = 20/80 solution. The insets are the resulting 2d-FFT power spectra, displaying an isotropic circular pattern, corresponding to the isotropic distribution of the PS-rich droplets.

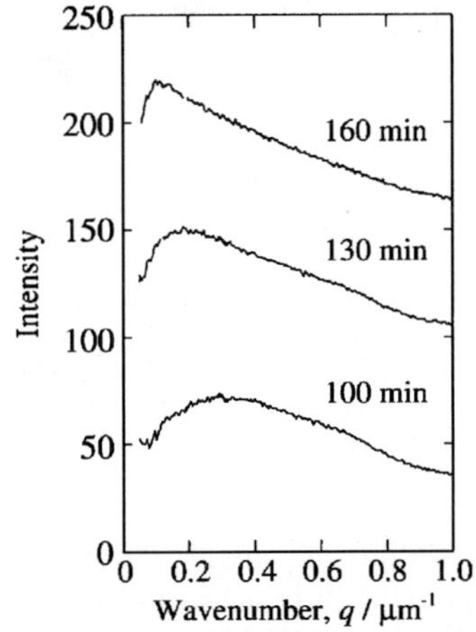

Fig. 10 One-dimensional FFT power spectra, as obtained from 2d-FFT patterns by conducting circular averaging.

by a Poiseuille flow, is relevant to $\alpha = 1$[1] and that the random (stochastic) collisions of droplets gives $\alpha = 1/3$.

As shown in Fig. 13, $\alpha \sim 1$ for $w_{PS} \leq 0.5$ and $\alpha \sim 1/3$ for $w_{PS} = 0.8$ were found. For the PS/PB = 20/80 blend solution, despite the cluster regime where the PS-rich droplets were observed, $\alpha \sim 1$ that is typical of the percolation regime was obtained. Moreover, the result is more likely pronounced for the larger ΔT. Bearing in mind that $\alpha \sim 1$ is relevant to the hydrodynamically controlled growth, it may be considered that the coarsening of the clusters under the temperature gradient is controlled by hydrodynamics. To understand this fact, we also conducted the PCOM observations with changing precisely the height of a focal plane, as shown in Fig. 14 (for the PS/PB = 20/80 solution after 35 min). As a result, the PS-rich droplets with about 10 μm size were observed in the lower position of the sample layer where the temperature is higher than the phase separation temperature, implying that no droplets should exist. Nevertheless, the PS-rich droplets were in fact observed. What does this mean? Presumably, the denser PS-rich droplets initiated near the top plate of the solution layer sink down while gradually growing. The PS-rich droplets with about 10 μm size sink down across the border into the phase dissolution region. However, once the droplets are formed, it is hard to dissolve them under a quiescent state even if the temperature is above the phase separation temperature. As a matter of fact, the PCOM observations (Fig. 14) revealed that the PS-rich droplets rather keep on growing while continuously sinking down deeper towards the bottom plate. Due to such self-sedimentation, the deterministic (or directional) collisions of the droplets take place. As schematically illustrated in Fig. 15, the directional (or

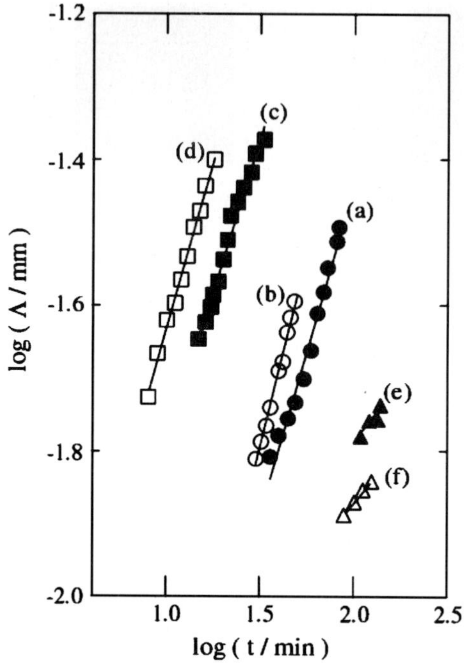

Fig. 11 Double logarithmic plots of Λ *vs.* time, as a function of the blend composition and the applied temperature difference (ΔT). Here, Λ denotes the characteristic length ($\Lambda = 2\pi/q^*$) which was evaluated from the peak position, q^*, of the 1d-FFT profile. (a) PS/PB = 20/80, ΔT = 10 °C, (b) PS/PB = 20/80, ΔT = 20 °C, (c) PS/PB = 50/50, ΔT = 10 °C, (d) PS/PB = 50/50, ΔT = 20 °C, (e) PS/PB = 80/20, ΔT = 10 °C, and (f) PS/PB = 80/20, ΔT = 20 °C.

multiple) collision of the droplets is a possible model of domain growth giving rise to $\alpha = 1$. Thus, the growth exponent $\alpha \sim 1$ for $w_{PS} \leq 0.5$ in Fig. 13 can be accounted for. For the larger ΔT, it is reasonable to consider that the effect is more pronounced. As for the PS/PB = 50/50 case, the growth exponent α is anyhow equal to 1 because this sample stayed in the percolation regime, giving rise to $\alpha = 1$ (Fig. 12(a)). On the other hand, for PS/PB = 80/20 the lighter PB-rich droplets initiated near the top plate of the solution can stay at the same position in the initial stage as long as the Soret-driven convection does not set in, so that the random (stochastic) collisions of the droplets govern the growth. Therefore, $\alpha \sim 1/3$ was obtained irrespectively for both ΔT = 10 °C and 20 °C.

Finally, the effects of convective flows upon the domain growth (which may play an important role at the reverse cluster-to-percolation transition and phase inversion, as shown in Fig. 6(b)–(c)) are discussed. For this purpose, we focused on the domain growth of the PS-rich droplets after the percolation-to-cluster transition in the PS/PB = 50/50 blend solution with 3.2 wt.% total polymer

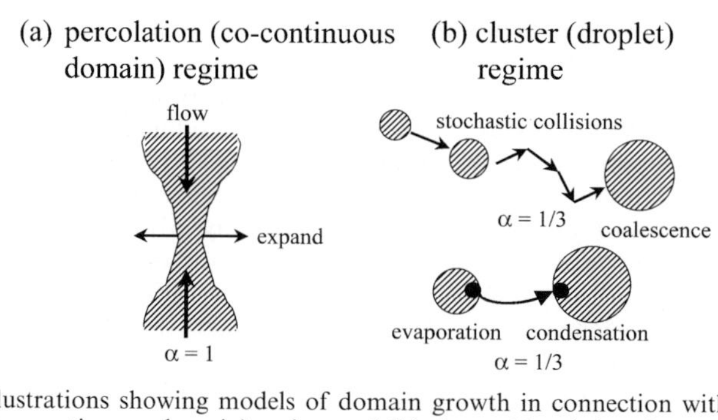

Fig. 12 Schematic illustrations showing models of domain growth in connection with the growth exponent α for (a) percolation (co-continuous domain) regime and (b) cluster (droplet) regime.

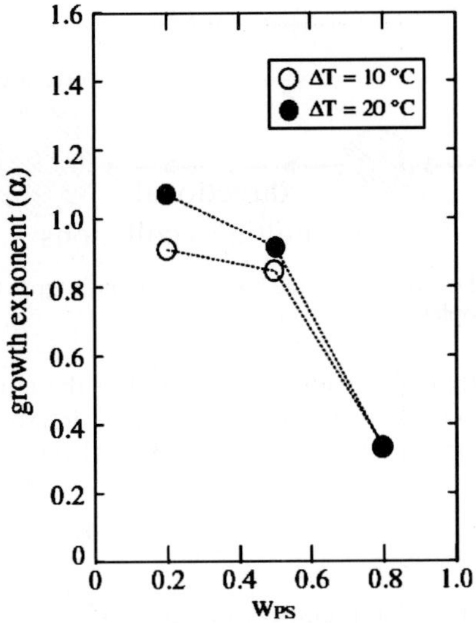

Fig. 13 Composition (w_{PS}) dependence of the growth exponent (α) for $\Delta T = 10$ or 20 °C.

concentration (phase diagram shown in Fig. 4(b)), where the effects of the convection appeared explicitly in the PCOM images. The double logarithmic plots of Λ *vs.* time, as a function of the applied temperature difference (ΔT) are shown in Fig. 16. For each data set, two straight lines with different slopes are drawn. The reason for the two-line fitting is that the growth regime is different in the earlier stage as far as the percolation (the co-continuous structure) was maintained and in the later stage after the percolation-to-cluster transition. Fig. 17 shows plots of the growth exponent α as a function of ΔT for both the percolation and the cluster regimes. The former gives $\alpha \sim 1$, irrespective of ΔT, which is again reasonable for the percolation growth regime and is consistent with the result presented for the PS/PB = 50/50 solution shown in Fig. 13. As for the cluster regime, the exponent exhibits pronounced ΔT dependence. The exponent monotonically increases with ΔT, starting from $\alpha \sim 1/3$ at $\Delta T = 0$ and approaching gradually to $\alpha \sim 1$ for $\Delta T > 20$ °C. This ΔT

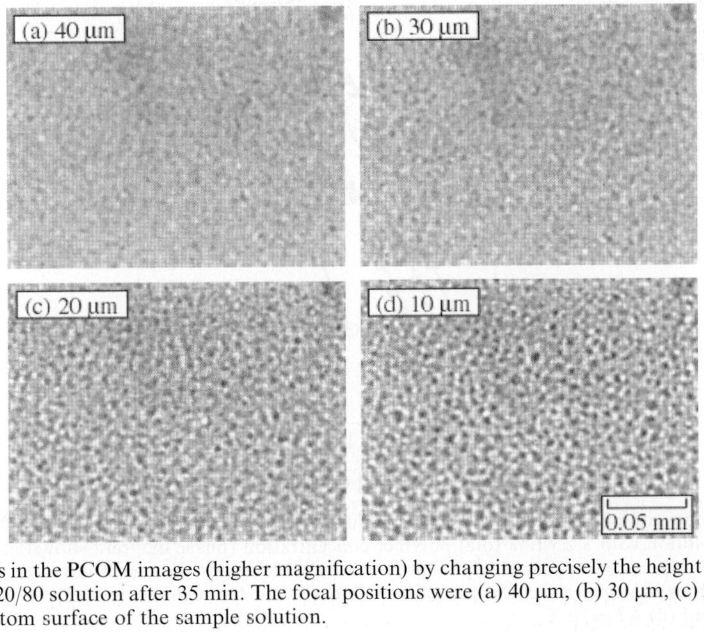

Fig. 14 Changes in the PCOM images (higher magnification) by changing precisely the height of the focal plane for the PS/PB = 20/80 solution after 35 min. The focal positions were (a) 40 μm, (b) 30 μm, (c) 20 μm, and (d) 10 μm from the bottom surface of the sample solution.

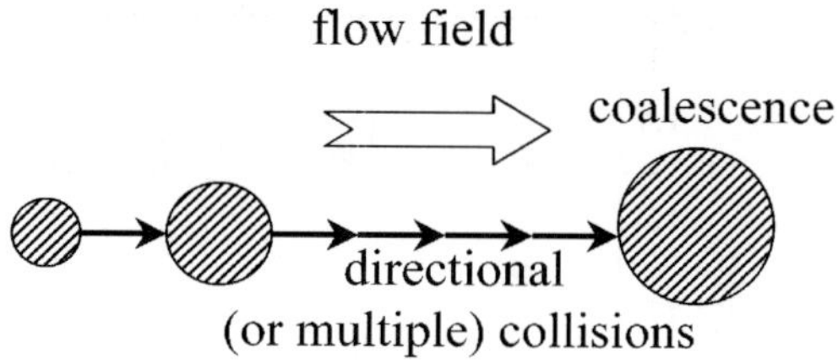

Fig. 15 Schematic illustration showing a possible model of domain growth in connection with the growth exponent $\alpha = 1$ for the cluster (droplet) regime.

dependence clearly indicates that the mechanism governing the cluster growth changes from the random (stochastic) collision to the convective-flow mediated deterministic (or directional) collision with an increase of ΔT. Thus, the result in Fig. 17 shows that the regime change is not abrupt but rather gentle.

Conclusions

We conducted phase separation under a temperature gradient for polymer blend solutions (PS/PB/DOP) to develop self-organized domain structures. For this purpose, a blend solution with various compositions (PS/PB = 20/80, 50/50 or 80/20) was prepared with a total polymer concentration (PS + PB) at 3.0 wt.%. The solution was confined in a layer (0.1 mm thick), which was further subjected to a 10 or 20 °C temperature difference ($T_{top} < T_{bot}$). By analyzing the PCOM images, self-similar domain coarsening was confirmed and the growth exponent could be evaluated. Since the growth mechanism can be rationalized with the growth exponent, effects of (convective) flows were also comprehensively discussed.

For all compositions, micro-convection was generated due to the Soret effect. An unusual phase separating process was observed as a result of dynamic coupling with convective flows. For the PS/PB = 20/80 solution, the PS-rich droplets were formed in the initial stage. Although the droplets

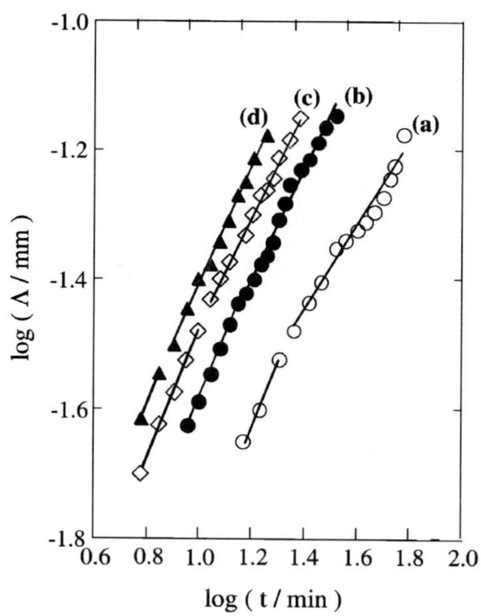

Fig. 16 Double logarithmic plots of Λ *vs.* time, as a function of the applied temperature difference (ΔT) for a PS/PB = 50/50 solution with 3.2 wt.% total polymer concentration (phase diagram shown in Fig. 4(b)), where the effects of the convection appeared explicitly in the PCOM images. For each data set, two straight lines with different slopes are drawn for the percolation and the cluster regimes. (a) $\Delta T = 5$ °C, (b) $\Delta T = 10$ °C, (c) $\Delta T = 15$ °C, and (d) $\Delta T = 20$ °C.

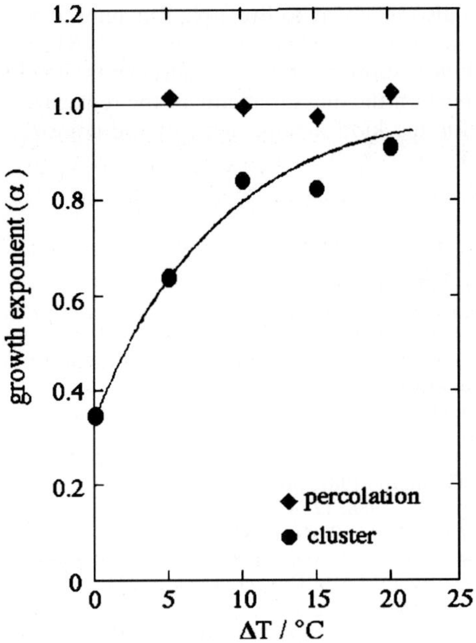

Fig. 17 Plots of the growth exponent α as a function of ΔT for both the percolation and the cluster regimes for the PS/PB = 50/50 solution with 3.2 wt.% total polymer concentration.

grow, their shape was not round and it seems that the linear boundary between two merging droplets stays for a while without spontaneous annihilation. Furthermore, the PS-rich domains were not round droplets but exhibit complicated shape and more likely were being faded out in Fig. 5(d). These features in Fig. 5(c) and (d) are unusual for the isothermal phase separation process. Although it seemed that such unusual phenomena were ascribed to convection, the PCOM observations with changing precisely the height of a focal plane revealed that the motion of sinking droplets is rather dominant. Namely, the denser PS-rich droplets being initiated near the top plate sink down. The motion of sinking droplets makes the collision of the droplets more deterministic as compared to the stochastic collision.

For the PS/PB = 50/50 solution, rich phenomena were observed, such as reverse cluster-to-percolation transition and phase inversion due to dynamic coupling of phase separation with the Soret-driven convection. It was found that the effect of convection on the domain coarsening becomes more pronounced gradually with an increase of ΔT. Especially, an interesting hierarchical structure was developed for the PS/PB = 50/50 solution. This structure is characterized by a doubly surrounding "sea-island". Namely, the "sea" contains large "islands" which involve tiny "islands". The large "islands" are spatially arranged and embedded in polygonal convection cells.

For the PS/PB = 80/20 solution, the most obvious convection pattern was observed because of the most powerful driving force of the convection due to the most evident Soret effect in the PS/PB = 80/20 blend solution among the three different compositions compared herewith.[19,20] Such powerful convections organize distribution of the PB-rich droplets so that the PB-rich droplets are localized near the dark strings (or meshes). The PCOM observations with changing precisely the height of a focal plane revealed that the lighter PB-rich droplets are floating near the top plate and they were accumulated by the convective flows near their sinking positions. For this case, however, no effect of convection on the domain coarsening was found in the initial stage.

Acknowledgements

This work was supported in part by the Grant-in-aid from the former Japanese ministry of education (09750993). The research grant from the Tokuyama Foundation is also gratefully acknowledged. The authors also indebted to Prof. N. Tsutsumi at Kyoto Institute of Technology (KIT) for his kind arrangement of the Abbe refractometer NAR-1T (Atago Co., Ltd.), Mr. K. Yokotani (KIT) for his technical support on the measurements of solution viscosity, Ms S. Kizaki (KIT) for her technical support on the measurements of the refractive index, and Mr. N. Hirano at Nihon SiberHegner K.K. for his kind arrangement of the compact rheometer MCR300 (Paar Physica).

References

1 E. D. Siggia, *Phys. Rev. A*, 1979, **20**, 595.
2 J. D. Gunton, M. san Miguel and P. S. Sahni, in *Phase Transition and Critical Phenomena*, eds. C. Domb and J. L. Lebowitz, Academic Press, London, 1983, vol. **8**.
3 H. Furukawa, in *Structure and Properties of Multiphase Polymeric Materials*, eds. T. Araki, Q. Tran-Cong and M. Shibayama, Marcel Dekker, 1998, p. 35.
4 A. Onuki, *Phase Transition Dynamics*, Cambridge University Press, 2002.
5 T. Nose, *Phase Transitions*, 1986, **8**, 245.
6 T. Hashimoto, in *Current Topics in Polymer Science-1984 Vol. 2 Polymer Physics*, ed. L.A. Utracki, Carl Hanser Verlag, West Germany, 1987, pp. 199–242.
7 T. Hashimoto, *Phase Transitions*, 1988, **12**, 47.
8 H. Tanaka, *J. Phys.: Condens. Matter*, 2000, **12**, R207.
9 A. I. Nakatani and C. C. Han, *Structure and Properties of Multiphase Polymeric Materials*, eds. T. Araki, Q. Tran-Cong and M. Shibayama, Marcel Dekker, 1998, p. 233.
10 S. R. de Groot, *L'Effet Soret*, North-Holland, Amsterdam, 1947.
11 H. J. V. Tyrrell, *Diffusion and Heat Flow in Liquids*, Butterworths, London, 1961.
12 P. Kolodner, H. Williams and C. Moe, *J. Chem, Phys.*, 1988, **88**, 6512.
13 W. Koehler, *J. Chem. Phys.*, 1993, **98**, 660; W. Koehler and B. Mueller, *J. Chem. Phys.*, 1995, **103**, 4367.
14 K. Lerman, E. Bodenschatz, D. S. Cannell and G. Ahlers, *Phys. Rev. Lett.*, 1993, **70**, 3572; K. Lerman, G. Ahlers and D. S. Cannell, *Phys. Rev. E*, 1996, **53**, R2041.
15 K. J. Zhang, M. E. Briggs, R. W. Gammon and J. V. Sengers, *J. Chem, Phys.*, 1996, **104**, 6881.
16 M. Assenheimer, B. Khaykovich and V. Steinberg, *Physica A*, 1994, **208**, 373.
17 M. C. Cross and P. C. Hohenberg, *Rev. Mod. Phys.*, 1993, **65**, 851.
18 S. Chandrasekhar, *Hydrodynamic and Hydromagnetic Stability*, Dover, New York, 1981.
19 S. Sakurai, Y.-c. Wang, T. Kushiro, T. Nambu and S. Nomura, *Chem. Phys. Lett.*, 2001, **348**, 242.
20 S. Sakurai and K. Yokotani, in preparation.

Shear-induced smectic ordering and crystallisation of isotactic polypropylene

Liangbin Li*† and Wim H. de Jeu‡

FOM-Institute for Atomic and Molecular Physics, Kruislaan 407, 1098, SJ Amsterdam, The Netherlands. E-mail: liangbin@amolf.nl

Received 23rd February 2004, Accepted 10th May 2004
First published as an Advance Article on the web 11th August 2004

Shear-induced smectic ordering and crystallisation of isotactic polypropylene (iPP) has been studied with *in-situ* small- and wide-angle X-ray scattering. Shear-induced smectic bundles with a periodicity of about 4 nm have been observed at temperatures below as well as above the melting point. This applies to iPP of different molecular weight and from different sources. An increase in the average molecular weight leads to a larger periodicity of the smectic layers. The smectic layers assemble in a fibrillar morphology with a length and a width up to 200 and 10 μm, respectively. After crystallisation, the smectic bundles show upon heating higher melting temperatures than their crystalline counterparts. In agreement with this behaviour, the correlation length along the smectic layer normal is of the order of tens of nanometers, much larger than the crystal thickness. We present an anisotropic drop model of smectic domains forming a conserved system in which the smectic layers can rotate. On the basis of this model we can explain the relative orientation of the smectic layers, the crystalline lamellae and the long axis of the drop, as well as the reversibility of the smectic periodicity during cooling and heating. In the supercooled melt, the smectic ordering is followed by crystallisation; during this process crystals grow epitaxially on the surface of the smectic bundles. This leads to a new picture of the shish-kebab structure in which smectic bundles rather than extended-chain crystals play the role of the shish. The crystallisation process of the smectic regions themselves indicates that the 'mesophase' previously reported in fast-quenched iPP, is a metastable state formed during the transition from the high-temperature smectic phase to a crystal. Moreover smectic domains rather than α crystallites form the nuclei for crystallisation of the β phase. The high-temperature smectic phase presents an ideal model system to study the coupling between density and conformational ordering under shear flow. Our results support a nucleation and growth process for polymer crystallisation, in which smectic bundles or other mesophases are the primary nuclei.

Introduction

In spite of considerable efforts by the polymer community during the past 60 years, the fundamental mechanisms of polymer crystallisation are still poorly understood.[1–6] For many years, nucleation

† Present address: Polymer Materials and Engineering, Delft University of Technology, Delft, The Netherlands.
‡ Also at the Dutch Polymer Institute, Eindhoven University of Technology, Eindhoven, The Netherlands.

and growth as a stepwise process has dominated the discussion.[4,5] In contrast to this view Strobl[7] proposed a multi-stage process to explain polymer crystallisation, while others concluded on the basis of X-ray scattering[8–16] a spinodal-assisted crystallisation process. As a result, during recent years polymer crystallisation has been subjected to an important and still open debate.

Using small- and wide-angle X-ray scattering (SAXS and WAXS), Imai *et al.*[8–11] reported density fluctuations before the onset of crystallisation of polyethylene terephthalate (PET) near the glass transition temperature. Time-resolved SAXS showed a peak indicating the growth of structures at the length scale of the semi-crystalline lamellar stacks but at times significantly before the appearance of crystal Bragg peaks in the WAXS. Other semi-rigid chain polymers, such as polyethylene naphthalate (PEN),[12] poly(ether ketone ketone) (PEKK)[17] and flexible chain polymers, such as polyethylene (PE) and polypropylene (PP), showed a similar pre-ordering process before the onset of crystallisation.[13–16] For polystyrene (sPS and iPS) conformation changes favouring the crystalline structure were reported before the onset of crystallisation[18,19] on the basis of Fourier-transformed infrared (FTIR) spectroscopy and depolarized light scattering.

The framework of early ideas by Flory can be used to interpret ordering of semi-rigid chain polymers.[20] Local parallel ordering of polymer chains can occur and be followed by an increase in the intermolecular interaction leading to crystallisation. The relative importance of these two steps is controlled by chain flexibility. If the chain flexibility is smaller than a critical value, the initial ordered state is more stable than the disordered one. Calculations for rigid-rod polymers indicate a critical concentration above which nematic ordering is more stable than an isotropic state.[21–23] However, these ideas cannot be directly applied to flexible chain polymers in which no liquid crystallinity is expected at all. Instead, Olmsted *et al.*[14] proposed that coupling between density and chain conformation induces a liquid–liquid binodal within the equilibrium liquid-crystalline solid coexistence region. This results in two liquid phases with different densities. In a certain temperature range the crystallisation of polymers is predicted to be preceded at an early stage by spinodal decomposition.

Based on similar experimental evidence, some authors suggested that nucleation and growth alone can explain polymer crystallisation. Amis *et al.* argued that SAXS data on the induction period of PE crystallisation could be attributed to polydispersity. Their results indicate simultaneous changes in the SAXS and WAXS intensities for samples with a narrow molecular-weight distribution ($M_w/M_n = 1.1$), while the SAXS peak appears before the WAXS in samples with a broad distribution.[24,25] Muthukumar *et al.*[26] proposed, on the basis of computer simulations, the concept of smectic pearls as a nucleation model. The same single chain connects many 'baby nuclei' with a size much smaller than the final lamellar thickness. The calculated structure factor is in agreement with experiments but not with the predictions of the spinodal decomposition model.

Interestingly pre-ordering was already implied in some rather early studies of polymer crystallisation, but did not get the present type of attention attached to this issue. As early as 1967 Katayama *et al.*[30] observed a SAXS peak obviously earlier than the corresponding crystal Brag peaks in WAXS. They proposed density fluctuations before the formation of any crystals. The idea of a multi-stage process dates back to 1967 by Yeh and Geil,[27,28] while Schultz introduced in 1981 a spinodal approach for an orientated system.[29] However, the essential question after the nature of pre-ordering still remained open. Moreover, the concept of 'pre-ordering before crystallisation' has not always been used in a precise way. A polymer system cannot develop such ordering because it wants to crystallise. The large enthalpy associated with crystallisation/melting prevents pre-transitional fluctuations as found for second-order or weakly first-order phase transitions. Hence pre-ordering before crystallisation in polymers is necessarily an independent effect that enables subsequent crystallisation. Such a pre-ordering can be the result of a long-living or transient metastable state.

Materials often crystallise into a metastable phase when cooled from melt. According to Ostwald's stage rule,[31] a phase transformation from one stable state to another one proceeds *via* metastable states—whenever these exist—in stages of increasing stability. Once formed, at large supercooling a metastable phase may remain stable indefinitely. However, nucleation can create a transient phase that converts into the stable form during some stage of growth. In such a case no temperature may exist at which the metastable phase remains indefinitely. Such transient states are not only intermediate forms on the path to the final state but can also be a key to understanding crystallisation behaviour.[32–37] In polymer systems so-called conformational disordered crystals frequently serve as a transient phase inducing nucleation.[38] Well-known examples are the rotator phase in paraffins[39] and the hexagonal phase in PE.[35,36] As discussed before, a liquid-crystal phase may also act to provide nuclei for the final crystalline state. Such a process still largely follows the route of nucleation and growth.

In the context of the spinodal approach,[14] external fields such as flow and shear may change the phase diagram and move the binodal temperature toward the crystallisation temperature. This may turn a metastable phase into a stable one and make it easier to detect the pre-ordering process. Moreover, in the case of shear-induced crystallisation oriented molecules serve as the precursor of primary nucleation and form the so-called shish.[39,40] After the formation of the shish, so-called kebabs grow epitaxially around it.[41] Such a two-step process potentially allows the exploration of primary nucleation and/or identifying precursors by investigating the structure of the shish.

In this work, we apply time-resolved SAXS and WAXS to study the shear-induced ordering of isotactic polypropylene (iPP). iPP is a typical semi-crystalline flexible-chain polymer. Three crystal modifications α, β and γ have been found that all show a three-fold helical conformation of the chains.[42] A mesosphase has been reported in rapidly quenched samples.[43] Though its nature is still a matter of controversy, it is generally accepted that it is metastable and readily transforms into a crystalline phase at temperatures above 80 °C.[44–49] Because flow-induced phase transitions are fundamental (but poorly understood) properties of non-equilibrium systems and flow fields are inevitably applied during polymer processing,[50] the ordering process under such conditions has attracted increasing interest. However, because no liquid-crystal phase is expected at high temperatures in a flexible-chain polymer like iPP, the focus has been on shear-induced crystallisation rather than on low-dimensional ordering.[51–61] In this work, we report that a smectic phase of iPP is induced by a step shear at temperatures above as well as below the melting point. The smectic ordering has a profound influence on any subsequent crystallisation.[62,63] We focus on the stability of the smectic phase and the relation between the smectic ordering and crystallisation. Our results show that the 'pre-ordering' before crystallisation can be liquid crystalline even in a flexible chain polymer melt and support a nucleation and growth process for the crystallisation.

Experimental

Isotactic polypropylene from three different sources was used; the molecular weight and its distribution, the melt flow index and the source are listed in Table 1. Simultaneously small- and wide-angle X-ray scattering (SAXS, WAXS) was recorded using an in-house setup with a rotating anode X-ray generator (Rigaku RU-H300, 18 kW). The system was equipped with two parabolic multi-layer mirrors (Bruker, Karlsruhe), giving a highly parallel beam (divergence about 0.012°) of monochromatic Cu Kα radiation (wavelength $\lambda = 0.1542$ nm). The SAXS intensity was collected with a two-dimensional gas-filled wire detector (Bruker Hi-Star). A semitransparent beamstop placed in front of the area detector allowed monitoring of the intensity of the direct beam. The WAXS intensity was recorded with a linear position-sensitive detector (PSD-50M, M. Braun), which could be rotated around the beam path to measure in either the meridional or the equatorial direction. The SAXS and WAXS intensities were normalized to the intensity of the direct beam. In most cases, the two-dimensional SAXS intensity was integrated azimuthally to obtain the scattering profile as a function of $q = (4\pi/\lambda)\sin\theta$, the modulus of the momentum transfer $\vec{q}$, 2θ being the scattering angle.

A Linkam CSS450 temperature-controlled shear system was employed as sample stage. The glass windows were replaced by two brass plates with apertures, covered by 50 μm thick kapton foil allowing transmission of the X-ray beam. The iPP sample was held in the gap between the two windows and sheared by a single rotation step of the bottom plate. A similar construction was

Table 1 Molecular characteristics of the iPP materials

Sample	M_n ($\times 10^5$)	M_w ($\times 10^5$)	M_w/M_n	Melt index/g (10 min)$^{-1}$	Source
iPP1	1.5	7.2	4.8	0.3	Sabic-Europe
iPP2	1.6	5.8	3.6	0.5	Aldrich
iPP3	1.0	3.7	3.7	1.1	Yanshan, China

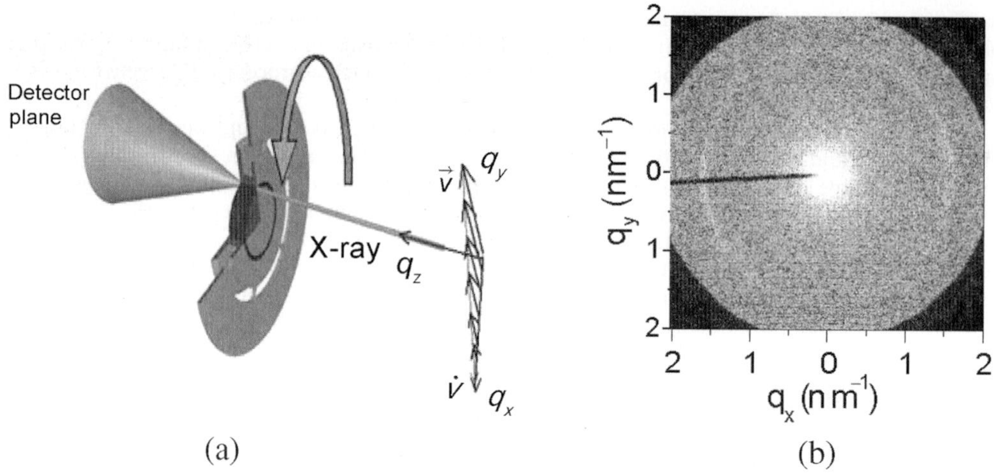

Fig. 1 (a) Reciprocal space picture of the scattering wave vector $\vec{q}$ in a shear geometry. (b) Two-dimensional SAXS pattern of iPP1 displaying smectic ordering at 210 °C after a steady shear with rate 1 s^{-1} for 3 min.

described in ref. 51. In addition to the standard two-plate shearing cell configuration, alternatively a Couette cell was incorporated. The original bottom plate of the Linkam shear cell was replaced by a brass cylinder with an inner diameter of 8 mm while the upper plate was changed into a rod with a diameter of 7 mm, both with a height of 4 mm. Setting the gap at 1 mm, all the other parameters remained the same. This allowed the X-rays to be incident parallel to $\vec{v} \times \dot{v}$ ($\vec{v}$ being the shear velocity) ensuring detection of oriented layer structures. The shear geometry in reciprocal space is sketched in Fig. 1a, where $\vec{q} = (q_x, q_y, q_z)$. The sample was first melted at 210 °C and then cooled or heated to the predetermined temperature for a shear experiment.

Infra red (IR) spectra were taken under nitrogen atmosphere with a FTIR spectrometer (Excalibur Series BIORAD, FTS3000MX) using the attenuated total reflection technique. The crystallisation behaviour in samples with thicknesses of about 100 to 200 μm was monitored with polarized optical microscopy (POM) using an OLYMPUS BX51, employing the Linkam shear system with the original glass windows. Differential scanning calorimetry was performed under nitrogen atmosphere using a Perkin-Elmer DSC7.

Results

Shear-induced smectic ordering

Fig. 1b gives a typical SAXS picture of a shear-induced smectic peak at 210 °C from iPP1. At this temperature no crystallisation is expected at all, as confirmed by the absence of any crystalline reflections in both SAXS and WAXS. Upon application of this mild shear, smectic ordering with a period around 4 nm can be observed in a wide temperature range up to 225 °C. The nature of the smectic layering was further investigated by considering the shear alignment. After a shear pulse at low temperatures, the smectic layers can be detected with the X-ray beam parallel to the direction of $\dot{v}$ (two-plate shear cell configuration),[62] but in this geometry a slight increase of the shear strain causes the smectic peak to disappear. Fig. 2 demonstrates that this is due to a change in the alignment of the smectic layers. Fig. 2a shows the weak initial smectic peak in the SAXS pattern, slightly oriented, as obtained by a shear pulse at 180 °C. After imposing a steady shear with a shear rate of 0.1 s^{-1} for 10 min, the smectic peak disappears completely (Fig. 2b). Rotating the sample over 90° aligns the X-ray beam parallel to $\vec{v} \times \dot{v}$ and a highly oriented smectic peak is visible again (Fig. 2c).[63] These alignment effects make it difficult to observe the smectic layering in a two-plate shear configuration, even though it can already be induced by pressing or any other deformation during the sample preparation.

The smectic ordering has also been observed in the other iPP samples, again both above and below the melting temperature. Fig. 3a gives typical one-dimensional SAXS curves of the three

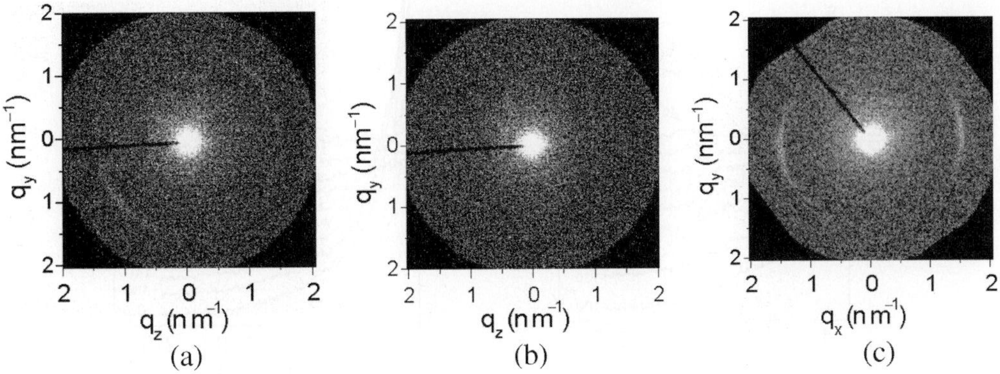

Fig. 2 (a) Two-dimensional SAXS pattern showing a weakly oriented smectic peak for iPP1 sample subjected to a shear pulse with a rate of 1 s^{-1} and a strain of 3000% at 180 °C. (b) Disappearance of the smectic peak after a steady shear of 0.1 s^{-1} for 10 min. (c) SAXS pattern with the incident X-ray beam parallel to $\vec{v} \times \dot{v}$ showing highly oriented smectic layers.

types of iPP samples at 190 °C. Fig. 3b shows the smectic spacing from the different materials *vs.* temperature. After imposing the same shear field, the smectic peak intensity of the three materials follows the relation: $I_1 : I_2 : I_3 \approx 1.6 : 1.0 : 1.2$ (iPP1 > iPP3 > iPP2). In all cases the smectic periodicity decreases with increasing temperature. Comparing results at the same temperature, the smectic periodicity is larger for iPP with a higher average molecular weight. At temperatures below 170 °C the shear-induced smectic ordering is followed by crystallisation.[62,63] In this situation both conventional crystalline lamellae with a large periodicity and crystallised smectic layers (with only an about 10% increased periodicity) are present. In the following we shall refer to the latter layering as the smectic peak, even though the underlying structure may have changed with temperature and crystallisation time.

Upon heating a crystallized sample, the smectic layers show a higher melting temperature than their crystalline counterparts. Figs. 4a and b display one-dimensional SAXS profiles of melt-sheared iPP3 (sheared above the melting temperature) during step cooling and heating, respectively. The smectic periodicity during this process is plotted *vs.* temperature in Fig. 4c. Between 150 to 180 °C, the behaviour of the smectic periodicity is reversible even though the sample has crystallised. The smectic periodicity of iPP1 during step cooling and heating after shear is plotted *vs.* temperature in Fig. 5c, again indicating a reversible process within the experimental errors. Note that during the step cooling crystallisation occurs around 160 °C, which does not result in any β crystals as indicated by the WAXS.

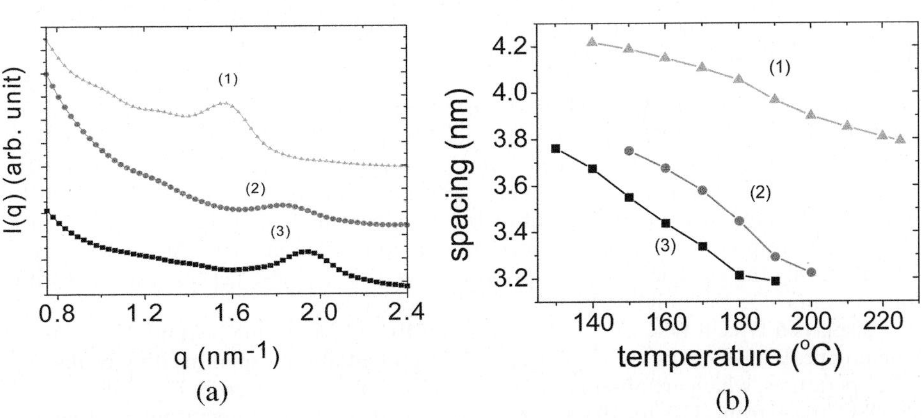

Fig. 3 (a) One-dimensional SAXS pattern of the three types of iPP after shearing at 190 °C with a rate of 1 s^{-1} for 3 min. (b) Corresponding variation of the smectic periodicity with temperature.

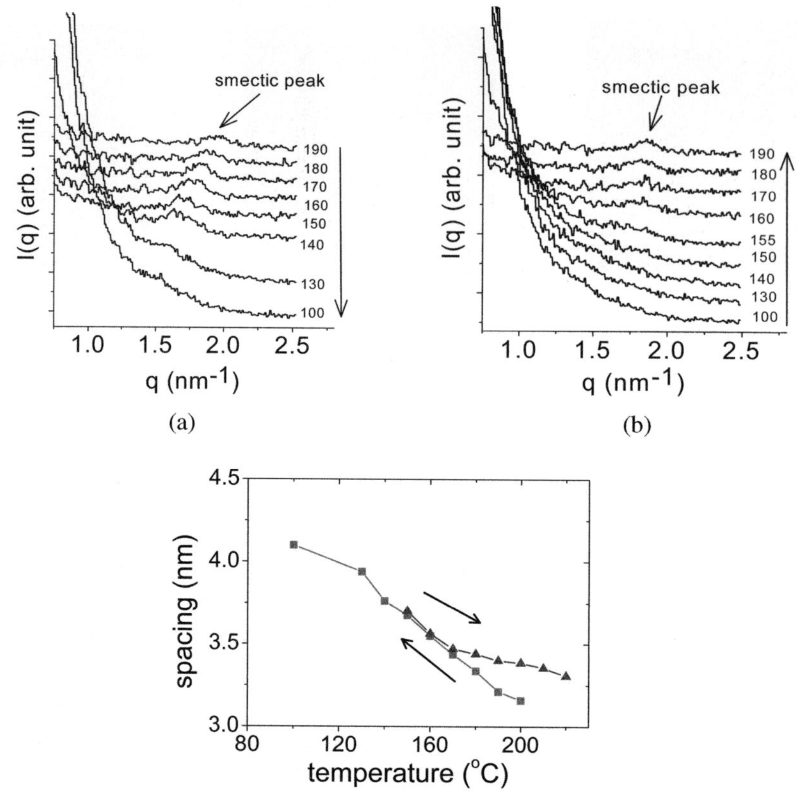

(c)

Fig. 4 One-dimensional SAXS curves of melt-sheared iPP3 during step cooling (a), during a heating scan (b), and plot of the smectic periodicity *vs.* temperature (c). The sample was originally sheared at 190 °C with a rate 1 s^{-1} for 3 min; ■ and ▲ indicate cooling and heating, respectively.

Another iPP1 sample was quenched down to 110 °C for crystallisation after being sheared at 180 °C. In addition to the α phase, β crystals were also found. After crystallisation a step-heating scan was monitored by SAXS and WAXS, see Figs. 5a and b. The crystalline peaks in both SAXS and WAXS disappear completely near 180 °C, while the smectic peak in the SAXS pattern remains up to a temperature of about 190 °C. The smectic periodicity *vs.* temperature during heating, plotted in Fig. 5c, shows a clear inflexion point around 140 °C. We attribute this discontinuous change to a transition crystal-smectic in the original smectic regions.

The iPP1 sample crystallised at 110 °C was also investigated with FTIR during a heating scan; the results are displayed in Fig. 6. As is well established for iPP, specific regularities (helix bands) can be related to a different critical length n of an isotactic sequence.[64] The minimum n-values for bands appearing at 973, 998, 841, and 1220 cm^{-1} are 5, 10, 12 and 14 monomer units in a 3/1 helical sequence, respectively. The band at 1220 cm^{-1}, corresponding to helical sequences of at least 14 monomers, is observed at temperatures up to 180 °C. Upon increasing the temperature to 190 °C, bands remain corresponding to helical sequences with up to 12 monomers. As the X-ray scattering indicates that only the smectic ordering survives above 180 °C, these helical sequences with up to 14 monomers must be the building blocks of the smectic layers.

Fig. 7 displays a typical DSC curve of a sheared iPP1 sample during heating. In addition to the large melting peak of iPP crystals around 165 °C, several additional small melting peaks appear at higher temperatures, which are absent in non-sheared samples. The insert in Fig. 7 shows a blow-up of the DSC trace from 170 to 195 °C. We attribute these high-temperature melting peaks to contributions from the smectic phase. Fig. 8 shows POM pictures of the sheared iPP1 sample during

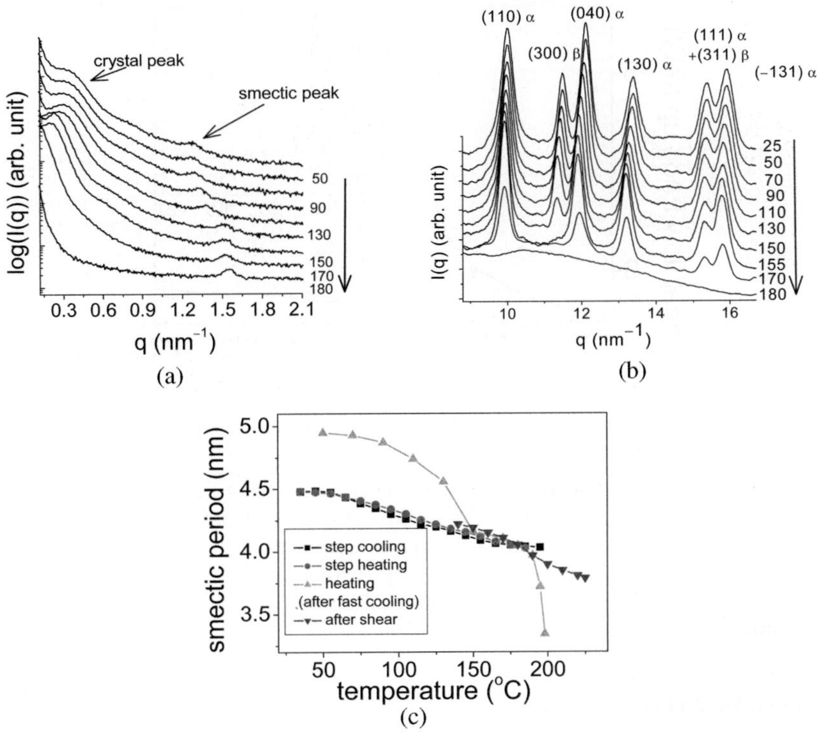

Fig. 5 One-dimensional SAXS (a) and WAXS (b) patterns of a sheared iPP1 sample during a heating scan; the numbers indicate the temperature in °C. (c) Corresponding smectic periodicity. ▼ and ▲ indicate the original shear-induced smectic spacing and the spacing during heating, respectively, of the fast quenched down sample used also in (a) and (b). ■ and ● indicate the smectic spacing during step cooling and during the heating process of a different iPP1 sample. Both samples were initially sheared at 190 °C for 3 min with a rate of 1 s^{-1}.

a step-heating scan. The smectic layers are seen to assemble into a fibrillar morphology. These fibres melt in the temperature range where X-ray scattering, DSC and FTIR also indicate melting of the smectic layers.

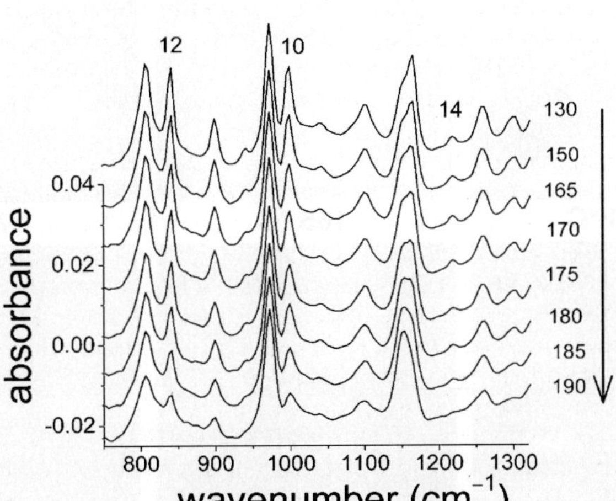

Fig. 6 IR spectra of a sheared iPP1 sample during a heating scan. At specific bands the monomer number associated with the helical sequence is indicated.

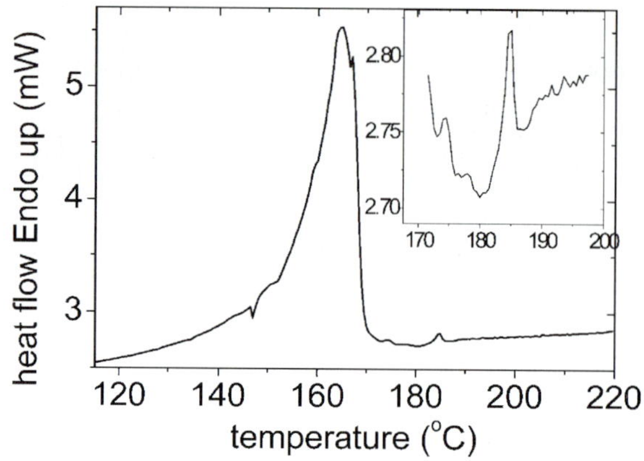

Fig. 7 DSC curves of a melt-sheared iPP1 sample during heating at a rate of 10 °C min^{-1}.

Upon heating directly after shearing the melt, the smectic structures can generally stand higher temperatures than in the situation of cooling and subsequent heating. However, even after the disappearance of the smectic peak at sufficiently high temperatures, upon lowering the temperature the smectic ordering can reappear without applying further additional shear. This is illustrated in Fig. 9, which shows the SAXS patterns of iPP1 initially sheared at 200 °C, then heated up to 260 °C and subsequently cooled.

Smectic-phase-induced crystallisation

The smectic phase can be induced by shear flow below and above the melting temperature of iPP crystals. This provides us with two possibilities to investigate the influence of smectic ordering on crystallisation. In Scheme 1 the smectic layering is induced by shearing the sample above the melting point and subsequently cooling down to the crystallisation temperature. In Scheme 2 the iPP sample is cooled down to the crystallisation temperature, then sheared and left to crystallise.[62]

Let us first consider the situation of Scheme 1.[63] Figs. 10a and b show one-dimensional SAXS patterns of iPP1 samples during crystallisation at 145 °C, with and without shearing above the

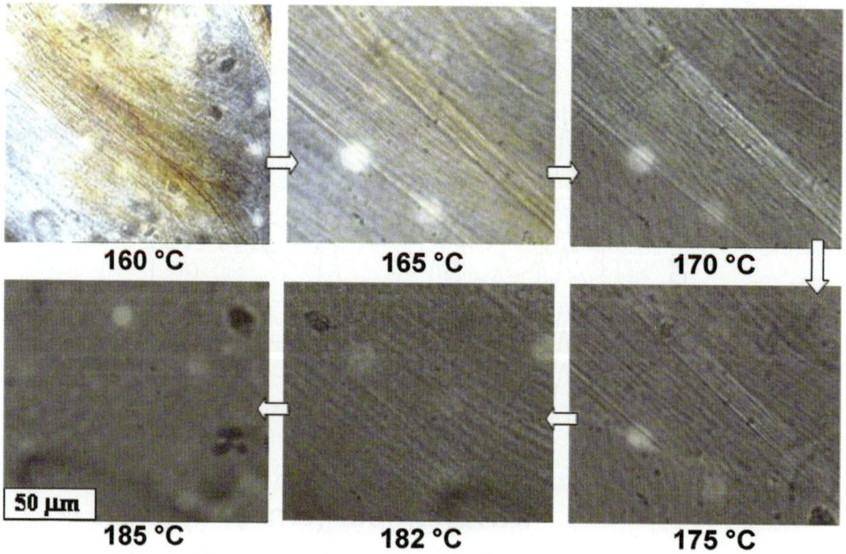

Fig. 8 POM pictures of a melt-sheared iPP1 sample during step heating.

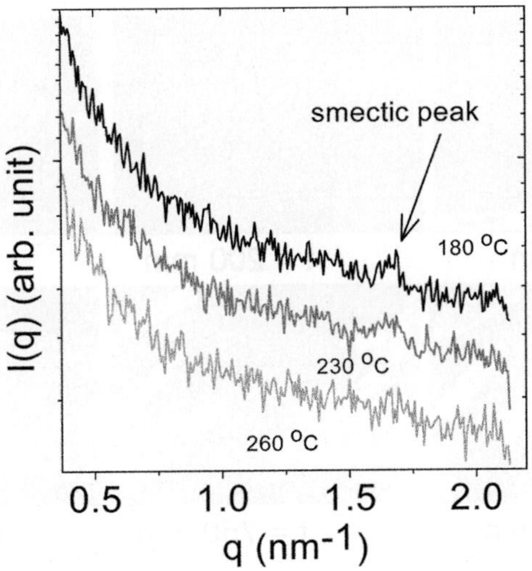

Fig. 9 One-dimensional SAXS patterns of iPP1 in the melt. The sample was initially sheared for 2 min at a rate of 1 s^{-1} at 200 °C, heated to 260 °C for several minutes and subsequently cooled to the temperatures indicated.

melting point. The relative SAXS intensities are plotted *vs.* crystallisation time in Fig. 10c. The sample with shear-induced smectic bundles crystallises faster than the other one. This suggests that the smectic bundles induce the nucleation of crystals. This is confirmed by optical microscopy as

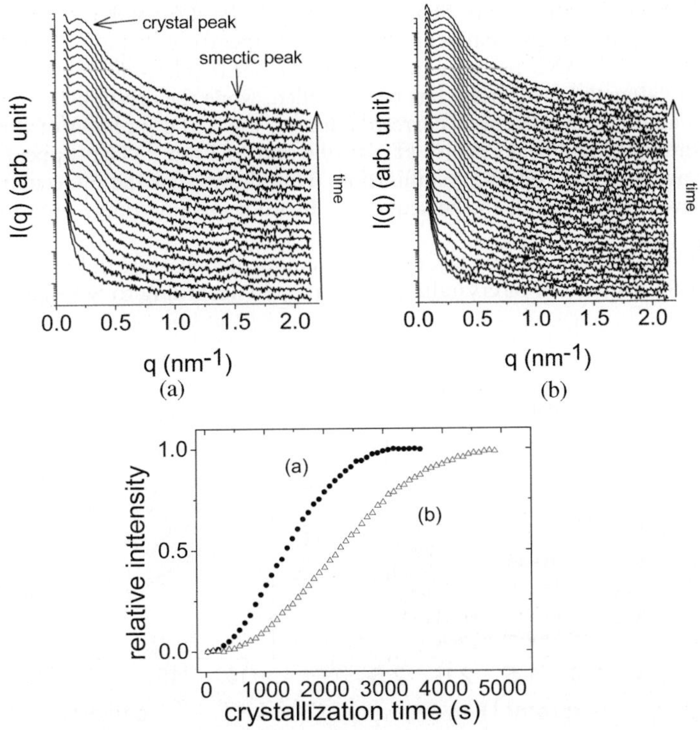

Fig. 10 One-dimensional SAXS curves of iPP1 with (a) and without (b) melt shearing, during crystallisation at 145 °C. (c) Relative intensities *vs.* crystallisation time.

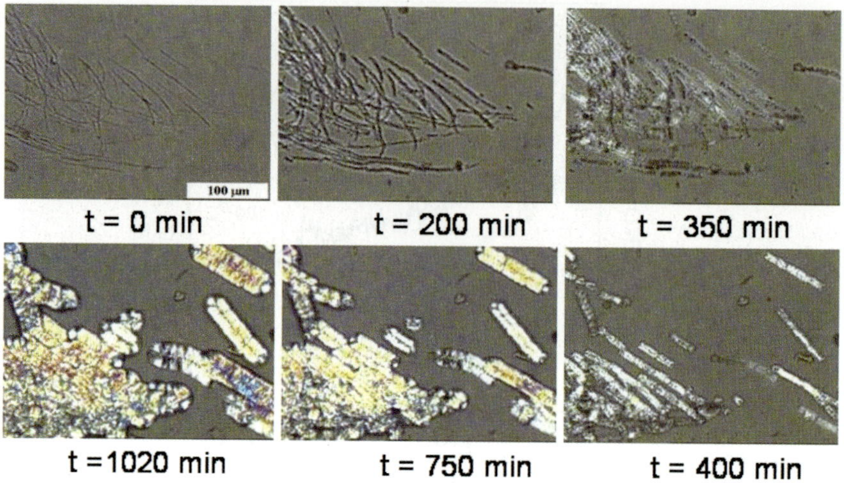

Fig. 11 POM pictures showing smectic-bundle-induced crystallisation at 160 °C. The scale bar for the pictures in the second row is twice that of the first row.

shown in Fig. 11. This figure shows a series of POM images of an iPP1 sample with shear-induced smectic bundles during crystallisation at 160 °C. The original smectic layers assemble into bundles with a diameter and length up to 10 and 200 μm, respectively. After about 100 min, crystals of iPP grow from the surface of these bundles. Upon increasing the crystallisation time so-called transcrystallites develop.

Let us now consider the situation of Scheme 2, in which the shear-induced smectic ordering and the crystallisation happen sequentially at the same temperature.[62] Fig. 12a and b show one-dimensional SAXS and WAXS patterns of iPP2 after a shear pulse at 150 °C. The smectic order forms immediately after the shear pulse has been applied, while the crystallisation takes more than 480 s to start. In addition to reflections corresponding to the α and β phases of iPP, in the WAXS patterns two additional small peaks evolve at $q \approx 10.6$ and $14.3\ nm^{-1}$. They appear after about 360 s but disappear again at the end of crystallisation. This shows that the molecular packing in the smectic layers develops according to the sequence: disorder → short-range ordering → long-range ordering.

The relative orientations of smectic layers and crystalline lamellae are different. Fig. 13a shows a two-dimensional SAXS pattern of crystallised iPP1 treated according to Scheme 1. Before cooling

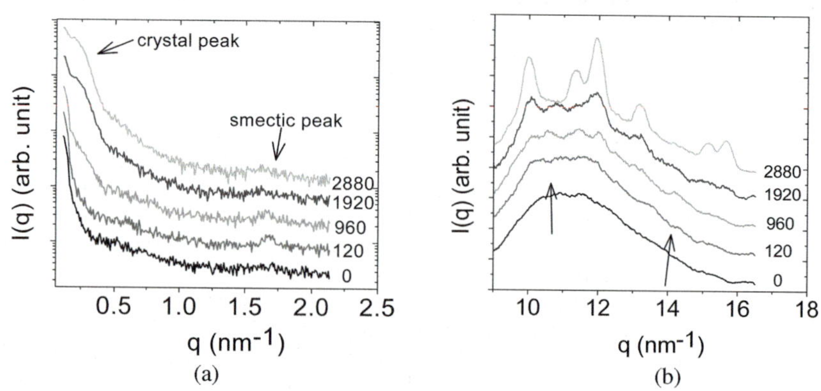

Fig. 12 (a) One-dimensional SAXS and WAXS patterns of iPP2 during crystallisation at 150 °C after a shear pulse with a rate of $1\ s^{-1}$ and strain of 1500%. The number indicates the crystallisation time in seconds.

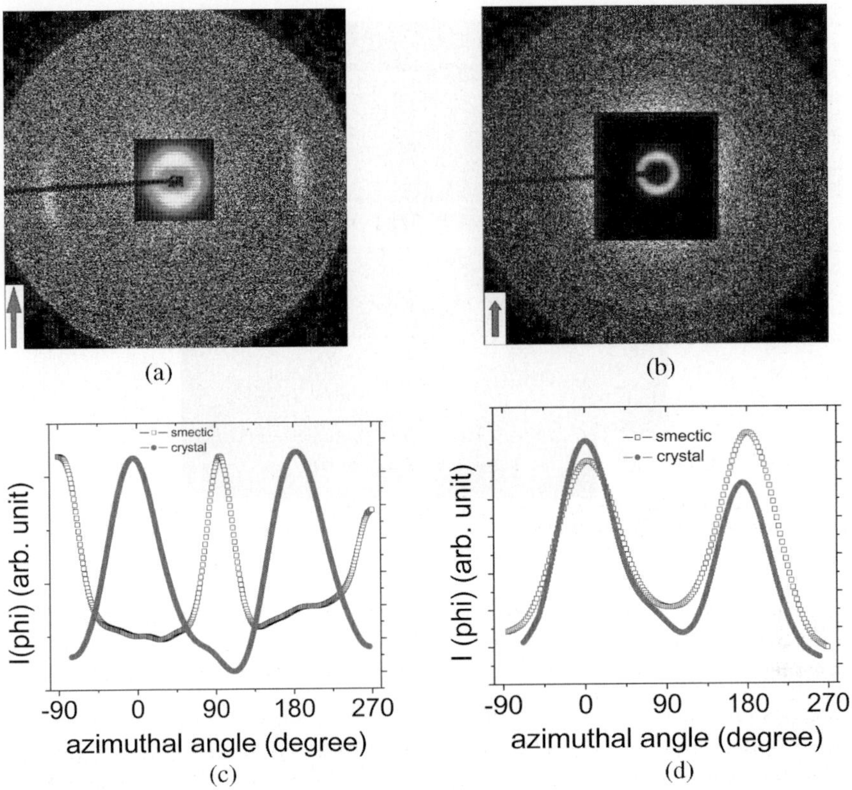

Fig. 13 Two-dimensional SAXS patterns of (a) an iPP1 sample crystallised at 140 °C after shearing in the melt (180 °C, a steady shear of 0.1 s^{-1} for 10 min) and (b) iPP3 sheared and crystallised at 140 °C with a shear pulse with a rate of 30 s^{-1} and a strain of 1500% applied. The corresponding azimuthal intensity distribution of the smectic and crystalline peaks are shown in (c) and (d), respectively. The arrows indicate the flow direction.

down to crystallization temperature, the smectic layers were aligned in the velocity gradient direction by weak steady shear. A similar pattern of iPP3 crystallising following Scheme 2, is displayed in Fig. 13b. The corresponding azimuthal intensity distributions of both the smectic peak and the crystalline peak are plotted in Fig. 13c and d, respectively. In Scheme 1, the smectic layers are parallel to the flow direction, and the crystal lamellae have developed perpendicular the smectic layers and thus also to the flow. The scattering from the smectic bundles (see Figs. 8 and 11) can also be observed and is shown in Fig. 14. It shows that the normal of the crystalline lamellae follows the direction of the bundle axis, consistent with the transcrystallite model. Here the smectic layer normal is perpendicular to the bundle axis. In Scheme 2, the initial smectic layers and the crystalline lamellae have the same preferred orientation, perpendicular to the direction of the shear flow. Here we only present two extreme cases; any intermediate situation can be obtained by different shear alignment of the smectic layers before crystallization.

Finally we also checked the smectic layering in the case of fibre spinning with an elongational flow. Fig. 15 presents a one-dimensional SAXS patterns from iPP1 fibres taken *ex situ* at room temperature. With the x-rays perpendicular to the fibre axis—the usual geometry for *in-situ* X-ray scattering on fibres—we observe the scattering peak from crystals rather than the smectic layering. The smectic peak appears after rotating the sample, positioning the X-ray beam parallel to the fibre axis. This indicates smectic layers parallel to the fibre axis. In principle this should still be seen in the perpendicular geometry as a two-dimensional azimuthal average. However, the averaging of an already small peak gives a rather low scattering that can be easily missed.

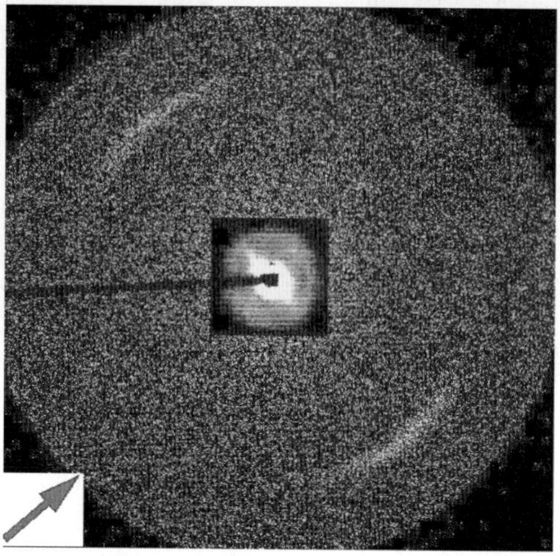

Fig. 14 Two-dimensional SAXS pattern as in Fig. 13a, but showing at small angles the perpendicular direction of the bundle axis with respect to the smectic layers.

Discussion

Smectic ordering

From the results presented, we conclude that shear-induced smectic ordering is a general intrinsic property of iPP. As materials with different molecular weight from different sources show similar features (see Fig. 3), speculations concerning any possible influence of impurities or additives can be excluded. Interestingly, tracing back the literature about flexible chain polymers, several reports already point to the possible existence of a high-temperature mesophase. Smith[70] studied the temperature dependence of the specific heat C_p of PE and iPP and found several reversible transitions in the melt at temperatures 50–100 °C above the respective melting points. A comparison of the values of ΔC_p with those of liquid crystalline compounds (higher aliphatic acid salts) leads to the suggestion of several types of smectic modifications being formed. The smectic layer spacings in PE and iPP were estimated to be 2–5 nm. More recently, two high-temperature melting peaks were observed in DSC of a Ziegler–Natta PE melt.[71] At present no model exists that can explain these observations.

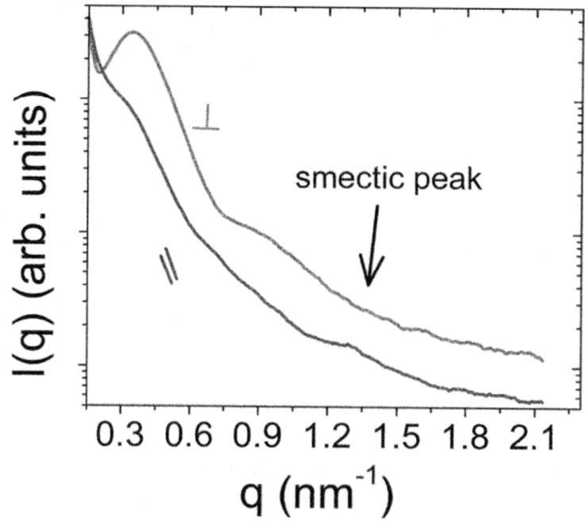

Fig. 15 One-dimensional SAXS curves of iPP1 fibres with the X-ray beam incident parallel and perpendicular to the fibre axis, respectively.

Why has the smectic ordering not been observed before? iPP is a very suitable polymer for studying shear-induced crystallisation and since Ziegler–Natta's invention a large number of reports have been published; see for a review ref. 59. Why has smectic ordering in the iPP melts not been observed earlier in the last half a century? There are two possible reasons: X-scattering instrumentation and the orientation of the smectic layers. First, *in-situ* SAXS measurements have only been possible in the last two decades, triggered off by synchrotron radiation and new X-ray optics and detection systems.[65] *In-situ* SAXS measurement on shear-induced crystallisation of iPP started only about five years ago. Moreover, the smectic peak at about 4 nm easily falls outside the detector area, especially at synchrotron beamlines with a large sample-to-detector distance. Second, most work so far on shear-induced crystallisation employs a shear cell with a two-plate configuration.[51–62] As the smectic layers are easily aligned parallel to the plates, in this geometry X-rays do not detect the layer structure (see Fig. 2). Our results from fibre samples indicate that the usual geometry with the X-rays perpendicular to the fibre axis can also easily miss the layer structure (see Fig. 15).[47] Finally, at room temperature even in the right geometry it is rather difficult to observe the smectic peak, due to the strong scattering from crystals and the shift of the smectic peak to low-q values. At room temperature, in many cases the original smectic peak merges into the high-q tail of the crystal peak (see Fig. 4).

How can a flexible chain polymer like iPP form a liquid-crystalline phase? A general recipe for synthesizing a main-chain liquid-crystalline polymer is to take a series of small rod-like molecules and join them end-to-end. The rod-like units often contain aromatic rings and may be connected *via* flexible 'spacers'.[66] The connection between liquid crystallinity and rod-like structures is well accepted and liquid-crystal theories are usually built on a rod-like molecular model. Some intrinsic stiffness is needed to achieve the straightness in the liquid-crystalline polymer that allows orientational ordering below the isotropic-liquid crystal transition temperature. Nevertheless, theories of flexible-chain polymers show that rigidity induced by a coil–helix transition could provide the necessary stiffness to meet the conditions of an isotropic–nematic transition.[67,68] Many examples can be found in biopolymers such as DNA and proteins, in which hydrogen bonds stabilize the helices.[69] A coil–helix transition seems to be the only possibility for iPP to gain stiffness. In that case the density modulation of the smectic layering must arise from alternating rigid helical sequences and flexible disordered segments. Following Doi–Edwards' theory, the critical persistence length l_p for an isotropic–nematic transition can be calculated from the following formula:[19–23]

$$l_p = \frac{4.19M_0}{bl_0\rho N_A}. \tag{1}$$

In this equation b represents the diameter of a polymer segment, ρ the density, N_A Avogadro's number, and M_0 and l_0 the mass and the length of a monomer, respectively. For the iPP melt, $M_0 = 42$ g mol^{-1}, $b = 0.665$ nm, $l_0 = 0.217$ nm and $\rho = 0.85$ g cm^{-3}. Inserting these values into eqn. (1) we find $l_p \approx 2.4$ nm, which is for a 3/1 helix equivalent to about 11 monomers.[64] Though the theory applies to rigid rods, it is gratifying that the rod length of iPP helices as obtained from IR meets this criterion (see Fig. 6). However, present theories predict a nematic phase rather than a smectic one.[19–23,67,68] A broad distribution of the helical sequence length should exist in the melt of flexible polymers in a dynamic equilibrium, which contributes a favourable entropy term.[67,68] If the conformational ordering and orientational ordering occur in sequence, again only a nematic phase is expected. It is not clear how the smectic periodicity enters the problem as a new length scale. The selection of the sequence lengths must happen during the formation of the smectic layers, thus the conformational ordering (or coil–helix transition) and the alignment of helices take place simultaneously. The coupling between conformational ordering and density determines the formation of the smectic layers.

Why is the smectic phase more stable than its crystalline counterpart? The smectic phase can not only be induced by shear flow in the melt, but also has a higher melting temperature than its crystalline counterpart (see Figs. 4–8). The smectic periodicity is reversible during heating and cooling in a temperature range from 150 to about 180 °C. Based on our experimental observations (see Figs. 13 and 14), we propose in Fig. 16 an anisotropic drop model to explain this reversible

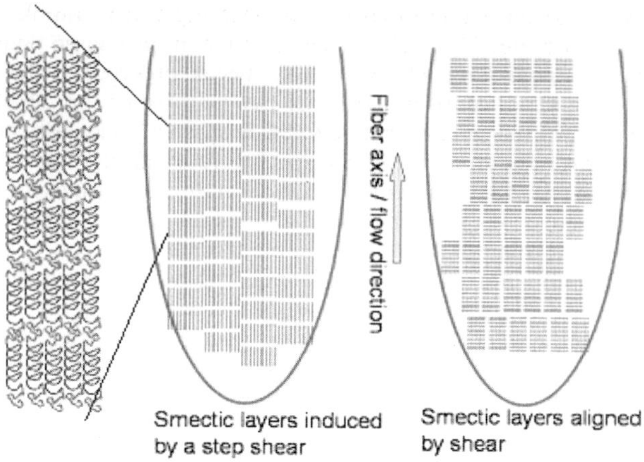

Fig. 16 Anisotropic drop model of the smectic domains based on the POM pictures of Figs. 8 and 10, and the scattering patterns of Figs. 13 and 14.

behaviour. X-ray scattering reveals that after initial subjection of the iPP melt to a shear pulse the smectic layer normal is parallel to the flow direction (Fig. 13b and d). As the smectic layers assemble into bundles, the smectic layer normal follows the direction of the bundle axis. Upon further imposing a shear field, the smectic layers rotate and finally align in the direction of the velocity gradient (Figs. 2, 13a and 13c). Now the question arises whether the bundle axis also rotates to follow the velocity gradient direction. In other words, do the smectic layers rotate independently from the bundles direction? The answer is 'yes': the smectic layers rotate within the fibrillar entities, while the bundle axis remains in the original direction following the flow, in agreement with POM observations. The relative orientation of smectic layers, crystalline lamellae and bundles, as revealed by X-ray scattering in Figs. 13 and 14, supports the answer. In the iPP1 sample with initially pre-aligned smectic layers above the melting point (Scheme 1), the smectic layers are parallel to the bundle axis and to the normal of the crystalline lamellae, the latter two following the flow direction. Based on the width of the smectic peak, we estimate the correlation length of the smectic layers to be of the order of tens of nanometers. This is two and three orders less than the width and the length, respectively, of the fibrillar entities. Aligning the smectic layers parallel to the bundle axis, hundreds of smectic domains with internally correlated layers are required to match the width of the bundles. Thus the fibrillar entities can be defined as an anisotropic drop containing smectic domains with a size of tens of nanometers (Fig. 16). The anisotropic drop model provides a conserved system and thus explains the reversible process of the smectic periodicity during heating and cooling.

Though the anisotropic drop model can provide the possibilities of superheating and super-cooling of the smectic bundles due to confinement effects, it still cannot explain by itself their stability. To evaluate the thermodynamic stability, it is more appropriate to look at the smectic domains with correlated layers. A model of formation and growth of smectic filaments (similar to the Hoffman–Lauritzen[5,6] model for polymer crystallisation) was proposed by Frenkel and Schilling,[72] and can be extended to iPP in a natural way. Following this model the free energy associated with the formation of a small circular smectic disk with radius R is given by

$$\Delta F_0(R) = \pi R^2(2\gamma_\perp - \rho d|\Delta\mu_s|) + 2\pi R d\gamma, \tag{2}$$

where d is the thickness of the smectic disk, $\gamma_\perp$ and γ represent the surfaces energy perpendicular and parallel to the molecular chains, respectively, $\Delta\mu_s = \mu_s - \mu_l$ is the difference in chemical potential between the smectic and the liquid phase, and ρ is the number density of the bulk smectic phase. For $|\Delta\mu_s| < 2\gamma_\perp/(\rho d)$ we find that $\Delta F_0(R)$ is positive, which means that a single smectic layer cannot grow. $\Delta F_0(R)$ can be decreased by adding another smectic disk with the same diameter. Repeating this process several times eventually a smectic filament with $n = 2R\ \gamma_\perp/(R\rho d|\Delta\mu_s| - 2d\ \gamma)$ layers becomes absolutely stable and can grow. A Gaussian fitting of the smectic peak gives a correlation length of the smectic layers up to 120 nm, which corresponds to about 30 smectic layers. As measured by DSC, X-ray scattering and POM during heating scans, crystals with a periodicity of about 4.5 nm, crystallised from the original smectic regions, transform around 140 °C back to smectic bundles that melt at about 185 °C. At the same time crystalline lamellae with a periodicity of about 25 nm melt at temperatures around 165 °C. DSC shows several small additional melting peaks that can be attributed to different correlation length of smectic filaments. Following the Gibbs–Thomson equation,[73] which is generally applied to estimate the relation between crystal thickness and melting temperature, we can write

$$T_{\mathrm{m}} = T_{\mathrm{m}}^{\mathrm{o}}\left(1 - \frac{2\gamma_{\perp}}{l\Delta H_{\mathrm{m}}}\right), \tag{3}$$

where T_{m} and $T_{\mathrm{m}}^{\mathrm{o}}$ are the melting points of a smectic filament with a thickness l and an infinitely long one, respectively, ΔH_{m} is the melting heat of the smectic phase per unit volume. Eqn. (3) cannot only explain the multi-melting-point behaviour of the smectic phase, but also helps to understand why it can stand higher temperatures than the crystalline one. iPP crystals have a higher melting heat per unit volume than the smectic bundles. Wunderlich estimated the latent heat from a transient state to a crystal to be about 0.6 kJ mol^{-1}. Comparing this value to the 148 kJ mol^{-1} of the melting heat of iPP crystals,[73] this suggests that the difference between the smectic melting heat and the crystal one is small. On the other hand, the length l of a smectic filament is up to 3 times larger than the thickness of a crystal, while the smectic surface free energy $\gamma_{\perp}$ is smaller. Based on this comparison, according to eqn. (3) a smectic filament with a length up to 120 nm can have a higher melting point than a crystal with a thickness of less than 20 nm. The variation of l is discrete because the growth of a smectic filament is a layer-by-layer process ($l = nd$). This discrete nature is consistent with the multi-melting-point behaviour.

How do the molecular weight and its distribution influence the smectic ordering? Though the smectic periodicities of three iPP materials show a similar temperature dependence, the one with the higher average molecular weight gives a larger value (see Fig. 3). Because the smectic layer spacing and the molecular chain length are at completely different length scales, a direct connection is not expected. However, different molecular weights correspond to different visco-elastic properties. From a rheological point of view, such differences can lead to a different orientation, stretching and relaxation of the molecular chains under external shear. The formation of smectic regions induced by shear flow is a non-equilibrium process. For the same temperature and external field, the effect of the stimuli may vary for different molecular chain length, leading to a different smectic layer periodicity. The molecular weight distribution may also be an important factor influencing the process, because only a small percentage of the iPP melt transforms into smectic regions after shear. As the influence of the smectic ordering on shear-induced crystallisation is significant[62] and high-molecular-mass species play an important role for the acceleration of polymer crystallisation under shear,[51–62] we believe that the high-molecular-mass part of the distribution also gives the main contribution to the smectic regions.

Crystallisation experiments on a blended bimodal system show that the ratio between long and short chain length, rather than the average chain length, controls the shear-induced enhancement of crystallisation.[56–59] Hence one may wonder whether a similar effect plays a role in the smectic ordering. The molecular weight distribution of the three materials follows the sequence: iPP1 > iP-P3 > iPP2 (see Table 1). Imposing the same shear field, the intensity of the smectic peak of iPP1 is much larger than that of iPP2, while iPP3 gives an intermediate intensity. Because the smectic peak intensity indicates the concentration of the smectic bundles, the relative amount of smectic regions in the three iPP materials shows the same sequence as the molecular weight distribution. We cannot establish quantitative relationships at this stage, but at least qualitatively the following statements seem to be correct. *A higher (average) molecular weight corresponds to a larger smectic periodicity and hence a higher stability; a larger molecular weight distribution (or larger ratio between the lengths of long and short chains) leads to an increased concentration of smectic regions*. The ratio between the lengths of long and short chains is even more crucial if we associate with the smectic ordering a phase separation between long and short chains, which may follow spinodal decomposition kinetics. These rules suggest that it may be easier to observe shear-induced smectic ordering in

Ziegler–Natta iPP than in metallocene-based iPP with a narrower molecular weight distribution. Further experiments are underway to test this quantitatively.

What is the nature of the low-temperature 'mesophase' in fast quenched iPP? In 1959, Natta *et al.*[43] pointed out a mesomorphic form of iPP when the polymer was quenched from the melt in ice water. The mesosphase was reported to have a degree of order intermediate between the amorphous phase and the crystalline phase. Natta *et al.* also concluded that the mesophase is 'smectic', containing bundles of parallel stretches of right- and left-handed 3-fold (3/1 and 3/2) helices. Though since then many studies on the nature of this 'mesophase' have been carried out, the experimental evidence for smectic ordering is still not conclusive. Miller[46] suggested that the ordering could be of the type described by Hosemann as 'paracrystalline'. Gailey and Ralston[45] proposed a partially ordered phase composed of small hexagonal crystals. Wunderlich and Grebowicz[38] suggested that the concept of conformationally disordered crystals was perhaps more appropriate for the mesomorphic form. Corradini *et al.*[44,45] considered models corresponding to pseudo-hexagonal crystals (as in the β form) and monoclinic ones (as in the α form) as well as several other disordered models combining character of both forms. They concluded that the mesomorphic form of iPP was not composed of small pseudo-hexagonal crystals but rather of disordered bundles of chains. Bruckner *et al.*[48] pointed out that the statistical structure of mesomorphic iPP eliminated the possibility of small crystals in the α form or the β form. Deformation studies confirmed that the mesomorphic form was quite disordered in relation to the typical crystalline unit cell symmetries.[47] The evidence of the 'mesophase' generally refers to two broad WAXS peaks at $q \approx 10.6$ and 14.3 nm^{-1}.[43–49] No direct layering comparable to a conventional smectic phase was observed.

Effectively all the above arguments can be understood on the basis of the process illustrated in Fig. 12. In the sheared high- temperature melt or at lower temperatures before crystallisation, the molecular chains in the smectic layers are still in the liquid state. During the crystallisation process of the smectic layers, the two broad WAXS peaks are observed (see arrows in Fig. 12b) similar to that usually obtained in the quenched samples and used as the evidence of the low-temperature mesophase.[43–49] Finally these two peaks disappear. However, through a fast quenching they can be frozen in. This strongly suggests that the structure observed in the quenched iPP samples is a transient metastable phase during the crystallisation of the (high-temperature) smectic phase. Different authors could easily report different observations because of possible differences in the thermal history.[43–49] From our results we believe that the low-temperature 'mesophase' has in most cases a structure of defective micro-crystals with short-range ordering.

Influence of smectic ordering on the crystallisation

How does the smectic ordering influence the crystallisation process? The nucleation effect is clearly demonstrated in Figs. 10–12. Samples containing smectic bundles show faster crystallisation kinetics than samples without smectic domains. The existence of high-temperature smectic ordering in iPP can explain several questions in polymer crystallisation under shear as well as at quiescent conditions. Though knowledge of flow-induced crystallisation in polymer melts is essential to control the final properties of products during industrial processing methods such as extrusion, injection, and blow moulding, no satisfactory molecular theory exists that can describe polymer crystallisation under flow conditions. During the past half a century, a great deal of work has been dedicated to this subject, featuring in particular iPP because of its practical importance and high-temperature stability.[51–63] On this basis several points regarding shear-induced crystallisation of polymers have been established.[58] (i) Shear accelerates the crystallisation kinetics and, under severe conditions, changes the semi-crystalline morphology from spherulites to crystallites oriented in the flow direction. (ii) The enhancement of the crystallisation kinetics is attributed to an increase in the nucleation rate caused by distortions of the polymer chains in the melt. (iii) The molecular weight and the molecular weight distribution have a pronounced effect. These three points can be considered as a phenomenological description of shear-induced crystallisation rather than as a molecular explanation. The essential question of how shear promotes nucleation in polymer crystallisation is still open.

What is the nature of the pre-ordering before crystallisation? So-called pre-ordering before crystallisation has been widely studied, the main observation being the appearance of intensity in SAXS prior to Bragg reflections of crystals in WAXS. Let us investigate whether this effect could be attributed to the assembled domains of smectic layers, which can be present at different length scales (see Figs. 8 and 11). Somani *et al.* reported a precursor with a length scale of about 43 nm in a sheared supercooled iPP melt, while Heeley *et al.* observed precursors with other length scales from 15 to 30 nm in different iPP materials. The experiments under shear flow (Somani *et al.*[54]) and with a spin-line (Heeley *et al.*[13–15]) imposed a similar deformation history on the samples as we did. Hence there can be little doubt that smectic ordering took place in their samples, even though it was not observed due to the reasons discussed above. In principle the heterogeneity of the density due to the smectic ordering can lead to a SAXS signal, giving a natural explanation of these observations. However, as discussed above, the length scale of the smectic layers is not understood at this stage, which also hinders a more quantitative connection with the dimensions observed in the early SAXS. The results at quiescent conditions can be ascribed to a memory effect[15] (see next section).

The observation of smectic ordering in the flexible chain polymer iPP puts some doubt on the spinodal decomposition model,[14] which originally comprises both flexible- and rigid-chain polymers. Can—taking induced rigidity into account[67,68]—the old framework of Flory[20–23] for the isotropic–nematic transition explain the pre-ordering observation in both flexible- and rigid-chain polymers? The essential structural difference between the spinodal model and the idea of an isotropic–nematic transition is orientational ordering. In other words, the spinodal model suggests that no orientational ordering is needed for the ordered chain segments or helices to pack closely (gain density). Thus a dense liquid rather than liquid crystal results. The observation of smectic ordering in the iPP melt indicates that in fact conformational ordering, orientational ordering and density couple. In the absence of a more quantitative model this can be considered as a multiple chicken–egg problem. If and how these three types of ordering can be separated in a flexible chain melt is still a fully open question.

Is it appropriate to consider the smectic ordering as pre-ordering before crystallisation? As the smectic layering can be obtained at temperatures above the melting point, the answer must be no. During heating, the crystallised but initially smectic layers transform back to smectic bundles and show a higher melting temperature than the conventional crystalline lamellae. Though the smectic domains induce epitaxial growth of crystals in the supercooled melt, they do not easily crystallise themselves, as indicated by the small change of the periodicity (see Figs. 4 and 5). Hence there should be no doubt that the smectic bundles are the result of an intrinsic ordering process. More generally, an intermediate phase may induce the nucleation of a final state and both states may have the same origin, being for example driven by conformational ordering, orientational ordering and close packing.[31–37] However, the occurrence of such an intermediate state is not necessarily connected to the final state.

What is the origin of the memory effect? The existence of a high temperature phase provides a natural explanation to the so-called memory effect in polymer crystallisation.[74] This term refers to the effect that after heating well above the melting temperature of the crystals, athermal nucleation is still found after subsequent cooling below the melting point. Because of a lack of reproducibility it has sometimes been considered as a somewhat mysterious effect. As the smectic ordering can stand much higher temperatures than the crystals, the melting of crystals does not lead to a homogeneous liquid. Instead the smectic domains still survive in the melt and can serve as the nuclei for crystals after subsequently cooling down to crystallisation temperature. Fig. 10 presents an illustration of the exact phenomena involved in the memory effect. It is due to the presence of the smectic domains well above the melting point, the persistence of which depends strongly on the history of the sample. To get rid of the smectic ordering prolonged annealing above 200 °C is necessary,[63] which is usually avoided because of the risk of degradation of the sample.

The memory effect brings an explanation to the observation of pre-ordering at quiescent conditions. First we notice that a small shear or deformation during sample preparation can be sufficient to induce smectic ordering. Second, we have seen that the smectic phase can be 'frozen in' upon cooling and be recovered upon subsequent heating. Commercial polymer materials including iPP are generally subjected to some form of shear flow during industrial production. This all suggests that smectic domains may already be present in raw material, and cannot easily be avoided. As we have seen, keeping the sample at temperatures above the crystal melting point is not sufficient to erase the 'memory' of the smectic ordering. In Fig. 9 we demonstrated that even after a strong reduction of the smectic peak in SAXS at sufficiently high temperatures, the smectic ordering recovers at lower temperatures without additional shear. *We conclude that many experiments thought to be done under quiescent conditions in fact have been carried out on pre-sheared samples displaying smectic ordering.* Hence the recovery and growth of the smectic ordering can provide an explanation for the evolution of the SAXS intensity during the early stages of crystallisation at low temperatures, even when no shear was explicitly imposed.[13–16]

Following the above discussion, we believe that polymer crystallisation follows a nucleation and growth process. In this view the debate about pre-ordering before crystallisation concerns arguments about the nature of the primary nuclei. The well-established folded-chain crystal model of polymers has pushed the polymer community to acceptance of a folded chain as the primary nucleus. Zachmann[75] calculated the effective free energy of folded-chain and fringed-micelle (or bundle nucleus) models[76–79] and found that the folded-chain primary nucleus is favoured. However, in the case of the smectic bundles of iPP or the hexagonal phase of PE,[35] the surface free energy of the bundle nucleus is greatly overestimated in Zachmann's approach.[75] In the smectic phase or other mesophases the excess interfacial energy caused by entropy constraints on the connection of liquid segments to the nuclei, is much lower than in a crystalline phase. Hence for nucleation through smectic filaments the energy barrier is low, favouring fringed-micelle nuclei. For some recent reconsiderations of this concept see refs. 78 and 79. This picture of bundle or row nuclei is also consistent with Muthukumar's computer simulations.[1] In the polymer melt, the 'baby nuclei' do not necessarily reach the final lamellar size by thickening alone. When adjacent 'baby nuclei' aggregate or align and reach the critical size of a nucleus, crystal growth can take place.

What is the nature of the shish or row nuclei? To understand the mechanism leading to the enhancement of crystallisation under shear, we consider the structure of the so-call shish or row nuclei, the primary nucleus induced by flow. In the common view the shish consists of extended-chain crystals that may be defective.[39–41] Structures with some degree of order are assumed to be induced by shear flow before the onset of crystallisation, which can be long living at shallow supercooling.[51–61] They are called shear-induced precursors, and have been described as bundles with stretched chains or to be due to amorphous liquid–liquid phase separation. From our work, we conclude that these primary row nuclei or precursors are assemblies of smectic layers. The fibrillar morphology, high temperature stability, and molecular weight dependence of the smectic domains, all meet the general description of the so-called precursors as well as account for the final shish-kebab (or transcrystalline) structure. The corresponding relation between average molecular weight (its distribution) and smectic periodicity (the concentration of smectic bundles) provides a molecular understanding of shear-induced crystallisation. *The average molecular weight determines the stability of the shear-induced primary nuclei (the smectic bundles), while the distribution of molecular weight (or the ratio between long and short chains) controls the density of nuclei.* On the basis of our experimental results we present a model, Fig. 17, in which the row nuclei are the smectic filaments and not extended-chain crystals. In the case of shear pulse, smectic layers are induced by the flow but are not further aligned. Hence the epitaxially grown crystalline lamellae have the same orientation as the smectic layers (Fig. 17a). After imposing the shear field for a longer time, the induced smectic layers are aligned along the velocity gradient direction (see Figs. 2, 13 and 14). Though we describe these processes separately, it should be realized that in most experimental situations the formation and the alignment of the smectic layers occurs simultaneously. Depending on the shear field the smectic layers may to some extent lose their initial orientation (Fig. 17b). As a result the epitaxial growth of crystals starts to deviate from the direction perpendicular to the axis of the smectic bundles. This phenomenon is also helpful to understand the formation of the shear-induced β phase of iPP (see next section).

What are the nuclei for β phase? It is generally accepted that in sheared samples of iPP the β phase is promoted. The common view is that the surfaces of oriented α cylindrites provide nucleation sites for the growth of β crystals.[80–82] However, the helices pack on a scale that is much smaller than the

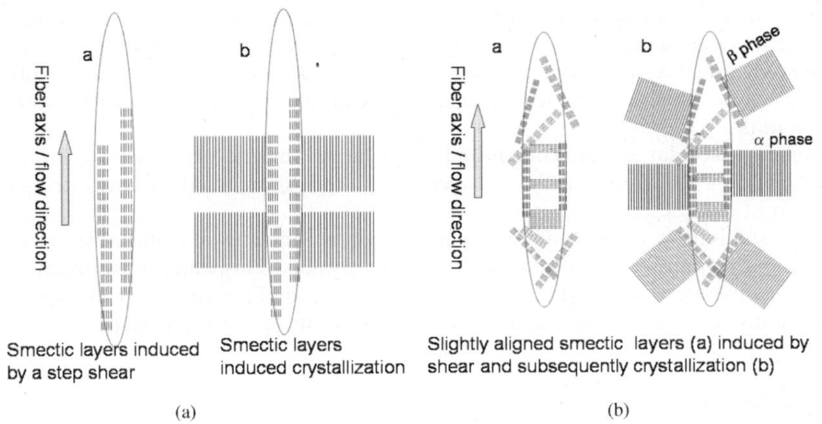

Fig. 17 Schematic model of crystallisation induced by smectic bundles; (a) and (b) indicate how the crystalline lamellae grow on the smectic bundles without and with shear-alignment, respectively.

cylindrites. Hence in this situation a shear field is not expected to have much influence on the growth of β crystals, in clear contradiction with the experimental results.[80–82] Instead we propose that the smectic bundles provide the nucleation sites for the growth of β crystals. During a heating scan on the sheared sample crystallised at 110 °C (Fig. 5) a discontinuous change of the smectic periodicity occurs around 140 °C, the melting temperature range of β crystals. This suggests that the crystal–smectic transition is from the β phase to smectic, rather than starting from the α phase. Obviously, then the reverse crystallisation process is a transformation of the smectic layers into the β modification. It has been reported that β crystals, unlike crystals in the α phase, show a random orientation even when induced by flow,[82] which is similar to the smectic ordering at this stage (see Fig. 17b). During a step cooling the sheared sample crystallises at about 160 °C and no β crystals are formed. In that case upon subsequent heating no discontinuous change of the smectic periodicity is observed (see Fig. 5c). This observation provides further support for a connection between smectic ordering and β crystals. From the viewpoint of helical selection,[83] it should be easier for smectic bundles (with a random helical handiness) to induce the growth of the β phase than for α crystals that have a regular right- and left-handed helical packing. We conclude that the view that smectic bundles provide the nucleus for growth of β crystals, gives a consistent explanation for a wide range of observations.[80–82]

Conclusions

On the basis of *in-situ* SAXS and WAXS, POM, DSC and FTIR of shear-induced ordering and crystallisation of isotactic polypropylene, we come to the following conclusions.

(1) Shear-induced smectic ordering is an intrinsic property of iPP that can be induced at temperatures above and below the melting point of the crystals. The smectic ordering with a periodicity of the order of 4 nm takes the form of filaments or bundles that crystallize upon cooling with only a small increase in periodicity. During heating these bundles show a higher melting temperature than their crystalline counterparts.

(2) An anisotropic drop model has been presented to explain the relative orientation of the smectic layers, the crystalline lamellae and the bundle axis. The anisotropic drop model forms a conserved system providing a basis for the reversibility of the smectic periodicity during cooling and heating.

(3) In the supercooled melt, the smectic ordering is followed by crystallisation, during which crystals grow epitaxially on the surface of the smectic bundles. This leads to a new picture of the shish-kebab structure in which smectic bundles rather than extended-chain crystals play the role of the shish. The notion of shear-induced pre-ordering concerns the formation of primary nuclei, which are fringed micelles rather than folded-chain lamellae.

(4) The survival of the smectic domains at a temperature well above the melting temperature of iPP crystals provides a natural explanation of the memory effect. The observation of SAXS intensity prior to WAXS signal at quiescent conditions can be attributed to 'frozen-in' smectic ordering, induced by shear or other deformations during sample preparation and/or during the production of the raw material.

(5) A higher molecular weight corresponds to a larger smectic periodicity; a larger molecular weight distribution (or larger ratio between long and short chains) leads to an increased concentration of smectic bundles.

(6) The crystallisation process of the smectic regions indicates a metastable state during the transition from the high-temperature smectic state to crystals. This transient state can be identified with the low-temperature 'mesosphase' of iPP reported in the literature after fast quenching.

(7) Smectic domains rather than α crystallites serve as nuclei for the β phase.

Acknowledgements

The authors thank D. Frenkel (Amsterdam) and J. Groenewold (Delft) for valuable discussions, Z. Chen (Eindhoven) for performing the IR measurements and B. Norder (Delft) for carrying out the DSC. L. Zhang (East. Polytech. China) and M. Soliman (SABIC-Europe) are acknowledged for providing iPP. This work is part of the Softlink research program of the 'Stichting voor Fundamenteel Onderzoek der Materie' (FOM), which is financially supported by the 'Nederlandse Organisatie voor Wetenschappelijk Onderzoek' (NWO).

References

1 M. Muthukumar, *Philos. Trans. R. Soc. London, Ser. A*, 2003, **361**, 539.
2 K. Armitstead and G. Goldbeck-Wood, *Adv. Polym. Sci.*, 1992, **100**, 221.
3 D. M. Sadler and G. H. Gilmer, *Phys. Rev. Lett.*, 1986, **56**, 2708.
4 J. J. Point, *Macromolecules*, 1979, **12**, 770.
5 J. I. Lauritzen and J. D. Hoffman, *J. Res. Natl. Bur. Stand.*, 1960, **64**, 73.
6 J. D. Hoffman and R. L. Miller, *Polymer*, 1997, **38**, 3151.
7 G. Strobl, *Eur. Phys. J. E*, 2000, **3**, 165.
8 M. Imai, K. Mori, T. Mizukami, K. Kaji and T. Kanaya, *Polymer*, 1992, **33**, 4451.
9 M. Imai, K. Kaji and T. Kanaya, *Phys. Rev. Lett.*, 1993, **71**, 4162.
10 M. Imai, K. Kaji and T. Kanaya, *Macromolecules*, 1994, **27**, 7103.
11 M. Iami, K. Kaji and T. Kanaya, *Phys. Rev. B*, 1995, **52**, 12 696.
12 G. Matsuba, T. Kanaya, M. Saito, K. Kaji and K. Nishida, *Phys. Rev. E*, 2000, **62**, R1497.
13 A. J. Ryan, A. Fairclough, N. J. Terrill, P. D. Olmstead and W. C. K. Poon, *Faraday Discuss*, 1999, **112**, 13.
14 P. D. Olmsted, W. C. K. Poon, T. C. B. Mcleish, N. J. Terrill and A. J. Ryan, *Phys. Rev. Lett.*, 1998, **81**, 373.
15 E. L. Heeley *et al.*, *Faraday Discuss*, 2003, **122**, 343.
16 E. L. Heeley, A. V. Maidens, P. D. Olmstead, W. Bras, I. P. Dolbnya, J. P. A. Fairclough, N. J. Terrill and A. J. Ryan, *Macromolecules*, 2003, **36**, 3656.
17 T. A. Ezquerra, E. Lopez-Cabarcos, B. S. Hsiao and F. J. Balta-Calleja, *Phys. Rev. E*, 1996, **54**, 989.
18 G. Matsuba, K. Kaji, K. Nishida, T. Kanaya and M. Imai, *Macromolecules*, 1999, **32**, 8932.
19 G. Matsuba, K. Kaji and K. Nishida, *Polym. J.*, 1988, **31**, 72.
20 P. J. Flory, *Proc. R. Soc., London, Ser. A*, 1956, **234**, 60.
21 T. Shimada, M. Doi and K. Okano, *J. Chem. Phys.*, 1988, **88**, 2815.
22 M. Doi, T. Shimada and K. Okano, *J. Chem. Phys.*, 1988, **88**, 4070.
23 M. Doi and S. F. Edwards, *The Theory of Polymer Dynamics*, Clarendon, Oxford, 1986.
24 Y. A. Akpalu and E. J. Amis, *J. Chem. Phys.*, 1999, **111**, 8686.
25 Y. A. Akpalu and E. J. Amis, *J. Chem. Phys.*, 2000, **113**, 392.
26 M. Muthukumar and P. Welch, *Polymer*, 2001, **41**, 8833.
27 G. S. Y. Yeh and P. H. Geil, *J. Macromol. Sci., Phys. B*, 1967, **1**, 235.
28 G. S. Y. Yeh and P. H. Geil, *J. Macromol. Sci., Phys. B*, 1967, **1**, 251.
29 J. M. Schultz, J. S. Lin, R. W. Hendricks, J. Peterman and R. M. Gohil, *J. Polym. Sci., Polym. Phys. Ed.*, 1981, **19**, 609.
30 K. Katayama, T. Amano and K. Nakamura, *Kolloid. Z. Z. Polym.*, 1968, **226**, 125.
31 W. Ostwald, *Z. Phys. Chem.*, 1897, **22**, 286.
32 M. Sutton, Y. S. Yang and J. Mainville, *Phys. Rev. Lett.*, 1989, **62**, 288.
33 P. R. Ten Wolde, M. J. Ruiz-Montero and D. Frenkel, *Phys. Rev. Lett.*, 1995, **75**, 2714.
34 Y. C. Shen and D. W. Oxtoby, *Phys. Rev. Lett.*, 1996, **77**, 3585.

35 M. Hikosaka *et al.*, *J. Macromol. Sci. B*, 1992, **31**, 87.
36 A. Keller and S. Z. D. Cheng, *Polymer*, 1998, **39**, 4461.
37 E. B. Sirota and A. B. Herhold, *Science*, 1999, **283**, 529.
38 B. Wunderlich, M. Moller, J. Grebowicz and H. Baur, *Conformational Motion and Disorder in Low and High Molecular Mass Crystals*, Springer-Verlag, Berlin, 1988.
39 A. J. Pennings, *Crystal Growth*, Pergamon, Oxford, 1967, p. 389.
40 R. L. Miller, *Flow-induced crystallisation in polymer systems*, Gordon and Breach Science Publishers, New York, 1979.
41 J. K. Hobbs, A. D. L. Homphris and M. J. Miles, *Macromolecules*, 2001, **34**, 5508.
42 B. Lotz, J. C. Wittmann and A. J. Lovinger, *Polymer*, 1996, **37**, 4979.
43 G. Natta, M. Peraldo and P. Corradini, *Rend. Accad. Naz. Lincei*, 1959, **26**, 14.
44 P. Corradini, V. Petraccone, C. De Rosa and G. Guerra, *Macromolecules*, 1986, **19**, 2699.
45 J. A. Gailey and P. H. Ralston, *Plast. Eng. Trans.*, 1964, **4**, 29.
46 R. L. Miller, *Polymer*, 1960, **1**, 135.
47 S. F. Ran, X. H. Zong, D. F. Fang, B. S. Hsiao, B. Chu and R. A. Philips, *Macromolecules*, 2001, **34**, 2569.
48 S. Bruckner, S. V. Meille, V. Petraccone and B. Pirozzi, *Prog. Polym. Sci.*, 1991, **16**, 361.
49 Z. G. Wang, B. S. Hsiao, S. Srinivas, G. M. Brown, A. H. Tsuo, S. Z. D. Cheng and R. S. Stein, *Polymer*, 2001, **42**, 7561.
50 A. I. Nakatani and M. D. Dadmun, *Flow-induced structure in polymers*, American Chemical Society, 1995.
51 R. H. Somani, B. S. Hsiao and A. Nogales *et al.*, *Macromolecules*, 2000, **33**, 9385.
52 P. Jerschow and H. Janeschitz-Kriegl, *Int. Polym. Proc.*, 1997, **12**, 72.
53 R. H. Somani, L. Yang, B.S. Hsiao, P. K. Agarwal, H. A. Fruitwala and A. H. Tsou, *Macromolecules*, 2002, **35**, 9096.
54 R. H. Somani, L. Yang and B. S. Hsiao, *Physica A*, 2002, **304**, 145.
55 A. Elmoumni, H. H. Winter, A. J. Eaddon and H. Fruitwala, *Macromolecules*, 2003, **36**, 6453.
56 G. Kumaraswamy, J. A. Kornfield, F. Yeh and B. S. Hsiao, *Macromolecules*, 2002, **35**, 1762.
57 M. Seki, D. Thurman, J. Oberhauser and J. A. Kornfield, *Macromolecules*, 2002, **35**, 2583.
58 G. Kumaraswamy, A. M. Issaian and J. A. Kornfield, *Macromolecules*, 1999, **32**, 7537.
59 J. A. Kornfield, G. Kumaraswamy and A. M. Issaian, *Ind. Eng. Chem. Res.*, 2002, **41**, 6383.
60 P. K. Agarwal, R. H. Somani, W. Q. Weng, A. Mehta, L. Yang, S. F. Ran, L. Z. Liu and B. S. Hsiao, *Macromolecules*, 2003, **36**, 5226.
61 M. C. Garcia Gutieerez, G. C. Alfonso, C. Riekel and F. Azzurri, *Macromolecules*, 2003, **36**, in press.
62 L. B. Li and W. H. de Jeu, *Macromolecules*, 2003, **36**, 4862.
63 L. B. Li and W. H. de Jeu, *Phys. Rev. Lett.*, 2004, **92**, 75 506.
64 X. Y. Zhu, D. Y. Yan and Y. P. Fang, *J. Phys. Chem. B*, 2001, **105**, 12 461.
65 B. Chu and B. S. Hsiao, *Chem. Rev.*, 2001, **101**, 1727.
66 *Polymer Liquid Crystals*, ed. A. Ciffrri, W. R. Krigraum and R. B. Meyer, Academic Press, New York, 1982.
67 R. Matheson and P. J. Flory, *J. Phys. Chem.*, 1984, **88**, 6606.
68 P. G. De Gennes and P. Pincus, *Polym. Preprints*, 1977, **18**, 161.
69 S. S. Plotkin and J. N. Onuchic, *Q. Rev. Biophys.*, 2002, **35**, 111.
70 P. P. A. Smit, *Kolloid Z. Z. Polym.*, 1972, **250**, 27.
71 I. A. Hussein and M. C. Williams, *Macromolecules*, 2000, **33**, 520.
72 D. Frenkel and T. Schilling, *Phys. Rev. E*, 2002, **66**, 41 606.
73 B. Wunderlich, *Macromolecular Physics*, Academic Press, New York, 1980, vol. **3**.
74 G. C. Alfonso and A. Ziabicki, *Colloid Polym. Sci.*, 1995, **273**, 317.
75 H. G. Zachmann, *Kolloid Z. Z. Polym.*, 1969, **231**, 504.
76 K. Hermann, O. Gerngross and W. Abitz, *Z. Phys. Chem. (Munich)*, 1930, **B10**, 371.
77 P. J. Flory, *J. Am. Chem. Soc.*, 1962, **84**, 2857.
78 G. Allegra and S. V. Meille, *Phys. Chem. Chem. Phys.*, 1999, **1**, 5179.
79 H. Kraack, M. Deutsch and E. B. Sirota, *Macromolecules*, 2000, **33**, 6174.
80 J. Varga and J. Karger-Koscic, *Polymer*, 1995, **36**, 4877.
81 S. V. Meille, D. R. Ferro, S. Bruckner, A. Lovinger and F. J. Padden, *Macromolecules*, 1994, **27**, 2615.
82 R. H. Somani, B. S. Hsiao, A. Nogales, H. Fruitwala, S. Srinivas and A. H. Tsou, *Macromolecules*, 2001, **34**, 5902.
83 B. Lotz, *Eur. Phys. J. E*, 2000, **3**, 185.
84 L. Cartier, T. Okihara and B. Lotz, *Macromolecules*, 1997, **30**, 6313.

General Discussion

Professor Schick opened the discussion of Professor Müller's paper: Because of the detection limit of isothermal DSC experiments you use a self-nucleation technique to enhance the crystallization rate of the PPDX block.

The crystallization from the isotropic melt could be studied by following the changes in heat capacity too. By using temperature modulated DSC to measure heat capacity during a quasi-isothermal run you would be able to follow infinite slow processes because one is independent on the release rate of the crystallization enthalpy.

Professor Müller replied: I agree. Another possibility is to employ the melting scans after isothermal crystallisation performed at different times to determine the crystallisation kinetics one step at a time ("isothermal step crystallisation"), see for instance V. Balsamo, N. Urdaneta, L. Pérez, P. Carrizales, V. Abetz and A. J. Müller, *Eur. Polym. J.*, 2004, **40**, 1033. However, in these two experiment types, we have to be careful since these materials are highly sensitive to degradation. The advantage of the self-nucleation technique is the reduction in the overall time needed for crystallization in the isothermal step.

Dr Bromley asked: None of the AFM images, either of the individual lamellae or of the growing spherulite, show the assumed planar aspect of the lamellae. This leaves the images looking extremely similar to those taken of amyloid fibrils. Do the lamellae ever appear flat in AFM or is this a surface effect? Can the aspect ratio of the lamellae be tuned such that they begin to approach the rod symmetry of fibrils (which also have the chain length running perpendicular to their longest dimension)?

Professor Müller responded: In the AFM images, the samples are thin films prepared by solvent casting or from the melt (typically 0.5 μm in thickness). In a bulk sample, the spherulites can have large diameters of up to 100 μm. So, the spherulites are rather flattened in the direction of the film thickness, in the case of thin films. Nevertheless, the main reason why more often edge on lamellae are seen rather than flat on is geometrical, if you take a thin sheet and cut sections through it at different angles, only a very few of these sections will show large flat areas. In many other AFM pictures, it was evident that some flat on lamellae were present. An example is presented in Fig. 1; the bottom right hand side represents a PCL-rich spherulite within $D_{23}C_{77}$ isothermally growing at 40 °C (a similar experiment to that shown in Fig. 2 of our paper with another sample of the same copolymer), many flat on lamellae are seen on the surface of the spherulite. Even in Fig. 2 of our paper, although it is difficult to see, some lamellae that looked rather thick are just seen from an angle and are partially flat on.

Professor Lodge said: It appears from the paper that there is ambiguity in the state of the materials above both melting temperatures; is it microphase separated, or not?

Have you tried rheology, particularly the frequency dependent dynamic moduli, which can be a very sensitive diagnostic of the state of order? Also, is it possible to deuterate one monomer? SANS should be more sensitive than SAXS, given the similarity in chemical structures.

Professor Müller replied: We have not been able to gather direct experimental evidence in the melt that lets us ascertain the segregation state of the copolymers. Nevertheless, we have seen two clear glass transition temperatures by DSC in the block copolymers during cooling from the melt or during heating depending on the composition. The values of these T_g are very similar to those of

DOI: 10.1039/b413663m *Faraday Discuss.*, 2005, **128**, 321–339 **321**

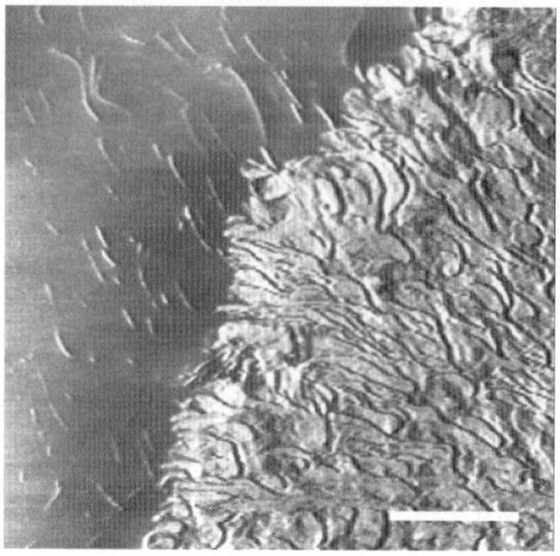

Fig. 1 AFM image of $D_{23}{}^{8}C_{77}{}^{27}$ diblock copolymer. The sample was first isothermally crystallized at 60 °C (only the PPDX could crystallize) and the picture was taken during isothermal crystallization at 40 °C. Scale bar represents 125 nm.

equivalent molecular weight homopolymers. Therefore, we conclude that the materials must be phase segregated in the melt. If they were to form one phase in the melt, even if crystallization could induced phase segregation during cooling, the two amorphous phases formed during cooling must according to thermodynamic rules of mixing be composed of a PPDX rich phase and a PCL rich phase. These phases, since they must contain at least a small amount of the other component are bound to have T_g values that are not equal to those corresponding to the homopolymers. Another indirect evidence is that we have performed separate blending experiments with PPDX and PCL homopolymers of very low molecular weights, down to a PCL of 600 g mol^{-1} and a PPDX of 5000 g mol^{-1} and no evidence of miscibility has been found. We have not been able to do rheological measurements because we have had limitations in the quantity of the block copolymers employed. The idea about deuterating one monomer is certainly worth exploring.

Dr Krebs asked: Can you explain the banding you observe in your spherulites? How do you see the lamellae twisting, given their 80 nm by 1 μm by 1 μm dimensions? What is the pitch of the twist and/or how does it relate to the spherulite and/or lamellae dimensions?

Professor Müller said: The lamellar thickness in PPDX is approximately 80 Å (or only 8 nm) and the exact value depends on the supercooling applied during crystallization. Therefore, the lamellae are ribbon like, so their lateral dimensions are usually shorter than their length. The length depends on crystallization conditions and can easily reach 1 μm or more. PPDX behaves very similar to many polyesters that exhibit banded spherulites when they crystallize from the melt. Banding is observed in PPDX and in PPDX-*b*-PCL block copolymers depending on the crystallization conditions.[1] The pitch has been determined and is an increasing function of the crystallisation temperature and resembles the behaviour reported for polyhydrohybutirate, PHB, poly(ε-caprolactone) and polyamides.[2]

The origin of ringed spherulites was studied in detailed by Keller[3] and by Keith and Padden.[4] The single ring extinction patterns in polymer spherulites have been interpreted as arising from a uniaxial indicatrix twisted about a normal to the optic axis. Doubled ringed spherulites, like those observed in PPDX, arise when the material possesses a biaxial indicatrix twisted about the optic normal leading to uneven spacing of the rings. Optically, the phenomenon is well understood, however, in morphological terms, its origin is controversial. The explanation that is compatible with the extinction pattern observed is that of ribbon like lamellae that grow radially within the spherulite and exhibit periodic twisting as they grow. Lamellar twisting has been observed for some polymers (including PPDX) by scanning electron microscopy and AFM.[5]

The classical interpretation of double banding or double-ringed spherulites in the literature is that of cooperative twisting of crystalline orientation about the radius or more specifically to twisting of crystalline units of optically biaxial character about their optic normals which are aligned precisely parallel to the radius. Keith has reinforced such a view by performing quantitative analyses of birefringence of ringed spherulites in poly(ethylene sebacate).[6] The reason behind the postulated regular twisting is controversial but one possible explanation is that lamellae twist to relieve surface stresses as they grow.[7]

1 M. A. Sabino, J. L. Feijoo and A. J. Müller, *Macromol. Chem. Phys.*, 2000, **201**, 2687; J. Albuerne, L. Márquez, A. J. Müller, J. M. Raquez, Ph. Degée, Ph. Dubois, V. Castelletto and I. Hamley, *Macromolecules*, 2003, **36**, 1633. M. A. Sabino, J. Albuerne, A. J. Müller, J. Brisson and R. E. Prud'homme, *Biomacromolecules*, 2004, **5**, 358.

2 H. D. Keith, F. J. Padden and T. P Russel (Jr.), *Macromolecules*, 1989, **22**, 666; P. J. Barham, A. Keller, E. L. Otum and P. A. Holmes, *J. Mater. Sci.*, 1984, **19**, 2781; M. A. Sabino, J. L. Feijoo and A. J. Müller, *Macromol. Chem. Phys.*, 2000, **201**, 2687.

3 A. Keller, *J. Polym. Sci.*, 1959, **39,** 151 and *J. Polym. Sci.*, 1955, **17,** 291.

4 H. D. Keith and F. J. Padden, *J. Polym. Sci.*, 1959, **39**, 101 and *J. Polym. Sci.*, 1959, **39**, 123.

5 Schultz, *J. Polym. Cryst.*, Oxford University Press, Washington, 2001 and Y. Jiang, J.-J. Zhou, L. Li, J. Xu, B.-H. Guo, Z.-M. Zhang, Q. Wu, G.-Q. Chen, L.-T. Weng, Z.-L. Cheung and C.-M. Chan, *Langmuir*, 2003, **19**, 7417. M. A. Sabino, J. Albuerne, A. J. Müller, J. Brisson and R. E. Prud'homme, *Biomacromolecules*, 2004, **5**, 358.

6 H. D. Keith, *Macromolecules*, 1982, **15**, 114.

7 U. W. Gedde, *Polymer Physics*, Chapman and Hall, London, 1995. Schultz, *J. Polym. Cryst.*, Oxford University Press, Washington, 2001.

Professor Schick asked: In Fig. 5 of your paper the inverse crystallization half time of the PCL block is shown for temperatures close to the melting temperature. The complex dependence on sample composition is discussed in terms of an increasing nucleation effect of the PPDX and increasing topological restrictions. Is this well supported by the measurements in that limited temperature range? In order to differentiate both effects it seems to be necessary to measure the overall crystallization rate in a wider temperature range at least including the maximum of crystallization rate.

Professor Müller replied: Extending the measurement range in Fig. 5, although desirable, is not easy in PCL or PCL components within diblock copolymers since the crystallization is very fast at lower crystallization temperatures. Furthermore, in the case of PCL (as well as in many other flexible polymers like polyethylene or polypropylene) a maximum in crystallization rate with T_c (in isothermal crystallization kinetic experiments as a function of T_c) will be extremely difficult to observe. The experimental limitation is that when the sample is quickly cooled to low isothermal crystallization temperatures (we employ 80 °C/min as a controlled cooling rate in our DSC) it starts to crystallize during the cooling period. In this case, one way to circumvent this limitation would be to use extremely high cooling rates to quench the samples from the melt to the crystallization temperature in order to prevent the crystallization upon cooling to T_c.

Nevertheless, we believe that the nucleation effects that are discussed regarding Fig. 5 are reasonable since other evidences that indicate a nucleation of PCL induced by PPDX have been presented in this publication and in a previous work by us (ref. 42 of the paper). They can be summarized as follows: (1) When a PPDX-*b*-PCL copolymer is cooled directly from the melt to a crystallization temperature low enough for both blocks to simultaneously crystallize, the crystallization kinetics is substantially slower than when PPDX is first crystallized and then the temperature is lowered so that PCL can crystallize (see ref. 42 of the present paper). (2) Figs. 5 and 7 in the present paper. (3) The behaviour observed after self-nucleation indicates that PPDX nucleates PCL, see Fig. 9 of the present paper and its discussion. (4) Evidence gathered by polarized optical microscopy that shows an enhancement in nucleation density of PCL when the PPDX has been previously crystallized.

Professor Hu asked: In your paper, you showed evidence for the enhancement of PCL crystallisation by the present crystallites of PPDX, particularly in the diblock copolymer of $D_{23}C_{77}$. But in Fig. 2 of your paper, we see PCL block initiating its crystallization from the area outside of that rich in the crystallites of PPDX. Does this picture belong to an accidental occurrence?

Professor Müller responded: We have obtained several evidences that indicate that the PPDX block is able to nucleate the PCL block in PPDX-*b*-PCL diblock copolymers (see also my previous response). All these evidences have been observed in bulk samples. Your question refers to AFM images obtained in thin films (approx. 0.5 μm thick) where surface effects may affect the nucleation of the material. In any case, we did not observed by AFM morphological evidences of this nucleation phenomenon as you rightly pointed out. However, in polarized optical microscopy experiments in thicker films (approx. 25–35 μm thick), very clear nucleation effects were observed whereby, after the crystallization of PPDX, there was a higher nucleation density for the PCL as compared to equivalent experiments with PCL homopolymer.

Professor Gido opened the discussion of Professor Hu's paper: (1) Since cutting the junction point between the coil and crystallizable blocks in your simulations is an artificial process that can be easily done in a simulation but not experimentally, I am not sure how you can say anything about how your observed switch from perpendicular to parallel crystalline chain stem alignment relates to experiments.

(2) At small undercoolings, the size of a stable nucleus may be quite large and have trouble fitting between the confining walls of the surrounding glassy layers. In this case, larger nuclei can form in the perpendicular chain orientation. This may favor the perpendicular orientation even for the case of clipped diblock junction points. Have you considered simulations of isothermal crystallization at small undercoolings?

Professor Hu replied: (1) In our simulations, we actually switched the confinement from the block junction dominant to the microphase interface dominant. This switch corresponds to a change of case from short chains to long chains in the diblock copolymer experiments. In the case of long chains, the effect of block junction becomes minor and the confinement by the microphase interface is dominant; therefore, parallel orientations of crystallites are often observed, see *e.g.* ref. 6 and 7 in our paper, and more recently reported by Huang *et al.*[1]

(2) The free energy barrier for crystal nucleation depends on both the surface area and the body volume of crystallites rather than on their shapes. The crystallites parallel to the lamellar compartment can grow in large and eventually approach at their critical size for nucleation. Therefore, the perpendicular preference seems not to be necessary. The time window of our current simulations was limited at the molecular level. It is far from the laboratory time-scale in waiting for the crystal nucleation under small supercoolings, so it seems formidable for us to simulate isothermal crystallization under the small supercoolings comparable to the experiments.

1 P. Huang, L. Zhu, Y. Guo, Q. Ge, A. J. Jing, W. Y. Chen, R. P. Quirk, S. Z. D. Cheng, E. L. Thomas, B. Lotz, B. S. Hsiao, C. A. Avila-Orta and I. Sics, *Macromolecules*, 2004, **37**, 3689–3698.

Dr Hamley asked: How will your conclusions depend on the volume fraction of the amorphous block? Whether parallel or perpendicular crystallisation is observed will depend in part on the interfacial area per junction. Also, have you considered the effect of crystallisation quench depth. Zhu, Cheng and co-workers have shown for PS-PEO diblocks that the stem orientation varies depending on supercooling. In particular, they observed tilted stems as well as parallel and perpendicular orientations for a wide range of undercoolings. Did you observe any tilted configurations in your simulations?

Professor Hu replied: Change in volume fractions of the amorphous block away from the symmetric case may lead to the cylindrical and spherical microphase compartments. These cases are under our investigation in the same way as the lamellar case. In principle, our conclusions here shall be applicable to these cases, since both the block junction and the microphase interfaces are still there.

We share the idea that the preferred orientation depends in part on the interfacial area per junction. We regard this dependence as mainly related with the whole chain length. The large interfacial area per junction corresponds to the case of long chains in which the confinement of microphase interfaces is dominant and hence the parallel orientations of crystallites are favored.

We have indeed checked the variable quench depth of crystallization in our simulations. Under small supercoolings, we observed the preferred perpendicular orientation in agreement with Zhu *et al.*'s experiments; but under large supercoolings, we only found random orientations without any preference on tilted or parallel.

Professor de Jeu said: In experiments of confined PEO crystallization in uniform lamellar block copolymer films, at small supercooling the crystalline stems are perpendicular to the block interfaces and want to fit the sublayer thickness by an integer or half-integer number of folds.[1] This even can lead to an expansion of the thickness of the crystalline layers (in this particular case in the rubbery surroundings of the amorphous block), in spite of the decrease in volume upon crystallization. Hence I would expect the results of the computer simulations with fixed boundary conditions to be extremely sensitive on whether the extended chain length chosen is close to a (half)multiple of the imposed layer thickness. If this is indeed true, any statement about the orientation of the crystal stems would also depend strongly on this condition.

1 R. Opitz, D. M. Lambreva and W. H. de Jeu, *Macromol.*, 2002, **35**, 6930.

Professor Hu replied: In Fig. 4a of our paper, the length of crystallizable blocks is 16 lattice sites and the lamellar microphase thickness is about 8. We indeed observed that the perpendicular crystalline blocks fit for the compartment thickness by once folding. However, we also found that the preferred perpendicular orientation was initiated at a very early stage of crystallization from the confined melt, far before the crystalline stems have reached the boundary of the microphase compartment. Therefore, we expect that at late stages of crystallization the integer folding may only serve as an additional thermodynamic stability to the crystallites, and this additional stability may not be able to drive the large crystallites rotating with hard restrictions from both the block junction and the narrow space.

Dr Hamley commented: I would like to emphasize that the observed result of perpendicular stem orientation cannot be universal, as indicated by experiments by ourselves and others. If the crystalline stem is longer than the preferred crystal domain thickness, parallel folding should be preferred, as noted in this discussion by de Jeu. Also, if the crystal stem is very long, parallel folding will be energetically/entropically preferable compared to multiple folds in a perpendicular orientation.

Professor Hu replied: The concerning question which is the same as in Professor de Jeu's comment is, at which stage of crystallization has the crystallite orientation been dictated? According to our paper, the answer is at the very early stage of crystallization, *i.e.* the primary crystal nucleation in the confined melt. The length of crystalline stems shall, however, be assigned to a consequence of the late stages of crystallization.

Professor Iwuoha asked: Isn't the simulation technique lacking in not accommodating a model involving equilibrium between the (i) stretched crystallisable chains and (ii) broken connection between block copolymer chains?

Professor Hu answered: If I understand your question correctly, I can say that we treated them as individual but parallel cases in our paper. In principle, the simulation technique allows us to model a thermodynamic equilibrium between stretched and broken connections, by assuming that the free chain ends make strong connections between different components when they meet at the microphase interfaces and this strong connection will be broken when the block coils become over-stretched. However, this situation belongs to another story beyond our current interest.

Professor Yamamoto commented: Polymer crystallization, especially its onset, is very sensitive to conditions, the system size, boundary conditions, chain length, distribution of chain-ends, *etc.* The authors are discussing chain orientations in the crystalline domains, for both the connected and the

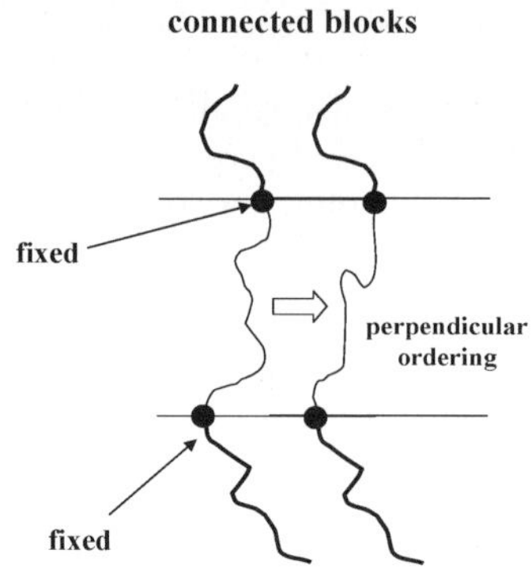

Fig. 2 Chain orientations for connected blocks.

disconnected blocks. The results for the connected blocks, *e.g.* the appearance of perpendicular orientation, are those readily expected (Fig. 2). However, it seems that the conclusions about the disconnected blocks depend on various assumptions implicitly made.

The general tendency of the chain ends toward segregation to the surfaces or interfaces between different phases is well recognized. The way how they segregate can act as decisive triggers to the later process of crystallization (Fig. 3); types a, b, c will lead to quite distinct chain orientations of the crystals.

The problem of the present work I noticed is that the way of chain-end segregation, including whether they segregate or not (type c in Fig. 3), can be affected by the starting molecular model; the present authors take the conventional lattice model. For example, our previous MD studies on the surface freezing of n-alkanes showed strong preference of the chain ends to be placed at the surfaces, which led to the predominance of perpendicularly orientated crystals. The chain-end accumulation on the surface was considered to be closely related with the lower free energy of the surfaces with abundant chain ends.

Professor Hu replied: We did not check the distribution of chain ends for the disconnected crystallisable blocks in the lamellar compartments. Even if the chain ends were rich at the interfaces, I would not expect any obvious correlation of this distribution with the observed parallel preference in crystallite orientations.

Professor Müller commented: The prediction shown in Fig. 1 of the paper where the lamellae are percolated seems to be in line with experiments in diblock copolymers where perfect confinement without any grain border contacts are very rarely encountered. This is the reason why Avrami indexes for strong segregated crystallisable components that form lamellae are nearly always much greater than unity.

Professor Hu opened the discussion of Professor Schick's paper: Is it possible to run temperature modulation in the fast-scan calorimeter?

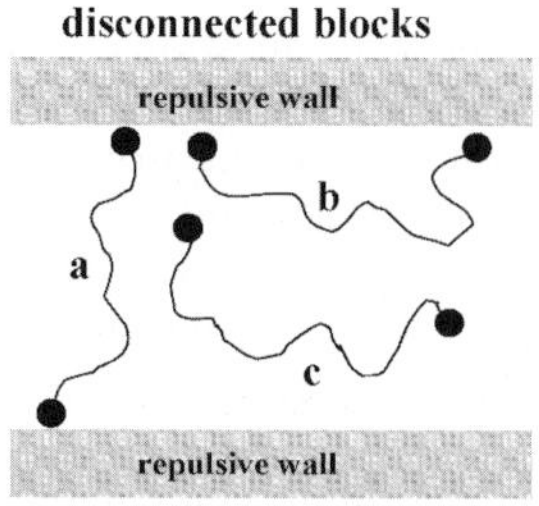

Fig. 3 Effect of type of segregation on chain orientation.

Professor Schick replied: Yes, temperature modulation is possible and successfully applied in two fields so far. Because of the small heat capacity of the sensor itself the thin film calorimeter allows temperature modulation at frequencies up to kilohertz. Combining the thin film calorimeter with conventional DSC we cover a frequency range from 10^{-4} Hz to 10^3 Hz. This allows heat capacity spectroscopy in a wide frequency range.

The other interesting application makes again use of the small heat capacity of the sensor. It allows us to study heat capacity of thin film samples. It is, as an example, possible to measure the glass transition in polystyrene films with thicknesses below 10 nm.

Professor Müller asked: (1) Have you studied the effect of superheating on the fast rate measurements?

(2) Why is the reorganisation process so fast as compared to crystallisation kinetics in a rather slow crystallising polymer like PET?

Professor Schick answered: (1) We see the superheating, *e.g.* Fig. 2B of the paper where the low melting peak shifts to higher temperature with increasing heating rate. The superheating of PET melting is in the order of 10 K at a heating rate of 2700 K s^{-1}. This shift is not due to thermal lag or other experimental artefacts because the glass transition temperature shifts in the expected range as discussed in ref. 13.

For isotactic polystyrene (iPS) superheating is even more pronounced as shown in Fig. 4. At 500 K s^{-1} superheating is in the order of 25 $\pm$ 5 K. In the case of semicrystalline iPS an independent check of the temperature scale is possible from the glass transition temperature. For heating rates of 0.17 K s^{-1} and 500 K s^{-1} T_g shifts for about 13 K as expected from the Vogel–Fulcher–Tamman (VFT) behaviour obtained from lower rates.[1] But for high heating rates there is still an uncertainty of the absolute temperature of about $\pm$5 K.

(2) Reorganisation requires a non-random melt. The remaining structures (melt memory) allow a different crystallization pathway compared to crystallization from an isotropic melt. Assuming a large fraction of *trans* isomers in the melt just after crystal melting the following reorganization step requires only local rearrangements, which seem to be very fast. This has to be checked with other more specific methods like IR spectroscopy. The thin film sensor may be very helpful in this respect because fast scanning is needed for the IR study too.

1 S. Weyer, M. Merzlyakov and C. Schick, *Thermochim. Acta*, 2001, **377**, 85–96

Professor Ryan commented: The effect of reorganisation after melting at T_c + 20 K is readily observable in a time resolved SAXS experiment (at typical 10 K min^{-1} heating rates) as the

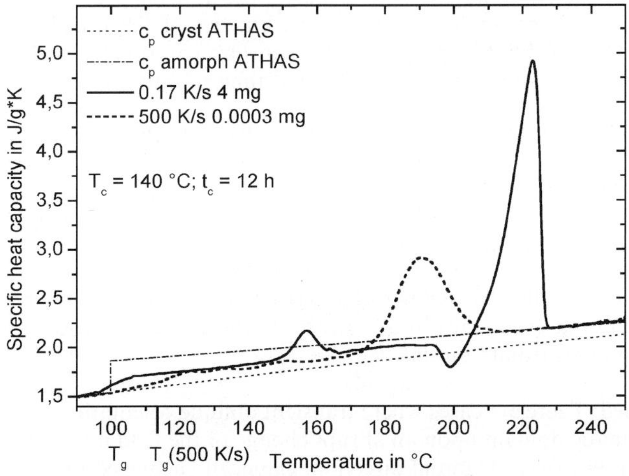

Fig. 4 Melting of iPS at different rates after isothermal crystallization at 140 °C for 12 h. Superheating is in the order of 25 $\pm$ 5 K.

d-spacing and lamellae thickness from correlation function increase, *i.e.* the crystals get thicker. This is readily understood as a transient self-seeding with thick crystals growing on the base of thinner.

Professor Schick replied: At low heating rates (10 K min^{-1}) a continuous melting–reorganisation process is observed. At fast heating we are able to prevent reorganisation during the heating scan. What we have studied, see Figs. 5 and 6 of the paper, is the reorganization of a not totally but mostly molten sample. Only a very few lamellae remain at 223 °C after fast heating. Therefore I believe the reorganization process after fast heating must be different from what you describe for the case of slow heating. If only the remaining lamellae grow large scale rearrangements are needed. Such a process would not be as fast as we see it. I see the reorganization process after fast heating as a crystallisation from a pre-ordered melt because there is not enough time to randomize the chain conformation after melting the crystals.

Professor Hu commented: The rigidity of PET may retard the process of disentanglement from the melt and hence make the crystallization slow. If the chains keep their disentangled states immediately after melting, the subsequent recrystallization shall be much faster. I recall a paper[1] published by Professor Strobl's group discussing this point.

1 J. Schmidtke, G. Strobl and T. Thurn-Albrecht, *Macromolecules*, 1997, **30**, 5804–5821.

Professor Schick said: I totally agree with Strobl's view. That is just what we want to prove by experiments, which have not been possible over such a broad temperature range before.

Professor Lodge asked: In the context of Fig. 7 in the paper , I am curious as to whether there is any "melt memory" above the equilibrium melting temperature? If you wait at T_{max} for long enough, doesn't the effect go away if $T_{max} \geq T_m^{\infty}$?

Professor Schick replied: I expect this but we have not been able to check it experimentally so far. It will be one of our next experiments.

Professor Ryan commented: One would expect the melt memory to have a characteristic time associated with the relaxation of straight unentangled chain segments. The pom-pom constitutive equation might be able to predict this characteristic time and differentiate between the effects of crystal thickness and the overall molecular weight.

Professor Schick answered: This is one of the interesting experiments one can try to realize with the fast heating calorimeter. If we are able to detect the vanishing of the melt memory above a certain temperature it will be possible to study the kinetics of the process and to compare with model predictions. One of the advantages of the thin film calorimeter is the short time constant which allow isothermal experiments with millisecond time resolution, see *e.g.* ref. 12 of the paper.

Professor Kabanov asked: Did you check the dependence of the melt memory on the degree of polymerization?

Professor Schick said: Not so far. But if we want to answer the questions by Professors Lodge and Ryan we have to study the influence of the molar mass too.

Dr Perez Mendez opened the discussion of Dr Davidson's paper: I find the rotation of the nematic director by 90°, resulting from only the increase of the magnetic field, not its direction, very interesting. Nothing intermediate but 90°.

Dr Davidson replied: There are cases where transient oblique orientations are observed during the realignment of a nematic domain upon an abrupt change of the field intensity. The reorientation of a nematic domain at 90° from its initial direction, classically achieved by a sudden change of field direction, does not proceed by a uniform rotation of the director but involves a transient hydrodynamic instability where domains are formed that slowly and uniformly rotate towards the new orientation. This process has been studied in detail with the model suspensions of the tobacco mosaic virus (TMV) by Meyer and Fraden.[1] The reorientation of the nematic goethite suspensions proceeds in the same way and then transient intermediate orientations can be observed on a time scale of hours.[2]

1 S. Fraden, A. J. Hurd, R. B. Meyer, M. Cahoon and D. L. D. Caspar, *J. Phys. Colloq.*, 1985, **46**, 85.
2 B. Lemaire, P. Davidson, D. Petermann, P. Panine, I. Dozov, D. Stoenescu and J. P. Jolivet, *Eur. Phys. J. E*, 2004, **13**, 309.

Dr Borsali asked: Do you expect similar results (magnetic field) when applying shear to the system, or both shear and magnetic field?

Dr Davidson answered: Nematic goethite suspensions have been submitted to shear-flow in a Couette cell and studied *in situ* at the ID02 experimental station of the European Synchrotron Radiation Facility in Grenoble. We observed the alignment of the nanorods (unpublished data) parallel to the shear velocity, as intuitively expected for quite elongated objects like these ones.[1] We did not observe any particular and unexpected orientation change upon shear-rate variation. The magnetic properties of goethite are more unusual and another type of experiment involving both shear flows and magnetic fields should therefore be devised. Such experiments are quite feasible and have sometimes been performed in the early studies of thermotropic liquid crystals.[1] Due care will have to be paid to precisely adjust the strengths of both the flow and the field so that an interesting balance can be reached.

1 P. G. De Gennes, *The Physics of Liquid Crystals*, Oxford University Press, Oxford, 1979, ch. 5.

Professor Penfold asked: Would you expect a larger effect than normally encountered in the magnetic alignment of nematic phases, as the magnetic moment/nanorod is higher than normally encountered? Polarised neutron small angle scattering could be used to identify the magnetic contribution; see published work on scattering for ferrofluids.

Dr Davidson replied: Yes, we definitely observe much larger magnetic effects than normally encountered with usual organic liquid crystals. For instance, the Fredericks thresholds beyond which magnetic-field alignment prevails over surface anchoring effects are about 50 times smaller than those reported for the model suspensions of TMV and about 30 times smaller than those observed with usual thermotropic liquid crystals. The magneto-optic coefficient that describes the susceptibility to field-induced alignment of the isotropic suspensions of goethite are at least 10^5 times larger than that of the TMV suspensions. These mineral colloids are indeed very unconventional in this respect and illustrate the interest of mineral entities in the field of soft matter.

Polarised neutron small angle scattering experiments could indeed be used to confirm our interpretations. This technique has already been successfully used many times to study the structure of ferrofluid suspensions.[1,2] More information about the orientation, with respect to the magnetic field, of the remanent magnetic moments of the particles and of their magnetic interactions could in principle be retrieved. Such experiments should certainly be performed; they would benefit from the use of heavy water (in the final stage of synthesis) to reduce incoherent scattering and to try contrast variation techniques.

1 A. Wiedenmann, *Physica B*, 2001, **297**, 226.
2 F. Gazeau, F. Boué, E. Dubois and R. Perzynski, *J. Phys.: Condens. Matter*, 2003, **15**, S1305.

Dr Semenov commented: The observed value of the orientational order parameter is significantly higher than the Onsager prediction for hard rods at the nematic transition, and also higher than the order parameter obtained in simulations of hard rods of finite length. Also the observed critical concentration at the nematic transition is about five times lower than that predicted for hard-core rods with the appropriate ratio of length to width.

The discrepancy may or may not be due to the Coulomb interactions between the rods. It might be useful to clarify the issue by a quantitative comparison of the experimental data with theories of nematic transitions in systems of charged rods, for example, by I. A. Nyrkova *et al.*[1]

1 I. A. Nyrkova, N. P. Shusharina and A. R. Khokhlov, *Macromol. Theory Simul.*, 1997, **6**, 965.

Dr Davidson responded: There are two sides to this question. Let us first focus on the values of the critical concentrations at the nematic transition. Directly applying Onsager's theory[1–3] with particle length $L \sim 150$ nm and diameter $D \sim 25$ nm would yield volume fractions at the transition of $\phi_I \sim 55\%$ and $\phi_N \sim 70\%$ for the isotropic and nematic phases, respectively. This is very much larger than the experimental values of 5.5% and 8.5%. However, goethite nanorods are charged and their surface charge density must also be considered. Electrostatic interactions have been incorporated into the Onsager model by a number of authors; these interactions usually tend to decrease the volume fraction at which nematic ordering takes place. In spite of a relatively large ionic strength ($I \sim 0.04$ M), the predicted volume fractions become $\phi_I \sim 34\%$ and $\phi_N \sim 46\%$, still much larger than the experimental ones. At this point, it should be recalled that the Onsager model and many related theories are virial expansions, generally truncated at the second term B_2. This approximation is valid only for very elongated particles, which is not the case for goethite rods. Then, comparisons should instead be made with numerical simulations[4] of hard rods, still keeping in mind electrostatic interactions by using an effective diameter (a rather rough approach). Consequently, predicted values of the volume fractions are around 15% (with a very small biphasic gap). Although these values are in much better agreement with the experimental ones, they are not quantitative yet. This may be due to two additional reasons: goethite particles are not cylindrical but lath-shaped and the suspensions are very polydisperse. In fact, compared to the model suspensions of viruses, suspensions of goethite nanorods are not very good systems to check the validity of theories of the nematic ordering. Their main interest lie in their magnetic properties. This discussion was already published in one of our previous publications,[5] but I agree that it should nevertheless be extended in even more detail.

Let us now discuss the large values of the nematic order parameter *S*. Onsager predicted a strongly first-order isotropic/nematic phase transition and therefore a large jump of *S* from 0 to 0.79.[1,2] This should be compared to experimental values of $S = 0.85 \pm 0.10$ obtained by SAXS and birefringence. The error bars more or less reflect the drastic assumptions involved in the data treatment performed to extract *S* by both techniques. However, the visual aspect of the SAXS diffuse scattering spots (from which *S* is obtained) qualitatively confirms the unusually high orientational order of these suspensions. There have been theoretical predictions[3] in literature of highly ordered nematic phases and the system should certainly be considered in this light. However, magnetic-field effects may also be at work here. We observed that *S* increases from 0.85 up to 0.95 at field intensities around 100 mT, which may be due to the freezing of fluctuations by the magnetic field. Similar effects have been reported for the influence of electric fields on thermotropic nematic phases. More detailed studies of such type of effects are made difficult by our inability to achieve proper nematic anchoring on cell surfaces.

1 L. Onsager, *Ann. N.Y. Acad. Sci.*, 1949, **51**, 627.
2 G. J. Vroege and H. N. W. Lekkerkerker, *Rep. Prog. Phys.*, 1992, **55**, 1241.
3 I. A. Nyrkova, N. P. Shusharina and A. R. Khokhlov, *Macromol. Theory Simul.*, 1997, **6**, 965.
4 P. Bolhuis and D. Frenkel, *J. Chem. Phys.*, 1997, **106**, 666.
5 B. Lemaire, P. Davidson, D. Petermann, P. Panine, I. Dozov, D. Stoenescu and J. P. Jolivet, *Eur. Phys. J. E*, 2004, **13**, 309.

Dr Perez Mendez asked: Is it a first order phase transition and is it discontinuous with temperature or is it changing in a broader range?

Dr Davidson replied: The isotropic/nematic phase transition is generally of first order and this is also the case of goethite suspensions in zero-field. Temperature has very little effect on the system, at least till 100 °C, and the phase transition remains first order. However, when a magnetic field is applied to the suspensions, it breaks the isotropic symmetry of the disordered phase (*i.e.* isotropic in zero-field) that is called "paranematic" and the coexistence region is limited. If the nematic aligns parallel to the field, then a continuous path links the paranematic to the usual nematic phase, as predicted by Khokhlov and Semenov (see Fig. 4a in our article).[1] If the nematic aligns perpendicular to the field, then it becomes biaxial under the influence of the field that breaks its uniaxial symmetry and there can be a second-order phase transition from the paranematic phase to the biaxial nematic phase (see Figs. 4b and 8 in our article).

1 A. R. Khokhlov and A. N. Semenov, *Macromolecules*, 1982, **15**, 1272.

Dr Fahmi asked: How are the sequences of the magnetic field applied, immediately or is there some time in between? Did you measure the magnetic properties by hysteresis? They may convert to superparamagnets!

Dr Davidson replied: (1) We have performed numerous experiments with these suspensions and we can distinguish two cases: (i) the isotropic phase at coexistence has a response time of about 10 ms. Therefore, there is no influence of magnetic-field history. (ii) the nematic phase, in contrast, may sometimes take hours to re-orient, depending on concentration, field intensity, surface anchoring, *etc*. Then, attention must be paid to the different sequences of magnetisation and rest, and it is useful to control (by SAXS or polarised light microscopy) that equilibrium has been reached.

(2) We investigated the magnetic properties of goethite suspensions with help of a SQUID magnetometer.[1,2] We could measure a remanent magnetic moment of about $1000\mu_B$ per particle by freezing the suspensions under field below 0 °C and switching off the field. The spin-flop transition of these particles is beyond at least 4 T at room temperature, that is, well beyond the field intensities that we used in our experiments. This means that the remanent magnetic moment is locked inside each particle. At temperatures above 0 °C, in zero field, the nanorods can reorient (at least in the isotropic phase) and no magnetisation can be detected. In the field of ferrofluids, this is called a case of "extrinsic superparamagnetism".

1 B. Lemaire, P. Davidson, J. Ferré, J. P. Jamet, D. Petermann, P. Panine, I. Dozov and J. P. Jolivet, *Eur. Phys. J. E*, 2004, **13**, 291.
2 B. Lemaire, P. Davidson, D. Petermann, P. Panine, I. Dozov, D. Stoenescu and J. P. Jolivet, *Eur. Phys. J. E*, 2004, **13**, 309.

Professor Hillmyer asked: Has there been any effort, published in something other than a thesis, on control of the nanorod size, shape, polydispersity?

Dr Davidson answered: I am not aware of much published work along these lines. The general principles of solution chemistry that need to be used for this purpose are now well established.[1] From the few conversations I had on this topic with chemists, I had the feeling that controlling nanorod size, shape and polydispersity is certainly feasible. However, it would be long, tedious and not very original from the chemical point of view. Therefore, I still need to convince the chemists that it could be worthwhile.

1 J. P. Jolivet, *Metal Oxide Chemistry and Synthesis. From Solution to Solid State*, Wiley, Chichester, 2000.

Dr Geoghegan opened the discussion of Professor Sakurai's paper: A temperature gradient exists during spin coating of polymer films. Could you comment on the parallels between your results and the spin coating of polymer blend films?

Professor Sakurai replied: Let me comment on the two points on effects of temperature gradients and in turn (convective) flows for the spin coating: (1) surface undulation or wrinkling and (2) bulk morphology of phase-separated domains.

(1) Surface undulation or wrinkling. The top surface of the sample solution is fixed by a glass plate in the experiment presented in the Faraday Discussion meeting. Therefore, no relevant insight can be extracted from this experiment. However, we have studied surface undulation of the polymer film cast from solution of which bottom surface was heated.[1–3] During solvent evaporation, convection takes place due to temperature difference, which is promoted by a removal of the latent heat as well as heating from the bottom. The convective flow on the free (top) surface wrinkles the top surface of the polymer solution. Surface tension distribution (inhomogeneity) on the free surface enhances convection and in turn wrinkling the top surface. When the solvent completely

evaporated, we observed a surface undulation of the as-cast polymer film due to modulation of polymer concentration formed in the preceding polymer solution.[1] The most commonly observed pattern of the surface undulation may be a polygonal mesh pattern, which seems to be an imprint of the convection pattern (looks like an hexagonal mesh).[1]

An exception is a peculiar concentric ring pattern, which is formed on the free surface of the ascast block copolymer film when the casting solution is heated from the bottom plate.[2,3] Although the relationship between morphology of the microphase separation and the peculiar pattern formation is not well understood, lamellar-forming block copolymers mostly exhibit the peculiar concentric ring pattern. This may be due to formation of a skin layer on the top surface of the polymer solution during solvent evaporation. Convections wrinkle the skin and the wrinkles of the skin are finally fixed as a surface undulation (exhibiting concentric ring pattern) of the as-cast polymer film.

(2) Bulk morphology of phase-separated domains. First of all, note that it becomes more difficult to induce convection in a thinner layer of fluid and a tremendously large temperature difference is required. Nevertheless, if a convection would be relevant in the spin coating, our results can be applied. Since the size of a convection cell is of the order of a thickness of the fluid layer, we can find that the convection governs the phase-separated domains, of which size is down to about one-tenth of a layer thickness of a convecting solution, in such a way that the convection controls domain size and organizes spatial distribution of domains. No influence may be found on the size below about one-tenth of the layer thickness of the convecting solution. As for the spin coating, since the thickness of a coated film is of the order of nanometer, it would be possible to organize nano-structures by convection, if it could be induced in such the thin layer of the solution.

1 S. Sakurai, C. Furukawa, A. Okutsu, A. Miyoshi and S. Nomura, *Polymer*, 2002, **43**, 3359–3364
2 S. Sakurai, K. Tanaka and S. Nomura, *Macromolecules*, 1992, **25**, 7066–7068.
3 S. Sakurai, N. Tameno and S. Nomura, *Forma*, 1993, **8**, 309–314.

Professor Gido said: The convection driven structure that you show seem to evolve continuously over time, increasing in length scale and sometimes changing the structural connectivity of the phases. In order to use this method to form a useful material it would perhaps be desirable to achieve a constant or fixed structure. This might be done with chemical methods, perhaps photochemical reactions that fix the structure. Or a constant structure might be achieved by continuously flipping the sample cell over thus reversing both the gravitational field and the temperature gradient on a time scale (perhaps about an hour) that achieves some sort of resonance with the physical processes generating the patterns. Can you comment on methods to achieve a stable structure?

Professor Sakurai responded: Although it is commented that the convection-driven structure seems to evolve and increase in length scale with sometimes changing the structural connectivity of the phases, I think it is only appropriate for the transient process before the formation of the dissipative structure (see Fig. 6(a) to (c) in the paper). When the dissipative structure is formed (see Fig. 6(d) in the paper), such the long-range structural change does not occur, as discussed in the main text. However, it is true that the coalescence of isolated polybutadiene (PB)-rich large domains across the bright string (the border between two neighboring convection cells) or spontaneous shrinking-and-annihilation of a PB-rich large domain (which looks like being swamped down in the bottom layer) was occasionally (not frequently) observed. Therefore, the structure cannot be referred to as a dissipative structure in a rigorous sense. Nevertheless, we dare to refer such the structure as shown in Fig. 6(d) as a dissipative structure in the current study, because the structure is temporarily steady within an hour or so. Since the comments raised are valuable from the viewpoint of preservation of structures formed in the less viscous solution, I would like to continue the discussion.

As pointed out in the comments, photochemical reactions are promising to fix in order to prevent an increase of the size and changes in morphology before reaching a steady state when a dissipative structure is formed. More important requirement to fix structure is ascribed to the fact that the solution viscosity is low so that the dissipative structure is delicate and can be easily spoiled by a perturbation due to an accidental swaying of the experimental set-up. Preservation of the dissipative structure is the first step of its utilization so that a possible photochemical reaction is promising.

The comment on the *resonance* of physical processes is worthy of deep consideration, although the *resonance* has nothing to do with preservation of domain structures. Actually, I give some conjectures for the case when the temperature difference is flipped continuously with a constant interval, namely;

$$T_{top}\text{: } T_1 \rightarrow T_2 \rightarrow T_1 \rightarrow T_2 \rightarrow T_1 \rightarrow \ldots \text{ (with a constant interval } \Delta t\text{)}$$

$$T_{bottom}\text{: } T_2 \rightarrow T_1 \rightarrow T_2 \rightarrow T_1 \rightarrow T_2 \rightarrow \ldots \text{ (with the same interval } \Delta t\text{)}$$

where T_{top} and T_{bottom} denote temperatures of the sample fluid at the top and bottom plates, respectively, and $\Delta T = T_2 - T_1$ (>0). Let us restrict the application of this procedure (temperature flipping) to the PS/PB = 50/50 blend solution; (1) when the initial co-continuous structure is formed (Fig. 6(a)), (2) when the convections are organizing the preceding phase-separated structures but the dissipative structure is not yet achieved (Fig. 6(c)), or (3) when the dissipative structure is already achieved (Fig. 6(d)).

(1) Application of the temperature flipping when the initial co-continuous structure is formed (Fig. 6(a)).

Since the *z* position (along the thickness (gravitational) direction), at which phase-separation sets in, changes upon the temperature flipping, the initial co-continuous domain structure (near the cooler plate, *i.e.*, the top plate at the initial state when $T_{top} = T_1$ and $T_{bottom} = T_2$ with $T_2 > T_1$) will stop growing and it will be going to disappear, and meanwhile a new co-continuous domain structure will be generated at a different *z*-position (near the cooler plate, *i.e.*, the bottom plate because the temperatures have been flipped, *viz.*, $T_{top} = T_2$ and $T_{bottom} = T_1$) and it will spontaneously grow. In case of a short duration of the interval Δt, such co-continuous structures will both survive because the domain dissolution is much slower than generation of phase-separated domains. At the second temperature flipping when the initial temperatures are retrieved (*i.e.*, $T_{top} = T_1$ and $T_{bottom} = T_2$), the initial co-continuous domain structure (near the top plate) retrieve ability of coarsening but it would take time due to reconstruction of modulation in the blend compositions, which is being homogenized at the first temperature flipping. On the other hand, the co-continuous structure having appeared upon the first temperature flipping (near the bottom plate) will stop growing and it will be going to disappear, but the domain dissolution would take time like that of the initial co-continuous structure would at the first temperature flipping. Thus, both of the co-continuous structures continuously survive, and the *resonance* will be achieved after a certain number of temperature flippings. On the contrary, if the interval Δt is too long, the initial co-continuous structure will completely disappear and meanwhile the co-continuous structure having appeared upon the first temperature flipping will alternatively prevail. After successive temperature flippings, such a cycle of alternative appearing of the different co-continuous domain structures will continue. For this particular case, the observed 2D images will look like oscillations (breathing), and the *resonance* will not be attained.

(2) Application of the temperature flipping when the convections are organizing the preceding phase-separated structures but the dissipative structure is not yet achieved (Fig. 6(c)).

For this particular case, the *resonance* is expected to attain. As discussed in the main text, the convections force PB-rich continuous phase (brighter phase) to be disconnected (and thus, PB clusters will be formed). However, if the temperatures are flipped before the disconnection of the PB-rich continuous phase takes place, the polystyrene (PS)-rich continuous phase will be going to be disconnected to retrieve stable morphological structure, which used to be existing before the convections set in, like the structure shown in Fig. 6(b), or even earlier one. This is because the convections cease upon the temperature flipping. Provided an appropriate interval Δt would balance those opposing tendencies of disconnections of the co-continuous domains, the *resonance* can be achieved and the co-continuous structure shown in Fig. 6(c) can be dynamically stabilized.

(3) Application of the temperature flipping when the dissipative structure is already achieved (Fig. 6(d)).

When the temperatures are flipped at the state of Fig. 6(d), convections cease and the bright string (mesh pattern) will disappear, because this is a convection pattern. Furthermore, tiny PS-rich droplets (dark droplets) will spontaneously appear and prevail inside of the isolated PB-rich large domains, and meanwhile tiny PB-rich droplets (bright droplets) will spontaneously appear and prevail in the matrix (outside of the isolated PB-rich large domains). Those droplets respectively undergo coarsening, and finally the original PB-rich large domains disappear and spatial regions that are originally enclosed by the interface of the PB-rich large domain may become ill defined. Thus, the temperature flipping always spoils the dissipative structure, irrespective of the duration of the interval of the temperature flipping (Δt). It is needless to say that the unfavorable situation stated above can be avoided by reducing the duration of interval Δt. However, the *resonance*, or in other words, more dynamically stable dissipative structure cannot be obtained by the temperature flipping for the case of the dissipative structure already formed at Fig. 6(d), because this structure is satisfactorily long living as far as the positive temperature difference ($\Delta T = T_2 - T_1 > 0$) is applied.

Dr Fahmi asked: Can you, through the temperature gradient between the two hot and cold plates, obtain the maximum orientation of block copolymer?

Professor Sakurai replied: The comment has nothing to do with the current study, because block copolymers were not used in the experiment. Nevertheless, let us consider the comment as "to what extent can the polymer chains be oriented by the convection?" An answer is as follows: The convective flow is so slow that it takes about half an hour to make a cyclic loop for this particular case of the polymer–DOP solution of 0.1 mm in thickness. Therefore, the orientation of the polymer chain may be negligible.

Professor Lodge asked: I wonder if you would care to comment on what you had hoped to see when you began these experiments, and where do you see this research going? As a possible contribution to the second question, have you considered using a combinatorial approach, for example by applying concentration and mean temperature gradients in the plane of the surface?

Professor Sakurai responded: As stated in the main text, one of the main purposes of this study was to form a dissipative structure in a polymeric system, in which it had been supposed to be impossible due to high viscosity. Furthermore, we hoped to conduct structural organization of the phase-separated domain structures of polymer blends with a smart fashion by utilizing self-organization ability of a polymer solution with the assistance of convection. Namely, it was tried to induce long-range spatial orders in the ordinary phase-separated domain structures. Normally, phase-separated polymer blends merely show a short-range order of domains such that the most probable distance of the nearest neighboring domains is almost constant as well as the almost constant size of the domains. We really hoped to construct hierarchies for the polymer blend phase separation! In this sense, we have succeeded in constructing such a dissipative structure as shown in Fig. 6(d) (looks like living (biological) tissue). In a rigorous sense, self-organization can only be attained in the non-equilibrium open system, therefore our system, continuously applying the temperature gradient across the sample fluid, can provide one of the rare examples of self-organization in polymeric systems.

Next, let me say a few words on a possible combinatorial approach for the study of convection-driven structural organization in a phase-separated polymer blend solution. For this approach, we should design an experiment to give results at once when some different values of a couple of control parameters are set simultaneously. For instance, in case of the measurement of a phase diagram of polymer blend, the experiment should be designed so as to simultaneously vary a combination of blend composition and temperature. When a gradient of the blend composition can be formed along an axis (let it be the x-axis) of a rectangular sample cell, apply a gradient of temperature along the y-axis. After holding this condition at a sufficient duration, the rectangular sample cell itself visually exhibits the phase diagram with opaque appearance for the phase separated region (transparent for the one phase region) in the x–y-plane, corresponding to the composition–temperature plot. To consider application of the combinatorial approach to the study of convection-driven structural organization, the problem is how the temperature gradient is applied. It is needless to say that the temperature gradient can only be imposed along the gravitational direction for this particular case to induce convection. However, with this situation, the composition gradient prepared for this experiment (along the x-axis, for instance) will be mixed up by thus induced convections. To begin with, what we would really like to observe is a two-dimensional (2d) pattern of the structural organization as well as the change of the 2d pattern as a function of the blend composition. However, a combination of the blend composition and the 2d pattern of the resulted structure cannot be in principle performed. Thus, it seems that the combinatorial approach is hardly applicable to the convection studies.

Finally, I would like to comment on the future of this study. We have succeeded in achieving self-organization of the polymer blend solutions. As a consequence, a dissipative structure on the micrometre scale such as is shown in Fig. 6(d) can be constructed. It is hierarchical and has different long-range orders on different length scales, reminiscent of living tissue. The important point here is that the structure formation is spontaneous so that it takes much less time than the manipulation approach where the parts of the hierarchical structure are assembled one by one (it takes astronomical time to produce a gram of material!). Thus, the self-organization process enables us to conduct one-step construction of a hierarchical structure from a simple mixture of its many different parts (*viz.* multicomponent mixture) by application of a gradient of any external field (*i.e.*, temperature gradient in this study but any other field may be possible). Without using special polymers having functional groups or characteristic chain architectures (like block copolymers), we could obtain a variety of hierarchical structures. However, what we have done is only a two-component system. In order to examine polymeric systems for such self-organization ability as is stated above, the study should be extended to a multicomponent polymer system. Extension to multicomponent system is aimed at not only achieving multifunctionality but also increasing the rank of the hierarchical structure. As a first trial, a ternary system (Poly1/Poly2/Poly3/solvent) is reasonable. The third component will be a homopolymer with a different chemical structure or a block copolymer (a ring block copolymer may be interesting!).

Professor Gido commented: In response to general curiosity among the assembled scholars as to a good application for the interesting microconvection patterns shown in this paper, I suggest that they could form the basis for the world's smallest lava lamp. A lava lamp is a device, invented in the 1960's, in which two immiscible, translucent and colored liquids with different coefficients of thermal expansion are simultaneously illuminated and driven in convective motion by a light bulb in the base of the lamp. See Fig. 5. It was generally agreed that a nano-lava lamp was beyond the current capabilities of the technology, but that a micro-lava lamp was certainly feasible.

Dr Semenov opened the discussion of Dr Li's paper: How good is the smectic order: for how many layers does the smectic structure persist?

Dr Li replied: Usually the correlation length is tens of nanometres (of the order of about 10 layers). In iPP1, it can be up to 20 layers. The situation depends somewhat on the type of material and the shear history.

Fig. 5 Lava lamps.

Dr Hamley said: You only observe one weak Bragg reflection in the so-called "smectic phase". Did you ever attempt to measure higher order reflections that would provide further evidence for the nature of the layered structure indicated by the shear alignment experiments?

Dr Li responded: As even the first-order smectic peak is weak, we did not expect to see the higher orders. Nevertheless we tried, but did not observe any scattering at higher-order positions. This is one of the reasons that we used shear alignment to demonstrate the layering.

Professor Kabanov commented: I don't understand your cartoon consisting of separated "rigid" helix and "flexible" coil segments. PP 3, helix becomes rigid only in the environment of other helices *i.e.* when it becomes a part of some three-dimensional ordered structure. Separated helical segments of PP chain cannot behave like rigid segments.

Dr Li replied: Indeed the cartoon might be somewhat misleading regarding this point. In our paper we stress that the conformational ordering (the formation of helices) and orientational ordering (alignment of helices) take place simultaneously.

Professor Lodge said: The "smectic" layering implies a particular length scale, is it the mean helical length from chain-to-chain, or from helical region to helical region on one chain? What sets this length scale? We are familiar with the persistence length, the entanglement length, and the chain radius of gyration; where does the new length scale come from?

Dr Li and **Professor de Jeu** responded: The length scale evidently stems from the sum of the average rigid helical sequence of at least 11 monomers (about 2.6 nm) and a random-coil part of about 1.4 nm. Apart from this experimental fact we do not understand its origin. We compared the various length scales you mentioned, but none of them matches to the smectic periodicity. The answer to this question may lie in the disordered state of the polymer melt and the non-thermodynamic nature of the shear-induced smectic ordering. Following the fringed-micelle model, local correlations between chain segments exist in the polymer melt. This provides a length scale (correlation length) depending on temperature which under normal conditions is not expressed in ordering.[1] The latter requires an external field like shear or a temperature change as in fast quenching. The X-ray and IR results show that upon shear the length of helical sequences in the smectic layers just reaches the critical persistence length for an isotropic–liquid crystal transition. Though this sets a length for the liquid crystal ordering, it does not explain why the rodlike structures do not grow further. Could the non-thermodynamic promotion of helices by shear provide a possible selection mechanism?

A second point is the large polydispersity of helical rods and random-coil sequences that must exist. Hence we have been rather surprised that a nematic-like phase is not preferred by the system. However, it is well known from low-molecular-weight smectic phases, that in mixtures the smectic layers form a weight average of the long dimensions of the respective components.[2,3] If we accept that the same applies to a polydisperse distribution of helical sequences and random-coil parts in iPP, the average helical/random-coil dimension (whatever their origin) will determine the smectic periodicity.

1 L. Li and W. H. de Jeu, *Adv. Polym. Sci.*, in press.
2 V. M. Kaganer, S. Diele, B. I. Ostrovskii and W. Haase, *Mol. Mater.*, 1997, **9** 59.
3 H. R. Keymeulen, W. H. de Jeu, J. T. Slatery and M. Veum, *Eur. Phys. J. E.*, 2002, **9**, 443.

Professor Müller asked: (1) In your paper it is concluded that the smectic phase is the responsible for the nucleation of your melt sheared iPP. Furthermore, it is speculated that since the smectic phase is induced by shear and since most commercial iPPs have been processed by extrusion, they will probably already contained these smectic precursors. Nevertheless, it is well known that iPP (as well as other semicrystalline polymers) is usually heterogeneously nucleated by catalyst residues and other heterogeneities sometimes of unknown origin (many droplet crystallization experiments have been designed and performed to prove this). The number of active heterogeneities for iPP depends on the cooling rate but has been measured and is on the order of 10^6 heterogeneities cm^{-3} (see for instance ref. 49 in Professor Schick's paper, *i.e.*, Fillon *et al.*). In your paper there is no mention about heterogeneous nucleation, what would be the relative importance of heterogeneities *versus* smectic precursors in commercial iPP?

(2) What is the exact thermal treatment that you have to do in order to make the smectic phase completely disappear?. According to Fig. 9, even after "several minutes" at 260 °C, the smectic phase reappears during cooling. Since you have employed commercial iPP, it would be a good idea to indicate if you had to apply a specially long heat treatment at high temperatures to get rid of the remaining smectic phase (for instance in Fig. 10b).

(3) Is the DSC trace reported in Fig. 7 reproducible? The endotherm attributed to the "melting" of the smectic phase is very small and noisy and has the same order of magnitude of several other small peaks present in the trace. If it is reproducible, what is the enthalpy associated with that melting along with the error of the measurement?, *i.e.*, ΔH_m and its associated error.

Dr Li replied: (1) Comparing the crystallization kinetics of samples with and without smectic domains (see Figs. 10 and 11 of the paper), the nucleation effect of the smectic domains on crystals is obvious. The heterogeneous nuclei certainly do exist but show a pronounced effect only at lower temperatures than the range we studied.

(2) The stability of the smectic phase is determined by its thermal and shearing history. As the only indicator is the smectic peak we cannot give a general rule, but it seems that in many cases annealing a couple of hours at high temperatures (say at least 230 °C) is necessary.

(3) The melting peaks from the smectic domains are well above the noise level, which has been checked systematically. Taking both the low enthalpy and the low volume fraction of the smectic domains into account, small additional melting peaks seem natural. We did not make a further quantitative evaluation.

Professor Ryan added: I take exception to the prior comment of Professor Müller. The effect of flow, which causes long-lived orientation of the inner portions of the high M_w fraction, is a much more efficient nucleating agent than the 10^6 particles cm^{-3} that he refers to. Shear flow (and chain orientation) is a very efficient nucleating agent even if there are no smectic phases.

Mr Follestad addressed Professor Ryan and Dr Li: Another comment about PP commercial samples already being oriented when arriving for analysis: Not only is there the effect of orientation by the extrusion in the commercial poly-propylene plant. Even before that, the polymerisation process itself makes all virgin PP with molecules with a lower degree of entanglement, corresponding to a lower entropy, than relaxed molecules.

Dr Al-Hussein commented: Smectic layers do not transform into crystalline lamellae but rather keep their initial structure, meanwhile they nucleate other "kebab" lamellae after the shear field is off.

Dr Li replied: The anisotropic droplets, assembled by the smectic layers, induce the growth of kebabs. Meanwhile, the initial smectic layers also transform to crystalline lamellae in a supercooled melt. Indeed the lamellae crystallized from the smectic layers only show a small increase of the layer spacing ($\sim$4.5 nm), which is still much smaller than the thickness in the kebab (10–20 nm).

Mr Follestad commented: In the paper, you have a discussion on how smectic filaments could be stable above the melting points of iPP crystals, based on the Gibbs–Thomson equation. This equation I tried to derive last evening, with the following results:

The equation is obviously based on the Gibbs free energy $\Delta G_{\text{melt}} = \Delta H_{\text{melt}} - T_{\text{melt}}\Delta S_{\text{melt}} = 0$, for the melting process.

Firstly: The Gibbs–Thomson equation is based on the assumption that the surface energy of the "crystal"'s sides can be neglected versus that of its ends. Understanding from your paper that the smectic filaments are longer than their width, I suggest reconsidering the relevance of this equation.

Secondly: The Gibbs–Thomson equation, at least as ordinarily used and as you use it, assumes that a molecule melting off from the crystal, with an entropy change of $\Delta S_{melt} = S_{melt} - S_{\text{“cryst”}}$, has an entropy S_{melt} which is that of the relaxed, equilibrium entangled molecule. However, this ideal state may not be reached. If the molecule is long, the heating fast and the crystal long in the chain direction, then for this molecule $S_{melt,real} > S_{melt} \Rightarrow \Delta S_{melt,real} > \Delta S_{melt} \Rightarrow T_{melt,real} > T_{melt}$.

This helps explain crystal-like structures above the regular melting point, *e.g.* smectic phase high temperature stability, also (low temperature) melting point increase with heating rate increase in Fig. 2 of the paper presented by Professor Schick. Trying experimental higher and lower heating rates exactly when passing a phase transition region could help pinpoint such an effect.

Dr Li responded: Concerning the first point, our description of filaments may be somewhat misleading. One may think of filaments being very thin, which of course results in a large lateral area. In fact, we do not have much information on the width of the filaments, which can be rather large. Moreover, in the Gibbs–Thomson equation the specific surface free energy rather than the total surface free energy matters. Generally we assume that the specific free energy of top and bottom surface is much larger than that of the lateral surface.

Second, the Gibbs–Thomson equation is a thermodynamic approach, while your argument is more a kinetic view. We agree that both kinetic and thermodynamic factors will influence the experimental melting point. It is well know that with increasing heating rate extended-chain crystals give a larger superheating than folded-chain crystals. Our X-ray experiments monitor a step heating process and each step takes more than 10 min, which is not expected to lead to superheating.

Professor Schick commented: You argue that the observed smectic bundles are the primary nuclei for iPP crystallization. Because all injection molded samples were subjected to shear they should contain such smectic bundles. I am wondering how to explain the observed temperature dependence of the overall crystallization rate on crystallization temperature as shown in Fig. 6 within the context of your discussion?

The observed very broad and double peaked maximum in overall crystallization rate requires at least two different mechanisms for structure formation for this iPP sample.

Dr Li responded: We agree there may be two different crystallization processes in the two temperature ranges. As we know a “mesophase” can be obtained following fast quenching, and crystallisation is not completed. This could lead to somewhat different processes for crystallisation at lower and higher temperatures. We speculate that at temperatures below around 35 °C one may see a melting–smectic ordering–crystallization process, while at higher temperatures the smectic ordering may be more suppressed and the direct melting–crystallization process could be more pronounced.

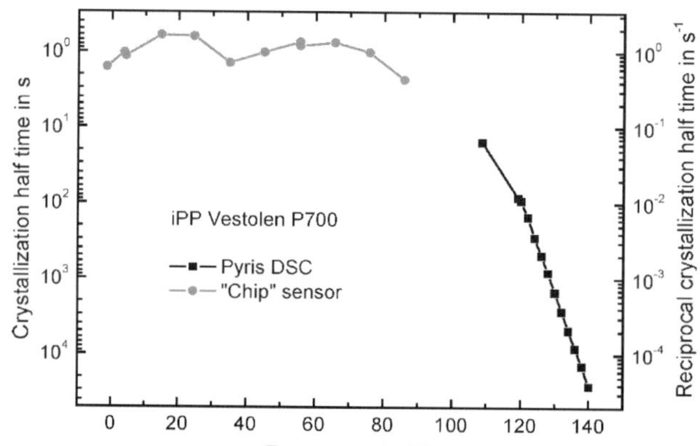

Fig. 6 Overall isothermal crystallization rate of iPP Vestolen P700 after cooling from the melt (210 °C) at 5000 K s^{-1} represented as crystallization half life (left axis) and reciprocal half life (right axis).

Dr Hobbs asked: If a previously unsheared sample of iPP is crystallized at high temperature with thick lamellae, do these thick lamellae 'melt' into a smectic phase on heating? Have you looked for this? In other words, can you use the previous crystallization as an alternative to shear to introduce alignment and, ultimately, smectic bundles?

Dr Li replied: We have not done such experiments systematically. But from our general knowledge it is unlikely that the smectic ordering in iPP can be obtained through melting normal thick lamellae. However, if the sample is highly stretched as the case of iPP fibres, this may be possible. Recent work[1] shows that a semi-molten iPP fibre can induce the growth of the beta phase. As stated in our paper, the smectic bundles are the nuclei for beta phase and this may imply that smectic ordering occurs during the melting of iPP fibres.

1 H. Li, S. Jiang, J. Wang, D. Wang and S. Yan, *Macromolecules*, 2004, **36**, 2802.

Professor Ryan made a general comment: The much maligned spinodal theory of precrystalline density fluctuations (refs. 13–16) of the paper requires conformation and density coupling. This is provided by the 12 monomer helices seen by IR at high T. So your data could be compatible with phase separation if high density "smectic" domains separated from the normal amorphous melt.

Dr Li responded: In the paper we write that the coupling between the conformational ordering, orientational ordering and density is a multiple chicken–egg problem. Clearly such a coupling is needed to obtain the smectic phase, and the random coil–rigid helix transition provides a mechanism. Note that smectic domains are characterized by a density modulation along the layer normal, rather than by a density difference with the surroundings.

Dr Hamley said: It would be an understatement to say that the spinodal decomposition model as a precursor to polymer crystallization is controversial. Could your data suggest an alternative input to the model by Olmsted, Ryan *et al.*,[1] *i.e.* could the smectic clusters correspond to the local regions of orientationally ordered polymer discussed in that model, replacing "nematic" type domains with "smectic" ones? Or is this model the nail in the coffin of the spinodal decomposition model for polymer crystallisation?

1 P. D. Olmstead, W. C. K. Poon, T. C. B. McLeish, N. J. Terrill and A. J. Ryan, *Phys. Rev. Lett.*, 1998, **81**, 373–376.

Dr Li and **Professor de Jeu** replied: The present work indicates unambiguously that in iPP smectic ordering arises as a result of local coil–helix transitions in the polymer chain. The resulting smectic filaments act as row nuclei for subsequent crystallisation that can be described by nucleation and growth. Of course this does not necessarily exclude any other mechanism, and perhaps one could argue that some form of spinodal decomposition contributes to the smectic ordering. However, the occurrence of the smectic peak at rather high temperatures in the melt of iPP makes it hard to draw any direct connection between the smectic ordering and a spinodal decomposition model.

Self-organization of cationic dendrimers in polyanionic hydrogels

Victor A. Kabanov,*[a] Alexander B. Zezin,[a] Valentina B. Rogacheva,[a] Tatyana V. Panova,[a] Elena V. Bykova,[a] Jacques G. H. Joosten[b] and Josephine Brackman[b]

[a] *Polymer Division, Faculty of Chemistry, Moscow Lomonosov State University, Leninskie Gori, 119992, Moscow, Russia. E-mail: kabanov@genebee.msu.su; Fax: +7(095)939 01 74*

[b] *DSM Research, P.O. Box 18, 6160, MD Geleen, The Netherlands*

Received 24th March 2004, Accepted 14th May 2004

First published as an Advance Article on the web 16th September 2004

Protonated poly(propylene imine) dendrimers (Astramol™) of five generations: DAB-dendr-$(NH_2)_x$ (where $x = 4$, 8, 16, 32 or 64) are sorbed by slightly cross-linked polyanionic hydrogels: poly(sodium acrylate) and poly(sodium 2-acrylamido-2-methylpropane sulfonate). As a result highly swollen original hydrogel transforms into compact cross-linked polyelectrolyte-dendrimer complexes. Sorption of dendrimers by the hydrogels is a chemically drawn frontal diffusion process. Driving force comes from the gain in the free energy of interpolyelectrolyte coupling reaction between the charged dendrimer molecules and the oppositely charged hydrogel network, accompanied with entropically favourable release of low molecular salt into environment. The amount of a simple salt released is equivalent to a number of intermolecular salt bonds, formed between protonated dendrimers and hydrogel networks. Apparently the mechanism of dendrimer uptake involves a "relay-race" transfer of dendrimer polycations from one fragment of polyelectrolyte network to the other *via* interpolyelectrolyte exchange reaction. As a result "core–shell" constructs consisting of outer weakly swollen complex shell and highly swollen hydrogel core are formed at intermediate stages of the process. The rate of sorption is determined by the rate of the interpolyelectrolyte exchange reaction that is the rate of the formation of free fragments of polyelectrolyte network (vacancies) on the inner complex-hydrogel boundary. The amount of vacancies depends on the area of this boundary. Consequently kinetics of dendrimer uptake could not be fitted in terms of Fickian diffusion (except DAB-dendr-$(NH_2)_4$), but expressed in terms of the kinetic equation derived for a frontal heterogeneous reaction. Sorbed dendrimers of all studied generations at pH values ensuring complete protonation of primary and tertiary amine groups are closely packed in hydrogel networks, so that all dendrimer cationic units form ion pairs with anionic units of hydrogels. In other words polyanionic network fragments are able to penetrate into the interior of fully protonated DAB-dendr-$(NH_2)_x$ species as it was earlier shown for flexible linear polyanions. In such case the ultimate amount of sorbed dendrimer molecules is always determined by the condition $n_a/N^- = 1$, where n_a is the total number of dendrimer amine groups, N^- is the number of the anionic hydrogel units. The latter is also true for the complex shell composition in the heterogeneous reacting samples formed at intermediate stages of dendrimers uptake. Variation of pH and sorption extent is an effective tool to control dendrimer distribution, self-organization and the final structure of dendrimer–hydrogel constructs.

Introduction

It is known that slightly cross-linked highly swollen polyelectrolyte hydrogels (#HG) absorb oppositely charged linear polyion chains[1–6] as well as compact protein globules[7–11] from water solutions to form cross-linked interpolyelectrolyte complexes (#IPEC). Such sorption is nothing but cooperative interpolyelectrolyte reaction resulting in formation of numerous salt bonds between the reagents and release of an equivalent number of simple counter ions originally neutralized the dissolved polyions and the #HG network. It is precisely the increase in entropy of released counter ions provides the gain in free energy, which drives the whole process. By and large the process represents a heterogeneous reaction propagating inward the piece of #HG followed with compaction of the reacting sample as shown in Scheme 1.

Its mechanism was established in our early studies.[3–5,11] Linear polyions apparently diffuse into the #HG by reptation, while compact protein globules migrate as a whole, if the cross-link density is sufficiently low. However, to our knowledge nothing was known about interaction of dendritic polyions with oppositely charged #HG.

This paper presents the results obtained for five generations (G1–G5) of poly(propylene imine) dendrimers, further abbreviated as DAB-dendr-$(NH_2)_x$, which contain external primary and internal tertiary amino groups. Here x is the number of primary amino groups, equal to 4, 8, 16, 32 and 64 for G1, G2, G3, G4 and G5 dendrimers, correspondingly. The total number of primary and ternary amino groups is $X = 2x - 2$. Radii of gyration of the dendrimer molecules, measured by small angle neutron scattering in water are: 0.44, 0.69, 0.93, 1.16 and 1.39 nm.[12] Scheme 2 shows the chemical structure of the representative of this dendrimer family (G3).

If amino groups are protonated, DAB-dendr-$(NH_2)_x$ molecules naturally transform into dendritic polycations. We studied their interaction with two slightly cross-linked polyanionic

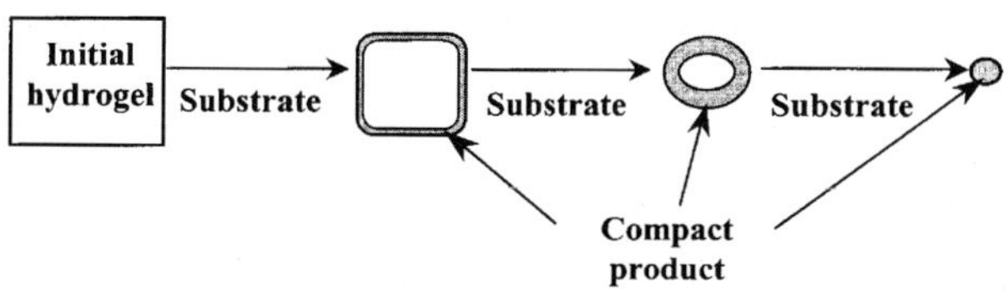

Scheme 1

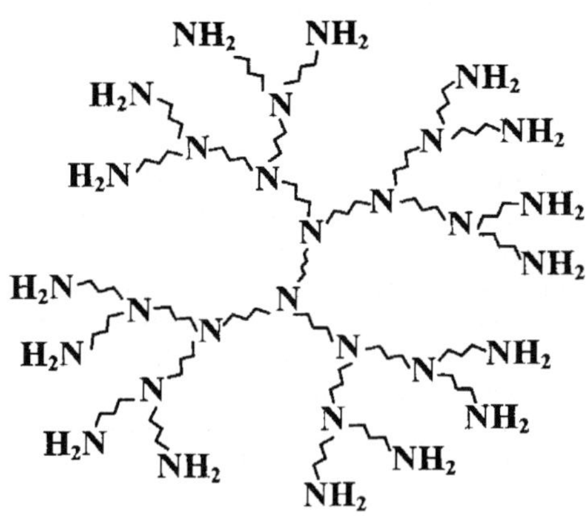

Scheme 2

#HGs: poly(sodium acrylate) (#PA-Na) and poly(sodium 2-acrylamide-2 methylpropane sulfonate) (#PAMPS-Na).

Earlier it was shown that fully protonated DAB-dendr-$(NH_2)_x$ (DAB-dendr-$(NH_2)_x \cdot HCl$) of all five generations complex with flexible linear polyanions so that each positively charged amino group forms ion pair (salt bond) with the negatively charged group of the polyanion.[13] In other words negatively charged polymer chains penetrate into the dendrimer interior. Therefore at the beginning it was not obvious that dendritic polycations especially those of higher generations adsorbed on the negative #HG surface would be able to diffuse with a measurable rate inward the macroscopic piece of #HG because of possible entangling dendrimer branches and the network. However, if they do it would provide a route to prepare a novel family of environmentally responsive polymer constructs consisting of well defined dendritic molecular species imbedded into polyelectrolyte network. Their composition, structure, morphology and behavior may be effectively controlled by such factors as pH, nature and concentration of external salts *etc.* We explored such possibilities in this work.

Experimental

DAB-dendr-$(NH_2)_X$ (Astramol™) of five generations were synthesized as described in ref. 14 and presented to us by DSM (Netherlands). #PA-Na and #PAMPS-Na were prepared by radical copolymerization of acrylic acid or sodium 2-acrylamido-2-methylpropanesulfonate with *N*,*N*-methylene-bis-acrylamide as a cross-linker (to the extent 1.0% and 1.5% of the monomer mass in 10% and 20% aqueous solutions correspondingly). A redox system composed of potassium persulfate and sodium metabisulfite was used as an initiator in the extent of 0.25 wt.% of the monomers content. The copolymerization was performed in sealed ampoules contained 10 ml of the co-monomer mixture at 40 °C for 1 day in the absence of air. The obtained #HGs were completely neutralized by NaOH, *i.e.* transformed into the salt form. Then highly swollen #HG samples were washed by distilled water to pH 8.5–9.0 in the case of #PA-Na and 6.0–7.0 in the case of #PAMPS-Na to purify them from the residual monomers, sol fractions and an excess of NaOH. The swelling behaviour of the original gels and the products of their reaction with the dendrimers was studied by weighting technique and characterized by the equilibrium degree of swelling $H = (m_s - m_d)/m_d$ (here m_s and m_d are the masses of the swollen and dry samples respectively). The H values for #PA-Na and #PAMPS-Na in pure water equalled to 750 and 350, while concentrations of $-COO^-$ and $-SO_3^-$ groups in equilibrium swollen gels were 0.012 and 0.014 base-mol l^{-1}, correspondingly. The water content measured in arbitrary chosen volume elements of the #HG samples was found equal indicating macroscopically uniform distribution of cross-links. The synthesized #HGs had in average not more than one cross-link per 200 (for #PA-Na) and 50 (for #PAMPS-Na) monomer units. So the average contour length of chain fragments between two neighbouring cross-links in #HG networks deliberately exceeded the diameters of all studied dendrimer molecules. The total mol amount of ionogenic groups in a swollen #HG sample was calculated as $N = m_s/[(H+1)M_u]$, where M_u is molecular mass of the #HG monomer unit. $N = N^-$ for fully ionized #HG.

Interaction of #HGs with the HCl protonated dendrimers was studied in a following manner. Equilibrium swollen cubic shaped monolithic #HG sample with a mass 1–3 g was immersed in the salt free solution of the dendrimer at 20 °C and given pH. Use of the large #HG specimen makes possible to monitor accurately change in its mass upon interaction with the dendrimer polycations if occurs, observe visually macroscopic changes in the reaction system and also cut and analyze separately different parts of the reacting sample. The advantages of such experimental approach have been strongly demonstrated earlier.[15] The shape of the reaction vessel and the solution volume were chose so that the #HG sample was fully submerged in the solution. Total concentration of dendrimer primary and tertiary amino groups in stock solutions was usually varied in the range 0.002–0.01 base-mol l^{-1}. The reassigned ratio, Z, between the molar amount of all amino groups (n_a) and ionogenic groups in #HG sample, N, in different experiments was varied from 3 to 0.2. The #HG samples were taken out of the solution at certain intervals, blotted to remove excess liquid from its surface and then weighted. At the same time concentrations of the residual dendrimer, as well as Cl^- and Na^+ counter ions released were measured. The composition of the products formed upon interaction between DAB-dendr-$(NH_2)_x \cdot HCl$ and #HGs was expressed as $\varphi = (n_a/N)_{IPEC}$. In some cases φ was determined by elemental analysis.

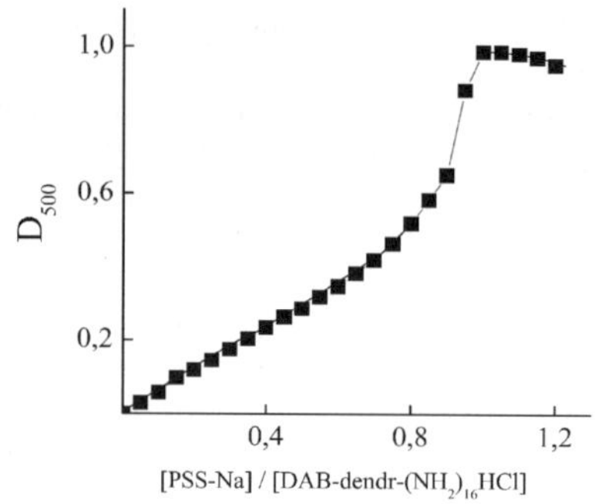

Fig. 1 Turbidimetric titration curve of 0.01 base-mol l^{-1} DAB-dendr-$(NH_2)_{16}\cdot$HCl solution with 0.01 base-mol l^{-1} PSS-Na solution; $T = 20\,°C$.

Earlier it has been shown[16] that potentiometric titration curve for DAB-dendr-$(NH_2)_4$ (G1) exhibits two rather sharp pH jumps corresponding to protonation (deprotonation) of its primary and tertiary amino groups. So in this work concentration of DAB-dendr-$(NH_2)_4$ was monitored by potentiometric titration accurate to ±0.02 pH unit using pH-meter RTS 882 Radiometer (Denmark). However, that is not the case of higher dendrimers generations which potentiometric titration curves do not exhibit distinct inflation points. Therefore concentration of fully protonated G2, G3, G4, and G5 was measured by turbidimetric titration ($\lambda = 500$ nm) of their diluted solutions with water solution of poly(sodium styrenesulfonate) (PSS-Na), $M_w = 100\,000$, (Aldrich). PSS-Na concentration was determined measuring UV-absorbance at $\lambda = 261$ nm ($\varepsilon = 450$ l mol^{-1} cm^{-1}) by spectrophotometer Hitachi 150-20 (Japan). Concentration of the dendrimer was found from the position of the equivalence point, which corresponds to the maximum on the turbidimetric titration curve. As an example, Fig. 1 presents the typical curve of turbidimetric titration of the solution of DAB-dendr-$(NH_2)_{16}\cdot$HCl (G3) with PSS-Na solution.

As is seen, the curve shows a well-pronounced maximum at an equimolar ratio between the components. Similar curves were also typical for the solutions of G2, G4 and G5 dendrimers. The titration data were well reproduced for the dendrimer solutions with concentrations varying from 0.001 to 0.02 base-mol l^{-1}. However, addition of PSS-Na to the dendrimer solutions caused visually detected turbidity even at dendrimer concentration as low as 0.0002 base-mol l^{-1}.

Concentration of Cl^- ions in aqueous solutions was estimated by Ag-metric titration on a pH 673 pH-meter with an ASL-51-07 glass pAg electrode and an AVL-1 M Ag/AgCl reference electrode. Concentration of Na^+ ions was measured by flame photometry using an FPA-2 automatic flame photometer (Russia).

The composition of the products obtained upon interaction of #HG with the dendrimers was determined by elemental analysis using C,N automatic analyser Carlo Erba 1106 (Italy).

Results and discussion

As was found, studied polyanionic #HGs are able to sorb dendrimers of all five generations from aqueous solutions. Sorption occurs in certain pH intervals, but only if concentration of external salts doesn't exceed some characteristic values. In particular, #PAA uptakes dendrimers from salt-free water solutions in interval $2.5 < pH < 11.5$, that is, under the conditions when at least part of dendrimer amino groups and swollen network units are ionized. No sorption is observed beyond this pH interval. #PAMPS-Na being a strong anionic polyelectrolyte is ionized over the whole pH range. So #PAMPS-Na uptakes the dendrimers in the entire region of $pH < 11.5$, where they are positively charged. It was also found that sorbed dendrimers release into surrounding solutions from the both #HGs on loose their charge, *i.e.* at pH exceeding 11.5. They also release from #PAA at pH lower than 2.5, where PAA degree of ionization becomes close to 0.

As a result of dendrimer uptake the dimensions of the originally transparent highly swollen #HG sample gradually decrease and, finally, the sample is transformed into a compact only slightly swollen ($H \approx 1$) piece of the reaction product. In other words, sorption of the dendrimers results in collapse of the original #HGs. Their volume decreases by two orders of magnitude and even more; the samples becomes opaque. At intermediate stages of dendrimer uptake the collapsing sample remains heterophase consisting of a newly formed compact shell separated by the detectable boundary from the unreacted highly swollen original #HG core.

Dendrimer uptake by hydrogels as interpolyelectrolyte reaction

Sorption of all DAB-dendr-$(NH_2)_x \cdot HCl$ by studied #HGs is accompanied by the release of Na^+ and Cl^-, counterions originally neutralized the #HG networks and the dendrimers, respectively. The simple salt concentration increases in the surrounding solution in parallel with the extent of dendrimer uptake. It apparently means that sorption of the dendrimer polycations with the swollen polyanionic network is nothing but cooperative interpolyelectrolyte coupling (addition) reaction similar to those proceeding upon interaction of linear polyelectrolytes, proteins or ionic surfactants with oppositely charged hydrogels. Such reactions are driven by an increase in translational entropy of the released counterions and result in formation of #IPECs. In other words the processes under study may be described by the following equilibria:

$$\#PA{-}COO^{\ominus}\,Na^{\oplus} + D_x{-}NH_3^{\oplus}Cl^{\ominus} \rightleftharpoons \#PA{-}COO^{\ominus}\,{}^{\oplus}NH_3{-}D_x + Na^{\oplus} + Cl^{\ominus} \quad (1a)$$

$$\#PA{-}COO^{\ominus}\,Na^{\oplus} + D_x{-}NH^{\oplus}Cl^{\ominus} \rightleftharpoons \#PA{-}COO^{\ominus}{-}{}^{\oplus}NH{-}D_x + Na^{\oplus} + Cl^{\ominus} \quad (1b)$$

for #PA-Na, and

$$\#PAMPS{-}SO_3^{\ominus}Na^{\oplus} + D_x{-}NH_3^{\oplus}Cl^{\ominus} \rightleftharpoons \#PAMPS{-}SO_3^{\ominus}\,{}^{\oplus}NH_3{-}D_x + Na^{\oplus} + Cl^{\ominus} \quad (1'a)$$

$$\#PAMPS{-}SO_3^{\ominus}Na^{\oplus} + D_x{-}NH^{\oplus}Cl^{\ominus} \rightleftharpoons \#PAMPS{-}SO_3^{\ominus}{-}{}^{\oplus}NH{-}D_x + Na^{\oplus} + Cl^{\ominus} \quad (1'b)$$

for #PAMPS-Na, where #PA, #PAMPS and D_x are the rests of the anion networks and the dendrimer polycations, correspondingly.

If at a given initial concentration the total amount of the dendrimer in the reaction mixture was not sufficient to saturate the whole #HG sample, the residual concentration of the dendrimers in surrounding solutions established when sorption completed was extremely small and did not exceed 0.0002 base-mol l^{-1} (*e.g.* at initial concentration of 0.01 base-mol l^{-1}). It means that the above equilibriums are almost fully shifted toward the formation of #IPECs for all studied dendrimers.

The composition, $\varphi = (n_a/N)_{IPEC}$, of the samples transformed into #IPECs was determined at the reassigned Z values equal to 1 and 3 in the conditions at which the dendrimers were fully protonated with hydrochloric acid and the original #HGs fully ionized. The φ values were calculated from decrease of DAB-dendr-$(NH_2)_x \cdot HCl$ in gel-surrounding solution taking into account that $N = N^- = m_s/[(H+1)M_u]$. For DAB-dendr-$(NH_2)_{16} \cdot HCl$ and DAB-dendr-$(NH_2)_{32} \cdot HCl$ φ values were also estimated by elemental analysis. Importantly φ was practically equal to unity both for $Z = 1$, and $Z = 3$. It shows, that at equimolar composition of the reaction mixture ($Z = 1$), the #HGs uptake nearly all dendrimer polycations from surrounding solution, so that #IPECs formed contain both interacting polyelectrolytes in almost equivalent ratio. Moreover, with increasing content of the dendrimers in the surrounding solution above equivalent ratio ($Z = 3$), no change in #IPEC composition occurs. In other words the #HG networks are unable to uptake the dendrimers in amounts exceeding 1 : 1 charge ratio. As was shown earlier,[13] interaction of DAB-dendr-$(NH_2)_x \cdot HCl$ with linear polyanions PA-Na and PSS-Na also leads to the formation of stoichiometric complexes.

For DAB-dendr-$(NH_2)_{16} \cdot HCl$ and DAB-dendr-$(NH_2)_{32} \cdot HCl$ interacting with either #PA-Na or #PAMPS-Na at $Z = 1$ the concentrations of Na^+ and Cl^- ions, released to the surrounding solution on approaching $\varphi = 1$, were determined using flame photometry and Ag-metric titration. The corresponding data are listed in Tables 1 and 2 in terms of mol amounts of counterions released: $\nu_{Cl} = [Cl^-]V$ and $\nu_{Na} = [Na^+]V$, where V is the final volume of the reaction system. The tables also list the total mol amounts ν^0_{Cl} and ν^0_{Na} of counter ions originally neutralized the interacting polymer

Table 1 Mole amounts and fractions of Cl^- counter ions as released into the solution upon the interaction of DAB-dendr-$(NH_2)_x \cdot HCl$ with #PA-Na or #PAMPS-Na at $Z = 1$

System	ν^0_{Cl}	ν_{Cl}	ν_{Cl}/ν^0_{Cl}
# PA-Na–DAB-dendr-$(NH_2)_{16} \cdot HCl$	6.01×10^{-5}	5.98×10^{-5}	0.96
# PA-Na–DAB-dendr-$(NH_2)_{32} \cdot HCl$	5.85×10^{-5}	5.75×10^{-5}	0.98
# PAMPS-Na–DAB-dendr-$(NH_2)_{16} \cdot HCl$	2×10^{-5}	1.97×10^{-5}	0.98
# PAMPS-Na–DAB-dendr-$(NH_2)_{32} \cdot HCl$	1.7×10^{-5}	1.68×10^{-5}	0.99

Table 2 Mole amounts and fractions of Na^+ ions as released into the solution upon the interaction of DAB-dendr-$(NH_2)_x \cdot HCl$ with #PA-Na or #PAMPS-Na at $Z = 1$

System	ν^0_{Na}	ν_{Na}	ν_{Na}/ν^0_{Na}
# PA-Na–DAB-dendr-$(NH_2)_{16} \cdot HCl$	6.01×10^{-5}	5.98×10^{-5}	0.96
# PA-Na–DAB-dendr-$(NH_2)_{32} \cdot HCl$	5.85×10^{-5}	5.75×10^{-5}	0.98
# PAMPS-Na–DAB-dendr-$(NH_2)_{16} \cdot HCl$	3.06×10^{-5}	3.03×10^{-5}	0.99
# PAMPS-Na–DAB-dendr-$(NH_2)_{32} \cdot HCl$	3.2×10^{-5}	3.15×10^{-5}	0.97

components and also fractions of counter ions ν_{Cl}/ν^0_{Cl} and ν_{Na}/ν^0_{Na} released through reactions (1) and (1′).

As is seen, almost all counterions originally belonged to #HGs (Na^+) and the dendrimers (Cl^-) are released to the surrounding solution after the interpolyelectrolyte reactions are completed. This fact conforms with the above conclusion that equilibria (1) and (1′) are completely shifted from left to right. Importantly it also shows that practically all ionic groups of the dendrimer polycations and anionic #HG networks form salt bonds with each other. In other words in addition to easily accessible peripheral primary amino groups, virtually all ternary amino groups located in the inner regions of the dendrimers are involved in the interaction with the network units. It follows that, as in the case of the interaction of flexible linear polyions,[13] dendrimer polycations also appear to be fully permeable for flexible chain fragments of slightly cross-linked polyianionic networks, as shown in Scheme 3.

Distribution of dendrimers in cross-linked interpolyelectrolyte gels

To quantify kinetics and clarify the mechanism of penetration of the dendrimers into polyanionic #HGs one should know, how the dendrimer polycations are distributed in the reacting sample at intermediate stages of transformation of highly swollen #HG into compact #IPEC. The experimental procedure was markedly simplified by using large-sized #HG samples. In fact the picture represented in Scheme 1 could be observed directly on the slices of the samples reacting with G2,

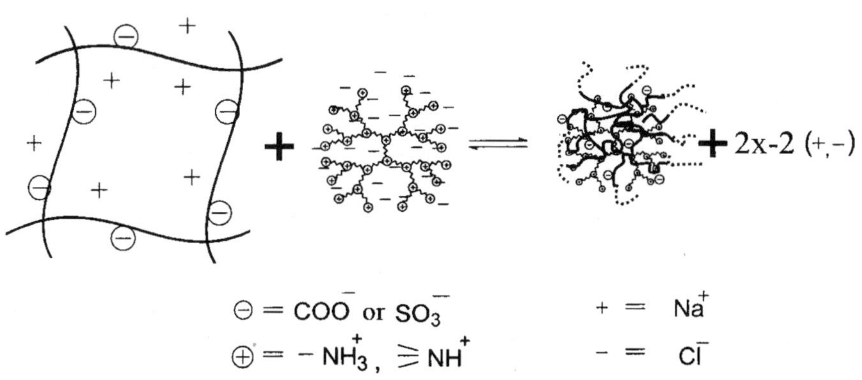

Scheme 3

G3, G4 and G5 dendrimers at intermediate degrees of conversion. Only in the case of G1 dendrimer the shell-core boundary was gradually blurred in time.

To solve the problem #PAMPS-Na samples were immersed in aqueous solutions of protonated dendrimers at $Z = 0.5$ so that the amount of the dendrimer was insufficient for complete conversion of #HG into #IPEC. After the added dendrimer was consumed a partially converted gel sample was allowed to stand for 2–3 days and then cut into pieces by using a razor blade to separate the peripheral part from the core. Independently it was found, that all studied dendrimers originally involved in #IPEC composition quantitatively released into the gel-surrounding solution at pH exceeding 11.5. So each piece of the cut sample was placed into 1 ml of 0.25 N NaOH solution, kept there for 24 h, the time wide margin of complete extraction of the sorbed dendrimer, then removed. The extract was acidified with 0.2 N HCl to pH *ca.* 3.5, and the concentration of the dendrimer released was measured by turbidimetric titration with 0.01 base-mol l^{-1} PSS-Na water solution. The concentration of released DAB-dendr-$(NH_2)_4 \cdot HCl$ and DAB-dendr-$(NH_2)_8 \cdot HCl$ was also estimated by potentiometric titration. The data obtained for #PAMPS-Na–DAB-dendr-$(NH_2)_x \cdot HCl$ systems are listed in Table 3, where m is mass of partly transformed reacting sample, m_1 and m_2 are masses of its separated outer and inner parts, ν_0 is the initial molar amount of the dendrimer in the solution; ν_1 and ν_2 are molar amounts of the dendrimer in inner and outer parts of the sample. As is seen only G1 dendrimer is uniformly distributed between outer and inner layers. Contrastingly, G2, G3 and G4 dendrimers are located in the compact outer shell to form stoichiometric #IPEC ($\varphi = 1$). The inner parts of partially converted samples are almost free of the dendrimers; at least, their contents in the highly swollen cores are below the experimentally detected level.

The similar behavior is also characteristic of the products formed upon interaction of DAB-dendr-$(NH_2)_x \cdot HCl$ with #PA-Na.

The above results allow visualizing transformation of original highly swollen #HG sample to compact dendrimer containing #IPEC as it is shown in Scheme 1, earlier established for slightly cross-linked polyelectrolyte hydrogels reacting with oppositely charged linear polyions,[1–5] globular proteins[7–9,11] and ionic surfactants.[17,18]

It is important that in the case of higher-generation dendrimers the boundary between compact peripheral #IPEC and inner highly swollen #HG phases in partly transformed samples then immersed in plane water, neither smears nor moves at least for the time exceeding one month. Correspondingly, the location of the dendrimer polycations within the sample also remains unchanged as it has been shown for DAB-dendr-$(NH_2)_8 \cdot HCl$, DAB-dendr-$(NH_2)_{16} \cdot HCl$, and DAB-dendr-$(NH_2)_{32} \cdot HCl$ using the above described experimental procedure. The similar behavior was also established earlier for the complexes of #HGs with oppositely charged linear polyions, globular proteins and ionic surfactants.[1–5,7–9,11,17,18]

Stable two-phase product can be prepared either *via* exhaustive sorption of reassigned deficient amount of higher generations' dendrimers or simple removing the sample from the dendrimer solution. Both approaches make it possible to easily control thickness of the #IPEC outer shell and may be used to design constructs consisting of a highly swollen charged hydrogel core coated with outer #IPEC layer. It is also important that sorption of dendrimer polycations, ceased at any intermediate stage, restarts and shell-core boundary continues to move inwards the partly converted sample, if transferred from water back to the dendrimer solution.

Dendrimer uptake is accompanied by compaction and a corresponding dramatic loss in mass of reacting #HG samples. The dependence of mass upon extent of conversion (F) of #HG to #IPEC provides additional important information on the sorption process. Indeed, Scheme 1 assumes that

Table 3 Core and shell composition of partially converted products of interaction of DAB-dendr-$(NH_2)_x \cdot HCl$ and #PAMPS-Na at $Z = 0.5$

	Core				Outer layer			
System	m_1/g	m_1/m	ν_1/mol	ν_1/ν_0	m_2/g	m_2/m	ν_2/mol	ν_2/ν_0
DAB-dendr-$(NH_2)_4 \cdot$ HCl-#PAMPS-Na	0.68	0.33	4.60×10^{-6}	0.3	1.4	0.67	9.85×10^{-6}	0.6
DAB-dendr-$(NH_2)_8 \cdot$ HCl-#PAMPS-Na	0.38	0.33	0	0	0.78	0.67	1.29×10^{-5}	1
DAB-dendr-$NH_2)_{16} \cdot$ HCl-#PAMPS-Na	0.50	0.61	0	0	0.32	0.39	1.32×10^{-5}	1
DAB-dendr-$(NH_2)_{32} \cdot$ HCl-#PAMPS-Na	0.58	0.62	0	0	0.35	0.38	1.25×10^{-5}	1

only ratio but not the composition of the coexisting phases changes in course of the heterogeneous reaction. In other words the collapse exhibited by the reacting hydrogel is of a local character as it was earlier established for linear polyelectrolytes, proteins and surfactants sorbed by oppositely charged #HGs.[1–5,8,9,11,18] Then the process should fit the relationship:

$$m = m_0 - F(m_0 - m_{\#IPEC}) \tag{2}$$

where m_0 and $m_{\#IPEC}$ are correspondingly initial mass of equilibrium swollen #HG sample and the mass of a sample fully converted to #IPEC. Because m_0 exceeds $m_{\#IPEC}$ by three orders of magnitude the above relationship can be replaced with fair accuracy by

$$m/m_0 = 1 - F \tag{3}$$

To achieve and fix progressively increasing F values in the experimental measurements the following procedure was used. #HG sample was immersed in the solution containing the dendrimer in amount from 5 to 20 mol% of the ultimate #HG capacity and allowed to stand for 48 h, the time with a safety margin for exhaustive uptake of the dissolved dendrimer. Then partially converted sample was washed with distilled water to remove released NaCl from the surface, weighted and immersed in the dendrimer solution again to repeat the above procedure. Fig. 2 presents reduced mass (m/m_0) values, plotted against F for #PA-Na (a) and #PAMPS-Na (b) samples, interacting with DAB-dendr-$(NH_2)_x \cdot HCl$. It is seen that the behavior of most studied systems is perfectly described by eqn. (3), except two of them containing DAB-dendr-$(NH_2)_4 \cdot HCl$.

Marked positive deviations of the exceptional curves from the universal straight lines in Fig. 2 correlate with above mentioned in time blurring of #IPEC-#HG interface and establishing uniform dendrimer distribution in partly converted #HG samples, when DAB-dendr-$(NH_2)_4 \cdot HCl$ is sorbed. However, the reduced mass of all fully converted collapsed samples including those containing DAB-dendr-$(NH_2)_4$ approaches the same small value, which corresponds to H around unity.

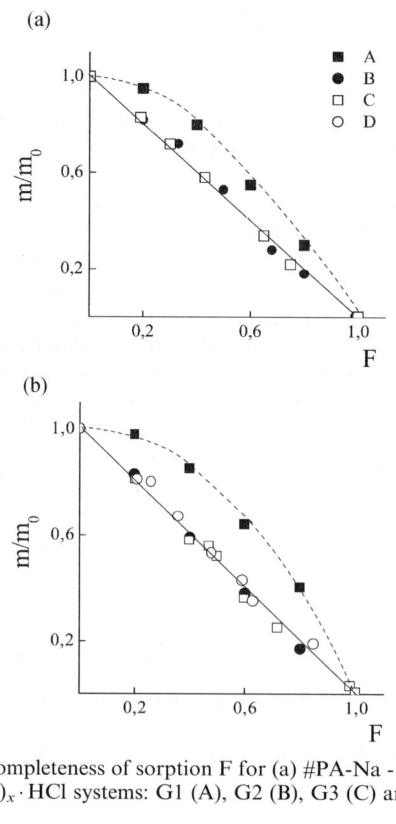

Fig. 2 m/m_0 ratio *vs.* degree of completeness of sorption F for (a) #PA-Na - DAB-dendr-$(NH_2)_x \cdot HCl$ and (b) #PAMPS-Na - DAB-dendr-$(NH_2)_x \cdot HCl$ systems: G1 (A), G2 (B), G3 (C) and G4 (D) dendrimers.

Local collapse states observed upon sorption of G2-G5 dendrimers by #HGs at each fixed F less than 1, most probably correspond to a series of equilibriums established in partly transformed reaction system. It can be understood taking into account that each dendrimer polycation sorbed by an oppositely charged hydrogel serves as $(2x-2)$-functional ionic cross-linker, which forms water insoluble #IPEC cluster within the reacted sample. Clusters are the larger and the more favourably disposed towards additional micro segregation the higher is the dendrimer generation. One may assume that in such case uniform distribution of higher generation dendrimer polycations, *i.e.* ionic cross-links within the #HG sample, would result in decrease of the total volume of the sample if to compare with integrated volume of the collapsed shell and highly swollen core. If this is the case the gain in entropy of mixing at homogeneous dendrimer distribution may not compensate an increase in osmotic pressure of simple counter ions within the #HG partly transformed into #IPEC, so that the system in local collapse state has a lower free energy. It may not be the case of DAB-dendr-$(NH_2)_4$ cations, each forms only six ionic bonds with anionic #HG network. Therefore uniform distribution of this ionic cross-linker can be reached with a smaller additional contraction of the whole sample, so that the gain in entropy of mixing dominates. In the latter case collapse occurring at F close to unity most likely is described within the general theories.[19]

Kinetics and mechanism of dendrimers uptake

The mechanism of transportation of dendrimer polyions from environmental solution inward the reacting hydrogel sample through the already formed #IPEC shell is of a primary importance. Kinetic measurements are helpful to solve this problem.

The kinetics of sorption of protonated G1–G5 dendrimers by #HG samples from dilute salt-free solutions was studied. As was revealed in special experiments, sorption rate becomes concentration independent at dendrimers concentration, C, in the surrounding solution, *i.e.* not controlled by external diffusion, at C exceeding 0.007 base-mol l^{-1}. All measurements were performed at $C = 0.01$ base-mol l^{-1}. Therefore, the data obtained correspond to the regime when sorption kinetics is not controlled by the external diffusion but the rate of transfer of dendrimer polycations in the interior of #HG phase.

The experiments were performed according to the following procedure. The swollen-to-equilibrium #HG sample with a mass of 1 g was dipped into the certain volume of 0.01 base-mol l^{-1} DAB-dendr-$(NH_2)_x \cdot HCl$ stock solution containing the total amount of the dendrimer required to neutralize all anionic groups in #HG network. After some time the sample was taken out, and the dendrimer concentration in the remaining solution was measured by either turbidimetric or potentiometric titration. Then the same sample was dipped again into a fresh portion of the stock solution. The above procedure was repeatedly performed until sorption was completed.

The extent of sorption was calculated as $F(t) = \nu_t/N^-$, where ν_t is the cumulative dendrimer mole amount as sorbed by #HG at time t.

Fig. 3 presents kinetic curves of sorption of fully protonated dendrimers by #PA-Na and #PAMPS-Na hydrogel samples. As it is seen, the sorption rate decreases in the series of dendrimers from G1 to G3 that seems to be a natural consequence of an increase in the molecular mass and, correspondingly, effective radii of the dendrimer polycations ($R_D \sim M_D^{1/3}$).

Surprisingly on passing from G3 to G4 dendrimers, sorption does not slow down: the kinetic curves for DAB-dendr-$(NH_2)_{16} \cdot HCl$ and DAB-dendr-$(NH_2)_{32} \cdot HCl$ by either #PA-Na or #PAMPS-Na hydrogels almost coincide. Presently, this fascinating and reproducible fact has no any justified explanation and requires further studies.

Another peculiarity was observed for G5 dendrimer consisting in dramatic retardation of its sorption rate with F(t) increase if to compare with lower generations. Moreover, in contrast to other dendrimers, complete conversion of the hydrogels into #IPECs can be hardly achieved in the case of fully ionized DAB-dendr-$(NH_2)_{64} \cdot HCl$. F(t)-t curves approach their limits not exceeding 0.3–0.5 (in various experiments). However, if such partly converted two-phase sample approached sorption shutdown was cut into two pieces then dipped into G5 dendrimer solution, sorption recommenced through the freshly formed surfaces. Repeating this procedure one could entirely transform the original #HG material into DAB-dendr-$(NH_2)_{64}$-containing #IPEC. A possible reason behind above kinetic shutdown may relate to the topology of a dendritic molecule and polyelectrolyte network. One may anticipate that there are certain traps in the network. Being entrapped in such

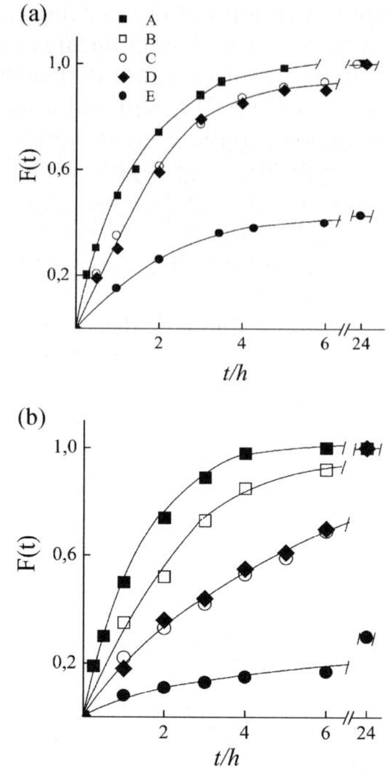

Fig. 3 Kinetic curves of sorption of DAB-dendr-$(NH_2)_x \cdot HCl$ by (a) #PA-Na and (b) #PAMPS-Na gels: G1 (A), G2 (B), G3 (C), G4 (D), and G5 (E) dendrimers; $m_0 = 1$ g; $C = 0.01$ base-mol l^{-1}; $T = 20$ °C.

trap, dendrimer molecule loses its ability for any further translational diffusion within the hydrogel phase either temporally or permanently. For example, a network junction may play the role of such a trap if inside the dendrimer polycation as is shown in Scheme 4a.

Then a dendrimer species embedded in #HG network due to topological entanglements is not only excluded from translational motion, but also restricts diffusion of neighboring species. Such events may, probably, happen to dendrimers of all generations, but the higher the generation is, the more is probability for a dendrimer species to become permanently embedded. Indeed, it is apparent that a chance to release from the trap is the lowest for G5 dendrimer polycation, because it requires

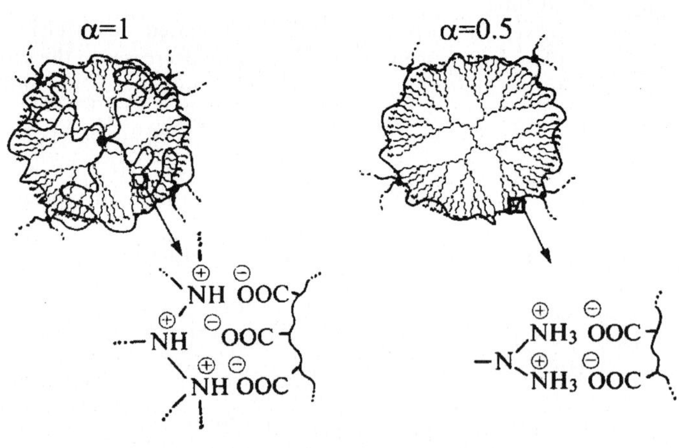

Scheme 4

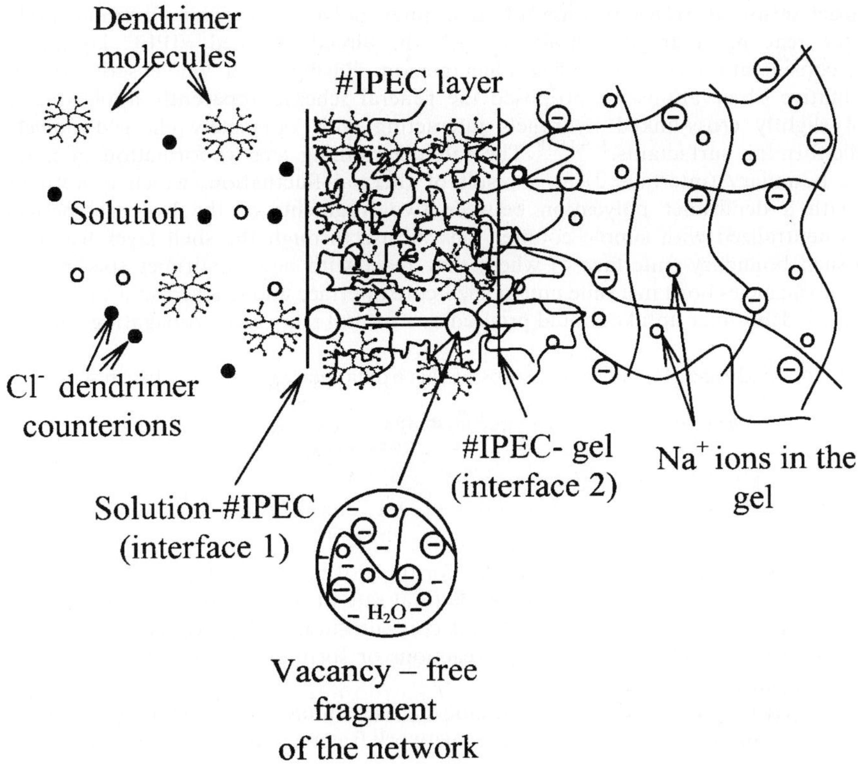

Scheme 5

coordinated thermal motions of the largest number of the dendrimer and network repeating units. In this respect, the behavior of the studied dendrimers in hydrogel networks is somewhat different from that of high molecular mass linear polyions, which diffuse *via* reptation mechanism. In the latter case, sorption shutdown does not occur even at rather high degrees of polymerization (10^3 and higher).[6]

Remarkably, kinetic hindrances retarding DAB-dendr-$(NH_2)_{64} \cdot HCl$ diffusion in #HG may be avoided if degree of protonation (α) of amine groups is reduced. In particular, at $\alpha = 0.5$, peripheral primary amino groups of higher basicity as compared with that of inner ternary amino groups are mostly protonated and positively charged.[16] It is precisely these amino groups that form salt bonds with negatively charged network units, which do not penetrate deep inside the neutral G5 dendrimer core as is shown in Scheme 4b. Therefore, such partly ionized G5 dendrimer species do not become permanently entrapped in the #HG so that sorption process proceeds without limitation.

To gain a deeper insight into the above phenomena, one should resort to the mechanism of transfer of charged polymer species in oppositely charged networks. The basic fundamentals of this mechanism were first stated,[3,4] where interaction of linear polyions with oppositely charged #HGs was considered in terms of a three-phase model. Such model adopted for the systems on study is presented in Scheme 5.

The system represented in Scheme 5 consists of three coexisting phases: dendrimer solution, anionic #HG and #IPEC. Dendrimer solution and #IPEC phases are separated by interface 1, while #HG and #IPEC phases by interface 2. Interpolyelectrolyte coupling reaction between dendrimer polycations and anionic network resulting in formation of electro neutral stoichiometric #IPEC proceeds at interface 1. It is precisely this reaction that provides the driving force for the whole process of dendrimer uptake. Thus, either adsorption of dendrimer polycations on #HG surface or their diffusion inward #HG sample are chemically facilitated processes consuming free energy of interpolyelectrolyte coupling reaction.

The mechanism of transportation of dendrimer polycations from environmental solution inward the reacting hydrogel sample through the already formed #IPEC layer is a crucial point in consideration of the studied phenomenon. Basing on a whole series of representative qualitative observations we proposed the general scheme apparently applicable to a wide range of slightly cross-linked hydrogels interacting with oppositely charged polyelectrolytes or micelle-forming surfactants.[3–5,11,18] The rate-determining step is formation of a vacancy at shell-core boundary (interface 2) as a result of thermal fluctuation, which is nothing but exchange with a dendrimer polycation between two fragments of the hydrogel network. Such vacancies neutralized with simple counter ions diffuse through the shell layer backward to the solution-shell boundary (interface 1) where they react with new dendrimer species. Then if the numbers of vacancies born in a time unit at shell-core interface and reached solution-core boundary are equal, the dendrimer uptake should proceed as a frontal relay race propagating a heterogeneous reaction.

For #PA-Na hydrogel the individual steps of this exchange reaction may be presented as follows:

$$\mathbf{\#PA}\text{—}\overset{\ominus}{COO}\overset{\oplus}{N}H_3\text{—}Dx + \mathit{\#PA}\text{—}\overset{\ominus}{COO}\overset{\oplus}{Na} \rightleftarrows \underbrace{\mathbf{\#PA}\text{—}\overset{\ominus}{COO}\overset{\oplus}{Na}}_{\text{vacancy unit}} + \mathit{\#PA}\text{—}\overset{\ominus}{COO}\overset{\oplus}{N}H_3\text{—}Dx \tag{4}$$

$$\mathbf{\#PA}\text{—}\overset{\ominus}{COO}\,\overset{\oplus}{N}H\text{—}Dx + \mathit{\#PA}\text{—}\overset{\ominus}{COO}\overset{\oplus}{Na} \rightleftarrows \underbrace{\mathbf{\#PA}\text{—}\overset{\ominus}{COO}\overset{\oplus}{Na}}_{\text{vacancy unit}} + \mathit{\#PA}\text{—}\overset{\ominus}{COO}\,\overset{\oplus}{N}H\text{—}Dx \tag{4'}$$

Precisely these individual steps are responsible for the movement of interface 2 inwards toward the reacting #HG sample. They also determine the total rate of the dendrimer uptake. As was shown earlier, the same factors, such as external salt concentration and polyelectrolyte charge density, control either the rate of similar exchange reactions or sorption of other polyions by oppositely charged hydrogels.[3,6,20]

In the connection with the above mechanism, one should note that the vacancy shown in Scheme 5 in addition to the negatively charged network fragment also contains either the equivalent number of Na^+ counterions or water. Virtually, it is these vacancies that serve as pump elements providing the facilitated efflux of Na^+ cations and water from still unreacted #HG phase through already formed #IPEC layer to external solution. Each vacancy approached interface 1 couples with DAB-dendr-$(NH_2)_x \cdot HCl$ species coming from external solution. As a result #IPEC fragment forms, while equivalent amounts of free Cl^- and Na^+ release to the solution phase *via* interpolyelectrolyte addition reactions (1) and (1′). Water is also delivered by the same vacancies from the inner heavily swollen phase 2 through phase 3 to the surrounding solution 1. This mechanism also responsible for water pumping out provides a decrease in both mass and volume of the reacting sample. As a result the above described local collapse occurs in course of dendrimer uptake.

The above "relay-race" mechanism of the facilitated transport of the dendrimer polycations into anionic hydrogels assumes that the rate of dendrimer uptake ($\mathrm{d}C/\mathrm{d}t$) is controlled by the rate of nucleation of vacancies ($\mathrm{d}N/\mathrm{d}t$) at interface 2

$$\frac{\mathrm{d}C}{\mathrm{d}t} \sim \frac{\mathrm{d}N}{\mathrm{d}t} \tag{5}$$

Under the steady-state regime all vacancies formed at interface 2 approaching interface 1 are eliminated *via* their interaction with dendrimer species. An increase in the number of vacancies ($\mathrm{d}N$) should be proportional to a decrease in the area of interface 2 ($-\mathrm{d}S$):

$$\mathrm{d}N = -\beta\,\mathrm{d}S \tag{6}$$

where β is the proportionality coefficient. Since even at low extent of conversion the samples assume almost spherical shape, the area of inner interface 2 is equal to $S = 4\pi r^2$, where r is the current radius of still unreacted #HG core. Therefore,

$$\mathrm{d}N = -\beta\,\mathrm{d}S = -\beta \mathrm{d}(4\pi r^2) = -8\pi\beta r\,\mathrm{d}r \tag{7}$$

$$\frac{\mathrm{d}N}{\mathrm{d}t} = \beta k_s S = \beta k_s 4\pi r^2 \tag{8}$$

Joint solution of eqns. (7) and (8) gives:

$$-2\frac{dr}{dt} = K_s r \tag{9}$$

where $K_s = k_s/\beta$ is the first order rate constant. Its integration leads to:

$$\ln\frac{r}{r_0} = -\frac{K_s t}{2} \tag{10}$$

where r_0 is the radius of an equivalent sphere with its volume equal to the volume of initial gel sample. Since, as was shown above the mass of the sample and, hence, its volume linearly change with the extent of #HG sample transformation into #IPEC, one may put down the following relationship:

$$F(t) = 1 - \left(\frac{r}{r_0}\right)^3 \tag{11}$$

By substituting eqn. (11) into eqn. (10), we arrive at the final expression:

$$\ln[1 - F(t)] = -\frac{3K_s}{2}t \tag{12}$$

Thus, under the assumption of the relay-race mechanism for dendrimer uptake experimental kinetic curves should be straightened in the $\ln[1 - F(t)] - t$ coordinates.

Fig. 4a and 4b presents the kinetic curves plotted in the above coordinates for DAB-dendr-$(NH_2)_4 \cdot HCl$, DAB-dendr-$(NH_2)_{16} \cdot HCl$, DAB-dendr-$(NH_2)_{32} \cdot HCl$ and DAB-dendr-$(NH_2)_{64} \cdot HCl$. As is seen, for G3 and G4 dendrimers the above dependences are linear; that is, they fit eqn. (12). The rate constant calculated for G3 and G4 dendrimers which sorption kinetics follow eqn. (12) is $8.2 \times 10^{-5} sec^{-1}$. Let's note that this value is close to the value earlier obtained for the system #PA-Na–cytochrome *c*,[11] although cytochrome *c* molecular mass (12 400) is considerably higher that those of the dendrimers (1687 and 3514 correspondingly).

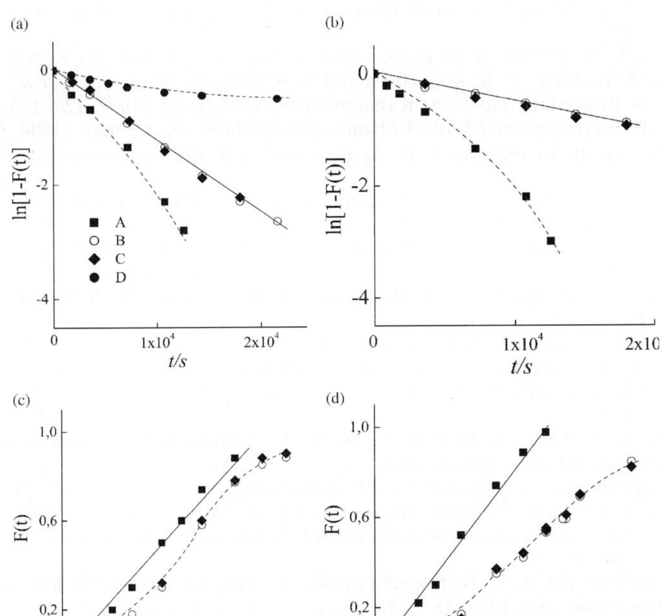

Fig. 4 Kinetic curves for sorption of DAB-dendr-$(NH_2)_x \cdot HCl$ by #PA-Na (a,c) and #PAMPS-Na (b,d) gels in the coordinates of eqn. (12) (a,b) and the Fick equation (c,d): G1 (A), G3 (B), G4 (C) and G5 (D) dendrimers.

However the $\ln[1 - F(t)] - t$ plots for G1 and G5 dendrimers markedly deviate from linearity. This behavior reflects the above mentioned different character of sorption of various dendrimer homologs. In particular, it fully agrees either with observed homogenation over time of partly transformed #HG samples when sorbing DAB-dendr-$(NH_2)_4 \cdot HCl$, and retardation followed with shutdown of DAB-dendr-$(NH_2)_{64} \cdot HCl$ uptake. Moreover, in contrast to G3 and G4 dendrimers kinetic curves of DAB-dendr-$(NH_2)_4 \cdot HCl$ sorption are well linearized if plotted in the coordinates of the diffusion Fick law, $F(t) - t^{1/2}$ (Figs. 4c and 4d).

Conclusions

Two different mechanisms of chemically facilitated sorption of dendritic DAB-dendr-$(NH_2)_x \cdot HCl$ polycations by oppositely charged slightly cross-linked anionic hydrogels were established. In one case it is a frontal heterogeneous reaction under which a sharp boundary between thickening #IPEC layer and diminishing highly swollen original hydrogel core retains till the end of the whole process. The kinetic equation was derived which permitted to quantify the above mechanism. In the case of the lowest generation dendrimer homolog (DAB-dendr-$(NH_2)_4 \cdot HCl$) kinetics of its facilitated diffusion inwards the hydrogel follows Fick's law and the process is accompanied with blurring of the shell–core interface. Both mechanisms confirmed by the scope of direct physical–chemical measurements exploring macroscopic hydrogel samples and kinetic analysis.

Acknowledgements

This work was supported by the Russian Foundation for Basic Research (project 02-03-32058) and NWO Foundation (The Netherlands, grant 99005725).

References

1 V. A. Kabanov, A. B. Zezin, V. B. Rogacheva and E. A. Litmanovich, *Dokl. Akad. Nauk SSSR*, 1986, **288**, 1408.
2 V. B. Rogacheva, V. A. Prevish, A. B. Zezin and V. A. Kabanov, *Polym. Sci. USSR*, 1988, **30**, 2262.
3 V. A. Kabanov, A. B. Zezin, V. B. Rogacheva and V. A. Prevish, *Macromol. Chem.*, 1989, **190**, 2211.
4 A. B. Zezin, V. B. Rogacheva and V. A. Kabanov, *Macromol. Symp.*, 1997, **126**, 123.
5 A. B. Zezin, V. B. Rogacheva and V. A. Kabanov, *Intell. Mater. Syst. Struct.*, 1994, **5**, 144.
6 A. M. Chupyatov, V. B. Rogacheva, A. B. Zezin and V. A. Kabanov, *Polym. Sci., Ser. A Ser. B*, 1994, **36**, 169.
7 V. B. Karabanova, V. B. Rogacheva, A. B. Zezin and V. A. Kabanov, *Polym. Sci., Ser. A Ser. B*, 1995, **37**, 1138.
8 V. B. Skobeleva, V. B. Rogacheva, A. B. Zezin and V. A. Kabanov, *Dokl. Acad. Nauk SSSR*, 1996, **347**, 207.
9 V. B. Skobeleva, A. V. Zinchenko, V. B. Rogacheva, A. B. Zezin and V. A. Kabanov, *Polym. Sci., Ser. A Ser. B*, 2001, **43**, 315.
10 A. B. Zezin, V. B. Rogacheva, V. B. Skobeleva and V. A. Kabanov, *Polym. Adv. Technol.*, 2002, **13**, 919.
11 V. A. Kabanov, V. B. Skobeleva, V. B. Rogacheva and A. B. Zezin, *J. Phys. Chem. B*, 2004, **108**, 1485.
12 R. Scherrenberg, B. Coussens, P. Van Vliet, G. Edouard, J. Brackman and E. De Brabander, *Macromolecules*, 1998, **31**, 456.
13 V. A. Kabanov, A. B. Zezin, V. B. Rogacheva, Zh. G. Gulyaeva, M. F. Zansochova, J. Joosten and J. Brackman, *Macromolecules*, 1999, **32**, 1904.
14 E. M. M. de Brabander, J. Brackman, M. Mure-Mak, H. de Man, M. Hogeweg, J. Keulen, R. Scherrenberg, B. Coussens, Y. Mengering and Sj. van der Wal, *Macromol. Symp.*, 1996, **102**, 9.
15 T. Tanaka, in *Frontiers of Macromolecular Science*, ed. T. Saegusa, T. Higashimura and A. Abe, Blackwell Science, Oxford, 1989, p. 325.
16 V. A. Kabanov, A. B. Zezin, V. B. Rogacheva, Zh. G. Gulyaeva, M. F. Zansochova, J. Joosten and J. Brackman, *Macromolecules*, 1998, **31**, 5142.
17 Yu. V. Khandurina, V. B. Rogacheva, A. B. Zezin and V. A. Kabanov, *Polymer Science*, 1994, **36**, 184.
18 V. A. Kabanov, A. B. Zezin, V. B. Rogacheva, Yu. V. Khandurina and O. A. Novoskoltseva, *Macromol. Symp.*, 1997, **126**, 79.
19 A. R. Khokhlov, S. G. Starodubtsev and V. V. Vasilevskaya, *Adv. Polym. Sci.*, 1993, **109**, 123.
20 K. N. Bakeev, V. A. Izumrudov, S. I. Kuchanov, A. B. Zezin and V. A. Kabanov, *Macromolecules*, 1992, **25**, 4249.

Design of chimaeric polymersomes

J. G. E. M. Fraaije, C. A. van Sluis, A. Kros, A. V. Zvelindovsky and G. J. A. Sevink

Leiden Institute of Chemistry, Leiden University, PO Box 9502, 2300 RA Leiden, The Netherlands

Received 1st March 2004, Accepted 19th May 2004
First published as an Advance Article on the web 29th October 2004

We discuss the development of hierarchical polymer particles, or variegated polymersome composites, in which at least two different components are phase separated within one polymersome chimaera. We briefly discuss the present status in experimental polymersome research, and then discuss a speculative design strategy, based on mesoscopic simulations with a dynamical variant of polymer self-consistent field theory (Mesodyn). The main conclusion is that the counter-intuitive co-assembly of demixing block copolymers is the key in controlling hierarchical structures on a mesoscopic scale. This is the classical paradox of a chimaera: the constituents live in the same scaffold, but apart. Block copolymers beyond a certain length will always split the assembly, and without further precautions, polymer based chimaerae are intrinsically unstable. To this end, we propose the application of a branched block copolymer as composite compatibilizer, glueing the separate domains together, and thereby stabilizing the chimaeric polymersome.

Introduction

The formation of vesicles is a well-established and familiar property of phospholipids. Such hollow structures have been applied in a variety of techniques in drug release or as micro-reactors. In recent years, it was shown various amphiphilic block copolymers also form vesicles, the so-called polymersomes. Polymersomes have been prepared of biodegradable,[1] thermal tunable,[2] conducting[3] or polyelectrolyte materials.[4] A number of papers, especially from the Eisenberg group, describe an entire 'Wonderland' of more complex polymersome architectures.[5–15] For example, vesicles with hollow rods running inside and parallel to the surface of the walls, or vesicular structures with protrusions or holes. Döbereiner and coworkers[16] published the properties of giant polymersomes (> 10 μm) with a wall formed either by a double-layer, which is connected by a lattice of passages, or a tubular network with hexagonal symmetry. In all cases the polymer is impure, sometimes deliberately so, and the structures are discovered by trial and error. 'Peptosomes', that is, polymersomes—or more general various tubes, micelles *etc.*—based on (conjugated) block polypeptides, were demonstrated very recently.[17–22] Assembling random peptides have in fact a much longer history. In the origin of life area, proteineous microspheres had been prepared already a few decades ago by Fox,[23,24] by simply baking amino acids, and exposing the residue to water. These structures lacked precisely defined compositions.

Polymersome research exploded following the publication of a few high-profile papers on the general principles[25] and detailed physical chemical characterization.[26,27] The interest is understandable from a technological point of view, since polymeric vesicles have many advantages over low-molecular weight vesicles, such as enhanced stability and flexibility in chemical modification.

The formation of one-component hollow polymersomes of low complexity is well understood by the formation of polymeric bilayers, guided by mixing and demixing rules set forth by standard Flory–Huggins theory. Also for amphiphile packing the theory seems straightforward. The essential ingredients are not so much different from lipid bilayers. The molecule comprises a hydrophilic 'head' and a hydrophobic 'tail'. The packing factor, derived from the sizes of head and tail, determines whether the amphiphile aggregates in spheres, cylinders or lamellae (bilayers). The only difference between polymersomes and lipid vesicles is the molecular conformation: random in the case of flexible polymers and more narrowly distributed in the case of the packed lipids. Simple concepts are adequate to design single-component polymersomes, but they need extension in designing compound aggregates.

We foresee the design of organelle mimicking phases, or the design of reactive genetic assemblies in which *in situ* polymer synthesis is coupled to a genetic coding string by self-assembly. These *polymersome chimaerae* would be much more complex than the spheres, cylinders or hollow sacs one observes with single-component polymersomes. Our envisioned chimaeric particles have three geometrical characteristics which set them aside: (1) they are not hollow, but filled with a microstructure (which, when continued, leads to a hierarchy of embedded structures), (2) they have a means to interact with the environment through protrusions, 'legs', or invaginations, 'holes', *etc.* and (3), when the polymers are produced *in situ*, the geometry is dynamical, dissipative and may grow and fission and potentially obey evolutionary rules.

Chimaeric complexity obviously offers enormous potential for the design of "intelligent" self-assembled structures, but at the same time the challenge for keeping control also increases. The apt question is: How to construct a plastic mitochondrion? Or an artificial *de novo* macromolecular chloroplast? Construction implies rational design, not just trial and error. And clearly, the 'trial-and error' approach, which has led to the plethora of complex polymersomes structures so far, is inadequate and unsatisfactory from an intellectual point of view. What we need is, in the clearest terms possible, general rules by which one can design a desired hierarchical polymersome structure from block polymer mixtures and possibly from biological components. This is nothing less than the ambitious development of *a theory for abiotic proto-cells*. But even a superficial comparison between an *in vivo* biological organelle and a current state-of-the-art *in vitro* polymer assembly teaches us humility. It makes us realize that chemical science is a very, very long way from attaining such goals. Here is where we start: the real, living, biological organelles serve as beacons, posing a distant goal. We will concentrate on a very basic subtask: design of a polymersome with two polymers, co-assembling in different domains within one vesicular body.

Theory and simulation

The computations require the repeated simulation of polymer vesicle formation and stability, given a certain mixture of block copolymers, so as to generate a microphase diagram of polymersome morphologies. The rules are based on Flory–Huggins χ-parameters to account for block demixing, and size ratios of the blocks for the induction of bilayer curvature. In previous work we studied the formation of single component polymersomes, demonstrating the delicate influence of soft confinement on the classical microphase structures.[28] Here, following the same simulation method, we study two-component polymersome chimaerae, a mixture of two diblock polymers, where the two hydrophobic blocks become demixed, and the hydrophilic blocks having different curvature preference. In this way, unstable domains or rafts are formed, which then lead to protrusions and invaginations.

In our calculations, we generate structures by quenching a homogeneous vesicular droplet of polymer surfactant in an aqueous bath. The vesicle relaxes by a non-linear diffusion process: $\dot{\rho}_I = D_I\nabla^2(\rho_{sI} - f_I)$ where f_I is the density functional of the polymer (or solvent) molecule, with the mean-field chemical potentials as variable (details of the model are in ref. 28; see the Appendix for the explicit formula of f_I).

It can be shown that the non-linear diffusion equations have the same kinetic coefficients as in collective Rouse dynamics, which is correct for the spinodal-like microphase separation internal to the vesicular structures. The simulation parameters are for diblock polymer surfactants in strongly selective solvent and strong segregation. They are: the architecture of the two polymers A13H7 and B13H17, where each chain is a necklace of "beads", and Flory–Huggins parameters $\chi_{HS} = 0.5$ (theta solvent for the hydrophilic blocks), $N\chi_{AH} = 40$, $N\chi_{BH} = 60$ (strong segregation within one polymer,

N the polymer length), $\chi_{AS} = \chi_{BS} = 2.0$, (strongly hydrophobic blocks with respect to solvent) and $\chi_{AB} = 0.5$ (weakly mutually phase separating hydrophobic blocks of the two polymers).

One should realize that in the mean-field model any polymer surfactant solution, with the same properly scaled interaction parameters, behaves in the same way. Taking into account that each bead or statistical unit represents at least three or four monomers, a possible system with the given parameters values would be a mixture of polybutadiene–polyethylene oxide, and polypropylene-oxide–polyethyleneoxide diblocks, with number of monomers each $O(100)$. The dimensionless time indicated in Figs. 1–4 is $\tau = Dt/a^2$; where a is the bead size. The diffusion coefficient of all components is identical in all simulations. The typical real time scale of the polymer composite phase separation is on a mesoscopic scale (seconds–hours)—depending on realistic values for the bead size and diffusion coefficient. Both polymer architectures affect the intrinsic stability of the assembly structure curvatures differently: the larger polymer (B13H17) is asymmetric, promotes high curvature and destabilizes the vesicle wall, the smaller polymer (A13H7) is more symmetric and forms stable vesicles (data not shown).

The free energy model is that for an nVT ensemble, and consequently we do not calculate the global equilibrium of an open system, but rather the local equilibrium morphology of an isolated vesicular structure. The situation is analogous to that of classic studies of the shape of isolated lipid vesicles, when interactions between vesicles are less relevant. With the selected values of the Flory–Huggins parameters and length of the polymers the polymers molecules are all insoluble, hence deformations of the vesicles occur at constant mass of polymer.

The simulations proceed by a sudden quench of a vesicle in a solvent bath. Following the quench, the assembly takes up or releases additional solvent locally and globally, depending on the particular morphology being formed. The unimer polymer concentration is zero, and this remains so during adaptations in morphology. We have found it advantageous to add white noise to the mean-field chemical potentials. The uncorrelated noise does not obey the fluctuation-dissipation theorem, but it helps to cross barriers in the free energy landscape. The vesicles are placed in the center of the box with sufficient space, $N/2$ cells, between the vesicular surface and the boundaries of the computational box, thereby avoiding artifacts resulting from the periodic boundary conditions. In all cases the droplets develop a fuzzy outer layer of solvophilic blocks. Since the confinement of the polymers is not restricted, the vesicular surface is not necessarily spherical, nor is the topology constant.

The results (see Figs. 1 and 2) clearly demonstrate demixing of the two block copolymers. In related simulations (data not shown) we decreased the chemical incompatibility between the two hydrophobic blocks slightly (by a few tenths of a kT unit to $\chi_{AB} = 0.25$) and observed no phase separation. Here, with slightly larger incompatibility, we observe phase separation into rafts, but the phase separation is already so strong the two polymers further demix into separate assemblies entirely, depending on the relative composition. This can be seen very clearly in Fig. 1a and 1b.

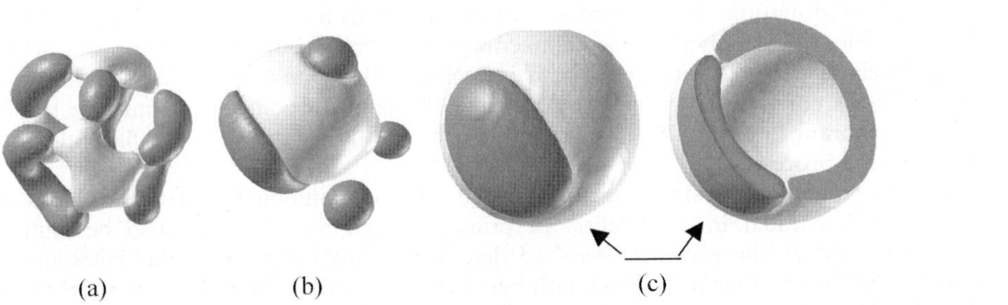

Fig. 1 Two-component polymersome chimaera, simulated with a dynamical variant of self-consistent field theory (Mesodyn), shapshots at $\tau = 5000$. Fractional composition $x = 0.25/0.50/0.75$ (left to right), x(A13H7) + (1 − x)(B13H17) (rightmost image is 75% composition, but in open view). Architectures of Gaussian chains (A and B, hydrophobic and H hydrophilic bead types):

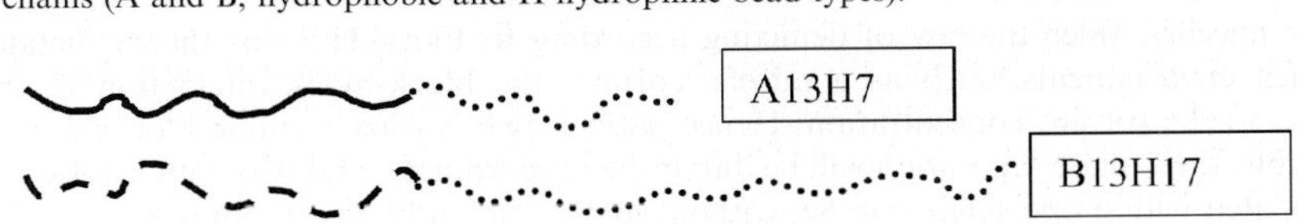

(1)

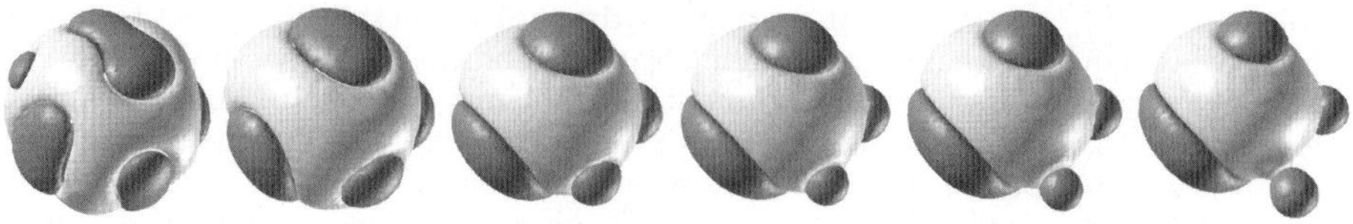

Fig. 2 Formation dynamics of 50% polymersome chimaera (Fig. 1b). Snapshots at (from left to right) $\tau = 500$, 1000, 2500, 3000, 3500, 4000, 4500 (5000 in Fig. 1b).

Fig. 2 illustrates the time evolution of the domains for one composition ratio (50%). Here, we also observe the disappearance of separate smaller domains, and the growth of larger domains (Ostwald ripening), but there is a peculiar topological alteration too: a small domain pinches off as a micelle, while the larger domains respond by buckling ($\tau = 4500$). Phase separation also depends on the initial composition ratio of the two polymers. In Fig. 1a, the majority phase is the asymmetric polymer, and the vesicles bursts apart, completely disrupting the assembly structure into separate islands. In Fig. 1c, however, the more symmetric polymer is the majority phase, the vesicle is stable, with at the end one domain of the minority phase. From the theory of membrane fluid–fluid coexistence,[29] it is known that in such a case budding will occur when the domain of the minority phase is above a certain critical size. We would therefore expect that a larger vesicle (not yet accessible to simulation at present), of same initial composition as in Fig. 1c would split apart too.

Demixing vesicles, even if the topology does not change, do not maintain a perfect smooth spherical structure. Rather, the domain-domain interface, here a *line*, has a reduced bilayer thickness. This is easy to understand: since when the two polymers demix, the contact between domains must be as small as possible, hence the evolution to a few large domains, and the decrease in domain-domain interface perpendicular to the membrane.

A theoretical description of fluid-fluid coexistence in multi-component (lipid) vesicular membranes has been the subject of intensive research for over ten years,[29,30] and was recently observed aslo in mixed lipid vesicles by advanced light microscopy.[31] Phenomenological approaches of different degree of sophistication have been applied to two types of complex membranes: lipid membranes anchored by macromolecules (*e.g.* proteins or synthetic flexible polymers), and domain-induced budding of lipid membranes.

The starting point of the theoretical analysis is usually a Helfrich Hamiltonian of a membrane or a membrane patch. Such an analysis is useful if the topology is constant. In addition to the analytical theory, membrane-budding was extensively studied by Monte Carlo simulation.[29] The evolution of a multi-component lipid membrane experiences the following distinct time regimes during the phase separation process: (1) formation and growth of intramembrane domains, (2) multiple bud formation, and (3) coalescence of small buds into larger ones. Although both the time and size scales of the polymersome chimaerae are different from lipid systems, our field simulation qualitatively agrees with this scenario. We observe both the in-plane phase separation of different types of block copolymers, as well as the escape of one of the components into the third dimension due to membrane flexibility. Coarse graining of the resulting aggregates is a natural consequence of the demixing process.

Another comparable system is a concentrated solution containing a mixture of block copolymers, microphase separated in co-micelles. In principle, such micelles may also be multi-component chimaerae, just as the polymersomes we described above, and so similar rules for mixing and demixing will apply. One may distinguish here between (1) the co-micellisation of block polymers of same chemical composition, but of different block lengths (such co-micellisation is usually accidental following the application of heterodisperse polymer samples and not further discussed here), and (2) the mixing of block copolymers, where either the core forming, or corona forming blocks are different. The case where the corona forming blocks are incompatible is less interesting for forming chimaerae micelles, since the law of demixing according to Flory–Huggins theory dictates that in solvent rich environments, such as a micelle corona, the block-block interaction is reduced, in proportion to the solvent concentration. Hence, even if two corona forming blocks are chemically incompatible, their phase separation will be diminished accordingly, and two such blocks may well be accommodated within one single micelle, without internal microphase separation.

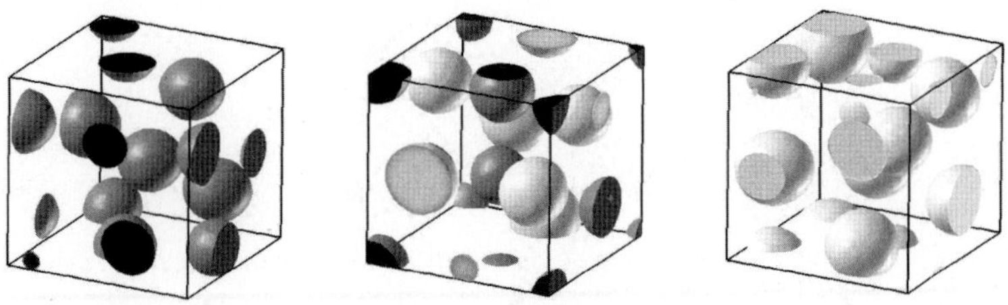

Fig. 3 Pure and mixed micelles at polymer fraction $x = 0/0.50/1.00$ (Fig. 1) at 70% total polymer concentration. Chimaeric phase is unstable.

The case of demixing core-forming blocks is very similar to our scenario.[32,33] Here, since the cores are dense, demixing again follows the strong phase separation principle we discussed before. When the core-forming blocks demix, the micelles will split into two, when the core-forming blocks mix, the micelles will house the two blocks together. But the internal mixing will then be homogeneous, without forming chimaeric micelles! Hence, such a system is either bimodal in micelle distribution, or unimodal, but rarely microphase separated within each individual micelle. The situation is clarified by considering the stability of three phases: chimaeric micelles, a mixture of micelles and homogeneous micelles. When we increase the interactions, the phase boundary is crossed between the homogeneous state and the mixed micelles state, and the chimaeric state is never reached. Essentially the same conclusion was reached for co-micellisation in block copolymer blends.[33]

Our simulations, applied to 70% polymer solutions (Fig. 3) precisely follow the proposed rule. The pure block copolymer solutions (0% or 100% of each polymer), are micellar, the mixed copolymer solution (50% of each) is completely demixed in micelles composition. Notice that the separate micellar fractions do *not* macrophase separate (as discussed in ref. 33), since the hydrophilic coronas of identical chemical nature ensure good mixing of the dissimilar micelles (this is not a general conclusion for all co-micelles but applies here due to the particular choice of the corona blocks under consideration).

Discussion

The simulations so far indicate that a design rule based on simply mixing diblock copolymers in *one* assembly, is inadequate to predict the formation of stable chimaeric polymersomes. When one mixes two block polymers, of same hydrophobic block, and same hydrophilic block, but different chain lengths, the polymer with the longer hydrophilic block accumulates in regions of higher curvature (this is already known from the Eisenberg experiments; for a discussion of the effect of polymer dispersity see for example ref. 34). When the two hydrophilic blocks have different chemical composition, lateral phase separation leads to the formation of raft structures.

For our purpose of designing polymersome chimaerae, the most favorable situation is when the two hydrophobic blocks have different *chemical* composition, so that the polymers phase separate, and in addition the hydrophilic blocks must be of dissimilar length, so that planar rafts are unstable and open. This is due to the fact that both the chemical composition (Flory–Huggins parameters χ) and block length (entropic effect) contribute to the effective bending moduli of structures. This seems an attractive scenario, because of its simplicity, but we believe the practical range of chemical incompatibility is too narrow. When the phase separation is too strong, the two polymers will pinch off and individually form assemblies. When the phase-separation is too weak, the hydrophilic blocks will not be forced to accumulate into rafts, the longer hydrophilic block remains dispersed homogeneously, and the wall is not destabilized. There may be a value range in parameter space $0 < \chi_{AB} < \chi^c$ where the chemical incompatibilities between the hydrophobic blocks is strong enough for phase separation, but not yet strong enough for splitting the assembly. But arguably, such region will be very small. Since we are dealing with polymers, not with low molecular weight surfactants or lipids, even a small discrepancy in chemical compatibility on a monomer scale, is amplified at the polymer level. The relevant interaction parameter is $N\chi_{AB}$, not χ_{AB}, and $\chi^c = O(1/N)$. Therefore, given a certain chemical composition of the monomers, $\chi^c \to 0$ as the block length increases, and long block copolymers, trapped in a bilayer, will either mix homogeneously, or

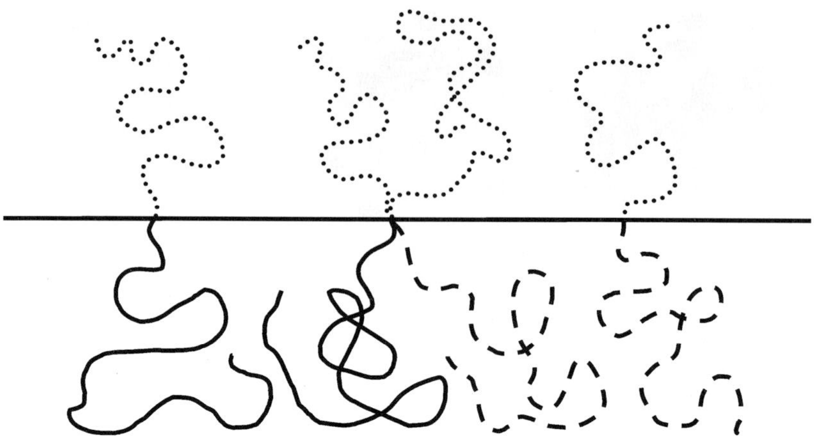

Fig. 4 Four-arm in-plane compatibilizer.

the assembly splits. There is no intermediate outcome: without further precautions chimaerae are intrinsically unstable.

Robust rules for polymersome composite design must somehow counteract the fission of the vesicle due to the built-in phase separation, that is, the polymers must separate individually to form hierarchical structures, but without splitting the overall assembly structure. This is the classical paradox of a chimaera: the polymers must "live in the same house, but apart". The biological strategy for dealing with the paradox is manifold, for example, (1) by generating the amphiphiles *in situ* (for example, specifically *inside* an assembly), in such a way that on the biological life-span, the transport is too limited by barriers for the assembly to fall apart, or, (2) by maintaining a dissipative process for the integrity of the assembly structure (that is, the assembly is far out of equilibrium), or (3) by the incorporation of additional *networking* or *skeleton* molecules for spanning and cross-linking the membrane. Except for covalent cross-linking, in man-made polymersome composites such strategies cannot be employed. Also, in a practical approach we are limited to rather crude macroscopic assembly techniques, such as by quench, film de-stabilization or sonication. These methods lack the inherent fine-graded 'nanoscopic' instrumentation by which biological entities can shape and mold individual assembly structures. The challenge is therefore reduced to the development of additional block copolymers, capable of spontaneously glueing and stitching together the phase separated domains within each single polymersome. We have speculated that multi-arm block copolymers may be useful for such purpose, but any experimental evidence is lacking at present. Fig. 4 displays a cartoon of the four-arm copolymer glue in action: a planar block copolymer bilayer is separated into two different domains, each pure in one block copolymer. The compatibilizer has four arms, two branches similar to the two hydrophobic blocks, and two branches similar to the hydrophilic blocks. In this way, the branched polymer stabilizes the *in-plane* domain–domain interface of the polymersome chimaerae, similar in action to diblock compatibilizers stabilizing mixed polymer phases in bulk industrial systems, on a much larger scale. In our naïve design, branches of identical composition and length act as the separate blocks of the constituting di-block copolymers. Perhaps, rather than *two* hydrophilic blocks, *one* bigger hydrophilic block may suffice: this would be a minimal three-arm compatibilizer. It may be interesting to test our hypothesis experimentally, while we calculate the phase diagram theoretically.

Appendix

The expanded formula for the density functional is:[28]

$$f_I(\boldsymbol{r}) = Z\sum_{s=1}^{N}\theta_{Is}\int \mathrm{d}\{\boldsymbol{R}\}\delta(\boldsymbol{r}-\boldsymbol{R}_s)\mathrm{e}^{-\beta(H+X)}$$

$$H = \frac{3}{2a^2}\sum_{s=2}^{N}(\boldsymbol{R}_s - \boldsymbol{R}_{s-1})^2$$

$$X = \sum_{s=1}^{N}\mu_s^{\mathrm{MF}}(\boldsymbol{R}_s)$$

$$\mu_I^{\mathrm{MF}}(\boldsymbol{r}) = \textstyle\sum_J \int \mathrm{d}\boldsymbol{r}'\, \varepsilon_{IJ} K(\boldsymbol{r}-\boldsymbol{r}')\rho_J(\boldsymbol{r}') + \kappa\nu\rho_J(\boldsymbol{r})$$

where Z is a normalization constant ($\int \mathrm{d}\boldsymbol{r} f_I = \int \mathrm{d}\boldsymbol{r}\, \rho_I$), θ_{Is} is 1 when bead s is of type I and 0 otherwise, H is the ideal Gaussian chain Hamiltonian, μ_s^{MF} is the mean-field chemical potential, the Flory–Huggins parameter is $\chi_{IJ} = \nu^{-1}\beta(\varepsilon_{IJ} + \varepsilon_{JI} - \varepsilon_{II} - \varepsilon_{JJ})$, ν is the bead volume and κ the Helfand compressibility parameter.

References

1 F. Meng, C. Hiemstra, G. H. M. Engbers and J. Feijen, *Macromolecules*, 2003, **36**, 3004–3006.
2 R. J. Thibault, P. J. Hotchkiss, M. Gray and V. M. Rotello, *J. Am. Chem. Soc.*, 2003, **125**, 11 249–11 252.
3 D. M. Vriezema, J. Hoogboom, K. Velonia, K. Takazawa, P. C. M. Cristianen, J. C. Maan, A. E. Rowan and R. J. M. Nolte, *Angew. Chem., Int. Ed. Engl.*, 2003, **42**, 772–776.
4 J. N. Cha, H. Birkedal, L. E. Euliss, M. H. Bartl, M. S. Wong, T. J. Deming and G. D. Stucky, *J. Am. Chem. Soc.*, 2003, **125**(27), 8285–8289.
5 N. S. Cameron, M. K. Corbierre and A. Eisenberg, *Can. J. Chem.*, 1999, **77**, 1311.
6 L. Zhang and A. Eisenberg, *Science*, 1995, **268**, 1728.
7 H. Shen and A. Eisenberg, *Macromolecules*, 2000, **33**, 2561.
8 H. Shen and A. Eisenberg, *J. Phys. Chem. B*, 1999, **103**, 9473.
9 Y. S. Yu and A. Eisenberg, *J. Am. Chem. Soc.*, 1997, **119**, 5980.
10 Y. Yu, L. Zhang and A. Eisenberg, *Macromolecules*, 1998, **31**, 1144.
11 L. Zhang, K. Yu and A. Eisenberg, *Science*, 1996, **272**, 1777.
12 L. Zhang and A. Eisenberg, *Macromolecules*, 1996, **29**, 8805.
13 L. Zhang and A. Eisenberg, *Macromolecules*, 1999, **32**, 2239.
14 L. Zhang, H. Shen and A. Eisenberg, *Macromolecules*, 1997, **30**, 1001.
15 K. Yu, L. Zhang and A. Eisenberg, *Langmuir*, 1996, **12**, 5980–5984.
16 C. K. Haluska, W. T. Gózdz, H.-G. Döbereiner, S. Förster and G. Gompper, *Phys. Rev. Lett.*, 2002, **89**(23), 238302.
17 S. Kimura, D. H. Kim, J. Sugiyama and Y. Imanishi, *Langmuir*, 1999, **15**(13), 4461–4463.
18 H. Kukula, H. Schlaad, M. Antonietti and S. Forster, *J. Am. Chem. Soc.*, 2002, **124**(8), 1658–1663.
19 S. Vauthey, S. Santoso, H. Y. Gong, N. Watson and S. G. Zhang, *Proc. Natl. Acad. Sci. USA*, 2002, **99**(8), 5355–5360.
20 S. S. Santoso, S. Vauthey and S. G. Zhang, *Curr. Opin. Colloid Interface Sci.*, 2002, **7**(5–6), 262–266.
21 S. Santoso, W. Hwang, H. Hartman and S. Zhang, *Nano Lett.*, 2002, **2**, 687–691.
22 M. S. Wong, J. N. Cha, K. -S. Choi, T. J. Deming and G. D. Stucky, *Nano Lett.*, 2002, **2**, 583–587.
23 S. W. Fox, *The Emergence of Life: Darwinian Evolution from the Inside*, Basic Books, New York, 1988.
24 S. W. Fox and K. Harada, *Science*, 1958, **128**, 1214.
25 D. E. Discher and A. Eisenberg, *Science*, 2003, **297**(5583), 967–973.
26 H. Bermudez, A. K. Brannan, D. A. Hammer, F. S. Bates and D. E. Discher, *Macromolecules*, 2002, **35**(21), 8203–8208.
27 J. C. M. Lee, M. Santore, F. S. Bates and D. E. Discher, *Macromolecules*, 2002, **35**(2), 323–326.
28 J. G. E. M. Fraaije and G. J. A. Sevink, *Macromolecules*, 2003, **36**(21), 7891–7893.
29 R. Lipowsky, *J. Biol. Phys.*, 2002, **28**(2), 195–210.
30 R. Lipowsky and R. Dimova, *J. Phys.-Condens. Matter*, 2003, **15**(1), S31–S45.
31 T. Baumgart, S. T. Hess and W. W. Webb, *Science*, 2003, **425**, 821–824.
32 D. F. K. Shim, C. Marques and M. E. Cates, *Macromolecules*, 1991, **24**(19), 5309–5314.
33 A. L. Borovinskii and A. R. Khokhlov, *Macromolecules*, 1998, **31**(4), 1180–1187.
34 O. Terreau, L. Luo and A. Eisenberg, *Langmuir*, 2003, **19**, 5601–5607.

Organisation in two series of low-dimensional polymer electrolytes with high ambient lithium salt conductivity†

Jianguo Liu, Yungui Zheng, Yen-Po Liao, Xiangbing Zeng, Goran Ungar and Peter V. Wright

Dept. of Engineering Materials, University of Sheffield, Mappin St., Sheffield, UK, S1 3JD

Received 26th February 2004, Accepted 18th May 2004
First published as an Advance Article on the web 14th September 2004

The amphiphilic polymers poly[2,5,8,11,14-pentaoxapentadecamethylene(5-hexadecyloxy-1,3-phenylene)] **I** (abbrev. C16O5), its homologues C16O*n*, $n = 1$–4, and its copolymers with C16O1 have been synthesised by two Williamson procedures. Method X gives polyether-esters; method Y gives pure polyethers. DSC, optical microscopy, small-to-wide angle X-ray diffraction and complex impedance spectroscopy have been used to investigate organisation and ionic conductivities of the C16O*n* and the C16O5–C16O1 copolymer series and their complexes with $LiClO_4$ and $LiBF_4$. Four-component mixtures of the latter systems (polymers **I**) with polyoxytetramethylene–decamethylene copolymer (**II**) and an octadecyl-terminated ABA block copolymer (**III**) (interfacial stabiliser) ('type D' systems) gave high ambient conductivities ($\geq 10^{-4}$ S cm^{-1}) with low temperature dependence. Highest conductivities (8×10^{-4} S cm^{-1} at 20 °C) were observed with a type D system based upon C12C18O5X.

Introduction

The current use of solvents in commercial lithium batteries for portable electronic applications are considered to represent a safety hazard for larger scale heavy-duty applications such as electrical traction or in aerospace where solvent evaporation may occur at low-pressures.[1] Recent work at Sheffield[2–6] has shown that the amphiphilic polymers **I** ($m = 12$, 16, 18; $n = 5$) when blended and de-blended with copolymers **II** and a lithium salt LiX (*e.g.* $LiBF_4$ or $LiClO_4$) self-organise in solvent-free low-dimensional polymer electrolytes having conductivities at ambient temperature (20 °C) $\sim 10^{-3}$ S cm^{-1} which are of the order required for the operation of a lithium battery. These results approach those of the 'gel' (solvent-containing) electrolytes[7] in current use in commercial Li batteries and exceed by up to 100-fold conductivities for 'conventional' solvent-free amorphous systems[8] typically based upon poly(ethylene oxide). We had previously shown[9,10] that the conductivities of complexes of **I** : LiX (1 : 1) ('type A' systems) and **I** : **R** : LiX (1 : 1 : 1), where **R** is a long-chain alkane or alkyl molecule such as $C_{16}H_{33}OMe$ ('type B' systems), could exceed those of amorphous systems in a liquid-crystalline (lc) phase above the melting temperatures of the side-chains of **I** ($T_{lc} = 35$ °C–55 °C). The lc phase arises from the mesogenic character of the **I**–Li salt

† Electronic supplementary information (ESI) available: Synthetic procedures, molecular dynamics models, tabulated X-ray and thermal data. See http://www.rsc.org/suppdata/fd/b4/b402949f/

adduct, the salt-free **I** being amorphous above side chain melting. However, on cooling below T_{lc} to ambient the conductivities fell steeply as the side chains crystallised, crystallite shrinkage introducing resistive grain boundaries. In the **I** : **II** : Li salt ('type C') systems, after de-blending by heat treatment at the liquid crystal–isotropic transition temperature ($T_{iso} \sim 110\,°C$), polymer **II** functions as an ion bridge or 'glue' between the crystallites.[6] The higher level of conductivity was thus maintained down to ambient and below ($\sim 8 \times 10^{-5}$ S cm^{-1} was observed[2] at $-5\,°C$) with low temperature dependence. The de-blending process, *i.e.* the separation of 1–3 μm lamellae of **I**–salt complex as lc and crystalline phases from **II**, was observed by optical microscopy and monitored by complex impedance spectroscopy. The conductivities from AC impedance data between blocking electrodes were supported by DC polarisations between Li electrodes. The agreement between DC and AC data is consistent with predominantly Li^+ transport. Using one of these systems, a galvanic cell discharging at 0.1 mA cm^{-2} at 20 °C was demonstrated[6] confirming the passage of ions across the cell.

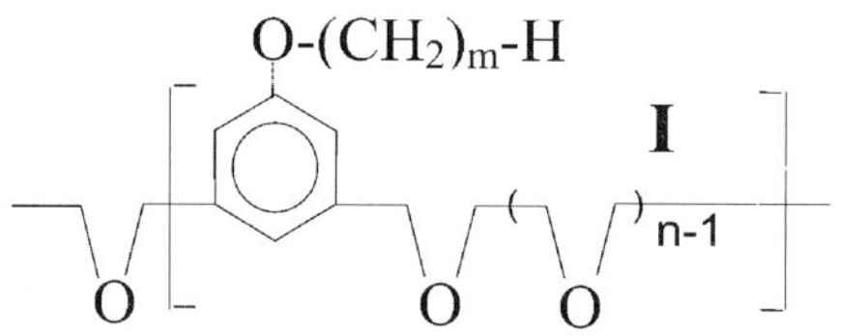

(1)

$-[(CH_2)_4\text{-}O\text{-}]_{23}\text{-}(CH_2)_{12}\text{-}O\text{-}$

II

Although ambient conductivities within the range $10^{-4} < \sigma < 10^{-2}$ S cm^{-1} have been repeatedly observed in type C systems having a variety of compositions, some variability and instability in conductivity measurements have been observed.[5,11] A significant observation was the apparent reversible dependence on the pressure generated within the measurement cell as the films were compressed between the electrodes. Release of the compression on the cell presumably created gaps at the interface between the polymers **I** and **II** inhibiting the passage of ions between the phases. Lack of consistency in the structure of polymer **I** was also an apparent source of poor reproducibility.

In this paper, two series of polymer **I** have been synthesised using the two methods X and Y shown in Scheme 1. In the first series the polymers C16O*n*, $n = 1$ to 5, have been prepared in order to investigate the association of the salts $LiBF_4$ and $LiClO_4$ with the polyether skeletal backbones. In the second series copolymers C16O1–C16O5 have been prepared in order to study the consequence of 'vacancies' in the pentaethoxy skeletal segments of **I** created by substitution of single oxygens. Equimolar complexes of both series of **I** with either $LiBF_4$ or $LiClO_4$, *e.g.* C16O*n* : $LiClO_4$ (1 : 1), ('type A') have been prepared and the pure polymers **I** and their complexes have been investigated using DSC, small-to-wide-angle X-ray diffraction and complex impedance spectroscopy.

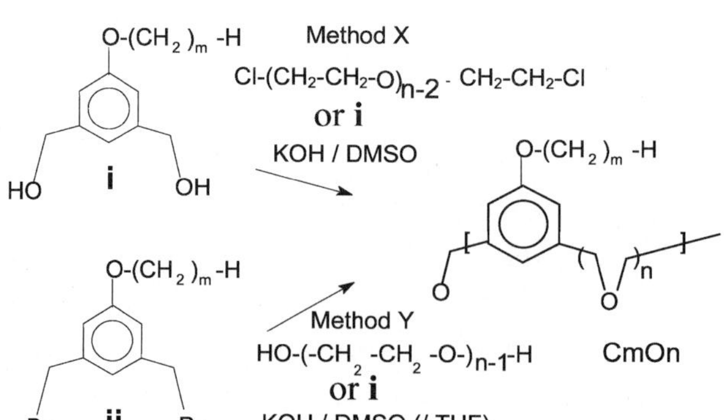

Scheme 1 Synthetic routes to C*m*O*n* polymers.

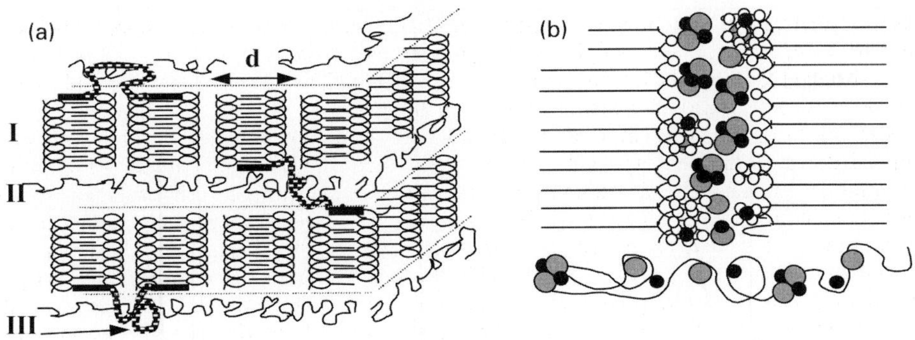

Fig. 1 A model for 'type D' systems. (a): In polymer **III**, spotted lines are [–$(CH_2)_4$–O–]– segments; solid lines are –$C_{18}H_{37}$ blocks; '*d*' is long spacing from SAXS. (b) C16O5–C16O1 copolymer; black circles are Li^+; grey circles are anions, open circles are oxygens.

Interfacial stability has been investigated by addition of polymer **III**, an ABA block copolymer, coded DC18PTHF $C_{18}H_{37}$–O–[–$(CH_2)_4$–O–]$_{\sim 50}C_{18}H_{37}$ ($\langle M_w \rangle \cong 4700$) in *four*-component systems **I**:**II**:**III**:Li salt. The ionophobic terminal –$C_{18}H_{37}$ blocks should more readily manœuvre alongside the alkyl side chains at the lamellar surfaces of **I** than would the main-chain –$(CH_2)_{12}$– blocks of **II** (see Fig. 1a). Polymer **III** was incorporated in the mixture so as to replace 20 wt.% of polymer **II**. Within the total system **III** amounts to ~8 wt.%. These 4-component systems are categorised as 'type D' systems and are designated **I**:**II**:**III**:Li salt (1:0.8:0.2:1.2) where the molar ratios refer to the repeat unit of **I**. The ionic species are located within the channels between lamellae created by side chain interdigitation as shown in Fig. 1(b). The roles of the ionic aggregates and 'free' ions in the conduction mechanism are discussed below. As before,[2–6] an equivalent 'molar mass' for the PTHF copolymers in the mixture was taken to be '500' after consideration of the heptamer sequence of PTHF required to traverse a unit cell of C16O5 in a hypothetical type C blend. The four-component blends were prepared from homogeneous dilute solutions of the mixtures in CH_2Cl_2/acetone by freeze drying to fine powders. The solids were then fused under vacuum at 50 °C.

Abbreviations

Polymers and copolymers prepared by method X are denoted C*m*O*n*X. This includes the polymer C16O1X prepared by dehydration polymerisation of monomer **i** in DMSO/KOH. Absence of 'X' in the abbreviation denotes preparation by method Y. The copolymers C16O1–C16O5 are denoted by the molar proportion of pentaethoxy segment in the skeletal backbone *e.g.* C16O5(40%), the remaining 60% of alkoxyphenylenes being linked by –CH_2–O–CH_2–.

Results and discussion

Synthetic methods

The amphiphilic nature of monomer **i** and the C*m*O*n* polymers and copolymers have a profound influence on the structure of the products in both methods X and Y. The polar solvent dimethylsulfoxide (DMSO) promotes the 'Williamson' condensation but it is a non-solvent for the polymers and **i** presumably exists within micellar aggregates.

The preparation of C16O1X by self-condensation of **i** in DMSO/KOH shows that a dehydration process, taking place within the phase-separated solvent-swollen product, occurs alongside the Williamson etherification in these systems. The use of pure DMSO for polymerisation in the presence of the oligoethoxy dihalide or glycol thus gives rise to C*m*O1–C*m*O*n* copolymers in which the incorporation of the oligoethoxy segment is limited by its permeation into the phase-separated polymeric medium. Thus an equimolar mixture of **ii** and tetraethylene glycol in pure DMSO yields C16O5(75%) by method Y; lower proportions of oligoethoxy segment maybe obtained by

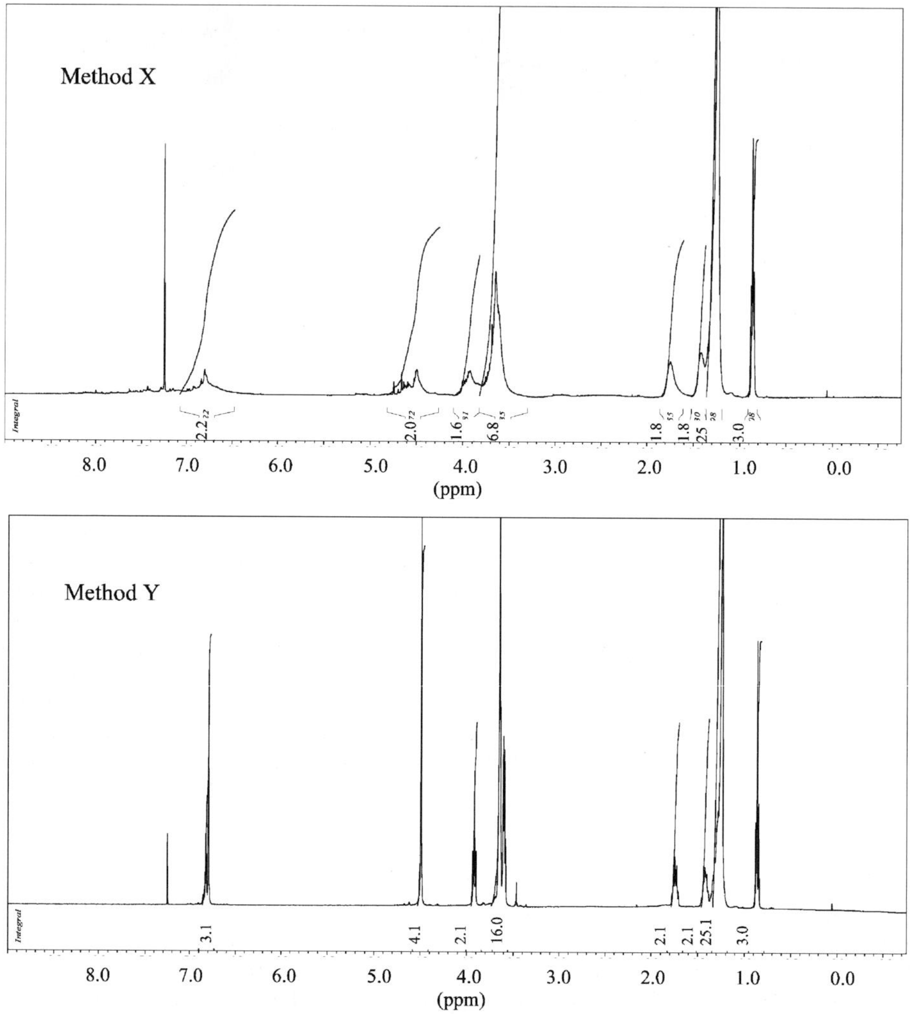

Fig. 2 ^{1}H NMR spectra for C16O5X(40%) synthesised by method X ($M_w = 257\,700$) and C16O5(100%) synthesised by method Y ($M_w = 38\,000$).

replacement of a proportion of the glycol by **i**; total replacement gives C16O1.The proportion of oligoethoxy segment is given by ^{1}H NMR from the integrated peak at $\delta = 3.65$ ppm (standard $SiMe_4$) as shown in Fig. 2 for C16O5X(40%) and C16O5(100%).

However, in method Y the benzylic halogens are significantly more reactive than the halogens of the tetraethyleneglycol dichloride. Thus, the pure C*m*O*n*(100%) homopolymer may be obtained by method Y using DMSO diluted (50/50) with a solvent for the polymeric products, such as tetrahydofuran.

Polymers obtained by both methods X and Y may be prepared to high molar mass. Alongside the Williamson condensation further chain extension takes place by dehydration of the benzylic alcohols to give C*m*O1 links. In method Y this process would follow the competing hydrolysis of benzylic halogens, **ii** + OH^- → **i** + Br^-. Thus molar mass distributions are generally broad. Samples with $\langle M_n \rangle = 5 \times 10^3$–$10^4$; $\langle M_w \rangle = 10^4$–$10^5$ and $\langle M_z \rangle = 10^5$–$10^6$ Da are readily prepared by both methods X and Y.

The products of method Y were found to be 'NMR pure' having ^{1}H NMR spectra consistent with the pure skeletal polyether **I** with side chain and aromatic nuclei assigned according to their expected proportions and the ethoxy hydrogens denoting the copolymer composition. (Fig. 2).

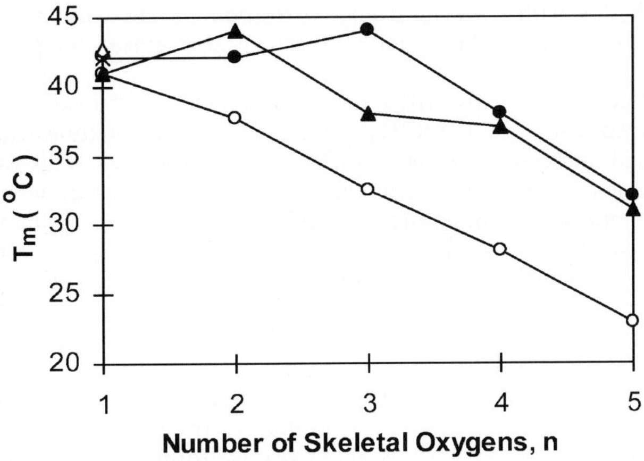

Fig. 3 Melting temperatures of C16O*n* systems: (○) pure polymer; (●) $LiBF_4$ complex of polymer; (▲) $LiClO_4$ complex of polymer; (×) C16O1X; (△) C16O1X : $LiClO_4$ (2 : 1).

However, whilst the products of method X had the expected side chain peaks in their ^1H NMR spectra, the bands in the region $\delta = 4.45$ corresponding to the benzyl α-CH_2 were broad and complex suggesting less than the required 4 hydrogens per repeating unit. The aromatic hydrogens were similarly complex and suggested less than 3 hydrogens at 6.85 ppm but having bands between 7.5 and *ca.* 8 ppm. FTIR spectra of the method X products generally feature a significant carbonyl band at 1720 cm^{-1}. The spectroscopic evidence is thus apparently consistent with the oxidation of a proportion (≤ *ca.* 50%) of the benzylic α-hydrogens to carbonyl ester. Weak OH bands at ~3500 cm^{-1} are observed in FTIR of the products from both methods and are assigned to the expected terminal units.

Homopolymers C16O*n* and complexes with Li salts

Thermal properties. Fig. 3 shows melting temperatures for polymers having full stoichiometric oligoethoxy content C16O*n*(100%) $n = 1$–5 and their equimolar complexes C16O*n*(100%) : Li salt (1 : 1) with $LiBF_4$ and $LiClO_4$. The melting temperatures of the pure polymers decrease almost linearly from *ca.* 41 °C to *ca.* 23 °C with the length of the oligoethoxy segments. Table 1 shows that the heats of fusion, $\Delta H_f \cong 30$ kJ mol^{-1} are constant within experimental error suggesting that the

Table 1 Thermal data for C16O*n* polymers and type A complexes with $LiClO_4$ and $LiBF_4$

Polymers or complexes	T_m/°C	ΔH/kJ mol^{-1}
C16O1	41	29
C16O1 : $LiBF_4$(1 : 1)	42	31.3
C16O1 : $LiClO_4$(2 : 1)	41	25.6
C16O2	38	29
C16O2 : $LiBF_4$(1 : 1)	42	32.3
C16O2 : $LiClO_4$(1 : 1)	44	22.4
C16O3	33	30.8
C16O3 : $LiBF_4$(1 : 1)	44	32.7
C16O3 : $LiClO_4$(1 : 1)	38	18.8
C16O4	28	30.2
C16O4 : $LiBF_4$(1 : 1)	38	25.8
C16O4 : $LiClO_4$(1 : 1)	37	28.75
C16O5(100%)	23	30
C16O5(100%) : $LiBF_4$(1 : 1)	32	23.9
C16O5(100%) : $LiClO_4$(1 : 1)	31	22.5

decrease in $T_m = \Delta H_f/\Delta S_f$ with n arising from a corresponding increase in the entropies of fusion with increase in the conformational probabilities of the flexible ethoxy links between hexadecyloxy-phenylene units.

The formation of complexes with $LiBF_4$ and $LiClO_4$ increases the side chain melting temperatures in all cases except for C16O1 : $LiClO_4$ (2 : 1) for which the melting temperature remains at *ca.* 41 °C. Only a small increase for C16O1 : $LiBF_4$ (1 : 1) is apparent. This complex and the $n = 2$ system probably dissolve less than an equimolar proportion of salt, as discussed further below, but the formation of complexes in these and the systems $n = 3$–5 with equimolar repeat unit: salt composition is confirmed by the observation of liquid crystal phases above side chain melting at 50 °C. In contrast with the pure, uncomplexed polymers which melt to form isotropic phases, all of the complexes C16On : Li salt form liquid crystal phases above their side chain melting temperatures at 50 °C as shown in Figs. 4c–h. (We have previously described the liquid crystal phases in C16O5 : Li salt complexes[9–12] and their mixtures with long chain alkane and alkyl compounds which insert into the side chain lattice layer). Cooling below side chain melting retains the morphology of the liquid crystal phase in all cases but the crystalline birefringence is generally more intense and coloured (Fig. 4a and b). C16O4 : $LiClO_4$ (1 : 1) readily formed a row-nucleated texture when sheared (Fig. 4b). Small increases in ΔH_f are apparent from Table 1 for $LiBF_4$ complexes with C16O1, C16O2 and C16O3 but the remainder have significantly lower ΔH_f than for the pure uncomplexed polymers, particularly in the cases of the $LiClO_4$ complexes. Thus the increases in melting temperatures for $n = 2$–5 may probably be ascribed to the reduction in ΔS_f as a consequence of the organisation in the liquid crystal phases following side chain melting. The lc phase derives from the mesogenic character imparted to the polyether backbone following its association with the Li salt. For $n = 3$–5 the increase in T_m on complex formation is substantial being 9–11 °C, whereas for $n = 1$ and $n = 2$ the increases are significantly less being 1 °C and 4–6 °C, respectively. The increases thus reflect the cumulative interactions of Li^+ with skeletal oxygens according to the length of the oligoethoxy segment. The cation may engage within a 'wrap-around' (helical) loop with 3 or 4 oxygens of the longer segments but for $n = 1$ and 2, unable to 'wrap-around', the interaction must be significantly weaker and in C16O1 has negligible influence on the side chain melting temperature. A significant endotherm at 115 °C in C16O1 : $LiBF_4$ (1 : 1) suggests 'excess' uncomplexed salt in this mixture.

Also shown in Fig. 3 are melting temperatures for the polymer and complexes of C16O1X synthesised by method X and giving a polyether-ester product, some of the benzylic α-hydrogens seemingly replaced by carbonyl oxygens. The corresponding melting temperatures (42.2 °C and 42.7 °C) are slightly higher than those of the C16O1 system but the increase on complex formation is similarly small. The formation of a complex in C16O1X systems is also confirmed by the striking textures at 20 °C and 50 °C (Fig. 4a and d), the uncomplexed C16O1X having an ill-defined granular texture at 20 °C and is isotropic at 50 °C. The formation of the textures of the complex required that the solution deposited film be first heated to *ca.* 120 °C (presumably to distribute the salt) before cooling.

X-ray diffraction in C16O*n* polymers and complexes with Li salts. X-ray traces of these materials generally show two major peaks below side chain melting corresponding to the long spacing, d, at low angles and the wide angle reflection in the region 4.2 Å corresponding to the hexagonal lattice of the side chains. The former may show higher order reflections (see Tables S1 and S2†) and usually remains sharp and intense in the liquid crystal phase following side chain melting; the latter becomes broader following side chain melting. The long-spacings, d, at 26 °C and 50 °C for the layers composed of the –O–$C_{16}H_{33}$ side chains, phenylene rings and skeletal ether oxygens in uncomplexed polymers and including the salt in the complexed forms are shown in Fig. 5. The spacings for the uncomplexed polymers $n = 1$ to 5 increase progressively from 32.0 Å to 33.7 Å, 37.6 Å, 38.4 Å, 37.6 Å, respectively, consistent with a general scheme of interdigitated bilayers shown for C16O5 in the molecular dynamics models.† The series in d reflects the spaces required to accommodate the different lengths of oligoethoxy loops as adjacent side chains coalesce although an apparent maximum for C16O4 at 26 °C (and its complex at 50 °C) may perhaps suggest more flexible tortuosity of the longer segment in C16O5 able to fold into a smaller space. For the uncomplexed polymers at 50 °C the d spacings are essentially constant with n at 32–34 Å. This presumably

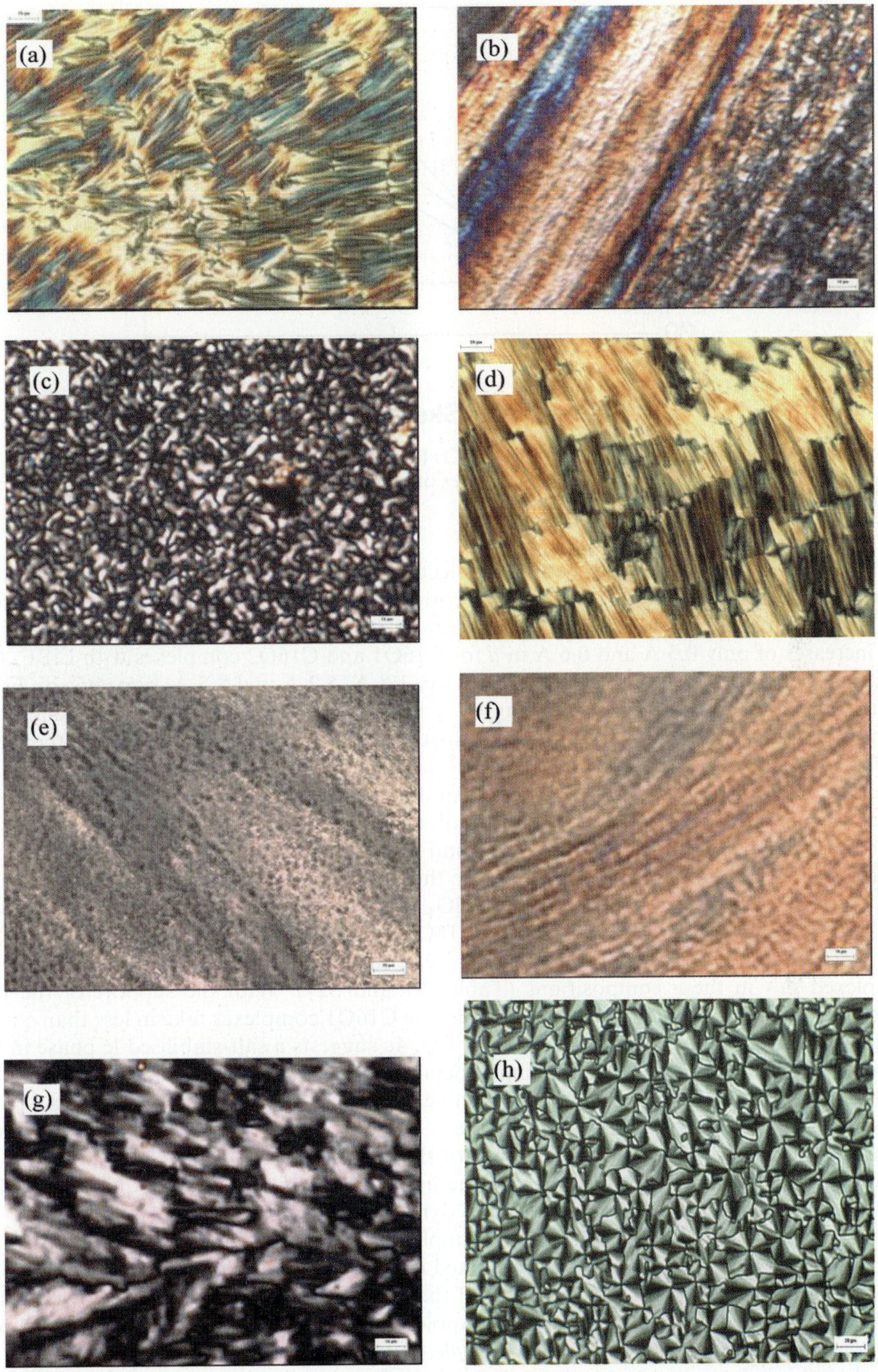

Fig. 4 Optical micrographs between crossed polars. (a) C16O1X : $LiClO_4$ (2 : 1) 25 °C; (b) C16O4(100%) : $LiBF_4$ (1 : 1) 20 °C (sheared film); (c) C16O1 : $LiClO_4$ (2 : 1) 50 °C; (d) C16O1X : $LiClO_4$ (2 : 1) 50 °C; (e) C16O2 : $LiClO_4$ (1 : 1) 50 °C; (f) C16O3 : $LiClO_4$ (1 : 1) 50 °C; (g) C16O4 : $LiBF_4$ (1 : 1) 50 °C; (h) C16O5(100%) : $LiClO_4$ (1 : 1) 50 °C (scale bars are 20 μm).

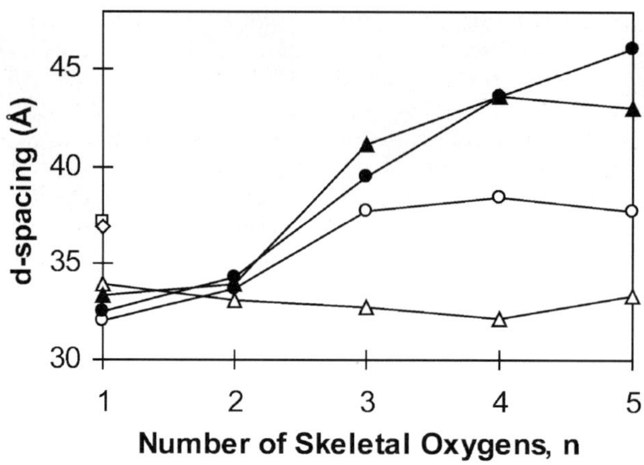

Fig. 5 Small angle X-ray data for C16O*n* systems. (○) Pure polymer at 26 °C, (△) pure polymer at 50 °C; (●) $LiBF_4$ complex of polymer at 26 °C; (▲) $LiBF_4$ complex of polymer at 50 °C; (◊) C16O1X; (□) C16O1X : $LiClO_4$ (2 : 1) at 26 °C.

indicates lateral thermal expansion between molecules. The wide angles show (Tables S1 and S2 †) that the sharp reflections in the region 4.2 Å corresponding to the hexagonal lattices of the side chains become broader above the melting temperatures with increased spacing to *ca.* 4.5.

The increases of only 0.5 Å and 0.6 Å in *d* for C16O1 and C16O2 complexes with $LiBF_4$ (Fig. 5) are minimal but for $n = 3$, 4 and 5 the increases are 1.8 Å, 5.2 Å and 8.5 Å, respectively. For $n = 4$ and 5 the increases in *d* spacings (reflecting the corresponding increases in the side chain melting temperatures) are consistent with the extra space which would be required for the 'inflated' oligoethoxy loops as they wrap around the Li^+ cations and accommodate rows of BF_4^- anions between the (helical) polyethoxy –Li^+ adducts of adjacent layers. For C16O1, C16O2 and C16O3 however, the increases in *d* do not reflect the anion diameter, even if only one mole of salt associates with *two* repeat units as surmised for C16O1 and C16O2. For C16O1 : $LiBF_4$ (1 : 1) the observed *d* spacing is 32.5 Å compared with 32.0 Å for the uncomplexed polymer but similar results for C16O1 : $LiBF_4$ (2 : 1) (32.7 Å) and C16O1 : $LiClO_4$ (2 : 1) (33.9 Å) are observed. The persistence of some sharp wide angle reflections at 50 °C in C16O1 : $LiBF_4$ (1 : 1) and C16O2 : $LiBF_4$ (1 : 1) in the region 3.9, 4.8 and 6.1 Å) are coincident with those of the pure salt and apparently indicates excess, uncomplexed salt in these compositions (Tables S1 and S2†). With the support of the thermal evidence mentioned above it may be inferred that the C16O1 complexes take in less than equimolar proportions of salt. Nonetheless, the texture of Fig. 4c suggests a salt-stabilised lc phase in C16O1 systems so that the small increase in long spacings for $n = 1$, 2 and 3 indicates that the salt is accommodated in spaces within the amphiphilic layered systems where it exerts a stabilising influence but makes little contribution to the *d* spacing. At 50 °C the complexes retain their crystalline *d*-spacings confirming the existence of the lc states for $n = 3$–5.

A more complete understanding of the structure of the complexes will require more detailed structural investigations. However, molecular dynamics modelling (shown in the ESI†) reveals striking differences between C16O5 : Li salt complex and the corresponding C16O1 : $LiBF_4$ complex. In the former, the salt is present as separated ions, the cation encapsulated by the oligoethoxy segment with the anion in attendance outside of the loop. In the C16O1 complex, however, the salt has coalesced into neutral (uncharged) quadrupole aggregates. Such aggregation, with minimum interaction with the polymer (essentially *decoupled*), would account for the small influence of the salt on the side chain melting temperature as well as the poor conductivities of C16O1 complexes described below. However, the aggregates are postulated to provide the 'relay' sites for Li^+ 'hopping' by a novel mechanism for high ambient conduction with low activation energies in copolymers having minor proportions of C16O5 as illustrated in Fig. 1b and discussed further at the end of this paper.

The C16O1X systems have larger *d* spacings (~36.9 Å and 37.2 Å for C16O1X and C16O1X : (LiClO4) (2 : 1), respectively) than the corresponding C16O1 systems even though the increases on complex formation are similarly almost negligible. The striking crystalline and liquid crystalline textures for the complexes in Fig. 4a and d contrast with the granular morphology of the uncomplexed C16O1X and confirm the formation of the lc state. However, the differences in the textures of the C16O1 and C16O1X systems alongside the significant differences in their *d* spacings are further evidence of structural differences in the products of synthetic methods X and Y, also apparent from ^{1}H NMR and FTIR. Whether the larger *d* spacings arise from less overlap in the interdigitation of side chains or from a larger interlayer gap in the skeletal heterochain interface, where the ions are assumed to be most stable, are issues which could have some significance for the mobilities of the ions in these systems.

Ionic conductivities of C16O*n* polymers and complexes with Li salts. $\log_{10} \sigma$ *versus* $1/T$ plots for the more conductive cooling cycles of C*m*O*n* : $LiClO_4$ (1 : 1) ('Type A') systems for $n = 2–5$ are shown in Fig. 6. The systems have similar conductivities at 110 °C but although C16O4 and C16O5 are similar at 20 °C the other systems increasingly diverge as the temperature decreases. The C16O4 system is the most and the C16O2 system is the least conductive. The impedance profiles (Fig. 6b) at temperatures well above the side chain melting temperatures (80–100 °C) show sharp 'ticks' typical of liquid-like mobility in amorphous or liquid crystal states. Below side chain melting at 20 °C the formation of new grain boundaries give rise to a semicircle at higher frequencies also seen, for example, in semicrystalline poly(ethylene oxide)–salt complexes.[13] The slight steepening of the $\log_{10} \sigma$ *versus* $1/T$ plots between 30 °C and 20 °C for $n = 3$, 4 and 5 and below 40 °C for $n = 1$ and 2 may be attributed to enlargement of grain boundaries following side chain crystallisation. The less complete arc at 50 °C is consistent with the persistence of the grain boundaries at temperatures immediately above side chain melting. It seems that in these two-component ('type A') materials, side chain melting to the liquid crystal state does not necessarily ensure a continuum of pathways for ion mobility. The corresponding data with $LiBF_4$ show a similar trend. The two component complexes of C16O1 with $LiClO_4$ and $LiBF_4$ appeared highly resistive giving impedance plots without clear minima but C16O1 : Li $(CF_3SO_2)_2N$ (1 : 1) gave the lowest plot shown in Fig. 6a. The anion of this salt ($TFSI^-$) is too large to achieve the minimum ion occupancy of the skeletal sites (half-filled[12]) to sustain a smectic texture in C16O5 systems and the material invariably appears optically transparent and has lower conductivities than the ClO_4^- and BF_4^- anions.[5,6] However, in C16O1 some solubility has apparently been achieved to give low but measurable conductivities.

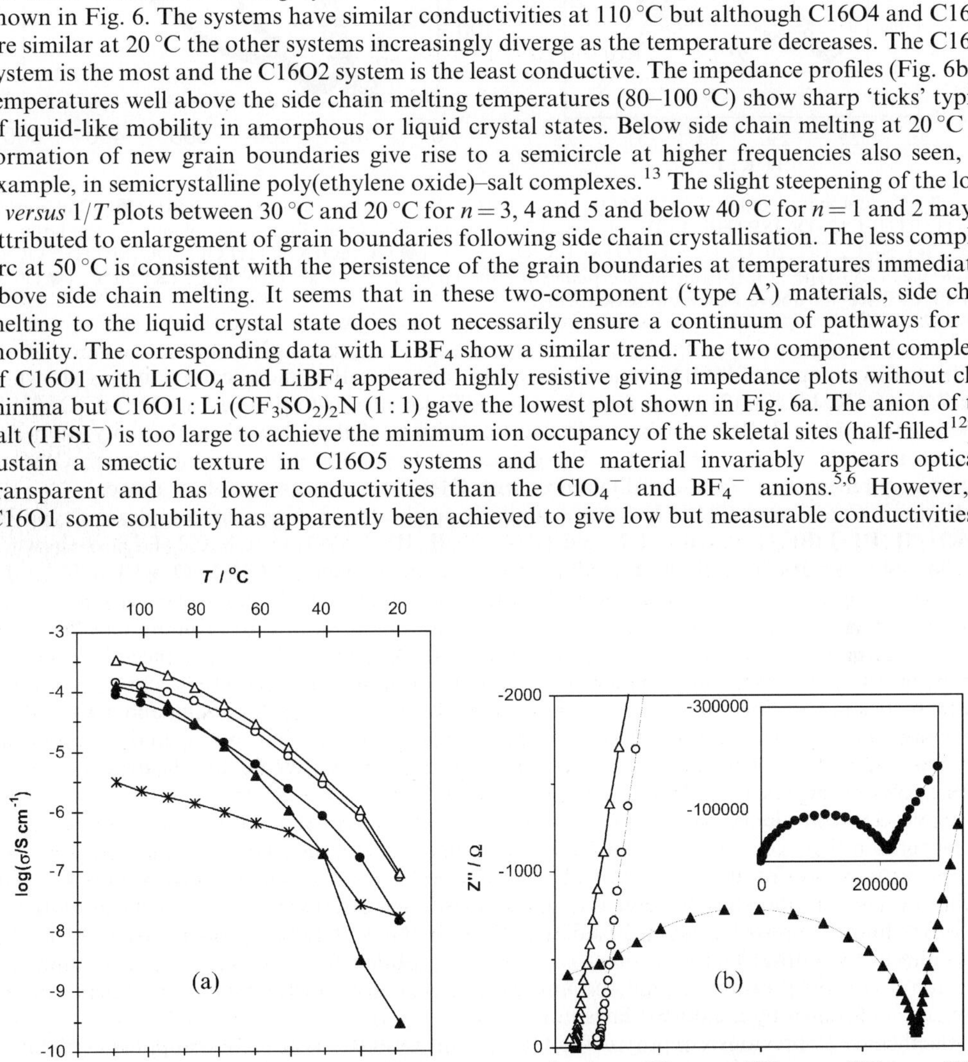

Fig. 6 Conductivity data for C16O*n* complexes. (a) (○) C16O5 : $LiClO_4$ (1 : 1); (△) C16O4 : $LiClO_4$ (1 : 1); (●) C16O3 : $LiClO_4$ (1 : 1); (▲) C16O2 : $LiClO_4$ (1 : 1); (*) C16O1 : LiTFSI (1 : 1). (b) Impedance data for C16O4 : $LiClO_4$ (1 : 1) (△) 110 °C; (○)80 °C; (▲)50 °C; (●)20 °C.

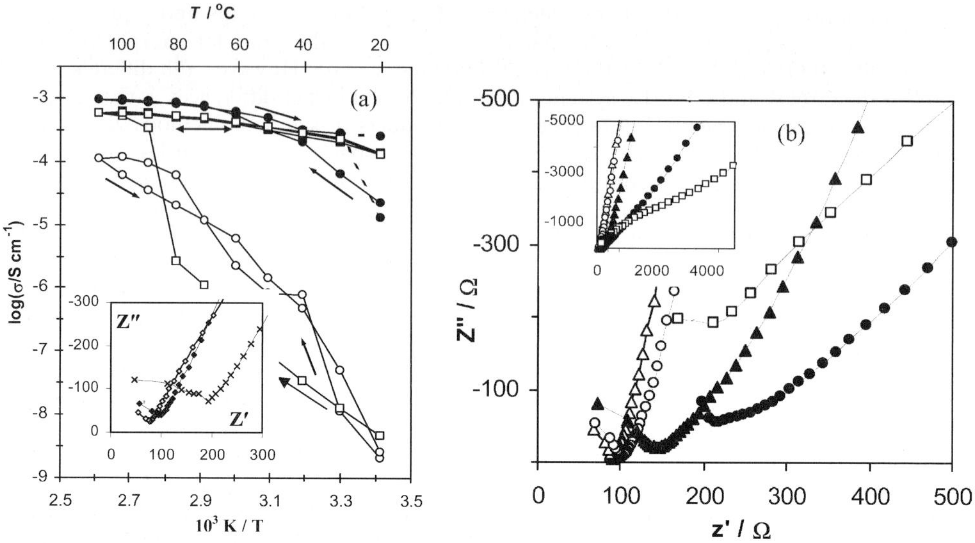

Fig. 7 Conductivity data for C16O1 and C16O1X systems. (a) (□) and (■ 2nd cycle) C16O1 : PTHFC12 : DC18PTHF : $LiBF_4$ (1 : 0.8 : 0.2 : 1.2) ('type D'). Inset: Impedance planes for the latter, (◊) 80 °C, (♦) 50 °C, (×) 20 °C. (○) C16O1X : $LiClO_4$ (2 : 1), (●) C16O1X : PTHFC12 : DC18PTHF : $LiClO_4$ (1 : 0.8 : 0.2 : 1.2) ('type D'). (b) Impedance planes for the latter, (△) 110 °C, (○) 80 °C, (▲) 50 °C, (●)30 °C, (□) 20 °C. Inset: the above with compressed scale.

The two-component C16O1X systems, however, appear rather more conductive as shown in Fig. 7. C16O1X : $LiClO_4$ (2 : 1) has $\log_{10} \sigma$ *versus* $1/T$ plot similar to that of C16O2 : $LiClO_4$ (1 : 1). (The corresponding $LiBF_4$ system is similar but with a slightly lower conductivity at higher temperatures (10^{-5} S cm^{-1} at 110 °C). The greater conductivities of the two-component C16O1X system may reflect their structural differences with C16O1 systems as described above.

However, C16O1 and C16O1X behave similarly in the *four*-component complexes ('type D') C16O1 : **II** : **III** : $LiBF_4$(1 : 0.8 : 0.2 : 1.2) and C16O1X : **II** : **III** : $LiClO_4$ (1 : 0.8 : 0.2 : 1.2), as shown in Fig. 7a and b. As described in the Introduction, **II** is the copolymer $-[(CH_2)_4-O-]_{\sim 23}-(CH_2)_{12}-O-$ having $\langle M_w \rangle = 3.9 \times 10^4$, and **III** is the ABA copolymer $H_{37}C_{18}-O-[(CH_2)_4-O-]_{\sim 50}-C_{18}H_{37}$ with $\langle M_w \rangle \cong 4700$. The latter is added as a minor component or 'surfactant' to promote the stability of the interface between the amphiphilic polymers **I** and **II** (see Fig. 1). The impedance planes for the C16O1-based system show single sharp 'tick' minima over the range 20–110 °C corresponding to conductivities in the range *ca.* 2×10^{-4} S cm^{-1}–10^{-3} S cm^{-1} for the first cooling and second heat–cool cycles. The latter is hardly discernible beneath the points for the first cooling in the log σ *vs.* $1/T$ plot showing that the 'transformation' (de-blending) that occurs after the first heating to T_{iso}, the liquid crystal–isotropic transformation temperature in the region 110 °C, is permanent and reversible. The impedance planes for the C16O1X-based system shown in Fig. 7b are less steep at temperatures in the range 30–50 °C becoming approximately $\pi/4$ degrees at 30 °C. This may suggest a tendency for diffusion control of the mobility as the temperature is lowered. This phenomenon, also observed in other type D systems,[14] was ascribed to the increasing limitation on ion mobilities imposed by temperature-dependent mobility in the polymer **II** phase. At 20 °C a minimum in the C16O1X-based system is less certain (at a high frequency of 500 kHz) and the inflexion at *ca.* 3000 Ω may signal the appearance of a second mininum and a more resistive interface, as denoted by the dotted lines in Fig. 7a.

Nonetheless, as previously reported for C16O5 and C18O5-based *three*-component mixtures,[6] these results for type D systems based on C16O1 and C16O1X suggest that highly conductive pathways may be created in either system at ambient temperatures where the two are in good agreement. In these cases the minima are observed at high frequencies (200–300 kHz) suggesting that DC measurements would encounter higher resistance barriers, of particular significance for the operation of batteries. However, DC polarisations of 3-component (type C) systems in cells with lithium electrodes show[2–4,6] stepwise increases in conductivity with time, eventually levelling at $\sim 10^{-3}$ S cm^{-1}. A current-induced 'self-tracking' process was proposed to account for this observation in which such higher impedance interfaces were progressively breached.

Copolymers of C16O5–C16O1 and their complexes

Thermal properties. When polymers C16O*n* (n = 2–5) are synthesised using KOH in pure DMSO solvent a dehydration reaction, which takes place within a separated phase of aggregated polymer and monomer, gives rise to C16O*n*–C16O1 copolymers. These are abbreviated C16O5(x%) where x is the percentage of $-(CH_2-CH_2-O-)_{(n-1)}-$ segments in the copolymer determined by ^{1}H NMR. Fig. 8 shows that side chain melting temperatures of the pure uncomplexed polymers decrease steadily from C16O1 at 41 °C as C16O1 links are replaced by the flexible tetraethoxy segments to 36.0 °C, 28.3 °C and 23.0 °C for C16O5(40%), C16O5(75%), and C16O5(100%), respectively. As noted in Fig. 3, the $LiClO_4$ and $LiBF_4$ complexes with C16O5(100%) melt 8–9 °C higher than the pure polymer and there is little change in the melting temperatures for C16O1 and C16O1X following complex formation. However, the intermediate copolymer systems show substantial increases following complex formation, particularly in the C16O5(75%) system for which increases of 13–15 °C are apparent. This presumably illustrates the influence of two mechanisms by which the skeletal polyether backbone is stiffened. The first is the stiffening influence of C16O1 links relative to the aliphatic ethers. The second, being the interaction of the Li salt with $-O-(CH_2CH_2O)_4-$ should have maximum effect in C16O5(100%) but apparently becomes almost negligible in pure C16O1 systems. The influence of both of these factors operating together is clearly apparent in the copolymer systems C16O5(75%) and C16O5(40%). Polymers and copolymers synthesised by method X generally decrease in a similar way but there is scatter in the data apparently depending on variations in the procedure and presumably reflecting some greater skeletal disorder. Both skeletal stiffening mechanisms may tend to promote the stabilities of the liquid crystal states this being reflected in the heats of fusion (Table 2). The latter might be expected to be lower than that for the pure polymers reflecting greater order in the lc state compared with the isotropic. This is generally true but exceptions are apparent in the C16O1 systems and in the half-stoichiometric C16O5(75%) : $LiBF_4$ (2 : 1). (Dias *et al.*[12] found that half-stoichiometric complexes with C*m*O5 systems contained the minimum salt for which lc phases were formed.) However, as noted above, smaller ΔH_f values are found following complex formation with $LiClO_4$. An indication of more stable lc states may derive from stronger $Li^+\cdots O_4Cl^-$ interactions than the corresponding ones with BF_4^-. This was also invoked[6] to account for the more laterally extensive lamellae observed with $LiClO_4$.

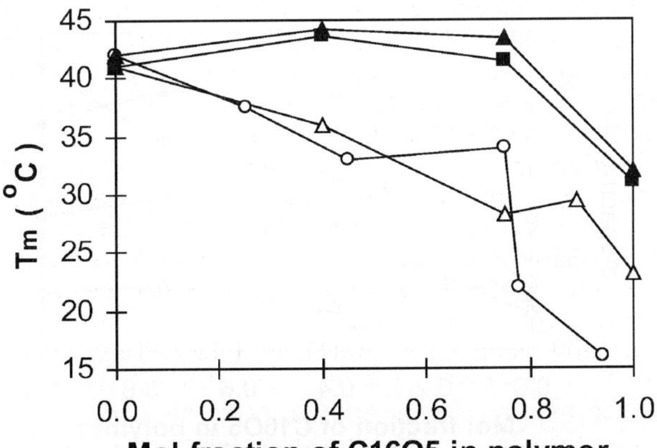

Fig. 8 Melting temperatures for C16O1–C16O5 copolymers and complexes. (○) Method X polymer; (△) method Y polymer; (■) $LiClO_4$ complex of polymer; (▲) $LiBF_4$ complex of polymer.

Table 2 Thermal data for polymers and copolymers of C16O5–C16O1

Polymer and complex	T_m/°C	ΔH/kJ mol^{-1}
C16O5(100%)	23	30
C16O5(100%) : $LiBF_4$(1 : 1)	32	23.9
C16O5(100%) : $LiClO_4$(1 : 1)	31	22.5
C16O5(75%)	28	30
C16O5(75%) : $LiBF_4$(1 : 1)	44	27.4
C16O5(75%) : $LiBF_4$(1 : 0.5)	41	35.8
C16O5(75%) : $LiClO_4$(1 : 1)	42	26.4
C16O5(40%)	36	33
C16O5(40%) : $LiBF_4$(1 : 1)	44	26.7
C16O5(40%) : $LiClO_4$(1 : 1)	44	24
C16O1	41	29
C16O1 : $LiBF_4$(1 : 1)	42	31.3
C16O1 : $LiClO_4$(2 : 1)	41	25.6
C1601X	42	21
C16O1X : $LiClO_4$(2 : 1)	43	11.3

X-ray diffraction of copolymers of C16O5–C16O1 and their complexes. Wide-angle X-ray data for the C16O5–C16O1 copolymers at 25 °C show sharp peaks at ~4.2 Å corresponding to the hexagonal lattices which become broader after melting, indistinguishable from the behaviour of the corresponding peaks in the homopolymers described above. This indicates that the $-C_{16}H_{33}$ side chains from the mixtures of the two monomers readily coalesce to form hexagonal lattices though at present we have no analysis of the sequences of monomers—random, blocked and/or mixtures of homopolymers. (Experiments with mixtures of pure homopolymer components C16O5(100%) and C16O1 suggest that mutual blending readily takes place.)

The long spacings, *d*, from small angle data are shown in Fig. 9. The *d* spacings for the C16O1–C16O5 copolymers follow similar trends to those of Fig. 5 for the C16O*n* homopolymers. For the uncomplexed salt-free materials, *d* rises to a shallow maximum at C16O5(75%) and in the melt at 50 °C the systems apparently undergo lateral thermal expansion and *d* remains approximately constant throughout the composition range. On complex formation, the *d* spacings increase as tetraethoxy segments are incorporated at 26 °C and in the lc state at 50 °C. Thus, the addition of skeletal oxygen atoms has a broadly similar influence in each class of materials.

Ionic conductivities of C16O1–C16O5 copolymers and complexes with Li salts. Investigations of type D systems based upon copolymers **I** within this series lead to the conclusion that optimum

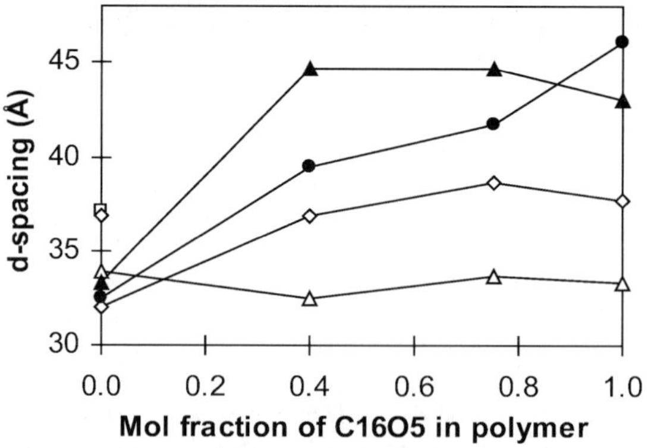

Fig. 9 X-ray data for copolymers C16O1–C16O5. (○) Pure polymer at 26 °C; (△) pure polymer at 50 °C; (●) $LiBF_4$ complex of polymer at 26 °C; (▲) $LiBF_4$ complex of polymer at 50 °C; (◊)C16O1X; (□) C16O1X : $LiClO_4$ (2 : 1) at 26 °C.

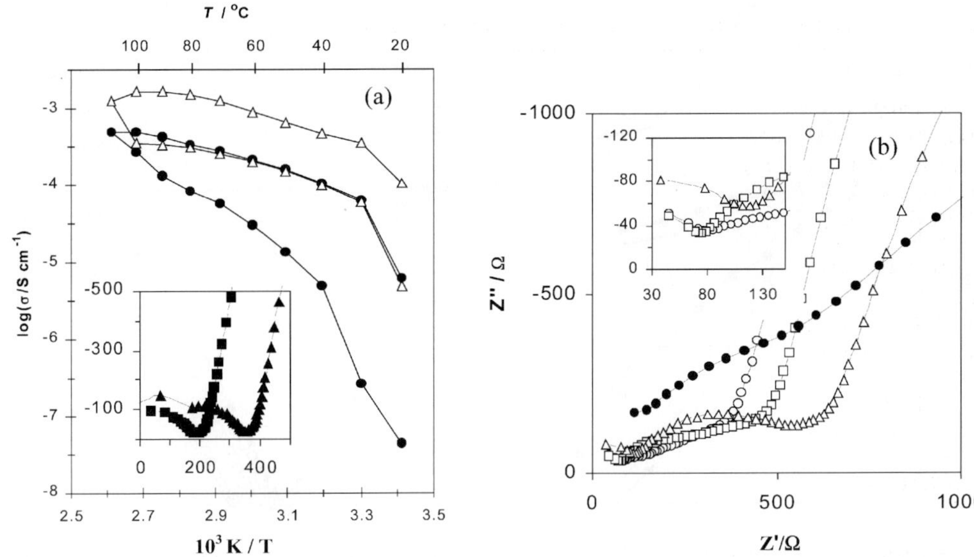

Fig. 10 Conductivity data for a type D system C16O5(40%) : **II** : **III** : $LiBF_4$(1 : 0.8 : 0.2 : 1.2). (a) (●) First thermal cycle; (△) second cycle. Inset: impedance planes for first cooling:(▲)50 °C; (■) 80 °C;. b) Impedance planes for second cooling: (○) 110 °C; (□) 80 °C; (△) 50 °C; (●) 20 °C.

conductivities tend to be observed with systems richer in C16O1 (*n.b.* Fig. 7). Several factors may conspire to account for this tendency. Firstly, the ion-containing interface is presumably *two*-dimensional, *i.e.* Li^+ ions are not confined to one-dimensional helical channels as might be the case in C16O5. This should render less critical the orientation of individual crystallites or organised domains in the creation of continuous pathways through the material. Furthermore, a C16O1-rich interface should be the more ionophobic engendering closer interactions with the alkylene and alkyl segments in polymers **II** and **III**. This should promote ion transfer across the interfaces. In a recent paper[14] we reported results for a type D system incorporating C16O5(75%). This system gave conductivities in the region 10^{-3} S cm^{-1} on cooling over the range 110 °C to 30 °C. However, at 20 °C the conductivity declined sharply by an order of magnitude. This reversible effect coincides with the crystallisation of sidechains and, presumably, with shrinkage of phase **I**. In the same paper, evidence was presented for different degrees of 'conversion' from the blended precursor to the more conductive morphology which takes place following cooling from the isotropic transition temperature in the region 110 °C. Data for a type D system incorporating C16O5(40%), which also demonstrates progressive conversion, is shown in Fig. 10. During the first cooling cycle the impedance planes are sharp 'ticks' giving conductivities $\geq$ *ca.* 10^{-4} S cm^{-1} at temperatures above 30 °C. On cooling to 20 °C the conductivity falls sharply. However, the second heating cycle follows the first cooling very closely until at 100 °C (in the region T_{iso}) there is an increase in conductivity. On cooling, the impedance data (Fig. 10b) indicates that *two* conductive phases are now present. The single ticks of the first cooling cycle have now resolved into two sets of minima, one indicating low impedances of bulk grains and the other perhaps arising from the new grain boundaries.

However, the type D system which undergoes conversion most completely and consistently that we have so far investigated is a mixed side chain copolymer, C12C18O5X(18%), also with low tetraethoxy content. This copolymer, which incorporates molecules of high molar mass ($\langle M_z \rangle \sim 10^6$ Da), has side chains $-C_{12}H_{25}$ and $-C_{18}H_{37}$ in equal proportion, presumably in random sequence. Within the type D mixture the polymer has an endotherm at 25 °C but with a 'shoulder' and 'tail' down to ~0 °C. The shoulder corresponds to crystallisation within polymer **II**. Fig. 11 shows conductivity data for one of a number of different preparations and conductivity runs for this material but which give almost identical results. The impedance planes in Fig. 11b are sharp ticks down to 20 °C giving conductivities ~10^{-3} S cm^{-1} over the temperature range. Significantly, the single tick minima occur at ~60 KHz which are almost an order lower than in Fig. 6. (In a separate

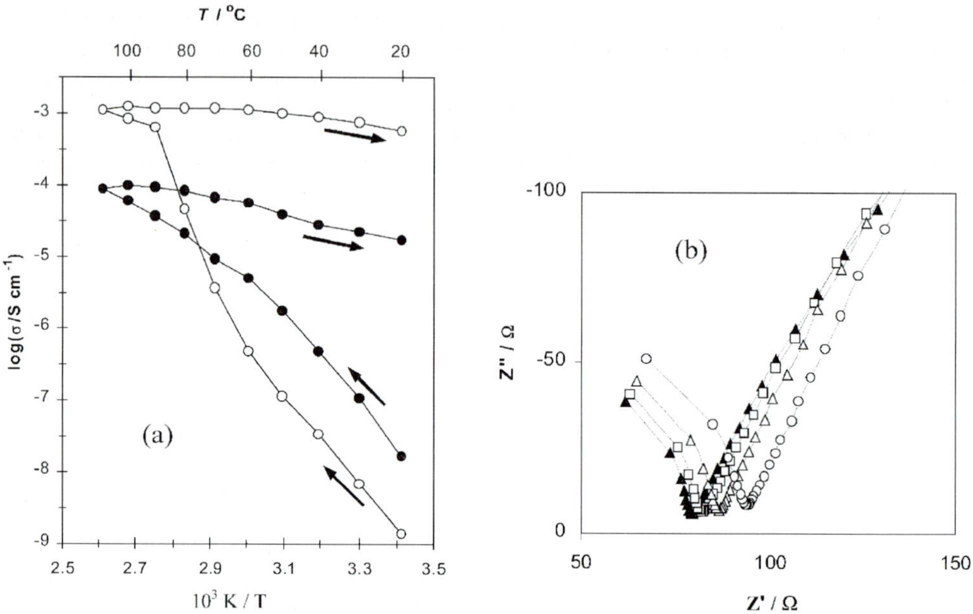

Fig. 11 Conductivity data for a Type D system. (a) (○) C12C18O5X(18%) : **II** : **III** : $LiBF_4$ (1 : 0.8 : 0.2 : 1.2); (●) PTHFC12 : $LiBF_4$ (1 : 1.2). (b) Impedance planes for the type D system cooling. (▲) 50 °C; (□) 40 °C; (△) 30 °C; (○) 20 °C.

experiment with this system, reported elsewhere,[14] the measurements were extended down to −10 °C and giving 5×10^{-5} S cm^{-1}. This result is in good agreement with that reported[2,4] earlier for a type C incorporating a side chain copolymer (C12C16O5X) which gave 8×10^{-5} S cm^{-1} at −5 °C.)

Also shown in Fig. 11a is a conductivity plot for polymer **II**. As in the type D systems, the conductivity undergoes an irreversible change to a more conductive state. This may also involve a morphological change, involving the coalescence of the decamethylene segments in polymer **II**. However, similar irreversibility is also observed in amorphous polytrimethylene oxide copolymers and blends which we have also investigated, but the effect is very much smaller in amorphous poly(ethylene oxide) systems. In the poorer solvent polyethers, the initial heating process may also serve to distribute salt throughout the system from isolated solution-deposited aggregates.

The morphologies of the type D systems based upon C16O5(40%) and C12C18O5X(18%) are compared in Fig. 12. Whereas the corresponding system based upon C16O5(75%) (reported

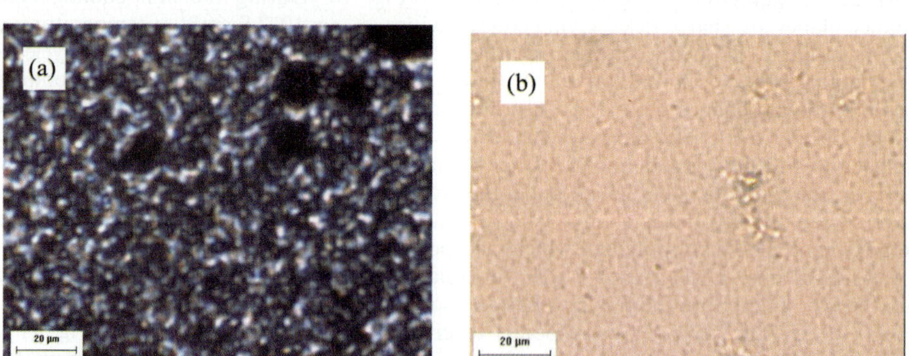

Fig. 12 Optical micrographs of type D systems. (a) C16O5(40%) : **II** : **III** : $LiBF_4$(1 : 0.8 : 0.2 : 1.2) between crossed polars; (b) C12C18O5X(18%) : **II** : **III** : $LiBF_4$ (1 : 0.8 : 0.2 : 1.2) (unpolarised light image). Scale bars are 20 µm.

elsewhere[14]) has a strongly birefringent morphology after cooling to ambient from 110 °C, the C16O5(40%) based system in Fig. 12a has a weaker, granular texture but crystalline domains of several micrometres in dimension are apparent in the polarising microscope. In the C12C18O5X(18%)- based system, however, the field is dark between crossed polars at 25 °C. At 10 °C a fine birefringent texture may be discerned but the texture may also be observed in the non-polarised image of Fig. 12b indicating domains on the micron scale. The optimum performance of this material may perhaps be attributed to a reduced tendency for phase separation owing to both mixed side chains and disorder within the ionophilic interface whilst the discrete ionophilic pathways and a low degree of coupling between ionic and polymer mobilities are retained. This copolymer has $\langle M_z \rangle \sim 10^6$ Da and makes tough, self-supporting films. Thus, high ambient conductivities may be observed in self-organised mixed complexes using polymers **I** prepared by either of methods X or Y and considering these findings alongside recently published work[5,6,14] it is apparent that the incorporation of the 'interfacial stabiliser', polymer **III** is beneficial in promoting consistency in conductivity behaviour.

The mechanism of ionic conduction. The conductivity plots in Figs. 7, 10 and 11 together with similar results published in refs. 6 and 14 testify to a remarkably weak temperature dependence and small activation energy for conduction in the 'converted material' following heating to 110 °C and cooling through the lc phase transition at the side chain melting temperatures over the range 25–45 °C. This virtual independence of temperature down to ambient temperatures and, at a lower conductivity (8×10^{-5} S cm^{-1}), over subambient temperatures[2–4] constrasts profoundly with the temperature dependence in conventional amorphous polymer electrolytes in which 'free-volume' (the VTF relationship) is considered to regulate the motions of the ions *coupled* to the polymer chains.[15] Consistent observations of the the high-conductivity–low temperature dependent regime in systems with copolymers **I** which are rich in the ionophobic component C16O1 and having only minor proportions (15–25%) of the more ionophilic C16O5 segments suggest a mechanism of conduction in which ion–polymer associations play a different role than in the conventional PEO-based amorphous systems. The molecular dynamics models for these components, shown in the ESI† indicate that, whilst the salt interacts with C16O5 to form separated ion pairs, in C16O1 ions coalesce to form non-associated or *decoupled* neutral aggregates, likely to be ion pairs and quadrupoles. This suggests a novel conduction mechanism in the copolymers of Li^+ hopping between neutral aggregates. The decoupled neutral aggregates and the temporary charged species (*e.g.* $Li_3A_2^+$) occupy channels between the lamellae bounded by the alkyl side chains with a separation largely determined by the tetraethoxy loops encapsulating a cation and its attendant anion. The latter then serve to 'dope' the neutral species, the cation migrating in the field *e.g.* $Li_3A_2^+\ Li_2A_2\ Li_2A_2 \rightarrow Li_2A_2\ Li_3A_2^+\ Li_2A_2$ This process is illustrated in the sketch of Fig. 1b. If the chain-decoupled cation exchange takes place with minimum activation energy and anion mobility is hindered by its size and space considerations within the channels such a mechanism could account for the observed temperature dependence and the high Li^+ transport suggested by DC polarisation experiments between Li metal electrodes.[2–4,6] In related polyesters with conventional mesogen side chains in glassy phases, Imrie and coworkers[16] have reported temperature–dependent ionic conductivities but which are independent of the glass transition and apparently decoupled from the motions of the chains.

Further investigations are in progress on the structure, the mechanism of ion mobility and the application of these systems to galvanic cells.

Experimental

General procedures

^{1}H NMR was performed using a Bruker AC250 instrument in the Dept. of Chemistry, University of Sheffield. Gel permeation chromatography was carried out using a Hewlett-Packard 1090 liquid chromatograph in the Dept. of Chemistry, University of Sheffield. Thermal analysis was performed on Du Pont Instruments 910 DSC. Infra-red spectroscopy was carried out using a Perkin-Elmer FTIR spectrum 2000 scanning between 4000 and 400 cm^{-1}. Optical microscopy was performed using an Olympus BX50 microscope fitted with a 'Coolsnap' digital camera. Elemental analysis was performed in the Dept. of Chemistry, University of Sheffield.

Simultaneous SAXS and WAXS measurements over ambient to 50 °C were carried out using graphite monochromatised Cu Kα radiation and a MAR Research Image plate with 5 or 10 min exposure times.

Conductivity measurements were carried out using a Solartron 1287A Electrochemical interface coupled to a 1250 Frequency Response Analyser. The conductivity cell was fabricated using indium–tin oxide (ITO) coated glass electrodes with the polymer sample sandwiched in a window within a cellulose acetate or polyethylene spacer. The cell was mounted on a heated block within an evacuated chamber. Impedance measurements were made over the range 1 Hz–1 MHz with an amplitude of 0.1 V. The resistance of the ITO electrodes was 40 Ω which was subtracted from the measured impedance minima.

Molecular dynamics modelling was performed with Cerius^2 molecular dynamics software. Modelling was done with 'annealing cycles' between room temperature and 600 K.

Two-component ('type A') **I**: salt complexes were prepared by dissolving the weighed amounts of salt and **I** in a mixed solvent (50/50) dichloromethane–acetone. The solvent was then removed under vacuum and the material annealed and dried at 50 °C under vacuum before storing in a dessicator. Four-component ('type D') **I** : **II** : **III** : Li salt systems were prepared by dissolving the weighed amounts of the components in (50/50) dichloromethane–acetone at a concentration of 5% (wt/vol). The clear solution was then frozen in liquid nitrogen, the coolant removed and the solvent removed under vacuum ('freeze drying'). The fine powdered mixture was then fused under vacuum to a melt at 50 °C. The melt was then vigorously stirred mechanically before finally drying under vacuum at 50 °C and storing in a dessicator.

Synthetic procedures are described in the ESI.†

Acknowledgements

We are indebted to the Engineering and Physical Sciences Research Council for support for JL, YZ, J-P.L in this work and to the European Space Agency for further support for JL.

References

1 J.-M. Tarascon and M. Armand, *Nature*, 2001, **414**, 359–367.
2 Y. Zheng, F. Chia, G. Ungar and P. V. Wright, *Chem. Commun.*, 2000, 1459–1460.
3 Y. Zheng, F. Chia, G. Ungar and P. V. Wright, *J. Power Sources*, 2001, **97–98**, 641–643.
4 Y. Zheng, F. Chia, G. Ungar and P. V. Wright, *Electrochim. Acta*, 2001, **46**, 1397–1405.
5 F. Chia, Y. Zheng, J. Liu, G. Ungar and P. V. Wright, *Solid State Ionics*, 2002, **147**, 275–280.
6 F. Chia, Y. Zheng, J. Liu, N. Reeves, G. Ungar and P. V. Wright, *Electrochim. Acta*, 2003, **48**, 1939–1951.
7 P. E. Stallworth, J. J. Fontanella, M. C. Wintersgill, C. D. Scheidler, J. J. Immel, S. G. Greenbaum and A. S. Gozdz, *J. Power Sources*, 1999, **81–82**, 739–747.
8 F. M. Gray, in *Polymer Electrolytes*, Royal Society of Chemistry, Cambridge, UK, 1997, ch. 5 and 6.
9 P. V. Wright, Y. Zheng, D. Bhatt, T. Richardson and G. Ungar, *Polym. Int.*, 1998, **47**, 34–42.
10 Y. Zheng, P. V. Wright and G. Ungar, *Electrochim. Acta*, 2000, **45**, 1161–1165.
11 F. Chia, PhD Thesis, University of Sheffield, 2002.
12 F. B. Dias, S. V. Batty, G. Ungar, J. P. Voss and P. V. Wright, *J. Chem. Soc., Faraday Trans.*, 1996, **92**, 2599–2606.
13 P. G. Bruce, in *Polymer Electrolytes 1*, eds. J. R. MacCallum and C. A. Vincent, Elsevier Applied Science, London and New York, 1987, ch. 8, p. 237.
14 Y. Zheng, J. Liu, Y.-P. Liao, G. Ungar and P. V. Wright, *Dalton Trans.*, 2004, 3053.
15 M. B. Armand, J. M. Chabagno and M. Duclot, in *Fast Ion Transport in Solids*, eds. P. Vashishta, J. N. Mundy and G. Shenoy, North Holland, Amsterdam, 1979, p.131.
16 C. T. Imrie, M. D. Ingram and G. S. McHattie, *Adv. Mater.*, 1999, **11**, 832; C. T. Imrie and M. D. Ingram, *Mol.Cryst. Liq. Cryst.*, 2000, **347**, 199.

Light-sensitive fusion between polymer-coated liposomes following physical anchoring of polymerisable polymers onto lipid bilayers by self-assembly

Kostas Kostarelos,*[ab] Dimitris Emfietzoglou[c] and Tharwat F. Tadros[d]

[a] *Centre for Drug Delivery Research, The School of Pharmacy, University of London, 29–39 Brunswick Square, London, UK, WC1N 1AX. E-mail: kostas.kostarelos@ulsop.ac.uk; Tel: +44 20 7835 2296*
[b] *Imperial College Genetic Therapies Centre, Flowers Building, South Kensington Campus, Imperial College London, London, UK, SW7 2AZ*
[c] *Department of Medical Physics, School of Medicine, University of Ioannina, Greece*
[d] *89 Nash Grove Lane, Wokingham, Berkshire, UK, RG40 4HE*

Received 4th March 2004, Accepted 25th May 2004
First published as an Advance Article on the web 17th September 2004

Delivery of therapeutics (drugs, radionuclides or genes) *in vivo* can be optimized when carried by a targeting delivery vehicle such as a surfactant vesicle, polymeric micelle or other polymer-coated colloidal particulate. In the present communication, we propose a general method based on self-assembly principles, to construct lipid-polymer bilayer vesicles whose featured characteristics may be altered according to the polymer molecule used, thus be easily designed along the needs of a particular delivery application. Polymer molecules containing non-polymerizable (polypropylene) and polymerisable (methacrylate) hydrophobic groups were used to construct lipid-polymer vesicles by following two different methods of preparation. In accord with our previous findings, when both types of polymer molecules are added to pre-formed liposomes, only weak adsorption onto the lipid surface occurs. Preparation of the vesicles by pre-mixing the lipid and polymer molecules has proved essential in order to allow the hydrophobic blocks of the copolymers to participate as integral parts of the bilayer. Anchoring of a polymerisable polymer onto the lipid bilayer by hydrophobic interactions, resulted in steric stabilization of the vesicles. When UV polymerization of the bilayer-incorporated [Methyl(PEG)$_{2000}$MA] polymer was induced, inter-vesicle fusion was triggered. Direct cryo-EM imaging of fusion between the PEG-coated liposomes has been observed. Such sterically stabilised fusogenic vesicles were constructed as potential triggered-release delivery systems, responsive to a variety of external stimuli depending on the type of polymerisable, hydrophobic group in the polymer molecule. By altering the properties of the incorporated hydrophobic group, liposomes able to fuse in response to initiators milder than UV light, such as green or red light, sound, temperature, oxygen or pH can be engineered.

Introduction

Engineering of liposome surfaces and bilayers is primarily aimed at improving the delivery of transported molecules and their controlled release onto targeted sites (*in vitro* or *in vivo*), for therapeutic or other purposes. There have been two major directions in liposome engineering research: (a) control of the rigidity and physicochemical stability of the liposome bilayer; and (b) rendering liposome surfaces recognisable by specific receptors at the target sites. Strategies whereby modification of the liposome surface is sought have been playing a leading role in the development of clinically relevant liposome systems during the last 15 years.[1,2] Placing an effective barrier (most commonly a neutral, hydrophilic macromolecular chain) protruding from the liposome surface to the external aqueous environment contributed to the rejuvenation of interest in the liposome field and led to an extremely rapid approval of liposome-based pharmaceuticals, clinically used in chemotherapeutic applications today.[3] Irrespective of the way in which the macromolecules are attached onto the liposome surface, the mechanism by which these systems function *in vitro* and *in vivo* is by steric stabilisation.[4,5] The term "Stealth"[6] has been coined to describe these systems, originating from their proposed *in vivo* behaviour leading to disguise from the blood serum macrophages and consequently, delayed recognition and sequestration by the RES (mainly the mononuclear phagocytic cells of the liver and spleen).[7,8]

Most of the ways to alter liposome surface characteristics in order to form sterically stabilized systems involve chemical conjugation of a polymer chain to the headgroup of a particular lipid (usually PE). An alternative methodology we developed almost a decade ago is aimed at engineering sterically stabilized liposomes by application of the concept of 'physical grafting'.[9–11] 'Physical grafting' constitutes one of the principal ways by which amphipathic proteins interact with the lipid bilayers to form plasma membranes. By mimicking this well documented grafting mechanism, abundantly found in nature, novel sterically stabilized liposomes were engineered by use of synthetic amphiphilic macromolecules.[9,12] Application of the 'physical grafting' method, eliminates the need for covalent chemical manipulation of the liposome-participating lipids, and can also be used as a generic strategy to engineer liposome surfaces by any macromolecule (synthetic or biological), provided its molecular structure allows for bilayer incorporation, therefore an effective physical anchor. However, the presence of polymeric macromolecules (chemically or physically grafted) at the liposome surfaces, pharmacologically leads only to prolonged blood circulation half-lives without incurring any means of controlling the release of encapsulated material at specific sites in the body. Even though the polymer-coated, steric stabilisation of liposomes is offering enhanced accumulation in the case of tissues exhibiting 'leaky' and 'faulty' vasculature (tumors, sites of infection) by a mechanism called 'passive targeting',[13] control of load release is still highly desirable. Inclusion of triggerable release functionalities has been attempted in the design of numerous liposome systems, particularly taking advantage of moderate temperature increases[14,15] and pH fluctuations.[16]

In the present study we extended application of 'physical grafting' of block copolymers onto lipid bilayers to include a polymerisable polymer containing a polymerisable hydrophobic group. Construction of photo-triggerably fusogenic sterically stabilised liposomes was therefore achieved, by incorporation of the polymer within the lipid bilayer by self-assembly, able to act both as the steric barrier (PEG domain) and the triggerable molecular switch (methacrylate domain) for interliposomal fusion.

Materials and methods

Liposome preparation and polymerisation

All liposomes were prepared using soybean lecithin (Sigma, UK) and doubly-distilled water. The tri-block copolymer, PF127 (PEO_{99}-PPO_{65}-PEO_{99}) was a gift from ICI Surfactants, Belgium and was used without further purification. The polymerizable polymer: methoxy-poly(ethylene glycol)-methacrylate [CH_3-O$(PEG)_{2000}$COCO(CH_3)] (M-$(PEG)_{2000}$MA); a gift from ICI Paints, Slough, UK) was used without further purification and was included in the lipid mixture following liposome preparation or during liposome formation. Azobisisobutyronitrile, AIBN (BDH, UK), was also included in the initial solvent lipid and polymer mixtures. This mixture of lipid, polymer and AIBN was dissolved in a chloroform : methanol (2 : 1) solvent phase, which was then rotary evaporated. A thin lipid or lipid-polymer film was rehydrated to form multilamellar vesicles (MLVs) and subsequently sonicated using a 50 Hz bath sonicator (Kerry) under nitrogen atmosphere and constant temperature (25 °C). Liposomes were filtered through Millipore 0.2 μm polycarbonate filters following sonication. A polymerization reaction was photochemically initiated using a tailor-made double jacketed quartz vessel with circulating water (Syngenta plc., formerly ICI Agrochemicals, Berkshire, UK), achieving UV illumination (125 W mercury lamp used as source of ultraviolet radiation; max output 240 nm light wavelength) of the sterically stabilized SUVs containing AIBN, by continuous stirring of the sample at constant temperature. The vesicle solutions were photo-irradiated while stirred constantly at 25 °C.

^{1}H-NMR

All spectra were recorded on a JEOL GX400 spectrometer, acquired with 45′ pulses and a 2 s repetition rate. All samples were prepared in D_2O (Aldrich, UK) at a fixed lipid concentration of 2% (w/w).

Optical microscopy

A Leitz Diaplan light microscope was used connected to a Cohu CCD camera. All optical microscopy images were obtained using a 40× objective lens under phase contrast.

Cryogenic transmission electron microscopy

Vesicles were imaged using a Zeiss EM 902 cryo-transmission electron microscope exactly as described elsewhere.[17] Briefly, samples were prepared using a custom-built controlled environment vitrification system, allowing immersion of the copper or carbon grids used to deposit a drop of the sample, into liquid ethane under controlled temperature and humidity. The accelerating voltage was 80 kV and the cryo-stage temperature was kept below 180 K.

Quasielastic light scattering

All measurements were carried out using Malvern 4700 (Malvern, UK) at a fixed 90° angle and 25 °C. The instrument cumulants analysis software was used for the mean diameter measurements. At least three independent series of ten measurements were carried out. The experimental error from the sizing measurements was never above 1% of the mean value.

Laser Doppler electrophoresis

The electrophoretic mobility of the liposomes was measured using the DELSA 440 Zetasizer instrument (Beckman-Coulter, UK) operating with a helium-neon laser beam (5 mW). Ten different measurements for each liposome system were carried out and all four different angles of detection were used to obtain the ζ-potential at the liposome surface by employing the Smoluchowski approximation on the electrophoretic mobility data obtained when a 5 V electric field was applied to the liposome suspension.

Results

In agreement with our previous findings, we observed that interaction of both the tri-block copolymer or polymerisable polymer molecules with the lipid bilayer by purely hydrophobic forces, resulted in polymer coating (steric stabilization) of the vesicles. This was done by preparation of liposomes containing PF127 and M-$(PEG)_{2000}$MA with the initial lipid mix in solvent, or addition of polymer molecules post-liposome formation. Fig. 1 shows the ^{1}H-NMR spectrum for a small unilamellar liposome dispersion (2% w/w) without addition of any polymer (A), following addition of 1% w/w PF127 molecules in the solvent mix (B) and after liposome preparation (C). Line broadening of the methyl peak in the hydrophobic PPO block was observed (black arrow), illustrating restriction of the PPO molecular movement when incorporated in the lipid bilayer

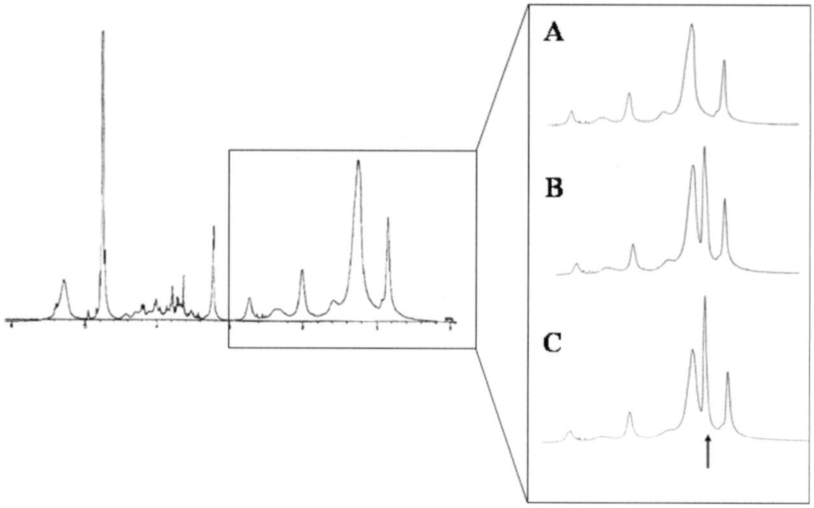

Fig. 1 ^{1}H-NMR spectrum of small unilamellar vesicles, and expansions of the spectral regions where the methyl signal from the PPO groups of the tri-block copolymer PF127 appears. (A) liposome alone; (B) liposome with PF127 mixed with the lipid molecules in solvent, prior to lipid hydration; (C) liposome with PF127 added post liposome formation.

(B). The structural characteristics of those multilamellar and unilamellar vesicles incorporating the copolymer were then studied by optical videomicroscopy and cryo-EM (Fig. 2). Interestingly, when PF127 molecules (1% w/w) were allowed to participate during liposome formation, structural changes occurred for both the MLV and SUV. Participation of the PPO blocks into the lipid bilayer led to elongation of the spherical MLVs (Fig. 2A) to rod-like structures (Fig. 2B - black arrows). The cryo-EM examination of the same samples following the high energy sonication process to

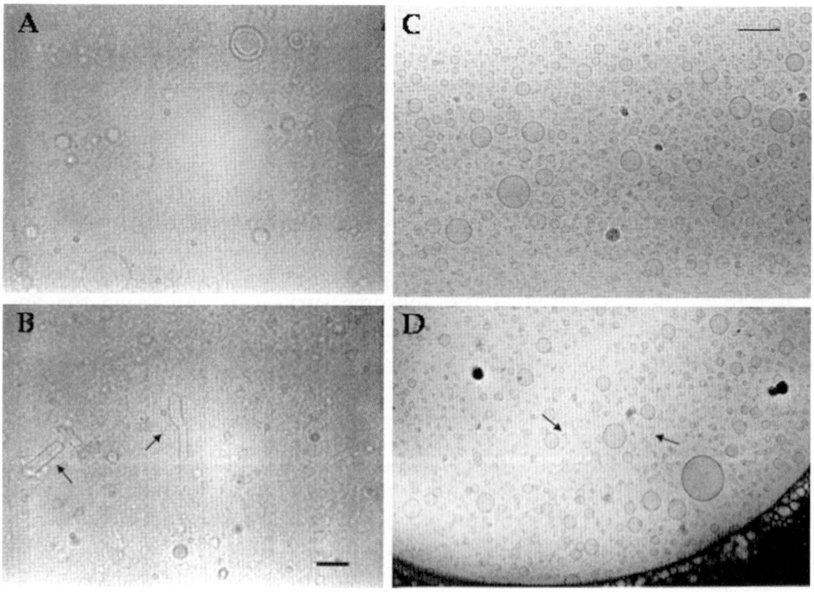

Fig. 2 Optical (A and B) and Cryo-EM (C and D) images of multilamellar (MLV) and small unilamellar (SUV) vesicles. (A) MLV alone; (B) MLV with PF127 added prior to lipid hydration; (C) SUV alone; (D) SUV with PF127 added prior to lipid hydration. Black arrows indicate the elongated structures formed in the presence of PF127. Scale bar for A and B is 1 μm; for C and D is 100 nm.

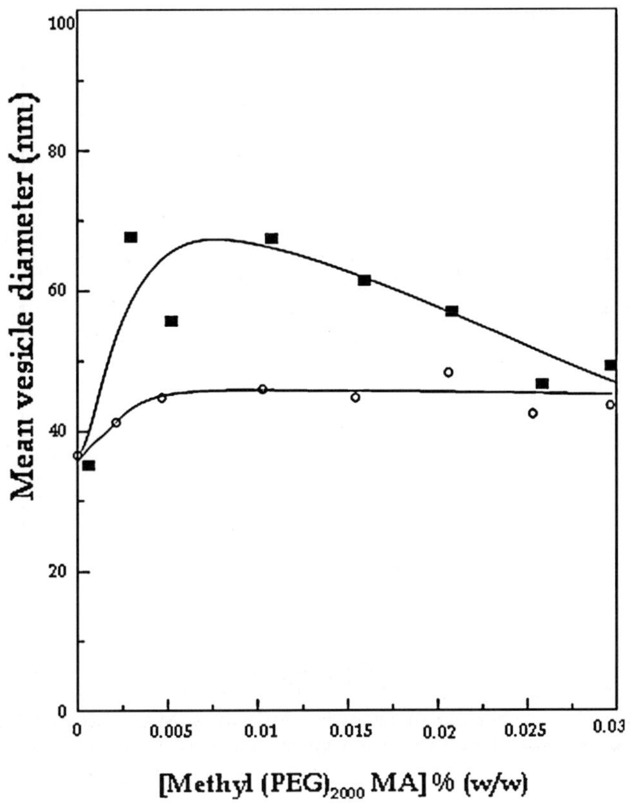

Fig. 3 Mean vesicle diameter by quasielastic light scattering for increasing $M(PEG)_{2000}MA$ concentrations. Filled squares are vesicles formed in the presence of polymer; open circles are vesicles with polymer added post-liposome formation.

form the SUVs, was indicative of elongated small flake-like structures only when F127 was allowed to participate in liposome formation at this molar ratio (approximately one copolymer molecule for every 30 lipids in the bilayer; Fig. 2D - black arrows). Both microscopic techniques indicated that incorporation of the tri-block copolymer molecule into the lipid bilayer was leading to structural alterations in the resultant lipid-polymer vesicles.

Fig. 3 shows QUELS mean particle diameter values following interaction between the $M\text{-}(PEG)_{2000}MA$ and lipids following the two different protocols described above for the tri-block copolymer. In the case of mixing the polymerisable polymer and the lipids in the chloform:methanol solvent followed by evaporation and hydration, the resultant SUV mean diameter peaked at a range of lipid:copolymer molar ratio. On the other hand, addition of the polymer molecules following liposome preparation led to vesicle surface adsorption evidenced by a moderate increase in the mean particle diameters. The molar ratio range between 6 : 8 lipids per polymer molecule was found to lead to maximum increases in mean vesicle size. Above this molar ratio, decreases in the mean particle diameter for the vesicles containing bilayer-anchored polymers are thought to be indicative of gradual lipid bilayer solubilisation towards the more preferred mixed micellar phase, with co-partitioning of the vesicle and mixed micelles in-between.

The effect on vesicle surface charge following the two types of interaction between the $M\text{-}(PEG)_{2000}MA$ and lipids were monitored by studying the ζ-potential at the vesicle surface (Fig. 4). Decreases in the ζ-potential of the liposome surface from −51.3 mV for the 'naked' liposomes to values between −30 and −40 mV for vesicles resulting from both types of polymer interaction with the lipids, indicates the presence of the hydrophilic polymer PEG layer at the liposome surface. The presence of polymer both prior and post liposome formation was therefore leading to polymer-coated, sterically stabilised vesicles.

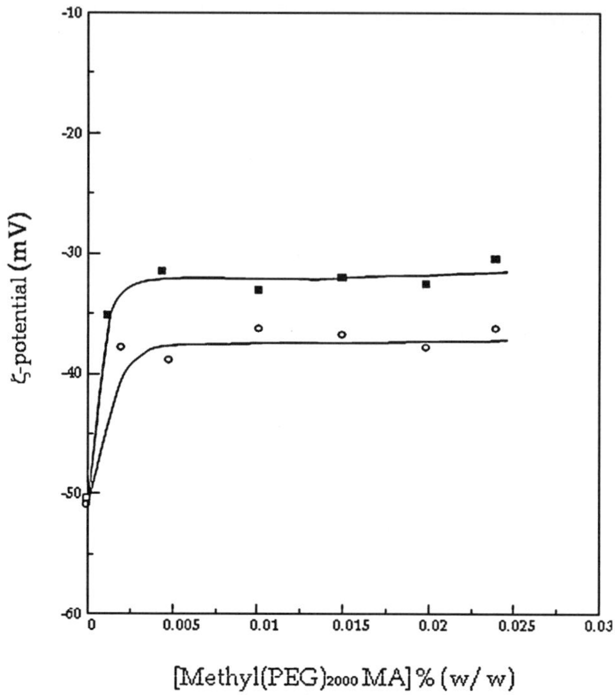

Fig. 4 Vesicle surface charge (ζ-potential) increasing $M(PEG)_{2000}MA$ concentrations. Filled squares are vesicles formed in the presence of polymer; open circles are vesicles with polymer added after liposome formation.

Vesicles (2% lipid w/w) that contained 'physically grafted' polymerisable polymer (*i.e.* bilayer incorporated) and the hydrophobic polymerisation initiator AIBN in the bilayer, were subsequently placed in a quartz polymerisation vessel and irradiated with UV light for different time points at 25 °C. Fig. 5 shows the cryo-EM microscopy of $M\text{-}(PEG)_{2000}MA$ incorporating vesicles before (Fig. 5A) and after 12 h of UV light irradiation (Fig. 5B). Inter-vesicle fusion leading to significantly larger closed vesicular structures can be observed. Determination of mean vesicle diameter for these vesicle systems by QUELS did not provide meaningful data because of the large, non-spherical, inhomogenous structures formed. Interestingly, five times dilution of this polymerised vesicle dispersion with double distilled water (Fig. 5C), led to rupture of the large vesicular structures providing an SUV dispersion of approximately the same mean vesicle size as the original sterically stabilised formulation prior to polymerisation (Fig. 5A). These images were indicative to us that vesicle fusion was triggered when a photochemical polymerization reaction between the hydrophobic blocks of the bilayer-buried polymer (methacrylate groups) was induced by UV light irradiation.

Discussion

In the present study we applied the principle of 'physically grafting' tri-block copolymer and polymerisable polymer molecules onto lipid bilayers to obtain polymer-coated liposomes. Extension of the 'physical grafting' principle was attempted by use of polymer molecules containing a polymerisable (methacrylate) group. On photochemical initiation, polymerisation of the methacrylate predictably residing within the bilayer led to vesicle destabilisation and extensive vesicle fusion, imaged directly by cryo-EM. The fused vesicle structures resembled cryo-EM images of recently reported intra-liposomal fusion following interaction with SNARE proteins.[18]

The self-assembly of ABA block copolymers with lipid molecules during the formation of vesicle bilayers has been proposed as an alternative novel method to sterically stabilize phospholipid

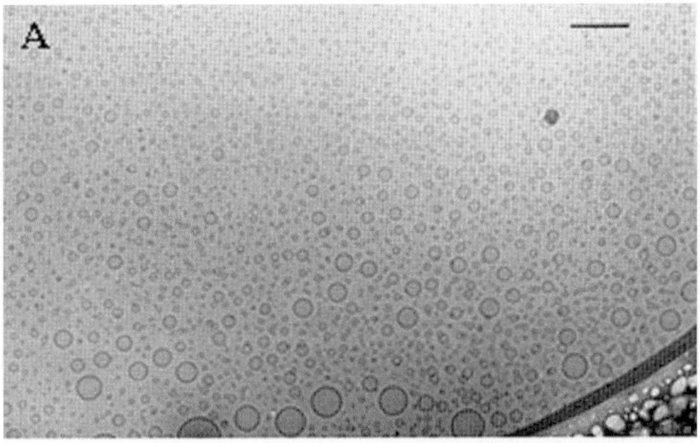

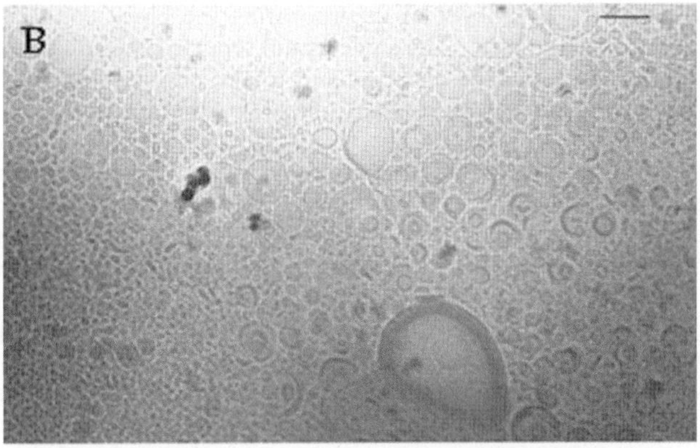

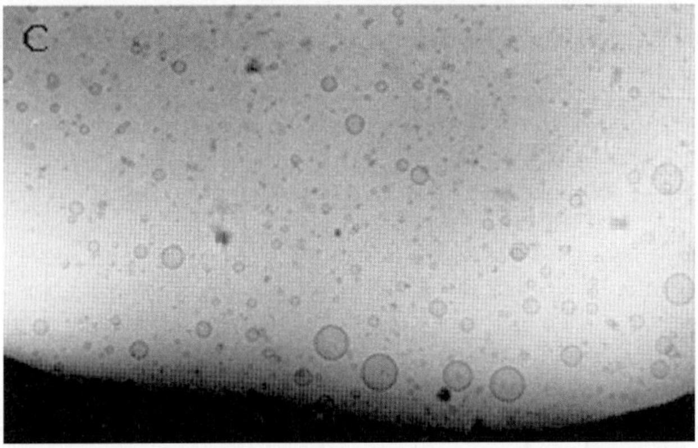

Fig. 5 Cryo-electron microscopy images of lipid:M(PEG)$_{2000}$MA vesicles formed in the presence of polymer. (A) lipid-M(PEG)$_{2000}$MA vesicles before polymerisation; (B) lipid:M(PEG)$_{2000}$MA vesicles following 12 h in the polymerisation quartz vessel; (C) sample shown in (B) diluted ×5 with double distilled water.

bilayer vesicles.[19] By allowing the anchoring B block of the copolymer to participate in the vesicle formation process a novel structure was obtained. The B block is physically grafted inside the bilayer, thus firmly anchoring the two A stabilizing chains dangling in solution. From the various experimental methods used previously[19,20] and in this study (Figs. 1–4) to elucidate the effect that the polymer molecules had on the vesicle systems, it was demonstrated that the polymer:lipid molar ratio is an extremely important factor determining aggregate structure. Interaction between polymer molecules and lipids in a vesicular form takes place at low to moderate molar ratios. For higher polymer:lipid ratios destruction of the vesicle structures occurs, with a concomitant formation of mixed micelles between the polymers and the lipids, similar to the phase behaviour reported for the poly(ethyleneglycol)-phosphatidylethanolamine (PEG-PE) conjugates.[21] In the present studies, both vesicle structural and morphological studies in the case of tri-block copolymers (PF127; Figs. 1 and 2) and vesicle particle size and surface characterization for the polymerisable polymer (M-$(PEG)_{2000}$MA; Figs. 3 and 4) revealed that 'physical grafting' of the hydrophobic group of the polymers and vesicle surface coating offered by the hydrophilic (PEO and PEG) chains was taking place.

The pronounced advantage of allowing association of the hydrophobic polymer groups with the lipid bilayer are evident, since the steric barriers are much more robust and less susceptible to desorption (caused by interparticle collisions, dilution effects, interactions with components of the biological milieu), therefore offering robust polymer coating to the vesicle. Interestingly, the concept of 'physically grafting' polymer molecules onto liposomes was subsequently reproduced independently by other groups.[22–24] More recently, in a complimentary development to our work, the thermosensitive properties of the liposomes incorporating the block copolymer molecule PF127 were explored in a series of reports.[24,25] The same group reported the enhanced *in vitro* and *in vivo* delivery of a hydrophilic dye molecule encapsulated in the internal aqueous phase of the block copolymer-incorporated liposomes using a local mild hyperthermia protocol on a subcutaneous tumor model.[26]

In an attempt to extend the concept and methodologies of 'physically grafting' macromolecules onto liposomes, the present study used a polymerisable polymer molecule containing a methacrylate (polymerizable) group as the hydrophobic domain and a methoxy-PEG group as its hydrophilic chain. By initiation of a polymerization reaction, a means for triggering intra-bilayer fusion was achieved (Fig. 5). The liposome structures were constructed in a way to combine small size (less than 100 nm; Fig. 3), surface polymer coating (*i.e.* steric stabilization; Fig. 4), and capability of inducible fusion (Fig. 5). A model mechanism describing the observed fusion between liposomes is proposed and schematically shown in Fig. 6. The methacrylate groups inside the lipid bilayer on initiation of the polymerisation reaction are thought to induce strain against the high bending curvature of the lipid bilayer towards more elongated structures as can be observed in Fig. 5B. As the reaction

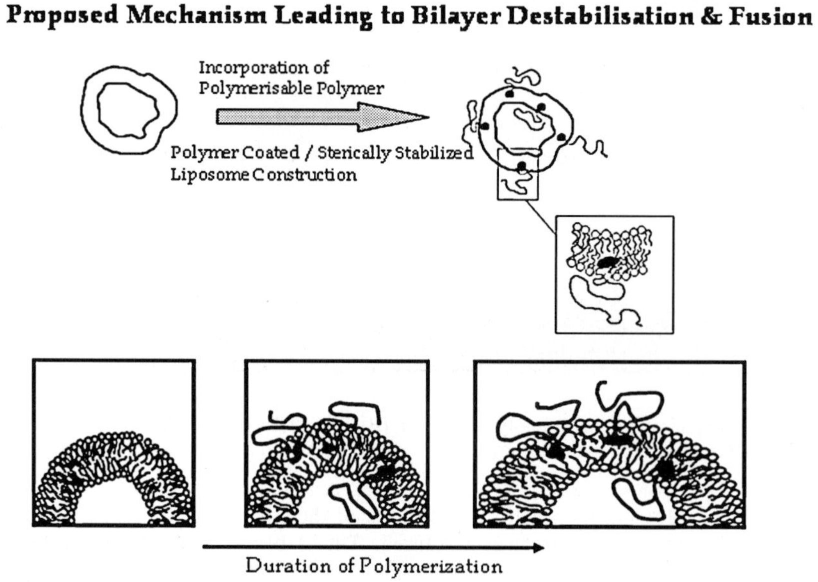

Fig. 6 Schematic representation of the 'physical anchoring' of $M(PEG)_{2000}MA$ polymer molecules into the lipid bilayer and a proposed scheme leading to intra-liposomal fusion on polymerisation.

proceeds, lateral phase separation at specific regions of the bilayer in combination with the high degree of Brownian motion (nanometer-range vesicles) leading to frequent close approach or collision between the lipid bilayers, results in fusion between the vesicles. Bilayer rigidity and lateral phase separation have been recognised as important parameters in the mechanism leading to content leakage from photopolymerised liposomes made of photochromic phospholipids,[27] however no fusion was observed or implicated in those studies.

The results obtained here also correlate well with the studies (rapid release of fluorescent marker) recently reported by O'Brien's group, by photo-inducing destabilization of sterically stabilized liposomes.[28,29] However, these investigators used newly synthesized polymerizable lipids (Bis SorbPC and analogues) and their pegylated forms instead of polymerizable polymers able to be buried within the liposome bilayer by self-assembly used in this work. Also, the perceived mechanism of bilayer destabilisation and consequent release of entrapped material from the Bis-SorbPC containing liposomes, is not through inter-vesicular fusion but due to shrinkage of the polymerised domains in the bilayer, leading to gaps through which materials release.[29] Various other triggerably fusogenic liposome systems have been engineered to release contents by acidic,[16,30] enzymatic,[31] thermal[14,15] or photochemical[32] triggering processes. Most commonly, novel DOPE analogues are synthesized, whereby external stimuli-triggered cleavage of a steric group covalently attached to the DOPE headgroup (acid-sensitivity most widely explored) exposes the cone-shaped lipid molecule, which tends to promote the bilayer-to-hexagonal phase transition facilitating membrane fusion.[33,34] Most relevant to the 'physically grafted' polymerisable polymer work we carried out herein, is the thermosensitive activation of liposomes by incorporation of *N*-isopropylacrylamide (NIPAM) polymers[12] and copolymers.[35,36] However, the inclusion of an effective steric stabilization barrier onto the NIPAM-thermosensitive liposomes without jeopardizing the temperature-responsive fusion and content release requires further work as recently described.[37] Moreover, in those studies the mechanism of thermosensitive content release has not drawn upon occurrence for inter-vesicle fusion, nor such fusion as evidenced in our studies has been previously observed. We herein propose a general method to engineer triggerably fusogenic, sterically stabilized vesicles that can be developed further by the self assembly between lipid and copolymer molecules responsive to other stimuli, as for example changes in oxidation,[38] constituting the next generation, of 'smarter' polymer–lipid hybrid vesicle systems.

Acknowledgements

We would like to acknowledge the late Prof. Dimitri Papahadjopoulos, University of California at San Francisco (UCSF), USA for enlightening discussions on the mechanisms of liposome fusion and useful advice at the initial steps of this project. Also, Prof. M. Almgren for providing the cryo-EM imaging of our samples and constructive comments. DE would like to acknowledge financial assistance from Grant 62/1293 registered with the University of Ioannina Committee of Research.

References

1 *Medical Applications of Liposomes*, ed. D. D. Lasic and D. Papahadjopoulos, Elsevier, Amsterdam, 1998.
2 T. M. Allen, P. Sapra, E. Moase, J. Moreira and D. Iden, *J. Liposome Res.*, 2002, **12**, 5.
3 A. Gabizon, H. Shmeeda and Y. Barenholz, *Clin. Pharmacokinet.*, 2003, **42**, 419.
4 M. C. Woodle and D. D. Lasic, *Biochim. Biophys. Acta*, 1992, **1113**, 171.
5 *Long Circulating Liposomes: Old Drugs, New Therapeutics*, ed. M. C. Woodle and G. Storm, Springer, Berlin, 1998.
6 Stealth™ is a registered trademark of Alza Corp., a Johnson & Johnson company.
7 T. M. Allen and C. Hansen, *Biochim. Biophys. Acta*, 1991, **1068**, 133.
8 *Stealth Liposomes*, ed. D. D. Lasic and F. J. Martin, CRC Press, Boca Raton, FL, 1995.
9 K. Kostarelos, P. F. Luckham and T. F. Tadros, *J. Liposome Res.*, 1995, **5**, 117.
10 K. Kostarelos, P. F. Luckham and T. F. Tadros, *J. Liposome Res.*, 1995, **5**, 443.
11 K. Kostarelos, in *Targeting of Drugs 6: Strategies for Stealth Therapeutic Systems*, ed. G. Gregoriadis and B. McCormack, Plenum, New York & London, 1998, p.139.
12 H. Ringsdorf, E. Sackmann, J. Simon and F. M. Winnik, *Biochim. Biophys. Acta*, 1993, **1153**, 335.
13 T. M. Allen, *Nat. Rev. Cancer*, 2002, **2**, 750.
14 D. Needham and M. W. Dewhirst, *Adv. Drug Deliv. Rev.*, 2001, **53**, 285.

15 K. Kono, *Adv. Drug Deliv. Rev.*, 2001, **53**, 307.
16 J. Shin, P. Shum and D. H. Thompson, *J. Controlled Release*, 2003, **91**, 187.
17 M. Almgren, K. Edwards and G. Karlsson, *Colloids Surf. A*, 2000, **174**, 3.
18 C. G. Schuette, K. Hatsuzawa, M. Margittai, A. Stein, D. Riedel, P. Kuster, M. Konig, C. Seidel and R. Jahn, *Proc. Natl. Acad. Sci. USA*, 2004, **101**, 2858.
19 K. Kostarelos, T. F. Tadros and P. F. Luckham, *Langmuir*, 1999, **15**, 369.
20 K. Kostarelos, M. Kipps, T. F. Tadros and P. F. Luckham, *Colloids Surf. A*, 1998, **136**, 1.
21 K. Edwards, M. Johnsson, G. Karlsson and M. Silvander, *Biophys. J.*, 1997, **73**, 258.
22 M. Johnsson, M. Silvander, G. Karlsson and K. Edwards, *Langmuir*, 1999, **15**, 6314.
23 J. D. Castile, K. M. G. Taylor and G. Buckton, *Int. J. Pharm.*, 2001, **221**, 197.
24 P. Chandaroy, A. Sen, P. Alexandridis and S. W. Hui, *Biochim. Biophys. Acta*, 2002, **1559**, 32.
25 P. Chandaroy, A. Sen and S. W. Hu, *J. Controlled Release*, 2001, **76**, 27.
26 J. Wells, A. Sen and S. W. Hui, *Int. J. Pharm.*, 2003, **261**, 105.
27 R. H. Bisby, C. Mead and C. G. Morgan, *FEBS Lett.*, 1999, **463**, 165.
28 B. Bondurant, A. Mueller and D. F. O'Brien, *Biochim. Biophys. Acta*, 2001, **1511**, 113.
29 T. Spratt, B. Bondurant and D. F. O'Brien, *Biochim.Biophys. Acta*, 2003, **1611**, 35.
30 X. Guo and F. C. Szoka, *Bioconjug. Chem.*, 2001, **12**, 291.
31 C. C. Pak, S. Ali, A. S. Janoff and P. Meers, *Biochim. Biophys. Acta*, 1998, **1372**, 13.
32 A. Mueller and D. F. O'Brien, *Chem. Rev.*, 2002, **102**, 727.
33 X. Guo, J. A. MacKay and F. C. Szoka, Jr., *Biophys. J.*, 2003, **84**, 1784.
34 S. Zalipsky, M. Qazen, J. A. Walker, 2nd, N. Mullah, Y. P. Quinn and S. K. Huang, *Bioconjug. Chem.*, 1999, **10**, 703.
35 H. Hayashi, K. Kono and T. Takagishi, *Bioconjug. Chem.*, 1999, **10**, 412.
36 E. Roux, R. Stomp, S. Giasson, M. Pezolet, P. Moreau and J. C. Leroux, *J. Pharm. Sci.*, 2002, **91**, 1795.
37 E. Roux, M. Lafleur, E. Lataste, P. Moreau and J. C. Leroux, *Biomacromolecules*, 2003, **4**, 240.
38 A. Napoli, M. Valentini, N. Tirelli, M. Muller and J. A. Hubbell, *Nat. Mater.*, 2004, **3**, 183.

Polyelectrolyte–surfactant complex: phases of self-assembled structures

C. von Ferber*[a] and H. Löwen[b]

[a] *Theoretical Polymer Physics, Freiburg University, 79104, Freiburg, Germany. E-mail: ferber@physik.uni-freiburg.de*
[b] *Institut für Theoretische Physik II, Heinrich-Heine-Universität Düsseldorf, 40225, Düsseldorf, Germany*

Received 29th March 2004, Accepted 28th April 2004
First published as an Advance Article on the web 27th September 2004

We study the structure of complexes formed between ionic surfactants (SF) and a single oppositely charged polyelectrolyte (PE) chain. For our computer simulation we use the "primitive" electrolyte model: while the polyelectrolyte is modeled by a tethered chain of charged hard sphere beads, the surfactant molecules consist of a single charged head bead tethered to a tail of tethered hard spheres. A hydrophobic attraction between the tail beads is introduced by assuming a Lennard-Jones potential outside the hard-sphere diameter. As a function of the strengths of both the electrostatic and the hydrophobic interactions, we find the following scenario: switching on and increasing the electrostatic forces first leads to a stretching of the PE and then by condensation of SF to the formation of a complex. For vanishing hydrophobic forces this complex has the architecture of a molecular bottle-brush cylindrically centered around the stretched PE molecule. Upon increasing the hydrophobic attraction between the SF tails, a transition occurs inverting this structure to a spherical micelle with a neutral core of SF tails and a charged corona of SF heads with the PE molecule wrapped around. At intermediate hydrophobicity there is a competition between the two structures indicated by a non-monotonic dependence of the shape as function of the Coulomb strength, favoring the cylindrical shape for weak and the spherical micellar complex for strong interaction.

I. Introduction

Polyelectrolyte–surfactant mixtures have proven to provide the basis for new materials with extraordinary properties that make them interesting for a wide range of applications.[1,2] In particular the Coulomb attraction between polyelectrolyte chains and oppositely charged ionic surfactant molecules in solution leads to aggregation and complex formation. The resulting complex differs in its conformational, structural and dynamical features from that of a solution containing only one of its constituents, the pure polyelectrolyte or the pure surfactant (for reviews see ref. 1 and 3). Much experimental work has been done on these systems and a rich variety of different complex structures was revealed as a function of chemical nature of the polyelectrolyte, surfactant and solvent molecules.[4–15] A theory that could give a systematic microscopic characterization and prediction of the different complexes on the other hand does not exist so far.

For complexes formed by a *single* polyelectrolyte chain with ionic and hydrophobic surfactants at low concentration a theoretical treatment that takes the long-ranged Coulomb interaction and the hydrophobic interactions fully into account is still to be developed. The case without hydrophobicity has recently been simulated by the present authors.[16] For these single polyelectrolyte complexes one expects structures either of bottle-brush shape for low hydrophobicity or, for strong hydrophobicity, possibly spherical micelles that aggregate together with the polyelectrolyte chain.[2,17] Theories so far in general make assumptions about the symmetry or the structure of the complex implying a complete adsorption of surfactant onto the polyelectrolytes for rigid[18,19] and flexible chains, see *e.g.* ref. 20, and cannot predict the complex structure or its symmetry[17] as far as it enters the theory as an input. For the special case of a system of rigid polyelectrolytes that form complexes *via* thermodynamic counterion condensation including the effect of added surfactant a theory has recently been developed by Kuhn, Levin and coworkers.[21–23] The collapse and partial collapse of semiflexible and flexible polymers in solutions with surfactants has been treated by a meanfield association theory by Diamant and Andelman.[24,25] However, this theory does not include explicitly the long range of the Coulomb interactions. A self consistent field theory with a spherically symmetric setup[17] on the other hand predicts the formation of spherically symmetric micelles with a neutral hydrophobic core and a surface layer where the charged surfactant heads and the oppositely charged polyelectrolyte are located. Based on this observation it has been proposed to simplify the model by replacing the ionic surfactant by large spherical counterions not treating the surfactant tails explicitly.[26] For a system of a polyelectrolyte in the presence of small counterions computer simulations[27–29] and theories[30,31] find a swelling and stretching of the polyelectrolyte and a subsequent collapse when increasing the Coulomb interaction. In this scenario a non-monotonic dependence of the polyelectrolyte conformation as a function of the Coulomb strength occurs: for weak Coulomb strength the PE stretches while in the strong Coulomb regime counterion condensation and a collapse to a coil conformation occurs. For large counterions representing micelles the polyelectrolyte may be wrapped around the charged sphere exhibiting different conformations depending on the chain flexibility.[26,32–46]

The bottle brush structure that is expected for the condensation of charged ionic surfactants onto a polyelectrolyte chain may be compared with corresponding neutral molecular bottle brushes in case the polarization effects in the complex are neglected. Such neutral bottle brush or comb polymers are built either by end linking the side chains to the backbone chain or by attaching them *via* strong hydrogen bonds.[47,48] Such configurations may be expected to become more rigid for higher densities of the side chains. A quantitative description, however, is still under debate since different simple scaling arguments[49–51] appear to predict different quantitative behavior for the sizes of the main and side chains depending on the model used and the limit that is considered.[52]

In the present paper we explore by computer simulation the phases of complex formation of a single polyelectrolyte chain with ionic surfactants in the phase space parameterized by the strengths of the Coulombic interactions and the hydrophobic attraction between the surfactant tails. As in our previous study[16] our simulation relies on the so-called "primitive" model[53] that treats the microscopic charges of the polyelectrolyte and the surfactant heads explicitly. This is in contrast to simulations on polymer–surfactant complexation which neglect the long-range nature of the Coulomb interactions.[54] The "primitive" model has recently also been applied to simulate the complexation of polyelectrolyte chains with charged spheres representing spherical micelles[55] and has proven to be reliable in different contexts of polyelectrolyte conformations.[27–29,56–59]

To treat surfactants with neutral or hydrophobic tails we model the surfactant by a tethered hard sphere chain with a charged head and a Lennard-Jones like effective attraction between the hydrophobic tail monomers allowing for the formation of micelles. As a result, we observe different phases of complexation depending on the strengths of the two interactions involved: the Coulomb and the hydrophobic interaction. An overview of these phases is given in Fig. 1 in terms of simulation snapshots. For vanishing hydrophobicity (leftmost column of Fig. 1) the mean square end–end distance R^2 of the polyelectrolyte is non-monotonic as a function of the Coulomb interaction. However, in contrast to the standard situation where a collapse of the polyelectrolyte due to counterion condensation occurs,[27–29] such a collapse is blocked by the steric interaction of the surfactant tails. As we have shown in ref. 16, for even longer surfactant tails the collapse is more sufficiently blocked, while the addition of salt weakens the complex as it replaces the surfactant in the complex. Increasing the hydrophobicity for vanishing Coulomb interaction on the other hand (bottom row of Fig. 1) leads to the aggregation of micellar clusters of surfactant molecules that are independent from the polyelectrolyte chain. Introducing a finite Coulomb strength in this latter

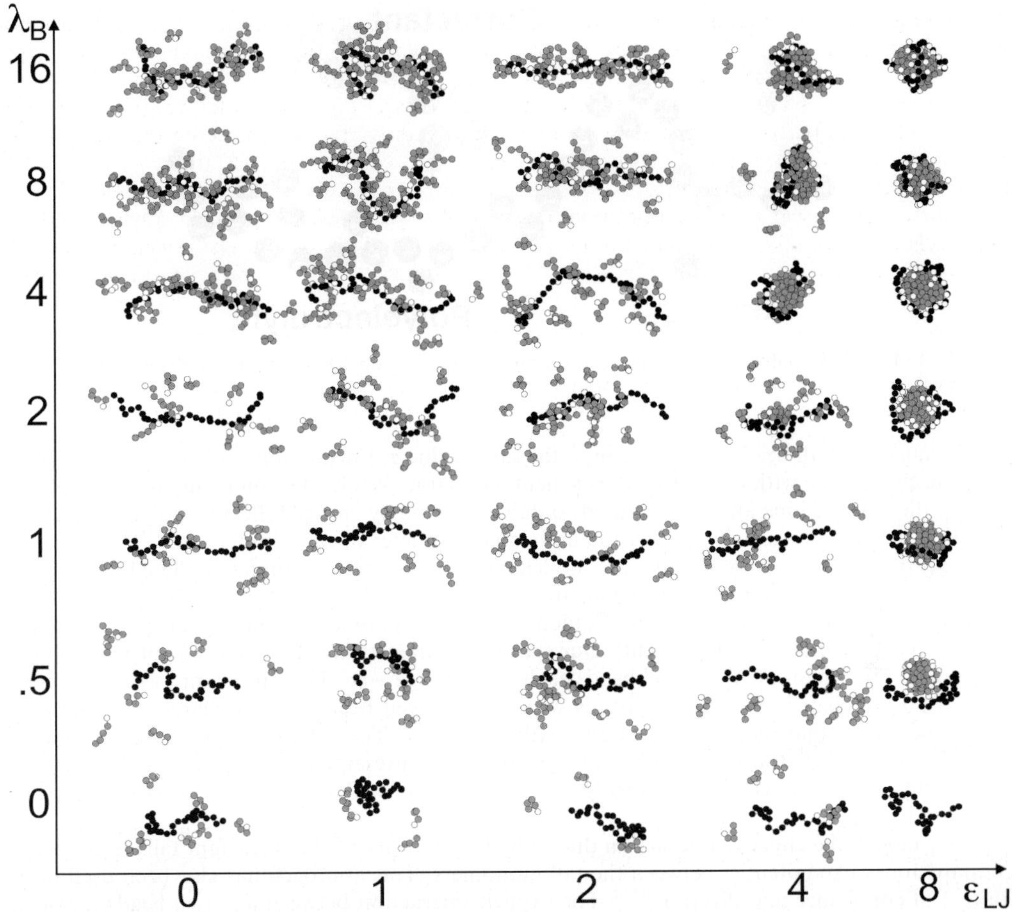

Fig. 1 Snapshots of the complex configurations showing the dependence on the two interactions: Coulombic (λ_B) and hydrophobic (ε_{LJ}).

situation favors the condensation of the surfactant at the polyelectrolyte. Thirdly we may start with a bottle brush conformation of the complex at high Coulomb but vanishing hydrophobic interactions (top left of Fig. 1). Gradually increasing the hydrophobicity (along the top row of Fig. 1) the tails start to aggregate and a transition occurs from the cylindrical structure with the tails on the outside to an inverted spherical structure with a core that is constituted by the surfactant tails. In our present resolution this transition appears to be sharp.

The paper is organized as follows: in section II, we describe our model in detail and we present details of the simulation procedure in section III. Results are given in section IV and we conclude in section V.

II. The model

In our simulation we model all molecules as chains of tethered hard spheres (beads) that represent Kuhnian segments of the molecule. Microions are modeled by hard sphere beads. Charges on these molecules and ions are represented by placing a point charges at the center of any charged bead. In this primitive model the solvent is not treated explicitly and only enters *via* its dielectric constant ε_r neglecting the discreteness of the solvent molecules. The microscopic interactions between the particles are given by the hard sphere excluded volume, the long-ranged Coulomb forces, and a short range hydrophobic interaction. We reduce the model parameter space by fixing the hard

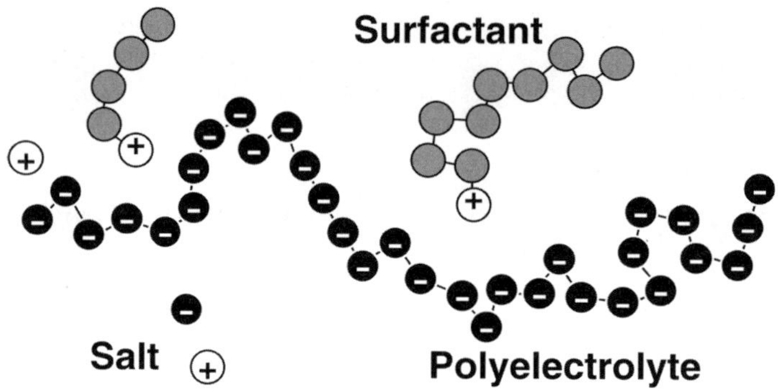

Fig. 2 Model of the polyelectrolyte–surfactant system with salt ions. Charges are indicated by "+" and "−" signs.

sphere diameter of all beads in the system to the same value σ. Furthermore, all charged beads have the same charge $\pm qe$ with e denoting the elementary charge. While this "minimal" model explicitly handles the counter ions and may be used to describe strongly charged, flexible polyelectrolytes,[27] analytical calculations usually neglect the individual character of the counter ions.[60–63]

As mentioned above, all molecules are modeled by chains of freely jointed, tethered hard sphere beads. Subsequent beads in the same chain are tethered. The extension of all bonds is limited by the tether to a fixed value b. For vanishing Coulomb and hydrophobic interactions the model is purely entropic (*i.e.* temperature-independent). Our specific parameters are such that the polyelectrolyte is represented by a chain of N charged beads each with charge $-qe$. The surfactant molecules on the other hand are chains of n tethered beads with one charged head bead of charge $+qe$ and $n-1$ neutral tail beads that may interact due to hydrophobicity. The full model may also contain salt which has been treated in ref. 16. The salt ions are then represented by single charged beads. The constituents of the model of polyelectrolytes, surfactant molecules and microions is schematically shown in Fig. 2.

In our model, the effective interaction due to hydrophobicity of the surfactant tails is given by a Lennard-Jones like potential between the tail monomers. This approach has also been used in the context of copolymer micellization.[64,65] The explicit interaction between any two beads i, j of the model at a distance r between their centers is given by a potential $V_{ij}(r)$. To denote chain connectivity we use an adjacency matrix T_{ij}. For beads i, j that are tethered $T_{ij}=1$ otherwise $T_{ij}=0$. With $q_i \in \{\pm 1, 0\}$ denoting the charge and $u_i \in \{1, 0\}$ the hydrophobicity number of bead i, the potential $V_{ij}(r)$ is given by

$$\frac{1}{k_{\mathrm{B}}T}V_{ij}(r) = q_i q_j \lambda_{\mathrm{B}}\frac{\sigma}{r} + u_i u_j \varepsilon_{\mathrm{LJ}}\left(\left(\frac{\sigma}{r}\right)^{12} - \left(\frac{\sigma}{r}\right)^{6}\right) + V_{\mathrm{H}}(\sigma - r) + T_{ij}V_{\mathrm{H}}(r-b). \tag{1}$$

Here, the dimensionless Coulomb strength λ_{B} is the Bjerrum length measured in bead diameters σ. The Bjerrum length $\sigma\lambda_{\mathrm{B}} = q^2e^2/\varepsilon_r k_{\mathrm{B}}T$ sets the length scale where the Coulomb pair interaction is comparable to the thermal energy $k_{\mathrm{B}}T$. The dimensionless hydrophobicity parameter $\varepsilon_{\mathrm{LJ}}$ controls the depth of the Lennard-Jones potential. Furthermore, the hard core and tether interactions are represented by the one dimensional hard wall potential

$$V_{\mathrm{H}}(x) = \begin{cases} 0 & \text{if } x<0 \\ \infty & \text{else} \end{cases}. \tag{2}$$

We note that in our notation the product $T_{ij}V_{\mathrm{H}}(r-b)$ is defined to be zero if T_{ij} is zero.

The key parameters of our model are: the number N of charged monomers of the polyelectrolyte chain, the number $n-1$ of neutral beads of the surfactant tails, the parameters λ_{B} and $\varepsilon_{\mathrm{LJ}}$ which measure the strengths of the Coulomb and hydrophobic interactions and finally the relative salt concentration c_s. In the following we shall explore the parameter space in particular in the two variables λ_{B} and $\varepsilon_{\mathrm{LJ}}$ while keeping fixed the particle hard-core diameters, the microion charges q_i, the monomer number $N=32$ and the maximal tether length $b=1.5\sigma$ and in general also the salt concentration $c_s = 0$ and the surfactant length $n = 5$. In our previous work[16] we also investigated the influence of surfactant tail variation as well as of finite salt concentration. A brief discussion of the corresponding results is also included in the following.

III. Simulation technique

We simulate by standard Monte Carlo (MC) methods a system of a single polyelectrolyte chain of N charged beads at finite concentration with polyelectrolyte charge density $\rho = N/L^3 = 0.001\sigma^{-3}$ which is kept fixed throughout all simulations. The finite concentration is taken into account by periodic boundary conditions of our finite cubic simulation box of length L and replicated images of the charges. Summations over the interactions with the periodic images are calculated by the Lekner sum method.[66] The Metropolis rates of the MC moves are determined by the interactions given by eqn. (1).

The relaxation of the system, especially in the fully assembled bottle brush configurations is very slow. This is also known from simulations of conventional bottle brush molecules.[52] To ensure sufficient relaxation we performed $\sim 5 \times 10^6$ attempted MC moves per particle at each state point. The acceptance ratio is roughly 0.8 in situations with an open structure. In the micellar cases however, caging occurs inside the densely packed core reducing the acceptance rate down to 0.3.

In all simulations the global charge of our system vanishes. We performed simulations for a system of a single polyelectrolyte chain together with $N = 32$ oppositely charged surfactant molecules. The case $n = 1$ of no tail beads at all serves as a natural reference for pure counter ions. Such a situation was already studied in ref. 27 and implemented in the frames of the present setup in ref. 16 for different Bjerrum lengths. Note, however, that chain connectivity was modeled differently in ref. 27 *via* a finite extension potential.

IV. Results

A. Zero hydrophobicity of the surfactant

In the case of vanishing hydrophobic interactions $\varepsilon_{\mathrm{LJ}} = 0$ the surfactant tails interact only due to their steric repulsion. This special case was treated by the present authors in ref. 16 in detail, so we only recall some of the basic results. We first discuss the case of no added salt. As a reference situation serves the surfactant without tail, *i.e.* $n = 1$. The averaged square end-to-end distance R^2 of the polyelectrolyte chain is defined *via*

$$R^2 = \langle(\vec{R}_1 - \vec{R}_N)^2\rangle \tag{3}$$

where $\langle\ldots\rangle$ denotes a statistical average and $\vec{R}_1$ and $\vec{R}_N$ are the actual positions of the two end-monomers of the polyelectrolyte.

R^2 is plotted as a function of Coulomb strength λ_{B} in Fig. 3. Here, the collapse scenario found in other simulations[27–29] and analytical treatments[67] is confirmed. This scenario for simple counterions is the following: At zero Bjerrum length $\lambda_{\mathrm{B}} = 0$ the interaction is purely excluded volume and the chain attains a polymer coil configuration. Increasing λ_{B} the chain stretches due to the repulsion of its charged beads. For low λ_{B} the counterion screening is weak and stretching continues until $\sigma\lambda_{\mathrm{B}}$ is of the order of the counterion diameter, *i.e.* $\lambda_{\mathrm{B}} \simeq 1$. For larger $\lambda_{\mathrm{B}} > 1$ the condensation of counterions induces a shrinking of the PE extension which for high λ_{B} may even become smaller than the neutral coil; an effect not yet seen in the range for λ_{B} that we simulated.

Substituting in the former system the counterions by charged surfactants with neutral tails the behavior as a function of the Coulomb strength λ_{B} changes in the following way: The PE chain stretches for increasing λ_{B} with the maximal extension attained for higher values of λ_{B} corresponding to the the larger effective diameter of the charged surfactant acting as a counterion. When counterion condensation occurs at higher λ_{B} the surfactant and the PE chain form a complex in which the surfactants aggregate with their heads near the PE chain and their tails pointing away from the PE chain in a manner resembling the structure of a molecular bottle brush. The subsequent collapse of the PE chain that is observed in the previous scenario now is effectively blocked due to the excluded volume interaction of the surfactant tails. The internal structure of this complex is discussed below. For the special case of vanishing hydrophobicity the influence of added salt on the

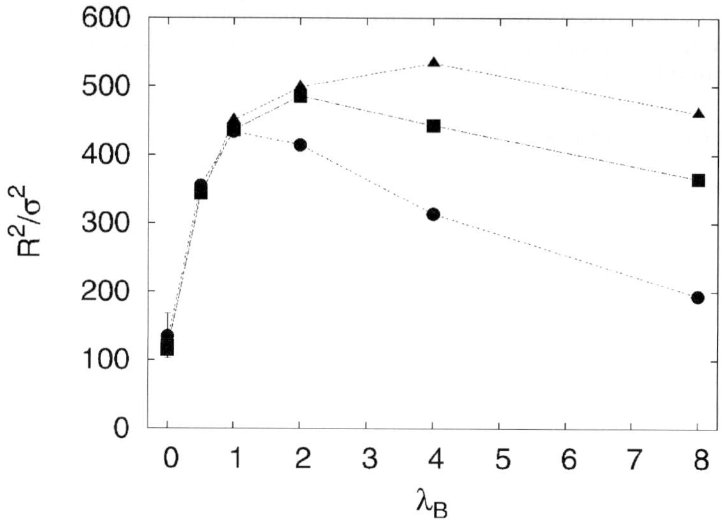

Fig. 3 The mean square end-to-end distance R^2 as a function of the Bjerrum length λ_B for $n = 1$(●), 5(■) and 10(▲) (from below) monomer surfactants for vanishing hydrophobicity parameter $\varepsilon_{LJ} = 0$. The statistical uncertainty is indicated by an error bar.

end-to-end distance R was investigated in ref. 16. Adding N_s additional 1 : 1 salt pairs to our system we define the relative concentration as $c_s = N_s/N$ where N is the number of charges on the polyelectrolyte. The results are shown in Fig. 4. The surfactant molecules in the complex are replaced in favor of salt ions. This is driven by entropy of mixing and corresponds to a chemical equilibrium. The replacement is therefore increasing with increasing salt concentration. This destroys the structure of the complex as seen from the reduction in the end-to-end distance of the polyelectrolyte.

In all our simulations we observe that the contour length L of the polyelectrolyte given by the sum of the bond lengths is very stable under all changes of the parameters. Varying the Bjerrum length in

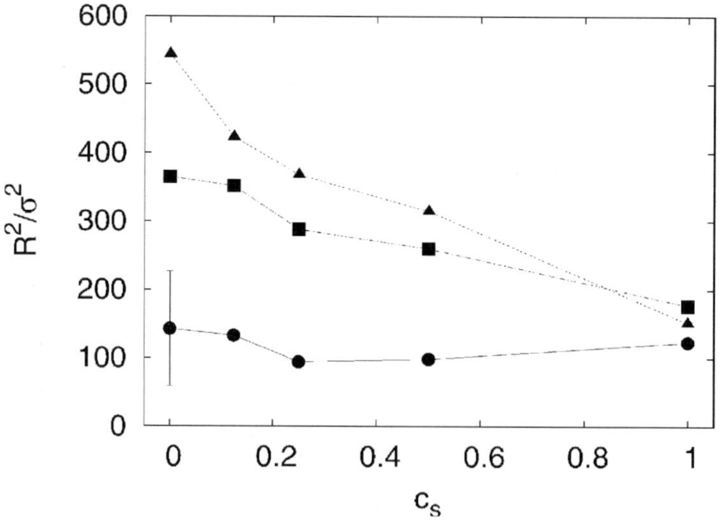

Fig. 4 The mean square end-to-end distance R^2 for Bjerrum length $\lambda_B = 8\sigma$ as a function of the relative salt concentration c_s for (from below) $n = 1$(●), 5(■) and 10(▲) monomer surfactants. The statistical uncertainty is indicated by an error bar.

the range $\lambda_B = 0 \ldots 8$ we find a corresponding increase from $L = 40$ to $L = 42$, a change that is within the statistical uncertainty. For this reason also the persistence length L_p does not give additional information. For a worm-like chain namely it is related to the end–end distance R and the contour length L by[68]

$$R^2 = 2LL_p - 2L_p^2(1 - \exp(-L/L_p)) \tag{4}$$

a relation also confirmed by simulations.[27]

B. Clusters and micelles

The results described in this and the following sections all refer to a system with a single polyelectrolyte chain of 32 charged beads and 32 oppositely charged 5-bead surfactant molecules with one charged head bead and four neutral possibly hydrophobic tail beads. For finite values of the Coulombic and hydrophobic interactions we observe in general one or more aggregates or clusters of molecules. To analyze these structures in detail we define the clusters by assuming that any two beads belong to the same cluster if their distance is less than 2σ. Among these clusters we identify the complex as the one that contains the polyelectrolyte chain. We monitor the number of clusters and the sizes of the complex and of the largest aggregates that are detached from the polyelectrolyte. These latter aggregates we identify with isolated micelles. The number of identified clusters is shown in Fig. 5 as a function of the Coulomb interaction parameter λ_B for different values of the hydrophobicity strength ε_{LJ}: For vanishing interactions clustering occurs only by chance and almost each cluster observed represents one of the 33 single molecules in the system. For increasing interactions the molecules condense to a few larger clusters with only one or two remaining in the strong interaction case. When the Coulomb interaction vanishes, *i.e.* $\lambda_B = 0$, the polyelectrolyte will always be detached from the micelles and thus a minimal number of two clusters remain even for strong hydrophobic attraction. For any finite Coulomb interaction and high hydrophobicity on the other hand all molecules condense to a single complex in the present setup. The sizes of the complex and of the largest detached micelle are shown in Fig. 6 where we plot the masses of these aggregates as a function of the Coulomb parameter λ_B for different values of the hydrophobicity ε_{LJ}. It is clearly seen from these plots that a large detached micelle appears only for small Coulombic strength $\lambda_B < 1$ and high hydrophobic attraction $\varepsilon_{LJ} > 5$. For these high $\varepsilon_{LJ} > 4$ the largest detached micelle shrinks rapidly with increasing λ_B while at the same time the mass of the complex grows such that the surfactant is nearly quantitatively condensed in the complex for $\lambda_B > 1$. For smaller hydrophobicity $\varepsilon_{LJ} < 4$ on the other hand no larger detached micelles appear at all while the

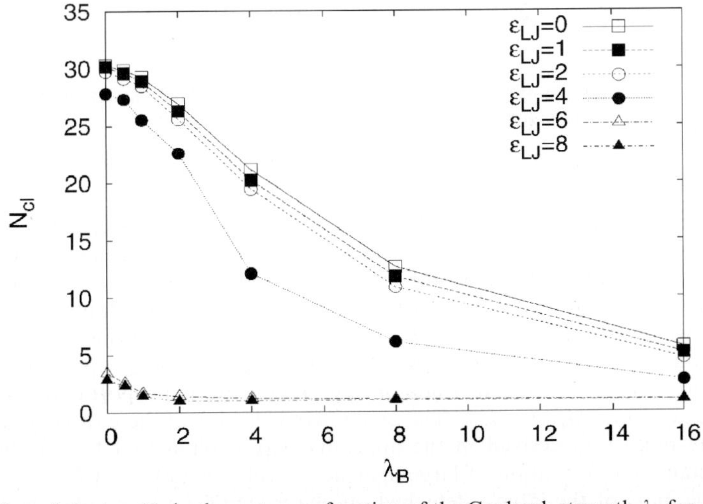

Fig. 5 The number of clusters N_{cl} in the system as function of the Coulomb strength λ_B for different values of the hydrophobicity parameter $\varepsilon_{LJ} = 0(\square)$, $1(\blacksquare)$, $2(\bigcirc)$, $4(\bullet)$, $6(\triangle)$, $8(\blacktriangle)$ from above.

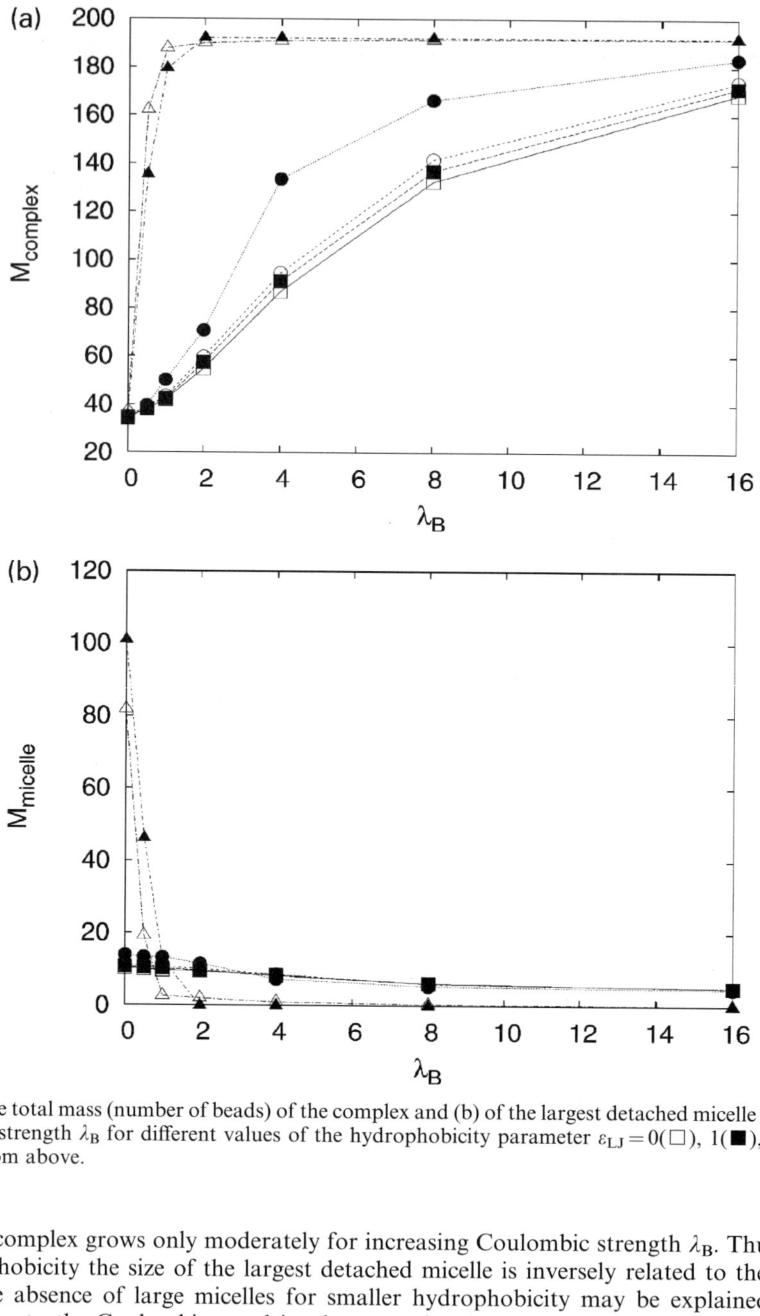

Fig. 6 (a) The total mass (number of beads) of the complex and (b) of the largest detached micelle as function of the Coulomb strength λ_B for different values of the hydrophobicity parameter $\varepsilon_{LJ} = 0(\square)$, $1(\blacksquare)$, $2(\bigcirc)$, $4(\bullet)$, $6(\triangle)$, $8(\blacktriangle)$ from above.

mass of the complex grows only moderately for increasing Coulombic strength λ_B. Thus, while for high hydrophobicity the size of the largest detached micelle is inversely related to the size of the complex, the absence of large micelles for smaller hydrophobicity may be explained with their instability due to the Coulombic repulsion between the charged surfactants in combination with entropy. As shown in the appendix the total binding energy of the surfactant tails in the micelle grows proportionally to its mass M_{mic} with a surface correction proportional to $M_{mic}^{2/3}$. The electrostatic energy of the surfactant heads on the other hand is proportional to $M_{mic}^{5/3}$ while the entropy term behaves like $M_{mic} ln\, M_{mic}$. The qualitative behavior of these three terms together with the approximate prefactors derived in the appendix may explain that for the finite number of available surfactants the formation of large detached micelles is unfavorable for $\varepsilon_{LJ} < 6$ and even for $\varepsilon_{LJ} = 6$, 8 the mass of the largest detached micelle is significantly smaller than the total mass $M_{mic}^{max} = 160$ of available surfactant.

C. Shapes of the complex

In the case of complex formation we characterize the shape of the complex in terms of its matrix of inertia

$$\Theta_{\alpha\beta} = \frac{1}{M}\sum_{j\in C}\left(\mathbf{r}_j^2\delta_{\alpha\beta} - r_j^{(\alpha)}r_j^{(\beta)}\right) \equiv \overline{r^2\delta_{\alpha\beta} - r^{(\alpha)}r^{(\beta)}} \tag{5}$$

where the sum is over all M beads j that are part of the complex and $r_j = (r_j^{(1)}, r_j^{(2)}, r_j^{(3)})$ is the position of bead j with respect to the center of mass of the complex. Diagonalizing $\Theta_{\alpha\beta}$ results in the three eigenvalues $\vartheta_x \geq \vartheta_y \geq \vartheta_z$ with respective axes x, y, z of the principal moments of inertia with the center of mass as the origin. Note that in these coordinates one has *e.g.*

$$\vartheta_z = \overline{x^2 + y^2} \tag{6}$$

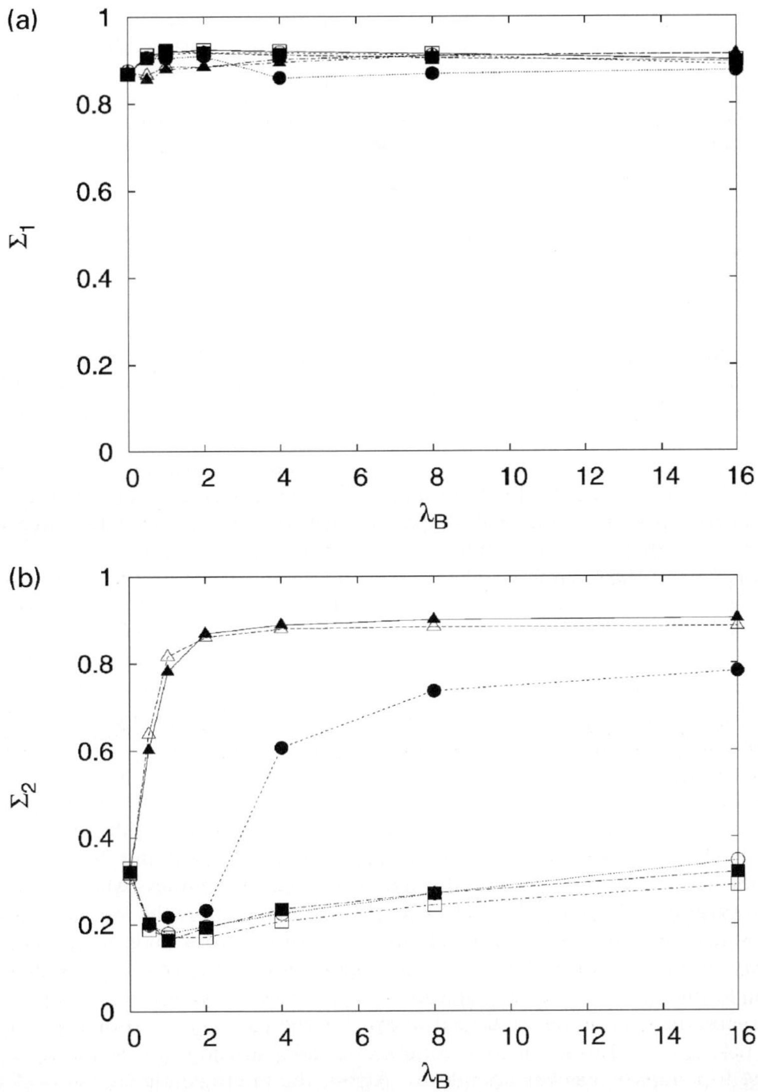

Fig. 7 (a) The ratio $\Sigma_1 = \vartheta_y/\vartheta_x$ of the second largest and largest principal moments of inertia and (b) the ratio $\Sigma_2 = \vartheta_z/\vartheta_x$ of the smallest and the largest principal moments of inertia as function of the Coulomb strength λ_B for different values of the hydrophobicity parameter $\varepsilon_{LJ} = 0(\square)$, $1(\blacksquare)$, $2(\bigcirc)$, $4(\bullet)$, $6(\triangle)$, $8(\blacktriangle)$ from below.

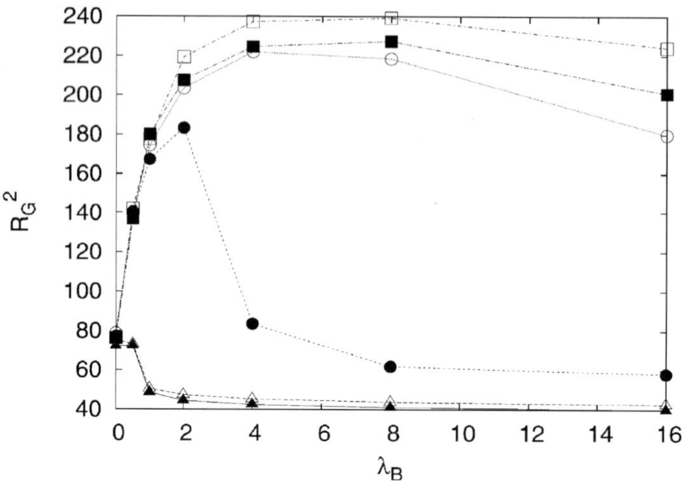

Fig. 8 The radius of gyration of the complex as a function of the Coulomb strength λ_B for different values of the hydrophobicity parameter $\varepsilon_{LJ} = 0(\square)$, $1(\blacksquare)$, $2(\bigcirc)$, $4(\bullet)$, $6(\triangle)$, $8(\blacktriangle)$ from above.

such that the radius R_G of gyration of the complex is given by

$$R_G^2 = (\vartheta_x + \vartheta_y + \vartheta_z)/2 = \tfrac{1}{2}\mathrm{Tr}\Theta$$

The relative values of the principal moments determine the global shape of the complex. The averaged ratios $\Sigma_1 = \langle\vartheta_y/\vartheta_x\rangle$ and $\Sigma_2 = \langle\vartheta_z/\vartheta_x\rangle$ are plotted as functions of the Coulomb strength λ_B for different hydrophobicities ε_{LJ} in Fig. 7. These serve as indicators for the transition of the complex from a cylindrical shape to an almost spherical one. The cylindrical shape is characterized by two large approximately equal inertia, *i.e.* $\Sigma_1 \approx 1$ and a third smaller moment such that $\Sigma_2 \approx 0.2\ldots0.3$. For an elongated cylindrical shape the z-axis corresponding to the smallest moment ϑ_z is the symmetry axis. The spherical shapes in turn are characterized by three almost equal moments of inertia with $\Sigma_2 \approx \Sigma_1 \approx 1$. We observe the following—see Fig. 7: At $\lambda_B = 0$, $\varepsilon_{LJ} = 0$ only the PE chain is part of the 'complex'. The corresponding values we find for the neutral chain are $\Sigma_1 = \langle\vartheta_y/\vartheta_x\rangle = 0.87$ and $\Sigma_2 = \langle\vartheta_z/\vartheta_x\rangle = 0.34$. These may be compared with $\langle\vartheta_y\rangle/\langle\vartheta_x\rangle = 0.87$ and $\langle\vartheta_z\rangle/\langle\vartheta_x\rangle = 0.27$ calculated from data found for self avoiding walks.[69] Note that the relation $\langle\vartheta_z/\vartheta_x\rangle > \langle\vartheta_z\rangle/\langle\vartheta_x\rangle$ is in accordance with the Schwartz inequality for averages. For weak hydrophobicity $\varepsilon_{LJ} < 3$ the shape of the complex remains cylindrical even for high Coulomb strength, while for strong hydrophobicity $\varepsilon_{LJ} > 5$ the transition from cylindrical to spherical shape occurs already for small Coulomb strengths $\lambda_B \approx 1$. For the intermediate hydrophobicity $\lambda_B = 4$ though, we observe a competition between the formation of the cylindrical bottle brush and the spherical micellar complex. This fact is expressed by the non-monotonic behavior of the shape parameter Σ_2 as a function of λ_B. While for small λ_B the cylindrical shape wins and Σ_2 decreases with respect to the value of the neutral situation, a transition occurs between $\lambda_B = 2$ and $\lambda_B = 4$ where the shape parameter indicates a transition towards the spherical micellar complex, with Σ_2 increasing to 0.6 and further towards 0.8 for $\lambda_B > 8$. To verify the transition we analyze the internal structure of these intermediate states in more detail below. A similar division into strong, weak and intermediate hydrophobicity we find for the behavior of the square radius of gyration R_G^2 which is plotted as function of λ_B in Fig. 8. For small hydrophobicity $\varepsilon_{LJ} < 3$ the behavior of R_G^2 parallels that of the square end-to-end distance R^2 as described above for the case $\varepsilon_{LJ} = 0$. For high hydrophobicity $\varepsilon_{LJ} > 5$ the radius of gyration of the complex shrinks already for small $\lambda_B \approx 1$ to values corresponding to compact micellar complexes. Again, the intermediate case $\varepsilon_{LJ} = 4$ behaves in a special way: For small Coulomb strengths $\lambda_B < 3$ the radius grows due to the stretching of the polyelectrolyte chain and only for higher $\lambda_B > 6$ it again shrinks below the polymer coil value but still remains well separated from the high hydrophobicity case.

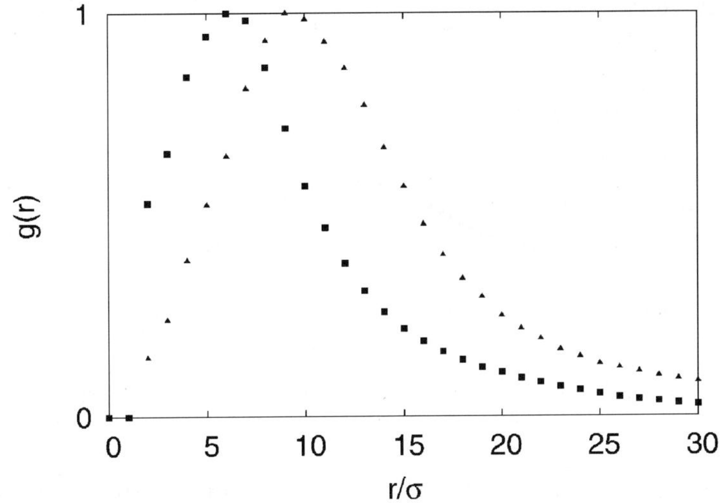

Fig. 9 The pair correlation function $g(r)$ for pairs of a polyelectrolyte monomer and a surfactant tail monomer at Bjerrum length $\lambda_B = 8\sigma$. The functions are scaled to a maximum value of 1. The maximal number of tail ends is found at $r = 6\sigma$ from the polyelectrolyte backbone in the case of $n = 5$(■) monomer surfactants. For $n = 10$(▲) this most probable distance is $r = 9\sigma$.

D. Internal structure of the complex

While the overall shape of the complex is characterized by the inertia matrix, its internal structure can be explored by either the internal monomer–monomer correlations or by the density distributions of the different complexes in a given reference frame. We have used both approaches to identify the key features of the self-assembled structures. To verify the idea of the molecular bottle brush like structures in the case of low hydrophobicity ε_{LJ} and high Coulomb strength λ_B, we have measured in particular the correlation between the polyelectrolyte monomers and the surfactant tail ends. As shown in Fig. 9 using also data from ref. 16 we indeed find that the surfactant tail end is found in this situation with highest probability at a distance from the polyelectrolyte chain that increases with the length of the tail in accordance with the bottle brush picture. To monitor, on the other hand, the internal reorganization along with the transition from the cylindrical bottle-brush to the spherical micellar aggregate, we chose as a reference frame a coordinate system defined by the directions of the principle moments of inertia of the complex with the origin at its center of mass. We define the x, y, z-directions such that the corresponding moments are ordered by $\vartheta_x \geq \vartheta_y \geq \vartheta_z$. As mentioned above, the z-axis then is the symmetry axis for the averaged cylindrical structures.

In Figs. 10–12 we show as examples for typical conformations of the complex the density distributions measured in this coordinate system. For vanishing hydrophobicity $\varepsilon_{LJ} = 0$ and $\lambda_B = 4$ the corresponding distributions are shown in Fig. 10. The densities as a function of the distance r_{xy} from the z-axis as plotted in Fig. 10(a) show clearly that the polyelectrolyte monomers and the surfactant heads are localized near the symmetry (z)-axis while the density of the surfactant-tail monomers surpasses that of the former two species at larger distances from the central axis. The distributions measured along the z-axis on the other hand show the remarkable feature of maxima near the ends of the cylinder. Surprising as it may seem, this effect is even exhibited by the equilibrium coil of a neutral polymer chain and in particular in our simulation for vanishing interactions.

The spherical micellar complexes, on the other hand, which we observe for sufficiently large parameter values of λ_B and ε_{LJ} are characterized by density distributions along and perpendicular to the z-axis that are nearly identical—see Fig. 11. Only the different differential volumes used for measurement lead to deviations. A more natural variable in the spherical case is of course the distance r from the center of mass. The density $\rho(r)$ as displayed in Fig. 11 parallels the corresponding results of Wallin and Linse obtained by a self-consistent spherical lattice field approach:[17] The core of the micelle is formed by almost densely packed surfactant tails surrounded

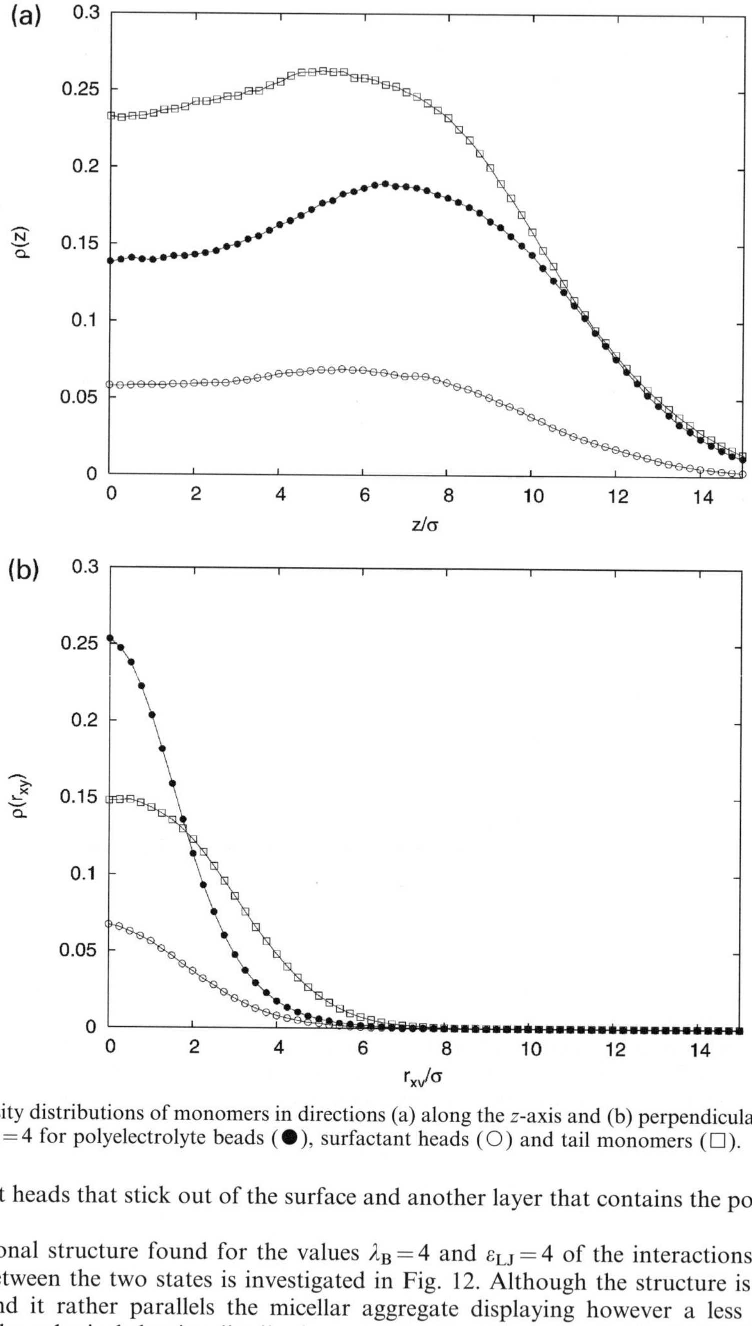

Fig. 10 Density distributions of monomers in directions (a) along the *z*-axis and (b) perpendicular to the *z*-axis for $\lambda_B = 0$, $\varepsilon_{LJ} = 4$ for polyelectrolyte beads (●), surfactant heads (○) and tail monomers (□).

by surfactant heads that stick out of the surface and another layer that contains the polyelectrolyte chain.

A transitional structure found for the values $\lambda_B = 4$ and $\varepsilon_{LJ} = 4$ of the interactions near to the boundary between the two states is investigated in Fig. 12. Although the structure is much more loosely bound it rather parallels the micellar aggregate displaying however a less pronounced layering in the spherical density distribution, see Fig. 12. The comparison of Fig. 12(a) and (b) however shows that the distributions display deviations from spherical symmetry. Nonetheless, the present data indicates that the transition between the two states of the complex is rather sharp.

V. Conclusions

In extensive MC simulations we have investigated the behavior of a single polyelectrolyte chain that interacts with oppositely charged surfactant molecules taking into account both long range

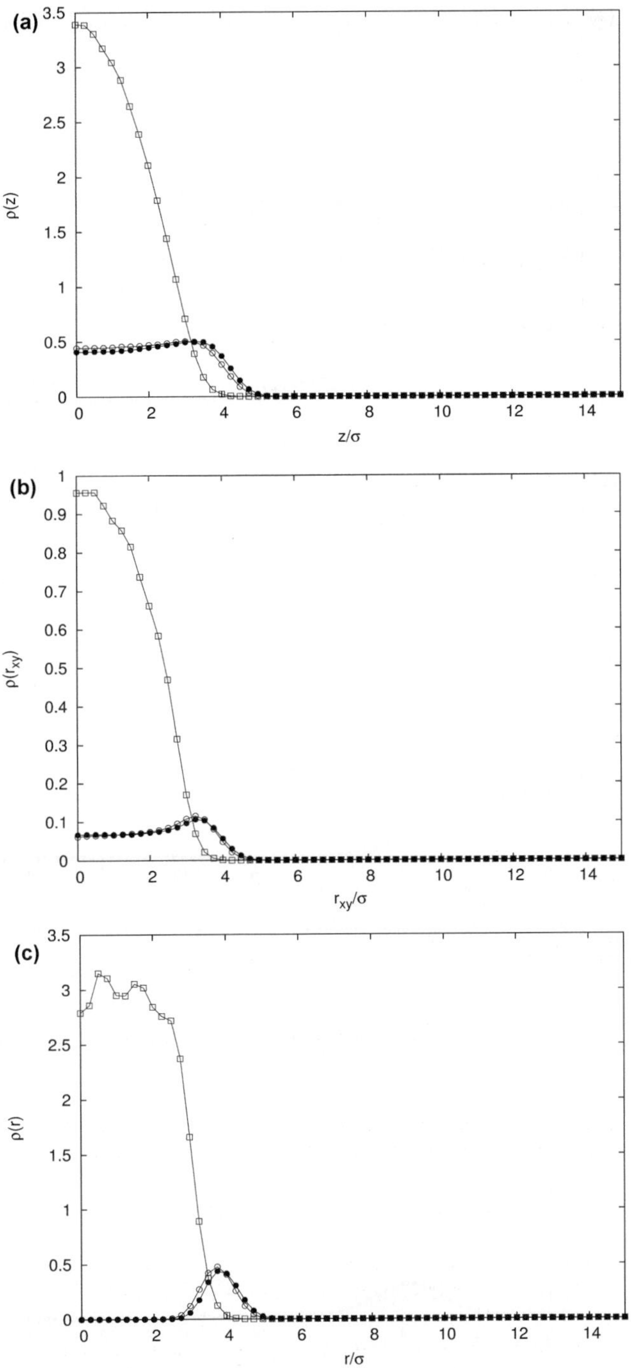

Fig. 11 Density distributions of monomers in directions (a) along the *z*-axis, (b) perpendicular to the *z*-axis, and (c) as function of the distance from the center of mass for $\lambda_B = 16$, $\varepsilon_{LJ} = 8$ for polyelectrolyte beads (●), surfactant heads (○) and tail monomers (□).

Coulomb and short range hydrophobic interactions. Exploring the parameter space in both of these interactions we have established a "phase diagram" for this system and several distinct regions ("phases") with characteristic structural behavior of the system as sketched in Fig. 1: for weak

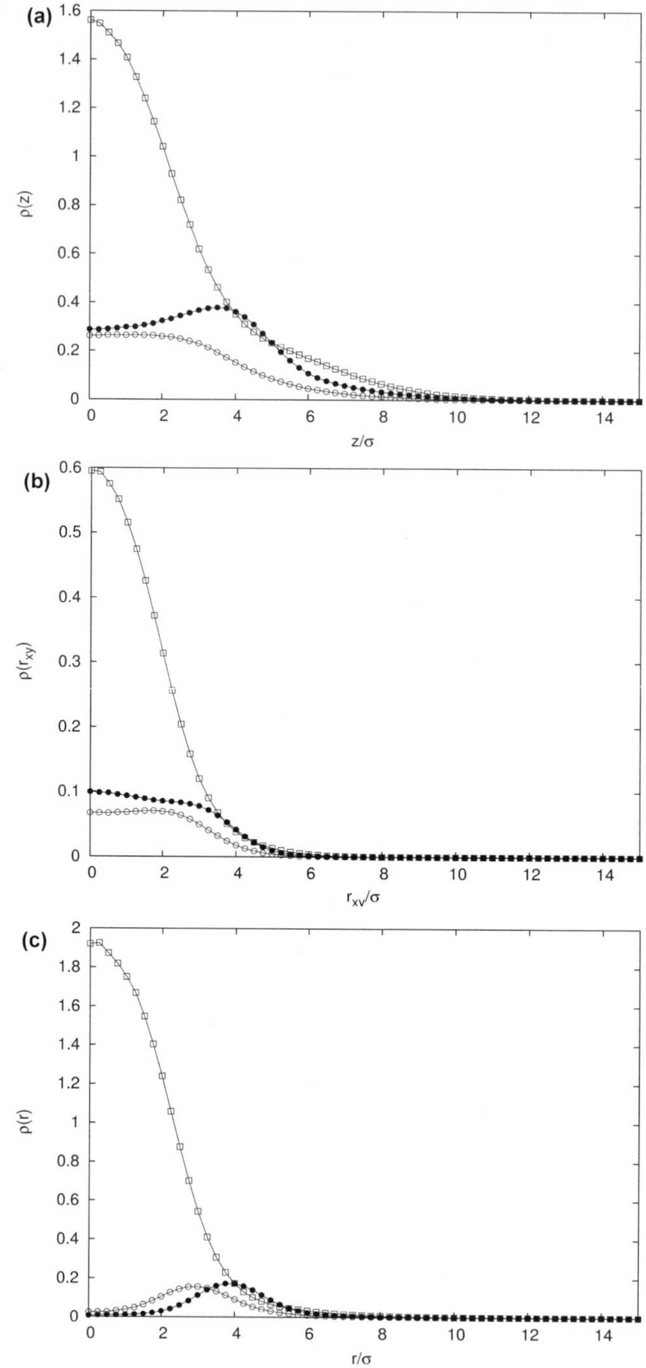

Fig. 12 Density distributions of monomers in directions (a) along the z-axis, (b) perpendicular to the z-axis, and (c) as function of the distance from the center of mass for $\lambda_B = 4$, $\varepsilon_{LJ} = 4$ for polyelectrolyte beads (●), surfactant heads (○) and tail monomers (□).

interactions both constituents behave nearly independently and the PE chain has a possibly stretched coil conformation. For high Coulomb strength but weak hydrophobicity we observe the formation of a complex with a cylindrical bottle brush structure in which the PE chain is stretched along the cylinder axis. If both interactions are strong, an inverted spherical micellar complex structure emerges with the SF tails constituting the core of a spherical micelle and the SF heads and the PE chain confined to the surface of this sphere. For intermediate values of the hydrophobicity there is a competition between the two structurally different complexation modes resulting in a non-trivial behavior of the shape as function of the Coulomb strength. For high hydrophobicity, but low Coulomb interaction the surfactant forms micelles to which the PE chain is only loosely attached. As we have reported in ref. 16 complexation in this system is in general weakened by the addition of salt. Our results can in principle be verified in experiments by systematically varying the surfactant tail hydrophobicity and the surfactant/polyelectrolyte charge and/or the dielectric permittivity of the solvent.

A number of extensions of our present approach are possible and would be interesting to follow. Of particular interest would be a study of a multi PE-chain system. For small numbers of chains a corresponding simulation appears to be feasible using our present approach. One might expect to observe in such a case a competition between the cooperative assembly of a multichain complex and single chain complexation. At high densities in systems with many chains one expects the formation of layered structures with many different possible phases.[70] However, high density and a large number of chains are out of reach of the present approach, due to the requirement of calculating the long range interaction between any pair of charged beads and the slowing down of the simulation by the caging effect at high densities. To simulate such systems either one has to release the hard sphere non-penetrability[71] or resort to lattice methods and simulate only with a screened shorter range Coulomb interaction.

Other situations that can lead to significant changes in the behavior might be found for a choice of different valency of the SF heads and the PE chains to study over-charging effects of the polyelectrolyte for high valency salt ions.[72] Also, in general, introducing neutral beads to the PE chain and varying the distribution of charges along the chain may lead to significant effects. In addition, the chain stiffness has been shown to be an important parameter. For a single micelle modeled as a charged sphere, many recent studies have revealed different topologies of the defect depending on the chain stiffness.[26,32–46]

Appendix

A crude argument for the stability of micelles built from our model surfactants can be derived from the following considerations: We assume that the core of a spherical micelle is given by m_t tail monomers at a packing fraction η such that the core radius r_C is given by

$$r_C(m_t) = \frac{\sigma}{2}\left(\frac{m_t}{\eta}\right)^{1/3}. \tag{7}$$

The number of beads in the surface layer is then given by

$$m_\sigma(m_t) = \left[r_C^3(m_t) - \left(r_C(m_t) - \frac{\sigma}{2}\right)^3\right]\eta. \tag{8}$$

Assuming that each bead inside the core has 12 neighbors at an optimal distance (*i.e.* at the minimum of the potential V_{LJ}), while the surface beads have only 6 neighbors we find the total Lennard-Jones energy of the micelle as

$$\frac{1}{k_B T} E_{LJ}(m_t) = -(m_t - m_\sigma(m_t)/2)\frac{3}{2}\varepsilon_{LJ}. \tag{9}$$

The total Coulomb energy of the SF heads distributed on the micellar surface may be derived from the numerical solution of the Thomson problem for distributing k charges on a spherical surface which is given by the fit formula[73]

$$\frac{1}{k_B T} E_C(k, m_t) = \frac{\lambda_B}{r_C(m_t)}\frac{k^2}{2}\left(1 - ak^{-1/2} + bk^{-3/2}\right), \tag{10}$$

with $a = 1.10461$ and $b = 0.137$. Finally, we may estimate the relevant entropy term by assuming that for every molecule it is given by restricting it to the volume of the micelle. This volume entropy term is thus

$$\frac{1}{k_{\mathrm{B}}} S(k, m_{\mathrm{t}}) = k \ln\left(r(m_{\mathrm{t}})^3 \big/ \Omega \right) \tag{11}$$

where Ω is the total system volume. For our system of 32 surfactant molecules with one charged bead and four tail beads we have $m_{\mathrm{t}} = 4k$ and a minimum of the free energy $F(k, m_{\mathrm{t}})$ with

$$\frac{1}{k_{\mathrm{B}} T} F(k, m_{\mathrm{t}}) = E_{\mathrm{C}}(k, m_{\mathrm{t}}) + E_{\mathrm{LJ}}(m_{\mathrm{t}}) - S(k, m_{\mathrm{t}}) \tag{12}$$

at $n \neq 0$ existing for a given Coulomb interaction only for sufficiently large hydrophobicity. Note, however that our estimate is too crude to give more than an order of magnitude result showing that $\varepsilon_{\mathrm{LJ}}$ should be at least of the order of λ_{B} and of the order of $\varepsilon_{\mathrm{LJ}} \sim 1$ for $\lambda_{\mathrm{B}} = 0$ to find micellar aggregation for our case of surfactant length $n = 5$, *i.e.* $m_{\mathrm{t}} = 4k$ and the given overall number of surfactant molecules $k \leq 32$.

Acknowledgements

We thank A. F. Thünemann, O. Farago, H. Diamant and R. Messina for helpful discussions. This work was supported by the Deutsche Forschungsgemeinschaft through grant No. LO 418/7.

References

1 J. C. T. Kwak, *Polymer-Surfactant Systems*, Marcel Dekker, New York, 1998.
2 M. Antonietti, C. Burger and A. Thünemann, *Trends Polym. Sci.*, 1997, **5**, 262.
3 P. Hansson and B. Lindman, *Curr. Opin. Colloid Interface Sci.*, 1996, **1**, 604.
4 E. Sokolov, F. Yeh, A. Khokhlov, V. Y. Grinberg and B. Chu, *J. Phys. Chem. B*, 1998, **102**, 7091.
5 S. Kosmella, J. Kotz, K. Shirahama and J. Liu, *J. Phys. Chem. B*, 1998, **102**, 6459.
6 P. M. Claesson, M. L. Fielden, A. Dedinaite, W. Brown and J. Fundin, *J. Phys. Chem. B*, 1998, **102**, 1270.
7 M. Tsianou and P. Alexandridis, *Langmuir*, 1999, **15**, 8105.
8 A. F. Thünemann, S. Kubowicz and U. Pietsch, *Langmuir*, 2000, **16**, 8562.
9 R. Dias, S. Mel'nikov, B. Lindman and M. G. Miguel, *Langmuir*, 2000, **16**, 9577.
10 V. G. Babak, E. A. Merkovich, J. Desbrieres and M. Rinaudo, *Polym. Bull.*, 2000, **45**, 77.
11 X. M. Liao and D. A. Higgins, *Langmuir*, 2001, **17**, 6051.
12 Y. Guan, Y. P. Cao, Y. X. Peng, J. Xu and A. S. C. Chen, *Chem. Commun.*, 2001, **17**, 1694.
13 D. J. F. Taylor and R. K. Thomas, *Langmuir*, 2002, **18**, 4748.
14 P. Hansson, S. Schneider and B. Lindman, *J. Chem. Phys. B*, 2002, **106**, 9777.
15 S. Guillot, D. M. Loughlin, N. Jain, M. Delsanti and D. Langevin, *J. Phys.: Condensed Matter*, 2003 **15**, S219.
16 C. von Ferber and H. Löwen, *J. Chem. Phys.*, 2003, **118**, 10 774.
17 T. Wallin and P. Linse, *Langmuir*, 1998, **14**, 2940.
18 K. Shirahama, H. Yuasa and S. Sugimoto, *Bull. Chem. Soc. Jpn.*, 1981, **54**, 375.
19 K. Shirahama and M. Tashiro, *Bull. Chem. Soc. Jpn.*, 1984, **57**, 377.
20 R. Sear, *J. Phys.: Condensed Matter*, 1998, **10**, 1677.
21 P. S. Kuhn, Y. Levin and M. C. Barbosa, *Chem. Phys. Lett.*, 1998, **298**, 51.
22 P. S. Kuhn, M. C. Barbosa and Y. Levin, *Physica A*, 2000, **283**, 113.
23 M. B. A. Silva, P. S. Kuhn and L. S. Lucena, *Physica A*, 2001, **296**, 31.
24 H. Diamant and D. Andelman, *Phys. Rev. E*, 2000, **61**, 6740.
25 H. Diamant and D. Andelman, *Macromolecules*, 2000, **33**, 8050.
26 T. Wallin and P. Linse, *Langmuir*, 1996, **12**, 305.
27 M. Stevens and K. Kremer, *J. Chem. Phys.*, 1995, **103**, 1669.
28 R. G. Winkler, M. Gold and P. Reineker, *Phys. Rev. Lett.*, 1998, **80**, 3731.
29 U. Micka, C. Holm and K. Kremer, *Langmuir*, 1999, **15**, 4033.
30 N. V. Brilliantov, D.V. Kuznetsov and R. Klein, *Phys. Rev. Lett.*, 1998, **81**, 1433.
31 P. S. Kuhn, *Physica A*, 2002, **311**, 50.
32 E. Gurovitch and P. Sens, *Phys. Rev. Lett.*, 1999, **82**, 339.
33 E. Mateescu, C. Jeppesen and P. Pincus, *Europhys. Lett.*, 1999, **46**, 493.
34 R. R. Netz and J.-F. Joanny, *Macromolecules*, 1999, **32**, 9026.
35 S. Y. Park, R. F. Bruinsma and W. M. Gelbart, *Europhys. Lett.*, 1999, **46**, 454.
36 K. K. Kunze and R. R. Netz, *Phys. Rev. Lett.*, 2000, **85**, 4389.
37 T. T. Nguyen and B. I. Shklovskii, *Physica A*, 2000, **293**, 324.

38 P. Welch and M. Muthukumar, *Macromolecules*, 2000, **33**, 6159.
39 H. Schiessel, R. F. Bruinsma and W. M. Gelbart, *J. Chem. Phys.*, 2001, **115**, 7245.
40 M. Jonsson and P. Linse, *J. Chem. Phys.*, 2001, **115**, 3406.
41 M. Jonsson and P. Linse, *J. Chem. Phys.*, 2001, **115**, 10 975.
42 A. Akinchina and P. Linse, *Macromolecules*, 2002, **35**, 5183.
43 P. Chodanowski and S. Stoll, *J. Chem. Phys.*, 2001, **115**, 4951.
44 P. Chodanowski and S. Stoll, *Macromolecules*, 2001, **34**, 2320.
45 M. Brynda, P. Chodanowski and S. Stoll, *Colloid Polym. Sci.*, 2002, **280**, 789.
46 K. Keren, Y. Soen, G. B. Yoseph, R. Gilad, E. Braun, U. Silvan and Y. Talmon, *Phys. Rev. Lett.*, 2002 **89**, 88 103.
47 G. ten Brinke and O. Ikkala, *Trends Polym. Sci.*, 1997, **5**, 213.
48 J. Ruokolainen, G. ten Brinke, O. Ikkala, M. Torkkeli and R. Serimaa, *Macromolecules*, 1996, **29**, 3409.
49 T. M. Birshtein, O. V. Borisov, Y. B. Zhulina, A. R. Khokhlov and T. A. Yurasova, *Polym. Sci. USSR*, 1987, **29**, 1293.
50 G. H. Fredrickson, *Macromolecules*, 1993, **26**, 2825.
51 Y. Rouault and O. V. Borisov, *Macromolecules*, 1996, **29**, 2605.
52 P. G. Khalatur, D. G. Shirvanyanz, N. Y. Starovoitova and A. R. Khokhlov, *Macromol. Theory Simul.*, 2000, **9**, 141.
53 J.-P. Hansen and H. Löwen, *Annu. Rev. Phys. Chem.*, 2000, **51**, 209.
54 R. D. Groot, *Langmuir*, 2000, **16**, 7493.
55 J. Rescic and P. Linse, *J. Phys. Chem. B*, 2000, **104**, 7852.
56 T. Wallin and P. Linse, *J. Phys. Chem. B*, 1997, **101**, 5506.
57 R. G. Winkler, M. O. Steinhauser and P. Reineker, *Phys. Rev. E*, 2002, **66**, 21 802.
58 A. Jusufi, C. N. Likos and H. Löwen, *Phys. Rev. Lett.*, 2002, **88**, 18 301.
59 R. Messina, C. Holm and K. Kremer, *Phys. Rev. E*, 2002, **65**, 41 805.
60 T. Odijk, *Macromolecules*, 1979, **12**, 688.
61 M. Fixman and J. Skolnik, *Macromolecules*, 1978, **11**, 863.
62 P. G. deGennes, P. Pincus and R. Velasco, *J. Phys. (Paris)*, 1976, **37**, 1461.
63 J.-L. Barrat and J.-F. Joanny, *Europhys. Lett.*, 1993, **3**, 343.
64 R. Connolly, E. G. Timoshenko and Y. A. Kuznetsov, *J. Chem. Phys.*, 2003, **119**, 8736.
65 E. G. Timoshenko and Y. A. Kuznetsov, *J. Chem. Phys.*, 2000, **112**, 8163.
66 J. Lekner, *Physica A*, 1991, **176**, 485.
67 H. Schiessel and P. Pincus, *Macromolecules*, 1998, **31**, 7953.
68 M. Doi and S. F. Edwards, *The Theory of Polymer Dynamics*, Oxford University Press, Oxford, 1986.
69 G. Zifferer, *J. Chem. Phys.*, 1998, **109**, 3691.
70 A. Thünemann, *Langmuir*, 1997, **13**, 6040.
71 R. D. Groot, *J. Chem. Phys.*, 2003, **118**, 11 265.
72 R. Messina, C. Holm and K. Kremer, *J. Chem. Phys.*, 2002, **117**, 2947.
73 J. R. Morris, D. M. Deaven and K. M. Ho, *Phys. Rev. B*, 1996, **53**, R1740.

General Discussion

Professor Ryan opened the discussion of Professor Kabanov's paper: Is there a critical mesh size (cross link density) below which a given dendrimer will not enter the gel? Does the molecular weight between crosslink have a predictable effect on the diffusion and condensation of dendrimers in oppositely charged gel?

Professor Kabanov responded: We didn't study highly cross-linked hydrogels with respect to sorption of dendrimers. However, I believe that such critical cross-link density should exist. In fact it does exist for hydrogels with respect to oppositely charged protein uptake. The effect of molecular weight between cross-links should be studied experimentally. The quantitative predictions cannot be made at the moment.

Dr Geoghegan asked: The hydrogels that you describe are very heterogeneous; there will be parts within the gels which are highly crosslinked and others that are only loosely connected to the rest of the network, indeed there may even by microgels contained within the networks. With this in mind, it may even be unhelpful to classify the network by means of some average mesh size. The distribution of charge around the network will be affected by this heterogeneity. As well as this, differences in rate coefficients for the crosslinking of the #PA-Na and #PAMPS-Na hydrogels will probably mean that the gels have very different structures. Could you therefore comment on the consequences of the nature of your networks for your sorption measurements?

Professor Kabanov replied: In the Experimental part it is mentioned that water content measured in arbitrary chosen volume elements of the hydrogel samples was found equal. It indicates that the macroscopic samples we used were macroscopically homogeneous. I don't think that microscopic inhomogenity in cross-links distribution, which may exist, would markedly change the results of the measurements performed. Moreover, such inhomogeneity is hardly significant because of close monomer–cross linker copolymerization constants.

Professor Ross-Murphy said: You mentioned that the charge on your protonated poly(propylene imine) dendrimers can be adjusted by pH. Your figures appear to show the kinetics of transport only for fully protonated systems. I am interested to know how this rate is effected for such partially protonated systems? Can this effect be factored into your proposed "relay-race" mechanism equations?

Professor Kabanov responded: This is a good question. Up to now precise kinetic measurements have been performed only for partly protonated (pH 9.0, α *ca.*. 0.5) DAB-dendr-$(NH_2)_{64}$ (G5). In contrast to fully protonated (pH 4.0) G5 species, whose uptake by the studied hydrogels is markedly retarded (see Fig. 3 in the paper) and doesn't meet eqn. (12) (see Fig. 4a), no retardation was observed in the case of partly protonated G5. Moreover, in the latter case sorption kinetics is perfectly described by eqn. (12), *i.e.* in terms of the relay-race mechanism. The apparent explanation is that charged groups of the partly protonated G5 are localized in the periphery of the dendrimer molecule, while the core imine groups are not charged and therefore not available for the charged hydrogel network fragments. Therefore each move of the dendrimer species from one vacancy to another requires less correlated molecular motions of the interacting oppositely charged repeating units and correspondingly proceeds with a higher probability (rate constant). It might be true also for the partly protonated DAB-dendr-$(NH_2)_X$ of lower generations. However, it should be checked in the future.

DOI: 10.1039/b413664k

Dr Perez Mendez asked: Did you entrap negatively charged DNA?

Professor Kabanov responded: Single stranded DNA interpenetrates in the positively charged slightly cross-linked PDADMAC hydrogels. Double helical DNA does not, but is sorbed at the interface in compacted form.[1]

1 V. G. Sergeyev, O. A. Novoskoltseva, O. A. Pyshkina, A. A. Zinchenko, V.B. Rogacheva, A. B. Zezin, K. Yoshikawa and V. A. Kabanov, *J. Am. Chem. Soc.*, 2002, **124**, pp. 11324–11333.

Dr Fahmi asked: How do impurities affect the cross-linked polyelectrolyte gels?

Professor Kabanov replied: I don't know a general answer to this question. It depends on what the impurities are.

Dr Semenov asked: What would be the time dependence of the dendrimer uptake rate in the model situation with infinite plain interface between the gel and the solution?

Professor Kabanov responded: This is a good question for a theoretician, but I am not one. In any case the effect of contraction should be taken into account, which may cause the formation of mechanical cracks in a real experiment. Therefore some relaxation mechanism should be involved in any model considerations.

Professor Gido asked: Is it possible to use this type of measurement of dendrimer mobility in gel as a type of chromatography to shed light on such issues as dendrimer conformation and core accessibility as a function of generation?

Professor Kabanov responded: The dendrimer core accessibility for charged network groups depends on whether it is charged or not. At least we know this for G5. However, at the moment I cannot say any more on this matter.

Professor Hamley opened the discussion of Professor Fraaije's paper: Could you explain how the phase separation process is initiated in your simulations (*i.e.* what is a "quench in a solvent bath") and what is the initial condition? Also, how can the conditions in the simulation be achieved in experiments?

I would also like to note the relationship between lateral phase separation in your simulations and the possibility for this in the cylindrical pore morphology formed by blends of AB and AC diblocks presented by Hillmyer *et al.* at this meeting.

Professor Fraaije answered: The phase separation is initiated by placing a hollow spherical droplet of polymer surfactant in the simulation box, where the droplet is surrounded internally and externally by solvent. The initial wall is homogenous with thickness about twice R_g (see my earlier response to Professor Semenov). At this point ($t = 0$) the interaction parameters are set as given in the text. Phase separation follows spontaneously, while at each time step a small amount of noise is added.

The structure formation observed by Hillmyer *et al.* concerns the pore formation in mixed composites of block copolymers AB + AC, where the core forming hydrophobic block A is the same for both polymers, and the hydrophilic B and C ideally are miscible (and where for example B is degradable). Hillmyer *et al.* use polystyrene for the A block and polylactic acid and polyethylene oxide for the B and C blocks. In our simulations, the hydrophilic block is the same in the two polymers, but the hydrophobic blocks are different and not miscible. Hence the two systems are conceptually very different: the B and C blocks in Hillmyer *et al*'s system are intentionally similar, and in our case they are intentionally dissimilar.

But if in Professor Hillmyer's system the compatibility between B and C would be too low—(polylactic acid is more hydrophobic than polyethylene oxide so care must be taken with chain lengths, (co)solvents, and temperature), the system would be more similar to ours, as pointed out by Professor Hamley. The two polymers would then indeed phase separate, perhaps accompanied by lateral phase separation in early stages, and then demixed assemblies would be formed, as observed in our block copolymer micelles simulations. In earlier papers however[1] it is demonstrated that in a blend pure polymers of poly(L) lactic acid and polyethylene oxide are miscible (in the absence of water), hence we must conclude that such lateral phase separation in Professor Hillmyer's system is unlikely.

1 A. J. Nijenhuis, E. Colhuis, D. W. Grijpma and A. J. Pennings, *Polymer*, 1996, **37**, 5849.

Dr Webber said: The presenter stated that polymersomes of short chain polymers do not from chimaeric structures such as that shown in Fig. 3 of the paper, whilst long chain polymers do. Is the theory developed by the authors able to predict the critical polymer length below which homogeneous structures form, but above which chimaeric structures are present? Additionally, is the theory able to predict structures when the mixed polymers are of different chain length?

Professor Fraaije responded: The paper states that chimaeric polymersomes of *long* block copolymers will be intrinsically unstable; such polymersomes will necessarily eventually (it may take a long time) split into separate and pure polymersomes. The argument is simply that the entropy of (lateral) mixing of long block copolymers, while confined in a plane or wall, is more than compensated eby the unfavourable enthalpy of mixing. This is the same argument as one uses traditionally to explain demixing of polymer blends in 3D. Here, we state that it also holds for oriented block copolymers in a vesicular wall. Similarly as with the traditional approach for blends it is possible to predict at which critical Flory–Huggins parameter the dissimilar block copolymers will demix laterally, for given chain lengths. Such a calculation has not yet been done. Also, such a calculation should be accompanied by estimates of the line tension, spinodal and binodals (for the in-plane phase diagrams) *etc.* The problem is complicated by the soft confinement of the vesicular walls, since the walls can bulge or bend.

The method we used to calculate the 3D structures is applicable to any mixture of (branched) block copolymers, whether of same length or not, provided the polymers are all long and flexible, so that the mean-field model is appropriate.

Professor Semenov commented: But you actually get different structures with the same initial conditions, *i.e.* the aim (to find the relationship between the terminal structure of the system and the molecular plus external parameters) does not seem to have been achieved yet.

Professor Fraaije responded: Yes, we have not achieved it in this paper, and indeed a full report seems more than a life-time's work. Yes, for different block copolymers, using the same quenching procedure, we get different terminal structures. These results—in themselves not very surprising—have been published before for single-component polymersomes.[1] Then we have many simulations (all unpublished) of same block copolymers, with different starting conditions but also different terminal structures! The documentation and classification of the temporal behaviour of polymersome composites is clearly at its infancy. For example, we have observed that when we start with a homogeneously filled droplet of a symmetric diblock polymer, the result is a multi-lamellar 'onion'. But when we start with a droplet with internal void—filled with solvent—and when the void is large enough, the result is a uni-lamellar vesicle. We explain the different outcomes by subtle symmetry-breaking effects and huge free energy barriers associated with breaking up a block copolymer lamella. We have not yet found the conditions by which an initial homogeneous droplet will transform spontaneously into a large unilammellar vesicle. For this reason, we start the simulation with a spherically imposed vesicular structure (a droplet with a large void filled with solvent), with a homogeneous wall thickness commensurate with the block copolymer length (see my earlier response to Professor Hamley). Following the quench (at $t = 0$) the wall then phase separates internally and laterally.

The range of molecular and processing parameters is enormous, and one can easily be led astray into an irrelevant area, and predict the outcome of an experiment which can never be done. But the experimental approaches we have available now are not always suitable for physical-chemical study of the process. For example, it is simply not possible to simulate the sequence of intricate steps in the simple dialysis method (where one first dissolves the block copolymer in a common good solvent, and subsequently performs a dialysis against a selective solvent). It would be most welcome to have an experimental approach such that we have the freedom to study the processing parameters, in the same detail as we now have to study the molecular interactions. The impingent jet method pursued by Prud'Homme from Princeton might offer such an approach. It seems important to have a means of rapidly making a sizable emulsion or droplet spray by a physical (rheological) force, which would then be far out of equilibrium initially, followed by slow internal self-organisation. In this way, one avoids complications due to free energy barrier effects in micellar aggregation, and one introduces effectively a newly controllable processing parameter, namely the initial droplet size.

1 J. G. E. M. Fraaije and G. J. A. Sevink, *Macromolecules*, 2003, **36**(21), 7891–7893.

Professor Gido commented: (1) There has been some experimental work done on block copolymer microphases separating within spherical droplets of nanometer to micron size. Accounts of this work and TEM images of the morphologies obtained can be found in ref. 1 and 2 here. The droplets are produced by spraying a dilute solution of block copolymer from an atomizer into a vertical column. As the droplets settle down the column they dry and self-assemble to produce structures some of which are similar to those shown in Professor Fraaije's simulations. Especially interesting are the raspberry-like morphologies of the TEM images of Thomas and Reffner.

(2) Your simulations indicate a preference for droplets to split up into separate droplets of incompatible components, as in Fig. 2 in your paper. However, this tendency must depend upon the balance between polymer–polymer interactions and the interactions with the surrounding solvent or medium. In the case of droplets forming while suspended in air, as in the Thomas *et al.* experiments, might the surface tension of the droplets be sufficient to favor the incompatible polymers staying together in one droplet?

1 E. L. Thomas, J. R. Reffner and J. Bellare, *Colloq. Phys.*, 1990, **C7**, 363.
2 J. R. Reffner, PhD Thesis, *Polymer Science and Engineering*, University of Massachusetts, 1992.

Professor Fraaije responded: (1) While the indicated papers unfortunately are unavailable to us, similar atomizer experiments have been carried out by several groups[1] interested in preparation of template-assisted precipitation of inorganics, and are in fact also actively pursued by pharmaceutical scientists, interested in making microparticles as drug carriers. A host of methods is available, differing in composition and detail of the particle formation process. In the template-assisted precipitation experiments, the interest is in the formation of structured but hard particles, typically by creating and drying an aerosol of a complex composition (including surfactants and metal oxide precursors). The ensuing particles (of the order of 1 micron) then show a 'frozen-in' mesoscopic structure, of the precipitated metal oxide given the soft template set by the (polymer) surfactant (often a pluronic). In our calculation scheme, in principle we could include reactions too[2] in order to mimick the interaction of the metal oxide precursor with the soft template, and also the solvent evaporation process. But in as far as the frozen-in mesoscopic metal oxide structure is a copy of the soft template, some of the observed structures already resemble the structures we have simulated. For example, the multi-layered 'ionion' or raspberry phases in the aforementioned Lu *et al.* and Andersson *et al.* papers are very similar to the structures we have observed in ref. 3. In all these cases, the soft template is thought to be formed by a single component surfactant. As far as we know though, no experiment has yet been published demonstrating the properties of the additional hierarchical structures in *mixed* or chimaeric polymer surfactant particles, as we have shown in this paper.

(2) The splitting of the droplets in a solvent free system, such as an aerosol is an interesting problem. Surely, one may consider 'air', or vacuum, a very bad solvent for any polymer, and in this respect it is extremely unlikely that an individual polymer molecule will be squeezed out as an unstable chimaeric polymer particle. But as Giodi correctly points out, splitting of droplets remains a possibility, and the better alternative is to consider the surface tensions of the pure droplets, and the corresponding surface energies, given the geometry of the mother droplet and daughter droplets. In principle the total surface energy of split spherical droplets always goes up, compared to a spherical mother droplet, and thus one may assume splitting in such an environment is unfavourable. However, the total free energy balance must include the unfavourable interactions of the immiscible polymers along the line of tension of the polymer patch in the mother droplet (this is

favourable for splitting). Also one must consider, perhaps pathological, that neither the mother chimaera nor the daughter droplets are necessarily spherical (a cylindrical chimaeric mother droplet when split into two pure spherical droplets, could actually result in decrease of surface energy, when the initial cylinder is stretched).

1 Y. F. Lu, H. Y. Fan, A. Stump, T. L. Ward, T. Rieker and C. J. Brinker, *Nature*, 1999, **398**(6724), 223–226; N. Andersson, P. C. A. Alberius, J. S. Pedersen and L. Bergstrom., *Microporous Mesoporous Mater.*, 2004, **72**(1–3), 175–183.
2 N. M. Maurits, G. J. A. Sevink, A. V. Zvelindovsky and J. G. E. M. Fraaije, *Macromolecules*, 1999, **32**(22), 7674–7681.
3 J. G. E. M. Fraaije and G. J. A. Sevink, *Macromolecules*, 2003, **36**(21), 7891–7893.

Dr Borsali asked: Has any work been done in predicting the morphological changes when comparing linear diblock copolymer with cyclic diblock copolymer?

Professor Fraaije responded: We have previously calculated the phase diagram according to RPA for cylindrical block copolymers in the weak segregation limit,[1] no such calculation was done for polymer particles. Also, the numerical evaluation of the polymer path integral in the present calculation scheme would have to be modified for cyclic polymers.

1 A. N. Morozov and J. G. E. M. Fraaije, *Macromolecules*, 2001, **34**(5), 1526–1528.

Professor McLeish said: Could you comment on the suggestion by some groups (*e.g.* Rao and Major in Bangalore) that structures such as the Golgi are maintained not by equilibrium separation of rafts, but by kinetic flux of lipid and membrane proteins? Might the polymersome case also produce new morphologies if maintained at a well-tuned flux of membrane components?

Professor Fraaije responded: Professor McLeish hints at the possibility of including a dissipative process in our structure formation model. We have already done so in the past[1] simply by adding a reactive term for block polymer formation (out of free blocks) to the diffusion–convective model. In addition we have unpublished results on growing polymersomes in 2D and 3D, where we have added a position dependent source term to the diffusion equation. There is obviously an almost overwhelming variety of choices for kinetic parameters and conditions, and at the same time an overwhelming lack of experiments to justify a specific model. We have not yet studied the case where there is kinetic steady-state, with a source and sink of block copolymers, thus mimicking the formation and breakdown of (bio)amphiphiles. Such a model would be quite interesting, and there can be no doubt in that case the flux intensity will be coupled to the structure; that is the flux is an additional *non-equilibrium* parameter in structure building. For example, when in a bilayer the local build-up of an asymmetric block polymer is faster than a certain critical lateral diffusion rate, build-up could easily destroy the bilayer locally (that is, form a pore). With some chair-cartoon thinking one can produce many scenarios like this. But at this point we feel such a model cannot be much more than a toy. For the application in biology, the mean field is not appropriate for lipids or proteins (or only in a very qualitative sense), let alone that it is not possible to take into account the many hundreds if not thousands of different chemicals making up a complex cell organelle. Sure, the computer does not complain when we integrate a few hundred (non-linear) diffusion equations simultaneously, but without good ideas for parameters such an approach would be less than useful, even confusing, and therefore inconclusive with respect to the suggestions put forward by Rao and Major (however appealing). For the application in the polymer area, there are perhaps more realistic possibilities for application of a reaction–diffusion–convection model in humble but important areas such as emulsion polymerisation and reactive processing of engineering plastics.

Despite my pessimism regarding quantitative application to biological cell-organelle problems, it would be very interesting to find out if there are stable points in an off-equilibrium amphiphile phase diagram (with steady-state flux intensity as a non-equilibrium thermodynamic parameter), and if there is universal behaviour. And if, and only if, there is universal behaviour, this might be of relevance to the cell organelle problem.

1 N. M. Maurits, G. J. A. Sevink, A. V. Zvelindovsky and J. G. E. M. Fraaije, *Macromolecules*, 1999, **32**(22), 7674 7681.

Professor Ryan asked: I have a graduate student who is currently trying to make such mixed vesicles. Should I tell him to stop because they cannot be made?

Professor Fraaije answered: No! The message is that chimearic vesicles of two demixing block copolymers are necessarily unstable thermodynamically. But one can easily envisage a process where a mixed emulsion at a certain stage of phase separation is quenched below a glass temperature, or halted by polymerisation (when using unsaturated monomers). This would effectively freeze the structure somewhere between homogeneous and completely separated polymersomes. Such structures will then necessarily be kinetically trapped, but if they are sufficiently long-lived for subsequent experiments this will not be a problem. The phase separation process is very slow anyway, and may easily take hours or even days for completion, more than enough time for study of separate stages. I am sure your student will write a wonderful thesis!

Dr Davidson opened the discussion of Professor Wright's paper: The conductivity of organized phases such as crystals and liquid-crystals often strongly depend on defect densities. (By defect, I mean here domain walls, dislocation lines...) Did you try to measure and control the defect densities in your samples? Did you try to reduce these densities by annealing and/or aligning the mesophase by applying a magnetic field or a shear flow?

Professor Wright responded: The most significant 'defects' in our systems are apparently the 'grain' boundaries between the crystalline/liquid crystalline lamellae of polymer **I** (see Fig. 1a in the paper) and the interlamellar phase, polymer **II** and **III**. The latter is added to promote interfacial contact (see ref. 14 in the paper). The requirement for an interlamellar ion bridge or 'glue' was perceived following investigations (see ref. 9 and 10 in the paper)) on polymer **I**–salt complexes alone and with intercalated long chain alkyls, in which the conductivity was observed to plunge steeply with reducing temperature. Above the crystal–liquid crystal transitions at 40–50 °C conductivities were observed at the levels reported in the present paper (*i.e.* approaching 10^{-3} S cm^{-1}) but they declined to *ca.* 10^{-7} S cm^{-1} at 20 °C. The introduction of polymers **II** (see ref. 2–6 in the paper), together with **III** more recently (see ref. 14 of the paper) has brought about the extrapolation of the higher level of conductivity to lower temperatures, with low temperature dependence (see Fig. 11 in the paper and Fig. 1 here), even though the side chains in **I** crystallise and the materials become tough, waxy solids. The conductivity declines as polymer **II** begins to crystallise below *ca.* 20 °C (as shown in the more comprehensive results (see ref. 14 of the paper) of Fig. 1 here for the same system as Fig. 11 in the paper) but still has a high value (4×10^{-5} S cm^{-1})

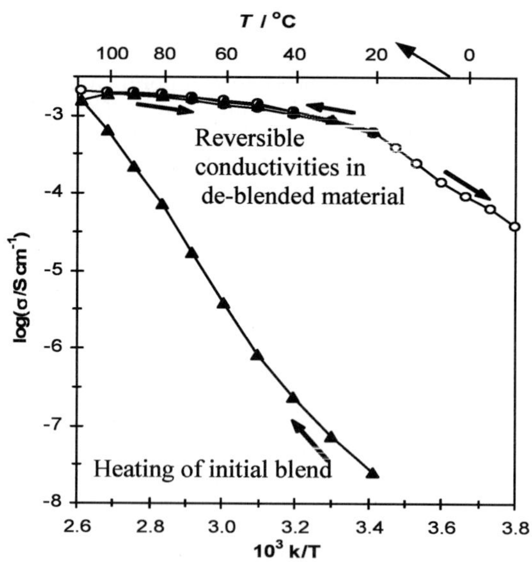

Fig. 1 C18C12O5X(18%):**II**:**III**:$LiBF_4$ AC impedance measurements. ITO electrodes.

at −10 °C. Thus, in polymer **I** ion and aggregate mobility is apparently *decoupled* from the polymer chain motion whereas in polymer **II** ion and aggregate mobility may be *coupled* to chain motion and is temperature-dependent. As is well understood in conventional amorphous polymer electrolytes (see ref. 8 in the paper) such motion is free-volume dependent. The formation of the lamellar morphology depicted in Fig. 1a in the paper follows slow cooling at 0.3 deg min^{-1} ('de-blending') from the liquid crystal/isotropic temperature (*ca.* 110 °C) after which the higher level of conductivity becomes permanent as also shown by the second thermal cycle in Fig. 1 here.

We have previously noted that alignment of the material before deblending may be beneficial (unpublished observations). We believe that this may reflect 'blockiness' in the structure of our C*m*O1–C*m*O5 copolymers. This may allow channels formed from C*m*O1 sequences, which engender small salt aggregates, to lie alongside C*m*O5 sequences which promote ion separation. Such close mutual alignment would allow facile charge balancing for cation hopping along the aggregate rows by small displacements of anions from the separated ions. This proposal is supported by conductivity measurements on synthesised block copolymers and is now under closer investigation.

The optimum ion mobility in phase **I** alone presumably depends critically on the concentration of the small salt aggregates within the channels. These themselves may presumably be mobile, requiring some free space, but must encounter each other in sufficiently close proximity for facile cation exchange. Their concentration, in turn, must also depend on the distribution between **I** and **II** which is under investigation.

These polymers only become mesogenic when alkali salts are complexed with them. The salt-free polymers and copolymers become isotropic at side chain melting temperatures 20–40 °C (see Figs. 3 and 8 in the paper). Their interaction with the salt introduces some chain rigidity at the same time causing increases in the side chain melting temperatures. We have not investigated the effect of magnetic fields on these systems. However, the passage of DC current between lithium electrodes at 30–40 °C brings about progressive increases in conductivities over several hours eventually reaching the same level as the AC impedance measurements ($\sim 10^{-3}$ S cm^{-1}) (and in some systems somewhat higher) without the requirement for higher temperature cycling as required in the AC experiments, as described in ref. 6 of the paper and shown in Fig. 2 here. We are investigating this phenomenon but at present we presume that it arises from current-driven morphological changes involving the disintegration of less stable salt-deficient lamellae and the formation of more stable salt containing ones, appropriately oriented with respect to the current direction.

Professor Iwuoha asked: Lithium hopping between neutral aggregates was proposed as the conduction mechanism in the copolymers. Is this not rather a case of charge propagation along and between a conducting copolymer chain? It is not clear how DC polarisation experiments 'suggested' the Li^+ transport. Cyclic voltammetric experiments should have given a clearer indication of charge propagation.

Professor Wright responded: If by the term 'conducting polymer chain' the questioner is referring to an intrinsically (electronically) conducting polymer system we have shown previously (see ref. 9 of the paper) that our polymers, when pure and salt-free are insulators with ambient conductivities $\sim 10^{-14}$ S cm^{-1}. The addition of salt brings about considerable increases in conductivity to the levels described in the paper. Our materials are therefore electrolytes. However, the essence of the proposed mechanism is that arrays or rows of small aggregates, such as ion pairs or quadrupoles, may themselves be mobile, but also critically serve as the substrates for cation hopping, *decoupled* from the polymer chains. This differs from the mechanism in conventional amorphous, elastomeric polymer electrolytes in which lithium ions coordinate to flexible polyether chains and their mobility requires local cooperation (coupling) of the polymer chain in bringing about the thermal disintegration of the leaving site and the formation of a new acceptor site.

DC polarisation between lithium metal electrodes in the cell

$$\text{Li}|\text{LiA electrolyte}|\text{Li}$$

is a well-recognised procedure for determination of the transference numbers of Li^+(see, for example, ref. 1 here). At zero time both ions may perhaps be mobile but over an extended period the anion contribution must decline with cell polarisation and the 'limiting current fraction' $= I_0/I_\infty$

reflects the net passage of cationic species across the cell in whatever state of association. Furthermore if the limiting DC current is in good agreement with that measured by AC impedance spectroscopy this is a good indication that cationic species are the predominant current carriers. Fig. 2 here shows a DC polarisation experiment for a low dimensional system closely related in structure to those described in the present paper. Typically, in these low-dimensional systems any decay of current due to anion polarisation is masked by the increase due the apparent formation of new conducting pathways (see my previous response to Dr Davidson). The high resistance over the first 8 h is believed to arise from the interfacial layer ('solid electrolyte interface') at the lithium–polymer electrolyte interface because the initial AC impedance result is in excellent agreement with the DC measurement after the first step. The further stepped increase in the DC measurement after 12 h presumably arises from the formation of new conducting pathways in the system and is also in agreement with the AC impedance result at the end of the experiment. (Further details of this experiment are given in ref. 6 of the paper.)

Cyclic voltammetry gives similar values for the conductivity but has revealed some dependence of the resistance on the applied field for fields > *ca.* 10 mV when decreases of up to 50% in resistance have been observed.

1 F. M. Gray in *Polymer Electrolytes*, Royal Society of Chemistry, Cambridge, UK, 1997, ch. 4, pp. 128–9.

Professor Imrie said: The conductivity plots reveal a remarkably weak temperature dependence on cooling from the isotropic phase. This behaviour is, however, dependent on the thermal history of the sample presumably implying, at least in part, a synergy between structure and mechanism. Do the authors consider that a different mechanism, perhaps a more conventional free-volume mechanism, operates in the more ordered structures?

Professor Wright responded: Please see my earlier response to Dr Davidson. The low temperature dependence on cooling from the isotropic phase undoubtedly depends on a morphological change, the new morphology developing on cooling through the liquid crystal phase *ca.* 110 °C > T > *ca.* 45 °C (see Fig. 12(a) of the paper). Furthermore, it is apparent that the high conductivity/low temperature dependence arises from a morphological change taking place *in situ* within the cell. If the material is subjected to the same thermal regime *outside* of the cell before its incorporation into the cell, the initial conductivity returns to that of the initial blend *ca.* 10^{-7} S cm^{-1}. The high conductivity/low temperature dependence regime can then be recovered in part by thermal cycling. This suggests structural alignment at the electrode surfaces and is consistent with the lamellae developing to lie along the electrodes and the conductive planes favourably oriented normal to them.

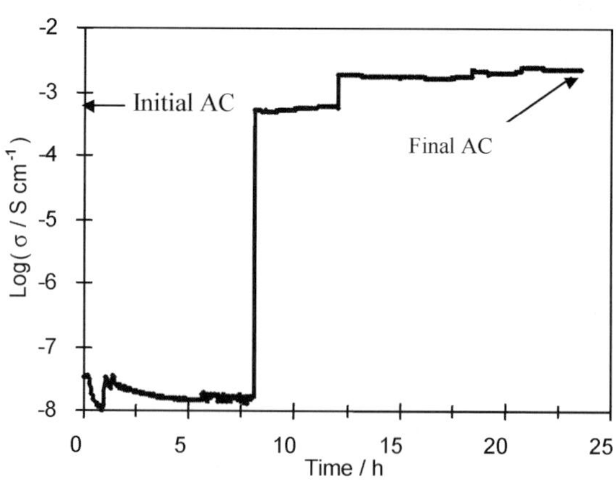

Fig. 2 DC polarisation of the cell Li | C18O5:**II**:$LiBF_4$ (1:1:1) | Li at 50 °C. (From ref. 6 of the paper, reproduced with kind permission from Elsevier.)

As we have explained above, we believe that the observation of new complex forms in the ionophobic C16O1-Li salt systems (Fig. 4 of the paper), their low conductivities (Fig. 6 of the paper) and molecular dynamics model (see ESI in the paper) indicate structures incorporating ionic aggregates. We are now pursuing further (spectroscopic) evidence for such adducts. Meanwhile, we are strongly inclined to the conclusion that the high conductivity/low activation energy regime in the systems which are predominantly ionophobic (Fig. 11 of the paper and Fig. 1 here, $T > 20$ °C) involves hopping of Li^+ between rows of aggregates *uncoupled* to the polymer chains in polymer **I**. (Analogies may perhaps reside in the protonation of water aggregates $(H_2O)nH^+$ and in the 'doping' of neutral conjugated polymers by electron-rich or electron-deficient centres, both of which lead to large increases in conductivity.) An 'ion-coupled free volume mechanism' (Vogel–Tamman–Fulcher) gives rise to the steeper curved plots of log σ *vs.* $1/T$, as seen, for example, in amorphous PEO systems (see ref. 8 in the paper) and is also tentatively postulated here for the mechanism operating in polymers **II**/**III**, the interlamellar 'glue' phase at $T < 20$ °C (Fig. 1 here) but mobile, decoupled aggregates are likely to be important components in this ionophobic (polytetramethylene oxide) phase also. However, the curved plots of Fig. 6(a) of the paper also conceal the formation of interlamellar grain boundaries at temperatures $T \leq$ *ca.* 50 °C (side chain crystallisation) in two-component systems as is evident from the impedance data of Fig. 6(b). The overall log σ *versus* $1/T$ profile in the **I**:**II**/**III**:Li salt systems is thus envisaged as a superposition of the mechanisms operating in each phase according to an appropriate mixing rule for the 'block-like' distribution of phase **I** in a continuum of **II** observed in $LiBF_4$ systems (see ref. 6 in the paper).

Professor Lodge asked: What other structural or morphological characterization tools, besides optical methods, support the detailed schematic diagrams in Fig. 1 of your paper?

Professor Wright replied: Small-to-wide angle X-ray data shown in Fig. 3 here are typical of the data obtained for lithium salt complexes with polymer **I**. At ambient temperatures the small angle peak shows multiple orders of reflection corresponding to d-spacings over the range 30–50 Å such as those shown in Figs. 5 and 9 of the paper. The wide angle peak at ambient corresponds to the hexagonal lattice (~4.1 Å) adopted by the alkyl side chains as they coalesce both intra- and intermolecularly. At 50 °C, above the side chain melting temperature (40–45 °C) the wide-angle peak has become diffuse but the intensity of the small-angle peak is retained signifying a mesogenic phase (see Figs. 4 c–h of the paper). In the absence of salt the small-angle region shows a peak of much diminished intensity and between crossed polars the field is dark. As the isotropic transformation region is approached (100–110 °C) the small angle peak in the complex also diminishes in intensity. The X-ray data are consistent with the molecular dynamics models for the molecular organisation shown in the ESI (see paper).

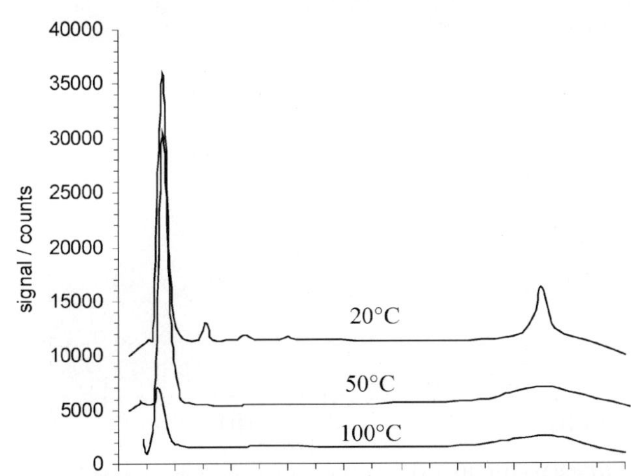

Fig. 3 Small-to-wide X-ray diffraction traces for C18O5–C18O1:$LiClO_4$.

The morphological organisation shown in Fig. 1a in the paper is clearly an adaptation of the well-established lamellar organisation in conventional semicrystalline polymers in which the inter-lamellar amorphous component is provided by polymers **II** and **III**. However, lamellae are typically much thicker than normally observed in flexible linear polymers and, as shown in ref. 6 in the paper, may be observed by high power optical microscopy to be 1–3 μm in thickness. Micrographs for a **I**:**II**:$LiBF_4$ system are reproduced in Fig. 4 here. Unlike the systems with $LiClO_4$ which spontaneously form well-defined lamellae with more extensive lateral dimensions, lamellae of the $LiBF_4$ systems are apparently slower to develop and in unsheared samples appear as 'blocks' growing from the blended mixture. However, shearing promotes lamellar growth normal to the shear direction as in the well-recognised 'shish-kebab' morphology. This observation gives good evidence for the orientation of molecular chains within the lamellae, in support of the model in Fig. 1(a) of the paper. The lateral mechanical rigidity of the lamellae should promote their development parallel to an electrode substrate *i.e.* with salt-containing channels appropriately oriented with respect to the direction of current flow. However, this is under investigation.

Dr Perez Mendez said: I did not realize the 2 discotic layers appearing in Fig. 1 in your paper were identical. I had thought these were two different lattices.

Professor Wright responded: Fig. 1a of the paper represents a continuum of an organised phase, polymer **I** plus complexed salt, interspersed with an amorphous phase polymer **II** and **III** plus complexed salt. Fig. 1b is intended to represent enlarged detail of the interface between these phases. (See my previous response to Professor Lodge.)

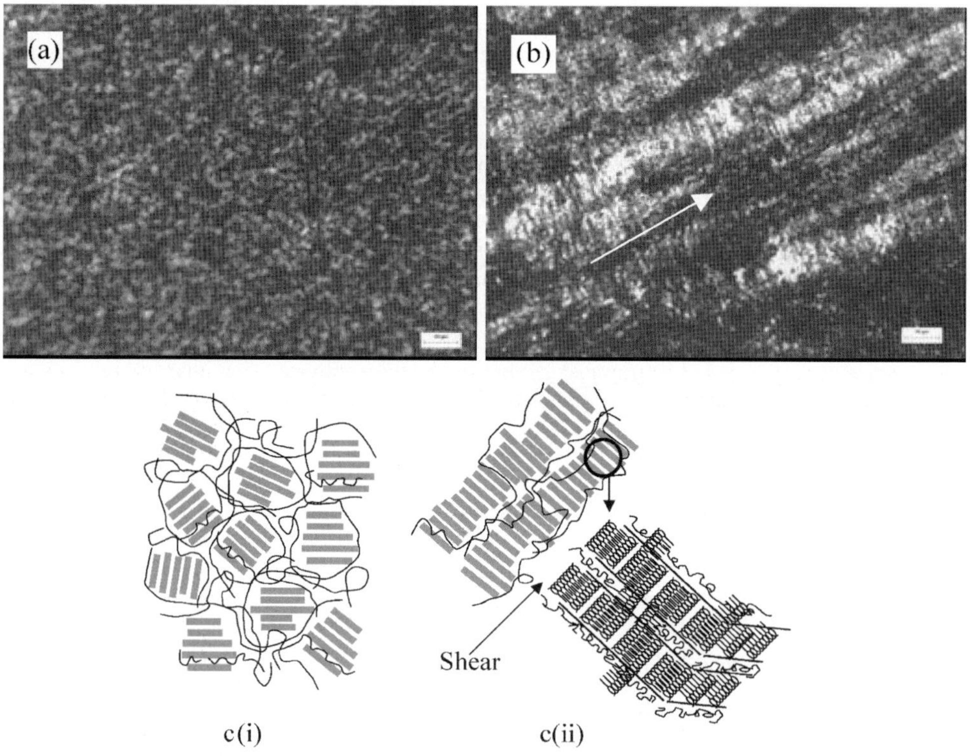

Fig. 4 Morphology of **I**(C18O5):II:LiBF4 (1:1:1) Optical micrographs viewed between crossed polars (scale bar = 20 μm). (a) Unsheared between glass slide, 20 °C. (b) Sheared at 50 °C and cooled to 20 °C. (c) Schematic models: (i) unsheared and (ii) sheared. (Reproduced from ref. 6 with kind permission from Elsevier.)

Dr Li commented: The conductivity during first heating and cooling is hugely different. What is the origin? Is the structure stable? Why does crystallisation not influence the conductivity?

Professor Wright replied: Conduction in these low-dimensional systems contrasts with that in conventional amorphous-phase ion conductors such as the semi-crystalline poly(ethylene oxide) (PEO)–Li salt complexes. In the latter, it is widely recognised (see ref. 8 in the paper) that conduction takes place predominantly through the amorphous phases of the complex. When the temperature falls below 60–65 °C the conductivity declines sharply to give ambient conductivities $< 10^{-5}$ S cm^{-1} as salt-free PEO crystallises.

However, in these low-dimensional systems the principal conduction phase, at least at temperatures ≥ 20 °C (see Fig. 11 of the paper and Fig. 1 here) is the *organised* phase polymer **I**:Li salt. As depicted in Fig. 1 an amorphous phase (polymers **II** and **III**) serves as an ion bridge or 'glue' between lamellae of **I** but the facile conduction in **I** dominates until polymer **II** begins to crystallise on cooling below 20 °C as shown in Fig. 1 here (see my earlier responses to Dr Davidson, Professor Iwuoha and Professor Lodge). Before the first heating in Fig. 11 of the paper and Fig. 1 here the blended, disordered components present only high resistance pathways to the passage of ions. On reaching the isotropic transition (~ 110 °C) and with slow cooling to ambient the organised lamellar morphology depicted in Fig. 1 in the paper forms whereupon new stable organised pathways for conduction are created and the higher level of conductivity is reversible and permanent as shown by the second cycle heating in Fig. 1 here for which the conductivity almost precisely follows that of the first cooling. Although some hysteresis in conductivity might be anticipated over the range -10 °C to $+20$ °C owing to the slow crystallisation of polymer **II**, this is apparently too small an effect to be perceptible. However, a non-crystallisable form of polymer **II** is clearly desirable for extrapolation of the higher conductivity regime of **I** to subambient temperatures and we are investigating this possibility. Crystallisation of the side chains in **I** at the crystal–liquid crystal transition temperature has no perceptible influence on the conductivity presumably because the conductive channels enclosing the salt aggregates remain largely unchanged. This is particularly noticeable in a system based upon C16O5:$LiBF_4$ as described in ref. 14 of the paper. In this case the log σ *versus* $1/T$ plot is virtually identical to Fig. 1 here and the side chain melt at approximately 45 °C. (See also my earlier response to Dr Davidson.)

Professor Fraaije opened the discussion of Dr Kostarelos's paper: Could you elaborate on the physical picture of the lipsome–polymer interaction? (1) Since the bilayer is fluid, the block copolymer isotherm will be strongly curved with respect to the polymer concentration in solution, so the important parameter to understand is the (percentage) coverage of liposomes with polymer, rather than the volume ratio of lipid/polymer. Do you have information on the polymer adsorption? (2) What is the size of the methacrylate block, and how does it compare with the thickness of the bilayer?

Dr Kostarelos responded: The physical picture of the liposome–polymer interactions, in general, differ dramatically depending on the way in which the components were allowed to interact. This is why two different preparation methodologies were developed and compared. In the case of polymers being present during vesicle formation, we believe that the hydrophobic groups of the polymer partition within the lipid bilayer anchoring the polymer molecule onto the vesicle, however, as the polymer:lipid molar ratio is increased, there are only a finite number of polymer molecules to be accommodated into the bilayer before bilayer solubilisation starts taking place. We believe that polymer:lipid molar ratio is reached when maximum peaks in the dynamic light scattering data are obtained (*e.g.* squares in Fig. 3 in the paper), indicating peak vesicle diameters and the beginning of vesicle and polymer micelle co-existence. In the case of polymers being included post-vesicle formation, then again at low polymer:lipid molar ratios the hydrophobic groups will tend to adsorb at the vesicle surface until maximum coverage of the liposome surface is achieved. At high polymer:lipid molar ratios, micelles will co-exist with the coated vesicles initially and will gradually exchange lipid molecules with the bilayer in a time and temperature dependent fashion leading to gradual bilayer solubilisation. This is the physical picture of the interactions we have in mind. However, these interactions will depend a lot on the lipids and polymers interacting. We have carried out numerous experimental studies on the interactions between zwitterionic phospholipids and a range of tri-block copolymer molecules (Pluronics). The physical interactions described above are conclusions we have reached over the years, primarily from these studies. It may well be shown that other polymers behave differently. For the polymerizable polymer we used in the present communication the experimental observations obtained were very similar to the ones using the block copolymers, so we speculate that similar physical interactions will be taking place between this polymer and the lipids. However, this still remains a speculation until further studies are carried out using this polymer.

Answering your specific questions now: (1) Your point refers to the methodology where polymers are added to pre-formed liposomes only. I agree that the degree of liposome surface coverage is the important physical parameter in this interaction. The lipid:polymer molar ratio described is an experimental indication of what happens at various concentration regiments and is not a physical parameter of importance, however it is important in designing and repeating experiments. We have previously carried out calculations of Γ (mg m^{-2} of liposome surface) from the adsoption curves obtained for various block copolymers (ref. 19 in our paper). For the polymer molecule used herein we do not have quantitative information of the percentage of liposome coverage obtained. (2) There is only one methacrylate group per polymer molecule and the lipid bilayer of *ca.* 4 nm thickness.

Dr Borsali asked: (1) Could you explain the physical meaning of the maximum observed in Fig. 3 in your paper? Indeed the mean vesicle diameter increases as a function of methyl PEG %w/w, reaches a maximum at about 70 nm and then decreases to 40 nm. (2) What is the polydispersity and error bars?

Dr Kostarelos responded: (1) I will refer you to the description of the physical interactions we believe are taking place in my earlier response to Professor Fraaije. We believe that the increase in mean size indicates vesicles incorporating the maximum number of polymer molecules in the bilayer, and that the decrease indicates that polymer–lipid mixed micelles are the predominant dispersed phase. (2) The polydispersity for vesicles prepared by ultrasonication (as is the case here) is generally quite high, at 0.1–0.15. The error in our experiments was never larger than 3% between five measurements of each sample. All sizing experiments were repeated at least twice for each of the polymer molecules exhibiting almost identical results.

Dr Lopes asked: As far as I remember only one polymer to phospholipid ratio was tried, 1 : 30. What is the influence of this ratio on the stability of the system (related to the degree of coverage)?

Professor Kostarelos responded: I refer you to my response to Professor Fraaije's question where exactly this point was addressed.

Dr Perez Mendez asked: Have you proved that the polymer interacts with the membrane by broadening its main transition temperature range (by steady state fluorescence(?))? Is this interaction stable in time?

Professor Kostarelos responded: We have shown that block copolymers interact with the lipid bilayer using a variety of experimental techniques (see refs. 9, 10, 19 and 20 of our paper) particularly when allowed to participate with the lipids during vesicle formation. Moreover, following our studies, other groups have monitored the pre-transition temperature of the bilayer using high-sensitivity differential scanning calorimetry (ref. 23 in our paper) verifying our observations and supporting our model of membrane-spanning PPO groups through the lipid bilayer. We have not carried out such studies for the PEG–MMA polymerisable polymer though. In light of our observation herein, it will be very interesting for someone to do so.

Professor Hillmyer asked: (1) Does the lipid that you employ have polymerizable groups? (2) Did you have a control experiment in which you initiated the AIBN in the absence of PEG–MMA?

Dr Kostarelos replied: (1) No, the lipids we have employed in these studies do not carry the polymerisable groups. Therefore, any effect following polymerisation is due to changes induced by the polymer moelcules. (2) Our controls included vesicles incorporating the PEG–MMA without any initiator in the bilayer, indicating no increase in vesicle size after 240 min irradiated in the polymerisation chamber.

Dr Perez Mendez asked: If liposomes with the polymer added "externally", after the membrane is formed are more stable, then the polymer must be covering the surface of the membrane. The entrance into the bilayer of the polymer added "internally", before the liposome is formed, could be through it. Perhaps the polymer is concentrated too much and a broader vesicle needs to be formed to really stabilize it?

Dr Kostarelos responded: If I understand your question correctly, your understanding is exactly what we believe is taking place. When the polymer is added 'internally' (as you describe in your question) the vesicles formed are becoming 'broader' to accommodate the bilayer-incorporated polymer molecules. That we believe is manifested as the observed increases in mean vesicle diameter up to peak values. As more polymer is added, mixed micelles are becoming the preferred aggregate to form between the polymers and the lipids, leading to decreases in size.

Professor Vamvakaki asked: Do you have any indication that the polymer remains in the lipid bilayer after polymerisation? I would expect that this will require too much stretching of the PEG2000 chains as there will be one PEG2000 chain for every two backbone carbon atoms.

Dr Kostarelos responded: Unfortunately we don't have any indication of where the polymer is after the polymerisation reaction is allowed to occur. Maybe the polymer is leaving the bilayer for a more preferred phase leading to this vesicle collapse. Or perhaps the polymer during polymerisation is concentrated in regions at the liposome surface leading to lateral phase separation locally, thus fusing with the neighbouring vesicle. In any case, we have no idea where the polymer molecules are located following polymerisation.

Professor Fraaije commented: I consider it likely that the polymers are distributed homogeneously over the polymerisable surface. Also likely is that the hydrophobic block embedded in the lipid bilayer does not induce later phase separation—otherwise the polymer rafts would pinch off the lipsomes.

Dr Borsali opened the discussion of Dr von Ferber's paper: Adding salt or surfactant to single chain polyelectrolyte will decrease the size of the chain—well known behaviour that is the consequence of electrostatic interaction screening . Does your model predict a phase separation of the polyelectrolyte/surfactant complex at high surfactant concentrations? Such behaviour is observed for the sodium hyaluronate/C_{10}TAB system (*e.g.* polyion/cationic surfactant).[1]

1 M. A. Villetti, R. Borsali, J. S. Crespo, V. Soldi and K. Fukada, *Macromol. Chem. Phys.*, 2004, **205**, 907–917.

Dr von Ferber responded: Unfortunately, our current simulation cannot answer this question as we are restricted to a very dilute system. It might be interesting to extend the simulations to many chain systems.

Professor Penfold asked: I was very interested in Fig. 1 of your paper which shows snapshots of the complex configurations as a function of electrostatic and hydrophobic interactions. The conformations of the polyelectrolyte wrapping around the surfactant micelles has been inferred from recent small angle neutron scattering studies (see ref. 1 here).

We have extensively studied the adsorption of polyelectrolyte–ionic surfactant complexes at the air–water interface by surface tension and neutron reflectivity (see ref. 2 here). At low surfactant concentrations we observe a mixed surfactant/polyelectrolyte monolayer at the interface which has a structure similar to the structures in your Fig. 1 at lower hydrophobic and high electrostatic interactions. At higher surfactant concentrations, for some of the systems, layers of complex with approximately that structure are formed. You might wish to comment, from your experience of the simulations, on the role of the interface in apparently suppressing the hydrophobic interaction.

Incidentally, your ref. 13 is not correct (see list).

1 R. D. Wesley, T. C. Cosgrove, L. Thompson, S. P. Armes and F. L. Baines, *Langmuir*, 2002, **18**, 5704; S. A. Couderc, J. Sidhu, T. Thurn, R. Xu, D. M. Bloor, J. Penfold, J. F. Holzwarth and E. Wyn-Jones, in preparation.
2 D. J. F. Taylor, R. K. Thomas, J. D. Hines, K. Humphreys and J. Penfold, *Langmuir*, 2002, **18**, 9783; D. J. F. Taylor, R. K. Thomas and J. Penfold, *Langmuir*, 2002, **18**, 4748; D. J. F. Taylor, R. K. Thomas, Z. X. Li and J. Penfold, *Langmuir*, 2003, **19**, 3712; J. Penfold, D. J. F. Taylor, R. K. Thomas, E. Staples, I. Tucker, L. J. Thomas and D. J. F. Taylor, *Langmuir*, 2002, **18**, 5147, *Langmuir*, 2002, **18**, 5139.

Dr von Ferber replied: (1) I am very grateful for your comments and for making me aware of your large collection of experimental evidence for the structures that we observe in our simulation.

(2) Note that the hydrophobic attraction that we imply between the surfactant tails is due to the fact that the hydrophobic tails need to minimize their contact with water. At the air–water interface this can be achieved by simply having the tails on the water surface while preserving configurational entropy and nothing more is gained in this geometry by close proximity of the tails. This could well explain the absence of the hydrophobic attraction between the tails at the interface.

However, in our current setup we do not have a surface and we would have to adjust the hydrophobic interaction at such a surface according to the above consideration.

Professor McLeish asked: What is your supposed physical cause behind the phenomenon of "necklace" formation indicated in your last slide? Are the length scales of "blob" diameter and phase separation consistent with the Flory-level calculations of Rubinstein and Dobrynin? Finally, can you say whether the transition from linear to spherical structures across your composition diagram possesses first-order or second-order character?

Dr von Ferber responded: (1) A. V. Dobrynin *et al.*[1] (DRO) consider an isolated charged polyelectrolyte chain in a bad solvent, implying a hydrophobic attractive force in competition with Coulomb repulsion. The transformations to a series of necklace structures in this case is related to the Rayleigh[2]instability of a charged droplet which occurs when the Coulomb energy exceeds the effective surface term. In our ongoing simulation sketched in my last slide we also observe necklace structures in our model of surfactant–polyelectrolyte complexation. In our case however, the overall charge is in general negligible. The cause for the instability of a globule consisting of a single large blob is rather related to the incompatibility of the blob radius with the length of the surfactant and the fact that the strong hydrophobic interaction of the surfactant tails pushes the hydrophilic heads and the polyelectrolyte to the blob surface.

Thus, we expect a strong dependence of the blob geometry on the surfactant structure and a differing impact of hydrophobicity than in the DRO model.

(2) The transition observed in Fig. 1 appears to be sharp in the current resolution of parameters. A higher resolution would be needed to confirm a first order transition, however.

1 A. V. Dobrynin, M. Rubinstein and S. P. Obukov, *Macromolecules*, 1996, **29**, 2974.
2 Rayleigh, *Philos. Mag.*, 1882, **14**, 184.

Dr Al-Hussein said: Would you expect any changes in the phase diagram by changing the size of the counterions?

Dr von Ferber responded: We have in fact changed the effective size of the counterions by changing the tail length of the surfactant with the effects seen in Figs. 3 and 4. We observe a shift of the maximal extension of the polyelectrolyte towards a Bjerrum length that is comparable to the effective size of the counterions. Another approach would be the change of the size of the charged surfactant head. For a moderate increase I would not expect any qualitative changes in terms of the structures we find. Only if the head parts become so large that they do not fit on the surface of the spherical micelle anymore, as we find for larger hydrophobicity, qualitatively different results may occur.

Concluding Remarks
Self-organising polymers

Anthony J. Ryan

Department of Chemistry, University of Sheffield, Sheffield, UK

Received 11th August 2004, Accepted 11th August 2004
First published as an Advance Article on the web 11th November 2004

I have always been impressed with the Faraday Discussions having formal concluding remarks on the schedule, but I had never realised the pressure it puts on the person who has to make them! Not only are they the last item on the longest morning, following an inevitably splendid banquet replete with the "Loving Cup", but also the remarks must be made only moments after the last paper has been discussed. It is, however, an honour to be asked to take on the responsibility and be part of such an established tradition.

There is an opportunity to integrate the main themes of the meeting. To ask where were the conflicts? Report whether arguments were resolved. And, if a consensus emerged, have the opportunity to be first to present the newly-received wisdom to the world. Well, this wasn't one of those sorts of Faraday Discussions. There weren't any fundamental disagreements, nor two groups working on the same subject and coming up with completely conflicting results. Neither was there a personality driven spat to provide entertainment for casual observers. Notwithstanding this, the meeting wasn't short of personalities—inspection of the recorded discussion will show that we had a lively debate over a broad ranging collection of subjects. Polymer science is a fortunate strand of physical chemistry in that it also lies in the domain of condensed matter physics, materials science, molecular biology and chemical engineering, to name but a few other disciplines. Faraday Discussions in this area attract a wide range of contributors from the diverse population interested in synthetic and natural polymers, and we were joined at this meeting by colleagues from the supramolecular chemistry community as well other colleagues from Leeds.

The title of the meeting "Self Organising Polymers" presented both a problem and an opportunity. It means different things to different people, and this was clearly demonstrated in the first session. In the opening lecture we were treated to a marvellous exposition of the subtle interplay between block polymer architecture and the self-organised structures that are formed therein. Followed by an equally striking analysis of the structures formed by proteins, illuminated by being given in the context of "conventional polymer physics". But the "different things" weren't just restricted to the polymers, we covered the whole gamut of our existence from the mundane commodities made from polypropylene right through to the meaning of life. Or at least how polymers self-organise to allow life to happen.

So was there a common meeting ground for this disparate community? We all converged at the nanoscale—that self-organising polymers operate over length scales covering the range from nanometres to microns was apparent throughout. And one of the things we agreed upon was that the polymer community had been doing nanoscale science for many decades. Polymer scientists are starting to adopt the nanotechnology with various levels of commitment: in some cases a fashionable rebranding; in others a complete change in research area; and those occupying the moral high-ground and having nothing to do with it.

Did a common theme emerge? Well the thing that struck me was how complex polymers are! Once one gets away from the properties that are dominated by topology (homogenous melt rheology, equilibrium phase behaviour, bulk diblock copolymer melts, the mechanical properties of glassy polymers) then the local effects of chemical structure become frighteningly complicated. It is the interaction between local chemical detail and topology-dominated coarse-graining that becomes important. For example, the understanding and prediction of the crystallisation of a simple helical homopolymer appeared to be extraordinarily complex.

So what have we learned? Everyone will have taken away their own education from the meeting and there was certainly a lot of material covered. In fact, one of the features was a lot of "show and tell", with many papers describing what had been made or measured. Whilst this was interesting and educational it didn't necessarily foster debate. From the evidence of this discussion polymer science and supra-molecular chemistry have converged sufficiently that a new discipline is starting to develop. There were lots of examples of complex, highly engineered structures held together by weak bonds, formed at room temperature and pressure, using molecules that have information written into them. In some cases the macromolecules are used as both templates and the components of devices. If Faraday Discussion 128 had been a Martian visitor's first experience of science he might have concluded that Biology is the progeny of supra-molecular chemistry and polymer science. And, in this case, the egg did come before the chickens!

Another interpretation of Faraday Disussion 128 was that three communities met: polymer crystallisation; self-organised (macro)molecular structures; and the biomimetic or biokleptic. The first group were interested in polymer crystallisation, the second thermodynamics and kinetics, and the latter the meaning of life! But did they really talk to each other, share ideas, find common ground? An inspection of the discussion will see whether this happened, but with the exception of a small number of individuals, I suspect they went their different ways. But one has to ask was there anything controversial? Were people breaking new ground? Where were the groups working on the same subjects but having data (or theories) that told different stories? Did we all agree with everything we heard or just choose to ignore what we did not like or understand.

At the start of the meeting Professor Lodge[1] pointed out the common theme of self-assembly in nanoscience. He also clearly defined some terms and it is useful to remind ourselves of them here. "The central goal of nanoscience is the control of structure (and function) on the 1–100 nm scale." The classes of materials we discussed were bulk materials, thin films and surfaces, with some mention of discrete objects and devices. "Nanotechnology is making something useful out of nanoscience." Given that the meeting was dominated by the making of self-assembled structures it is clear that the various strategies for control are well established.

Of paramount importance is the design of the constituent molecules and the meeting had a surfeit of synthetic detail. The obvious control factor, the molecular weight, wasn't explicitly the subject of discussion but the composition of the polymer, especially the incorporation of responsive groups and the polymer architecture (ABAB . . . , ABC, linear, branched, cycle) were obviously important. The subtlety of their effects was clearly brought out for block copolymers in the papers by Lodge[1] (ABC and branching), Gido[2] (stars) and Borsali[3] (cyclics). The nature of the interactions between polymers (dispersive, polar, H-bond, ionic) was illustrated by the theoretical discussions by Semenov[4] and Fraaije[5] for block copolymers in micelles and vesicles respectively and from many of the contributions to the discussion. The influence of conformational directors (α-helix, β-sheet, mesogens) borrowed from classical polymer and liquid crystal science to understand either biological molecules, for example the work on β-lactoglobulin presented by Donald,[6] or synthetic molecules containing biological motifs, including the block copolymer containing peptides discussed by Klok[7] and Lecommandoux.[8] And of course there was the self-contained discussion about polymers with crystallizable components, but more of that later.

Many of the papers presented complex systems with multiple components and the composition of the mixture is another factor controlling self-organised polymers. The most obvious example is selective solvation in block copolymers that leads to micellisation and the formation of vesicles. But a recurring theme was the response of these systems to external stimuli. We also heard about complementary interactions: specifically ionic interactions from Kabanov[9] and Wright[10] on how they can be used to design and control self-organised systems and a theoretical discourse of related systems from von Ferber.[11] I was particularly interested in the special case of AB + BC block copolymers, having worked a little in this area, and the paper by Hillmyer[12] illustrated the kinds of complex objects that can be made by rational design. It turns out that getting block copolymers to microphase separate before they macrophase separate is notoriously difficult. Whilst Hillmyer[12] was successful in this regard, Fraaije[5] was only able to generate theoretical chimera, by making vesicles of AB + BC block copolymers in the computer, which showed internal phase separation before they disappeared into a separated equilibrium. The blurring of the spectrum between supramolecular and polymer chemistry was demonstrated by the specific binding of metals and ligands in the work described by Rowan,[13] and this work was related (in the discussion) to the complementary interactions of hydrogen bonds in natural and synthetic systems as discussed above. All of which confirmed my prejudice that structural biology is essentially conventional polymer physics with the addition of hydrogen bonds!

Much was heard about responsive systems, this was over and above the normal response of polymers to changes in temperature pressure and flow, as exemplified in papers by Schick[14] and Li,[15] but in the specific design of self-organised systems that responded to specific external stimuli. The specific response of a phase separating system to a temperature gradient was brought to our attention by Sakurai.[16] Whilst this had intriguing phenomenology, one was left with the question: why? In a different vein, Davidson[17] showed the phase behaviour of suspended particles could be manipulated by a magnetic field and asked some deep theoretical questions.

One of the features of biological systems is that they operate essentially isothermally, and, more often then not isopiestically. The essential feature of the biomimetic, or, depending on how you are feeling, biokleptic, systems is that they operate in chemical potential (concentration) gradients. At this meeting the chemical potential gradient of choice was pH and the papers by the authors Vamvakaki,[18] Lecommandoux[8] and Webber[19] all used $[H^+]$ to stimulate a change in molecular shape or self-organisation. We used pH to generate force in a reciprocating manner, whereas Lecommandoux[8] and Webber[19] changed the organization of the polymer with pH and ionic strength. Vamvakaki[18] was the sole champion of pH changes as a processing route, using micelles to make metallic nanoparticles. It was intriguing that there were no advocates of concentration gradients of other species, Ca^{2+} comes to mind as an appropriate gradient in a biomimetic sense. But this question did emerge in the discussion.

The opening speech was dominated by Faraday Discussion 68, and the Faraday President even quoted from it after the banquet. Twenty five years have elapsed since then but given the connections that were made to Faraday Discussion 68 we must ask the question: Is crystallisation in polymers a uniquely tough problem? We know that it cannot be equilibrium but there is no agreement in the literature about the most appropriate kinetic model. The use of classical nucleation and growth models was evident from a number of the papers presented, particularly those of Müller[20] and Li.[15] Whilst it would be "throwing the baby out with the bath water" to suddenly disregard the nucleation and growth theories that have been so well used over the years, it is clear that a new theoretical framework is needed now. A number of the papers referred to work presented by the author at Faraday Discussions 112 and 122, and the literature on crystallisation shows the prevalence of precrystalline mesophases in a wide range of materials. In the specific case of isotactic polyproplylene, Li and de Jeu[15] showed there is a smectic precursor that has a specific relationship to shear and concluded that this subsequently nucleated the β-phase. Schick[14] showed us, with a clever advance in instrumentation, the dynamic nature of semi-crystalline polymers with their ability to reorganize to the smallest change in conditions. This further emphasises the non-equilibrium nature of the semi-crystalline state. Further theoretical contributions in this area used numerical modelling and Hu[21] was able to predict the orientation of stems in crystallisable block copolymers, whereas Yamamoto[22] had captured the beginings of helicity. But is there a consensus for us to reach? Well, not really given the limited number of viewpoints available at this meeting. That the theory of polymer crystallisation needs a whole new framework was apparent from the contributors to this meeting and I hope that somebody out there is looking for universality in this problem. At least then we will stop hiding behind the facile argument that a useful model should incorporate all the foibles of each specific polymer system. And finally, the use of Hoffmann's regimes in the analysis of block copolymer crystallisation? I don't think so!

Often a question to a presenter would would start "Your cartoon shows . . . . " and Hillmyer treated us to a hilarious quote from a referees report explaining the difference between a cartoon and a schematic diagram. *The American Heritage® Dictionary of the English Language* offers the following definitions (amongst others) of the word cartoon:

(1) A drawing depicting a humorous situation, often accompanied by a caption.

(2) A drawing representing current public figures or issues symbolically and often satirically: *a political cartoon.*

(3) A ridiculously oversimplified or stereotypical representation.

(4) A preliminary sketch similar in size to the work, such as a fresco, that is to be copied from it.

We all appreciate the first two definitions and the initial question in this Discussion is an example of a political cartoon in words! Hillmyers' point was that when we say cartoon we mean schematic diagram but the third definition is all too often a very accurate description. We only have to look through the papers here to see that our structural schematics are often so ridiculously oversimplified and stereotypical that they only serve to mislead. The "cartoon" rarely looks as fuzzy and disordered as the electron microscope or AFM image. The "cartoon problem" is at its worst in high impact factor journals that specialise in short reports or letters. We all have our favourite examples of a high profile publication which is long on conjecture, short on data and illustrated beautifully with molecular cartoons. I include the fourth definition of the word cartoon to point out that its proper usage was brought to our attention by the organising committee's ever-erudite Tom McLeish!

There is a serious point to be made here. The cartoons routinely used in newspapers and on the TV to describe nanotechnology, an enduring if not explicit theme of this meeting, are causing serious damage to our credibility. The mainstream media, with a few exceptions, have taken a remarkably responsible and informed position on nanotechnology, seeking out sensible commentary from well-regarded scientists. But even the most responsible journalists have been let down by the images that accompany the articles. There is a lack of visually arresting but scientifically accurate images of nanotechnology. Aside from the technical difficulty of finding pictures of things that are literally too small to be visualised, one of the biggest problems is the prevalence of the "Fantastic Voyage" theme. Many excellent media pieces on nanotechnology are let down by being illustrated with library pictures of nanoscale submarines and devices that are scientifically quite absurd, but, because of their high visual impact appeal to picture editors. We have a collective responsibility to prepare accurate and informative schematic diagrams and forgo the glib cartoon!

So, in conclusion, what was missing from our Faraday Discussion? That we can make all sorts of things was demonstrated by the community and is discussed above. But do we have a good enough theoretical framework? For single phase polymer systems we know that the theory of polymer dynamics is basically OK, both diffusion and flow behaviour can be predicted as a function of molar mass and concentration using the appropriate constitutive equations. Similarly, the statics of multiphase polymers systems are well understood, though there are a number of outstanding problems in their dynamics. Host–guest interactions are also under control, even if they rely on simulation methods rather than quantum theory, and the hydrophobic effect is a quantifiable though appropriate parameterisation. However many of the systems we have discussed here have long-ranged electrostatics involved and its interplay with polymer topology is proving difficult even in the theoretical description of the static (equilibrium) properties of polyelectrolytes. For me the biggest question is: how do we put all these theoretical concepts together into a theory of nanotechnology? Or even more difficult, a theory of biology? This, I believe, is the grand challenge for the self-organised polymers community as we move forward.

Finally, I must commend the staff of the RSC for taking care of all the arrangements, we "physical chemists" appreciate the special efforts you make to preserve the unique nature of the Discussions. I think you will all join me in thanking Ian Hamley for proposing this subject and assembling an excellent meeting, and the staff of Devonshire Hall for looking after us so well.

References

1 T. Lodge, *Faraday Discuss.*, 2005, **128**, 1.
2 X. Hu, Y. Zhu, S. P. Gido, T. P. Russell, H. Iatrou, N. Hadjichristidis, F. M. Abuzaina and B. A. Garetz, *Faraday Discuss.*, 2005, **128**, 103.
3 J.-L. Putaux, E. Minatti, C. Lefebvre, R. Borsali, M. Schappacher and A. Deffieux, *Faraday Discuss.*, 2005, **128**, 163.
4 I. A. Nyrkova and A. N. Semenov, *Faraday Discuss.*, 2005, **128**, 113.
5 J. G. E. M. Fraaije, C. A. van Sluis, A. Kros, A. V. Zvelindovsky and G. J. A. Sevink, *Faraday Discuss.*, 2005, **128**, 355.
6 E. H. C. Bromley, M. R. H. Krebs and A. M. Donald, *Faraday Discuss.*, 2005, **128**, 13.
7 H.-A. Klok, G. W. M. Vandermuelen, H. Nuhn, A. Rösler, I. W. Hamley, V. Castelletto, H. Xu and S. S. Sheiko, *Faraday Discuss.*, 2005, **128**, 29.

8 J. Babin, J. Rodriguez-Hernandez, S. Lecommandoux, H.-A. Klok and M.-F. Achard, *Faraday Discuss.*, 2005, **128**, 179.
9 V. A. Kabanov, A. B. Zezin, V. B. Rogacheva, T. V. Panova, E. V. Bykova, J. G. H. Joosten and J. Brackman, *Faraday Discuss.*, 2005, **128**, 341.
10 J. Liu, Y. Zheng, Y. P. Liao, X. Zheng, G. Ungar and P. V. Wright, *Faraday Discuss.*, 2005, **128**, 363.
11 C. von Ferber and H. Löwen, *Faraday Discuss.*, 2005, **128**, 389.
12 H. Mao, P. L. Arrechea, T. S. Bailey, B. J. S. Johnson and M. A. Hillmyer, *Faraday Discuss.*, 2005, **128**, 149.
13 S. J. Rowan and J. B. Beck, *Faraday Discuss.*, 2005, **128**, 43.
14 A. A. Minakov, D. A. Mordvintsev and C. Schick, *Faraday Discuss.*, 2005, **128**, 261.
15 L. Li and W. H. de Jeu, *Faraday Discuss.*, 2005, **128**, 299.
16 T. Nambu, Y. Yamauchi, T. Kushiro and S. Sakurai, *Faraday Discuss.*, 2005, **128**, 285.
17 B. J. Lemaire, P. Davidson, J. Ferré, J.-P. Jamet, D. Peterman, P. Panine, I. Dozov, D. Stoenescu and J.-P. Jolivet, *Faraday Discuss.*, 2005, **128**, 271.
18 M. Vamvakaki, L. Papoutsakis, V. Katsamanis, T. Afchoudia, P. G. Fragouli, H. Iatrou, N. Hadjichristidis, S. P. Armes, S. Sidorov, D. Zhirov, V. Zhirov, M. Kostylev, L. M. Bronstein and S. H. Anastasiadis, *Faraday Discuss.*, 2005, **128**, 129.
19 G. B. Webber, E. J. Wanless, S. P. Armes and S. Biggs, *Faraday Discuss.*, 2005, **128**, 193.
20 A. J. Müller, J. Albuerne, L. Marquez, J.-M. Raquez, P. Degée, P. Dubois, J. Hobbs and I. W. Hamley, *Faraday Discuss.*, 2005, **128**, 231.
21 W. Hu and D. Frenkel, *Faraday Discuss.*, 2005, **128**, 253.
22 T. Yamamoto, N. Orimi, N. Urakami and K. Sawada, *Faraday Discuss.*, 2005, **128**, 75.

List of Posters

Orientation of cylindrical microdomains of block copolymers under the magnetic field **S. Sakurai, E. Funai, S. Munakata, S. Okamoto, M. Yamato** and **T. Kimura**, *Kyoto Insitutute of Technology, Japan*

Nanotriosensors systems for organophosphate pesticides 1: Template synthesized polyaniline nanocomposites on mercaptobenzothiazole-modified gold electrodes **V. S. Somerset, T. Mabula, P. Baker** and **E. Iwuoha**, *University of the Western Cape, South Africa*

Evidence of pre-order in polymer melts revealed from dielectric spectroscopy **A. Wurm, R. Soliman, J. P. G Goossens, W. Bras** and **C. Schick**, *University of Rostock, Germany*

Rheological and structural characterization of hydrophobically modified polyacrylamide solutions **V. Castelletto, I. W. Hamley, W. Xue, C. Sommer, J. S. Pedersen** and **P. D. Olmsted**, *University of Leeds, UK*

Phase behavior of hydrophobically modified cationic polysaccharides and oppositely charged surfactants **N. Nichifor, S. Lopes, M. J. Capitan Aranda** and **A. Lopes**, *Instituto de Tecnologia Químca e Biológica, Portugal*

Nanoscopic ordering of a semifluorinated liquid crystalline diblock copolymer **M. Al-Hussein** and **W. H. de Jeu**, *FOM Institute for Atomic and Molecular Physics, The Netherlands*

Adsorption of linear and star polymers on surfaces studied by atomic force microscopy **E. Glynos, V. Koutsos, G. Petekidis** and **J. Roovers**, *University of Edinburgh, UK*

Water-soluble nano-fibrous aggregation structure of β-sheet-forming (Lys-Leu) oligopeptide **T. Sakurai, M. Koga, M. Takafuji** and **H. Ihara**, *Kumamoto University, Japan*

Temperature effect on self-assembly of linear and cyclic block copolymers **C. Lefebvre, M. Schappacher, A. Deffieux** and **R. Borsali**, *ENSCPB, France*

Light scattering and SAXS from self-assembled properties of linear and cyclic A-B and A-B-A-B block poly(styrene-block-isoprene) copolymers **C. Houga, C. Lefebvre, S. Lecommandoux, M. Schappacher, A. Deffieux** and **R. Borsali**, *ENSCPB, France*

Small-angle neutron scattering study of stabilized pluronic micelles **J. Pleštil, H. Pospíšil, Z. Sedlakova, I. Grillo** and **P. Strunz**, *Institute of Macromolecular Chemistry, 16206 Prague, Czech Republic*

Micropatterned poly(organosiloxane) nanofilms: Template for area-selective arrangement of charged microparticles **T. Koga, M. Morita, H. Otsuka** and **A. Takahara**, *Kyushu University, Japan*

Phase behavior of a new type of weakly segregated block copolymers: Poly(acrylate)-*b*-polystyrene block copolymers **K. Yamamoto, Y. Sugino, K. Usami, S. Okamoto, S. Shimada** and **S. Sakurai**, *Nagoya Institute of Technology, Japan*

The fold density effects on the crystalline structure of long-chain alkanes: Molecular dynamics study **N.-Y. Tanabe** and **S. Hanna**, *University of Bristol, UK*

Crystallization of block copolymer in confined geometries **C. Vasilev** and **G. Reiter**, *University of Sheffield, UK*

Study on δ to γ to α form transitions of syndiotactic polystyrene by time-resolved SAXS and WAXS simultaneous measurement **S. Okamoto, M. Suzuki, Y. Yamamoto, H. Yoshimizu, Y. Tsujita, K. Yamamoto** and **S. Sakurai**, *Nagoya Institute of Technology, Japan*

Spatially-correlated metallic nanostructures on self-assembled functional polymeric templates **A. W. Fahmi, P. Moriarty** and **M. Stamm**, *University of Nottingham, UK*

Fabrication of layer-by-layer assembly of TiO_2/polymer film: Photochemical characterization for reduction of viologen **T. Sagawa, J. H. G. Steinke, H. Ihara** and **K. Ohkubo**, *Kyoto University, Japan and Imperial College London, UK*

Self-assembly of lipid A-phosphate exhibiting ordered structures on the nanoscale **C. A. Faunce, H. Reichelt, H. Brown, D. K. Ross** and **H. H. Paradies**, *The University of Salford, UK*

pH induced conformational behaviour of polyelectrolytes in bulk solution and grafted to surfaces: Fluorescence and neutron reflectrometry studies **L. Ruiz-Perez, A. J. Parnell, S. J. Martin, J. R. Howse, P. Topham, A. Pryke, L. Swanson, I. Soutar, C. Crook, A. J. Ryan, R. A. Jones, M. Geoghegan** and **A. Menelle**, *University of Sheffield, UK*

Aggregation and gelation in randomly charged ionomer solutions **E. De Luca, T. A. Waigh** and **J. S. Kim**, *University of Leeds, UK*

Adsorption of a cholesteric liquid-crystal polyester on silver and gold nanoparticles and films studied by surface-enhanced Raman scattering (SERS) **Sánchez-Cortes, R. Marsal-Berenguel** and **M. Pérez-Méndez**, *CSIC, Madrid, Spain*

Structure and dynamics of poly(γ)benzyl-$_L$-glutamate)-*b*-polyglycine block copolymers **P. Papadopoulos, G. Floudas, I. Schnell, T. Aliferis, H. Iatrou** and **N. Hadjichristidis**, *University of Ioannina, Greece*

List of Participants

Dr D. Adams, *Unilever, UK*
Dr S. Adams, *Unilever, UK*
Dr M. Al-Hussein, *FOM-Institute AMOLF, The Netherlands*
Dr A. Berman, *Ben Gurion University, Israel*
Professor S. Biggs, *University of Leeds, UK*
Dr R. Borsali, *ENSCPB-University Bordeaux, France*
Dr R. Brazil, *Royal Society of Chemistry, UK*
Dr E. Bromley, *University of Cambridge, UK*
F. Case, *Case Scientific, USA*
Dr V. Castelletto, *University of Leeds, UK*
Miss K. Daniel, *RSC, UK*
Dr P. Davidson, *Université Paris Sud CNRS, France*
Dr K. Davies, *Royal Society of Chemistry, UK*
Dr R. Daw, *Nature, UK*
Professor W. de Jeu, *FOM Institute for Atomic & Molecular Physics, The Netherlands*
Dr E. De Luca, *University of Leeds, UK*
Professor A. Donald, *University of Cambridge, UK*
Dr H. Eickenbusch, *VDI Technologiezentrum GmbH, Germany*
Dr A. Fahmi, *University of Nottingham, UK*
Mr R. Folkes, *University of Durham, UK*
Mr A. Follestad, *Borealis, Norway*
Professor H. Fraaije, *Leiden Institute of Chemistry, Leiden University, The Netherlands*
Professor N. Freeman, *Farfield Sensors Ltd., UK*
Dr M. Geoghegan, *University of Sheffield, UK*
Professor S. Gido, *University of Massachusetts, USA*
Mr E. Glynos, *The University of Edinburgh, UK*
Dr W. Gosal, University of Leeds, *UK*
Miss C. Hall, *Royal Society of Chemistry, UK*
Dr I. Hamley, *University of Leeds, UK*
Mr P. Hemström, *Umea University, Sweden*
Mr I. Henderson, *University of Durham, UK*
Professor M. Hillmyer, *University of Minnesota, USA*
Professor J. Hobbs, *University of Sheffield, UK*
Mr R. Horner, *Proctor & Gamble Technical Centres Ltd., UK*
Professor W. Hu, *Nanjing University, China*
Professor C. Imrie, *University of Aberdeen, UK*
Professor E. Iwuoha, *University of the Western Cape, South Africa*
Professor V. Kabanov, *Moscow State University, Russia*
Professor H. Klok, *École Polytechnique Fédérale De Lausanne, Switzerland*
Mr T. Koga, *Kyushu University, Japan*
Dr K. Kostarelos, *Centre for Drug Delivery Research, UK*
Dr M. Krebs, *University of Cambridge, UK*
Dr S. Lecommandoux, *ENSCPB, University of Bordeaux 1, France*
Miss C. Lefebvre, *LCPO, France*
Dr L. Li, *FOM-Institute AMOLF, The Netherlands*
Mr Y. Liao, *Department of Engineering Materials, Sheffield, UK*
Mr J. Liu, *University of Sheffield, UK*
Professor T. Lodge, *University of Minnesota, USA*
Dr A. Lopes, *Instituto De Tecnologia Quimica E Biologica, Portugal*
Professor T. McLeish, *University of Leeds, UK*
Professor A. Müller, *Universidad Simon Bolivar, Venezuela*
Mr M. Murray, *I.C.I. Strategic Technology Group, UK*
Miss A. Nordborg, *Umea University, Sweden*
Professor S. Okamoto, *Nagoya Institute of Technology, Japan*
Mr P. Papadopoulos, *University of Ioannina, Greece*
Professor H. Paradies, *The University of Salford, UK*

Mr P. Parras, *University of Leeds, UK*
Professor J. Penfold, *CCLRC Rutherford Appleton Laboratory, UK*
Dr M. Perez Mendez, *Instituto De Ciencia Y Technologia Polimeros, Spain*
Dr J. Plestil, *Institute of Macromolecular Chemistry, Czech Republic*
Ms R. Quine, *Royal Society of Chemistry, UK*
Professor S. Ross-Murphy, *King's College London, UK*
Professor S. Rowan, *Case Western Reserve University, USA*
Miss L. Ruiz-Pérez, *University of Sheffield, UK*
Professor A. J. Ryan, *University of Sheffield, UK*
Dr T. Sagawa, Kyoto University, *Japan*
Professor S. Sakurai, *Kyoto Institute of Technology, Japan*
Dr T. Sakurai, *Kumamoto University, Japan*
Professor C. Schick, *University of Rostock, Germany*
Dr A. Semenov, *Institut Charles Sadron, France*
Professor J. Simpson, *University of Otago, New Zealand*
Professor T. Spalding, *University College Cork, Republic of Ireland*
Mr N. Tanabe, *University of Bristol, UK*
Dr R. Thompson, *University of Durham, UK*
Professor M. Vamvakaki, *Foundation for Research and Technology-Hellas, Greece*
Mr C. Vasilev, *University of Sheffield, UK*
Dr C. Von Ferber, *Universität Freiburg, Germany*
Dr G. Webber, *University of Leeds, UK*
Ms G. Wilkins, *University of Leeds, UK*
Professor P. Wright, *University of Sheffield, UK*
Dr A. Wurm, *University of Rostock, Germany*
Dr K. Yamamoto, *Nagoya Institute of Technology, Japan*
Professor T. Yamamoto, *Yamaguchi University, Japan*
Miss Z. Yang, *University of Cambridge, UK*

Index of Contributors*

* The page numbers in **bold** type indicate papers submitted for discussions.